Linear Multivariable Control Systems

This comprehensive and thorough yet accessible textbook provides broad and systematic coverage of linear multivariable control systems, including several new approaches to design. In addition to standard state-space theory, it provides a new measurement-based approach to linear systems, including a generalization of Thevenin's Theorem, a new single-input single-output approach to multivariable control, and analytical design of PID controllers developed by the authors. Each result is rigorously proved and applied to control systems applications, such as the servomechanism problem, the fragility of high-order controllers, multivariable control, and PID controllers. Illustrative examples solved using MATLAB® and Simulink®, with easily reusable programming scripts, are included throughout. Numerous end-of-chapter homework problems enhance understanding.

Based on course-tested material, this textbook is ideal for a single or two-semester graduate course on linear multivariable control systems in aerospace, chemical, electrical, and mechanical engineering.

Shankar P. Bhattacharyya is a professor of Electrical Engineering at Texas A&M University. He is a Fellow of the IEEE and IFAC, and a Foreign Member of The Brazilian Academy of Sciences and The Brazilian National Academy of Engineering. His contributions to control theory include the first solution of the multivariable servomechanism problem, an algorithm for eigenvalue assignment, robustness and fragility, and a modern analytical approach to the design of PID controllers.

Lee H. Keel is a professor of Electrical and Computer Engineering at Tennessee State University. His areas of expertise include linear systems, parametric robust control, computational methods, and control applications.

Linear Multivariable Control Systems

SHANKAR P. BHATTACHARYYA
Texas A&M University

LEE H. KEEL
Tennessee State University

CAMBRIDGE
UNIVERSITY PRESS

CAMBRIDGE
UNIVERSITY PRESS

University Printing House, Cambridge CB2 8BS, United Kingdom

One Liberty Plaza, 20th Floor, New York, NY 10006, USA

477 Williamstown Road, Port Melbourne, VIC 3207, Australia

314–321, 3rd Floor, Plot 3, Splendor Forum, Jasola District Centre, New Delhi – 110025, India

103 Penang Road, #05–06/07, Visioncrest Commercial, Singapore 238467

Cambridge University Press is part of the University of Cambridge.

It furthers the University's mission by disseminating knowledge in the pursuit of education, learning, and research at the highest international levels of excellence.

www.cambridge.org
Information on this title: www.cambridge.org/highereducation/isbn/9781108841689
DOI: 10.1017/9781108891561

First published 2022

Printed in the United Kingdom by TJ Books Limited, Padstow, Cornwall

A catalogue record for this publication is available from the British Library.

Library of Congress Cataloging-in-Publication Data
Names: Bhattacharyya, S. P. (Shankar P.), 1946- author. | Keel, L. H. (Lee H.), author.
Title: Linear multivariable control systems / Shankar K. Bhattacharyya, Lee H. Keel.
Description: Cambridge ; New York, NY : Cambridge University Press, 2022. |
    Includes bibliographical references and index.
Identifiers: LCCN 2021034422 (print) | LCCN 2021034423 (ebook) | ISBN
    9781108841689 (hardback) | ISBN 9781108841689 (ebook)
Subjects: LCSH: Linear control systems–Textbooks. | Linear control
    systems–Data processing–Textbooks.
Classification: LCC TJ220 .B435 2022 (print) | LCC TJ220 (ebook) | DDC 629.8/32–dc23
LC record available at https://lccn.loc.gov/2021034422
LC ebook record available at https://lccn.loc.gov/2021034423

ISBN 978-1-108-84168-9 Hardback

Additional resources for this publication at www.cambridge.org/bhattacharyya-keel

S. P. Bhattacharyya dedicates this book to his parents Hem Nalini and Nil Kantha Bhattacharyya and his ancestors who were Sanskrit scholars

L. H. Keel dedicates this book to his beloved wife, Kuisook

Contents

Preface

This textbook on linear control theory is an outcome of the authors' combined experience of 85 years teaching and researching this rich and beautiful subject. While we have covered all standard topics, we have also included new results and presented the subject from the unique perspective gained from our own research.

The text is directed at graduate students in Electrical, Mechanical, Aerospace, Chemical, and Biomedical Engineering, and would serve the needs of the first two graduate courses in Control Systems in these fields. It would also be valuable as a reference for control scientists and engineers involved in the design of conventional and modern control systems. The latter includes UAVs, driverless cars, robots, and learning or intelligent control systems. The material presented here has been class-tested at our respective schools.

The field of control is conveniently divided into pre-1960 classical control and post-1960 modern control. Roughly speaking, classical control consists of the study of single-input single-output (SISO) dynamic systems using Laplace transforms, the Nyquist criterion, and frequency response methods, whereas modern control concentrates on the analysis and design of multi-input multi-output (MIMO) systems using time domain-based state-variable models and design by optimization. Each approach has its advantages and limitations with respect to applicability. In order to be effective, the control engineer needs to be thoroughly grounded in both classical and modern control methods. We hope to have provided this balance in the present textbook. The Epilogue contains additional comments regarding this perspective, the role of mathematics in an engineering field, and the tussle between robustness and optimality that the control field witnessed.

The contents of this book can adequately serve two graduate courses, for example: Linear Mutivariable Systems (LMS) and Robust and Optimal Control (ROC). The first course (LMS) would cover an overview of classical control (Chapters 1 and 2), state-variable models (Chapters 3 and 4), controllability, observability, and realization theory (Chapter 5), state feedback and observers (Chapter 6), and linear servomechanisms (Chapter 8). The second course (ROC), which would require LMS as a prerequisite, could then deal with linear quadratic regulators (Chapter 7), H_2 and H_∞ optimal control (Chapter 8), robust control (Chapter 10), modern PID control (Chapter 11), and a SISO approach to MIMO systems (Chapter 12).

Our textbook, while covering traditional theory such as state feedback, observers, and minimal realizations thoroughly, contains some innovative and novel approaches to design that bear elaboration. (A) In Chapter 11 we develop an approach to the design of proportional integral derivative (PID) controllers based on computation of the complete stabilizing set. This allows for the imposition of multiple design specifications such as gain margin, phase margin, settling time, and H_∞ norm optimization in an analytical and rational manner, extending classical control design theory. (B) A new approach to multivariable control based on the Smith–McMillan diagonal representation of the plant is developed in Chapter 12. This reduces the multivariable controller design problem to a number of independent SISO problems to which classical design methods for designing PID and low-order controllers can be applied. We also show in this chapter how a multivariable feedback stability problem can be reduced to that of a scalar feedback loop to which the classical Nyquist criterion can then be readily applied. This procedure avoids state-space realizations. (C) In Chapter 1 we describe a measurement-based, as opposed to model-based, solution to the problem of adding design elements to an unknown system. Our results are based on showing that Cramer's rule can be used to extend Thevenin's Theorem of classical circuit theory to arbitrary linear systems, making it applicable to mechanical, chemical, economic, and other systems. (D) Our treatment of robust control in Chapter 10 includes both the norm-bounded H_∞ theory as well as the extremal results of the parametric approach, which are quite distinct. (E) We have emphasized the controller fragility of high-order controllers even though they may be plant-robust. (F) The servomechanism problem has been given a position of central importance. We believe that these results can help close the theory–practice gap in the field and should therefore be taught to future engineers. In this sense our writing of this extensive book is amply justified, as the material significantly extends that covered in the existing textbooks in the field and enhances the control curriculum. The smooth and efficient editorial support and assistance of Sarah Strange at Cambridge University Press throughout the writing process is gratefully acknowledged.

We take this opportunity to express deep gratitude to: our mentors Professors J. B. Pearson, W. M. Wonham, C. A. Desoer, and Academician Yakov Tsypkin; former students P. M. G. Ferreira, Afonso Celso, Eurice de Souza, Dal Yong Ohm, H. Hwang, Herve Chapellat, Kevin Moore, Guillermo Silva, Ming-Tzu Ho, Navid Mohsenizadeh, Sangjin Han, Ivan Jesus Diaz-Rodrigues, Samir Ahmad, Olawale Adetona, Richard Tantaris, Carlos Beane, Pavana Sirisha Kallakuri, Esther Amullen, and Mohammad Rahman; colleagues Antonio Vicino, Alberto Tesi, B. T. Polyak, V. L. Kharitonov, Swaroop Darbha, Jo W. Howze, Aniruddha Datta, and Young Chol Kim. What we have learnt from them is priceless and cannot be quantified or even described.

1   A Measurement-Based Approach to Linear Systems

In this chapter, we develop a model-free measurement-based approach to the analysis, synthesis, and design of linear systems described by linear algebraic equations of the form

$$\mathbf{A}(\mathbf{p})\mathbf{x} = \mathbf{Bu},$$
$$\mathbf{y} = \mathbf{Cx} + \mathbf{Du},$$

(1.1)

where $\mathbf{A}, \mathbf{B}, \mathbf{C}, \mathbf{D}$ are matrices, $\mathbf{x}, \mathbf{u}, \mathbf{y}$ are state, input, and output vectors, and $\mathbf{p}$ is a vector of parameters. The inputs or parameters are to be determined or designed to control the output $\mathbf{y}$. This framework encompasses direct current (DC) circuits, alternating current (AC) circuits operating in sinusoidal steady state at a fixed frequency, spring mass systems, hydraulic networks, and civil engineering truss structures. It also applies to block diagrams where $\mathbf{x}$, $\mathbf{u}$, $\mathbf{y}$ represent Laplace transforms and the "parameters" are transfer functions.

First, we determine how the output can be assigned a prescribed value by adjusting some sources in an unknown linear system, based on measured data obtained by applying test inputs and measuring the corresponding outputs.

Next we consider the problem of controlling or assigning states or outputs to prescribed values by adjusting all or some of the components of the parameter vector $\mathbf{p}$. The main result establishes that some or all components of such a design parameter can be determined from a small set of strategic measurements made on the system without knowledge of the system model. This result, it turns out, is a generalization and extension of *Thevenin's Theorem* of classical circuit theory, to arbitrary linear systems. Examples of this method applied to DC and AC circuits, mechanical and hydraulic systems are given by way of illustration.

1.1   Output Control of Unknown Systems Using Prescribed Inputs

In this section, we consider the problem of controlling the output of an unknown linear system using specific inputs. We begin with a simple example.

EXAMPLE 1.1   Consider the unknown linear DC circuit shown in Figure 1.1.

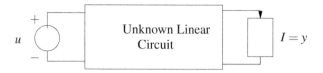

Figure 1.1　Scalar output control (Example 1.1)

Our problem is to determine the scalar source voltage u, so that the scalar output y, of the unknown linear system is a prescribed value, say y^*.

Measurement-Based Solution
In general, y will be an affine function of all the independent sources. Since all sources except u are fixed, we must have the following affine relationship:

$$y = \alpha_0 + \alpha_1 u, \tag{1.2}$$

where α_0, α_1 are fixed constants. Now conduct the following experiment:

$$\text{Set } u = u^1 \text{ and measure the current response, say } y^1.$$
$$\text{Set } u = u^2 \text{ and measure the current response, say } y^2. \tag{1.3}$$

From this data, form the measurement equation below:

$$\begin{bmatrix} y^1 \\ y^2 \end{bmatrix} = \begin{bmatrix} \alpha_0 \\ \alpha_0 \end{bmatrix} + \begin{bmatrix} u^1 \\ u^2 \end{bmatrix} \alpha_1, \tag{1.4}$$

from which we can extract α_0, α_1:

$$\alpha_1 = \frac{y^1 - y^2}{u^1 - u^2}, \tag{1.5a}$$

$$\alpha_0 = y^1 - \alpha_1 u^1 = \frac{y^2 u^1 - y^1 u^2}{u^1 - u^2}. \tag{1.5b}$$

With α_0, α_1 determined, the input u^* necessary to assign to the output y a desired value y^* can be determined:

$$u^* = \frac{y^* - \alpha_0}{\alpha_1}. \tag{1.6}$$

We consider a second example where two outputs y_1 and y_2 are to be assigned to prescribed values using two independent sources u_1 and u_2 in an unknown linear network shown in Figure 1.2.

Again, because of linearity and since there may be additional unknown sources in the network which are fixed, we must have

$$\mathbf{y} := \underbrace{\begin{bmatrix} y_1 \\ y_2 \end{bmatrix}}_{} = \underbrace{\begin{bmatrix} y_1^0 \\ y_2^0 \end{bmatrix}}_{\mathbf{y}^0} + \underbrace{\begin{bmatrix} g_{11} & g_{12} \\ g_{21} & g_{22} \end{bmatrix}}_{\mathbf{G}} \underbrace{\begin{bmatrix} u_1 \\ u_2 \end{bmatrix}}_{\mathbf{u}}, \tag{1.7}$$

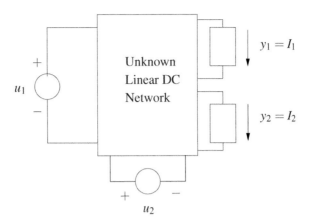

Figure 1.2 Vector output control

with y_1^0, y_2^0, g_{ij} for $i \in \mathbf{2} := \{1,2\}, j \in \mathbf{2} := \{1,2\}$ being constants. These constants can be determined from three experiments, namely setting the vector inputs to three different values $\mathbf{u} = 0, \mathbf{u} = \mathbf{u}^1, \mathbf{u} = \mathbf{u}^2$ and measuring the corresponding outputs:

$$\mathbf{y} = \mathbf{y}^0 = \begin{bmatrix} y_1^0 \\ y_2^0 \end{bmatrix}, \quad \mathbf{y} = \mathbf{y}^1, \quad \mathbf{y} = \mathbf{y}^2, \tag{1.8}$$

as illustrated in the numerical example below. Clearly $\mathbf{y}^0$, corresponding to the input $\mathbf{u} = 0$, may be called the **zero-input response** of the system.

EXAMPLE 1.2 (Assignment of two currents using two sources)   In the unknown circuit of Figure 1.2, the measurements shown in Table 1.1 are made.

Table 1.1 Measurements (Example 1.2)

Measurement No.	u_1	u_2	y_1	y_2
1	0 V	0 V	2 A	3 A
2	2 V	1 V	3 A	4 A
3	3 V	4 V	2 A	1 A

Therefore, we must have

$$y_1^0 = 2, \quad y_2^0 = 3, \tag{1.9a}$$

$$\begin{bmatrix} 3 \\ 4 \end{bmatrix} = \begin{bmatrix} 2 \\ 3 \end{bmatrix} + \begin{bmatrix} g_{11} & g_{12} \\ g_{21} & g_{22} \end{bmatrix} \begin{bmatrix} 2 \\ 1 \end{bmatrix}, \tag{1.9b}$$

$$\begin{bmatrix} 2 \\ 1 \end{bmatrix} = \begin{bmatrix} 2 \\ 3 \end{bmatrix} + \begin{bmatrix} g_{11} & g_{12} \\ g_{21} & g_{22} \end{bmatrix} \begin{bmatrix} 3 \\ 4 \end{bmatrix}. \tag{1.9c}$$

Solving Equations (1.9b) and (1.9c), we obtain

$$
\begin{bmatrix} g_{11} & g_{12} \\ g_{21} & g_{22} \end{bmatrix} = \begin{bmatrix} \dfrac{4}{5} & -\dfrac{3}{5} \\ \dfrac{6}{5} & -\dfrac{7}{5} \end{bmatrix}.
\tag{1.10}
$$

Now suppose we need to assign the vector output current as follows:

$$
y_1^* = 5\text{ A}, \quad y_2^* = 5\text{ A}
\tag{1.11}
$$

by choosing the vector control input (u_1^*, u_2^*). The solution is obtained by substituting $\mathbf{y}^*, \mathbf{u}^*$, and $\mathbf{y}^0$ into Equation (1.7):

$$
\mathbf{y}^* = \mathbf{y}^0 + \mathbf{G}\mathbf{u}^*.
\tag{1.12}
$$

This leads to

$$
\begin{bmatrix} 5 \\ 5 \end{bmatrix} - \begin{bmatrix} 2 \\ 3 \end{bmatrix} = \begin{bmatrix} \dfrac{4}{5} & -\dfrac{3}{5} \\ \dfrac{6}{5} & -\dfrac{7}{5} \end{bmatrix} \begin{bmatrix} u_1^* \\ u_2^* \end{bmatrix}
\tag{1.13}
$$

and to the final solution

$$
\begin{bmatrix} u_1^* \\ u_2^* \end{bmatrix} = \begin{bmatrix} 7.5 \\ 5 \end{bmatrix}\text{ V}.
\tag{1.14}
$$

In the above example, although applying $\mathbf{u} = 0$ was convenient to determine $\mathbf{y}^0$, the zero-input response, it was certainly not necessary and any three vector inputs could have been applied and the corresponding outputs measured. This would give us the six equations necessary to determine all the unknowns in Equation (1.7). The solution of the above two examples is readily generalized to the assignment of an m-dimensional vector output (m scalar outputs) using an r-dimensional vector input (r scalar inputs). In general, $r + 1$ experiments must be conducted by setting the input r-vector to $r + 1$ different values and measuring the corresponding m-dimensional output vector $r + 1$ times. From this data, $m(r+1)$ measurement-based equations may be formed and solved for the same number of unknowns. When less than m inputs are available ($r < m$), it is easy to see that arbitrary assignment is impossible. When more than m inputs are available ($r > m$), the output can in general be assigned arbitrarily and one may even find a $\mathbf{u}^*$ of minimum norm assigning $\mathbf{y}^*$. In general, the set of achievable outputs is of the form $\mathbf{y}^0 + \mathcal{G}$, where $\mathcal{G}$ is the subspace spanned by the columns of the matrix $\mathbf{G}$.

1.2   Current and Voltage Assignment Using Resistors in DC Circuits

In this section we consider a different problem, namely the control of current or voltage in a DC circuit using resistors. Consider an unknown DC circuit shown in Figure

1.2, consisting of an arbitrary connection of m resistors R_i and independent current or voltage sources $u_1, u_2, \ldots, u_r$. Let I_i and V_i denote the current through and voltage across $R_i, i \in \mathbf{m} := [1, 2, \ldots, m]$. The system is described by two sets of equations, one structural and the other parametric.

The structural equations are, from Kirchoff's laws:

$$\sum (\text{currents at a junction}) = 0,$$
$$\sum (\text{voltages around each loop}) = 0.$$

(1.15)

The parametric equations are, from Ohm's law:

$$V_i = R_i I_i, \quad \text{for } i \in \mathbf{m} = \{1, 2, \ldots, m\}.$$

(1.16)

There are m equations in each set and thus Equations (1.15) and (1.16) constitute $2m$ equations in the $2m$ variables V_i, I_i for $i \in \mathbf{m}$. Define

$$J := \begin{bmatrix} I_1 \\ I_2 \\ \vdots \\ I_m \end{bmatrix}, \quad V := \begin{bmatrix} V_1 \\ V_2 \\ \vdots \\ V_m \end{bmatrix}, \quad \mathbf{u} := \begin{bmatrix} u_1 \\ u_2 \\ \vdots \\ u_r \end{bmatrix}, \quad \mathbf{R} := \begin{bmatrix} R_1 & & \\ & \ddots & \\ & & R_m \end{bmatrix}$$

(1.17)

and write Equation (1.15) as

$$\mathbf{S}_1 J + \mathbf{S}_2 V = B_1 \mathbf{u}$$

(1.18)

and Equation (1.16) as

$$V = \mathbf{R} J.$$

(1.19)

Therefore, letting $\mathbf{I}$ denote the $m \times m$ identity matrix, we have

$$\underbrace{\begin{bmatrix} \mathbf{R} & -\mathbf{I} \\ \mathbf{S}_1 & \mathbf{S}_2 \end{bmatrix}}_{\mathbf{A}} \underbrace{\begin{bmatrix} J \\ V \end{bmatrix}}_{\mathbf{x}} = \underbrace{\begin{bmatrix} 0 \\ B_1 \end{bmatrix}}_{\mathbf{b}} \mathbf{u} = \mathbf{b}.$$

(1.20)

EXAMPLE 1.3   Consider the DC circuit in Figure 1.3.

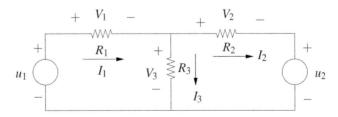

Figure 1.3 A DC circuit (Example 1.3)

The structural equations are

$$I_1 = I_2 + I_3, \quad u_1 = V_1 + V_3, \quad V_3 = V_2 + u_2$$

(1.21)

and the parametric equations are

$$V_i = R_i I_i, \quad \text{for } i \in \mathbf{3} := \{1, 2, 3\}. \tag{1.22}$$

These can be organized as

$$\underbrace{\begin{bmatrix} R_1 & 0 & 0 & -1 & 0 & 0 \\ 0 & R_2 & 0 & 0 & -1 & 0 \\ 0 & 0 & R_3 & 0 & 0 & -1 \\ 1 & -1 & -1 & 0 & 0 & 0 \\ 0 & 0 & 0 & 1 & 0 & 1 \\ 0 & 0 & 0 & 0 & -1 & 1 \end{bmatrix}}_{\mathbf{A}} \underbrace{\begin{bmatrix} I_1 \\ I_2 \\ I_3 \\ V_1 \\ V_2 \\ V_3 \end{bmatrix}}_{\mathbf{x}} = \underbrace{\begin{bmatrix} 0 & 0 \\ 0 & 0 \\ 0 & 0 \\ 0 & 0 \\ 1 & 0 \\ 0 & 1 \end{bmatrix}}_{\mathbf{B}} \underbrace{\begin{bmatrix} u_1 \\ u_2 \end{bmatrix}}_{\mathbf{u}} = \underbrace{\begin{bmatrix} 0 \\ 0 \\ 0 \\ 0 \\ u_1 \\ u_2 \end{bmatrix}}_{\mathbf{b}}. \tag{1.23}$$

To proceed, we assume that $\mathbf{u}$ is fixed since our focus is now on control using components of the parameter vector and not the inputs. Introduce the parameter vector

$$\mathbf{p} = \begin{bmatrix} p_1 \\ p_2 \\ \vdots \\ p_m \end{bmatrix} := \begin{bmatrix} R_1 \\ R_2 \\ \vdots \\ R_m \end{bmatrix} \tag{1.24}$$

and note that the matrix $\mathbf{A}$ depends on $\mathbf{p}$. This is denoted explicitly by writing

$$\mathbf{A} := \mathbf{A}(\mathbf{p}). \tag{1.25}$$

Solving for the ith component, x_i of $\mathbf{x}$ from Equation (1.20), using Cramer's rule (see Appendix A.1), we have

$$x_i = \frac{\det[\mathbf{T}_i(\mathbf{p})]}{\det[\mathbf{A}(\mathbf{p})]}, \quad \text{for } i \in \mathbf{m}, \tag{1.26}$$

where $\mathbf{T}_i(\mathbf{p})$ is the matrix obtained by replacing the ith column of $\mathbf{A}(\mathbf{p})$ by $\mathbf{b}$.

The solution for x_i given above can be generalized to the solution of any scalar output y_i of the form

$$y_i = \mathbf{c}_i^T \mathbf{x} + \underbrace{\mathbf{d}_i^T \mathbf{u}}_{d}. \tag{1.27}$$

Write the system equations, assuming $\mathbf{p}$ is fixed, as

$$\mathbf{A}\mathbf{x} = \mathbf{B}\mathbf{u} = \mathbf{b}_1 u_1 + \cdots + \mathbf{b}_r u_r =: \mathbf{b}. \tag{1.28}$$

Let

$$\mathbf{z} := \begin{bmatrix} \mathbf{x} \\ y_i \end{bmatrix} \tag{1.29}$$

and write Equations (1.27) and (1.28) as

$$\begin{bmatrix} \mathbf{A} & 0 \\ -\mathbf{c}_i^T & 1 \end{bmatrix} \mathbf{z} = \begin{bmatrix} \mathbf{b} \\ d \end{bmatrix} = \begin{bmatrix} \mathbf{b}_1 \\ d_{i1} \end{bmatrix} u_1 + \begin{bmatrix} \mathbf{b}_2 \\ d_{i2} \end{bmatrix} u_2 + \cdots + \begin{bmatrix} \mathbf{b}_r \\ d_{ir} \end{bmatrix} u_r. \qquad (1.30)$$

Then using Cramer's rule on Equation (1.30), and using Equations (1.27) and (1.28),

$$y_i = \frac{\det \begin{bmatrix} \mathbf{A} & \mathbf{b} \\ -\mathbf{c}_i^T & d \end{bmatrix}}{\det[\mathbf{A}]} \qquad (1.31)$$

$$= \underbrace{\frac{\det \begin{bmatrix} \mathbf{A} & \mathbf{b}_1 \\ -\mathbf{c}_i^T & d_{i1} \end{bmatrix}}{\det[\mathbf{A}]}}_{G_{i1}} u_1 + \underbrace{\frac{\det \begin{bmatrix} \mathbf{A} & \mathbf{b}_2 \\ -\mathbf{c}_i^T & d_{i2} \end{bmatrix}}{\det[\mathbf{A}]}}_{G_{i2}} u_2 + \cdots + \underbrace{\frac{\det \begin{bmatrix} \mathbf{A} & \mathbf{b}_r \\ -\mathbf{c}_i^T & d_{ir} \end{bmatrix}}{\det[\mathbf{A}]}}_{G_{ir}} u_r$$

$$= G_{i1} u_1 + G_{i2} u_2 + \cdots + G_{ir} u_r.$$

Equation (1.31) shows that the "transfer function" or "gain" between output i and input j with all other inputs set to zero is

$$G_{ij} = \frac{\det \begin{bmatrix} \mathbf{A} & \mathbf{b}_j \\ -\mathbf{c}_i^T & d_{ij} \end{bmatrix}}{\det[\mathbf{A}]}. \qquad (1.32)$$

Equation (1.31) is also a statement of the **Principle of Superposition**, applicable to linear systems, which asserts that the output resulting from all inputs being active simultaneously is the sum of the outputs in response to each input acting individually with all other inputs switched off.

Remark 1.1   We note that the Principle of Superposition applies only to linear systems where the zero-input response is zero.

In the next section we consider the following two synthesis problems in DC circuits:

(A) Control current I_i by adjusting resistor R_j, for $j \neq i$.
(B) Control current I_i by adjusting resistor R_i.

1.2.1   Current Control Using a Single Resistor

Consider the unknown linear DC circuit shown in Figure 1.4. Suppose that we wish to control the current in the ith branch, denoted by I_i, by adjusting the resistor R_j at an arbitrary location of the circuit.

The following lemma will be central to the development of our solution.

LEMMA 1.1

(a) $\det[\mathbf{A}(\mathbf{p})]$ is a multilinear polynomial in p_i, for $i \in \mathbf{m}$.
(b) $\det[\mathbf{T}_i(\mathbf{p})]$ is a multilinear polynomial in p_j, for $j \in \mathbf{m}$ and $j \neq i$.

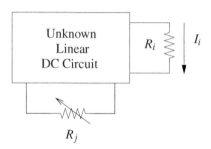

Figure 1.4 An unknown linear DC circuit

Proof   The statement of the lemma follows from the fact that

$$\mathbf{A}(\mathbf{p}) = \mathbf{A}_0 + p_1\mathbf{A}_1 + \cdots + p_m\mathbf{A}_m \tag{1.33}$$

and since $\mathbf{R}$ is diagonal, it follows from Equation (1.20) that

$$\text{rank}[\mathbf{A}_i] = 1, \quad i \in \mathbf{m}. \tag{1.34}$$

Therefore, $\det[\mathbf{A}(\mathbf{p})]$ is a polynomial of degree at most one in each $p_i, i \in \mathbf{m}$. This proves (a). To prove (b), note that the ith component of $\mathbf{p}$ is absent from $\mathbf{T}_i(\mathbf{p})$ and therefore $\det[\mathbf{T}_i(\mathbf{p})]$ is a multilinear polynomial in only the remaining components of $\mathbf{p}$. ∎

Following Example 1.3, in general, the resistor R_j will appear in the matrix $\mathbf{A}$ in Equation (1.23) with coefficient matrix of rank 1. This is because R_j is associated only with the branch current I_j and will therefore only appear in column j. In a later example, we will consider a gyrator resistance, in which case the parameter dependency is of rank 2. The rank 1 case however occurs generically and is assumed unless stated explicitly otherwise.

From the above observations we have the following result which is an extension of *Thevenin's Theorem* of classical circuit theory.

THEOREM 1.1   In a linear DC circuit, the functional dependency of any current I_i on any resistance R_j (which appears with rank 1 dependency) can be determined by at most three measurements of the current I_i obtained for three different settings of R_j.

Proof   Let us consider the two cases $i \neq j$ and $i = j$ separately.
Case 1: ($i \neq j$)   In this case, the matrices $\mathbf{T}_i(\mathbf{p})$ and $\mathbf{A}(\mathbf{p})$, in Equation (1.26), are both of rank 1 with respect to R_j. According to Lemma 1.1, the functional dependency of I_i on R_j can be expressed as

$$I_i(R_j) = \frac{\tilde{\alpha}_0 + \tilde{\alpha}_1 R_j}{\tilde{\beta}_0 + \tilde{\beta}_1 R_j}, \tag{1.35}$$

where $\tilde{\alpha}_0, \tilde{\alpha}_1, \tilde{\beta}_0, \tilde{\beta}_1$ are constants. If $\tilde{\beta}_0 = \tilde{\beta}_1 = 0$, then, I_i goes to ∞, for any value of the resistance R_j, which is physically meaningless as it implies a short circuit at the output terminals. Therefore, we rule out this case. Assuming that $\tilde{\beta}_1 \neq 0$, one can divide the numerator and denominator of Equation (1.35) by $\tilde{\beta}_1$ and obtain

$$I_i(R_j) = \frac{\alpha_0 + \alpha_1 R_j}{\beta_0 + R_j},$$ (1.36)

where $\alpha_0, \alpha_1, \beta_0$ are constants. In order to determine $\alpha_0, \alpha_1, \beta_0$ one conducts three experiments by setting three different values for the resistance R_j, namely R_{j1}, R_{j2}, R_{j3}, and measuring the corresponding currents I_i, namely I_{i1}, I_{i2}, I_{i3}. Then, the following set of measurement equations can be formed:

$$\underbrace{\begin{bmatrix} 1 & R_{j1} & -I_{i1} \\ 1 & R_{j2} & -I_{i2} \\ 1 & R_{j3} & -I_{i3} \end{bmatrix}}_{\mathbf{M}} \underbrace{\begin{bmatrix} \alpha_0 \\ \alpha_1 \\ \beta_0 \end{bmatrix}}_{\mathbf{t}} = \underbrace{\begin{bmatrix} I_{i1} R_{j1} \\ I_{i2} R_{j2} \\ I_{i3} R_{j3} \end{bmatrix}}_{\mathbf{m}}.$$ (1.37)

The set of equations in Equation (1.37) can be uniquely solved for the constants $\alpha_0, \alpha_1, \beta_0$, if and only if $\det[\mathbf{M}] \neq 0$. If $\det[\mathbf{M}] = 0$, the last column of the matrix $\mathbf{M}$ can be expressed as a linear combination of the first two columns because by assigning different values to the resistance R_j, the first two columns of $\mathbf{M}$ can be forced to become linearly independent. In such a case, the functional dependency of I_i on R_j can be expressed as

$$I_i(R_j) = \alpha_0 + \alpha_1 R_j,$$ (1.38)

where α_0, α_1 are constants that can be determined from any two of the experiments conducted earlier. The functional dependency in Equation (1.38) corresponds to the case where $\tilde{\beta}_1 = 0$ in Equation (1.35), and the numerator and denominator of Equation (1.35) are divided by $\tilde{\beta}_0$.

Case 2: ($i = j$) In this case, the matrix $\mathbf{A}(\mathbf{p})$ is of rank 1 with respect to R_i; however, the matrix $\mathbf{T}_i(\mathbf{p})$ is of rank 0 with respect to R_i. According to Lemma 1.1, the functional dependency of I_i on R_i can therefore be expressed as

$$I_i(R_i) = \frac{\tilde{\alpha}_0}{\tilde{\beta}_0 + \tilde{\beta}_1 R_i},$$ (1.39)

where $\tilde{\alpha}_0, \tilde{\beta}_0, \tilde{\beta}_1$ are constants. Assuming that $\tilde{\beta}_1 \neq 0$, and dividing the numerator and denominator of Equation (1.39) by $\tilde{\beta}_1$, we get

$$I_i(R_i) = \frac{\alpha_0}{\beta_0 + R_i},$$ (1.40)

where α_0, β_0 are constants that can be determined by conducting two experiments, by setting two different values for the resistance R_i, namely R_{i1}, R_{i2}, and measuring the corresponding currents I_i, namely I_{i1}, I_{i2}. The following set of measurement equations can then be formed:

$$\underbrace{\begin{bmatrix} 1 & -I_{i1} \\ 1 & -I_{i2} \end{bmatrix}}_{\mathbf{M}} \underbrace{\begin{bmatrix} \alpha_0 \\ \beta_0 \end{bmatrix}}_{\mathbf{t}} = \underbrace{\begin{bmatrix} I_{i1} R_{i1} \\ I_{i2} R_{i2} \end{bmatrix}}_{\mathbf{m}}.$$ (1.41)

The set of equations in Equation (1.41) can be uniquely solved for the constants α_0, β_0, provided $\det[\mathbf{M}] \neq 0$. If $\det[\mathbf{M}] = 0$ in Equation (1.41), it can be concluded that I_i is a constant:

$$I_i(R_i) = \alpha_0, \tag{1.42}$$

which corresponds to a current source in branch i in series with R_i, and can be determined from any of the experiments conducted earlier. In this case, the functional dependency in Equation (1.42) corresponds to the situation where $\tilde{\beta}_1 = 0$ in Equation (1.39), and the numerator and denominator of Equation (1.39) are divided by $\tilde{\beta}_0$. ∎

Remark 1.2   Suppose that $i \neq j$ and $\det[\mathbf{M}] \neq 0$ in Equation (1.37), then the derivative of $I_i(R_j)$ in Equation (1.36), with respect to R_j, can be calculated as

$$\frac{dI_i}{dR_j} = \frac{\alpha_1\beta_0 - \alpha_0}{(\beta_0 + R_j)^2}. \tag{1.43}$$

If $\beta_0 \geq 0$, we have the following:

1. The function in Equation (1.36) is monotonic in R_j. That is, $I_i(R_j)$ monotonically increases or decreases as R_j increases from 0 to large values. The limiting values of this function are: $I_i(0) = \frac{\alpha_0}{\beta_0}$ and $I_i(\infty) = \alpha_1$. If $\frac{\alpha_0}{\beta_0} > \alpha_1$, then Equation (1.36) will monotonically decrease, and if $\frac{\alpha_0}{\beta_0} < \alpha_1$, then Equation (1.36) will monotonically increase.

2. The range of I_i achievable by varying R_j in the interval $[0, \infty)$ is

$$\min\left\{\frac{\alpha_0}{\beta_0}, \alpha_1\right\} < I_i < \max\left\{\frac{\alpha_0}{\beta_0}, \alpha_1\right\}. \tag{1.44}$$

3. In a current control problem of this type, this monotonic behavior allows us to uniquely determine a range of values of the design parameter R_j, for $R_j^- \leq R_j \leq R_j^+$, for which the current I_i lies within a desired prescribed range, $I_i^- \leq I_i \leq I_i^+$, which of course must be within the achievable range in Equation (1.44).

These observations are also clear from the graph of Equation (1.36). For instance, if $\beta_0 > 0$, $\alpha_0 < 0$, and $\alpha_1 > 0$, the graph of Equation (1.36) has the general shape depicted in Figure 1.5.

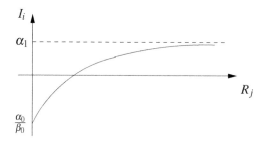

Figure 1.5 Graph of Equation (1.36) for $\beta_0 > 0$, $\alpha_0 < 0$, and $\alpha_1 > 0$

If $\beta_0 < 0$, we have

1. The function in Equation (1.36) is monotonic in R_j, in the intervals $[0, -\beta_0)$ and $(-\beta_0, \infty)$. If $\alpha_1 \beta_0 - \alpha_0 > 0$, then I_i starts at $\frac{\alpha_0}{\beta_0}$ and monotonically increases to $+\infty$ as R_j goes to $-\beta_0^-$, then as R_j goes to $-\beta_0^+$, it starts from $-\infty$ and monotonically increases to α_1 as R_j goes to ∞. If $\alpha_1 \beta_0 - \alpha_0 < 0$, I_i starts at $\frac{\alpha_0}{\beta_0}$ and monotonically decreases to $-\infty$ as R_j goes to $-\beta_0^-$, then, as R_j goes to $-\beta_0^+$, it starts from $+\infty$ and monotonically decreases to α_1 as R_j goes to ∞.
2. The range for I_i, achievable by varying R_j in the interval $[0, \infty)$, is

$$I_i \in \left(-\infty, \min\left\{ \frac{\alpha_0}{\beta_0}, \alpha_1 \right\} \right) \cup \left(\max\left\{ \frac{\alpha_0}{\beta_0}, \alpha_1 \right\}, +\infty \right). \tag{1.45}$$

3. Similarly, one can uniquely determine a range of values of the design parameter R_j, for $R_j^- \leq R_j \leq R_j^+$, for which the current I_i lies within a desired prescribed range, $I_i^- \leq I_i \leq I_i^+$, which of course must be within the achievable range in Equation (1.45).

The graph of Equation (1.36) for this case is shown in Figure 1.6.

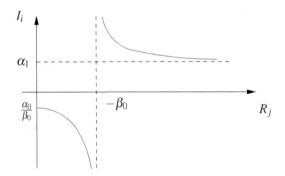

Figure 1.6 Graph of Equation (1.36) for $\beta_0 < 0$, $\alpha_0 > 0$, and $\alpha_1 > 0$

Thevenin's Theorem (the special case $i = j$)

Thevenin's Theorem of circuit theory follows as a special case of the results developed here. To see this, consider the current functional dependency given in Equation (1.40). From this relationship, it is clear that the short-circuit current I_{sc} is given by $I_{sc} = \frac{\alpha_0}{\beta_0}$, which is obtained by setting $R_i = 0$. Similarly, the open-circuit voltage V_{oc} is obtained by multiplying both sides of Equation (1.40) by R_i and taking the limit as R_i goes to ∞. This yields $V_{oc} = V_{Th} = \alpha_0$. Thus, the Thevenin resistance is given by $R_{Th} = \frac{V_{oc}}{I_{sc}} = \beta_0$, so that Equation (1.40) becomes

$$I_i(R_i) = \frac{V_{Th}}{R_{Th} + R_i}, \tag{1.46}$$

which is exactly Thevenin's Theorem. We point out that in our approach, it is *not* necessary to measure short-circuit current or open-circuit voltage; indeed, two *arbitrary* measurements suffice. This has practical and useful implications in circuits where short-circuiting and open-circuiting may sometimes be impossible.

Current Assignment Problem

Once the functional dependency of interest, Equations (1.36), (1.38), (1.40), or (1.42) is obtained, a synthesis problem can be solved. For instance, suppose that it is required to assign $I_i = I_i^*$, where I_i^* is a desired prescribed value of the current in the ith branch of the unknown circuit. Let us assume that the design variable is the resistance R_j, and $i \neq j$. How can one find a value of R_j for which $I_i = I_i^*$? Based on Theorem 1.1, since $i \neq j$, one conducts three experiments by setting three different values for the resistance R_j, and measuring the corresponding currents I_i. The matrix $\mathbf{M}$ in Equation (1.37) can then be evaluated from the measurements. If $\det[\mathbf{M}] \neq 0$, then the functional dependency of interest will be of the form given in Equation (1.36), and if $\det[\mathbf{M}] = 0$, it will be of the form obtained in Equation (1.38). Suppose that $\det[\mathbf{M}] \neq 0$ is the case; hence, the functional dependency of interest is as given in Equation (1.36). In order to determine the value of R_j, for which the desired current I_i^* is attained, one may solve Equation (1.36) for R_j, with $I_i = I_i^*$:

$$R_j(I_i^*) = \frac{\alpha_0 - I_i^* \beta_0}{I_i^* - \alpha_1}. \tag{1.47}$$

Interval Design Problem

Suppose now that the current I_i is to be controlled to stay within the following range (which is inside the achievable range in Equation (1.44)):

$$I_i^- \leq I_i \leq I_i^+, \tag{1.48}$$

by adjusting the design resistance R_j, $i \neq j$. Also, assume that after conducting three experiments, we found $\det[\mathbf{M}] \neq 0$ in Equation (1.37) and $\beta_0 \geq 0$. Therefore, the functional dependency of I_i on R_j is of the form in Equation (1.36) and is monotonic. Thus, one may find a unique corresponding interval for R_j values where Equation (1.48) is satisfied. Supposing that I_i, in Equation (1.36), monotonically increases as R_j increases, one gets

$$R_j^- \leq R_j \leq R_j^+, \tag{1.49}$$

where

$$R_j^- = \frac{\alpha_0 - I_i^- \beta_0}{I_i^- - \alpha_1}, \qquad R_j^+ = \frac{\alpha_0 - I_i^+ \beta_0}{I_i^+ - \alpha_1}. \tag{1.50}$$

If I_i, in Equation (1.36), monotonically decreases as R_j increases, then R_j^- and R_j^+ in Equation (1.49) can be calculated from

$$R_j^- = \frac{\alpha_0 - I_i^+ \beta_0}{I_i^+ - \alpha_1}, \qquad R_j^+ = \frac{\alpha_0 - I_i^- \beta_0}{I_i^- - \alpha_1}. \tag{1.51}$$

Following the same strategy, one may solve a synthesis problem for the case $i = j$. The problem of maintaining several currents in the circuit within prescribed intervals can be solved similarly.

Remark 1.3 (Generalization of Thevenin's Theorem)   Theorem 1.1 and the subsequent results in this chapter represent generalizations of Thevenin's Theorem. In Thevenin's Theorem, the current in a resistor/impedance connected to an arbitrary network can be obtained by representing the network by a voltage source in series with a resistance/impedance and these can be determined from short-circuit and open-circuit measurements made at these terminals. We have shown (1) that the resistor can be connected at a point different from the point where measurements are made and (2) that the current can be predicted from arbitrary measurements, not necessarily from short-circuiting or open-circuiting the system. The results given in the subsequent sections can be thought of as Thevenin-like results for mechanical, hydraulic, truss, and other systems. Note that in the circuit example above, the Thevenin resistance which equals β_0 may be negative in some cases, leading to the discontinuous behavior of the current at the value of $R_i = R_{Th}$. This atypical situation will not arise in regular cases.

1.2.2   Control Using Two Resistors

Consider the unknown linear DC circuit shown in Figure 1.7.

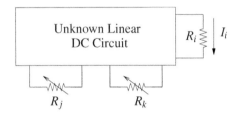

Figure 1.7 An unknown linear DC circuit

Suppose we want to control the current in the ith branch, denoted by I_i, by adjusting any two resistors R_j and R_k at arbitrary locations of the circuit. Assume, as before, that R_j and R_k are not gyrator resistances.

THEOREM 1.2   In a linear DC circuit, the functional dependency of any current I_i on any two resistances R_j and R_k (which appear with rank 1 dependency) can be determined by at most seven measurements of the current I_i obtained for seven different sets of values (R_j, R_k).

Proof   Let us consider the two cases $i \neq j, k$ and $i = j$ or $i = k$.
Case 1: $(i \neq j, k)$ In this case, the matrices $\mathbf{T}_i(\mathbf{p})$ and $\mathbf{A}(\mathbf{p})$, in Equation (1.26), are both of rank 1 with respect to R_j and R_k. Based on Lemma 1.1, the functional dependency of I_i on R_j and R_k can be expressed as

$$I_i(R_j, R_k) = \frac{\tilde{\alpha}_0 + \tilde{\alpha}_1 R_j + \tilde{\alpha}_2 R_k + \tilde{\alpha}_3 R_j R_k}{\tilde{\beta}_0 + \tilde{\beta}_1 R_j + \tilde{\beta}_2 R_k + \tilde{\beta}_3 R_j R_k}, \tag{1.52}$$

where $\tilde{\alpha}_0, \tilde{\alpha}_1, \tilde{\alpha}_2, \tilde{\alpha}_3, \tilde{\beta}_0, \tilde{\beta}_1, \tilde{\beta}_2, \tilde{\beta}_3$ are constants. Assuming that $\tilde{\beta}_3 \neq 0$ and dividing the numerator and denominator of Equation (1.52) by $\tilde{\beta}_3$, yields

$$I_i(R_j, R_k) = \frac{\alpha_0 + \alpha_1 R_j + \alpha_2 R_k + \alpha_3 R_j R_k}{\beta_0 + \beta_1 R_j + \beta_2 R_k + R_j R_k}, \tag{1.53}$$

where $\alpha_0, \alpha_1, \alpha_2, \alpha_3, \beta_0, \beta_1, \beta_2$ are constants. In order to determine these constants, one conducts seven experiments by assigning seven different sets of values to the resistances (R_j, R_k), and measuring the corresponding currents I_i. The following set of measurement equations will be obtained:

$$\underbrace{\begin{bmatrix} 1 & R_{j1} & R_{k1} & R_{j1}R_{k1} & -I_{i1} & -I_{i1}R_{j1} & -I_{i1}R_{k1} \\ 1 & R_{j2} & R_{k2} & R_{j2}R_{k2} & -I_{i2} & -I_{i2}R_{j2} & -I_{i2}R_{k2} \\ 1 & R_{j3} & R_{k3} & R_{j3}R_{k3} & -I_{i3} & -I_{i3}R_{j3} & -I_{i3}R_{k3} \\ 1 & R_{j4} & R_{k4} & R_{j4}R_{k4} & -I_{i4} & -I_{i4}R_{j4} & -I_{i4}R_{k4} \\ 1 & R_{j5} & R_{k5} & R_{j5}R_{k5} & -I_{i5} & -I_{i5}R_{j5} & -I_{i5}R_{k5} \\ 1 & R_{j6} & R_{k6} & R_{j6}R_{k6} & -I_{i6} & -I_{i6}R_{j6} & -I_{i6}R_{k6} \\ 1 & R_{j7} & R_{k7} & R_{j7}R_{k7} & -I_{i7} & -I_{i7}R_{j7} & -I_{i7}R_{k7} \end{bmatrix}}_{\mathbf{M}} \underbrace{\begin{bmatrix} \alpha_0 \\ \alpha_1 \\ \alpha_2 \\ \alpha_3 \\ \beta_0 \\ \beta_1 \\ \beta_2 \end{bmatrix}}_{\mathbf{t}} = \underbrace{\begin{bmatrix} I_{i1}R_{j1}R_{k1} \\ I_{i2}R_{j2}R_{k2} \\ I_{i3}R_{j3}R_{k3} \\ I_{i4}R_{j4}R_{k4} \\ I_{i5}R_{j5}R_{k5} \\ I_{i6}R_{j6}R_{k6} \\ I_{i7}R_{j7}R_{k7} \end{bmatrix}}_{\mathbf{m}}.$$

$$\tag{1.54}$$

This set of equations can be uniquely solved for the constants α_0, α_1, α_2, α_3, β_0, β_1, β_2, if and only if $\det[\mathbf{M}] \neq 0$ in Equation (1.54). In the case where $\det[\mathbf{M}] = 0$, one can follow the same procedure used in Section 1.2.1 to derive the corresponding functional dependency of I_i on R_j and R_k.

Case 2: ($i = j$ or $i = k$) Suppose that $i = j$ and recall Equation (1.26). In this case, the matrix $\mathbf{A}(\mathbf{p})$ is of rank 1 with respect to R_i and R_k; however, the matrix $\mathbf{T}_i(\mathbf{p})$ is of rank 0 with respect to R_i and is of rank 1 with respect to R_k. According to Lemma 1.1 and these rank conditions, the functional dependency of I_i on R_i and R_k can thus be expressed as

$$I_i(R_i, R_k) = \frac{\tilde{\alpha}_0 + \tilde{\alpha}_1 R_k}{\tilde{\beta}_0 + \tilde{\beta}_1 R_i + \tilde{\beta}_2 R_k + \tilde{\beta}_3 R_i R_k}, \tag{1.55}$$

where $\tilde{\alpha}_0, \tilde{\alpha}_1, \tilde{\beta}_0, \tilde{\beta}_1, \tilde{\beta}_2, \tilde{\beta}_3$ are constants. Assuming that $\tilde{\beta}_3 \neq 0$, one can divide the numerator and denominator of Equation (1.55) by $\tilde{\beta}_3$ and obtain

$$I_i(R_i, R_k) = \frac{\alpha_0 + \alpha_1 R_k}{\beta_0 + \beta_1 R_i + \beta_2 R_k + R_i R_k}, \tag{1.56}$$

where $\alpha_0, \alpha_1, \beta_0, \beta_1, \beta_2$ are constants that can be determined by conducting five experiments, by assigning five different sets of values to the resistances (R_i, R_k), and measuring the corresponding currents I_i. The following set of measurement equations can then be formed:

$$
\underbrace{\begin{bmatrix}
1 & R_{k1} & -I_{i1} & -I_{i1}R_{j1} & -I_{i1}R_{k1} \\
1 & R_{k2} & -I_{i2} & -I_{i2}R_{j2} & -I_{i2}R_{k2} \\
1 & R_{k3} & -I_{i3} & -I_{i3}R_{j3} & -I_{i3}R_{k3} \\
1 & R_{k4} & -I_{i4} & -I_{i4}R_{j4} & -I_{i4}R_{k4} \\
1 & R_{k5} & -I_{i5} & -I_{i5}R_{j5} & -I_{i5}R_{k5}
\end{bmatrix}}_{\mathbf{M}}
\underbrace{\begin{bmatrix}
\alpha_0 \\ \alpha_1 \\ \beta_0 \\ \beta_1 \\ \beta_2
\end{bmatrix}}_{\mathbf{u}}
=
\underbrace{\begin{bmatrix}
I_{i1}R_{j1}R_{k1} \\
I_{i2}R_{j2}R_{k2} \\
I_{i3}R_{j3}R_{k3} \\
I_{i4}R_{j4}R_{k4} \\
I_{i5}R_{j5}R_{k5}
\end{bmatrix}}_{\mathbf{m}}.
\tag{1.57}
$$

Again, this set of equations can be uniquely solved for the constants $\alpha_0, \alpha_1, \beta_0, \beta_1, \beta_2$, provided $\det[\mathbf{M}] \neq 0$ in Equation (1.57). If $\det[\mathbf{M}] = 0$, following the same strategy used in Section 1.2.1, one can derive the corresponding functional dependency of I_i on R_i and R_k. ∎

In this problem, the current I_i can be plotted as a surface in a 3D graph. In a synthesis problem of this type, any constraint on the current I_i results in a corresponding region in the R_j, R_k plane, if the solution set for that constraint is not empty.

1.2.3   Control Using m Resistors

As a generalization of the above results, consider the unknown linear DC circuit shown in Figure 1.8.

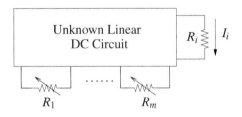

Figure 1.8 An unknown linear DC circuit

Suppose that the objective is to control the current in the ith branch of the circuit, denoted by I_i, by adjusting the m resistors R_j, $j = 1, 2, \dots, m$, at arbitrary locations of the circuit. Assume that the resistances R_j, $j = 1, 2, \dots, m$, are not gyrator resistances.

THEOREM 1.3   In a linear DC circuit, the functional dependency of any current I_i on any m resistances R_j, $j = 1, 2, \dots, m$, can be determined by at most $2^{m+1} - 1$ measurements of the current I_i obtained for $2^{m+1} - 1$ different sets of values of the vector $(R_1, R_2, \dots, R_m)$.

Proof   Let us consider the two cases $i \neq j$ for $j = 1, 2, \dots, m$ and $i = j$ for some $j = 1, 2, \dots, m$.

Case 1: ($i \neq j$, $j = 1, 2, \dots, m$)   In this case, the matrices $\mathbf{T}_i(\mathbf{p})$ and $\mathbf{A}(\mathbf{p})$, in Equation (1.26), are both of rank 1 with respect to R_j, $j = 1, 2, \dots, m$. Hence, based on Lemma

1.1, the functional dependency of I_i on $R_1, R_2, \ldots, R_m$ can be written as

$$I_i(R_1, R_2, \ldots, R_m) = \frac{\sum_{i_m=0}^{1} \cdots \sum_{i_2=0}^{1} \sum_{i_1=0}^{1} \tilde{\alpha}_{i_1 i_2 \cdots i_m} R_1^{i_1} R_2^{i_2} \cdots R_m^{i_m}}{\sum_{i_m=0}^{1} \cdots \sum_{i_2=0}^{1} \sum_{i_1=0}^{1} \tilde{\beta}_{i_1 i_2 \cdots i_m} R_1^{i_1} R_2^{i_2} \cdots R_m^{i_m}}, \tag{1.58}$$

where $\tilde{\alpha}_{i_1 i_2 \cdots i_m}$ and $\tilde{\beta}_{i_1 i_2 \cdots i_m}$ are constants. Assuming that $\tilde{\beta}_{11 \cdots 1} \neq 0$ and dividing the numerator and denominator of Equation (1.58) by $\tilde{\beta}_{11 \cdots 1}$, results in

$$I_i(R_1, R_2, \ldots, R_m) = \frac{\sum_{i_m=0}^{1} \cdots \sum_{i_2=0}^{1} \sum_{i_1=0}^{1} \alpha_{i_1 i_2 \cdots i_m} R_1^{i_1} R_2^{i_2} \cdots R_m^{i_m}}{\sum_{i_m=0}^{1} \cdots \sum_{i_2=0}^{1} \sum_{i_1=0}^{1} \beta_{i_1 i_2 \cdots i_m} R_1^{i_1} R_2^{i_2} \cdots R_m^{i_m}}, \tag{1.59}$$

where $\beta_{11 \cdots 1} = 1$, and $\alpha_{i_1 i_2 \cdots i_m}$ and $\beta_{i_1 i_2 \cdots i_m}$ are $2^{m+1} - 1$ constants. In order to determine these constants, one conducts $2^{m+1} - 1$ experiments, by setting $2^{m+1} - 1$ different sets of values for the resistances $(R_1, R_2, \ldots, R_m)$, and measuring the corresponding currents I_i. The obtained set of measurement equations has a unique solution for the constants if and only if $\det[\mathbf{M}] \neq 0$. If $\det[\mathbf{M}] = 0$ is the case, then one can follow the same procedure explained for the previous problems to derive the corresponding functional dependency.

Case 2: $(i = j$ for some $j = 1, 2, \ldots, m)$ Without loss of generality, suppose that $i = m$ and recall Equation (1.26). In this case, the matrix $\mathbf{A}(\mathbf{p})$ is of rank 1 with respect to R_j, $j = 1, 2, \ldots, m$; however, the matrix $\mathbf{T}_i(\mathbf{p})$ is of rank 0 with respect to R_m and is of rank 1 with respect to R_j, $j = 1, 2, \ldots, m-1$. According to these rank conditions and based on Lemma 1.1, the functional dependency of I_i on $R_1, R_2, \ldots, R_m$ will be

$$I_i(R_1, R_2, \ldots, R_m) = \frac{\sum_{i_{m-1}=0}^{1} \cdots \sum_{i_2=0}^{1} \sum_{i_1=0}^{1} \tilde{\alpha}_{i_1 i_2 \cdots i_{m-1}} R_1^{i_1} R_2^{i_2} \cdots R_{m-1}^{i_{m-1}}}{\sum_{i_m=0}^{1} \cdots \sum_{i_2=0}^{1} \sum_{i_1=0}^{1} \tilde{\beta}_{i_1 i_2 \cdots i_m} R_1^{i_1} R_2^{i_2} \cdots R_m^{i_m}}, \tag{1.60}$$

where $\tilde{\alpha}_{i_1 i_2 \cdots i_{m-1}}$ and $\tilde{\beta}_{i_1 i_2 \cdots i_m}$ are constants. Supposing $\tilde{\beta}_{11 \cdots 1} \neq 0$, one can divide the numerator and denominator of Equation (1.60) by $\tilde{\beta}_{11 \cdots 1}$ to get

$$I_i(R_1, R_2, \ldots, R_m) = \frac{\sum_{i_{m-1}=0}^{1} \cdots \sum_{i_2=0}^{1} \sum_{i_1=0}^{1} \alpha_{i_1 i_2 \cdots i_{m-1}} R_1^{i_1} R_2^{i_2} \cdots R_{m-1}^{i_{m-1}}}{\sum_{i_m=0}^{1} \cdots \sum_{i_2=0}^{1} \sum_{i_1=0}^{1} \beta_{i_1 i_2 \cdots i_m} R_1^{i_1} R_2^{i_2} \cdots R_m^{i_m}}, \tag{1.61}$$

where $\beta_{11 \cdots 1} = 1$, and the αs and βs are constants. These constants, say t in number, can be determined by conducting as many experiments, by assigning t different sets of values to the resistances $(R_1, R_2, \ldots, R_m)$, and measuring the corresponding currents I_i. The obtained system of measurement equations has a unique solution for the constants if and only if $\det[\mathbf{M}] \neq 0$. If $\det[\mathbf{M}] = 0$, following the same strategy presented for the previous problems, one can determine the corresponding functional dependency.   ∎

EXAMPLE 1.4 (A DC circuit example)   In this example, we show how the method proposed in Section 1.2.1 can be used toward control design problems in unknown linear DC circuits. Consider the unknown circuit shown in Figure 1.9.

In this example, it is desired to find the functional dependency of the current I_1 on the resistance R_9. Based on the results given in Section 1.2.1, one conducts three experiments, by setting three different values for R_9, and measuring the corresponding values

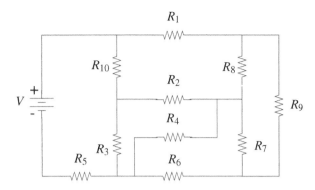

Figure 1.9 A DC circuit with unknown resistors (Example 1.4)

of currents I_1. Suppose that the experiments are done and let Table 1.2 summarize the numerical values, for this example, obtained from the three experiments.

Table 1.2 Numerical values of the measurements for the DC circuit (Example 1.4)

Exp. No.	R_9 (Ω)	I_1 (A)
1	1	0.054
2	5	0.056
3	10	0.058

Substituting the numerical values obtained from the experiments into the matrix **M** in Equation (1.37) resulted in $\det[\mathbf{M}] \neq 0$. Therefore, Equation (1.37) can be uniquely solved for the constants α_0, α_1, and β_0. This yields the following functional dependency which is plotted in Figure 1.10:

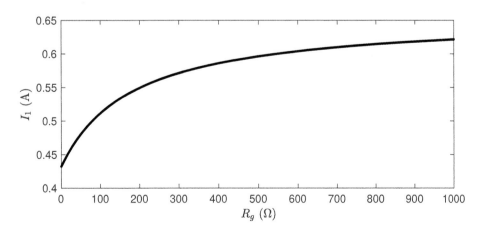

Figure 1.10 I_1 vs. R_9 (Example 1.4)

$$I_1(R_9) = \frac{78.4 + 0.66R_9}{181.3 + R_9}. \tag{1.62}$$

Remark 1.4

1. The current I_1 monotonically increases as R_9 increases.
2. By varying R_9 in the range $[0, \infty)$, the achievable range for I_1 becomes

$$\left[\frac{\alpha_0}{\beta_0}, \alpha_1\right] = [0.43, 0.66]. \tag{1.63}$$

3. In a synthesis problem where the current I_1 is to be controlled to stay within an acceptable interval, since I_1 is monotonic in R_9, one can find a corresponding interval for R_9 values for which the current I_1 stays within the acceptable range.

Suppose that we wish to design R_9 such that I_1 lies within the following achievable range:

$$0.5 \le I_1 \le 0.6 \text{ (A)}. \tag{1.64}$$

Using Equation (1.62), or Figure 1.10, the corresponding range for the design resistor R_9 can be obtained as

$$79 \le R_9 \le 550 \text{ (}\Omega\text{)}. \tag{1.65}$$

EXAMPLE 1.5   Consider the same circuit as above (Figure 1.9). Suppose now that the power levels within R_1, R_3, and R_9, denoted by P_1, P_3, and P_9, respectively, must remain in the following ranges:

$$6 \text{ (W)} \le P_1 \le 7 \text{ (W)}, \tag{1.66a}$$

$$7 \text{ (W)} \le P_3 \le 8 \text{ (W)}, \tag{1.66b}$$

$$3 \text{ (W)} \le P_9 \le 3.5 \text{ (W)}. \tag{1.66c}$$

Assume that the design resistor is R_9. One conducts three experiments by assigning three different values to R_9, and measuring the corresponding currents I_1, I_3, and I_9, passing through the resistors R_1, R_3, and R_9, respectively. In this problem, one also needs to measure the voltage across R_1 and R_3 from one of the experiments. Suppose that the experiments are done and let Table 1.3 summarize the numerical values for this example, obtained from the experiments.

Substituting the numerical values from Table 1.3 into the matrix $\mathbf{M}$ in Equation (1.37), for the currents I_1 and I_3, and into the matrix $\mathbf{M}$ in Equation (1.41), for the current I_9, yields $\det[\mathbf{M}] \ne 0$, for all cases. Therefore, the functional dependencies of

Table 1.3 Numerical values of the measurements for the DC circuit (Example 1.5)

Exp. No.	R_9 (Ω)	I_1 (A)	I_3 (A)	I_9 (A)
1	1	0.437	0.964	0.301
2	5	0.438	0.972	0.295
3	10	0.444	0.982	0.287

Exp. No.	R_9 (Ω)	V_1 (V)	V_3 (V)
1	1	8.67	4.82

P_1, P_3, and P_9 on R_9 will be

$$P_1(R_9) = \frac{8.67}{0.437}\left(\frac{78.4 + 0.66R_9}{181.3 + R_9}\right)^2, \tag{1.67a}$$

$$P_3(R_9) = \frac{4.82}{0.964}\left(\frac{174.4 + 1.34R_9}{181.3 + R_9}\right)^2, \tag{1.67b}$$

$$P_9(R_9) = R_9\left(\frac{54.9}{181.3 + R_9}\right)^2. \tag{1.67c}$$

Figure 1.11 shows the plots of the power levels P_1, P_3, and P_9 obtained above.

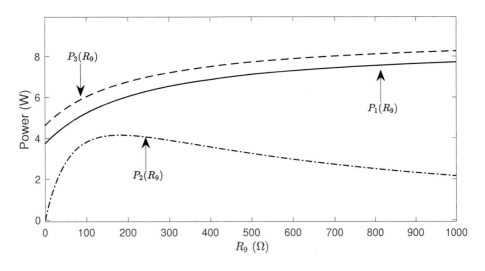

Figure 1.11 P_1, P_3, P_9 vs. R_9 (Example 1.5)

Using Equations (1.67a)–(1.67c), shown graphically in Figure 1.11, one imposes the power-level constraint Equations (1.66a)–(1.66c) to find the corresponding ranges of R_9 values. A necessary condition for the existence of a solution is that the constraint Equations (1.66a)–(1.66c) must be within their corresponding achievable ranges. For this example, the power-level constraints are within the achievable ranges; hence, we can find the following ranges for R_9 values:

$$190 \ (\Omega) \leq R_9 \leq 450 \ (\Omega), \tag{1.68a}$$

$$250 \ (\Omega) \leq R_9 \leq 690 \ (\Omega), \tag{1.68b}$$

$$(60 \ (\Omega) \leq R_9 \leq 80) \cup (420 \ (\Omega) \leq R_9 \leq 580) \ (\Omega), \tag{1.68c}$$

corresponding to the power-level constraints in Equations (1.66a), (1.66b), and (1.66c), respectively. Therefore, the range for R_9 values where Equations (1.66a), (1.66b), and (1.66c) are achieved simultaneously is the intersection of the ranges calculated above, that is

$$420 \ (\Omega) \leq R_9 \leq 450 \ (\Omega). \tag{1.69}$$

EXAMPLE 1.6   Consider the unknown linear DC circuit shown in Figure 1.12.

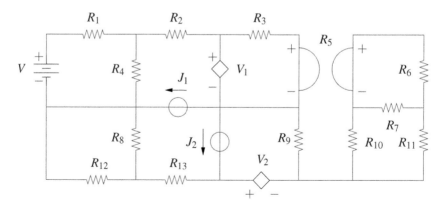

Figure 1.12 An unknown linear DC circuit (Example 1.6)

In this example, R_i, $i = 1, 2, \ldots, 13$, $i \neq 5$ are resistors, R_5 is a gyrator resistance, V, J_1, J_2 are independent sources, and V_1, V_2 are dependent sources. Our goal is to control the power levels in R_3, R_6, and R_{11}, denoted by P_3, P_6, and P_{11}, respectively, to be within the following prescribed ranges:

$$40 \ (W) \leq P_3 \leq 60 \ (W), \tag{1.70a}$$

$$1 \ (W) \leq P_6 \leq 8 \ (W), \tag{1.70b}$$

$$0.5 \ (W) \leq P_{11} \leq 5 \ (W). \tag{1.70c}$$

Assume that the design elements are the resistances R_1 and R_6. Therefore, we need to find the region in the R_1, R_6 plane where the constraints in Equations (1.70a), (1.70b), and (1.70c) are satisfied. In order to find the functional dependency of any power level in terms of any two resistances, one needs to do at most seven measurements of current and one measurement of voltage. Let us treat each power-level problem separately, as follows:

(a) P_3 vs. R_1 and R_6

In order to find the functional dependency of P_3 on R_1 and R_6, one needs to conduct seven experiments by setting seven different sets of values for the resistances (R_1, R_6), and measuring the corresponding values for current I_3. In addition to these measurements, one measurement of the voltage, across the resistor R_3, is needed to determine the functional dependency of the power on the resistors of interest. Suppose that this measurement is taken from the first experiment and denoted as V_{31}. Suppose that experiments are done and let Table 1.4 summarize the numerical values assigned to the resistances R_1 and R_6 along with the corresponding measurements of I_3 and V_{31}.

Table 1.4 Numerical values of the measurements for the DC circuit (Example 1.6)

Exp. No.	$R_1(\Omega)$	$R_6(\Omega)$	$I_3(A)$
1	7	1	3.31
2	13	8	2.7
3	21	19	2.46
4	35	26	2.56
5	40	32	2.52
6	52	45	2.47
7	59	56	2.44

Exp. No.	$R_1(\Omega)$	$R_6(\Omega)$	$V_{31}(V)$
1	7	1	33.1

Substituting the numerical values of Table 1.4 into the matrix $\mathbf{M}$, in Equation (1.54), it can be verified that $\det[\mathbf{M}] \neq 0$. Thus, the functional dependency of interest will be of the form

$$P_3(R_1, R_6) = \frac{V_{31}}{I_{31}} \underbrace{\left(\frac{\alpha_0 + \alpha_1 R_1 + \alpha_2 R_6 + \alpha_3 R_1 R_6}{\beta_0 + \beta_1 R_1 + \beta_2 R_6 + R_1 R_6}\right)^2}_{I_3^2(R_1, R_6)}, \quad (1.71)$$

where the constants $\alpha_0, \alpha_1, \alpha_2, \alpha_3, \beta_0, \beta_1, \beta_2$ can be determined by solving Equation (1.54), using the numerical values of Table 1.4. For this example, the constants are obtained as: $\alpha_0 = 96.3$, $\alpha_1 = 36$, $\alpha_2 = 6.4$, $\alpha_3 = 2.4$, $\beta_0 = 58.5$, $\beta_1 = 5$, $\beta_2 = 11.7$.

Hence, the functional dependency of P_3 on R_1 and R_6 will be

$$P_3(R_1, R_6) = \frac{33.3}{3.33} \left(\frac{96.3 + 36R_1 + 6.4R_6 + 2.4R_1R_6}{58.5 + 5R_1 + 11.7R_6 + R_1R_6} \right)^2. \tag{1.72}$$

Figure 1.13(a) shows the plot of the surface P_3 as a function of the design elements R_1 and R_6, obtained in Equation (1.72). Applying the constraint in Equation (1.70a) on P_3, one may extract the region in the R_1, R_6 plane, shown in black in Figure 1.13(b), where this constraint is satisfied.

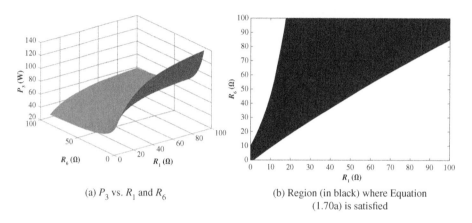

(a) P_3 vs. R_1 and R_6

(b) Region (in black) where Equation (1.70a) is satisfied

Figure 1.13 P_3 vs. R_1 and R_6 (Example 1.6)

(b) P_6 vs. R_1 and R_6

The functional dependency of P_6 on R_1 and R_6 can be determined by at most five measurements of current and one measurement of voltage. The plot of the surface P_6 as a function of R_1 and R_6 is shown in Figure 1.14(a). Applying the constraint in Equation (1.70b) on P_6, one may obtain the region in the R_1, R_6 plane, shown in black in Figure 1.14(b), where this constraint is valid.

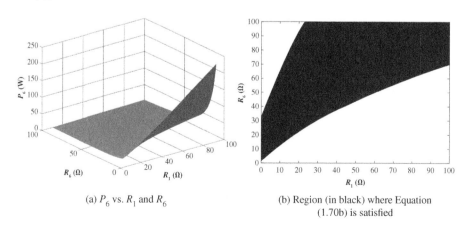

(a) P_6 vs. R_1 and R_6

(b) Region (in black) where Equation (1.70b) is satisfied

Figure 1.14 P_6 vs. R_1 and R_6 (Example 1.6)

(c) P_{11} vs. R_1 and R_6

Following the same procedure used to determine the functional dependency of P_3 on R_1 and R_6, one can determine the dependency of P_{11} on R_1 and R_6. The plot of the surface P_{11} as a function of R_1 and R_6 is shown in Figure 1.15(a). Applying the constraint in Equation (1.70c) on P_{11}, one finds the region in the R_1,R_6 plane, shown in black in Figure 1.15(b), where this constraint is satisfied.

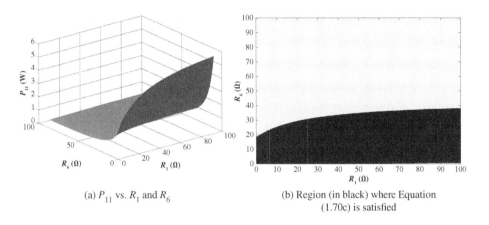

(a) P_{11} vs. R_1 and R_6

(b) Region (in black) where Equation (1.70c) is satisfied

Figure 1.15 P_{11} vs. R_1 and R_6 (Example 1.6)

In order to satisfy the constraints given in Equations (1.70a), (1.70b), and (1.70c), simultaneously, one has to intersect the regions shown in Figures 1.13(b), 1.14(b), and 1.15(b). Figure 1.16 shows the region (in black) in the R_1,R_6 plane where constraints in Equations (1.70a), (1.70b), and (1.70c) are satisfied, simultaneously.

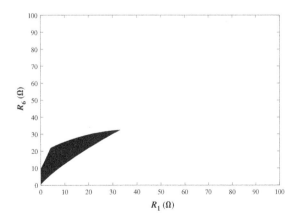

Figure 1.16 Region (in black) where Equations (1.70a), (1.70b), and (1.70c) are satisfied, simultaneously (Example 1.6)

1.3   AC Circuits

In this section we consider synthesis problems in unknown-model AC circuits, operating in steady state at a fixed frequency. The current, voltage, or power phasor in any branch of the circuit is to be controlled by adjusting the design elements, typically impedances, at arbitrary locations. We consider several cases of design elements and provide the results for each case.

The governing steady-state equations of a linear AC circuit, operating at a fixed frequency ω, can be represented in the following matrix form:

$$\mathbf{A}(\mathbf{p}(j\omega))\mathbf{x}(j\omega) = \mathbf{b}(j\omega), \tag{1.73}$$

where $\mathbf{A}(\mathbf{p}(j\omega))$ is the circuit characteristic matrix containing the circuit impedances, $\mathbf{x}(j\omega)$ is the vector of unknown current and voltage phasors, and $\mathbf{b}$ represents the vector of independent voltage and current sources which are fixed. Suppose that we want to control the current phasor in the ith branch of the circuit, denoted by $I_i(j\omega)$. Applying Cramer's rule to Equation (1.73), $I_i(j\omega)$ can be calculated from

$$x_i(j\omega) = I_i(j\omega) = \frac{\det\left[\mathbf{T}_i(\mathbf{p}(j\omega))\right]}{\det[\mathbf{A}(\mathbf{p}(j\omega))]}, \tag{1.74}$$

where $\mathbf{T}_i(\mathbf{p}(j\omega))$ is the matrix obtained by replacing the ith column of the characteristic matrix $\mathbf{A}(\mathbf{p}(j\omega))$ by the vector $\mathbf{b}(j\omega)$. We assume that the circuit is unknown, so that the matrices $\mathbf{T}_i(\mathbf{p}(j\omega),(j\omega))$, and $\mathbf{A}(\mathbf{p}(j\omega))$ are unknown. In the following subsections, for each choice of the design elements, a general rational function for the current phasor $I_i(\omega)$, in terms of the design elements, will be derived. For the sake of simplicity, we drop the argument $(j\omega)$ in writing the equations from now on.

1.3.1   Control Using a Single Impedance

Consider the unknown-model linear AC circuit shown in Figure 1.17.

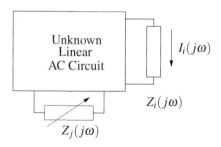

Figure 1.17 An unknown linear AC circuit

We want to control the current phasor in the ith branch, I_i, by adjusting any impedance Z_j at an arbitrary location of the circuit. Assume, for the moment, that Z_j is not a gyrator resistance so that the rank of the dependency is one.

THEOREM 1.4   In a linear AC circuit, the functional dependency of any current phasor I_i on any impedance Z_j (appearing with rank 1 dependency) can be determined by at most three measurements of the current phasor I_i obtained for three different complex values of Z_j.

Proof   The proof is similar to its DC circuit counterpart provided in Section 1.2.1. The main difference is that the circuit signals/variables and the constants appearing in the functional dependencies will be complex quantities rather than real numbers. Therefore, we provide the results and leave the details of the proof to the reader.

Case 1: $(i \neq j)$ Because of the rank 1 dependency of A on Z_j, the functional dependency of I_i on Z_j will be of the form

$$I_i(Z_j) = \frac{\alpha_0 + \alpha_1 Z_j}{\beta_0 + Z_j}, \tag{1.75}$$

where $\alpha_0, \alpha_1, \beta_0$ are complex constants that can be uniquely determined by solving the measurement equations,

$$\underbrace{\begin{bmatrix} 1 & Z_{j1} & -I_{i1} \\ 1 & Z_{j2} & -I_{i2} \\ 1 & Z_{j3} & -I_{i3} \end{bmatrix}}_{\mathbf{M}} \underbrace{\begin{bmatrix} \alpha_0 \\ \alpha_1 \\ \beta_0 \end{bmatrix}}_{\mathbf{t}} = \underbrace{\begin{bmatrix} Z_{j1}I_{i1} \\ Z_{j2}I_{i2} \\ Z_{j3}I_{i3} \end{bmatrix}}_{\mathbf{m}}, \tag{1.76}$$

provided $\det[\mathbf{M}] \neq 0$. These complex quantities can be written as

$$\begin{aligned} \alpha_0(j\omega) &= \alpha_{0r}(\omega) + j\alpha_{0i}(\omega), \\ \alpha_1(j\omega) &= \alpha_{1r}(\omega) + j\alpha_{1i}(\omega), \\ \beta_0(j\omega) &= \beta_{0r}(\omega) + j\beta_{0i}(\omega). \end{aligned} \tag{1.77}$$

If $\det[\mathbf{M}] = 0$ in Equation (1.76), the functional dependency of I_i on Z_j can be expressed as

$$I_i(Z_j) = \alpha_0 + \alpha_1 Z_j, \tag{1.78}$$

where the complex quantities α_0, α_1 can be determined from any two of the experiments conducted earlier.

Case 2: $(i = j)$ In this case, the functional dependency of I_i on Z_i can be represented as

$$I_i(Z_i) = \frac{\alpha_0}{\beta_0 + Z_i}, \tag{1.79}$$

where α_0, β_0 are complex quantities that can be calculated by solving the measurement equations

$$\underbrace{\begin{bmatrix} 1 & -I_{i1} \\ 1 & -I_{i2} \end{bmatrix}}_{\mathbf{M}} \underbrace{\begin{bmatrix} \alpha_0 \\ \beta_0 \end{bmatrix}}_{\mathbf{t}} = \underbrace{\begin{bmatrix} I_{i1}Z_{i1} \\ I_{i2}Z_{i2} \end{bmatrix}}_{\mathbf{m}}, \tag{1.80}$$

if and only if $\det[\mathbf{M}] \neq 0$. In a situation where $\det[\mathbf{M}] = 0$ in Equation (1.80), the current phasor I_i will be a constant:

$$I_i(Z_i) = \alpha_0, \tag{1.81}$$

where the complex quantity α_0 can be obtained from one of the experiments conducted earlier.   ∎

As noted earlier, since the main difference between the results of this section and their DC circuit counterparts covered earlier is that in AC circuits the variables are complex quantities, rather than real numbers, the proofs of the following theorems are omitted.

1.3.2   Control Using Two Impedances

Consider the unknown linear AC circuit shown in Figure 1.18.

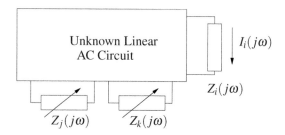

Figure 1.18 An unknown-model linear AC circuit

In this problem, we want to control the current phasor in the ith branch, I_i, by adjusting any two impedances Z_j and Z_k at arbitrary locations of the circuit. Assume that Z_j and Z_k are not gyrator impedances.

THEOREM 1.5   In a linear AC circuit, the functional dependency of any current phasor I_i on any two impedances Z_j and Z_k (each appearing with rank 1 dependency) can be determined by at most seven measurements of the current phasor I_i obtained for seven different sets of complex values (Z_j, Z_k).

Remark 1.5   If m impedances are considered as the design elements, one can use the results provided in Section 1.2.3 to derive the corresponding functional dependencies. Also the results given here extend to voltage control with little modification.

1.3.3   Current Control Using Gyrator Resistance

In this problem the design element is the resistance of a gyrator; thus, our objective is to control the current phasor I_i by a gyrator resistance, denoted by R_g, at an arbitrary location of the circuit.

THEOREM 1.6   In a linear AC circuit, the functional dependency of any current phasor I_i on any gyrator resistance R_g can be determined by at most five measurements of the current phasor I_i obtained for five different values of R_g.

Proof The gyrator resistance appears in the A matrix with rank 2 since it is associated with two currents. Thus, $\det[A(R_g)]$ and $\det[T_i(R_g)]$ are polynomials of degree 2 or less in R_g. The result now follows upon normalizing or dividing the numerator and denominator by the coefficient of the second-degree term in $\det[A(R_g)]$. ∎

1.3.4 Control Using m Independent Sources

Consider the problem of controlling the current phasor I_i using the independent current and voltage sources, denoted by $q_1, q_2, \ldots, q_m$, at arbitrary locations of the circuit (Figure 1.19).

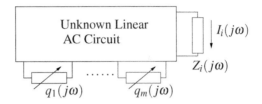

Figure 1.19 An unknown-model linear AC circuit

THEOREM 1.7 In a linear AC circuit, the functional dependency of any current phasor I_i on the independent sources can be determined by m measurements of the current phasor I_i obtained for m linearly independent sets of values of the source vector $\mathbf{q} = [q_1, q_2, \ldots, q_m]^T$.

EXAMPLE 1.7 (Example of AC circuit design) This example shows how the approach developed for the AC circuits can be used to solve a synthesis problem. Consider the unknown linear AC circuit shown in Figure 1.20, where L_4 and L_5 are inductively coupled.

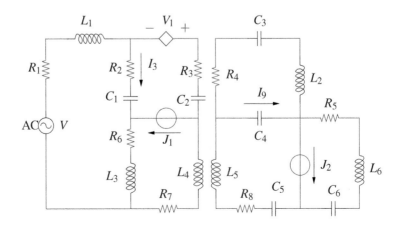

Figure 1.20 An unknown linear AC circuit (Example 1.7)

Let us assume that the operating frequency of this AC circuit is $f = 60$ (Hz) and thus the angular frequency will be $\omega_0 = 2\pi f = 120\pi$. Suppose that we want to control the current phasors I_3 and I_9 to be within the following ranges:

$$0 \text{ (A)} \le |I_3| \le 4 \text{ (A)}, \tag{1.82a}$$

$$10° \le \angle I_3 \le 30°, \tag{1.82b}$$

$$0 \text{ (A)} \le |I_9| \le 2.5 \text{ (A)}, \tag{1.82c}$$

$$-30° \le \angle I_9 \le -10°. \tag{1.82d}$$

Assume that the design elements are the inductor L_1 and the capacitor C_2. Therefore, we need to find the region in the L_1, C_2 plane where the constraints in Equations (1.82a)–(1.82d) are satisfied. Based on the approach provided in Section 1.3.2, one can determine the functional dependency of any current phasor in terms of any two impedances by taking seven measurements of the current phasor. Let us treat each current phasor problem separately as follows:

(a) I_3 vs. L_1 and C_2

To determine the functional dependency of I_3 on L_1 and C_2, one needs to do seven measurements of current phasor I_3 for seven different sets of values (L_1, C_2). Suppose that the measurements are taken and let Table 1.5 summarize the numerical values assigned to L_1 and C_2 and the corresponding measurements of I_3.

Table 1.5 Numerical values of the measurements for the AC circuit example (Example 1.7)

Exp. No.	L_1 (mH)	C_2 (μF)	I_3 (A)
1	13	10	$3.3 - j2.9$
2	25	20	$2.7 - j3.2$
3	32	23	$2.3 - j3.4$
4	45	29	$1.4 - j3.6$
5	54	33	$0.7 - j3.5$
6	68	40	$-0.5 - j2.9$
7	90	47	$-1.4 - j1.3$

For this case, the general functional dependency can be written as

$$I_3(L_1, C_2) = \frac{\alpha_0 + \alpha_1 L_1 j\omega_0 + \alpha_2/(C_2 j\omega_0) + \alpha_3 L_1/C_2}{\beta_0 + \beta_1 L_1 j\omega_0 + \beta_2/(C_2 j\omega_0) + L_1/C_2}, \tag{1.83}$$

where the complex constants $\alpha_0, \alpha_1, \alpha_2, \alpha_3, \beta_0, \beta_1, \beta_2$ can be determined by solving the set of seven measurement equations, which are now complex. Substituting the numerical values given in Table 1.5 into the measurement equations and solving for the unknown complex constants, yields

$$
\begin{aligned}
&\alpha_0 = -1502 - 2772j, &&\alpha_1 = 173 + 74j,\\
&\alpha_2 = 106 + 151j, &&\alpha_3 = 0,\\
&\beta_0 = -481 - 316j, &&\beta_1 = 13 + 13j,\\
&\beta_2 = 30 + 15j.
\end{aligned}
\tag{1.84}
$$

Thus, the functional dependency of I_3 on L_1 and C_2 will be

$$
I_3(L_1, C_2) = \frac{(-1502 - 2772j) + (173 + 74j)L_1 j\omega_0 + (106 + 151j)/(C_2 j\omega_0)}{(-481 - 316j) + (13 + 13j)L_1 j\omega_0 + (30 + 15j)/(C_2 j\omega_0) + L_1/C_2}.
\tag{1.85}
$$

Figures 1.21(a) and (b) show the plots of the magnitude and the phase of I_3 as a function of the design elements L_1 and C_2, as obtained in Equation (1.85).

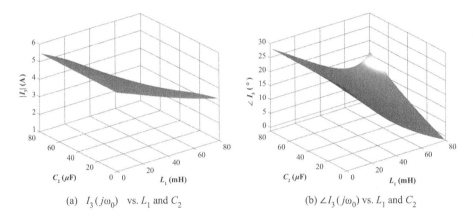

(a) $I_3(j\omega_0)$ vs. L_1 and C_2 (b) $\angle I_3(j\omega_0)$ vs. L_1 and C_2

Figure 1.21 $I_3(j\omega_0)$ vs. L_1 and C_2 (Example 1.7)

Applying the constraints in Equations (1.82a) and (1.82b) on I_3, one may obtain the region in the L_1, C_2 plane, shown in black in Figure 1.22, where these constraints are satisfied.

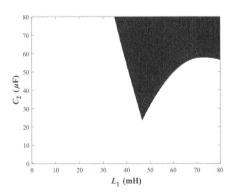

Figure 1.22 Region (in black) where Equations (1.82a) and (1.82b) are satisfied (Example 1.7)

(b) I_9 vs. L_1 and C_2

Following the same procedure as above, one may obtain the dependency of I_9 on L_1 and C_2. Plots of the magnitude and the phase of I_9 as a function of L_1 and C_2 are shown in Figures 1.23(a) and (b), respectively.

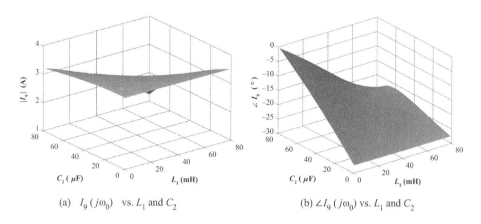

(a) $I_9(j\omega_0)$ vs. L_1 and C_2 (b) $\angle I_9(j\omega_0)$ vs. L_1 and C_2

Figure 1.23 $I_9(j\omega_0)$ vs. L_1 and C_2 (Example 1.7)

Applying the constraints in Equations (1.82c) and (1.82d) on I_9, one can find the region in the L_1, C_2 plane, shown in black in Figure 1.24(a), where these constraints are valid.

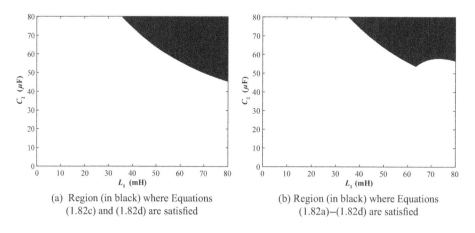

(a) Region (in black) where Equations   (b) Region (in black) where Equations
(1.82c) and (1.82d) are satisfied   (1.82a)–(1.82d) are satisfied

Figure 1.24 Regions satisfying all the conditions (Example 1.7)

In order to satisfy the constraints given in Equations (1.82a)–(1.82d), simultaneously, one has to intersect the regions shown in Figures 1.22–1.24. Figure 1.24(b) shows the region in the L_1, C_2 plane where the constraints in Equations (1.82a)–(1.82d) are satisfied, simultaneously.

1.4   Mass–Spring Systems

In this section we consider synthesis problems in an unknown mass–spring system where the displacements of the masses are to be controlled by adjusting the spring stiffness constants. Consider the unknown-model linear mass–spring system shown in Figure 1.25.

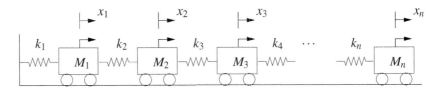

Figure 1.25 An unknown general mass–spring system

Suppose that we want to control the displacement of the ith mass, denoted by x_i, by adjusting the spring stiffness k_j at an arbitrary location. Assume that the spring k_j is composed of piezoelectric materials such that its stiffness can be controlled by applying an electrical field. The displacements can be measured using a variety of sensors such as potentiometers or linear variable differential transformers (LVDTs). In this problem the system of governing linear equations can be constructed in the form

$$
\underbrace{\begin{bmatrix}
k_1 + k_2 & -k_2 & 0 & \cdots & 0 & 0 \\
-k_2 & k_2 + k_3 & -k_3 & \cdots & 0 & 0 \\
0 & -k_3 & k_3 + k_4 & \cdots & 0 & 0 \\
\vdots & \vdots & \vdots & & \vdots & \vdots \\
0 & 0 & 0 & \cdots & -k_{n-1} & k_n
\end{bmatrix}}_{\mathbf{A(p)}}
\underbrace{\begin{bmatrix}
x_1 \\ x_2 \\ x_3 \\ \vdots \\ x_n
\end{bmatrix}}_{\mathbf{x}}
=
\underbrace{\begin{bmatrix}
F_1 \\ F_2 \\ F_3 \\ \vdots \\ F_n
\end{bmatrix}}_{\mathbf{b(q)}}, \quad (1.86)
$$

where $\mathbf{p} = [k_1, k_2, \ldots, k_n]^T$, $\mathbf{x}$ represents the vector of unknown displacements, and $\mathbf{q} = [F_1, F_2, \ldots, F_n]^T$ is the vector of fixed external forces. Applying Cramer's rule to Equation (1.86) to calculate x_i gives

$$
x_i(\mathbf{p}, \mathbf{q}) = \frac{|\mathbf{T}_i(\mathbf{p})|}{|\mathbf{A}(\mathbf{p})|}, \quad i = 1, 2, \ldots, n, \quad (1.87)
$$

where $\mathbf{T}_i(\mathbf{p})$ is the matrix obtained by replacing the ith column of $\mathbf{A}(\mathbf{p})$ by $\mathbf{b}$. We can now state the following theorem.

THEOREM 1.8   In a linear mass–spring system, with fixed external forces and fixed spring constants except k_j, the functional dependency of any mass displacement x_i on any spring stiffness k_j can be determined by three measurements of the displacement x_i obtained for three different values of k_j.

Proof   Note that the matrices $\mathbf{T}_i(\mathbf{p})$ and $\mathbf{A}(\mathbf{p})$, in Equation (1.87), are both of rank 1 with respect to k_j. According to Lemma 1.1, the functional dependency of x_i on k_j can be expressed as

$$x_i(k_j) = \frac{\tilde{\alpha}_0 + \tilde{\alpha}_1 k_j}{\tilde{\beta}_0 + \tilde{\beta}_1 k_j}, \tag{1.88}$$

where $\tilde{\alpha}_0, \tilde{\alpha}_1, \tilde{\beta}_0, \tilde{\beta}_1$ are constants. If $\tilde{\beta}_0 = \tilde{\beta}_1 = 0$, then $x_i \to \infty$, for any value of the spring stiffness k_j, which is physically impossible. Hence, we rule out this case. Assuming that $\tilde{\beta}_1 \neq 0$, one can divide the numerator and denominator of Equation (1.88) by $\tilde{\beta}_1$ and obtain

$$x_i(k_j) = \frac{\alpha_0 + \alpha_1 k_j}{\beta_0 + k_j}, \tag{1.89}$$

where $\alpha_0, \alpha_1, \beta_0$ are constants. In order to determine α_0, α_1, β_0, one conducts three experiments by setting three different values for the spring stiffness k_j, namely k_{j1}, k_{j2}, k_{j3} and measuring the corresponding displacements x_i, namely x_{i1}, x_{i2}, x_{i3}. The following set of measurement equations can then be formed:

$$\underbrace{\begin{bmatrix} 1 & k_{j1} & -x_{i1} \\ 1 & k_{j2} & -x_{i2} \\ 1 & k_{j3} & -x_{i3} \end{bmatrix}}_{\mathbf{M}} \underbrace{\begin{bmatrix} \alpha_0 \\ \alpha_1 \\ \beta_0 \end{bmatrix}}_{\mathbf{t}} = \underbrace{\begin{bmatrix} x_{i1}k_{j1} \\ x_{i2}k_{j2} \\ x_{i3}k_{j3} \end{bmatrix}}_{\mathbf{m}}. \tag{1.90}$$

The set of measurement equations in Equation (1.90) can be uniquely solved for the unknown constants $\alpha_0, \alpha_1, \beta_0$ provided that $\det[\mathbf{M}] \neq 0$. If $\det[\mathbf{M}] = 0$, the last column of the matrix $\mathbf{M}$ can be expressed as a linear combination of the first two columns because by assigning different values to the spring stiffness k_j, the first two columns of $\mathbf{M}$ become linearly independent. In this case, the functional dependency of x_i on k_j will be

$$x_i(k_j) = \alpha_0 + \alpha_1 k_j, \tag{1.91}$$

where α_0, α_1 are constants that can be determined from any two of the experiments conducted earlier. The functional dependency in Equation (1.91) corresponds to the case where $\tilde{\beta}_1 = 0$ in Equation (1.88), and the numerator and denominator of Equation (1.88) are divided by $\tilde{\beta}_0$. ∎

Remark 1.6 Suppose now that the design parameters are the external forces applied to the system. In such a case, the displacement x_i can be expressed as

$$x_i(F_1, F_2, \ldots, F_n) = \beta_1 F_1 + \beta_2 F_2 + \cdots + \beta_n F_n, \tag{1.92}$$

and the constants $\beta_1, \beta_2, \ldots, \beta_n$ can be determined by applying n different sets of vector forces $(F_1, F_2, \ldots, F_n)$ to the system and measuring the corresponding displacements x_i. This is the well-known Superposition Principle for mechanical systems. In addition, if the external forces vary in the intervals $F_j^- \leq F_j \leq F_j^+$, $j = 1, 2, \ldots, n$, then the displacement x_i will vary in a convex hull whose vertices can be computed using the vertices (F_j^-, F_j^+), $j = 1, 2, \ldots, n$.

EXAMPLE 1.8 (A mass–spring system example) Consider the three-story building frame shown in Figure 1.26(a). Suppose that the mechanical properties of the building components are unknown and the building is modeled as a mass–spring system shown is Figure 1.26(b) with unknown parameters.

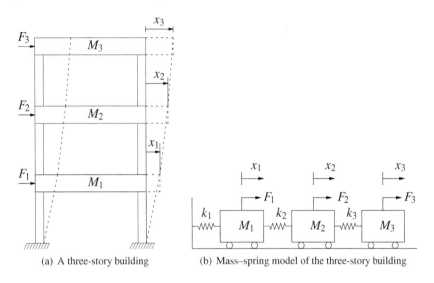

(a) A three-story building   (b) Mass–spring model of the three-story building

Figure 1.26 A mass–spring system (Example 1.8)

Assume that the links connecting the first and second floors are composed of piezo-electric materials and thus their stiffness can be controlled by applying an electrical field. Suppose that the design objective is to control the displacement of the second floor (mass M_2), denoted by x_2, by adjusting the stiffness constants of the links connecting the first and second floors (spring constant k_2), to be within the range

$$-0.05 \le x_2 \le -0.03 \text{ (m)}. \tag{1.93}$$

Hence, we need to find an interval of k_2 values for which Equation (1.93) is satisfied. Based on Theorem 1.8, the functional dependency of x_2 on k_2 can be written as

$$x_2 = \frac{\alpha_0 + \alpha_1 k_2}{\beta_0 + k_2}, \tag{1.94}$$

where $\alpha_0, \alpha_1, \beta_0$ are constants that can be determined by conducting three experiments, by setting three different values for the spring constant k_2, namely k_{21}, k_{22}, k_{23}, measuring the corresponding displacements x_2, namely x_{21}, x_{22}, x_{23}, and then solving the following system of measurement equations:

$$\begin{bmatrix} 1 & k_{21} & -x_{21} \\ 1 & k_{22} & -x_{22} \\ 1 & k_{23} & -x_{23} \end{bmatrix} \begin{bmatrix} \alpha_0 \\ \alpha_1 \\ \beta_0 \end{bmatrix} = \begin{bmatrix} k_{21}x_{21} \\ k_{22}x_{22} \\ k_{23}x_{23} \end{bmatrix}. \tag{1.95}$$

Suppose that the measurements are taken and let Table 1.6 show the numerical values for the experiments performed for this example.

Table 1.6 Numerical values for the experiments of the mass–spring example (Example 1.8)

Exp. No.	k_2 (N/m)	x_2 (m)
1	2×10^5	−0.035
2	3×10^5	−0.030
3	5×10^5	−0.026

Substituting these numerical values into Equation (1.95) and solving for the constants, yields

$$x_2 = \frac{-3000 - 0.02k_2}{k_2}, \tag{1.96}$$

which is plotted in Figure 1.27.

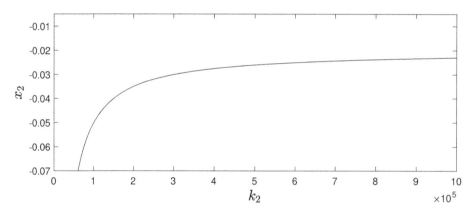

Figure 1.27 x_2 vs. k_2 for the mass–spring example (Example 1.8)

Applying the design constraint given in Equation (1.93) yields the following range of k_2 values:

$$10^5 \leq k_2 \leq 3 \times 10^5 \ (\text{N/m}) \tag{1.97}$$

as the solution.

1.5   Hydraulic Networks

In this section, we extend our measurement-based approach to linear hydraulic networks. Suppose that, in a hydraulic network, all the flows are in the laminar state, which implies that the resulting steady-state equations are linear. Here the objective is to control the flow rates in some set of pipes.

In a laminar flow, the pressure drop occurring in a pipe can be obtained from

$$\Delta P = \frac{8\mu L Q}{\pi r^4},\tag{1.98}$$

where μ is the dynamic viscosity of the fluid, L is the length of the pipe, Q is the volume flow rate, and r is the inner radius of the pipe. Let us rewrite Equation (1.98) as

$$\Delta P = RQ,\tag{1.99}$$

where

$$R = \frac{8\mu L}{\pi r^4}\tag{1.100}$$

is called the pipe resistance constant, which is a function of the mechanical properties (length L and radius r) of the pipe.

To illustrate the approach, we begin by considering an unknown-model general hydraulic network as shown in Figure 1.28.

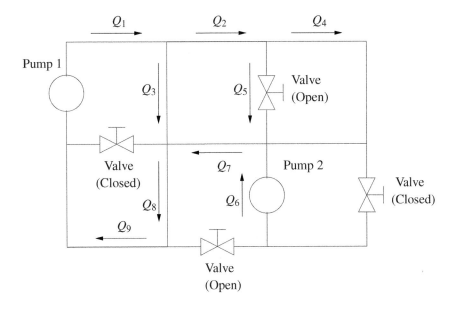

Figure 1.28 An unknown-model general hydraulic network

Similar to linear circuits, applying Kirchhoff's laws to a linear hydraulic network

(Figure 1.28) gives the set of governing linear equations shown below in matrix form:

$$
\underbrace{\begin{bmatrix}
1 & -1 & -1 & 0 & 0 & 0 & 0 & 0 \\
0 & 1 & 0 & -1 & -1 & 0 & 0 & 0 \\
0 & 0 & 0 & 1 & 1 & 1 & -1 & 0 \\
0 & 0 & 1 & 0 & 0 & 0 & 1 & -1 \\
R_1 & 0 & R_3 & 0 & 0 & 0 & 0 & R_8 \\
0 & -R_2 & R_3 & 0 & -R_5 & R_6 & 0 & R_8 \\
0 & 0 & 0 & -R_4 & R_5 & 0 & 0 & 0 \\
0 & -R_2 & R_3 & -R_4 & 0 & 0 & -R_7 & 0
\end{bmatrix}}_{\mathbf{A}(\mathbf{p})}
\underbrace{\begin{bmatrix}
Q_1 \\ Q_2 \\ Q_3 \\ Q_4 \\ Q_5 \\ Q_6 \\ Q_7 \\ Q_8
\end{bmatrix}}_{\mathbf{x}}
=
\underbrace{\begin{bmatrix}
0 \\ 0 \\ 0 \\ 0 \\ P_1 \\ P_2 \\ 0 \\ 0
\end{bmatrix}}_{\mathbf{b}}, \qquad (1.101)
$$

where $\mathbf{p} = [R_1, R_2, \ldots, R_8]$ is the vector of the pipe resistances (R_i, $i = 1, 2, \ldots, 8$ are the resistances of the set of pipes through which Q_i, $i = 1, 2, \ldots, 8$ flow), $\mathbf{x}$ denotes the vector of unknown flow rates, and $\mathbf{b}$ represents the vector of input parameters including the pump pressures. The flow rate Q_i can be calculated from Equation (1.101) using Cramer's rule:

$$
Q_i = x_i(\mathbf{p}) = \frac{|\mathbf{T}_i(\mathbf{p})|}{|\mathbf{A}(\mathbf{p})|}, \quad i = 1, 2, \ldots, n, \qquad (1.102)
$$

where $\mathbf{T}_i(\mathbf{p})$ is the matrix $\mathbf{A}(\mathbf{p})$ with the ith column replaced by $\mathbf{b}$.

Remark 1.7   Upon an application of Kirchhoff's laws we see that each pipe resistance R_j appears in only one column of the characteristic matrix $\mathbf{A}(\mathbf{p})$.

In hydraulic networks, it is common to carry out the measurements using a pressure-driven test. In the following design cases, assume that the set of design pipes are composed of piezoelectric materials, and thus the radii of these pipes can be controlled upon applying an electrical field. In case the radius can not be controlled by applying electrical fields, different pilot pipe sections have to be used for the experiments.

1.5.1   Flow Rate Control Using a Single Pipe Resistance

Assume that the design parameter is the resistance of one pipe, denoted by R_j, at an arbitrary location of the network. Therefore, we want to control the flow rate at some location of the network, denoted by Q_i, by adjusting the pipe resistance R_j. We can state the following theorem.

THEOREM 1.9   In a linear hydraulic network, with fixed pipe resistances except R_j, the functional dependency of any flow rate Q_i on the pipe resistance R_j can be determined by at most three measurements of the flow rate Q_i obtained for three different values of R_j.

Proof   Let us consider the two cases $i \neq j$ and $i = j$.

Case 1: ($i \neq j$)   Based on Remark 1.7, the matrices $\mathbf{T}_i(\mathbf{p})$ and $\mathbf{A}(\mathbf{p})$, in Equation (1.102), are both of rank 1 with respect to R_j. Therefore, according to Lemma 1.1, the functional dependency of Q_i on R_j can be expressed as

$$Q_i(R_j) = \frac{\tilde{\alpha}_0 + \tilde{\alpha}_1 R_j}{\tilde{\beta}_0 + \tilde{\beta}_1 R_j}, \tag{1.103}$$

where $\tilde{\alpha}_0, \tilde{\alpha}_1, \tilde{\beta}_0, \tilde{\beta}_1$ are constants. Assume that $\tilde{\beta}_1 \neq 0$; hence, one can divide the numerator and denominator of Equation (1.103) by $\tilde{\beta}_1$ and obtain

$$Q_i(R_j) = \frac{\alpha_0 + \alpha_1 R_j}{\beta_0 + R_j}, \tag{1.104}$$

where $\alpha_0, \alpha_1, \beta_0$ are constants that can be determined by conducting three experiments. The measurement equations can be written as

$$\underbrace{\begin{bmatrix} 1 & R_{j1} & -Q_{i1} \\ 1 & R_{j2} & -Q_{i2} \\ 1 & R_{j3} & -Q_{i3} \end{bmatrix}}_{\mathbf{M}} \underbrace{\begin{bmatrix} \alpha_0 \\ \alpha_1 \\ \beta_0 \end{bmatrix}}_{\mathbf{u}} = \underbrace{\begin{bmatrix} Q_{i1}R_{j1} \\ Q_{i2}R_{j2} \\ Q_{i3}R_{j3} \end{bmatrix}}_{\mathbf{m}}, \tag{1.105}$$

which can be uniquely solved for the unknown constants $\alpha_0, \alpha_1, \beta_0$, provided $\det[\mathbf{M}] \neq 0$. For the situations where $\det[\mathbf{M}] = 0$, corresponding to $\tilde{\beta}_1 = 0$ in Equation (1.103), one may use a similar strategy to that presented in Section 1.2.1 to derive the corresponding functional dependency. In this case:

$$Q_i(R_j) = \alpha_0 + \alpha_1 R_j. \tag{1.106}$$

Case 2: $(i = j)$ Recalling Equation (1.102) and based on Remark 1.7, the matrix $\mathbf{T}_i(\mathbf{p})$ is of rank 0 with respect to R_i; however, the matrix $\mathbf{A}(\mathbf{p})$ is of rank 1 with respect to R_j. According to Lemma 1.1, the functional dependency of Q_i on R_i will be

$$Q_i(R_i) = \frac{\tilde{\alpha}_0}{\tilde{\beta}_0 + \tilde{\beta}_1 R_i}, \tag{1.107}$$

where $\tilde{\alpha}_0, \tilde{\beta}_0, \tilde{\beta}_1$ are constants. Assuming that $\tilde{\beta}_1 \neq 0$, and dividing the numerator and denominator of Equation (1.107) by $\tilde{\beta}_1$, results in

$$Q_i(R_i) = \frac{\alpha_0}{\beta_0 + R_i}, \tag{1.108}$$

where α_0, β_0 are constants that can be determined by conducting two experiments. The following set of measurement equations can then be formed:

$$\underbrace{\begin{bmatrix} 1 & -Q_{i1} \\ 1 & -Q_{i2} \end{bmatrix}}_{\mathbf{M}} \underbrace{\begin{bmatrix} \alpha_0 \\ \beta_0 \end{bmatrix}}_{\mathbf{t}} = \underbrace{\begin{bmatrix} Q_{i1}R_{i1} \\ Q_{i2}R_{i2} \end{bmatrix}}_{\mathbf{m}}, \tag{1.109}$$

and uniquely solved for the unknown constants α_0, β_0, if and only if $\det[\mathbf{M}] \neq 0$. If $\det[\mathbf{M}] = 0$ in Equation (1.109), it can be concluded that Q_i is a constant:

$$Q_i(R_i) = \alpha_0. \tag{1.110}$$

This case corresponds to $\tilde{\beta}_1 = 0$ in Equation (1.107).   ∎

1.5.2   Flow Rate Control Using Two Pipe Resistances

Suppose that the design parameters are any two pipe resistances, denoted by R_j and R_k, at arbitrary locations of the network, and the flow rate Q_i, at some location of the network, is to be controlled by adjusting these two pipe resistances.

THEOREM 1.10   In a linear hydraulic network, with fixed pipe resistances except R_j and R_k, the functional dependency of any flow rate Q_i on the pipe resistances R_j and R_k can be determined by at most seven measurements of the flow rate Q_i obtained for seven different sets of values of (R_j, R_k).

Proof   Again, let us consider the two cases $i \neq j, k$ and $i = j$ or $i = k$.

Case 1: $(i \neq j, k)$ In this case, based on Remark 1.7, the matrices $\mathbf{T}_i(\mathbf{p})$ and $\mathbf{A}(\mathbf{p})$, in Equation (1.102), are both of rank 1 with respect to R_j and R_k. According to Lemma 1.1, the functional dependency of Q_i on R_j and R_k becomes

$$Q_i(R_j, R_k) = \frac{\tilde{\alpha}_0 + \tilde{\alpha}_1 R_j + \tilde{\alpha}_2 R_k + \tilde{\alpha}_3 R_j R_k}{\tilde{\beta}_0 + \tilde{\beta}_1 R_j + \tilde{\beta}_2 R_k + \tilde{\beta}_3 R_j R_k}, \tag{1.111}$$

where $\tilde{\alpha}_0, \tilde{\alpha}_1, \tilde{\alpha}_2, \tilde{\alpha}_3, \tilde{\beta}_0, \tilde{\beta}_1, \tilde{\beta}_2, \tilde{\beta}_3$ are constants. Assuming that $\tilde{\beta}_3 \neq 0$, dividing the numerator and denominator of Equation (1.111) by $\tilde{\beta}_3$ yields

$$Q_i(R_j, R_k) = \frac{\alpha_0 + \alpha_1 R_j + \alpha_2 R_k + \alpha_3 R_j R_k}{\beta_0 + \beta_1 R_j + \beta_2 R_k + R_j R_k}, \tag{1.112}$$

where $\alpha_0, \alpha_1, \alpha_2, \alpha_3, \beta_0, \beta_1, \beta_2$ are constants. In order to determine these constants, one conducts seven experiments by setting seven different sets of values for the pipe resistances (R_j, R_k), and measuring the corresponding flow rates Q_i. The following set of measurement equations can then be formed:

$$\underbrace{\begin{bmatrix} 1 & R_{j1} & R_{k1} & R_{j1}R_{k1} & -Q_{i1} & -Q_{i1}R_{j1} & -Q_{i1}R_{k1} \\ 1 & R_{j2} & R_{k2} & R_{j2}R_{k2} & -Q_{i2} & -Q_{i2}R_{j2} & -Q_{i2}R_{k2} \\ 1 & R_{j3} & R_{k3} & R_{j3}R_{k3} & -Q_{i3} & -Q_{i3}R_{j3} & -Q_{i3}R_{k3} \\ 1 & R_{j4} & R_{k4} & R_{j4}R_{k4} & -Q_{i4} & -Q_{i4}R_{j4} & -Q_{i4}R_{k4} \\ 1 & R_{j5} & R_{k5} & R_{j5}R_{k5} & -Q_{i5} & -Q_{i5}R_{j5} & -Q_{i5}R_{k5} \\ 1 & R_{j6} & R_{k6} & R_{j6}R_{k6} & -Q_{i6} & -Q_{i6}R_{j6} & -Q_{i6}R_{k6} \\ 1 & R_{j7} & R_{k7} & R_{j7}R_{k7} & -Q_{i7} & -Q_{i7}R_{j7} & -Q_{i7}R_{k7} \end{bmatrix}}_{\mathbf{M}} \underbrace{\begin{bmatrix} \alpha_0 \\ \alpha_1 \\ \alpha_2 \\ \alpha_3 \\ \beta_0 \\ \beta_1 \\ \beta_2 \end{bmatrix}}_{\mathbf{u}} = \underbrace{\begin{bmatrix} Q_{i1}R_{j1}R_{k1} \\ Q_{i2}R_{j2}R_{k2} \\ Q_{i3}R_{j3}R_{k3} \\ Q_{i4}R_{j4}R_{k4} \\ Q_{i5}R_{j5}R_{k5} \\ Q_{i6}R_{j6}R_{k6} \\ Q_{i7}R_{j7}R_{k7} \end{bmatrix}}_{\mathbf{m}}.$$

$$\tag{1.113}$$

This set of equations can be uniquely solved for the constants $\alpha_0, \alpha_1, \alpha_2, \alpha_3, \beta_0, \beta_1, \beta_2$, provided $\det[\mathbf{M}] \neq 0$. If $\det[\mathbf{M}] = 0$, one can follow a similar procedure as presented in Section 1.2.2 to develop the corresponding functional dependency of Q_i on R_j and R_k.

Case 2: $(i = j$ or $i = k)$ Suppose that $i = j$ and recall Equation (1.102). Based on Remark 1.7, in this case, the matrix $\mathbf{A}(\mathbf{p})$ is of rank 1 with respect to R_i and R_k; however, the matrix $\mathbf{T}_i(\mathbf{p})$ is of rank 0 with respect to R_i and is of rank 1 with respect to R_k. According to Lemma 1.1 and these rank conditions, the functional dependency of

Q_i on R_i and R_k will be

$$Q_i(R_i, R_k) = \frac{\tilde{\alpha}_0 + \tilde{\alpha}_1 R_k}{\tilde{\beta}_0 + \tilde{\beta}_1 R_i + \tilde{\beta}_2 R_k + \tilde{\beta}_3 R_i R_k}, \qquad (1.114)$$

where $\tilde{\alpha}_0, \tilde{\alpha}_1, \tilde{\beta}_0, \tilde{\beta}_1, \tilde{\beta}_2, \tilde{\beta}_3$ are constants. Assuming that $\tilde{\beta}_3 \neq 0$, one can divide the numerator and denominator of Equation (1.114) by $\tilde{\beta}_3$, and obtain

$$Q_i(R_i, R_k) = \frac{\alpha_0 + \alpha_1 R_k}{\beta_0 + \beta_1 R_i + \beta_2 R_k + R_i R_k}. \qquad (1.115)$$

The constants $\alpha_0, \alpha_1, \beta_0, \beta_1, \beta_2$, can be determined by conducting five experiments by assigning five different sets of values to the pipe resistances (R_i, R_k), measuring the corresponding flow rates Q_i, and solving the obtained set of equations. ∎

EXAMPLE 1.9 (An example hydraulic network) Consider the unknown-model hydraulic network shown in Figure 1.29 and suppose that the flow is laminar, which results in the governing steady-state equations being linear.

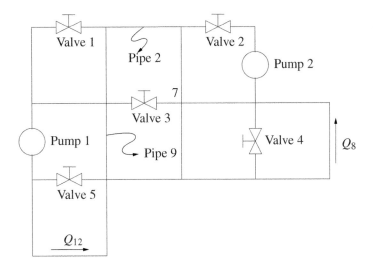

Figure 1.29 An unknown-model hydraulic network (Example 1.9)

Assume that the objective is to control the flow rates Q_8 and Q_{12} (as shown in Figure 1.29) to stay within the following ranges:

$$0.045 \leq Q_8 \leq 0.055 \ (\mathrm{m}^3/\mathrm{s}), \qquad (1.116)$$
$$0.01 \leq Q_{12} \leq 0.03 \ (\mathrm{m}^3/\mathrm{s}), \qquad (1.117)$$

by adjusting the radii of the pipes numbered 2 and 9, denoted by r_2 and r_9, respectively. Assume that these two pipes are made of piezoelectric materials and their radii can then be adjusted by applying an electrical field. In case the radius cannot be controlled by applying electrical fields, seven different pilot pipe sections (for this problem) have to be used for the experiments. Therefore, the design objective is to find regions in the r_2, r_9 plane for which the desired flow rates in Equations (1.116) and (1.117) are met.

Based on the results obtained in Section 1.5.2, one can determine the functional dependency of any flow rate on any two pipe resistances by at most seven measurements. Suppose that the measurements are done and let Table 1.7 show the numerical values of the measurements taken to find the functional dependency of Q_8 on r_2 and r_9.

Table 1.7 Numerical values for the experiments of the hydraulic network example (Example 1.9)

Exp. No.	r_2 (m)	R_2 (Pa.s/m³)	r_9 (m)	R_9 (Pa.s/m³)	Q_8 (m³/s)
1	0.05	408	0.05	408	0.038
2	0.07	107	0.08	62	0.043
3	0.09	39	0.11	17	0.049
4	0.1	26	0.13	9	0.051
5	0.12	12	0.15	5	0.054
6	0.14	6	0.17	3	0.055
7	0.17	3	0.2	1.6	0.056

Substituting these values into Equation (1.113) and solving for the unknown constants yields the following functional dependency (also recall Equation (1.100)):

$$Q_8(r_2, r_9) = \frac{8.7 \times 10^7 + \frac{1600}{r_2^4} + \frac{3500}{r_9^4} + \frac{0.034}{r_2^4 r_9^4}}{1.5 \times 10^9 + \frac{48{,}000}{r_2^4} + \frac{75{,}000}{r_9^4} + \frac{1}{r_2^4 r_9^4}}, \tag{1.118}$$

which is plotted in Figure 1.30(a). Applying the constraint in Equation (1.116) to Equation (1.118) yields the region shown (in black) in Figure 1.30(b) in the r_2, r_9 plane.

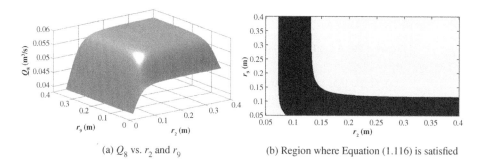

(a) Q_8 vs. r_2 and r_9 (b) Region where Equation (1.116) is satisfied

Figure 1.30 Q_8 vs. r_2 and r_9 (Example 1.9)

Similarly one can find the functional dependency of Q_{12} on r_2 and r_9, as plotted in

Figure 1.31(a). Figure 1.31(b) shows the region (in black) in the r_2, r_9 plane where the constraint in Equation (1.117) is satisfied.

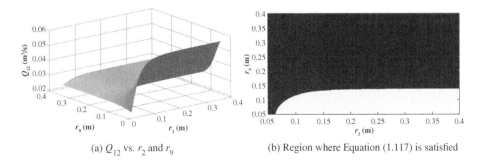

(a) Q_{12} vs. r_2 and r_9

(b) Region where Equation (1.117) is satisfied

Figure 1.31 Q_{12} vs. r_2 and r_9 (Example 1.9)

Intersecting the regions in Figures 1.30(b) and 1.31(b), one finds the region where both the constraints of Equations (1.116) and (1.117) are satisfied (see Figure 1.32).

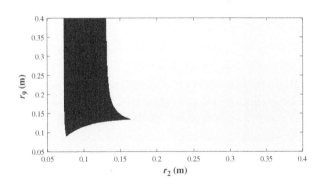

Figure 1.32 Region where Equations (1.116) and (1.117) are satisfied (Example 1.9)

1.6 Exercises

1.1 Measurements made at two pairs of terminals i and j of an unknown linear DC network in Figure 1.33 give the data shown in Table 1.8.
(a) Find the function $I_i(R_j)$.
(b) Determine the maximum and minimum values of I_i as R_j ranges from 0 to ∞.

1.2 Measurements made at a terminal of an unknown linear DC network given in Figure 1.34 give the data shown in Table 1.9. Determine:
(a) The function $I(R)$.
(b) The maximum power that can be extracted from the network by adjusting R.

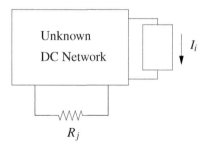

Figure 1.33 Unknown DC network (Exercise 1.1)

Table 1.8 Measurements (Exercise 1.1)

R_j	I_i
5 Ω	2 A
10 Ω	3 A
2 Ω	1 A

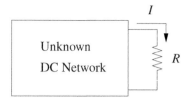

Figure 1.34 Unknown DC network (Exercise 1.2)

Table 1.9 Measurements (Exercise 1.2)

R	I
2 Ω	5 A
4 Ω	7 A

(c) The Thevenin equivalent of the network.
(d) The Norton equivalent of the network.
(e) The short-circuit current at the terminal.
(f) The open-circuit voltage.

1.3   Two sources u_1 and u_2 are to be used to control the scalar output y in an unknown linear system. The measurements in Table 1.10 are obtained.

Table 1.10 Measurements (Exercise 1.3)

Experiment No.	u_1	u_2	y
1	1	−1	1
2	1	1	2
3	2	1	−1

(a) Find an input $\begin{bmatrix} u_1 \\ u_2 \end{bmatrix}$ if possible so that $y = y^* = 5$.

(b) Find an input $\mathbf{u} = \begin{bmatrix} u_1 \\ u_2 \end{bmatrix}$ of minimum 2-norm so that $y = y^* = 5$.

1.4   A single source u is to be used to control two outputs y_1 and y_2 in an unknown linear system. The measurements in Table 1.11 are obtained from the system.
(a) Find the outputs for $u = 6, 10$, respectively.

Table 1.11 Measurements (Exercise 1.4)

Experiment No.	u	y_1	y_2
1	1	1	2
2	2	3	4

(b) Find the achievable set of output vectors **y** by adjusting $u \in [-10, 10]$.
(c) Find the achievable set of **y** for $u \in [-\infty, \infty]$.

1.5 An unknown linear AC network operating in steady state at a fixed frequency is shown in Figure 1.35 and the measurements obtained are shown in Table 1.12.

$y = I$

Table 1.12 Measurements (Exercise 1.5)

Experiment No.	u	y
1	$5\angle 60°$	$2\angle 45°$
2	$10\angle 30°$	$3\angle 60°$

Figure 1.35 Unknown AC network (Exercise 1.5)

Determine u^* so that $y^* = 20\angle 30°$.

1.6 Consider the unknown linear AC network in Figure 1.36 and the measurements obtained from it, shown in Table 1.13.

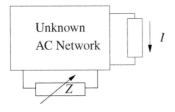

I

Table 1.13 Measurements (Exercise 1.6)

Experiment No.	Z	I
1	$2\angle 0°$	$5\angle 30°$
2	$4\angle 30°$	$4\angle 0°$
3	$6\angle 60°$	$2\angle 30°$

Figure 1.36 Unknown AC network (Exercise 1.6)

(a) Determine the function $I(Z)$.
(b) Find the impedance Z^* so that $I^* = 10\angle 30°$.

1.7 Measurements made on the unknown circuit in Figure 1.37, with $R_1 = 10\ \Omega$ and $R_2 = 20\ \Omega$, are tabulated in Table 1.14.
(a) Find the power P_0 consumed by the system as a function of u.
(b) Find the power P_1 consumed in R_1 as a function of u.
(c) Find the power P_2 consumed in R_2 as a function of u.

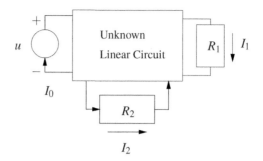

Figure 1.37 Unknown linear circuit (Exercise 1.7)

Table 1.14 Measurements (Exercise 1.7)

Experiment No.	u	I_0	I_1	I_2
1	2 V	1 A	6 A	5 A
2	5 V	7 A	3 A	6 A

(d) Find $P_1 + P_2$ as a function of u.

(e) Find the efficiency of power transfer $\frac{P_1 + P_2}{P_0}$ as a function of u.

In each of the above cases, display the result as a graph with $u \in [0, 100 \text{ V}]$.

1.8 Measurements made on the unknown DC circuit in Figure 1.38 are tabulated in Table 1.15.

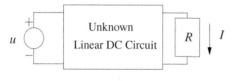

Figure 1.38 Unknown linear DC circuit (Exercise 1.8)

Table 1.15 Measurements (Exercise 1.8)

Experiment No.	R	I
1	2 Ω	10 A
2	10 Ω	5 A
3	4 Ω	8 A

Determine from these measurements the answer to the following:

(a) Are the measurements consistent?

(b) Find the function $I(R)$ and plot it for $R \in [0, 100 \text{ Ω}]$.

(c) Find the power $P(R)$ in R as a function of R and plot it for $R \in [0, 100 \text{ Ω}]$.

(d) Find, if possible, the "function" $R(P)$ and plot it for $P \in [0, 5000 \text{ W}]$.

1.7   Notes and References

The main results of this chapter were developed in Layek, Datta, and Bhattacharyya (2011), Layek et al. (2012), Mohsenizadeh et al. (2012), and Bhattacharyya, Keel, and

Mohsenizadeh (2014). A functional dependency of linear fractional form was first introduced in DeCarlo and Lin (1995). Also, some related works in the area of symbolic network functions can be found in Lin (1973) and Anderson and Lin (1973). Applications of Kirchhoff's laws in solving circuit analysis problems wherein the circuit model is available are given in Kirchhoff (1847). Thevenin's Theorem of circuit theory can be found in Thévenin (1883). See also Brittain (1990) and Johnson (2003).

For information about piezoelectric materials, see Tichý et al. (2010). Also, Nyce (2003) provides some information about sensors such as potentiometers and linear variable differential transformers (LVDTs). An application of finite-element methods in the analysis of truss structures can be found in Reddy (2006). For more details about the pressure-driven tests which are used to carry out measurements in hydraulic networks, see Giustolisi, Kapelan, and Savic (2008).

A two-story building frame example similar to the one in Example 1.8 can be found in Exercise 5.15 of Rao (2000, p. 260).

2 Classical Control Theory: A Brief Overview

In this chapter, we provide an introduction to classical control theory. This theory was developed mainly by Black, Nyquist, and Bode during the period 1925–1950 and deals with single-input single-output (SISO) linear time-invariant (LTI) systems, using Laplace transforms and frequency response methods. We have also included a description, with proof, of the Routh–Hurwitz criterion developed in 1877, and the Gauss–Lucas Theorem developed also in the 1800s as they are useful for determining stabilizing sets in parameter space. A good understanding of the above results is essential since these concepts dominate most applications. Moreover, a thorough understanding of SISO system theory is necessary for the understanding of multi-input multi-output (MIMO) systems.

2.1 Introduction

Progress in classical control theory was spearheaded by the development of the feedback amplifier by Black in 1926, and the Nyquist criterion in 1932. Black demonstrated how high-gain feedback could be used to build reliable systems using unreliable components. The Nyquist criterion developed a test, based on frequency response measurements on an open-loop system, to predict whether the closed-loop system would be stable or not. It also established gain and phase margins as quantitative measures of degree of stability for stable closed-loop systems. These were further developed by Bode and others into practical and efficient design methods for SISO feedback control systems. In the following sections, we briefly review these approaches beginning with Black's amplifier, operational amplifiers, and the unity feedback servomechanism.

2.2 Reliable Systems Using Unreliable Components: Black's Amplifier

In 1926 H. S. Black invented the feedback amplifier, as a solution to the problem of obtaining a reliable gain using unreliable components. Black's amplifier was the precursor to the operational amplifier (op-amp) which provides reliable addition, subtraction, multiplication, division, integration, and differentiation, and forms the backbone of modern signal processing technology. In this section, we describe the main concept underlying Black's amplifier.

Consider an amplifier shown, with nominal gain $A = A^0 = 100$, for example, built with components which have 50% reliability. Then A varies in the uncertainty interval

$$A \in [50, 150].$$ (2.1)

To overcome this uncertainty, Black proposed the high-gain feedback system (Figure 2.1), in which A_c is built with the same unreliable components, with high nominal gain, say $A_c^0 = 10,000$.

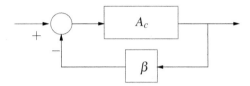

Figure 2.1 A feedback system: $A_c^0 = 10,000$; $\beta = 0.01$

The closed-loop system gain is, nominally:

$$A_{cl}^0 := \frac{A_c^0}{1 + A_c^0 \beta} = \frac{10,000}{1 + 100} = 99.0099.$$ (2.2)

Under perturbations, A_c varies in the interval

$$A_c \in [5000, 15,000].$$ (2.3)

However, A_{cl} varies in the relatively small interval $\left[\underline{A}_{cl}, \bar{A}_{cl}\right]$:

$$\underline{A}_{cl} = \frac{5000}{1 + 50} = 98.0392, \qquad \bar{A}_{cl} = \frac{15,000}{1 + 150} = 99.3377.$$ (2.4)

This is a remarkable improvement over Equation (2.1) considering that the same unreliable components make up A_c. To explain this, observe that

$$A_{cl} = \frac{A_c}{1 + A_c \beta} = \frac{1}{\frac{1}{A_c} + \beta},$$ (2.5a)

$$\lim_{A_c \to \infty} A_{cl} = \frac{1}{\beta} = 100.$$ (2.5b)

This shows that A_{cl} becomes independent of the unreliable gain A_c as $A_c \to \infty$. On the other hand, A_{cl} is completely determined in the limit $A_c \to \infty$, by the feedback gain β. Thus, the high-gain feedback system transfers the "gain sensitivity" of A_{cl} from A_c to β. It is thus appropriate to regard Black's amplifier as the very first result on robust control. In modern control systems, the same principle is used in integral control, which has infinite DC gain, and thus reliably zeros the tracking error asymptotically for step inputs.

2.3   Op-amp Circuits

Operational amplifiers (op-amps) allow us to carry out addition, subtraction, multiplication, division, integration, and differentiation, *accurately and reliably*. How do they do this? The answer is, again, *high-gain feedback* as we explain below.

First, consider an ordinary voltage divider shown in Figure 2.2.

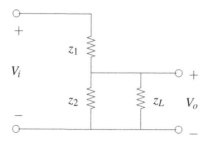

Figure 2.2 A circuit with load z_L

It is easy to see that

$$\frac{V_o}{V_i} = \frac{1}{1 + \frac{z_1}{z_2} + \frac{z_1}{z_L}} \tag{2.6}$$

and thus the "gain" varies from 0 to $\frac{1}{1 + \frac{z_1}{z_2}}$ as the "load" z_L varies from 0 to ∞. An op-amp circuit can be used to render the gain independent of the load as we show below.

An Op-amp Circuit
Now consider the op-amp circuit in Figure 2.3, where a high-gain amplifier has been inserted between the input and output.

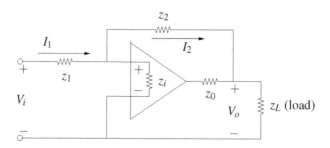

Figure 2.3 Op-amp circuit

If A denotes the amplifier gain, the system equations are:

$$V_i - z_1 I_1 - z_2 I_2 = V_o, \tag{2.7a}$$

$$A(I_1 - I_2) z_i + \left(I_2 - \frac{V_o}{z_L} \right) z_0 = V_o, \tag{2.7b}$$

$$V_i - z_i(I_1 - I_2) - z_1 I_1 = 0. \tag{2.7c}$$

From Equations (2.7a) and (2.7c):

$$z_i(I_1 - I_2) - z_2 I_2 = V_o. \tag{2.8}$$

Using Equations (2.7b) and (2.8), we get

$$V_o = \underbrace{\left(\frac{z_0 + A z_2}{1 - A + \dfrac{z_0}{z_L}} \right)}_{G(A)} I_2. \tag{2.9}$$

Note that

$$G(A) = \frac{\dfrac{z_0}{A} + z_2}{\dfrac{1}{A} - 1 + \dfrac{z_0}{A z_L}} \tag{2.10}$$

so that

$$\lim_{A \to \infty} G(A) = -z_2. \tag{2.11}$$

Therefore, as A approaches ∞:

$$V_o = -z_2 I_2. \tag{2.12}$$

From Equations (2.12) and (2.8), we obtain

$$I_1 = I_2 \tag{2.13}$$

and from Equation (2.7c):

$$V_i = z_1 I_1. \tag{2.14}$$

Finally, Equations (2.12) and (2.14) yield

$$\frac{V_o}{V_i} = -\frac{z_2}{z_1}. \tag{2.15}$$

Equation (2.15) shows that the "gain" is independent of the load z_L. By choosing $z_1 = R_1$ and $z_2 = R_2$, we obtain a multiplier; with $z_1 = R_1$ and $z_2 = \frac{1}{sC}$, we obtain an integrator; and with $z_1 = R$ and $z_2 = sL$, we obtain a differentiator. In each case, the mathematical operation is carried out with extreme precision since the gain $G(A)$ is independent of the load impedance z_L and the input and output impedances z_1 and z_0, as A approaches ∞. Equation (2.10) shows that A multiplies the load impedance z_L and divides the output impedance z_0.

2.4   Robust Feedback Linearization

In many instances, a physical element may have a nonlinear characteristic $y = f(u)$
(see Figure 2.4) and it may be desirable, for design or measurement purposes, to have a
linear input–output relationship.

Figure 2.4 A nonlinear relationship

This could be achieved by the high-gain feedback system shown in Figure 2.5 without
detailed knowledge of the nonlinear characteristic.

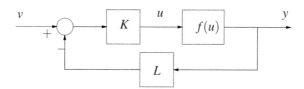

Figure 2.5 A high-gain "linear" feedback system

Since

$$v = Ly + \frac{1}{K}f^{-1}(y) \tag{2.16}$$

it follows that as K approaches ∞, v approaches Ly.

EXAMPLE 2.1   Consider a nonlinear system with

$$f(u) = 10u^3 = y. \tag{2.17}$$

Choose $L = 10$ and $K = 100$ in Figure 2.5. Then

$$v = 10y + \frac{1}{100}\sqrt[3]{\frac{y}{10}}. \tag{2.18}$$

If y varies between 0 and 10, we have v varying between 0 and 100 linearly with y with
an error less than 1%.

This example shows clearly that a nonlinear system can be rendered linear in the
closed loop by high-gain feedback. Thus an integral feedback controller (high gain at
$s = 0$) driving a nonlinear system will exhibit "linear" behavior. This is discussed next.

2.5   High Gain in Control Systems

A typical feedback control system is represented by the block diagram in Figure 2.6.

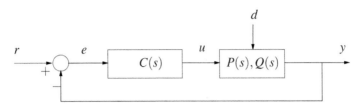

Figure 2.6 A typical feedback control system

Assuming all systems are linear and time-invariant and using transfer function matrix representations, we have

$$Y(s) = P(s)U(s) + Q(s)D(s), \qquad (2.19a)$$
$$U(s) = C(s)E(s), \qquad (2.19b)$$
$$E(s) = R(s) - Y(s), \qquad (2.19c)$$

where $P(s)$ and $Q(s)$ are the transfer functions between inputs u and d and output y, respectively. So

$$E(s) = \underbrace{[I+P(s)C(s)]^{-1}R(s)}_{E_r(s)} - \underbrace{[I+P(s)C(s)]^{-1}Q(s)D(s)}_{E_d(s)}. \qquad (2.20)$$

Equation (2.20) may be represented as in Figure 2.7.

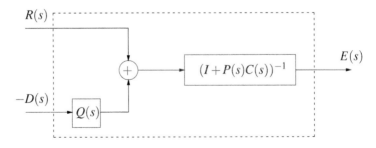

Figure 2.7 Diagram of Equation (2.20)

In this representation we see that the task of the control system is to maintain $e(t)$ "close" to zero even though persistent and unstable signals $r(t)$ and $d(t)$ may be working against it. The figure also shows that $e(t)$ may be maintained "close" to zero as long as $(I+P(s)C(s))^{-1}$ is "small." Since $P(s)$ is given, the latter condition means that $C(s)$ must be "large" or equivalently that

$$P(s)C(s) =: L(s), \qquad (2.21)$$

called the *loop gain*, must be "large" in order to achieve "small" errors with respect to reference inputs and disturbance simultaneously and independently. Since signals may be represented by their Fourier transforms it follows that $L(j\omega)$ must have large magnitude over the frequency band where r and d have significant "energy." In general this requirement runs into conflict with the requirement of closed-loop stability for most practical systems. Indeed, mastering this balancing act between performance and stability constitutes the major challenge in control system design. A compromise can often be reached by making the loop gain high pointwise. In this regard note that if $s = \alpha$ is a pole of $C(s)$ we have $C(\alpha) = \infty$ and thus $(I + P(\alpha)C(\alpha))^{-1} = 0$, which means that exponentials signals with exponent α are asymptotically tracked with zero steady-state error. The special case $\alpha = 0$ occurs when the signals to be tracked or rejected are constants and corresponds to the requirement that the controller be an integral controller. These ideas are explored in the next section.

2.6   Unity Feedback Servomechanisms

Most control systems exist to automatically execute the commands given by humans as exemplified by robots. A central problem in control engineering is the design of a controller C, that makes the output y of the plant track reference commands r and reject disturbances d, automatically. This is usually accomplished by driving the controller C with the tracking error $r - y$, which it must process to generate the plant input u to force the plant output y into correspondence with r. This structure is represented by the unity feedback block diagram shown in Figure 2.8.

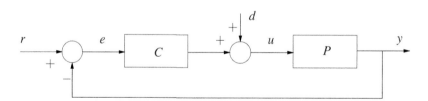

Figure 2.8 Unity feedback servomechanism

When the plant and controller are LTI systems, they may be represented by their respective transfer functions $P(s)$ and $C(s)$ and all signals may be represented by their Laplace transforms. Usually $P(s)$ and $C(s)$ are real, rational, proper (numerator degree $\leq$ denominator degree) and can be written as ratios of polynomials:

$$P(s) = \frac{n_p(s)}{d_p(s)}, \quad C(s) = \frac{n_c(s)}{d_c(s)} \qquad (2.22)$$

in the Laplace transform or s domain, and thus Figure 2.8 becomes Figure 2.9.

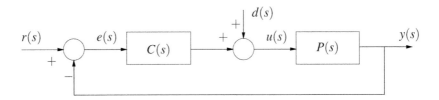

Figure 2.9 Transfer function representation of Figure 2.8

From Figure 2.9, we have

$$y(s) = P(s)u(s), \tag{2.23a}$$
$$u(s) = C(s)e(s) + d(s), \tag{2.23b}$$
$$e(s) = r(s) - y(s). \tag{2.23c}$$

Eliminating $u(s)$ and $y(s)$ from the above, we obtain

$$e(s) = \frac{1}{1 + P(s)C(s)}r(s) - \frac{P(s)}{1 + P(s)C(s)}d(s). \tag{2.24}$$

Substituting for $P(s), C(s)$ from Equation (2.22):

$$e(s) = \frac{d_p(s)d_c(s)}{d_p(s)d_c(s) + n_p(s)n_c(s)}r(s) - \frac{n_p(s)d_c(s)}{d_p(s)d_c(s) + n_p(s)n_c(s)}d(s). \tag{2.25}$$

2.7   Closed-Loop Stability

The denominator in Equation (2.25) clearly represents the *characteristic polynomial* of the closed-loop system, defined therefore as $d_{cl}(s)$:

$$d_{cl}(s) := d_p(s)d_c(s) + n_p(s)n_c(s). \tag{2.26}$$

The closed-loop system is asymptotically stable if and only if $d_{cl}(s)$ has all its roots, called closed-loop characteristic roots, with negative real parts, that is, in $\mathbb{C}^-$, the open left-half plane. This property will sometimes be referred to as Hurwitz stability of $d_{cl}(s)$ or alternatively, internal stability of the closed-loop system.

    Internal stability of the closed-loop system means that when the external signals r and d are switched off (set to zero), the response of all signals in the loop to arbitrary initial conditions in the plant and controller decay asymptotically to zero. The rate of decay is determined by the exponentials $e^{-\sigma_i t}$, where σ_i are the distances of the characteristic roots from the imaginary axis. Closed-loop stability will be a standing basic requirement of the unity feedback system above.

2.8   Classes of Reference and Disturbance Signals

In typical applications, the specific reference or disturbance signals are unknown a priori, but the types of reference and disturbance signals to be handled by the control system are specified during the design process to belong to classes such as steps, ramps, and sinusoids. That is, the signals are persistent (steps, sinusoids) and/or are "unstable" (ramps, for example). Signals of this type can be generated by placing initial conditions on a "signal generator," that is, a differential equation of the type

$$m(D)r(t) = 0, \tag{2.27a}$$
$$m(D)d(t) = 0, \tag{2.27b}$$

where $D := \frac{d}{dt}$ is the differentiation operator and $m(D)$ is a polynomial in D with obvious corresponding meaning. For example, if $r(t)$ can be arbitrary steps or ramps and $d(t)$ an arbitrary sine wave of frequency ω_0 rad/s, we choose

$$m(D) = D^2 \left(D^2 + \omega_0^2 \right). \tag{2.28}$$

In general, the requirement of asymptotic tracking and disturbance rejection in a servomechanism means that arbitrary signals $r(t)$ satisfying Equation (2.27a) must be tracked and arbitrary disturbances $d(t)$ satisfying Equation (2.27b) must be rejected independently of each other.

Taking Laplace transforms in Equations (2.27a) and (2.27b), we have

$$r(s) = \frac{r_0(s)}{m(s)}, \tag{2.29a}$$
$$d(s) = \frac{d_0(s)}{m(s)}, \tag{2.29b}$$

where $(r_0(s), d_0(s))$ are polynomials. For example, if

$$m(D) = D^2, \tag{2.30}$$

we have

$$\ddot{r}(t) = 0 \tag{2.31}$$

so that

$$s^2 r(s) - s r(0) - \dot{r}(0) = 0 \tag{2.32}$$

and

$$r(s) = \frac{\dot{r}(0) - s r(0)}{s^2}. \tag{2.33}$$

By suitable choices of $r(0)$ and $\dot{r}(0)$ in Equation (2.33), all steps and ramps can be generated. Similarly, if

$$m(D) = D^2 + \omega_0^2, \tag{2.34}$$

we have

$$\ddot{r}(t) + \omega_0^2 r(t) = 0 \tag{2.35}$$

so that

$$s^2 r(s) - sr(0) - \dot{r}(0) + \omega_0^2 r(s) = 0 \tag{2.36}$$

and

$$r(s) = \frac{\dot{r}(0) + sr(0)}{s^2 + \omega_0^2}. \tag{2.37}$$

Again, by suitable choice of $r(0)$ and $\dot{r}(0)$, any sinusoid of frequency ω_0 rad/s and arbitrary amplitude and phase may be generated.

2.9 Asymptotic Tracking and Disturbance Rejection

We rewrite Equations (2.24) and (2.25) as

$$e(s) = \underbrace{E_r(s)r(s)}_{e_r(s)} - \underbrace{E_d(s)d(s)}_{e_d(s)}, \tag{2.38}$$

where

$$E_r(s) := \frac{1}{1 + P(s)C(s)} = \frac{d_p(s)d_c(s)}{d_{cl}(s)}, \tag{2.39a}$$

$$E_d(s) := \frac{P(s)}{1 + P(s)C(s)} = \frac{n_p(s)d_c(s)}{d_{cl}(s)}, \tag{2.39b}$$

$$r(s) = \frac{r_0(s)}{m(s)}, \qquad d(s) = \frac{d_0(s)}{m(s)}, \tag{2.39c}$$

and assume that the zeros of $m(s)$ lie in $\mathbb{C}^+$, the closed right-half plane (RHP).

For asymptotic tracking and disturbance rejection, we must have

$$\lim_{t \to \infty} e_r(t) = 0,$$
$$\lim_{t \to \infty} e_d(t) = 0, \tag{2.40}$$

for all initial conditions in the plant, controller, and signal generator. Since

$$\mathscr{L}[e_r(t)] = e_r(s) = \frac{d_p(s)d_c(s)}{d_{cl}(s)} \cdot \frac{r_0(s)}{m(s)},$$
$$\mathscr{L}[e_d(t)] = e_d(s) = \frac{n_p(s)d_c(s)}{d_{cl}(s)} \cdot \frac{d_0(s)}{m(s)}, \tag{2.41}$$

where $d_{cl}(s)$ is Hurwitz, and $m(s)$ is anti-Hurwitz (all roots in $\mathbb{C}^+$), it follows that Equation (2.40) is satisfied if and only if

$$m(s) \mid d_c(s) \quad \text{(i.e., } m(s) \text{ divides } d_c(s)), \tag{2.42}$$

that is, the unstable poles in Equation (2.41) are cancelled perfectly by the zeros in Equation (2.41) assigned by $d_c(s)$. Thus, the controller $C(s)$ must be of the form

$$C(s) = \frac{\alpha(s)}{m(s)\beta(s)} =: \frac{n_c(s)}{d_c(s)},$$ (2.43)

where $\alpha(s), \beta(s)$ must be chosen so that

$$d_{cl}(s) = m(s)\beta(s)d_p(s) + \alpha(s)n_p(s)$$ (2.44)

is Hurwitz. Clearly, this requires that $n_p(s)$ and $m(s)d_p(s)$ be coprime over $\mathbb{C}^+$. For $C(s)$ to be proper, which is a common requirement in most applications, we may need in addition that

$$\deg[\alpha(s)] \le \deg[m(s)] + \deg[\beta(s)].$$ (2.45)

It is not difficult to show that by choosing $\beta(s)$ to be of degree $n-1$ all closed-loop characteristic roots, that is, roots of $d_{cl}(s)$, could be assigned arbitrarily and hence $d_{cl}(s)$ could, in particular, be rendered Hurwitz. The lowest-degree $\beta(s)$ satisfying Equation (2.45) and rendering $d_{cl}(s)$ Hurwitz is, in general, unknown.

EXAMPLE 2.2 (Pole assignment servocontroller)   Suppose

$$P(s) = \frac{s-1}{(s+1)^2}$$ (2.46)

and all steps and ramps must be tracked or rejected asymptotically. Then the controller

$$C(s) = \frac{\alpha_0 + \alpha_1 s + \alpha_2 s^2 + \alpha_3 s^3}{s^2(s+\beta_0)}$$ (2.47)

will provide asymptotic tracking and disturbance rejection provided the α_i, β_0 are chosen to stabilize

$$d_{cl}(s) = s^2(s+\beta_0)(s+1)^2 + (\alpha_0 + \alpha_1 s + \alpha_2 s^2 + \alpha_3 s^3)(s-1).$$ (2.48)

This may be done by choosing any monic Hurwitz polynomial $h(s)$ of degree 5 and setting

$$d_{cl}(s) = h(s).$$ (2.49)

Thus, all five characteristic roots are arbitrarily assignable by this controller, which also provides asymptotic tracking and disturbance rejection for the class of external reference and disturbance signals specified, that is steps and ramps in this case. This type of controller is therefore called a *pole assignment servocontroller*.

2.10 Integral Control: PI and PID Controllers

In many, if not most, applications, it is sufficient to design the controller to track and reject constant references and disturbances. In this case

$$\dot{r}(t) = 0, \quad \dot{d}(t) = 0 \tag{2.50}$$

and

$$r(s) = \frac{r_0}{s}, \quad d(s) = \frac{d_0}{s}. \tag{2.51}$$

Then the servomechanism controller must be of the form

$$C(s) = \frac{n_c(s)}{s\bar{d}_c(s)}. \tag{2.52}$$

A common choice of $\bar{d}_c(s)$ and $n_c(s)$ is

$$\bar{d}_c(s) = 1, \quad n_c(s) = sK_p + K_i, \tag{2.53}$$

resulting in

$$C(s) = \frac{sK_p + K_i}{s} = K_p + \frac{K_i}{s}, \tag{2.54}$$

which is known as a *proportional integral* (PI) controller because its action amounts to

$$u(t) = \underbrace{K_p e(t)}_{\text{proportional term}} + \underbrace{K_i \int_0^t e(\tau)d\tau}_{\text{integral term}}. \tag{2.55}$$

It is sometimes also possible to choose

$$\bar{d}_c(s) = 1, \quad n_c(s) = K_i + sK_p + s^2 K_d, \tag{2.56}$$

resulting in the *proportional integral derivative* (PID) controller:

$$C(s) = K_p + \frac{K_i}{s} + sK_d. \tag{2.57}$$

Note that $C(0) = \infty$ for both PI and PID controllers and thus when the closed-loop system is stable, both provide perfect steady-state tracking and disturbance rejection of arbitrary step inputs.

It is worth noting that PI and PID controllers account for 99% of all controllers in use in the world, across all industries.

EXAMPLE 2.3   Consider the plant and PI controller pair

$$P(s) = \frac{1}{s^2 + s + 1}, \quad C(s) = \frac{s+1}{s}. \tag{2.58}$$

This example illustrates via Simulink that the controller is designed such that the closed-loop system is stable and asymptotically tracks a step reference input beginning at $t = 0$ and also rejects a step disturbance that occurs at $t = 50$ s. We begin by opening Simulink

Model. The resources to construct Simulink models are found here: for example, the blocks representing systems in "continuous," signal generators such as step input and time in "source," the data collection blocks in "sink," and adders in "math operations." Figure 2.10 shows the Simulink model used to solve this example problem. The input parameter values for the plant and controller are shown in Figure 2.11.

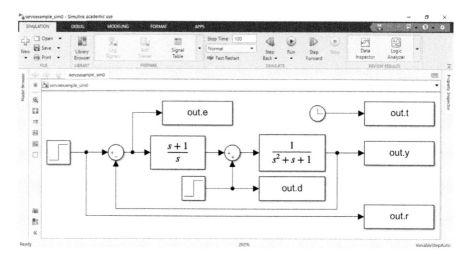

Figure 2.10 Simulink model (Example 2.3)

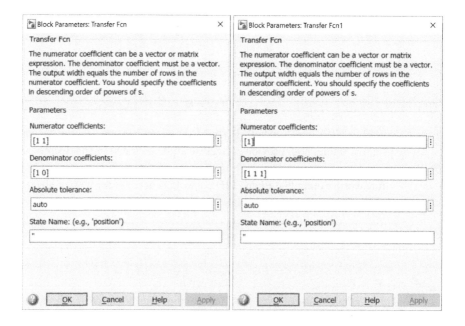

Figure 2.11 Parameters for Simulink plant/controller blocks (Example 2.3)

Similarly, the input parameters of the reference input and the disturbance blocks are shown in Figure 2.12.

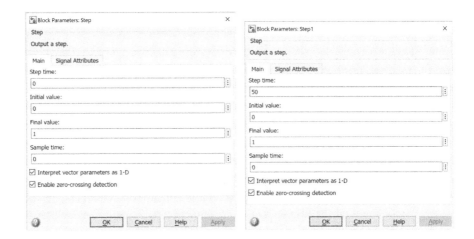

Figure 2.12 Parameters for Simulink reference/disturbance blocks (Example 2.3)

Figure 2.13 shows the output response $y(t)$ and the error $e(t)$ of the closed-loop system.

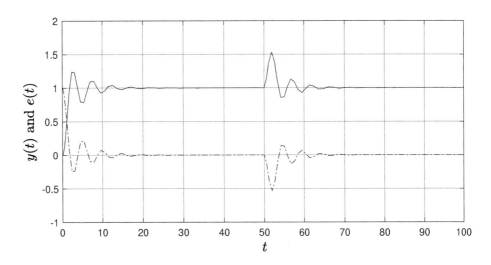

Figure 2.13 Output $y(t)$ and errot $e(t)$ (Example 2.3)

These results clearly show that the closed-loop system asymptotically tracks the step reference input and also rejects the disturbance occurring at $t = 50$ s, and continues to track the step reference input.

2.11   The Nyquist Criterion

The Nyquist criterion considers the system in Figure 2.14, where $G_{OL}(G_{CL})$ refers to the open-loop (closed-loop) system.

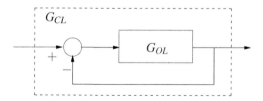

Figure 2.14 Open and closed-loop systems

The problem considered by Nyquist was the prediction of closed-loop stability or instability from measurements made on the open-loop system G_{OL}. Nyquist's solution was based on the Principle of the Argument from complex variable theory. This principle states that if a rational function $F(z)$ is evaluated as z moves, say, clockwise on a simple closed curve Γ in the complex plane, the resulting plot denoted Γ_F encircles the origin of the complex plane $z - p$ times in the clockwise direction, where $z(p)$ is the number of zeros (poles) of $F(z)$ inside Γ.

If G_{OL} is represented by the rational, proper transfer function

$$G_0(s) = \frac{N(s)}{D(s)}, \tag{2.59}$$

the closed-loop characteristic polynomial is

$$D_{CL}(s) = D(s) + N(s) = D(s)[1 + G_0(s)]. \tag{2.60}$$

If the $(D(s), N(s))$ pair is coprime, we can easily show that

$$D_{CL}(s_0) = 0 \quad \Longleftrightarrow \quad 1 + G_0(s_0) = 0, \tag{2.61}$$

that is, the characteristic roots are identical to the zeros of

$$F(s) := 1 + G_0(s) = \frac{D_{CL}(s)}{D(s)}. \tag{2.62}$$

Now let us bring in the Nyquist contour Γ shown in Figure 2.15 enclosing the entire closed right-half plane ($\mathbb{C}^+$). To apply the Nyquist criterion, the following steps are carried out:

A. Evaluate $G_0(s^*)$ as s^* moves on Γ clockwise and executes one circuit. Call Γ_{G_0} the resulting closed curve, known as the Nyquist plot of G_0. Note that for s^* on the segment ab, $s^* = j\omega$, and $G_0(s^*)$ is thus the *frequency response* $G_0(j\omega)$ for $\omega \in [0, \infty)$, which can often be obtained experimentally by measuring the response of the system to sinusoidal inputs. If $G_0(s)$ is proper (strictly proper), $G_0(\infty) = d(0)$, that is, a constant for s^* on bcd. Finally, if $G_0(s)$ has real coefficients, $G_0(-j\omega) = G_0^*(j\omega)$ so, as s^* moves from d to a, we obtain the mirror image of $G_0(j\omega)$, $\omega \geq 0$ about the real axis.

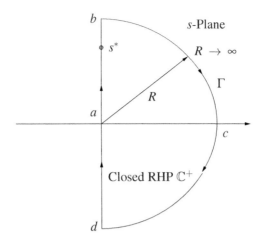

Figure 2.15 Nyquist contour

B. Count the number of clockwise encirclements of the point $-1 + j0$ by the Nyquist plot Γ_{G_0} and call it N. Counterclockwise encirclements are counted negatively. Thus N could be positive, zero, or negative.

C. Letting P_{CL}^+ (P_{OL}^+) denote the number of zeros of $D_{CL}(s)$ ($D(s)$) in $\mathbb{C}^+$, we must have

$$N = P_{CL}^+ - P_{OL}^+. \tag{2.63}$$

Equation (2.63) follows from the fact that $F(s)$ evaluated on Γ encircles the origin N times in the clockwise direction. By the Principle of the Argument and Equation (2.62), it follows that Equation (2.63) must hold.

D. For closed-loop stability, we require that

$$P_{CL}^+ = 0 \tag{2.64}$$

and therefore we must have

$$N = -P_{OL}^+. \tag{2.65}$$

The condition in Equation (2.65), which is a property of the open-loop system, is the statement of the Nyquist criterion for closed-loop stability. For an open-loop stable system ($P_{OL}^+ = 0$), there must in particular be no encirclements of $-1 + j0$, by the Nyquist plot of G_0.

EXAMPLE 2.4   Suppose

$$G_0(s) = \frac{K}{s(s+1)^2}. \tag{2.66}$$

Consider the two cases: (a) $K = 1$, (b) $K = 10$. The Nyquist plots of G_0 for the two cases are sketched in Figure 2.16. The pole at $s = 0$ is infinitesimally displaced into the left-half plane, to facilitate construction of this plot. Thus, the open-loop plant is stable.

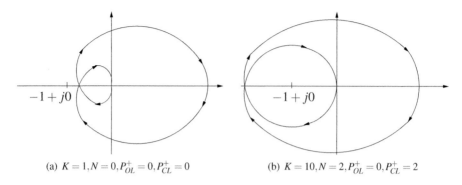

(a) $K = 1, N = 0, P_{OL}^+ = 0, P_{CL}^+ = 0$ (b) $K = 10, N = 2, P_{OL}^+ = 0, P_{CL}^+ = 2$

Figure 2.16 Nyquist plots: $P_{CL}^+ = 0$ with $K = 1$ and $P_{CL}^+ = 2$ with $K = 10$ (Example 2.4)

In case (a), the Nyquist plot Γ_{G_0} does not encircle $-1 + j0$, thus $N = 0$, and so $P_{CL}^+ = 0$, that is, the closed-loop system is stable. In case (b), the Nyquist plot Γ_{G_0} encircles $-1 + j0$ twice in the clockwise direction, thus $N = 2$ and so $P_{CL}^+ = 2$. Note also that because of symmetry about the real axis, only half of the Nyquist plot is needed.

MATLAB Solution
It is noted that the command "nyquist" does not provide sufficient information for the case when the plant has one or more poles on the imaginary axis. This is because $|G_0(j\omega)|$ becomes infinity at those frequencies. In this case, a third-party function "nyqlog" we use below is recommended (see Figure 2.17).

```
K=1;
n=K;
d=conv([1 0],conv([1 1],[1 1]));
G0=tf(n,d);
nyqlog(G0)
```

Figure 2.17 $K = 1, N = 0, P_{OL}^+ = 0, P_{CL}^+ = 0$ (Example 2.4)

For more information, including downloading the m file, the reader should refer to the *file exchange page* located at http://mathworks.com.

EXAMPLE 2.5   Consider

$$G_0(s) = \frac{s-1}{s^2(s+1)}.$$ (2.67)

The Nyquist plot in Figure 2.18 is obtained from the MATLAB script below.

```
n=[1 -1];
d=conv([1 0 0],[1 1]);
G0=tf(n,d);
nyqlog(G0)
```

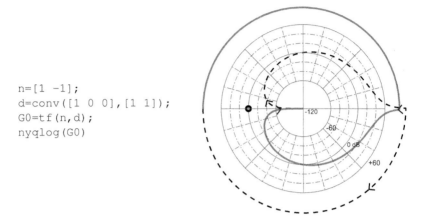

Figure 2.18 $N = 1, P_{OL}^+ = 0, P_{CL}^+ = 1$ (Example 2.5)

The Nyquist plot Γ_{G_0} encircles $-1 + j0$ once in the clockwise direction, thus $N = 1$ and so $P_{CL}^+ = 1$. Therefore, the closed-loop system is unstable with one RHP pole.

2.12   Counting Encirclements by the Nyquist Plot Using Bode Frequency Response

The Nyquist plot is a complex plane plot of $G(j\omega)$, the frequency response of the open-loop system. Once the plot is generated, the number of clockwise encirclements N of the point $-1 + j0$ needs to be found. It is possible, as an alternative, to determine N from the frequency response magnitude and phase plots (Bode plots) of $G(j\omega)$, as we illustrate in Figure 2.19.

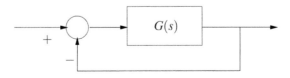

Figure 2.19 Unity feedback system

Consider, for example, the Nyquist plot displayed for non-negative frequencies in Figure 2.20.

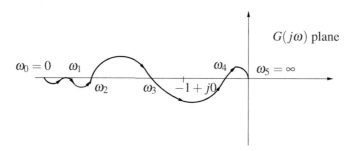

Figure 2.20 A typical Nyquist plot

In this example, we see that the number of *clockwise* encirclements of $-1 + j0$ can be counted as follows. There is one counterclockwise encirclement at $\omega_0 = 0$, none at $\omega = \omega_1$, two clockwise encirclements at $\omega = \pm\omega_2$ to account for the symmetry of the Nyquist plot about the real axis, similarly two counterclockwise encirclements at $\omega = \pm\omega_3$, and none at ω_4 and ω_5, leading to $N = -1$. Generalizing this example, we can write the following procedure (algorithm) to determine N from Bode plots when $G(s)$ has no $j\omega$ axis poles:

1. From the phase plot $\angle G(j\omega)$, determine the frequencies $\omega_0, \omega_1, \ldots, \omega_{l-1}, \omega_l = \infty$ where

$$|G(j\omega_c)| > 1 \quad \text{and}$$
$$\phi(\omega_i) := \angle G(j\omega_i) = \pm n\pi, \qquad \text{for } n = 1, 3, 5, \ldots. \tag{2.68}$$

2. Define

$$\Omega := \{\omega_0, \omega_1, \ldots, \omega_{l-1}, \omega_l\}. \tag{2.69}$$

Note that $\omega_0 = 0$ or $\omega_l = \infty$ in particular are included in Ω only if they satisfy Equation (2.68).

3. Map Ω into $\{+1, 0, -1\}$ as follows:

$$T(\omega_j) = i_j = \begin{cases} +1, & \text{if } \dot{\phi}(\omega_j) < 0 \\ -1, & \text{if } \dot{\phi}(\omega_j) > 0 \quad \text{for } j \in \{0, 1, 2, \ldots, l-1, l\}. \\ 0, & \text{if } \dot{\phi}(\omega_j) = 0 \end{cases} \tag{2.70}$$

4. $N = i_0 + \sum_{j=1}^{l-1} 2i_j + i_l$. When the system has a plant and controller, $G(s) = P(s)C(s)$ and

$$|G(j\omega)| = |P(j\omega)||C(j\omega)|, \tag{2.71a}$$
$$\angle G(j\omega) = \angle P(j\omega) + \angle C(j\omega), \tag{2.71b}$$
$$\dot{\phi}_G(j\omega) = \dot{\phi}_P(j\omega) + \dot{\phi}_C(j\omega), \tag{2.71c}$$

and the above algorithm can be suitably modified to obtain conditions on $C(s)$ for it to stabilize a given $P(s)$. This is discussed in the following sections.

2.13 Stabilizing Gains from the Nyquist Plot

Consider a unity feedback system with $KG(s)$ in the forward path. The Nyquist plot of $G(s)$ can be used to determine the set of all stabilizing gains. Assume for simplicity that $G(s)$ has no imaginary axis poles. Let

$$G(j\omega) = G_r(\omega) + jG_i(\omega) \tag{2.72}$$

determine all frequencies $\omega = \omega_j$ positive, zero, and negative for which

$$G_i(\omega_j) = 0, \quad \text{for } j = 0, 1, 2, \ldots. \tag{2.73}$$

Let

$$x_t := G_r(\omega_t), \quad \text{for } t = 1, 2, \ldots, l-1 \tag{2.74}$$

and order the real numbers x_t as

$$+\infty =: x_0 > x_1 > x_2 > \cdots > x_{l-1} > -\infty =: x_l. \tag{2.75}$$

Define the real axis segments

$$S_i := (x_{i+1}, x_i), \quad \text{for } i = 0, 1, 2, \ldots, l-1 \tag{2.76}$$

and the corresponding segments K_i:

$$K_i := \left(-\frac{1}{x_{i+1}}, -\frac{1}{x_i} \right), \quad \text{for } i = 0, 1, 2, \ldots, l-1. \tag{2.77}$$

Finally, let n_i denote the net number of *counterclockwise* encirclements of S_i by the Nyquist plot of $G(s)$. Call $\mathscr{S}_0$ the set of segments satisfying

$$\mathscr{S}_0 = \{S_i \ : \ n_i = P_{OL}^+, i = 1, 2, \ldots, l-1\}, \tag{2.78}$$

where P_{OL}^+ is the number of open-loop RHP poles and let $\mathscr{K}_0$ denote the union of the corresponding K segments

$$\mathscr{K}_0 = \{\cup K_i \ : \ S_i \in \mathscr{S}_0\}. \tag{2.79}$$

THEOREM 2.1 $\mathscr{K}_0$ is the complete set of stabilizing gains for the feedback system.

Proof The proof follows from the Nyquist criterion applied to $KG(s)$ and observing that $KG(j\omega)$ encircles $-1 + j0$ the same number of times as $G(j\omega)$ encircles the $-\frac{1}{K} + j0$ point. ∎

EXAMPLE 2.6 Consider the Nyquist plot of a system with $P_{OL}^+ = 1$ in Figure 2.21. Clearly, the only stabilizing segment of gains is $\left(-\dfrac{1}{x_3}, -\dfrac{1}{x_2} \right)$.

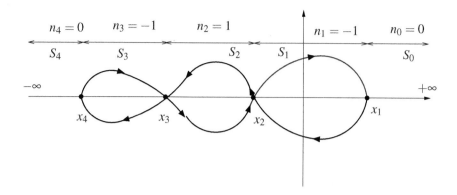

Figure 2.21 Nyquist plot of $G(s)$ (Example 2.6)

2.14   Gain and Phase Margin

Consider the stabilizing segment determined from the Nyquist plot shown in Figure 2.22.

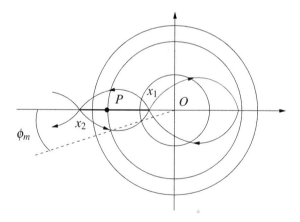

Figure 2.22 Nyquist plot with stabilizing segment

The endpoints of the segments have coordinates x_1, x_2 and this corresponds to the gain segment

$$(K_1, K_2) \quad \text{where} \quad K_1 := -\frac{1}{x_2}, \quad K_2 := -\frac{1}{x_1}. \tag{2.80}$$

For an operating gain, corresponding to the point P with coordinate z:

$$K = -\frac{1}{z}. \tag{2.81}$$

The gain margin, which is the distance to instability on the real axis, is *defined* as

$$g_m(K) = \min\{|K - K_1|, |K - K_2|\} \tag{2.82}$$

and the phase margin, which is the angular distance to instability, is $\phi_m(K)$ as shown. By drawing a family of circles, we can determine the gain and phase margins corresponding to each gain on the segment in Equation (2.80), as displayed in Figure 2.23.

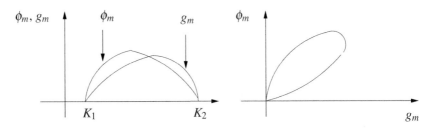

Figure 2.23 Gain margin and phase margin

2.15 A Frequency Response Parametrization of All Stabilizing Controllers

Let us begin by considering a finite-dimensional rational proper open-loop system G with $P_0 L^+$ open RHP poles, in a unity feedback configuration as shown in Figure 2.24.

Figure 2.24 A unity feedback system

Let $G(j\omega)$ be the frequency response of the open-loop system and let ω_i, $i = 0, 1, 2,$ $\ldots, k+1$ with $\omega_0 = 0$ and $\omega_{k+1} = \infty$ denote the non-negative frequencies where the Nyquist plot of $G(s)$ cuts the negative real axis of the complex plane to the left of $-1 + j$. In other words, these frequencies are the solutions of the following equation:

$$|G(j\omega_c)| > 1 \quad \text{and}$$
$$\phi(\omega_i) := \angle G(j\omega_i) = \pm n\pi, \quad \text{for } n = 1, 3, 5, \ldots. \tag{2.83}$$

Define the set of non-negative frequencies

$$\Omega = \{\omega_0, \omega_1, \ldots, \omega_k, \omega_{k+1}\} \tag{2.84}$$

where

$$0 =: \omega_0 < \omega_1 < \omega_2 < \cdots < \omega_k < \omega_{k+1} := \infty \tag{2.85}$$

and ω_0 and ω_{k+1} as well as the ω_j are included only if they satisfy the above angle and magnitude conditions. To count the number N of clockwise encirclements of $-1 + j0$ introduce as before the corresponding sequence of integers $\{i_0, i_1, i_2, \ldots, i_k, i_{k+1}\}$ corresponding encirclements at the respective frequencies ω_k:

$$i_t = 0, \quad \text{if } |G(j\omega_t)| < 1 \tag{2.86}$$

and otherwise,

$$i_t = \begin{cases} -1, & \text{if } \dfrac{d}{d\omega}\angle G(j\omega)\Big|_{\omega=\omega_t} > 0, \\[2mm] 0, & \text{if } \dfrac{d}{d\omega}\angle G(j\omega)\Big|_{\omega=\omega_t} = 0, \\[2mm] +1, & \text{if } \dfrac{d}{d\omega}\angle G(j\omega)\Big|_{\omega=\omega_t} < 0, \end{cases} \tag{2.87}$$

for $t = 0, 1, 2, \ldots, k+1$.

Remark 2.1   It is easy to see that Nyquist plot $G(j\omega)$ evaluated at $s = j\omega_t$ cuts the negative real axis when Equation (2.83) holds and it cuts it to the left of $-1 + j0$ when $|G(j\omega_t)| > 1$. The conditions in Equation (2.87) along with the condition $|G(j\omega_t)| > 1$ indicate that when the Nyquist plot cuts the negative real axis to the left of $-1 + j0$ downward, corresponding to a counterclockwise encirclement, $i_t = -1$ and when the plot cuts the negative real axis to the left of $-1 + j0$ upward, $i_t = +1$ corresponding to a clockwise encirclement of $-1 + j0$.

2.15.1   A Bode Equivalent of the Nyquist Criterion

We first assume that $G(s)$ has no imaginary axis poles and P_{OL}^+ open RHP poles. As usual we suppose that the Nyquist contour shown in Figure 2.15 is traversed in the clockwise direction, that is with ω increasing.

LEMMA 2.1   Under the assumption that $G(s)$ has no imaginary axis poles, let N denote the number of clockwise encirclements of $-1 + j0$ by the Nyquist plot of $G(s)$. Then

$$N = i_0 + \sum_{t=1}^{k} 2i_t + i_\infty =: i(G). \tag{2.88}$$

Proof   Consider the negative real axis cuts of the Nyquist plot of $G(s)$. The number of clockwise encirclements of $-1 + j0$ is equal to the net count of (upward less downward) cuts of the plot $G(j\omega)$ which occur on the negative real axis to the left of $-1 + j0$. Such a cut must satisfy the following condition:

$$|G(j\omega_k)| > 1. \tag{2.89}$$

For $\omega \in (-\infty, \infty)$, the $G(j\omega)$ plot passes through these points twice, corresponding to positive and negative frequencies or to the symmetry of the complete Nyquist plot about the real axis, whereas $G(j0)$ and $G(j\infty)$ can only induce one cut, positive or negative. Therefore, the expression of $i(G)$ in Equation (2.88) is nothing but the number of net clockwise encirclements around the point $-1 + j0$ by the Nyquist plot.   ∎

Using this lemma, we can now state the condition for stability of the feedback system.

THEOREM 2.2   Under the assumption that the plant has no imaginary axis poles, the unity feedback system in Figure 2.24 is stable if and only if

$$i(G) = -P_{OL}^+. \tag{2.90}$$

Proof From the Nyquist criterion, the feedback system is stable if and only if the complex plane plot of $G(j\omega)$ produces P_{OL}^+ net counterclockwise encirclements around the point $-1 + j0$ and therefore the theorem is evident from Lemma 2.1.   ■

Now consider the case when $G(s)$ has one or more poles at the origin. This addresses the class of systems where one or more integrators are required in the open-loop system, for the closed-loop system to track steps, ramps, or higher-order polynomial inputs.

Systems with Open-Loop Poles at the Origin

Let m_0 be the number of poles of $G(s)$ at the origin, and let i_0 denote the corresponding number of encirclements in the clockwise direction of the Nyquist plot of $G(s)$ at $s = 0$. Note that here

$$G(0^+) \neq G(j0^+). \tag{2.91}$$

As typically done in Nyquist theory, we use right indentation of the Nyquist Γ-contour when the contour approaches imaginary axis poles (see Figure 2.25).

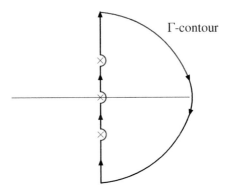

Figure 2.25 Γ-contour for Nyquist plot

Since the cases of odd and even numbers of poles at the origin are different, we separately state the results for these cases.

m_0 is odd

For the case of a plant with an odd number of poles at the origin, the Nyquist plot starts from the negative or positive imaginary axis as ω increases from zero, depending on the values of m_0. Furthermore, the Nyquist plot turns $180°$ clockwise for each pole at the origin. The first clockwise half circle is located in the LHP or RHP depending on the sign of $G(0^+)$. For simplicity, let us consider the case when $m_0 = 1$. If $G(0^+) < 0$, the clockwise half circle is located in the LHP and it results in a negative real axis cut to the left of $-1 + j0$. Since this cut is upward, we must set $i_0 = +1$. On the other hand, the clockwise half circle is located in the RHP for $G(0^+) > 0$ and this results in no negative real axis cut, that is, we must set $i_0 = 0$. From such considerations, we derive the following general formulas. For m_0 odd:

$$
i_0 = \begin{cases} \left(\dfrac{m_0 - 1}{2}\right), & \text{if } (-1)^{\frac{m_0-1}{2}} G(0^+) > 0, \\[3mm] \left(\dfrac{m_0 + 1}{2}\right), & \text{if } (-1)^{\frac{m_0+1}{2}} G(0^+) > 0. \end{cases}
\tag{2.92}
$$

m_0 is even

For the case of a plant with an even number of poles at the origin, the Nyquist plot begins from the negative or positive real axis as ω increases from zero, depending on the value of m_0. With considerations similar to the previous case we derive the following general conditions. For m_0 even:

$$
i_0 = \begin{cases} \left(\dfrac{m_0}{2} - 1\right), & \text{if } (-1)^{\frac{m_0}{2}-1} G(0^+) > 0 \text{ and } \left.\dfrac{d}{d\omega}\angle G(j\omega)\right|_{\omega=0^+} > 0, \\[3mm] \left(\dfrac{m_0}{2}\right), & \text{otherwise.} \end{cases}
\tag{2.93}
$$

Systems with Poles on the Imaginary Axis

In servomechanisms the controller may need to have $\pm j\omega_0$ poles to reject or track sinusoidal disturbances at the frequency ω_0. To address this situation, let the denominator of the transfer function $G(s)$ be

$$
(s^2 + u_1^2)^{m_1}(s^2 + u_2^2)^{m_2} \cdots (s^2 + u_k^2)^{m_k},
\tag{2.94}
$$

that is, $G(s)$ has m_i pairs of poles at $\pm j u_i$ for $i = 1, 2, \ldots, k$. Now define integer quantities j_i for $j = 1, 2, \ldots, k$ which denote the number of corresponding clockwise encirclements by the Nyquist plot of the point $-1 + j0$ at the corresponding frequency u_i, as follows. The verification of these is left to the reader and is based on arguments outlined in the previous cases.

$G(ju_i^-)$ is complex and m_i is odd:

$$
j_i = \begin{cases} (m_i - 1), & \text{if } \angle G(ju_i^-) \in (0, \pi), \\ (m_i + 1), & \text{if } \angle G(ju_i^-) \in (\pi, 2\pi). \end{cases}
\tag{2.95}
$$

$G(ju_i^-)$ is complex and m_i is even, $j_i = -m_i$.

$G(ju_i^-)$ is real and m_i is odd:

$$
j_i = \begin{cases} (m_i - 1), & \text{if } G(ju_i^-) > 0 \text{ and } \left.\dfrac{d}{d\omega}\angle G(j\omega)\right|_{\omega=u_i^+} > 0, \\[3mm] (m_i + 1), & \text{otherwise.} \end{cases}
\tag{2.96}
$$

$G(ju_i^-)$ is real and m_i is even:

$$j_i = \begin{cases} m_i, & \text{if } G(ju_i^-) > 0 \text{ or} \\ & \left(G(ju_i^-) < 0 \text{ and } \dfrac{d}{d\omega} \angle G(j\omega)\Big|_{\omega=u_i^+} > 0 \right), \\ (m_i + 2), & \text{otherwise.} \end{cases} \quad (2.97)$$

Theorem 2.2 can now be restated without restrictions on the location of poles of the plant.

THEOREM 2.3   The unity feedback system in Figure 2.24 is stable if and only if

$$i(G) := i_0 + \sum_{k=1}^{l} 2i_k + i_\infty + 2 \sum_{r=1}^{k} j_r = -P_{OL}^+, \quad (2.98)$$

where P_{OL}^+ is the number of open RHP poles of $G(s)$.

2.15.2   Arbitrary Order Controllers

We now consider a finite-dimensional rational proper controller with frequency response $C(j\omega)$ and with c^+ RHP poles and ask when it can stabilize a finite-dimensional rational, proper plant with frequency response $P(j\omega)$ with p^+ RHP poles.

Define the set of distinct non-negative frequencies ω_k

$$\Omega^+(\phi) := \{\omega_0, \omega_1, \ldots, \omega_l, \omega_{l+1}\} \quad (2.99)$$

with

$$0 =: \omega_0 < \omega_1 < \cdots < \omega_l < \omega_{l+1} := \infty \quad (2.100)$$

satisfying the phase condition

$$\angle C(j\omega) = \angle P^{-1}(j\omega) \pm n\pi, \ n = \pm 1, \pm 3, \pm 5, \pm 7, \ldots \quad (2.101)$$

and the magnitude condition

$$|C(j\omega_k)| > |P^{-1}(j\omega_k)|. \quad (2.102)$$

Note that these are the frequencies where the Nyquist plot of $P(s)C(s)$ intersects the negative real axis to the left of $-1 + j0$.

For the case when $P(s)C(s)$ has no imaginary axis poles, we introduce the integers i_k for $k = 0, 1, \ldots, l+1$:

$$i_k = \begin{cases} -1, & \text{if } \dfrac{d}{d\omega} \angle C(j\omega)\Big|_{\omega=\omega_k} > \dfrac{d}{d\omega} \angle P^{-1}(j\omega)\Big|_{\omega=\omega_k}, \\ 0, & \text{if } \dfrac{d}{d\omega} \angle C(j\omega)\Big|_{\omega=\omega_k} = \dfrac{d}{d\omega} \angle P^{-1}(j\omega)\Big|_{\omega=\omega_k}, \\ +1, & \text{if } \dfrac{d}{d\omega} \angle C(j\omega)\Big|_{\omega=\omega_k} < \dfrac{d}{d\omega} \angle P^{-1}(j\omega)\Big|_{\omega=\omega_k}. \end{cases} \quad (2.103)$$

The above count is obtained by replacing $G(j\omega)$ by $P(j\omega)C(j\omega)$ in Equation (2.87). Consider the first condition

$$\frac{d}{d\omega}\angle P(j\omega)C(j\omega)\bigg|_{\omega=\omega_k} > 0, \tag{2.104}$$

which is equivalent to

$$\frac{d}{d\omega}\angle P(j\omega)\bigg|_{\omega=\omega_k} + \frac{d}{d\omega}\angle C(j\omega)\bigg|_{\omega=\omega_k} > 0 \tag{2.105}$$

or

$$\frac{d}{d\omega}\angle C(j\omega)\bigg|_{\omega=\omega_k} > -\frac{d}{d\omega}\angle P(j\omega)\bigg|_{\omega=\omega_k} = \frac{d}{d\omega}\angle P^{-1}(j\omega)\bigg|_{\omega=\omega_k}. \tag{2.106}$$

Similarly, we can also restate Equations (2.92)–(2.97) in terms of the controller.

Plant–Controller Pairs with Poles at the Origin
Let m_0 be the number of poles at the origin, and i_0 the corresponding number of encirclements in the clockwise direction of $-1+j0$ by the Nyquist plot of $P(s)C(s)$ at $s=0$. Note that

$$P(0^+)C(0^+) \neq P(j0^+)C(j0^+). \tag{2.107}$$

Similar to the previous section, we can introduce the following:

m_0 **is odd**

$$i_0 = \begin{cases} \left(\dfrac{m_0-1}{2}\right), & \text{if } (-1)^{\frac{m_0-1}{2}}\text{sgn}[C(0^+)] = \text{sgn}[P(0^+)], \\[2mm] \left(\dfrac{m_0+1}{2}\right), & \text{if } (-1)^{\frac{m_0+1}{2}}\text{sgn}[C(0^+)] = -\text{sgn}[P(0^+)]. \end{cases} \tag{2.108}$$

m_0 **is even**

$$i_0 = \begin{cases} \left(\dfrac{m_0}{2}-1\right), & \text{if } (-1)^{\frac{m_0}{2}-1}\text{sgn}[P(0^+)] = \text{sgn}[C(0^+)] \text{ and} \\[1mm] & \quad \frac{d}{d\omega}\angle C(j\omega)\big|_{\omega=0^+} > \frac{d}{d\omega}\angle P^{-1}(j\omega)\big|_{\omega=0^+}, \\[2mm] \left(\dfrac{m_0}{2}\right), & \text{otherwise.} \end{cases} \tag{2.109}$$

Plant–Controller Pairs with Poles on the Imaginary Axis
Let the denominator of the transfer function $G(s)$ be

$$(s^2+u_1^2)^{m_1}(s^2+u_2^2)^{m_2}\cdots(s^2+u_k^2)^{m_k}. \tag{2.110}$$

Such an expression assumes that the controller and plant has m_i pairs of poles at $\pm ju_i$ for $i=1,2,\ldots,k$. Define integer quantities j_i for $j=1,2,\ldots,k$ as follows.

$P(ju_i^-)C(ju_i^-)$ is complex and m_i is odd:

$$j_i = \begin{cases} (m_i-1), & \text{if } \angle P^{-1}(ju_i^-) < \angle C(ju_i^-) < \angle P^{-1}(ju_i^-)+\pi, \\ -(m_i+1), & \text{if } \angle P^{-1}(ju_i^-)+\pi < \angle C(ju_i^-) < \angle P^{-1}(ju_i^-)+2\pi. \end{cases} \tag{2.111}$$

$P(ju_i^-)C(ju_i^-)$ is complex and m_i is even:

$$j_i = +m_i. \tag{2.112}$$

$G(ju_i^-)$ is real and m_i is odd:

$$j_i = \begin{cases} (m_i - 1), & \text{if } \mathrm{sgn}[P(ju_i^-)] = \mathrm{sgn}[C(ju_i^-)] \text{ and} \\ & \quad \frac{d}{d\omega}\angle C(j\omega)\big|_{\omega=u_i^+} < \frac{d}{d\omega}\angle P^{-1}(j\omega)\big|_{\omega=u_i^+}, \\ (m_i + 1), & \text{otherwise.} \end{cases} \tag{2.113}$$

$G(ju_i^-)$ is real and m_i is even:

$$j_i = \begin{cases} m_i, & \text{if } \mathrm{sgn}[P(ju_i^-)] = \mathrm{sgn}[C(ju_i^-)] \text{ or} \\ & \quad \left(\begin{array}{l} \mathrm{sgn}[P(ju_i^-)] = -\mathrm{sgn}[C(ju_i^-)] \text{ and} \\ \frac{d}{d\omega}\angle C(j\omega)\big|_{\omega=u_i^+} > \frac{d}{d\omega}\angle P^{-1}(j\omega)\big|_{\omega=u_i^+} \end{array} \right), \\ (m_i + 2), & \text{otherwise.} \end{cases} \tag{2.114}$$

Define

$$i(C) := i_0 + \sum_{k=1}^{l} 2i_k + i_\infty + 2\sum_{r=1}^{k} j_r. \tag{2.115}$$

THEOREM 2.4   The controller C stabilizes the plant P if and only if

$$i(C) = -(p^+ + c^+). \tag{2.116}$$

Remark 2.2   The Nyquist criterion can be roughly interpreted as follows: clockwise encirclements are "destabilizing" and counterclockwise encirclements are "stabilizing." Closed-loop stability is achieved when the net number of counterclockwise encircle-ments equals the number of RHP poles of the open-loop system. The presence of imag-inary axis poles increases the number of clockwise encirclements depending on their multiplicity at these frequencies. For closed-loop stability this must be compensated by additional counterclockwise encirclements in the rest of the Nyquist plot.

2.15.3   Performance Measures

The performance of a controller is often determined by the closed-loop stability margins it provides. The gain margin is such a performance measure. To compute it, define the distinct frequencies:

$$\{u : \angle C(ju) = \angle P^{-1}(ju) \pm n\pi, \ n = 1,3,5,\ldots\} = \{u_1, u_2, \ldots, u_m\} \tag{2.117}$$

and

$$\Omega(\phi) := \{u_0, u_1, \ldots, u_m, u_{m+1}\}, \tag{2.118}$$

where $u_0 = 0$ and $u_{m+1} = \infty$. Let us denote magnitudes measured in decibels as

$$C(\omega)_{\mathrm{dB}} := 20\log_{10}|C(j\omega)|, \tag{2.119a}$$

$$P^{-1}(\omega)_{\mathrm{dB}} := 20\log_{10}\left|P^{-1}(j\omega)\right|. \tag{2.119b}$$

Define

$$\Omega^+(\phi) := \left\{ \omega_k \in \Omega(\phi) \; : \; C(\omega_k)_{dB} > P^{-1}(\omega_k)_{dB} \right\}, \tag{2.120a}$$

$$\Omega^-(\phi) := \left\{ \omega_k \in \Omega(\phi) \; : \; C(\omega_k)_{dB} < P^{-1}(\omega_k)_{dB} \right\}. \tag{2.120b}$$

The upper (lower) gain margin is the smallest increase (decrease) in gain measured in decibels that destabilizes the closed loop.

THEOREM 2.5   If C is a stabilizing controller, the upper gain margin denoted K_{dB}^+ is

$$K_{dB}^+ = \min_{\omega_k \in \Omega^+(\phi)} \left\{ C(\omega_k)_{dB} - P^{-1}(\omega_k)_{dB} \right\}. \tag{2.121}$$

The lower gain margin denoted K_{dB}^- is

$$K_{dB}^- = \min_{u_k \in \Omega^-(\phi)} \left\{ P^{-1}(u_k)_{dB} - C(u_k)_{dB} \right\}. \tag{2.122}$$

The phase margin is also an important performance measure. To compute it, introduce the distinct frequencies v_i:

$$\Omega(g) = \left\{ v \; : \; |C(jv)| = \left| P^{-1}(jv) \right| \right\} = \{v_1, \, v_2, \, \ldots, \, v_m\}. \tag{2.123}$$

Similarly, we also define

$$\Omega^+(g) := \left\{ v_k \in \Omega(g) \; : \; \angle C(jv_k) > \angle P^{-1}(jv_k) + n\pi \right\}, \tag{2.124a}$$

$$\Omega^-(g) := \left\{ v_k \in \Omega(g) \; : \; \angle C(jv_k) < \angle P^{-1}(jv_k) + n\pi \right\}. \tag{2.124b}$$

The positive (negative) phase margin Φ^+ (Φ^-) is the minimum phase decrease (increase) that destabilizes the loop.

THEOREM 2.6   If C is a stabilizing controller, the positive phase margin is

$$\Phi^+ = \min_{n \text{ odd}} \min_{v_k \in \Omega^+(g)} \left\{ \angle C(jv_k) - \angle P^{-1}(jv_k) - n\pi \right\}. \tag{2.125}$$

The negative phase margin is

$$\Phi^- = \min_{n \text{ odd}} \min_{v_k \in \Omega^-(g)} \left\{ \angle P^{-1}(ju_k) - \angle C(ju_k) + n\pi \right\}. \tag{2.126}$$

The proofs of Theorems 2.5 and 2.6 follow from interpreting the Nyquist criterion in terms of the Bode magnitude and phase conditions.

We now illustrate the usefulness of Theorems 2.4–2.6 for controller design.

EXAMPLE 2.7   Consider a plant with four RHP poles and known frequency response $P(j\omega)$ as shown in Figure 2.26.

Let us examine the conditions for a stable controller to stabilize the given plant. From the frequency response data given, we have

$$\angle P^{-1}(j\omega)\big|_{\omega=0} = 0 \quad \text{and} \quad \angle P^{-1}(j\omega)\big|_{\omega=\infty} = \frac{\pi}{2} \left(\text{or} \; -\frac{3\pi}{2} \right). \tag{2.127}$$

Consider the following conditions:

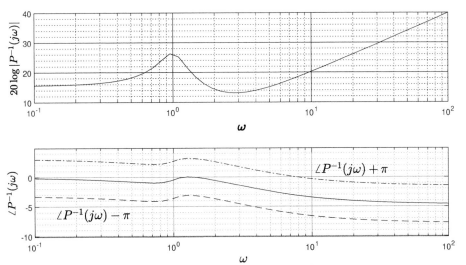

Figure 2.26 Frequency response of the plant considered (Example 2.7)

(1) $\angle C(j0) = \pm n\pi$ for any integer n and $|C(j0)| > |P^{-1}(j0)|$;

(2) $\angle C(j\infty) = (1 \pm 2n)\dfrac{\pi}{2}$ and $|C(j\infty)| > |P^{-1}(j\infty)|$;

(3) $|C(j\omega_k)| > |P^{-1}(j\omega_k)|$ and $\left.\dfrac{d}{d\omega}\angle C(j\omega)\right|_{\omega=\omega_k} > \left.\dfrac{d}{d\omega}\angle P^{-1}(j\omega)\right|_{\omega=\omega_k}$.

In order to achieve $i(C) = -4$, the frequency response of a stabilizing controller must satisfy the following:

A. Conditions (1) and (2) are satisfied and condition (3) is satisfied at one frequency, or
B. Conditions (1) and (2) are violated and (3) is satisfied at two frequencies.

Since the magnitude of $P^{-1}(j\omega)$ is unbounded as ω tends to ∞ and the controller is proper, it follows that $i_\infty = 0$. Thus i_0 must also be zero in order to generate an even number of encirclements. It is also easy to see that a constant gain cannot stabilize the plant. This is because the phase of a constant gain (zero angle) can only intersect the phase of $P^{-1}(j\omega)$ once at a nonzero frequency.

We can apply similar considerations for a first-order controller. If a controller is stable and minimum phase, then the maximum number of positive phase crossover frequencies is 3. Of these, only one contributes counterclockwise encirclement and therefore, the required $i(C) = -4$ cannot be attained.

We now consider verifying stabilizability by a given controller using the above arguments.

Suppose the controller to be tested is

$$C(s) = \frac{15.7091s^4 + 6.8889s^3 + 83.8916s^2 + 171.2964s + 62.5847}{s^4 + 4.2909s^3 - 2.1069s^2 - 7.1366s - 0.4308} \tag{2.128}$$

which has one RHP pole.

Since

$$\frac{d}{d\omega}\left[\angle C(j\omega)\right]_{\omega=0} < \frac{d}{d\omega}\left[\angle P^{-1}(j\omega)\right]_{\omega=0}, \tag{2.129}$$

we have $i_0 = 1$. Similarly, we have $i_\infty = 0$. From Figure 2.27:

$$i(C) = 1 - 2(1) + 2(0) - 2(1) + 2(0) - 2(1) + 0 = -5, \tag{2.130}$$
$$p^+ + c^+ = 4 + 1 = 5.$$

Thus the closed-loop system is stable.

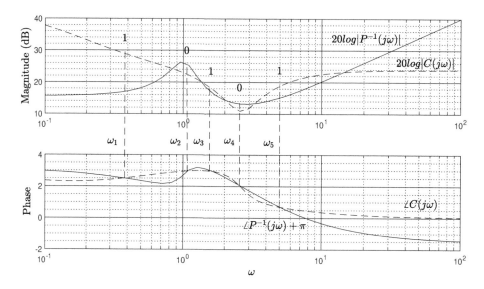

Figure 2.27 No poles on the imaginary axis (Example 2.7)

EXAMPLE 2.8　Consider a plant with three poles at the origin and the test PID controller is

$$C(s) = \frac{39.88s^2 + 50s + 49.71}{s}. \tag{2.131}$$

The poles of the open-loop plant are $\{0, 0, 0, -5, -2 \pm j4\}$. It is easy to see that $m_0 = 3 + 1 = 4$ (even) and $P(0^+)C(0^+) > 0$. Thus, from the condition in Equation (2.109):

$$i_0 = -\left(\frac{m_0}{2}\right) = -\frac{4}{2} = -2. \tag{2.132}$$

Now let us observe Figure 2.28. Since $|C(\infty)| < |P^{-1}(\infty)|$, $i_\infty = 0$. Therefore:

$$i(C) = i_0 + 2i_1 + i_\infty = -2 + 2(1) + 0 = 0 \quad \text{and} \quad p^+ + c^+ = 0 + 0 = 0. \tag{2.133}$$

We conclude that the closed-loop system is stable.

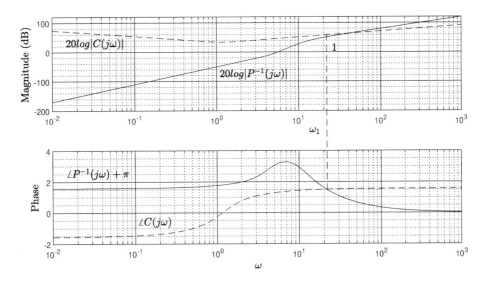

Figure 2.28 Poles at the origin (Example 2.8)

The following example illustrates the gain and phase margin computations.

EXAMPLE 2.9 Consider a plant with two RHP poles and let the frequency response of the system $P(j\omega)$ be known. Consider the test controller

$$C(s) = \frac{36s - 12}{s + 9}. \tag{2.134}$$

Since $C(0) < P^{-1}(0)$ and $C(\infty) < P^{-1}(\infty)$, we have $i_0 = i_\infty = 0$. From Figure 2.29, we have

$$i(C) = 0 - 2(1) + 0 = -2 \quad \text{and} \quad p^+ + c^+ = 2 + 0 = 2 \tag{2.135}$$

and the closed-loop system is stable. The gain and phase margins are shown in Figure 2.29 and by the combined Bode plots in Figure 2.30.

2.16 Gain, Phase, and Time-Delay Margins

The Nyquist criterion also allows for the quantitative determination of stability margins of the closed-loop system provided it is stable.

We consider the perturbed closed-loop system and assume that the nominal closed-loop system with $\Delta = 1$ is stable (see Figure 2.31).

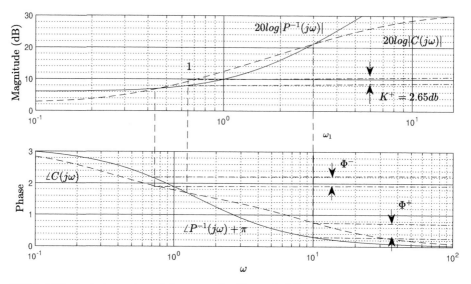

Figure 2.29 Illustrating gain and phase margins (Example 2.9)

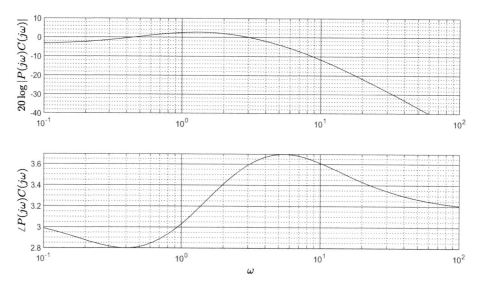

Figure 2.30 Bode plot of $P(s)C(s)$ (Example 2.9)

The perturbation Δ is usually allowed to be gain, phase, or time-delay:

$$\Delta = K, \tag{2.136a}$$

$$\Delta = e^{-j\phi}, \tag{2.136b}$$

$$\Delta = e^{-sT_d}. \tag{2.136c}$$

In each case, we ask: What is the "largest" perturbation that preserves closed-loop stability. We illustrate below how the Nyquist plot of $G_0(s)$ contains this information.

Figure 2.31 Perturbed closed-loop system

A typical Nyquist plot of $G_0(s)$ with the closed loop assumed to be stable is shown in Figure 2.32.

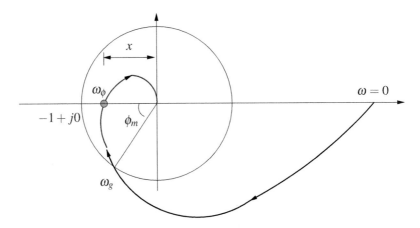

Figure 2.32 Illustrating gain and phase margin: ω_g = gain crossover frequency, ω_ϕ = phase crossover frequency

The perturbation Δ amounts to replacing $G_0(s)$ by $\Delta G_0(s)$ or successively by $KG_0(s)$, $e^{-j\phi}G_0(s)$, and $e^{-sL}G_0(s)$. In each case, this leaves unchanged the number P_{OL}^+ of open-loop unstable poles. Therefore, closed-loop stability is preserved as long as the perturbed Nyquist plot of $\Delta G_0(s)$ has the same number of encirclements of $-1 + j0$ as the nominal G_0. From this reasoning, we see that the gain margin GM, phase margin PM, and time-delay margin T_d^* are given by

$$\text{GM} = \frac{1}{x} = \frac{1}{|G(j\omega_\phi)|},\qquad (2.137a)$$

$$\text{PM} = \phi_m \text{ (rad)} = \angle G(j\omega_g) - \pi,\qquad (2.137b)$$

$$T_d^* = \frac{\phi_m}{\omega_g}.\qquad (2.137c)$$

EXAMPLE 2.10   Consider the unity feedback system with the plant and controller

$$P(s) = \frac{s+5}{s^2 + 2s + 3}, \quad C(s) = \frac{1}{s+1}.\qquad (2.138)$$

The Bode plots in Figure 2.33 are obtained from the MATLAB script below. Moreover, MATLAB command "margin" calculates phase crossover frequency (ω_{cg}), gain crossover frequency (ω_{cp}), gain margin (GM), and phase margin (PM). Caution should be exercised when using the command "margin" since it does not verify nominal closed-loop stability. In other words, GM and PM values computed by the program no longer represent stability margins if the nominal closed-loop system is not stable.

MATLAB script

```
C=tf(1,[1 1]);
P=tf([1 5],[1 2 3]);
sys=C*P;
bode(sys), grid
[Gm,Pm,Wcg,Wcp]=margin(sys)
```

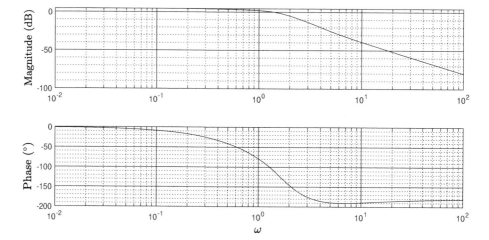

Figure 2.33 GM $= 6.0002$ dB, PM $= 70.5509°$, $\omega_{cg} = 3.3167$ rad/s, $\omega_{cp} = 1.4139$ rad/s (Example 2.10)

These margins may also be used to design a controller to obtain, if possible, prescribed stability margins as we show in the next section for a PI controller.

2.17   Stability Margin-Based Design of PI Controllers

Consider the PI controller

$$C(s) = k_p + \frac{k_i}{s} \tag{2.139}$$

for a given plant $P(s)$. Let $\mathscr{S}^0$ denote the set in the (k_p, k_i) plane for which the closed loop is stable. Every controller in $\mathscr{S}^0$ stabilizes the closed-loop system with the plant $P(s)$, and provides asymptotic tracking of reference step inputs and asymptotic rejection of step disturbances. To robustify the closed-loop system, we need to search for subsets of $\mathscr{S}^0$ where prescribed stability margins may be attained.

To proceed, we note that

$$|C(j\omega_g)| = \frac{1}{|P(j\omega_g)|} =: m_g \tag{2.140}$$

and

$$\angle C(j\omega_\phi) = n\pi - \angle P(j\omega_\phi) =: \phi, \tag{2.141}$$

where ω_g and ω_ϕ are the gain and phase crossover frequencies, respectively. If ω_g is known, so is m_g and so Equation (2.140) holds for all (k_p, k_i) satisfying

$$\left| \frac{j\omega_g k_p + k_i}{j\omega_g} \right| = m_g, \tag{2.142}$$

that is

$$\frac{\omega_g^2 k_p^2 + k_i^2}{\omega_g^2} = m_g^2 \tag{2.143}$$

or

$$\frac{k_p^2}{a^2} + \frac{k_i^2}{b^2} = 1 \tag{2.144}$$

with

$$a = m_g, \quad b = \omega_g m_g. \tag{2.145}$$

The above represents an ellipse with principal axes a and b. The phase margin ϕ_m is related to $\angle P(j\omega_g)$ and $\angle C(j\omega_g)$ by

$$\angle P(j\omega_g) + \angle C(j\omega_g) - \pi = \phi_m \tag{2.146}$$

and therefore

$$\angle C(j\omega_g) = \pi + \phi_m - \angle P(j\omega_g) =: \theta_g. \tag{2.147}$$

Therefore:

$$\tan \theta_g = -\frac{k_i}{k_p \omega_g} \tag{2.148}$$

or

$$k_i = \underbrace{-(\omega_g \tan \theta_g)}_{c} k_p, \tag{2.149}$$

a straight line in the (k_i, k_p) plane.

To summarize, the above calculations show that the intersection of the ellipse in Equation (2.144) and the straight line in Equation (2.149) defines a design point (K_p^*, K_i^*) with gain crossover frequency ω_g, and phase margin ϕ_m, provided only that it lies in $\mathscr{S}^0$ as shown in Figure 2.34.

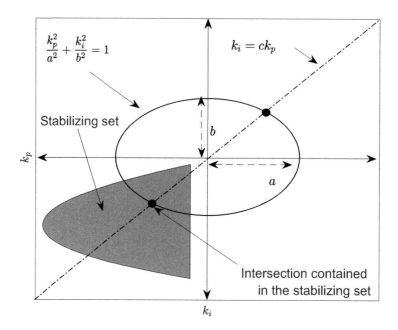

Figure 2.34 Ellipse and straight line intersecting with a stabilizing set

This suggests that the stabilizing set $\mathscr{S}^0$ can in fact be mapped into the space of achievable performance parameters ω_g and ϕ_m. This is illustrated in the example below.

EXAMPLE 2.11   Consider the plant

$$P(s) = \frac{s-1}{s+1} \tag{2.150}$$

with the PI controller

$$C(s) = K_p + \frac{K_i}{s}. \tag{2.151}$$

The open-loop transfer function

$$G_0(s) = \frac{(sK_p + K_i)(s-1)}{s(s+1)} \tag{2.152}$$

and the closed-loop characteristic polynomial is

$$\begin{aligned} d_{cl}(s) &= s(s+1) + (sK_p + K_i)(s-1) \\ &= s^2(1+K_p) + s(1+K_i - K_p) - K_i. \end{aligned} \tag{2.153}$$

Using the Routh–Hurwitz criterion, the stabilizing set $\mathscr{S}^0$ is described by

$$1 + K_p > 0, \quad 1 + K_i - K_p > 0, \quad -K_i > 0. \tag{2.154}$$

This is depicted in Figure 2.35.

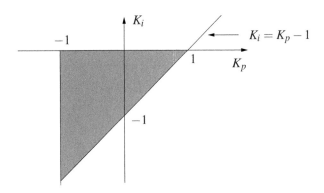

Figure 2.35 Stabilizing set $\mathscr{S}^0$ (Example 2.11)

To proceed, write the equation

$$|G_0(j\omega_g)| = 1 \tag{2.155}$$

in terms of K_p, K_i as

$$K_p^2 + \frac{K_i^2}{\omega_g^2} = 1. \tag{2.156}$$

If we choose $\omega_g = 1$, we obtain the circle

$$C: \quad K_p^2 + K_i^2 = 1 \tag{2.157}$$

which is shown superimposed on $\mathscr{S}^0$ in Figure 2.36.

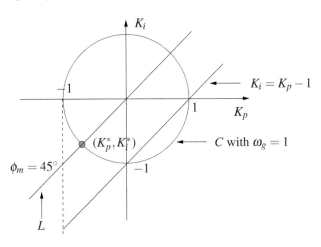

Figure 2.36 Design parameter selection (Example 2.11)

For a desired closed-loop phase margin $\phi_m = \frac{\pi}{4}$ (rad) $= 45°$, we obtain the line L

shown ($K_i = K_p$). The intersection of C and L gives the design point

$$(K_p^*, K_i^*) = \left(-\frac{1}{\sqrt{2}}, -\frac{1}{\sqrt{2}} \right) \tag{2.158}$$

which lies in $\mathcal{S}^0$. Therefore, the controller

$$C(s) = -\frac{1}{\sqrt{2}} \left(\frac{s+1}{s} \right) \tag{2.159}$$

gives a gain crossover frequency $\omega_g = 1$ and the phase margin 45°. Other crossover frequencies and phase margins can be explored in a similar manner. The time-delay margin can be determined similarly. In this example:

$$T_d^* = \frac{\phi_m}{\omega_g} = \frac{\pi}{4} \text{ s.} \tag{2.160}$$

2.18 Stability Theory for Polynomials

Stability analysis using polynomials is an important component of classical control theory. In this section we give a brief account of some useful results on stability analysis via polynomials. The stability of a finite-dimensional linear time-invariant system can always be characterized by its characteristic polynomial whose roots must lie in prescribed regions of the complex plane such as the open left-half plane for continuous-time systems or the open unit disc for discrete-time systems. The results described here are the Boundary Crossing Theorem, the Hermite–Biehler Theorem, the Routh–Hurwitz criterion, and the Gauss–Lucas Theorem.

2.18.1 The Boundary Crossing Theorem

We begin with the well-known Principle of the Argument of complex variable theory. Let $\mathscr{C}$ be a simple closed contour in the complex plane and $w = f(z)$ a function of the complex variable z, which is analytic on $\mathscr{C}$. Let Z and P denote the number of zeros and poles, respectively, of $f(z)$ contained in $\mathscr{C}$. Let $\Delta_C \arg[f(z)]$ denote the net change of argument (angle) of $f(z)$ as z transverses the contour $\mathscr{C}$.

THEOREM 2.7 (Principle of the Argument)

$$\Delta_C \arg[f(z)] = 2\pi(Z - P). \tag{2.161}$$

An important consequence of this result is the well-known theorem of Rouché.

THEOREM 2.8 (Rouché's Theorem)   Let $f(z)$ and $g(z)$ be two functions which are analytic inside and on a simple closed contour $\mathscr{C}$ in the complex plane. If

$$|g(z)| < |f(z)| \tag{2.162}$$

for any z on $\mathscr{C}$, then $f(z)$ and $f(z) + g(z)$ have the same number (multiplicities included) of zeros inside $\mathscr{C}$.

Proof Since $f(z)$ cannot vanish on $\mathscr{C}$, because of Equation (2.162), we have

$$\Delta_C \arg[f(z) + g(z)] = \Delta_C \arg\left[f(z)\left(1 + \frac{g(z)}{f(z)}\right)\right]$$
$$= \Delta_C \arg[f(z)] + \Delta_C \arg\left[1 + \frac{g(z)}{f(z)}\right]. \tag{2.163}$$

Moreover, since

$$\left|\frac{g(z)}{f(z)}\right| < 1 \tag{2.164}$$

for all $z \in \mathscr{C}$, the variable point

$$w = 1 + \frac{g(z)}{f(z)} \tag{2.165}$$

stays in the disc $|w - 1| < 1$ as z describes the curve $\mathscr{C}$. Therefore, w cannot wind around the origin, which means that

$$\Delta_C \arg\left[1 + \frac{g(z)}{f(z)}\right] = 0. \tag{2.166}$$

Combining Equations (2.163) and (2.166), we find that

$$\Delta_C \arg[f(z) + g(z)] = \Delta_C \arg[f(z)]. \tag{2.167}$$

Since $f(z)$ and $g(z)$ are analytic in and on $\mathscr{C}$, the theorem now follows as an immediate consequence of the argument principle. ∎

Note that the condition $|g(z)| < |f(z)|$ on $\mathscr{C}$ implies that neither $f(z)$ nor $f(z) + g(z)$ may have a zero on $\mathscr{C}$. Theorem 2.8 is just one formulation of Rouché's Theorem but it is sufficient for our purposes. The next theorem is a simple application of Rouché's Theorem. It is, however, most useful since it applies to polynomials.

THEOREM 2.9 Let

$$P(s) = p_0 + p_1 s + \cdots + p_n s^n = \prod_{j=1}^{m}(s - s_j)^{t_j}, \text{ for } p_n \neq 0,$$
$$Q(s) = (p_0 + \varepsilon_0) + (p_1 + \varepsilon_1)s + \cdots + (p_n + \varepsilon_n)s^n, \tag{2.168}$$

and consider a circle $\mathscr{C}_k$, of radius r_k, centered at s_k which is a root of $P(s)$ of multiplicity t_k. Let r_k be fixed in such a way that

$$0 < r_k < \min|s_k - s_j|, \quad \text{for} \quad j = 1, 2, \ldots, k-1, k+1, \ldots, m. \tag{2.169}$$

Then there exists a positive number ε such that $|\varepsilon_i| \leq \varepsilon$ for $i = 0, 1, \ldots, n$, implies that $Q(s)$ has precisely t_k zeros inside the circle $\mathscr{C}_k$.

Proof $P(s)$ is nonzero and continuous on the compact set $\mathscr{C}_k$ and therefore it is possible to find $\delta_k > 0$ such that

$$|P(s)| \geq \delta_k > 0, \quad \text{for all } s \in \mathscr{C}_k. \tag{2.170}$$

On the other hand, consider the polynomial $R(s)$, defined by

$$R(s) = \varepsilon_0 + \varepsilon_1 s + \cdots + \varepsilon_n s^n. \tag{2.171}$$

If s belongs to the circle $\mathscr{C}_k$, then

$$|R(s)| \leq \sum_{j=0}^{n} |\varepsilon_j||s^j| \leq \sum_{j=0}^{n} |\varepsilon_j| (|s - s_k| + |s_k|)^j \leq \varepsilon \underbrace{\sum_{j=0}^{n} (r_k + |s_k|)^j}_{M_k}. \tag{2.172}$$

Thus, if ε is chosen so that $\varepsilon < \frac{\delta_k}{M_k}$, it is concluded that

$$|R(s)| < |P(s)| \quad \text{for all } s \text{ on } \mathscr{C}_k, \tag{2.173}$$

so that by Rouché's Theorem, $P(s)$ and $Q(s) = P(s) + R(s)$ have the same number of zeros inside $\mathscr{C}_k$. Since the choice of r_k ensures that $P(s)$ has just one zero of multiplicity t_k at s_k, we see that $Q(s)$ has precisely t_k zeros in $\mathscr{C}_k$. ∎

COROLLARY 2.1 Fix m circles $\mathscr{C}_1, \ldots, \mathscr{C}_m$ that are pairwise disjoint and centered at $s_1, s_2, \ldots, s_m$, respectively. By repeatedly applying the previous theorem, it is always possible to find an $\varepsilon > 0$ such that for any set of numbers $\{\varepsilon_0, \ldots, \varepsilon_n\}$ satisfying $|\varepsilon_i| \leq \varepsilon$, for $i = 0, 1, \ldots, n$, $Q(s)$ has precisely t_j zeros inside each of the circles $\mathscr{C}_j$.

Note, that in this case, $Q(s)$ always has $t_1 + t_2 + \cdots + t_m = n$ zeros and must remain therefore of degree n, so that necessarily $\varepsilon < |p_n|$. The above theorem and corollary lead to our main result, the Boundary Crossing Theorem.

Let us consider the complex plane C and let $\mathscr{S} \subset C$ be any given open set. We know that $\mathscr{S}$, its boundary $\partial\mathscr{S}$ together with the interior $\mathscr{U}^o$ of the closed set $\mathscr{U} = C - \mathscr{S}$ form a partition of the complex plane, that is

$$\mathscr{S} \cup \partial\mathscr{S} \cup \mathscr{U}^o = C, \quad \mathscr{S} \cap \mathscr{U}^o = \mathscr{S} \cap \partial\mathscr{S} = \partial\mathscr{S} \cap \mathscr{U}^o = \emptyset. \tag{2.174}$$

Assume moreover that each one of these three sets is non-empty. These assumptions are very general. In stability theory one might choose for $\mathscr{S}$ the open left-half plane $\mathscr{C}^-$ (for continuous-time systems) or the open unit disc $\mathscr{D}$ (for discrete-time systems) or suitable subsets of these, as illustrated in Figure 2.37.

Consider a family of polynomials $P(\lambda, s)$ satisfying the following assumptions.

ASSUMPTION 2.1 $P(\lambda, s)$ is a family of polynomials of

(1) fixed degree n (invariant degree),
(2) continuous with respect to λ on a fixed interval $I = [a, b]$.

In other words, a typical element of $P(\lambda, s)$ can be written as

$$P(\lambda, s) = p_0(\lambda) + p_1(\lambda)s + \cdots + p_n(\lambda)s^n, \tag{2.175}$$

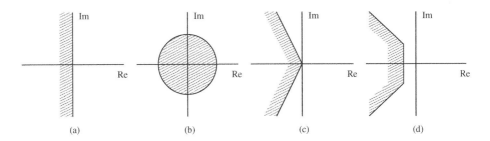

Figure 2.37 Some typical stability regions

where $p_0(\lambda)$, $p_1(\lambda)$, ..., $p_n(\lambda)$ are continuous functions of λ on I and where $p_n(\lambda) \neq 0$ for all $\lambda \in I$. From the results of Theorem 2.9 and its corollary, it is immediate that in general, for any open set $\mathcal{O}$, the set of polynomials of degree n that have all their roots in $\mathcal{O}$ is itself open. In the case above, if for some $t \in I$, $P(t,s)$ has all its roots in $\mathscr{S}$, then it is always possible to find a positive real number α such that

$$\text{for all } t' \in (t - \alpha, t + \alpha) \cap I, P(t',s) \text{ also has all its roots in } \mathscr{S}. \tag{2.176}$$

This leads to the following fundamental result.

THEOREM 2.10 (Boundary Crossing Theorem) Under Assumption 2.1, suppose that $P(a,s)$ has all its roots in $\mathscr{S}$ whereas $P(b,s)$ has at least one root in $\mathscr{U}$. Then there exists at least one ρ in $(a,b]$ such that:

(a) $P(\rho,s)$ has all its roots in $\mathscr{S} \cup \partial\mathscr{S}$, and
(b) $P(\rho,s)$ has at least one root in $\partial\mathscr{S}$.

Proof To prove this result, let us introduce the set E of all real numbers t belonging to $(a,b]$ and satisfying the following property:

$$\mathscr{P}: \qquad \text{for all } t' \in (a,t), \quad P(t',s) \text{ has all its roots in } \mathscr{S}. \tag{2.177}$$

By assumption, we know that $P(a,s)$ itself has all its roots in $\mathscr{S}$, and therefore as mentioned above, it is possible to find $\alpha > 0$ such that

$$\text{for all } t' \in [a, a+\alpha) \cap I, \quad P(t',s) \text{ also has all its roots in } \mathscr{S}. \tag{2.178}$$

From this it is easy to conclude that E is not empty since, for example, $a + \frac{\alpha}{2}$ belongs to E.

Moreover, from the definition of E the following property is obvious:

$$t_2 \in E, \text{ and } a < t_1 < t_2, \text{ implies that } t_1 \text{ itself belongs to } E. \tag{2.179}$$

Given this, it is easy to see that E is an interval and if

$$\rho := \sup_{t \in E} t \tag{2.180}$$

then it is concluded that $E = (a, \rho]$.

(A) On the one hand it is impossible that $P(\rho,s)$ has all its roots in $\mathscr{S}$. If this were the case then necessarily $\rho < b$, and it would be possible to find an $\alpha > 0$ such that $\rho + \alpha < b$ and

$$\text{for all } t' \in (\rho - \alpha, \rho + \alpha) \cap I, \; P(t',s) \text{ also has all its roots in } \mathscr{S}. \tag{2.181}$$

As a result, $\rho + \frac{\alpha}{2}$ would belong to E and this would contradict the definition of ρ in Equation (2.180).

(B) On the other hand, it is also impossible that $P(\rho,s)$ has even one root in the interior of $\mathscr{U}$, because a straightforward application of Theorem 2.9 would grant the possibility of finding an $\alpha > 0$ such that

$$\text{for all } t' \in (\rho - \alpha, \rho + \alpha) \cap I, \; P(t',s) \text{ has at least one root in } \mathscr{U}^0, \tag{2.182}$$

and this would contradict the fact that $\rho - \varepsilon$ belongs to E for ε small enough.

From (A) and (B) it is thus concluded that $P(\rho,s)$ has all its roots in $\mathscr{S} \cup \partial\mathscr{S}$, and at least one root in $\partial\mathscr{S}$. ∎

The above result is in fact very intuitive and just states that in going from one open set to another open set disjoint from the first, the root set of a continuous family of polynomials $P(\lambda,s)$ of fixed degree must intersect at some intermediate stage the boundary of the first open set. If $P(\lambda,s)$ loses degree over the interval $[a,b]$, that is if $p_n(\lambda)$ in Equation (2.175) vanishes for some values of λ, then the Boundary Crossing Theorem does *not* hold as we show in the next example.

EXAMPLE 2.12　Consider the Hurwitz stability of the polynomial

$$a_1 s + a_0, \quad \text{where } \mathbf{p} := [a_0 \; a_1]. \tag{2.183}$$

Referring to Figure 2.38, we see that the polynomial is Hurwitz stable for $\mathbf{p} = \mathbf{p}_0$. Now let the parameters travel along the path C_1 and reach the unstable point $\mathbf{p}_1$. Clearly no polynomial on this path has a $j\omega$ root for finite ω and thus boundary crossing *does not* occur. However, observe that the assumption of constant degree does not hold on this path because the point of intersection between the path C_1 and the a_0 axis corresponds to a polynomial where loss of degree occurs. On the other hand, if the parameters travel along the path C_2 and reach the unstable point $\mathbf{p}_2$, there is no loss of degree along the path C_2 and indeed a polynomial on this path has $s = 0$ as a root at $a_0 = 0$ and thus boundary crossing does occur. We illustrate this point in Figure 2.39(a). Along the path C_2, where no loss of degree occurs, the root passes through the stability boundary ($j\omega$ axis). However, on the path C_1 the polynomial passes from stable to unstable without its root passing through the stability boundary.

The above example shows that the invariant degree assumption is important. Of course we can eliminate the assumption regarding invariant degree and modify the statement of the Boundary Crossing Theorem to require that any path connecting $P_a(s)$ and $P_b(s)$ contains a polynomial which has a root on the boundary or which drops in degree.

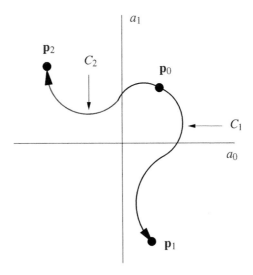

Figure 2.38 Degree loss on C_1, no loss on C_2 (Example 2.12)

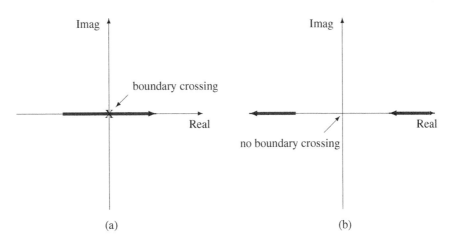

(a) (b)

Figure 2.39 (a) Root locus corresponding to the path C_2 (b) Root locus corresponding to the path C_1 (Example 2.12)

If degree dropping does occur, it is always possible to apply the result on sub-intervals over which $p_n(\lambda)$ has a constant sign. In other words, if the family of polynomials $P(\lambda, s)$ does not have a constant degree then of course Theorem 2.10 cannot be directly applied but that does not complicate the analysis terribly and similar results can be derived.

The following result gives an example of a situation where the assumption on the degree can be relaxed. As usual, let $\mathscr{S}$ be the stability region of interest.

THEOREM 2.11 Let $\{P_n(s)\}$ be a sequence of stable polynomials of *bounded degree* and assume that this sequence converges to a polynomial $Q(s)$. Then the roots of $Q(s)$ are contained in $\mathscr{S} \cup \partial \mathscr{S}$.

In words, the above theorem says that the limit of a sequence of stable polynomials of bounded degree can only have unstable roots which are on the boundary of the stability region.

Proof   By assumption, there exists an integer N such that degree$[P_n] \leq N$ for all $n \geq 0$. Therefore, we can write for all n:

$$P_n(s) = p_{0,n} + p_{1,n}s + \cdots + p_{N,n}s^N. \tag{2.184}$$

Since the sequence $\{P_n(s)\}$ converges to $Q(s)$, then $Q(s)$ itself has degree less than or equal to N so that we can also write

$$Q(s) = q_0 + q_1 s + \cdots + q_N s^N. \tag{2.185}$$

Moreover

$$\lim_{n \to +\infty} p_{k,n} = q_k, \quad \text{for } k = 0, 1, \ldots, N. \tag{2.186}$$

Now, suppose that $Q(s)$ has a root s^* which belongs to $\mathscr{U}^o$. We show that this leads to a contradiction. Since $\mathscr{U}^o$ is open, one can find a positive number r such that the disc $\mathscr{C}$ centered at s^* and of radius r is included in $\mathscr{U}^o$. By Theorem 2.9, there exists a positive number ε, such that for $|\varepsilon_i| \leq \varepsilon$, for $i = 0, 1, \ldots, N$, the polynomial

$$(q_0 + \varepsilon_0) + (q_1 + \varepsilon_1)s + \cdots + (q_N + \varepsilon_N)s^N \tag{2.187}$$

has at least one root in $\mathscr{C} \subset \mathscr{U}^o$. Now, according to Equation (2.186) it is possible to find an integer n_0 such that

$$n \geq n_0 \implies |p_{k,n} - q_k| < \varepsilon, \quad \text{for } k = 0, 1, \ldots, N. \tag{2.188}$$

But then Equation (2.188) implies that for $n \geq n_0$:

$$(q_0 + p_{0,n} - q_0) + (q_1 + p_{1,n} - q_1)s + \cdots + (q_N + p_{N,n} - q_N)s^N = P_n(s) \tag{2.189}$$

has at least one root in $\mathscr{C} \subset \mathscr{U}^o$, and this contradicts the fact that $P_n(s)$ is stable for all n. ∎

2.18.2   The Hermite–Biehler Theorem

We first present the Hermite–Biehler Theorem, sometimes referred to as the Interlacing Theorem. For the sake of simplicity we restrict ourselves to the case of polynomials with real coefficients. The corresponding result for complex polynomials will be stated separately. We deal with the Hurwitz case first and then the Schur case.

Hurwitz Stability

Consider a polynomial of degree n:

$$P(s) = p_0 + p_1 s + p_2 s^2 + \cdots + p_n s^n. \tag{2.190}$$

$P(s)$ is said to be a *Hurwitz* polynomial if and only if all its roots lie in the open left-half of the complex plane. We have the following two properties.

PROPERTY 1 If $P(s)$ is a real Hurwitz polynomial then all its coefficients are non-zero and have the same sign, either all positive or all negative.

Proof Follows from the fact that $P(s)$ can be factored into a product of first- and second-degree real Hurwitz polynomials for which the property obviously holds. ■

PROPERTY 2 If $P(s)$ is a Hurwitz polynomial of degree n, then $\arg[P(j\omega)]$, also called the phase of $P(j\omega)$, is a continuous and strictly increasing function of ω on $(-\infty, +\infty)$. Moreover, the net increase in phase from $-\infty$ to $+\infty$ is

$$\arg[P(+j\infty)] - \arg[P(-j\infty)] = n\pi. \tag{2.191}$$

Proof If $P(s)$ is Hurwitz then we can write

$$P(s) = p_n \prod_{i=1}^{n}(s - s_i), \text{ with } s_i = a_i + jb_i \text{ and } a_i < 0. \tag{2.192}$$

Then we have

$$\arg[P(j\omega)] = \arg[p_n] + \sum_{i=1}^{n} \arg[j\omega - a_i - jb_i]$$
$$= \arg[p_n] + \sum_{i=1}^{n} \arctan\left[\frac{\omega - b_i}{-a_i}\right] \tag{2.193}$$

and thus $\arg[P(j\omega)]$ is a sum of a constant plus n continuous, strictly increasing functions. Moreover, each of these n functions has a net increase of π in going from $\omega = -\infty$ to $\omega = +\infty$, as shown in Figure 2.40. ■

The even and odd parts of a real polynomial $P(s)$ are defined as:

$$P^{\text{even}}(s) := p_0 + p_2 s^2 + p_4 s^4 + \cdots,$$
$$P^{\text{odd}}(s) := p_1 s + p_3 s^3 + p_5 s^5 + \cdots. \tag{2.194}$$

Define

$$P^e(\omega) := P^{\text{even}}(j\omega) = p_0 - p_2 \omega^2 + p_4 \omega^4 - \cdots,$$
$$P^o(\omega) := \frac{P^{\text{odd}}(j\omega)}{j\omega} = p_1 - p_3 \omega^2 + p_5 \omega^4 - \cdots. \tag{2.195}$$

$P^e(\omega)$ and $P^o(\omega)$ are both polynomials in ω^2 and as an immediate consequence their root sets will always be symmetric with respect to the origin of the complex plane.

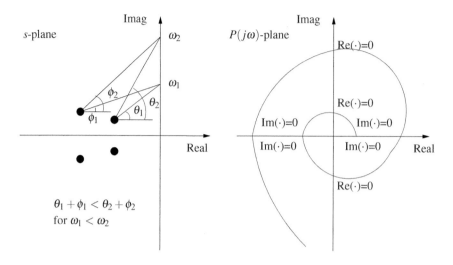

Figure 2.40 Monotonic phase-increase property for Hurwitz polynomials

Suppose now that the degree of the polynomial $P(s)$ is even, that is $n = 2m$, $m > 0$. In that case we have

$$P^e(\omega) = p_0 - p_2\omega^2 + p_4\omega^4 - \cdots + (-1)^m p_{2m}\omega^{2m},$$
$$P^o(\omega) = p_1 - p_3\omega^2 + p_5\omega^4 - \cdots + (-1)^{m-1} p_{2m-1}\omega^{2m-2}. \tag{2.196}$$

DEFINITION 2.1 A real polynomial $P(s)$ satisfies the interlacing property if

(a) p_{2m} and p_{2m-1} have the same sign.
(b) All the roots of $P^e(\omega)$ and $P^o(\omega)$ are real and distinct and the m positive roots of $P^e(\omega)$ together with the $m - 1$ positive roots of $P^o(\omega)$ interlace in the following manner:

$$0 < \omega_{e,1} < \omega_{o,1} < \omega_{e,2} < \cdots < \omega_{e,m-1} < \omega_{o,m-1} < \omega_{e,m}. \tag{2.197}$$

If, on the contrary, the degree of $P(s)$ is odd, then $n = 2m + 1, m \geq 0$, and

$$P^e(\omega) = p_0 - p_2\omega^2 + p_4\omega^4 - \cdots + (-1)^m p_{2m}\omega^{2m},$$
$$P^o(\omega) = p_1 - p_3\omega^2 + p_5\omega^4 - \cdots + (-1)^m p_{2m+1}\omega^{2m}, \tag{2.198}$$

and the definition of the interlacing property, for this case, is then naturally modified to

(a) p_{2m+1} and p_{2m} have the same sign.
(b) All the roots of $P^e(\omega)$ and $P^o(\omega)$ are real and the m positive roots of $P^e(\omega)$ together with the m positive roots of $P^o(\omega)$ interlace in the following manner:

$$0 < \omega_{e,1} < \omega_{o,1} < \cdots < \omega_{e,m-1} < \omega_{o,m-1} < \omega_{e,m} < \omega_{o,m}. \tag{2.199}$$

An alternative description of the interlacing property is as follows: $P(s) = P^{\text{even}}(s) + P^{\text{odd}}(s)$ satisfies the interlacing property if and only if

(a) the leading coefficients of $P^{\text{even}}(s)$ and $P^{\text{odd}}(s)$ are of the same sign, and

(b) all the zeros of $P^{even}(s) = 0$ and of $P^{odd}(s) = 0$ are distinct, lie on the imaginary axis, and alternate along it.

We can now enunciate and prove the following theorem.

THEOREM 2.12 (Interlacing or Hermite–Biehler Theorem)   A real polynomial $P(s)$ is Hurwitz if and only if it satisfies the interlacing property.

Proof   To prove the necessity of the interlacing property, consider a real Hurwitz polynomial of degree n:

$$P(s) = p_0 + p_1 s + p_2 s^2 + \cdots + p_n s^n. \tag{2.200}$$

Since $P(s)$ is Hurwitz it follows from Property 1 that all the coefficients p_i have the same sign, thus part (a) of the interlacing property is already proven and one can assume without loss of generality that all the coefficients are positive. To prove part (b) it is assumed arbitrarily that $P(s)$ is of even degree so that $n = 2m$. Now, we also know from Property 2 that the phase of $P(j\omega)$ strictly increases from $-n\pi/2$ to $n\pi/2$ as ω runs from $-\infty$ to $+\infty$. Due to the fact that the roots of $P(s)$ are symmetric with respect to the real axis, it is also true that $\arg(P(j\omega))$ increases from 0 to $+n\pi/2 = m\pi$ as ω goes from 0 to $+\infty$. Hence, as ω goes from 0 to $+\infty$, $P(j\omega)$ starts on the positive real axis $(P(0) = p_0 > 0)$, circles strictly counterclockwise around the origin $m\pi$ radians before going to infinity, and never passes through the origin since $P(j\omega) \neq 0$ for all ω. As a result it is very easy to see that the plot of $P(j\omega)$ has to cut the imaginary axis m times so that the *real* part of $P(j\omega)$ becomes zero m times as ω increases, at the positive values

$$\omega_{\mathscr{R},1}, \quad \omega_{\mathscr{R},2}, \quad \ldots, \quad \omega_{\mathscr{R},m}. \tag{2.201}$$

Similarly the plot of $P(j\omega)$ starts on the positive real axis and cuts the real axis another $m - 1$ times as ω increases, so that the imaginary part of $P(j\omega)$ also becomes zero m times (including $\omega = 0$) at

$$0, \quad \omega_{\mathscr{I},1}, \quad \omega_{\mathscr{I},2}, \quad \ldots, \quad \omega_{\mathscr{I},m-1} \tag{2.202}$$

before growing to infinity as ω goes to infinity. Moreover, since $P(j\omega)$ *circles* around the origin, we obviously have

$$0 < \omega_{\mathscr{R},1} < \omega_{\mathscr{I},1} < \omega_{\mathscr{R},2} < \omega_{\mathscr{I},2} < \cdots < \omega_{\mathscr{R},m-1} < \omega_{\mathscr{I},m-1} < \omega_{\mathscr{R},m}. \tag{2.203}$$

Now the proof of necessity is completed by simply noticing that the real part of $P(j\omega)$ is nothing but $P^e(\omega)$, and the imaginary part of $P(j\omega)$ is $\omega P^o(j\omega)$.

For the converse, assume that $P(s)$ satisfies the interlacing property and suppose for example that $P(s)$ is of degree $n = 2m$ and that p_{2m}, p_{2m-1} are both positive. Consider the roots of $P^e(\omega)$ and $P^o(\omega)$:

$$0 < \omega_{e,1}^p < \omega_{o,1}^p < \cdots < \omega_{e,m-1}^p < \omega_{o,m-1}^p < \omega_{e,m}^p. \tag{2.204}$$

From this, $P^e(\omega)$ and $P^o(\omega)$ can be written as

$$P^e(\omega) = p_{2m} \prod_{i=1}^{m} \left(\omega^2 - \omega_{e,i}^{p\,2} \right), \qquad P^o(\omega) = p_{2m-1} \prod_{i=1}^{m-1} \left(\omega^2 - \omega_{o,i}^{p\,2} \right). \qquad (2.205)$$

Now, consider a polynomial $Q(s)$ that is known to be stable, of the same degree $2m$, and with all its coefficients positive. For example, take $Q(s) = (s+1)^{2m}$. In any event, write

$$Q(s) = q_0 + q_1 s + q_2 s^2 + \cdots + q_{2m} s^{2m}. \qquad (2.206)$$

Since $Q(s)$ is stable, it follows from the first part of the theorem that $Q(s)$ satisfies the interlacing property, so that $Q^e(\omega)$ has m positive roots $\omega_{e,1}^q, \ldots, \omega_{e,m}^q$ and $Q^o(\omega)$ has $m-1$ positive roots $\omega_{o,1}^q, \ldots, \omega_{o,m-1}^q$, and

$$0 < \omega_{e,1}^q < \omega_{o,1}^q < \cdots < \omega_{e,m-1}^q < \omega_{o,m-1}^q < \omega_{e,m}^q. \qquad (2.207)$$

Therefore, we can also write

$$Q^e(\omega) = q_{2m} \prod_{i=1}^{m} \left(\omega^2 - \omega_{e,i}^{q\,2} \right), \qquad Q^o(\omega) = q_{2m-1} \prod_{i=1}^{m-1} \left(\omega^2 - \omega_{o,i}^{q\,2} \right). \qquad (2.208)$$

Consider now the polynomial $P_\lambda(s) := P_\lambda^{even}(s) + s P_\lambda^{odd}(s)$ defined by

$$P_\lambda^e(\omega) := \left((1-\lambda) q_{2m} + \lambda p_{2m} \right) \prod_{i=1}^{m} \left(\omega^2 - [(1-\lambda)(\omega_{e,i}^q)^2 + \lambda (\omega_{e,i}^p)^2] \right), \qquad (2.209)$$

$$P_\lambda^o(\omega) := \left((1-\lambda) q_{2m-1} + \lambda p_{2m-1} \right) \prod_{i=1}^{m-1} \left(\omega^2 - [(1-\lambda)(\omega_{o,i}^q)^2 + \lambda (\omega_{o,i}^p)^2] \right).$$

Obviously, the coefficients of $P_\lambda(s)$ are polynomial functions in λ which are therefore continuous on $[0,1]$. Moreover, the coefficient of the highest-degree term in $P_\lambda(s)$ is $(1-\lambda) q_{2m} + \lambda p_{2m}$ and always remains positive as λ varies from 0 to 1. For $\lambda = 0$ we have $P_0(s) = Q(s)$ and for $\lambda = 1$, $P_1(s) = P(s)$. Suppose now that $P(s)$ is not Hurwitz. From the Boundary Crossing Theorem it is then clear that there necessarily exists some λ in $(0,1]$ such that $P_\lambda(s)$ has a root on the imaginary axis. However, $P_\lambda(s)$ has a root on the imaginary axis if and only if $P_\lambda^e(\omega)$ and $P_\lambda^o(\omega)$ have a common real root. But, obviously, the roots of $P_\lambda^e(\omega)$ satisfy

$$\omega_{e,i}^{\lambda\,2} = (1-\lambda) \omega_{e,i}^{q\,2} + \lambda \omega_{e,i}^{p\,2}, \qquad (2.210)$$

and those of $P_\lambda^o(\omega)$,

$$\omega_{o,i}^{\lambda\,2} = (1-\lambda) \omega_{o,i}^{q\,2} + \lambda \omega_{o,i}^{p\,2}. \qquad (2.211)$$

Now, take any two roots of $P_\lambda^e(\omega)$ in Equation (2.210). If $i < j$, from Equation (2.204) $\omega_{e,i}^{p\,2} < \omega_{e,j}^{p\,2}$, and similarly from Equation (2.207) $\omega_{e,i}^{q\,2} < \omega_{e,j}^{q\,2}$, so that

$$\omega_{e,i}^{\lambda\,2} < \omega_{e,j}^{\lambda\,2}. \qquad (2.212)$$

In the same way, it can be seen that the same ordering as in Equations (2.204) and (2.207) is preserved between the roots of $P_\lambda^o(\omega)$, and also between any root of $P_\lambda^e(\omega)$

and any root of $P_\lambda^o(\omega)$. In other words, part (b) of the interlacing property is invariant under such convex combinations so that we also have for every λ in $[0,1]$:

$$0 < \omega_{e,1}^{\lambda}{}^2 < \omega_{o,1}^{\lambda}{}^2 < \cdots < \omega_{e,m-1}^{\lambda}{}^2 < \omega_{o,m-1}^{\lambda}{}^2 < \omega_{e,m}^{\lambda}{}^2. \tag{2.213}$$

But this shows that, whatever the value of λ in $[0,1]$, $P_\lambda^e(\omega)$ and $P_\lambda^o(\omega)$ can never have a common root, and this therefore leads to a contradiction which completes the proof. ∎

It is clear that the interlacing property is equivalent to the monotonic phase-increase property. If the stability region $\mathscr{S}$ is such that a stable polynomial does not have the monotonic phase-increase property, the interlacing of the real and imaginary parts will in general fail to hold. However, even in the case of such a region $\mathscr{S}$, the boundary crossing property must hold. This means that the transition from stability to instability can only occur if the real and imaginary parts simultaneously become zero at some boundary point.

A Frequency Domain Plot for Hurwitz Stability
The interlacing property of a polynomial can be verified by plotting either the graphs of $P^e(\omega)$ and $P^o(\omega)$ or the polar plot of $P(j\omega)$ as shown below.

EXAMPLE 2.13

$$P(s) = s^9 + 11s^8 + 52s^7 + 145s^6 + 266s^5 + 331s^4 + 280s^3 + 155s^2 + 49s + 6. \tag{2.214}$$

Then

$$P(j\omega) := P^e(\omega) + j\omega P^o(\omega) \tag{2.215}$$

with

$$P^e(\omega) = 11\omega^8 - 145\omega^6 + 331\omega^4 - 155\omega^2 + 6,$$
$$P^o(\omega) = \omega^8 - 52\omega^6 + 266\omega^4 - 280\omega^2 + 49. \tag{2.216}$$

The plots of $P^e(\omega)$ and $P^o(\omega)$ are shown in Figure 2.41 by using the MATLAB script below. They show that the polynomial $P(s)$ is Hurwitz because it satisfies the interlacing property.

MATLAB script

```
clear
p=[1 11 52 145 266 331 280 155 49 6];
if mod(length(p),2)==0,     % odd polynomial
  pe=p(2:end); po=p(1:end-1);
else
  pe=p(1:end); po=p(2:end-1);
end,
pe(2:2:end)=0; po(2:2:end)=0;
w=0:0.01:3;
pew=polyval(pe,j*w); pow=polyval(po,j*w);
plot(w,pew,w,pow),axis([0 3 -400 400]);
```

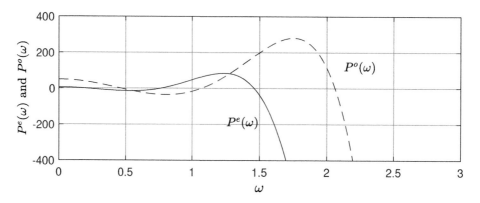

Figure 2.41 Interlacing property for Hurwitz polynomials (Example 2.13)

EXAMPLE 2.14

$$P(s) = s^9 + 21s^8 + 52s^7 + 145s^6 + 266s^5 + 331s^4 + 280s^3 + 155s^2 + 49s + 6. \quad (2.217)$$

Then

$$P(j\omega) := P^e(\omega) + j\omega P^o(\omega) \quad (2.218)$$

where

$$P^e(\omega) = 21\omega^8 - 145\omega^6 + 331\omega^4 - 155\omega^2 + 6,$$
$$P^o(\omega) = \omega^8 - 52\omega^6 + 266\omega^4 - 280\omega^2 + 49. \quad (2.219)$$

The plots of $P^e(\omega)$ and $P^o(\omega)$ are shown in Figure 2.42 using the MATLAB script given in Example 2.13. They show that the polynomial $P(s)$ is not Hurwitz because it fails to satisfy the interlacing property.

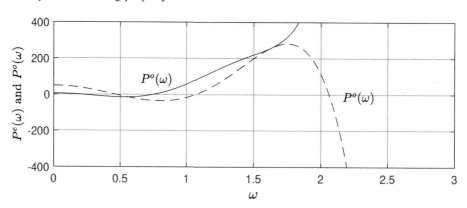

Figure 2.42 Interlacing fails for non-Hurwitz polynomials (Example 2.14)

Both the plots in the above examples are unbounded as ω tends to ∞. A bounded plot containing the same information can be constructed as follows. For a polynomial

$$P(s) = p_0 + p_1 s + \cdots + p_n s^n, \quad p_n > 0 \tag{2.220}$$

write as usual

$$P(j\omega) = P^e(\omega) + j\omega P^o(\omega) \tag{2.221}$$

and let $S(\omega)$ and $T(\omega)$ denote arbitrary continuous positive functions on $0 \le \omega < \infty$. Let

$$x(\omega) := \frac{P^e(\omega)}{S(\omega)}, \quad y(\omega) := \frac{P^o(\omega)}{T(\omega)}. \tag{2.222}$$

LEMMA 2.2   A real polynomial $P(s)$ is Hurwitz if and only if the frequency plot $z(\omega) := x(\omega) + jy(\omega)$ moves strictly counterclockwise and goes through n quadrants in turn.

Proof   The Hermite–Biehler Theorem and the monotonic phase property of Hurwitz polynomials shows that the plot of $P(j\omega)$ must go through n quadrants if and only if $P(s)$ is Hurwitz. Since the signs of $P^e(\omega)$ and $x(\omega)$, $\omega P^o(\omega)$ and $y(\omega)$ coincide for $\omega > 0$, the lemma is true.   ∎

Although the $P(j\omega)$ plot is unbounded, the plot of $z(\omega)$ can always be bounded by choosing the functions $T(\omega)$ and $S(\omega)$ appropriately. For example, $T(\omega)$ and $S(\omega)$ can be chosen to be polynomials with degrees equal to those of $P^e(\omega)$ and $P^o(\omega)$, respectively. A similar result can be derived for the complex case. Lemma 2.2 is illustrated with the following example.

EXAMPLE 2.15   Taking the same polynomial as in Example 2.13:

$$P(s) = s^9 + 11s^8 + 52s^7 + 145s^6 + 266s^5 + 331s^4 + 280s^3 + 155s^2 + 49s + 6 \tag{2.223}$$

and writing

$$P(j\omega) := P^e(\omega) + j\omega P^o(\omega) \tag{2.224}$$

we have

$$
\begin{aligned}
P^e(\omega) &= 11\omega^8 - 145\omega^6 + 331\omega^4 - 155\omega^2 + 6, \\
P^o(\omega) &= \omega^8 - 52\omega^6 + 266\omega^4 - 280\omega^2 + 49.
\end{aligned}
\tag{2.225}
$$

We choose

$$
\begin{aligned}
S(\omega) &= \omega^8 + \omega^6 + \omega^4 + \omega^2 + 1, \\
T(\omega) &= \omega^8 + \omega^6 + \omega^4 + \omega^2 + 1.
\end{aligned}
\tag{2.226}
$$

The frequency plot of $z(\omega)$ is shown in Figure 2.43 using the MATLAB script below.

MATLAB script

```
clear
p=[1 11 52 145 266 331 280 155 49 6];
if mod(length(p),2)==0,      % odd polynomial
  pe=p(2:end); po=p(1:end-1);
else
  pe=p(1:end); po=p(2:end-1);
end,
pe(2:2:end)=0;    po(2:2:end)=0;
w=0:0.001:3;
s=[1 0 -1 0 1 0 -1 0 1];
t=s;
xw=polyval(pe,j*w)./polyval(s,j*w);
yw=polyval(po,j*w)./polyval(t,j*w);
plot(xw,yw)
axis([-15 15 -20 50])
grid, xlabel('Real'), ylabel('Imag')
```

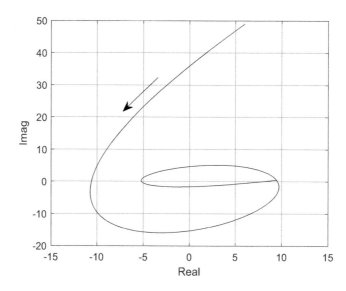

Figure 2.43 Frequency plot of $z(\omega)$ (Example 2.15)

The function $z(\omega)$ in Figure 2.43 turns strictly counterclockwise and goes through nine quadrants; this shows that the polynomial $P(s)$ is Hurwitz according to Lemma 2.2.

Hurwitz Stability for Complex Polynomials

The Hermite–Biehler Theorem for complex polynomials is given below. Its proof is a straightforward extension of that of the real case and will not be given. Let $P(s)$ be a complex polynomial

$$P(s) = (a_0 + jb_0) + (a_1 + jb_1)s + \cdots + (a_{n-1} + jb_{n-1})s^{n-1} + (a_n + jb_n)s^n. \quad (2.227)$$

Define

$$P_R(s) = a_0 + jb_1 s + a_2 s^2 + jb_3 s^3 + \cdots,$$
$$P_I(s) = jb_0 + a_1 s + jb_2 s^2 + a_3 s^3 + \cdots, \quad (2.228)$$

and write

$$P(j\omega) = P^r(\omega) + jP^i(\omega) \quad (2.229)$$

where

$$P^r(\omega) := P_R(j\omega) = a_0 - b_1\omega - a_2\omega^2 + b_3\omega^3 + \cdots,$$
$$P^i(\omega) := \frac{1}{j}P_I(j\omega) = b_0 + a_1\omega - b_2\omega^2 - a_3\omega^3 + \cdots. \quad (2.230)$$

The Hermite–Biehler Theorem for complex polynomials can then be stated as follows.

THEOREM 2.13 The complex polynomial $P(s)$ in Equation (2.227) is a Hurwitz polynomial if and only if

(1) $a_{n-1}a_n + b_{n-1}b_n > 0$.
(2) The zeros of $P^r(\omega)$ and $P^i(\omega)$ are all simple and real and interlace, as ω runs from $-\infty$ to $+\infty$.

Note that condition (1) follows directly from the fact that the sum of the roots of the polynomial $P(s)$ in Equation (2.227) is equal to

$$-\frac{a_{n-1} + jb_{n-1}}{a_n + jb_n} = -\frac{a_{n-1}a_n + b_{n-1}b_n + j(b_{n-1}a_n - a_{n-1}b_n)}{a_n^2 + b_n^2}, \quad (2.231)$$

so that if $P(s)$ is Hurwitz, then the real part of the above complex number must be negative.

Schur Stability

In fact it is always possible to derive results similar to the Interlacing Theorem with respect to any stability region $\mathscr{S}$ which has the property that the phase of a stable polynomial evaluated along the boundary of $\mathscr{S}$ increases monotonically. In this case the stability of the polynomial with respect to $\mathscr{S}$ is equivalent to the interlacing of its real and imaginary parts evaluated along the boundary of $\mathscr{S}$. Here we concentrate on the case where $\mathscr{S}$ is the open unit disc. This is the stability region for discrete-time systems.

DEFINITION 2.2 A polynomial

$$P(z) = p_n z^n + p_{n-1} z^{n-1} + \cdots + p_1 z + p_0 \quad (2.232)$$

is said to be a Schur polynomial if all its roots lie in the open unit disc of the complex plane. A necessary condition for Schur stability is $|p_n| > |p_0|$.

A Frequency Plot for Schur Stability

$P(z)$ can be written as

$$P(z) = p_n(z - z_1)(z - z_2) \cdots (z - z_n) \tag{2.233}$$

where the z_is are the n roots of $P(z)$. If $P(z)$ is Schur, all these roots are located inside the unit disc $|z| < 1$, so that when z varies along the unit circle, $z = e^{j\theta}$, the argument of $P(e^{j\theta})$ increases monotonically. For a Schur polynomial of degree n, $P(e^{j\theta})$ has a net increase of argument $2n\pi$, and thus the plot of $P(e^{j\theta})$ encircles the origin n times. This can be used as a frequency domain test for Schur stability.

EXAMPLE 2.16   Consider the stable polynomial

$$\begin{aligned} P(z) &= 2z^4 - 3.2z^3 + 1.24z^2 + 0.192z - 0.1566 \\ &= 2(z + 0.3)(z - 0.5 + 0.2j)(z - 0.5 - 0.2j)(z - 0.9). \end{aligned} \tag{2.234}$$

Let us evaluate $P(z)$ when z varies along the unit circle. The plot obtained in Figure 2.44 encircles the origin four times, which shows that this fourth-order polynomial is Schur stable.

MATLAB script

```
p=[2 -3.2 1.24 0.192 -0.1544];
theta=0:0.01:2*pi;
ptheta=polyval(p,exp(j*theta));
plot(real(ptheta), imag(ptheta))
grid, xlabel('Real'), ylabel('Imag')
```

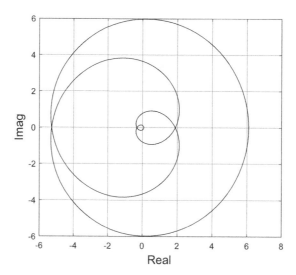

Figure 2.44 Plot of $P(e^{j\theta})$ (Example 2.16)

A simplification can be made by considering the *reverse polynomial* $z^n P(z^{-1})$:

$$z^n P(z^{-1}) = p_0 z^n + p_1 z^{n-1} + \cdots + p_n$$
$$= p_n(1 - z_1 z)(1 - z_2 z) \cdots (1 - z_n z). \tag{2.235}$$

$z^n P(z^{-1})$ becomes zero at $z = z_i^{-1}$, $i = 1, \ldots, n$. If $P(z)$ is Schur, the z_is have modulus less than one, so that the z_i^{-1} are located outside the unit disc. If we let $z = e^{j\theta}$ vary along the unit circle, the net increase of argument of $e^{jn\theta} P(e^{-j\theta})$ must therefore be zero. This means that for Schur stability of $P(z)$ it is necessary and sufficient that the frequency plot, $e^{jn\theta} P(e^{-j\theta})$ of the reverse polynomial *not encircle the origin*.

EXAMPLE 2.17 Consider the polynomial in the previous example. The plot of $z^n P(z^{-1})$ when z describes the unit circle is shown in Figure 2.45.

MATLAB script
```
clear
p=[2 -3.2 1.24 0.192 -0.1544];
theta=0:0.01:2*pi;
zn=exp(j*theta*(length(p)-1));
pitheta=zn.*polyval(p,exp(-j*theta));
plot(real(pitheta), imag(pitheta),'-k')
grid, xlabel('Real'), ylabel('Imag')
```

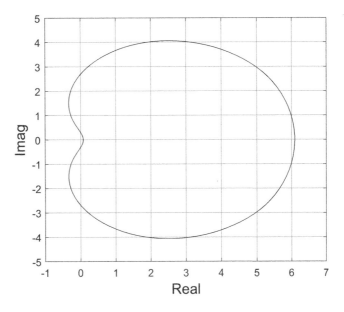

Figure 2.45 Plot of $e^{j4\theta} P(e^{-j\theta})$ (Example 2.17)

As seen, the plot does not encircle the origin and thus we conclude that $P(z)$ is stable.

We see that when using the plot of $P(z)$ we must verify that the plot of $P(e^{j\theta})$ en-circles the origin the correct number of times n, whereas using the reverse polynomial $R(z) = z^n P(z^{-1})$ we need only check that the plot of $R(e^{j\theta})$ excludes the origin. This result holds for real as well as complex polynomials.

For a real polynomial, it is easy to see from the above that the stability of $P(z)$ is equivalent to the interlacing of the real and imaginary parts of $P(z)$ evaluated along the upper-half of the unit circle. Writing $P(e^{j\theta}) = R(\theta) + jI(\theta)$, we have

$$R(\theta) = p_n \cos(n\theta) + \cdots + p_1 \cos(\theta) + p_0 \tag{2.236}$$

and

$$I(\theta) = p_n \sin(n\theta) + \cdots + p_1 \sin(\theta). \tag{2.237}$$

LEMMA 2.3   A real polynomial $P(z)$ is Schur with $|p_n| > |p_0|$ if and only if

(a) $R(\theta)$ has exactly n zeros in $[0, \pi]$,
(b) $I(\theta)$ has exactly $n+1$ zeros in $[0, \pi]$, and
(c) the zeros of $R(\theta)$ and $I(\theta)$ interlace.

EXAMPLE 2.18   Consider the polynomial

$$P(z) = z^5 + 0.2z^4 + 0.3z^3 + 0.4z^2 + 0.03z + 0.02. \tag{2.238}$$

As seen in Figure 2.46, the polynomial $P(z)$ is Schur since $\mathrm{Re}[P(e^{j\theta})]$ and $\mathrm{Im}[P(e^{j\theta})]$ have, respectively, five and six distinct zeros for $\theta \in [0, \pi]$, and the zeros of $\mathrm{Re}[P(e^{j\theta})]$ interlace with the zeros of $\mathrm{Im}[P(e^{j\theta})]$.

MATLAB script

```
clear
p=[1 0.2 0.3 0.4 0.03 0.02];
theta=0:0.01:2*pi;
ptheta=polyval(p,exp(j*theta));
plot(theta,real(ptheta),'-k',theta,imag(ptheta),'--k')
axis([0 6.3 -2 2])
```

EXAMPLE 2.19   Consider the polynomial

$$P(z) = z^5 + 2z^4 + 0.3z^3 + 0.4z^2 + 0.03z + 0.02. \tag{2.239}$$

Since $\mathrm{Re}[P(e^{j\theta})]$ and $\mathrm{Im}[P(e^{j\theta})]$ each do not have $2n = 10$ distinct zeros for $0 \le \theta < 2\pi$, as shown in Figure 2.47, the polynomial $P(z)$ is not Schur.

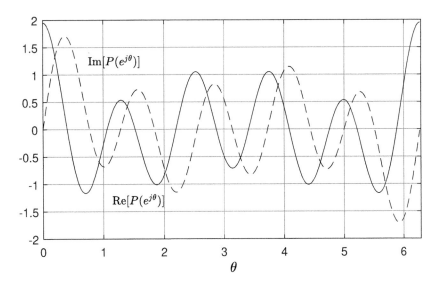

Figure 2.46 $\text{Re}[P(e^{j\theta})]$ and $\text{Im}[P(e^{j\theta})]$ (Schur case) (Example 2.18)

MATLAB script

```
clear
p=[1 2 0.3 0.4 0.03 0.02];
theta=0:0.01:2*pi;
ptheta=polyval(p,exp(j*theta));
plot(theta,real(ptheta),theta,imag(ptheta))
axis([0 6.3  -4 4])
```

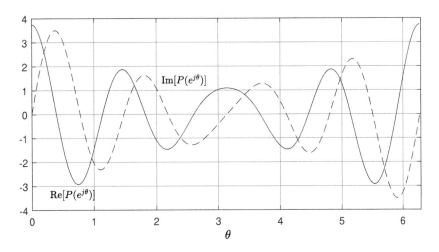

Figure 2.47 $\text{Re}[P(e^{j\theta})]$ and $\text{Im}[P(e^{j\theta})]$ (non-Schur case) (Example 2.19)

These conditions can in fact be further refined to the interlacing on the unit circle of the two polynomials $P_s(z)$ and $P_a(z)$ which represent the *symmetric* and *asymmetric* parts of the real polynomial $P(z) = P_s(z) + P_a(z)$:

$$P_s(z) = \frac{1}{2}\left[P(z) + z^n P\left(\frac{1}{z}\right)\right], \quad \text{and } P_a(z) = \frac{1}{2}\left[P(z) - z^n P\left(\frac{1}{z}\right)\right]. \tag{2.240}$$

THEOREM 2.14 A real polynomial $P(z)$ is Schur if and only if $P_s(z)$ and $P_a(z)$ satisfy the following:

(a) $P_s(z)$ and $P_a(z)$ are polynomials of degree n with leading coefficients of the same sign.

(b) $P_s(z)$ and $P_a(z)$ have only simple zeros which belong to the unit circle.

(c) The zeros of $P_s(z)$ and $P_a(z)$ interlace on the unit circle.

Proof Let $P(z) = p_0 + p_1 z + p_2 z^2 \cdots + p_n z^n$. Condition (a) is equivalent to $p_n^2 - p_0^2 > 0$, which is clearly necessary for Schur stability. Now we apply the bilinear transformation of the unit circle into the left-half plane and use the Hermite–Biehler Theorem for Hurwitz stability. It is known that the bilinear mapping

$$z = \frac{s+1}{s-1} \tag{2.241}$$

maps the open unit disc into the open left-half plane. It can be used to transform a polynomial $P(z)$ into $\hat{P}(s)$ as follows:

$$(s-1)^n P\left(\frac{s+1}{s-1}\right) = \hat{P}(s). \tag{2.242}$$

Write

$$\hat{P}(s) = \hat{p}_0 + \hat{p}_1 s + \cdots + \hat{p}_{n-1} s^{n-1} + \hat{p}_n s^n \tag{2.243}$$

where each $\hat{p}_i$ is a function which depends on the coefficients of $P(z)$. It follows that if the transformation is degree-preserving then $P(z)$ is Schur stable if and only if $\hat{P}(s)$ is Hurwitz stable. It is easy to verify that the transformation described above is degree-preserving if and only if

$$\hat{p}_n = \sum_{i=0}^{i=n} p_i = P(1) \neq 0 \tag{2.244}$$

and that this holds is implied by condition (c).

The transformation of $P(z)$ into $\hat{P}(s)$ is a linear transformation T. That is, $\hat{P}(s)$ is the image of $P(z)$ under the mapping T. Then $TP(z) = \hat{P}(s)$ may be written explicitly as

$$(s-1)^n P\left(\frac{s+1}{s-1}\right) = TP(z) = \hat{P}(s). \tag{2.245}$$

For example, for $n = 4$, expressing $P(z)$ and $\hat{P}(s)$ in terms of their coefficient vectors:

$$
\begin{bmatrix}
1 & 1 & 1 & 1 & 1 \\
-4 & -2 & 0 & 2 & 4 \\
6 & 0 & -2 & 0 & 6 \\
-4 & 2 & 0 & -2 & 4 \\
1 & -1 & 1 & -1 & 1
\end{bmatrix}
\begin{bmatrix}
p_0 \\ p_1 \\ p_2 \\ p_3 \\ p_4
\end{bmatrix}
=
\begin{bmatrix}
\hat{p}_0 \\ \hat{p}_1 \\ \hat{p}_2 \\ \hat{p}_3 \\ \hat{p}_4
\end{bmatrix}. \tag{2.246}
$$

Consider the symmetric and antisymmetric parts of $P(z)$, and their transformed images, given by $TP_s(z)$ and $TP_a(z)$, respectively. A straightforward computation shows that

$$
\begin{aligned}
TP_s(z) &= \hat{P}^{\text{even}}(s), \quad TP_a(z) = \hat{P}^{\text{odd}}(s), \quad \text{for } n \text{ even,} \\
TP_s(z) &= \hat{P}^{\text{odd}}(s), \quad TP_a(z) = \hat{P}^{\text{even}}(s), \quad \text{for } n \text{ odd.}
\end{aligned} \tag{2.247}
$$

Conditions (b) and (c) now follow immediately from the interlacing property for Hurwitz polynomials applied to $\hat{P}(s)$. ∎

The functions $P_s(z)$ and $P_a(z)$ are easily evaluated as z traverses the unit circle. Interlacing may be verified from a plot of the zeros of these functions as in Figure 2.48.

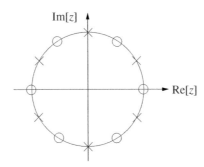

Figure 2.48 Interlacing of the symmetric and antisymmetric parts of a polynomial on the unit circle

2.19   Root Counting and the Routh Table

The Routh table is known as a procedure to determine the root distribution of a polynomial about the imaginary axis. In this section we describe how it does this. Let

$$
P(s) = p_n s^n + p_{n-1} s^{n-1} + \cdots + p_1 s + p_0 \tag{2.248}
$$

and

$$
\mu := \frac{p_n}{p_{n-1}}. \tag{2.249}
$$

We introduce the line segment of polynomials $Q(\lambda, s)$ defined below.
For $n = 2m$:

$$Q(\lambda, s) = \underbrace{\left(P^{\text{even}}(s) - \lambda\mu s P^{\text{odd}}(s)\right)}_{Q^{\text{even}}(\lambda, s)} + \underbrace{P^{\text{odd}}(s)}_{Q^{\text{odd}}(\lambda, s)}. \tag{2.250}$$

For $n = 2m + 1$:

$$Q(\lambda, s) = \underbrace{P^{\text{even}}(s)}_{Q^{\text{even}}(\lambda, s)} + \underbrace{\left(P^{\text{odd}}(s) - \lambda\mu s P^{\text{odd}}(s)\right)}_{Q^{\text{odd}}(\lambda, s)}. \tag{2.251}$$

More generally:

$$\begin{aligned} Q(\lambda, s) = (p_n - \lambda\mu p_{n-1}) s^n + p_{n-1} s^{n-1} + (p_{n-2} - \lambda\mu p_{n-3}) s^{n-2} \\ + p_{n-3} s^{n-3} + (p_{n-4} - \lambda\mu p_{n-5}) s^{n-4} + \cdots \end{aligned} \tag{2.252}$$

and we see that

$$Q(0, s) = P(s), \qquad Q(1, s) = Q(s) \tag{2.253}$$

and $Q(\lambda, s)$ has degree n for $\lambda \in [0, 1)$, and loses degree only at $\lambda = 1$. If we assume that $P(j\omega) \neq 0$ for $\omega \in [0, \infty)$, it follows from Equations (2.250) and (2.251) that $Q(\lambda, j\omega) \neq 0$ for $\lambda \in [0, 1)$ and $\omega \in [0, \infty)$. Thus, the root distribution about the imaginary axis of $Q(\lambda, s)$ remains invariant and equal to that of $P(s)$. As $\lambda \to 1$, the leading coefficient

$$p_n - \lambda\mu p_{n-1} = p_n - \lambda\frac{p_n}{p_{n-1}} p_{n-1} \tag{2.254}$$

of $Q(\lambda, s)$ tends to zero. As a result, one root of $Q(\lambda, s)$ tends to infinity along the positive or negative real axis. It is easy to see that for large $|s|$, we may approximate

$$\begin{aligned} Q(\lambda, s) &\approx (p_n - \lambda\mu p_{n-1}) s^n + p_{n-1} s^{n-1} \\ &\approx s^{n-1} \left[(p_n - \lambda\mu p_{n-1}) s + p_{n-1}\right] \end{aligned} \tag{2.255}$$

so that $n - 1$ roots of $Q(\lambda, s)$ are finite while the one root tending to infinity can be estimated by

$$s^* = \frac{-p_{n-1}}{p_n - \lambda\mu p_{n-1}} = -\frac{p_{n-1}}{p_n(1 - \lambda)}. \tag{2.256}$$

Equation (2.256) demonstrates that the real root that is lost in $Q(\lambda, s)$ as $\lambda \to 1$, tends to infinity along the negative real axis if p_n and p_{n-1} have the same sign and along the positive real axis if p_n and p_{n-1} are of opposite sign. This is exactly Routh's theorem, where the number of sign changes in the first column equals the number of RHP roots.

2.20   The Gauss–Lucas Theorem

The following classical result, known as the Gauss–Lucas Theorem, was implied in a note of Carl Friedrich Gauss dated 1836, and it was stated explicitly and proved by Félix Lucas in 1874.

THEOREM 2.15 (Gauss–Lucas Theorem)   Let $f(s)$ be a polynomial with real or complex coefficients. All the zeros of the derivative $f'(s) := \frac{df(s)}{ds}$ lie in the convex hull $\mathcal{H}$ of the set of zeros of $f(s)$. If the zeros of $f(s)$ are not co-linear, no zeros of the derivative $f'(s)$ lie on the boundary of $\mathcal{H}$ unless it is a multiple zero of $f(s)$.

Proof   Write an nth-order complex polynomial with α being the leading coefficient, as

$$f(s) = \alpha \Pi_{i=1}^{n} (s - a_i) \tag{2.257}$$

where the complex numbers a_i for $i = 1, 2, \ldots, n$ are all the (not necessarily distinct) zeros of the polynomial $f(s)$. It is easy to verify then that

$$\frac{f'(s)}{f(s)} = \sum_{i=1}^{n} \frac{1}{s - a_i}. \tag{2.258}$$

If z is a zero of $f'(s) = 0$ and $f(z) \neq 0$, then

$$f'(z) = f(z) \left(\sum_{i=1}^{n} \frac{1}{z - a_i} \right) = 0 \quad \text{or} \quad \sum_{i=1}^{n} \frac{1}{z - a_i} = \sum_{i=1}^{n} \frac{z^* - a_i^*}{|z^* - a_i^*|^2} = 0 \tag{2.259}$$

where "*" denotes the complex conjugate. The above equation may also be written as

$$\left(\sum_{i=1}^{n} \frac{1}{|z - a_i|^2} \right) z = \left(\sum_{i=1}^{n} \frac{1}{|z - a_i|^2} \right) a_i. \tag{2.260}$$

We see that z is a weighted sum with positive coefficients that sum to one, of the complex numbers a_i.

If z is a zero of both $f^*(s)$ and $f(s)$, then

$$z = (1)a_i + \left(\sum_{j=1, j \neq i}^{n} (0)a_j \right), \quad \text{for some } i \tag{2.261}$$

and so is still a convex combination of the zeros of $f(s) = 0$.   ■

The Gauss–Lucas Theorem can provide useful information on stabilizing sets for control systems by reducing it to lower-order problems, as shown below.

2.20.1   Application to Control Systems

In this section, we apply the Gauss–Lucas Theorem to derive results on the stability and performance of closed-loop control systems.

Consider the feedback system in Figure 2.49.

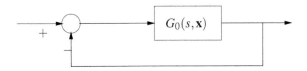

Figure 2.49 Closed-loop system $S_0(\mathbf{x})$

Let

$$G_0(s, \mathbf{x}) = \frac{n(s, \mathbf{x})}{d(s, \mathbf{x})}, \tag{2.262}$$

where $n(s, \mathbf{x})$ and $d(s, \mathbf{x})$ are polynomials of degree m and $n (\geq m)$, respectively, with real coefficients, and $\mathbf{x}$ is a real parameter vector.

The closed-loop characteristic polynomial of the system $S_0(\mathbf{x})$ is

$$p(s, \mathbf{x}) = d(s, \mathbf{x}) + n(s, \mathbf{x}), \tag{2.263}$$

and $S_0(\mathbf{x})$ is said to be *stable* if all roots of Equation (2.263) lie in an open convex set $\mathscr{C}$, of the complex plane, denoted the *stability region*.[1]

Let $n^i(s, \mathbf{x})$, $d^i(s, \mathbf{x})$, $G_i(s, \mathbf{x})$, and $S_i(\mathbf{x})$ (see Figure 2.50) be defined as follows:

$$n^i(s, \mathbf{x}) = \begin{cases} \dfrac{d^i n(s, \mathbf{x})}{ds^i}, & \text{for } i = 1, 2, \ldots, m, \\ 1, & \text{for } i = m+1, \ldots, n-1, \end{cases}$$

$$d^i(s, \mathbf{x}) = \begin{cases} \dfrac{d^i d(s, \mathbf{x})}{ds^i}, & \text{for } i = 1, 2, \ldots, m, \\[2mm] \dfrac{d^i d(s, \mathbf{x})}{ds^i} - 1, & \text{for } i = m+1, \ldots, n-1, \end{cases} \tag{2.264}$$

$$G_i(s, \mathbf{x}) = \frac{n^i(s, \mathbf{x})}{d^i(s, \mathbf{x})}, \quad \text{for } i = 1, 2, \ldots, n-1.$$

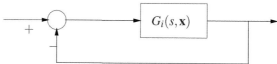

Figure 2.50 Closed-loop system $S_i(\mathbf{x})$

Let $\mathscr{X}_i$ denote the stabilizing set of parameters for $S_i(\mathbf{x})$ for $i = 0, 1, 2, \ldots, n-1$.

THEOREM 2.16

(a) $S_0(\mathbf{x})$ is stable only if $S_i(\mathbf{x})$, $i = 1, 2, \ldots, n-1$ are stable.
(b) The stabilizing sets $\mathscr{X}_i$, $i = 0, 1, 2, \ldots, n-1$ satisfy

$$\mathscr{X}_0 \subset \mathscr{X}_1 \subset \mathscr{X}_2 \subset \cdots \subset \mathscr{X}_{n-1}. \tag{2.265}$$

(c) The gain margin of $S_0(\mathbf{x})$ is upper bounded by the gain margins of $S_i(\mathbf{x})$, $i = 1, 2, \ldots, n-1$.
(d) The phase margin of $S_0(\mathbf{x})$ is upper bounded by the phase margins of $S_i(\mathbf{x})$, $i = 1, 2, \ldots, n-1$.

---

[1]Note that special cases of $\mathscr{C}$ are the open left-half plane and the open unit disc corresponding to Hurwitz and Schur stability, respectively.

Proof The proof of (a) and (b) follows immediately from the Gauss–Lucas Theorem and the convexity of $\mathscr{C}$. The proof of (c) and (d) follows from applying the Gauss–Lucas Theorem to

$$d(s,\mathbf{x}) + Kn(s,\mathbf{x}) \tag{2.266}$$

and

$$d(s,\mathbf{x}) + e^{-j\theta}n(s,\mathbf{x}), \tag{2.267}$$

respectively. ∎

In the rest of this chapter, we provide some applications of this result to control systems. The main idea is that one may reduce the system order by using the Gauss–Lucas Theorem.

EXAMPLE 2.20 An example is provided to illustrate Theorem 2.16. Let the plant be

$$P(s) = \frac{s^3 + 6s^2 - 2s + 1}{s^5 + 3s^4 + 29s^3 + 15s^2 - 3s + 60} \tag{2.268}$$

and a controller be

$$C(s) = \frac{sk_p + k_i}{s}. \tag{2.269}$$

Then

$$\begin{aligned}
p(s) = s^6 + 3s^5 + (k_p + 29)\,s^4 + (k_i + 6k_p + 15)\,s^3 \\
+ (6k_i - 2k_p - 3)\,s^2 + (k_p - 2k_i + 60)\,s + k_i.
\end{aligned} \tag{2.270}$$

Its first- and second-derivative polynomials after division by the leading coefficient are

$$\begin{aligned}
p'(s) = s^5 + \frac{5}{2}s^4 + \left(\frac{2}{3}k_p + \frac{58}{3}\right)s^3 + \left(\frac{1}{2}k_i + 3k_p + \frac{15}{2}\right)s^2 \\
+ \left(2k_i - \frac{2}{3}k_p - 1\right)s - \frac{1}{3}k_i + 1\frac{1}{6}k_p + 10
\end{aligned} \tag{2.271}$$

and

$$p''(s) = s^4 + 2s^3 + \left(\frac{2}{5}k_p + \frac{58}{5}\right)s^2 + \left(\frac{1}{5}k_i + \frac{6}{5}k_p + 3\right)s + \frac{2}{5}k_i - \frac{2}{15}k_p - \frac{1}{5}. \tag{2.272}$$

The set of stabilizing parameters in the controller parameter space for the actual system is found and plotted (see Figure 2.51).

Additionally, the stabilizing parameters for the lower-order systems are calculated corresponding to two derivatives of the given characteristic polynomial, and show outer approximations of the original stabilizing set. By continuing this procedure down to a second- and first-order characteristic polynomial more and more relaxed outer approximations can be obtained. The last two approximations would be convex sets, as may easily be verified.

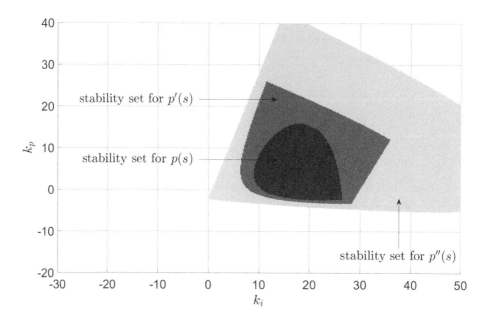

Figure 2.51 The Hurwitz stability sets of Example 2.20 © 2013 IEEE. Reproduced from Knap, Keel, and Bhattacharyya (2013) with permission.

2.21   Exercises

2.1   Consider the discrete-time system shown in Figure 2.52.

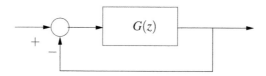

Figure 2.52 Discrete-time system (Exercise 2.1)

$G(z)$ is a real rational transfer function. The plot of $G(e^{j\theta})$ for $\theta \in [0, 2\pi]$ with θ increasing from θ to 2π encircles the $-1 + j0$ point N times in the clockwise direction. Let P_{CL}^o (P_{OL}^o) denote the number of closed-loop (open-loop) characteristic roots outside the unit circle. Show that

$$P_{CL}^o - P_{OL}^o = N \qquad (2.273)$$

and thus that closed-loop stability holds if and only if

$$N = -P_{OL}^o. \qquad (2.274)$$

2.2 Sketch the Nyquist plots of the following open-loop transfer functions. Assume each $T_i > 0$. Also verify your answers by using MATLAB (use the command "nyquist" or "nyqlog"):

(1) $\dfrac{1}{1+sT_1}$,

(2) $\dfrac{1}{s(1+sT_1)}$,

(3) $\dfrac{1}{(1+sT_1)(1+sT_2)}$,

(4) $\dfrac{1}{s(1+sT_1)(1+sT_2)}$,

(5) $\dfrac{1}{s^2(1+sT_1)(1+sT_2)(1+sT_3)}$,

(6) $\dfrac{1+sT_1}{s^2(1+sT_2)}$, for $T_1 > T_2$,

(7) $\dfrac{1+sT_1}{s^2(1+sT_2)}$, for $T_1 < T_2$.

In each case, determine the number of closed-loop characteristic roots in the right-half plane.

2.3 Show that each plant with transfer function of one of the forms

$$P(s) = \frac{K}{s+a}, \qquad P(s) = \frac{s-z}{s-p} \tag{2.275}$$

can be stabilized by a PI controller

$$C(s) = K_p + \frac{K_i}{s}. \tag{2.276}$$

Consider positive and negative values of all plant parameters.

2.4 For the plants in Exercise 2.3, determine the complete stabilizing set in the (K_p, K_i) plane in each case, in terms of the plant parameters.

2.5 Choose specific numerical values for the plant parameters in Exercise 2.4 and determine the PI stabilizing set. Choose a square inside this set, set up a grid of 25 points, and map them into the achievable gain and phase margin plane indexed by the gain crossover frequency ω_g.

2.6 Prove that the PID controller

$$C(s) = \frac{K_i + sK_p + s^2 K_d}{s(s+a)} \tag{2.277}$$

can stabilize any plant of the form

$$P(s) = \frac{b_0 + b_1 s}{s^2 + a_1 s + a_0}, \quad \text{with } b_0 \neq 0. \tag{2.278}$$

2.7 The Nyquist plot of a stable system with transfer function $KG(s)$ is shown in Figure 2.53 with $K = 1$: determine the values of K for which the closed-loop system is stable.

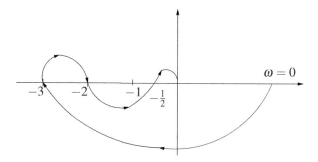

Figure 2.53 Nyquist plot of a stable system $KG(s)$ with $K = 1$ (Exercise 2.7)

2.8 Frequency response measurement data for a stable open-loop system is shown in Table 2.1.

Table 2.1 Measurement data (Exercise 2.8)

ω	0.1	1	10	100	1000
$\|G(j\omega)\|$ (dB)	10	5	2	0	-10
$\angle G(j\omega)$	$-10°$	$-120°$	$-145°$	$-150°$	$-180°$

Is the closed-loop system stable? If so, what are the gain and phase margins of the closed-loop system?

2.9 Consider the unity feedback stable closed-loop system with

$$G(s) = \frac{N(s)}{D(s)} \tag{2.279}$$

in the forward path. Assume that the leading coefficients of $D(s)$ and $N(s)$ are of the same sign. Prove that if z_i^+ (p_j^+) denote the RHP real distinct zeros and poles of $G(s)$, then:

(a) $N(p_j^+)$, $D(z_i^+)$ for $i = 1, 2, \ldots; j = 1, 2, \ldots$ all have the same signs, positive or negative.

(b) The numbers of zeros counting multiplicities, to the right of every pole p_j^+, and the numbers of poles to the right of every zero z_j^+ are all even or all odd.

Hint: Use the fact that

$$D(\sigma) + N(\sigma) \neq 0, \quad \text{for all } \sigma \geq 0. \tag{2.280}$$

2.10 Suppose that the Nyquist plot of G encircles an arbitrary point $s^* \in \mathbb{C}$, n^* times in the counterclockwise direction. Then the *open-loop* system has

$$p_{OL}^+ \geq \max_{s^* \in \mathbb{C}} n^*. \tag{2.281}$$

Hint: Replace $G(s)$ by $ke^{j\theta}G(s)$.

2.11 Consider the first-order plant with delay

$$P(s) = \frac{Ke^{-sL}}{1+sT} \tag{2.282}$$

and PID controller

$$C(s) = K_p\left(1 + \frac{1}{\tau_i s} + \tau_d s\right). \tag{2.283}$$

The Ziegler–Nichols design specifies that

$$K_p = \frac{1.2T}{L}, \qquad \tau_i = 2L, \qquad \tau_d = \frac{1}{2}L. \tag{2.284}$$

Choose various $K \in [1,5]$, $T \in [1,5]$, $L \in [1,5]$ and determine in each case the corresponding gain margin, phase margin, and gain crossover frequencies.

2.12 For the plant

$$P(s) = \frac{s-1}{(s-L)(s-3)}, \tag{2.285}$$

design a controller to track steps and ramps, and place closed-loop poles at $-1 \pm j$ and the remaining poles at $s = -2$.

2.13 Consider the open-loop system

$$G(s) = \frac{K}{s(s+a_1)(s+a_2)} \quad \text{with } K, a_1, a_2 > 0. \tag{2.286}$$

Using the Nyquist plot, determine conditions on K, a_1, a_2 for closed-loop stability, and the gain margin of the closed-loop system.

Answer: $K < a_1 a_2 (a_1 + a_2)$.

2.14 In Exercise 2.13, choose numerical values for $a_1 \in [1,5], a_2 \in [1,5]$ and appropriate stabilizing Ks. In each case, find the gain margin, phase margin, and gain crossover frequency.

2.15 The open-loop system transfer function is

$$G(s) = \frac{K(s-z)}{s(s+a_1)(s+a_2)}, \quad \text{with } z, a_1, a_2 > 0. \tag{2.287}$$

Show that the phase crossover frequency ω_ϕ is given by

$$\omega_\phi^2 = \frac{za_1a_2}{a_1 + a_2 + z}. \tag{2.288}$$

Determine conditions on K for closed-loop stability and the gain margin as a function of K.

2.16 Give op-amp circuit realizations of the following controller transfer functions, using RLC components:

(a) $C(s) = \dfrac{2s + 1}{s}$,

(b) $C(s) = \dfrac{1 + 2s + 3s^2}{s}$,

(c) $C(s) = \dfrac{K(1 + sT_1)}{1 + sT_2}$,

(d) $C(s) = \dfrac{s - 1}{s}$.

2.17 Consider the plant and PI controller

$$P(s) = \frac{1}{s^2 + 2s + 2}, \qquad C(s) = \frac{K_i + K_p s}{s}. \tag{2.289}$$

Choose appropriate values of K_i and K_p so that the closed-loop system asymptotically tracks a step reference and also rejects a step disturbance by using Simulink. Try at least 10 different sets of (K_i, K_p) parameters that make the closed-loop system stable and choose the parameter set that results in "good response." As in Example 2.3, assume that the disturbance begins after the output response reaches its steady state.

2.22   Notes and References

The Nyquist stability criterion was formulated by Harry Nyquist in Nyquist (1932). It remains as one of the cornerstone results of control theory. The Bode plot was originally introduced and developed by Hendrik Wade Bode in Bode (1945). Harold Stephen Black discovered the negative feedback amplifier in 1926; it is described in Black (1934). The relationship between Nyquist plots and Bode plots, including the counting of encirclements using Bode plots described in Section 2.12, was developed in Keel and Bhattacharyya (2010a). The frequency response characterization of all stabilizing controllers was reported in Keel, Shafai, and Bhattacharyya (2009). The Ziegler–Nichols design formulas used in Exercise 2.11 were developed in Ziegler and Nichols (1942). The proof of the Routh–Hurwitz criterion using the Hermite–Biehler Theorem was given in Bhattacharyya, Chapellat, and Keel (1995) and Ho, Datta, and Bhattacharyya (1998). The Gauss–Lucas Theorem was used to develop approximations of stabilizing sets for control systems in Knap, Keel, and Bhattacharyya (2013).

3 The $[\mathbf{A}, \mathbf{B}, \mathbf{C}, \mathbf{D}]$ State-Variable Model

This chapter introduces the standard state-variable representation of linear time-invariant dynamic systems. This standard model is an important and basic tool for representing, simulating, and designing continuous-time and discrete-time dynamic systems. It offers a unified framework and platform for the analysis and design of dynamic systems in various fields such as electrical, mechanical, aerospace, and chemical engineering. The state-space model $[\mathbf{A}, \mathbf{B}, \mathbf{C}, \mathbf{D}]$ is also the basic mathematical model on which optimal control, optimal estimation, and various applications in signal processing, circuit and control theory are based.

3.1 Introduction

A dynamic or static system is driven by input signals and processes them via the internal state to produce the corresponding outputs. This generic model holds not only in engineering systems, but also in population, weather forecasting, social, economic, biological, and other systems.

We regard a *system* as consisting of external variables, labeled, *inputs* $u_1(t)$, $u_2(t)$, ..., $u_r(t)$, *outputs* denoted $y_1(t)$, $y_2(t)$, ..., $y_m(t)$, and *internal* or *state* variables $x_1(t)$, $x_2(t)$, ..., $x_n(t)$. Using vector notation

$$\mathbf{y}(t) := \begin{bmatrix} y_1(t) \\ y_2(t) \\ \vdots \\ y_m(t) \end{bmatrix}, \quad \mathbf{u}(t) := \begin{bmatrix} u_1(t) \\ u_2(t) \\ \vdots \\ u_r(t) \end{bmatrix}, \quad \mathbf{x}(t) := \begin{bmatrix} x_1(t) \\ x_2(t) \\ \vdots \\ x_n(t) \end{bmatrix} \tag{3.1}$$

denote the m-output vector, r-input vector, and n-state vector, respectively, at time t (see Figure 3.1). For example, in a circuit or power system $\mathbf{y}(t)$ and $\mathbf{u}(t)$ might represent terminal voltages and input or source voltages, and $\mathbf{x}(t)$ the vector of internal currents and voltages in all branches. Considering the human body, the inputs may be food intake, beverages consumed, and medication administered, and the output $\mathbf{y}(t)$ may be body temperature, blood pressure, and pulse rate, while the internal variables $\mathbf{x}(t)$ may be heart rate, blood sugar concentration, brain activity level, and various hormone levels.

For linear time-invariant (LTI) systems it will be possible, in many cases, to write the

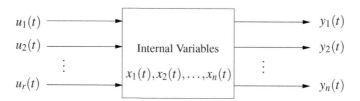

Figure 3.1 A conceptual model

system equations in the standard form

$$\dot{\mathbf{x}}(t) = \mathbf{A}\mathbf{x}(t) + \mathbf{B}\mathbf{u}(t), \tag{3.2a}$$
$$\mathbf{y}(t) = \mathbf{C}\mathbf{x}(t) + \mathbf{D}\mathbf{u}(t), \tag{3.2b}$$

where $\mathscr{X}$, $\mathscr{U}$, $\mathscr{Y}$ are vector spaces over the real or complex fields, $\mathbf{x}(t) \in \mathscr{X}$, $\mathbf{y}(t) \in \mathscr{Y}$, $\mathbf{u}(t) \in \mathscr{U}$, and the maps $\mathbf{A} : \mathscr{X} \to \mathscr{X}$, $\mathbf{B} : \mathscr{U} \to \mathscr{X}$, $\mathbf{C} : \mathscr{X} \to \mathscr{Y}$, $\mathbf{D} : \mathscr{U} \to \mathscr{Y}$ are represented by matrices with appropriate dimensions. This representation will also be called the [A, B, C, D] model.

In general the *internal* or *state* variables should be chosen to provide a "complete" description of the system. If the equations describing a system are defined for all time, the system is called a *continuous-time* system. When the variables occur at discrete instants of time, it is called a *discrete-time system*. Continuous-time systems are described by differential equations while discrete-time systems are described by difference equations.

In the following, we give some examples of physical systems that can be modeled in state-variable form.

EXAMPLE 3.1 (*RLC* circuits)   Consider the series *RLC* circuit shown in Figure 3.2.

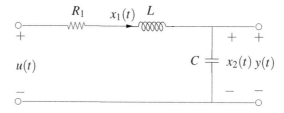

Figure 3.2 An *RLC* circuit (Example 3.1)

Choosing the states $x_1(t)$ and $x_2(t)$ as the inductor current and capacitor voltage, respectively, we see that the vector $(x_1(t), x_2(t))$ provides a complete description of all currents and voltages in the circuit. The system equations are obtained from Kirchhoff's laws:

$$x_1(t) = C\dot{x}_2(t) \qquad \text{(current summation)}, \tag{3.3a}$$

$$u(t) = Rx_1(t) + L\dot{x}_1(t) + x_2(t) \qquad \text{(voltage summation)}, \tag{3.3b}$$

$$y(t) = x_2(t) \qquad \text{(output)}. \tag{3.3c}$$

These can be rendered into the standard form

$$\begin{bmatrix} \dot{x}_1(t) \\ \dot{x}_2(t) \end{bmatrix} = \underbrace{\begin{bmatrix} -\dfrac{R}{L} & -\dfrac{1}{L} \\ \dfrac{1}{C} & 0 \end{bmatrix}}_{A} \begin{bmatrix} x_1(t) \\ x_2(t) \end{bmatrix} + \underbrace{\begin{bmatrix} \dfrac{1}{L} \\ 0 \end{bmatrix}}_{B} u(t),$$

$$\tag{3.4}$$

$$y(t) = \underbrace{[0 \quad 1]}_{C} \begin{bmatrix} x_1(t) \\ x_2(t) \end{bmatrix} + \underbrace{[0]}_{D} u(t).$$

EXAMPLE 3.2 (Two-input two-output *RLC* circuit)   Consider the state-space model of the circuit in Figure 3.3.

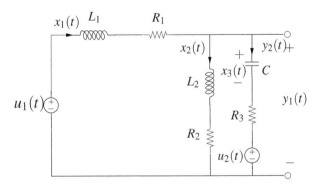

Figure 3.3 Two-input two-output *RLC* circuit (Example 3.2)

Choosing inductor currents $x_1(t)$, $x_2(t)$ and capacitor voltage $x_3(t)$ as state variables, we can write Kirchhoff's equations:

$$x_1(t) = x_2(t) + C\dot{x}_3(t) \quad \text{(current summation)}, \tag{3.5a}$$

$$u_1(t) = L_1\dot{x}_1(t) + R_1x_1(t) + x_3(t) + R_3C\dot{x}_3(t) + u_2(t) \quad \text{(voltage summation)}, \tag{3.5b}$$

$$L_2\dot{x}_2(t) + R_2x_2(t) = x_3(t) + R_3C\dot{x}_3(t) + u_2(t), \tag{3.5c}$$

$$y_1(t) = x_3(t) + R_3C\dot{x}_3(t) + u_2(t) \quad \text{(outputs)}, \tag{3.5d}$$

$$y_2(t) = x_1(t) - x_2(t). \tag{3.5e}$$

These can be rewritten in the standard state-variable form

$$
\begin{bmatrix} \dot{x}_1(t) \\ \dot{x}_2(t) \\ \dot{x}_3(t) \end{bmatrix} = \underbrace{\begin{bmatrix} -\dfrac{(R_1+R_3)}{L_1} & \dfrac{R_3}{L_1} & -\dfrac{1}{L_1} \\[2mm] -\dfrac{R_3}{L_2} & -\dfrac{(R_2+R_3)}{L_2} & \dfrac{1}{L_2} \\[2mm] \dfrac{1}{C} & -\dfrac{1}{C} & 0 \end{bmatrix}}_{\mathbf{A}} \begin{bmatrix} x_1(t) \\ x_2(t) \\ x_3(t) \end{bmatrix} + \underbrace{\begin{bmatrix} \dfrac{1}{L_1} & -\dfrac{1}{L_1} \\[2mm] 0 & \dfrac{1}{L_2} \\[2mm] 0 & 0 \end{bmatrix}}_{\mathbf{B}} \begin{bmatrix} u_1(t) \\ u_2(t) \end{bmatrix},
$$

$$
\begin{bmatrix} y_1(t) \\ y_2(t) \end{bmatrix} = \underbrace{\begin{bmatrix} R_3 & -R_3 & 1 \\ 0 & 1 & 0 \end{bmatrix}}_{\mathbf{C}} \begin{bmatrix} x_1(t) \\ x_2(t) \\ x_3(t) \end{bmatrix} + \underbrace{\begin{bmatrix} 0 & 1 \\ 0 & 0 \end{bmatrix}}_{\mathbf{D}} \begin{bmatrix} u_1(t) \\ u_2(t) \end{bmatrix}. \tag{3.6}
$$

Mechanical Systems

The common elements of mechanical systems are masses, springs, and dampers. If a linear spring is stretched (compressed) from its relaxed position by a certain amount x, it produces a restoring force Kx in the opposite direction. A *linear* damper as shown in Figure 3.4 produces a force $B\dot{x}$ opposing the motion, with $\dot{x}(t)$ being the relative velocity of the plunger with respect to the housing in the positive direction.

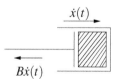

Figure 3.4 A linear damper (dash-pot)

The main law of physics governing the behavior of mechanical systems is Newton's equation of motion, which can be stated as follows: The algebraic sum of forces acting on a mass m, in the positive $x(t)$ direction $= m\ddot{x}(t)$, where $\ddot{x}(t)$ is the acceleration of m in the positive x direction.

A similar law holds for rotational motion with force, mass, and linear acceleration replaced by torque, moment of inertia, and angular acceleration, respectively.

EXAMPLE 3.3 (Electromechanical system) An armature-controlled, separately excited DC motor is shown in Figure 3.5 driving a rotational load with a resistive torque T_L.

The armature resistance, inductance, and current are denoted R_a, L_a, and $i_a(t)$, respectively, $e_b(t)$ denotes the back electromotive force (emf), T_m the motor torque, θ and ω denote the motor shaft's angular position and velocity, respectively, J denotes the moment of inertia of the load (reflected to the motor drive shaft), and B the viscous damping constant.

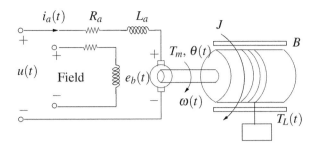

Figure 3.5 Separately excited DC motor (Example 3.3)

The motor equations are

$$T_m(t) = K_t i_a(t),$$
$$e_b(t) = K_b \omega(t), \tag{3.7}$$

where K_t, K_b are constants provided by the manufacturer. We write the system equations by summing voltages around the loop (electrical balance) and sum of mechanical torques resulting in acceleration (mechanical balance):

$$u(t) = R_a i_a(t) + L_a \dot{i}_a(t) + e_b(t) \qquad \text{(electrical balance)}, \tag{3.8a}$$
$$T_m(t) = T_L(t) + B\dot{\theta}(t) + J\ddot{\theta}(t) \qquad \text{(mechanical balance)}. \tag{3.8b}$$

Assuming that angular position and velocity are the outputs of interest, we have

$$y_1(t) = \theta(t),$$
$$y_2(t) = \omega(t) := \dot{\theta}(t). \tag{3.9}$$

We may now choose $x_1(t) := \theta(t)$, $x_2(t) := \omega(t)$, and $x_3(t) := i_a(t)$ as the state variables to cast the DC motor model in the standard form

$$\begin{bmatrix} \dot{x}_1(t) \\ \dot{x}_2(t) \\ \dot{x}_3(t) \end{bmatrix} = \underbrace{\begin{bmatrix} 0 & 1 & 0 \\ 0 & -\dfrac{B}{J} & \dfrac{K_t}{J} \\ 0 & -\dfrac{K_b}{L_a} & -\dfrac{R_a}{L_a} \end{bmatrix}}_{A} \begin{bmatrix} x_1(t) \\ x_2(t) \\ x_3(t) \end{bmatrix} + \underbrace{\begin{bmatrix} 0 & 0 \\ 0 & -\dfrac{1}{J} \\ \dfrac{1}{L_a} & 0 \end{bmatrix}}_{B} \begin{bmatrix} u(t) \\ T_L(t) \end{bmatrix},$$

$$\tag{3.10}$$

$$\begin{bmatrix} y_1(t) \\ y_2(t) \end{bmatrix} = \underbrace{\begin{bmatrix} 1 & 0 & 0 \\ 0 & 1 & 0 \end{bmatrix}}_{C} \begin{bmatrix} x_1(t) \\ x_2(t) \\ x_3(t) \end{bmatrix} + \underbrace{\begin{bmatrix} 0 & 0 \\ 0 & 0 \end{bmatrix}}_{D} \begin{bmatrix} u(t) \\ T_L(t) \end{bmatrix}.$$

EXAMPLE 3.4 (Inverted pendulum on a cart)   Consider the inverted pendulum shown in Figure 3.6.

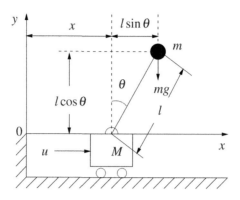

Figure 3.6 Inverted pendulum (Example 3.4)

There are two equations, respectively force balance and torque balance. First we consider force balance. Since the only force in the horizontal direction is the force on the cart, we have

$$M \frac{d^2}{dt^2} x(t) + m \frac{d^2}{dt^2} x_m(t) = u(t), \qquad (3.11)$$

where the coordinate of the point mass m is given by (x_m, y_m) in the reference coordinate system. Then we have

$$\begin{aligned} x_m(t) &= x(t) + l \sin \theta(t), \\ y_m(t) &= l \cos \theta(t). \end{aligned} \qquad (3.12)$$

Thus, we can rewrite Equation (3.11) as

$$M \frac{d^2}{dt^2} x(t) + m \frac{d^2}{dt^2} (x(t) + l \sin \theta(t)) = u(t). \qquad (3.13)$$

Noting that

$$\frac{d^2}{dt^2} \sin \theta(t) = -(\dot{\theta}(t))^2 \sin \theta(t) + \ddot{\theta}(t) \cos \theta(t), \qquad (3.14)$$

we get

$$(M + m)\ddot{x}(t) - ml(\dot{\theta}(t))^2 \sin \theta(t) + ml\ddot{\theta}(t) \cos \theta(t) = u(t). \qquad (3.15)$$

Next, we write equations for torque balance. Refer to Figure 3.7 for the notation. Since the torque due to the acceleration force is balanced by the torque due to the gravity force, we must have

$$F_{x_m} \cos \theta(t) - F_{y_m} \sin \theta(t) = mg \sin \theta(t), \qquad (3.16)$$

where the forces are

$$\begin{aligned} F_{x_m} &= m \frac{d^2}{dt^2} x_m(t) = m \left(\ddot{x}(t) - l(\dot{\theta}(t))^2 \sin \theta(t) + l\ddot{\theta}(t) \cos \theta(t) \right), \\ F_{y_m} &= m \frac{d^2}{dt^2} y_m(t) = -m \left(l(\dot{\theta}(t))^2 \cos \theta(t) + l\ddot{\theta}(t) \sin \theta(t) \right). \end{aligned} \qquad (3.17)$$

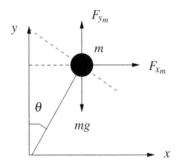

Figure 3.7 Torque directions (Example 3.4)

Substituting Equation (3.16) in Equation (3.17) gives

$$
\begin{aligned}
mg\sin\theta(t) &= F_{x_m}\cos\theta(t) - F_{y_m}\sin\theta(t) \\
&= m\left(\ddot{x}(t) - l(\dot{\theta}(t))^2\sin\theta(t) + l\ddot{\theta}(t)\cos\theta(t)\right)\cos\theta(t) \\
&\quad - m\left(l(\dot{\theta}(t))^2\cos\theta(t) + l\ddot{\theta}(t)\sin\theta(t)\right)\sin\theta(t) \\
&= m\ddot{x}\cos\theta(t) + ml\ddot{\theta}(t).
\end{aligned}
\tag{3.18}
$$

Using a small signal approximation about $\theta(t) = 0$, we have $\cos\theta(t) \cong 1$, $\sin\theta(t) \cong \theta(t)$, and $\dot{\theta}(t)^2 \cong 0$, and the inverted pendulum equations become

$$
(M+m)\ddot{x}(t) + ml\ddot{\theta}(t) = u(t), \tag{3.19}
$$

$$
m\ddot{x}(t) + ml\ddot{\theta}(t) = mg\theta. \tag{3.20}
$$

Now we choose the state vector as

$$
\mathbf{z}(t) := \begin{bmatrix} \theta(t) \\ \dot{\theta}(t) \\ x(t) \\ \dot{x}(t) \end{bmatrix}. \tag{3.21}
$$

Since $\dot{\mathbf{z}}(t)$ has elements $\ddot{x}(t)$ and $\ddot{\theta}(t)$ in the vector, we derive these expressions using Equations (3.19) and (3.20). First, from Equation (3.20) we have

$$
ml\ddot{\theta}(t) = mg\theta(t) - m\ddot{x}(t) \tag{3.22}
$$

and substituting this into Equation (3.19):

$$
\begin{aligned}
(M+m)\ddot{x}(t) + ml\ddot{\theta}(t) = u(t) &\Leftrightarrow (M+m)\ddot{x}(t) + mg\theta(t) - m\ddot{x}(t) = u(t) \\
&\Leftrightarrow M\ddot{x}(t) + mg\theta(t) = u(t) \\
&\Leftrightarrow \ddot{x}(t) = -\frac{m}{M}g\theta(t) + \frac{1}{M}u(t).
\end{aligned}
\tag{3.23}
$$

Next, substituting Equation (3.23) back into Equation (3.22), we get

$$ml\ddot{\theta}(t) = mg\theta(t) - m\ddot{x}(t) \Leftrightarrow l\ddot{\theta}(t) = g\theta(t) + \left(\frac{m}{M}g\theta(t) - \frac{1}{M}u(t)\right)$$

$$\Leftrightarrow \ddot{\theta}(t) = \frac{g}{l}\theta(t) + \frac{1}{l}\left(\frac{m}{M}g\theta(t) - \frac{1}{M}u(t)\right) \tag{3.24}$$

$$\Leftrightarrow \ddot{\theta}(t) = \frac{g}{l}\left(1 + \frac{m}{M}\right)\theta(t) - \frac{1}{lM}u(t).$$

Finally, we can write the state-space representation in the form

$$\dot{\mathbf{z}}(t) = \mathbf{A}\mathbf{z}(t) + \mathbf{B}u(t),$$
$$\mathbf{y}(t) = \mathbf{C}\mathbf{z}(t) + \mathbf{D}u(t). \tag{3.25}$$

$$\begin{bmatrix} \dot{z}_1(t) \\ \dot{z}_2(t) \\ \dot{z}_3(t) \\ \dot{z}_4(t) \end{bmatrix} = \underbrace{\begin{bmatrix} 0 & 1 & 0 & 0 \\ \frac{g}{l}\left(1 + \frac{m}{M}\right) & 0 & 0 & 0 \\ 0 & 0 & 0 & 1 \\ -\frac{m}{M}g & 0 & 0 & 0 \end{bmatrix}}_{\mathbf{A}} \begin{bmatrix} z_1(t) \\ z_2(t) \\ z_3(t) \\ z_4(t) \end{bmatrix} + \underbrace{\begin{bmatrix} 0 \\ -\frac{1}{lM} \\ 0 \\ \frac{1}{M} \end{bmatrix}}_{\mathbf{B}} u(t), \tag{3.26}$$

$$\begin{bmatrix} y_1(t) \\ y_2(t) \end{bmatrix} = \underbrace{\begin{bmatrix} 1 & 0 & 0 & 0 \\ 0 & 0 & 1 & 0 \end{bmatrix}}_{\mathbf{C}} \begin{bmatrix} z_1(t) \\ z_2(t) \\ z_3(t) \\ z_4(t) \end{bmatrix} + \underbrace{\begin{bmatrix} 0 \\ 0 \end{bmatrix}}_{\mathbf{D}} u(t),$$

where

$$\begin{bmatrix} \dot{z}_1(t) \\ \dot{z}_2(t) \\ \dot{z}_3(t) \\ \dot{z}_4(t) \end{bmatrix} = \dot{\mathbf{z}}(t) = \begin{bmatrix} \dot{\theta}(t) \\ \ddot{\theta}(t) \\ \dot{x}(t) \\ \ddot{x}(t) \end{bmatrix} \quad \text{and} \quad \begin{bmatrix} z_1(t) \\ z_2(t) \\ z_3(t) \\ z_4(t) \end{bmatrix} = \mathbf{z}(t) = \begin{bmatrix} \theta(t) \\ \dot{\theta}(t) \\ x(t) \\ \dot{x}(t) \end{bmatrix}. \tag{3.27}$$

3.2   State-Variable Models from Differential Equations

In this section, we show how a high-order differential equation can be converted to state-variable form.

3.2.1 Single-Input Single-Output Systems

Consider first a *single-input single-output (SISO)* system described by the differential equation

$$\frac{d^n y(t)}{dt^n} + a_{n-1}\frac{d^{n-1}y(t)}{dt^{n-1}} + \cdots + a_0 y(t) = c_0 u(t) + c_1\frac{du(t)}{dt} + \cdots + c_m\frac{d^m u(t)}{dt^m}. \quad (3.28)$$

A standard $\{\mathbf{A},\mathbf{B},\mathbf{C},\mathbf{D}\}$ representation for Equation (3.28) can be derived if and only if

$$m \leq n. \quad (3.29)$$

If $m > n$, the standard state-space model has to be extended to the form

$$\begin{aligned}\dot{\mathbf{x}}(t) &= \mathbf{A}\mathbf{x}(t) + \mathbf{B}\mathbf{u}(t),\\ \mathbf{y}(t) &= \mathbf{C}\mathbf{x}(t) + \mathbf{D}\mathbf{u}(t) + \mathbf{D}_1\dot{\mathbf{u}}(t) + \mathbf{D}_2\ddot{\mathbf{u}}(t) + \cdots.\end{aligned} \quad (3.30)$$

To proceed, we assume Equation (3.29) and treat the two cases $m < n$ and $m = n$.

Case 1: $(m < n)$ In this case, Equation (3.28) has the form

$$\begin{aligned}\frac{d^n y(t)}{dt^n} &+ a_{n-1}\frac{d^{n-1}y(t)}{dt^{n-1}} + \cdots + a_0 y(t)\\ &= c_0 u(t) + c_1\frac{du(t)}{dt} + \cdots + c_{n-1}\frac{d^{n-1}u(t)}{dt^{n-1}}.\end{aligned} \quad (3.31)$$

Define the differential operators

$$D := \frac{d}{dt}, \quad D^k := \frac{d^k}{dt^k}, \quad \text{for } k = 0,1,2,\ldots, \quad (3.32a)$$

$$A(D) := D^n + a_{n-1}D^{n-1} + \cdots + a_1 D + a_0, \quad (3.32b)$$

$$C(D) := c_0 + c_1 D + \cdots + c_{n-1}D^{n-1}, \quad (3.32c)$$

and write Equation (3.31) as

$$A(D)\{y(t)\} = C(D)\{u(t)\}. \quad (3.33)$$

Now introduce $z(t)$, defined by

$$u(t) = A(D)\{z(t)\}. \quad (3.34)$$

Then if we set

$$y(t) = C(D)\{z(t)\}, \quad (3.35)$$

we have (using $A(D)C(D) = C(D)A(D)$):

$$\begin{aligned}A(D)\{y(t)\} &= A(D)C(D)\{z(t)\}\\ &= C(D)A(D)\{z(t)\}\\ &= C(D)\{u(t)\},\end{aligned} \quad (3.36)$$

which proves that $(y(t), u(t))$ satisfy Equation (3.31).

Now define

$$z(t) =: x_1(t), \qquad \frac{dz(t)}{dt} =: x_2(t), \qquad \dots, \qquad \frac{d^{n-1}z(t)}{dt^{n-1}} =: x_n(t) \tag{3.37}$$

so that Equation (3.34) can be rewritten as

$$
\begin{aligned}
\dot{x}_1(t) &= x_2(t), \\
\dot{x}_2(t) &= x_3(t), \\
&\vdots \\
\dot{x}_n(t) &= -a_0 x_1(t) - a_1 x_2(t) - \cdots - a_{n-1} x_n(t) + u(t),
\end{aligned}
\tag{3.38}
$$

and Equation (3.35) can be rewritten as

$$y(t) = c_0 x_1(t) + c_1 x_2(t) + \cdots + c_{n-1} x_n(t). \tag{3.39}$$

Thus, a state-space representation of Equation (3.34) is obtained, with

$$\mathbf{x}(t) := \begin{bmatrix} x_1(t) \\ x_2(t) \\ \vdots \\ x_n(t) \end{bmatrix} \tag{3.40}$$

and

$$\dot{\mathbf{x}}(t) = \underbrace{\begin{bmatrix} 0 & 1 & 0 & \cdots & 0 \\ 0 & 0 & 1 & & 0 \\ & & & \ddots & \\ & & & & 1 \\ -a_0 & -a_1 & -a_2 & \cdots & -a_{n-1} \end{bmatrix}}_{\mathbf{A}} \mathbf{x}(t) + \underbrace{\begin{bmatrix} 0 \\ 0 \\ \vdots \\ 0 \\ 1 \end{bmatrix}}_{\mathbf{B}} u(t), \tag{3.41a}$$

$$y(t) = \underbrace{\begin{bmatrix} c_0 & c_1 & c_2 & \cdots & c_{n-1} \end{bmatrix}}_{\mathbf{C}} \mathbf{x}(t) + \underbrace{\begin{bmatrix} 0 \end{bmatrix}}_{\mathbf{D}}. \tag{3.41b}$$

Case 2: (m = n)   In this case, Equation (3.28) has the form

$$\frac{d^n y(t)}{dt^n} + a_{n-1} \frac{d^{n-1} y(t)}{dt^{n-1}} + \cdots + a_0 y(t) = b_0 u(t) + b_1 \frac{du(t)}{dt} + \cdots + b_n \frac{d^n u(t)}{dt^n}. \tag{3.42}$$

Define

$$y_o(t) := y(t) - b_n u(t) \tag{3.43}$$

and rewrite Equation (3.42) as

$$
\begin{aligned}
\frac{d^n y_o(t)}{dt^n} &+ a_{n-1} \frac{d^{n-1} y_o(t)}{dt^{n-1}} + \cdots + a_0 y_o(t) \\
&= c_0 u(t) + c_1 \frac{du(t)}{dt} + \cdots + c_{n-1} \frac{d^{n-1} u(t)}{dt^{n-1}},
\end{aligned}
\tag{3.44}
$$

with

$$c_i = b_i - a_i b_n, \quad \text{for } i = 0, 1, \ldots, n-1. \tag{3.45}$$

Since Equation (3.44) is of the form in Equation (3.28), we obtain the state-variable realization

$$\dot{\mathbf{x}}(t) = \underbrace{\begin{bmatrix} 0 & 1 & 0 & \cdots & 0 \\ 0 & 0 & 1 & & 0 \\ & & & \ddots & \\ & & & & 1 \\ -a_0 & -a_1 & -a_2 & \cdots & -a_{n-1} \end{bmatrix}}_{\mathbf{A}} \mathbf{x}(t) + \underbrace{\begin{bmatrix} 0 \\ 0 \\ \vdots \\ 0 \\ 1 \end{bmatrix}}_{\mathbf{B}} u(t), \tag{3.46}$$

$$y(t) = \underbrace{\begin{bmatrix} c_0 & c_1 & c_2 & \cdots & c_{n-1} \end{bmatrix}}_{\mathbf{C}} \mathbf{x}(t) + \underbrace{\begin{bmatrix} b_n \end{bmatrix}}_{\mathbf{D}} u(t),$$

from Equations (3.41b) and (3.45). This realization is called a *controllable companion realization*.

An equivalent procedure can be developed in terms of a transfer function model. Let $g(s)$ denote a rational transfer function with denominator degree n and numerator degree m. For arbitrary n and $m - n = r \geq 0$, $g(s)$ has the general form

$$g(s) = \frac{c_0 + c_1 s + \cdots + c_{n-1} s^{n-1}}{s^n + a_{n-1} s^{n-1} + \cdots + a_1 s + a_0} + d_0 + d_1 s + \cdots + d_r s^r. \tag{3.47}$$

If $m < n$:

$$d_0 = 0, \quad d_1 = 0, \quad \ldots, \quad d_r = 0. \tag{3.48}$$

From the previous results, it follows that a state-variable representation of Equation (3.47) is

$$\dot{\mathbf{x}}(t) = \underbrace{\begin{bmatrix} 0 & 1 & 0 & \cdots & 0 \\ 0 & 0 & 1 & & 0 \\ & & & \ddots & \\ & & & & 1 \\ -a_0 & -a_1 & -a_2 & \cdots & -a_{n-1} \end{bmatrix}}_{\mathbf{A}} \mathbf{x}(t) + \underbrace{\begin{bmatrix} 0 \\ 0 \\ \vdots \\ 0 \\ 1 \end{bmatrix}}_{\mathbf{B}} u(t), \tag{3.49a}$$

$$y(t) = \underbrace{\begin{bmatrix} c_0 & c_1 & c_2 & \cdots & c_{n-1} \end{bmatrix}}_{\mathbf{C}} \mathbf{x}(t) + \underbrace{d_0}_{\mathbf{D}} u(t) + \underbrace{d_1}_{\mathbf{D}_1} \frac{du(t)}{dt} + \cdots$$

$$+ \underbrace{d_r}_{\mathbf{D}_r} \frac{d^r u(t)}{dt^r}. \tag{3.49b}$$

When $r = 0$ $(m = n)$, the transfer function is *proper* and when $r < 0$ $(m < n)$, it is said to be *strictly proper*.

3.2.2   Multi-Input Multi-Output Systems

Consider now the *multi-input multi-output (MIMO)* case. There are m outputs $y_i(t)$ and r inputs $u_j(t)$, described by a matrix transfer function $\mathbf{H}(s)$ with each entry being proper or strictly proper. Then

$$\lim_{s \to \infty} \mathbf{H}(s) =: \mathbf{D} \tag{3.50}$$

exists and

$$\mathbf{H}(s) = \mathbf{G}(s) + \mathbf{D}, \tag{3.51}$$

where each entry of $\mathbf{G}(s)$ is strictly proper. Moreover, rewrite $\mathbf{G}(s)$ introducing common factors as necessary so that each entry of column i has the same denominator $a_i(s)$ for each $i = 1, 2, \ldots, r$. Thus

$$\mathbf{G}(s) = \begin{bmatrix} \dfrac{c_{11}(s)}{a_1(s)} & \dfrac{c_{12}(s)}{a_2(s)} & \cdots & \dfrac{c_{1r}(s)}{a_r(s)} \\ & \vdots & & \vdots \\ \dfrac{c_{m1}(s)}{a_1(s)} & \dfrac{c_{m2}(s)}{a_2(s)} & \cdots & \dfrac{c_{mr}(s)}{a_r(s)} \end{bmatrix}. \tag{3.52}$$

Let

$$a_i(s) = s^{n_i} + a_{n_i-1}^i s^{n_i-1} + \cdots + a_1^i s + a_0^i, \quad \text{for } i = 1, 2, \ldots, r \tag{3.53}$$

and let c_{ij} denote the row vector of coefficients of $c_{ij}(s)$ in ascending order. Then it is easy to see from Equation (3.46) that a state-variable representation of $\mathbf{H}(s)$ is given by $[\mathbf{A}, \mathbf{B}, \mathbf{C}, \mathbf{D}]$, where

$$\mathbf{A} = \begin{bmatrix} \mathbf{A}_1 & & \\ & \ddots & \\ & & \mathbf{A}_r \end{bmatrix}, \ \mathbf{B} = \begin{bmatrix} \mathbf{B}_1 & & \\ & \ddots & \\ & & \mathbf{B}_r \end{bmatrix}, \ \mathbf{C} = \begin{bmatrix} c_{11} & c_{12} & \cdots & c_{1r} \\ \vdots & & & \vdots \\ c_{m1} & c_{m2} & \cdots & c_{mr} \end{bmatrix}, \tag{3.54}$$

with

$$\mathbf{A}_i = \begin{bmatrix} 0 & 1 & & & & \\ & \ddots & \ddots & & & \\ & & \ddots & & \ddots & \\ & & & \ddots & & \ddots \\ & & & & 0 & 1 \\ -a_0^i & -a_1^i & \cdots & \cdots & -a_{n_i-2} & -a_{n_i-1} \end{bmatrix}, \ \mathbf{B}_i = \begin{bmatrix} 0 \\ 0 \\ \vdots \\ 0 \\ 1 \end{bmatrix} \tag{3.55}$$

and $\mathbf{D}$ is as in Equation (3.50).

It is important to emphasize that the above methods of developing a state-variable model are by no means unique. Different procedures for developing state-variable models, also called *realizations*, are discussed in greater detail in Chapter 5, where the important problem of obtaining a *minimal-order realization* is also discussed.

EXAMPLE 3.5 (MIMO case)

$$
\begin{bmatrix} \dfrac{1}{s+1} & \dfrac{1}{s+3} \\[2mm] \dfrac{1}{s+2} & \dfrac{s}{s+4} \end{bmatrix} = \begin{bmatrix} \dfrac{s+2}{s^2+3s+2} & \dfrac{s+4}{s^2+7s+12} \\[2mm] \dfrac{s+1}{s^2+3s+2} & \dfrac{-4s-12}{s^2+7s+12} \end{bmatrix} + \begin{bmatrix} 0 & 0 \\ 0 & 1 \end{bmatrix}.
\tag{3.56}
$$

Then we have

$$
\mathbf{A} = \left[\begin{array}{cc:cc} 0 & 1 & 0 & 0 \\ -2 & -3 & 0 & 0 \\ \hdashline 0 & 0 & 0 & 1 \\ 0 & 0 & -12 & -7 \end{array}\right], \quad \mathbf{B} = \left[\begin{array}{c:c} 0 & 0 \\ 1 & 0 \\ \hdashline 0 & 0 \\ 0 & 1 \end{array}\right],
$$

$$
\mathbf{C} = \left[\begin{array}{cc:cc} 2 & 1 & 4 & 1 \\ 1 & 1 & -12 & -4 \end{array}\right], \quad \mathbf{D} = \left[\begin{array}{c:c} 0 & 0 \\ 0 & 1 \end{array}\right].
\tag{3.57}
$$

3.3 Solution of Continuous-Time State Equations

The standard state-variable model equations can be solved for $\mathbf{x}(t)$ and $\mathbf{y}(t)$ in quasi-closed form. Consider the system

$$
\dot{\mathbf{x}}(t) = \mathbf{A}\mathbf{x}(t) + \mathbf{B}\mathbf{u}(t),
\tag{3.58a}
$$

$$
\mathbf{y}(t) = \mathbf{C}\mathbf{x}(t) + \mathbf{D}\mathbf{u}(t),
\tag{3.58b}
$$

where $\mathbf{x}(t)$ is an n-vector. If a system has r inputs and m outputs, $\mathbf{u}(t)$ and $\mathbf{y}(t)$ are vectors of dimension r and m, respectively.

Taking the Laplace transform of Equation (3.58a), we have

$$
s\mathbf{X}(s) - \mathbf{x}(0) = \mathbf{A}\mathbf{X}(s) + \mathbf{B}\mathbf{U}(s),
\tag{3.59}
$$

where $\mathbf{X}(s) = \mathscr{L}[\mathbf{x}(t)]$, $\mathbf{U}(s) = \mathscr{L}[\mathbf{u}(t)]$. Hence

$$
(s\mathbf{I} - \mathbf{A})\mathbf{X}(s) = \mathbf{x}(0) + \mathbf{B}\mathbf{U}(s)
\tag{3.60}
$$

or, since $(s\mathbf{I} - \mathbf{A})$ is always invertible (i.e., $\det[s\mathbf{I} - \mathbf{A}] = s^n + \cdots$):

$$
\mathbf{X}(s) = (s\mathbf{I} - \mathbf{A})^{-1}\mathbf{x}(0) + (s\mathbf{I} - \mathbf{A})^{-1}\mathbf{B}\mathbf{U}(s).
\tag{3.61}
$$

Note that

$$
(s\mathbf{I} - \mathbf{A})^{-1} = \frac{\mathbf{I}}{s} + \frac{\mathbf{A}}{s^2} + \frac{\mathbf{A}^2}{s^3} + \cdots
\tag{3.62}
$$

and thus

$$
\mathscr{L}^{-1}\left[(s\mathbf{I} - \mathbf{A})^{-1}\right] = \mathbf{I} + \mathbf{A}t + \frac{\mathbf{A}^2 t^2}{2!} + \frac{\mathbf{A}^3 t^3}{3!} + \cdots =: e^{\mathbf{A}t}.
\tag{3.63}
$$

The solutions, $\mathbf{x}(t)$ and $\mathbf{y}(t)$, of the state equations are thus given by

$$\mathbf{x}(t) = \mathscr{L}^{-1}\left[(s\mathbf{I} - \mathbf{A})^{-1}\right]\mathbf{x}(0) + \mathscr{L}^{-1}\left[(s\mathbf{I} - \mathbf{A})^{-1}\mathbf{B}\mathbf{U}(s)\right]$$

$$= e^{\mathbf{A}t}\mathbf{x}(0) + \int_0^t e^{\mathbf{A}(t-\tau)}\mathbf{B}\mathbf{u}(\tau)d\tau \tag{3.64}$$

and

$$\mathbf{y}(t) = \mathbf{C}e^{\mathbf{A}t}\mathbf{x}(0) + \mathbf{C}\int_0^t e^{\mathbf{A}(t-\tau)}\mathbf{B}\mathbf{u}(\tau)d\tau + \mathbf{D}\mathbf{u}(t). \tag{3.65}$$

From Equation (3.61) we have

$$\mathbf{Y}(s) = \mathbf{C}(s\mathbf{I} - \mathbf{A})^{-1}\mathbf{x}(0) + \left[\mathbf{C}(s\mathbf{I} - \mathbf{A})^{-1}\mathbf{B} + \mathbf{D}\right]\mathbf{U}(s). \tag{3.66}$$

For $u(t) = 0$:

$$\mathbf{y}(t) := \mathscr{L}^{-1}\left[\mathbf{C}(s\mathbf{I} - \mathbf{A})^{-1}\mathbf{x}(0)\right] =: \mathbf{y}_o(t) \tag{3.67}$$

is called the *zero-input response* or the initial condition response. For $\mathbf{x}(0) = 0$:

$$\mathbf{y}(t) = \mathscr{L}^{-1}\left[\left(\mathbf{C}(s\mathbf{I} - \mathbf{A})^{-1}\mathbf{B} + \mathbf{D}\right)\mathbf{U}(s)\right] =: \mathbf{y}_u(t) \tag{3.68}$$

is called the *zero-state response* or the forced response. Also

$$\mathbf{Y}_u(s) = \underbrace{\left[\mathbf{C}(s\mathbf{I} - \mathbf{A})^{-1}\mathbf{B} + \mathbf{D}\right]}_{\mathbf{G}(s)}\mathbf{U}(s) \tag{3.69}$$

and $\mathbf{G}(s)$ an $m \times r$ rational proper matrix is called the *system transfer function matrix*. We show below that

$$\mathbf{x}(t) = e^{\mathbf{A}t}\mathbf{x}(0) + \int_0^t e^{\mathbf{A}(t-\tau)}\mathbf{B}\mathbf{u}(\tau)d\tau \tag{3.70}$$

is a solution of

$$\dot{\mathbf{x}}(t) = \mathbf{A}\mathbf{x}(t) + \mathbf{B}\mathbf{u}(t) \tag{3.71}$$

without using Laplace transforms.

THEOREM 3.1 (A fundamental theorem of calculus)   Let $f(t)$ be a continuous function on $[a, b]$, then for any $x \in (a, b)$

$$\frac{d}{dx}\int_a^x f(t)dt = f(x). \tag{3.72}$$

Now from Equation (3.70), using Theorem 3.1, we have

$$\dot{\mathbf{x}}(t) = \frac{d}{dt}\mathbf{x}(t)$$

$$= \mathbf{A}e^{\mathbf{A}t}\mathbf{x}(0) + \frac{d}{dt}\left(e^{\mathbf{A}t}\right)\int_0^t e^{-\mathbf{A}\tau}\mathbf{B}\mathbf{u}(\tau)d\tau + e^{\mathbf{A}t}\frac{d}{dt}\int_0^t e^{-\mathbf{A}\tau}\mathbf{B}\mathbf{u}(\tau)d\tau$$

$$= \mathbf{A}e^{\mathbf{A}t}\mathbf{x}(0) + \mathbf{A}e^{\mathbf{A}t}\int_0^t e^{-\mathbf{A}\tau}\mathbf{B}\mathbf{u}(\tau)d\tau + e^{\mathbf{A}t}e^{-\mathbf{A}t}\mathbf{B}\mathbf{u}(t) \tag{3.73}$$

$$= \mathbf{A}\underbrace{\left(e^{\mathbf{A}t}\mathbf{x}(0) + \int_0^t e^{\mathbf{A}(t-\tau)}\mathbf{B}\mathbf{u}(\tau)d\tau\right)}_{\mathbf{x}(t)} + \mathbf{B}\mathbf{u}(t) = \mathbf{A}\mathbf{x}(t) + \mathbf{B}\mathbf{u}(t).$$

EXAMPLE 3.6 If

$$\mathbf{A} = \begin{bmatrix} 0 & 1 \\ -2 & -3 \end{bmatrix},$$

$$e^{\mathbf{A}t} = \mathscr{L}^{-1}\left[(s\mathbf{I} - \mathbf{A})^{-1}\right] = \mathscr{L}^{-1}\left[\begin{pmatrix} s & -1 \\ 2 & s+3 \end{pmatrix}^{-1}\right]$$

$$= \mathscr{L}^{-1}\begin{bmatrix} \dfrac{2}{s+1} - \dfrac{1}{s+2} & \dfrac{1}{s+1} - \dfrac{1}{s+2} \\ \dfrac{-2}{s+1} + \dfrac{2}{s+2} & \dfrac{-1}{s+1} + \dfrac{2}{s+2} \end{bmatrix} = \begin{bmatrix} 2e^{-t} - e^{-2t} & e^{-t} - e^{-2t} \\ -2e^{-t} + 2e^{-2t} & -e^{-t} + 2e^{-2t} \end{bmatrix}.$$

MATLAB Solution

```
A=[0 1; -2 -3];
syms t
expm(A*t)
```

EXAMPLE 3.7 Consider the system

$$\dot{\mathbf{x}}(t) = \begin{bmatrix} 0 & 1 & 0 \\ 0 & 0 & 1 \\ 0 & -3 & -2 \end{bmatrix}\mathbf{x}(t) + \begin{bmatrix} 1 & 0 \\ 0 & 0 \\ 0 & 1 \end{bmatrix}\mathbf{u}(t),$$

$$\mathbf{y}(t) = \begin{bmatrix} 1 & 0 & 0 \\ 0 & 0 & 1 \end{bmatrix}\mathbf{x}(t) + \begin{bmatrix} 0 & 1 \\ 0 & 0 \end{bmatrix}\mathbf{u}(t),$$ (3.74)

with initial condition

$$\mathbf{x}(0) = \begin{bmatrix} 1 \\ 0 \\ -1 \end{bmatrix}$$ (3.75)

and input

$$\mathbf{u}(t) = \begin{bmatrix} \delta(t) \\ \sin t \end{bmatrix}.$$ (3.76)

Determine the following:

(a) The zero-input state and output responses.
(b) The zero-state or forced response for state and output.
(c) The transfer function.

Solution

$$\mathbf{X}_0(s) = (s\mathbf{I} - \mathbf{A})^{-1}\mathbf{X}(0) = \begin{bmatrix} s & -1 & 0 \\ 0 & s & -1 \\ 0 & 3 & s+2 \end{bmatrix}^{-1} \mathbf{x}(0)$$

$$= \begin{bmatrix} \dfrac{1}{s} & \dfrac{s+2}{s^3 + 2s^2 + 3s} & \dfrac{1}{s^3 + 2s^2 + 3s} \\[3mm] 0 & \dfrac{s+2}{s^2 + 2s + 3} & \dfrac{1}{s^2 + 2s + 3} \\[3mm] 0 & -\dfrac{3}{s^2 + 2s + 3} & \dfrac{s}{s^2 + 2s + 3} \end{bmatrix} \begin{bmatrix} 1 \\ 0 \\ -1 \end{bmatrix}, \tag{3.77}$$

$$\mathbf{X}_u(s) = (s\mathbf{I} - \mathbf{A})^{-1}\mathbf{B}U(s) = \begin{bmatrix} \dfrac{1}{s} & \dfrac{1}{s^3 + 2s^2 + 3s} \\[3mm] 0 & \dfrac{1}{s^2 + 2s + 3} \\[3mm] 0 & \dfrac{1}{s^2 + 2s + 3} \end{bmatrix} \begin{bmatrix} 1 \\ \dfrac{1}{s^2 + 1} \end{bmatrix}. \tag{3.78}$$

Therefore, the zero-input state response is

$$\mathbf{x}_0(t) = \mathscr{L}^{-1}\left[\mathbf{X}_0(s)\right] = \begin{bmatrix} \left(\dfrac{1}{3} - \dfrac{1}{3}e^{-t}\cos\sqrt{2}t - \dfrac{1}{\sqrt{2}}e^{-t}\sin\sqrt{2}t\right)u(t) \\[3mm] -\dfrac{1}{\sqrt{2}}e^{-t}\sin\sqrt{2}t\,u(t) \\[3mm] -e^{-t}\left(\cos\sqrt{2}t - \sin\sqrt{2}t\right)u(t) \end{bmatrix} \tag{3.79}$$

and the zero-input, output response is

$$\mathbf{y}_0(t) = \begin{bmatrix} \left(\dfrac{1}{3} - \dfrac{1}{3}e^{-t}\cos\sqrt{2}t - \dfrac{1}{3}e^{-t}\sin\sqrt{2}t\right)u(t) \\[3mm] e^{-t}\left(\cos\sqrt{2}t + \sin\sqrt{2}t\right)u(t) \end{bmatrix}. \tag{3.80}$$

The zero-state or forced response of the state is

$$\mathbf{x}_u(t) = \mathscr{L}^{-1}\left[\mathbf{X}_u(s)\right]$$

$$= \begin{bmatrix} \left(1 + \dfrac{1}{4}\left(\sin t - \cos t + e^{-t}\cos\sqrt{2}t\right)\right)u(t) \\[3mm] \dfrac{1}{4}\left(\sin t - \cos t + e^{-t}\cos\sqrt{2}t\right)u(t) \\[3mm] \left(\dfrac{1}{4}(\cos t + \sin t) - \dfrac{1}{4}e^{-t}\cos\sqrt{2}t - \dfrac{1}{\sqrt{2}}e^{-t}\sin\sqrt{2}t\right)u(t) \end{bmatrix} \tag{3.81}$$

and the zero-state or forced output response is

$$
\mathbf{y}_u(t) =
\begin{bmatrix}
\left(1 + \dfrac{1}{4}\left(\sin t - \cos t + e^{-t}\cos\sqrt{2}t\right) + \sin t\right)u(t) \\[2mm]
\left(\dfrac{1}{4}\left(\cos t + \sin t - e^{-t}\cos\sqrt{2}t\right) - \dfrac{1}{\sqrt{2}}e^{-t}\sin\sqrt{2}t\right)u(t)
\end{bmatrix}.
\tag{3.82}
$$

The transfer function is

$$
\mathbf{G}(s) = \mathbf{C}(s\mathbf{I} - \mathbf{A})^{-1}\mathbf{B} + \mathbf{D} =
\begin{bmatrix}
\dfrac{1}{s} & 1 \\[2mm]
0 & \dfrac{s}{s^2 + 2s + 3}
\end{bmatrix}.
\tag{3.83}
$$

The system is both internally and externally unstable because of the eigenvalue and pole at $s = 0$.

MATLAB & Simulink: Modeling and Simulation
Modeling, simulation, and analysis of dynamic systems may be done conveniently by using Simulink, MATLAB's graphical environment. Figure 3.8 shows how to enter $\mathbf{A}, \mathbf{B}, \mathbf{C}, \mathbf{D}$.

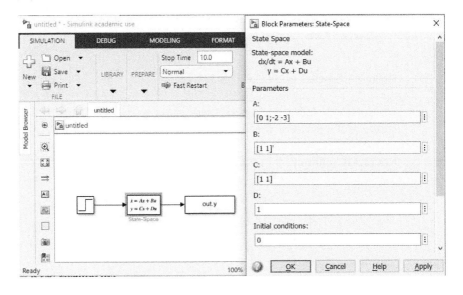

Figure 3.8 A Simulink interactive window

EXAMPLE 3.8 In this example, we show how Simulink can be used to obtain the time-domain response of the following system:

$$
\mathbf{x}(t) =
\begin{bmatrix}
0 & 1 \\
-2 & -2
\end{bmatrix}
\mathbf{x}(t) +
\begin{bmatrix}
1 & 0 \\
1 & 1
\end{bmatrix}
\mathbf{u}(t), \quad \mathbf{y}(t) = \mathbf{I}_2\mathbf{x}(t),
\tag{3.84}
$$

with

$$u_1(t) = \text{unit step} \quad \text{and} \quad u_2(t) = \begin{cases} 10, & \text{for } 0 \le t \le 0.1, \\ 0, & \text{otherwise.} \end{cases} \qquad (3.85)$$

Note that $u_2(t)$ is a simulated impulse. Figure 3.9(a) shows the Simulink model representing the example and Figure 3.9(b) shows the plant model.

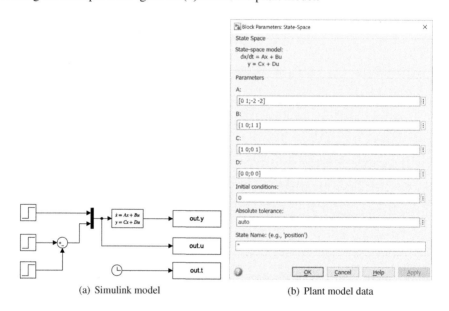

(a) Simulink model   (b) Plant model data

Figure 3.9 Simulink model (Example 3.8)

Figure 3.10 show the inputs and outputs. As seen in the figure, the output responses converge. If the matrix **A** were unstable, the output would not be bounded.

MATLAB Solution

The following MATLAB script also solves the problem and produces Figure 3.10.

```
A=[0 1; -2 -2];
B=[1 0; 1 1];
C=eye(2);
D=zeros(2,2);
sys=ss(A,B,C,D);
t=0:0.01:6;
u1=heaviside(t);
u2=10*(heaviside(t)-heaviside(t-0.1));
u=[u1' u2'];
[y, t]=lsim(sys,u,t);
plot(t,u1,t,u2,t,y(:,1),t,y(:,2));
axis([0 6 -2 10]), grid
```

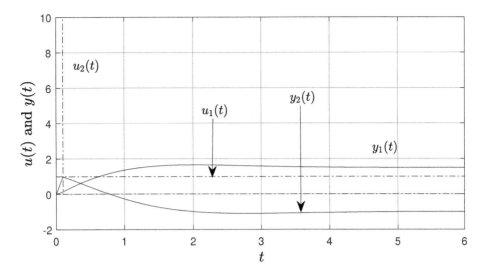

Figure 3.10 Input and output responses (Example 3.8)

EXAMPLE 3.9 With the same system used in Example 3.8, the second input is changed to $u_2(t) = \sin 2t$. The Simulink model used and its response are depicted in Figures 3.11 and 3.12.

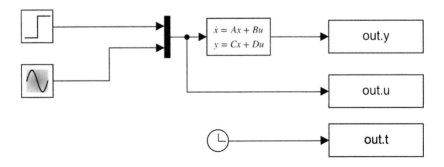

Figure 3.11 Simulink model (Example 3.9)

MATLAB Solution

The following MATLAB script also solves the problem and produces Figure 3.12.

```
A=[0   1;-2   -2];
B=[1   0; 1   1];
C=eye(2);
D=zeros(2,2);
sys=ss(A,B,C,D);
t=0:0.01:10;
u=[heaviside(t)'   sin(2*t)'];
[y, t]=lsim(sys,u,t);
plot(t,y(:,1),t,y(:,2))
```

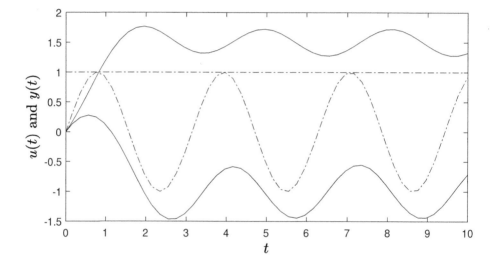

Figure 3.12 Simulink simulation (Example 3.9)

3.4   Realization of Systems from State-Variable Models

One important use of a state-variable model is that it indicates how the system may be
constructed or realized by interconnecting simple blocks carrying out addition, subtraction, multiplication, division, integration, or differentiation. The latter operations can be
carried out with great accuracy using high-gain operational amplifier (op-amp) circuits
for instance, as we showed in Section 2.3.

3.4.1   Integrators or Differentiators

The state-space models contain derivatives. We may physically obtain derivatives of the
state using a differentiation block or an integration block as depicted in Figure 3.13.
    In implementations, the use of integrators is preferred to that of differentiators be-

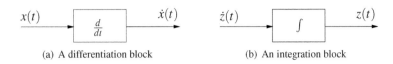

(a) A differentiation block   (b) An integration block

Figure 3.13 Differentiation and integration

cause differentiators *amplify* high-frequency noise whereas integrators *attenuate* the latter (Figure 3.14).

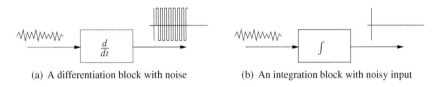

(a) A differentiation block with noise   (b) An integration block with noisy input

Figure 3.14 Differentiation and integration with noisy inputs

Since high-frequency noise is always present in physical systems, it is impractical to realize them with differentiators. This is why integrators and not differentiators are used to realize state-space systems.

3.4.2 Synthesis of State-Space System Using Integrators, Adders, and Multipliers

As discussed in Section 2.3, op-amp circuits can be used to build adders, multipliers, and integrators, which are standard integrated circuit components nowadays. We show below how a state-variable model can be synthesized using these standard components or blocks.

Consider the equation

$$\begin{bmatrix} \dot{x}_1(t) \\ \dot{x}_2(t) \end{bmatrix} = \begin{bmatrix} a_{11} & a_{12} \\ a_{21} & a_{22} \end{bmatrix} \begin{bmatrix} x_1(t) \\ x_2(t) \end{bmatrix} + \begin{bmatrix} b_1 \\ b_2 \end{bmatrix} u(t), \tag{3.86}$$

$$y(t) = \begin{bmatrix} c_1 & c_2 \end{bmatrix} \begin{bmatrix} x_1(t) \\ x_2(t) \end{bmatrix} + \begin{bmatrix} d \end{bmatrix} u(t). \tag{3.87}$$

Its realization is shown in Figure 3.15.

Similarly, it is easy to see that the state-space system in Equation (3.46) has the realization shown in Figure 3.16.

We also note that any circuit consisting of integrators, multipliers, and adders can be represented by a state-space $[\mathbf{A}, \mathbf{B}, \mathbf{C}, \mathbf{D}]$ model by defining the output of each integrator as an independent state variable, and the input of the same integrator as the derivative of this state variable. Thus, we see that realizations and $[\mathbf{A}, \mathbf{B}, \mathbf{C}, \mathbf{D}]$ representations have a one-to-one correspondence (modulo relabeling of the states). Moreover, the entries of the matrices $\mathbf{A}, \mathbf{B}, \mathbf{C}, \mathbf{D}$ are precisely the numerical values of the gains appearing in the realization.

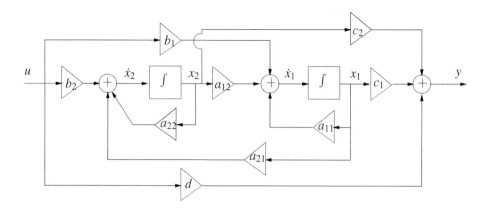

Figure 3.15 A synthesis using adders, multipliers, and integrators

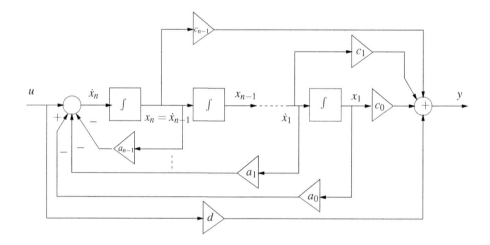

Figure 3.16 Realization of the system in Equation (3.46)

3.5   A Formula for the Characteristic Equation

We see that the computation of $(s\mathbf{I} - \mathbf{A})^{-1}$ is the main hurdle in the solution of state-space systems. Denoting adjoint by "Adj" and determinant by "det," we have

$$(s\mathbf{I} - \mathbf{A})^{-1} = \frac{\text{Adj}(s\mathbf{I} - \mathbf{A})}{\det(s\mathbf{I} - \mathbf{A})}. \tag{3.88}$$

We first develop in this section, a formula for $\det(s\mathbf{I} - \mathbf{A})$, the characteristic polynomial of $\mathbf{A}$.

Notation   Let $\mathbf{n} := \{1, 2, \dots, n\}$ and let $\mathbf{A} \in \mathbb{R}^{n \times n}$ denote a real $n \times n$ matrix with elements a_{ij}, $i \in \mathbf{n}, j \in \mathbf{n}$. Let $m_j(\mathbf{A})$ denote the sum of the $j \times j$ leading principal minors of A for $j \in \mathbf{n}$. Thus, if $n = 3$:

$$\mathbf{A} = \begin{bmatrix} a_{11} & a_{12} & a_{13} \\ a_{21} & a_{22} & a_{23} \\ a_{31} & a_{32} & a_{33} \end{bmatrix} \tag{3.89}$$

and we have

$$m_1(\mathbf{A}) = a_{11} + a_{22} + a_{33}, \tag{3.90}$$

$$m_2(\mathbf{A}) = \det \begin{bmatrix} a_{11} & a_{12} \\ a_{21} & a_{22} \end{bmatrix} + \det \begin{bmatrix} a_{11} & a_{13} \\ a_{31} & a_{33} \end{bmatrix} + \det \begin{bmatrix} a_{22} & a_{23} \\ a_{32} & a_{33} \end{bmatrix}$$

$$= (a_{11}a_{22} - a_{12}a_{21}) + (a_{11}a_{33} - a_{13}a_{31}) + (a_{22}a_{33} - a_{23}a_{32}), \tag{3.91}$$

$$m_3(\mathbf{A}) = \det[\mathbf{A}]$$

$$= a_{11}(a_{22}a_{33} - a_{23}a_{32}) + a_{12}(a_{31}a_{23} - a_{21}a_{33}) + a_{13}(a_{21}a_{32} - a_{22}a_{31}). \tag{3.92}$$

We show below that the coefficients of the characteristic polynomial of A can be expressed in terms of $m_j(A)$. We first need the following lemma.

LEMMA 3.1 If $\mathbf{T}$ is an $n \times n$ matrix:

$$\det[\mathbf{I} + \mathbf{T}] = 1 + t, \tag{3.93}$$

where

$$t = m_1(\mathbf{T}) + m_2(\mathbf{T}) + \cdots + m_n(\mathbf{T}). \tag{3.94}$$

Proof The proof is a direct calculation and is derived in Appendix A.2. ∎

THEOREM 3.2

$$\det[s\mathbf{I} - \mathbf{A}] := s^n + b_1 s^{n-1} + \cdots + b_{n-2}s + b_{n-1}, \tag{3.95}$$

where

$$b_j = (-1)^j m_j(\mathbf{A}) \quad \text{for } j \in \mathbf{n}. \tag{3.96}$$

Proof We have

$$\det[s\mathbf{I} - \mathbf{A}] = \det \left[s \left(\mathbf{I} - \frac{\mathbf{A}}{s} \right) \right] = s^n \det \left[\mathbf{I} - \frac{\mathbf{A}}{s} \right]. \tag{3.97}$$

Applying Lemma 3.1 with $\mathbf{T} = -\dfrac{\mathbf{A}}{s}$, we get

$$\det \left[\mathbf{I} - \frac{\mathbf{A}}{s} \right] = 1 + \sum_{j \in \mathbf{n}} m_j \left(-\frac{\mathbf{A}}{s} \right) = 1 + \sum_{j \in \mathbf{n}} (-1)^j m_j(\mathbf{A})s^{-j}, \tag{3.98}$$

so that

$$\det[s\mathbf{I} - \mathbf{A}] = s^n + \sum_{j \in \mathbf{n}} (-1)^j m_j(\mathbf{A})s^{n-j}, \tag{3.99}$$

proving the theorem.   ∎

EXAMPLE 3.10   Consider

$$\dot{\mathbf{x}}(t) = \begin{bmatrix} 0 & -1 & 1 \\ 1 & -2 & 1 \\ -3 & 1 & -4 \end{bmatrix} \mathbf{x}(t). \tag{3.100}$$

Then we have

$$m_1(\mathbf{A}) = -6, \qquad m_2(\mathbf{A}) = 11, \qquad m_3(\mathbf{A}) = -6 \tag{3.101}$$

and thus the characteristic polynomial of A is

$$\delta(s) = s^3 + 6s^2 + 11s + 6. \tag{3.102}$$

EXAMPLE 3.11 (Time-delay systems)   The formula derived above can also be used to calculate the characteristic equation of time-delay systems. As an example, consider the following time-delay system:

$$\dot{\mathbf{x}}(t) = \begin{bmatrix} 0 & -1 & 1 \\ 1 & -2 & 1 \\ -3 & 1 & -4 \end{bmatrix} \mathbf{x}(t) + \begin{bmatrix} 0.5 & -1.0 & 1.0 \\ 1.0 & -1.5 & 1.0 \\ -3.0 & 1.0 & -3.5 \end{bmatrix} \mathbf{x}(t - T_1)$$

$$+ \begin{bmatrix} 0 & 0 & 0 \\ 0 & 1 & 0 \\ 1 & 1 & -1 \end{bmatrix} \mathbf{x}(t - T_2). \tag{3.103}$$

Now let

$$\mathbf{A}(s) := \begin{bmatrix} 0 & -1 & 1 \\ 1 & -2 & 1 \\ -3 & 1 & -4 \end{bmatrix} + \begin{bmatrix} 0.5 & -1.0 & 1.0 \\ 1.0 & -1.5 & 1.0 \\ -3.0 & 1.0 & -3.5 \end{bmatrix} e^{-sT_1} + \begin{bmatrix} 0 & 0 & 0 \\ 0 & 1 & 0 \\ 1 & 1 & 1 \end{bmatrix} e^{-sT_2} \tag{3.104}$$

so that

$$m_1(\mathbf{A}(s)) = -6 - 4.5e^{-sT_1} + 2e^{-sT_2},$$

$$m_2(\mathbf{A}(s)) = 11 + 16e^{-sT_1} + 5.75e^{-2sT_1} - 4e^{-sT_2} - e^{-2sT_2} - 4e^{-s(T_1+T_2)},$$

$$m_3(\mathbf{A}(s)) = -6 - 12.5e^{-sT_1} + 4e^{-sT_2} - 8.5e^{-2sT_1} - e^{-2sT_2} - 1.875e^{-3sT_1}$$

$$\qquad + 6e^{-s(T_1+T_2)} - 1.5e^{-s(T_1+2T_2)} + 2e^{-s(2T_1+T_2)}, \tag{3.105}$$

and thus the characteristic equation of $\mathbf{A}(s) := \det[s\mathbf{I} - \mathbf{A}(s)]$ is the quasi-polynomial

$$
\begin{aligned}
\delta(s) =\, & s^3 + \left(4.5e^{-sT_1} + 6 + 2e^{-sT_2}\right)s^2 \\
& + \left(11 + 16e^{-sT_1} + 5.75e^{-2sT_1} - 4e^{-sT_2} - e^{-2sT_2} - 4e^{-s(T_1+T_2)}\right)s \\
& + \left(12.5e^{-sT_1} + 8.5e^{-2sT_1} - 4e^{-sT_2} + 1.875e^{-3sT_1} + e^{-2sT_2}\right. \\
& \left. - 6e^{-s(T_1+T_2)} + 1.5e^{-s(T_1+2T_2)} - 2e^{-s(2T_1+T_2)} + 6\right).
\end{aligned}
\tag{3.106}
$$

The roots of this quasi-polynomial need to be in the open left-half plane for the stability of the time-delay system.

3.6   Calculation of $(s\mathbf{I} - \mathbf{A})^{-1}$: Faddeev–LeVerrier Algorithm

We describe an algorithm that determines $(s\mathbf{I} - \mathbf{A})^{-1}$ without symbolic calculations. This algorithm is known as the Faddeev–LeVerrier algorithm. The algorithm was first published in 1840 by LeVerrier and its present form is due to Faddeev and Sominsky.

First, write

$$
(s\mathbf{I} - \mathbf{A})^{-1} = \frac{\mathrm{Adj}[s\mathbf{I} - \mathbf{A}]}{\det[s\mathbf{I} - \mathbf{A}]} = \frac{\mathbf{T}(s)}{a(s)},
\tag{3.107}
$$

where

$$
a(s) = \det[s\mathbf{I} - \mathbf{A}] = s^n + a_{n-1}s^{n-1} + \cdots + a_1 s + a_0,
\tag{3.108}
$$

$$
\mathbf{T}(s) = \mathrm{Adj}[s\mathbf{I} - \mathbf{A}] = \mathbf{T}_{n-1}s^{n-1} + \cdots + \mathbf{T}_1 s + \mathbf{T}_0, \quad \text{for } \mathbf{T}_i \in \mathbb{R}^{n \times n}.
\tag{3.109}
$$

The Faddeev–LeVerrier algorithm is a recursive formula to calculate the a_is and $\mathbf{T}_i$s as follows:

$$
\begin{aligned}
\mathbf{T}_{n-1} &= \mathbf{I}, & a_{n-1} &= -\mathrm{Trace}\left(\mathbf{A}\mathbf{T}_{n-1}\right), \\
\mathbf{T}_{n-2} &= \mathbf{A}\mathbf{T}_{n-1} + a_{n-1}\mathbf{I}, & a_{n-2} &= -\frac{1}{2}\mathrm{Trace}\left(\mathbf{A}\mathbf{T}_{n-2}\right), \\
\mathbf{T}_{n-3} &= \mathbf{A}\mathbf{T}_{n-2} + a_{n-2}\mathbf{I}, & a_{n-3} &= -\frac{1}{3}\mathrm{Trace}\left(\mathbf{A}\mathbf{T}_{n-3}\right), \\
&\;\;\vdots \\
\mathbf{T}_1 &= \mathbf{A}\mathbf{T}_2 + a_2\mathbf{I}, & a_1 &= -\frac{1}{n-1}\mathrm{Trace}\left(\mathbf{A}\mathbf{T}_1\right), \\
\mathbf{T}_0 &= \mathbf{A}\mathbf{T}_1 + a_1\mathbf{I}, & a_0 &= -\frac{1}{n}\mathrm{Trace}\left(\mathbf{A}\mathbf{T}_0\right).
\end{aligned}
\tag{3.110}
$$

The proof of the algorithm is given in Appendix A.10. The computation sequence above constitutes the Faddeev–LeVerrier algorithm.

EXAMPLE 3.12

$$\mathbf{A} = \begin{bmatrix} 2 & -1 & 1 & 2 \\ 0 & 1 & 1 & 0 \\ -1 & 1 & 1 & 1 \\ 1 & 1 & 1 & 0 \end{bmatrix}. \quad \text{Let} \quad (s\mathbf{I} - \mathbf{A})^{-1} = \frac{\mathbf{T}_3 s^3 + \mathbf{T}_2 s^2 + \mathbf{T}_1 s + \mathbf{T}_0}{s^4 + a_3 s^3 + a_2 s^2 + a_1 s + a_0},$$

$$\mathbf{T}_3 = \mathbf{I} \quad \Rightarrow \quad a_3 = -\text{Trace}[\mathbf{A}] = -4,$$

$$\mathbf{T}_2 = \mathbf{A}\mathbf{T}_3 + a_3\mathbf{I} = \mathbf{A} - 4\mathbf{I} = \begin{bmatrix} -2 & -1 & 1 & 2 \\ 0 & -3 & 1 & 0 \\ -1 & 1 & -3 & 1 \\ 1 & 1 & 1 & -4 \end{bmatrix} \Rightarrow a_2 = -\frac{1}{2}\text{Trace}[\mathbf{A}\mathbf{T}_2] = 2,$$

$$\mathbf{T}_1 = \mathbf{A}\mathbf{T}_2 + a_2\mathbf{I} = \begin{bmatrix} -1 & 4 & 0 & -3 \\ -1 & 0 & -2 & 1 \\ 2 & 0 & 0 & -5 \\ -3 & -3 & -1 & 5 \end{bmatrix} \Rightarrow a_1 = -\frac{1}{3}\text{Trace}[\mathbf{A}\mathbf{T}_1] = 5,$$

$$\mathbf{T}_0 = \mathbf{A}\mathbf{T}_1 + a_1\mathbf{I} = \begin{bmatrix} 0 & 2 & 0 & -2 \\ 1 & 5 & -2 & -4 \\ -1 & -7 & 2 & 4 \\ 0 & 4 & -2 & -2 \end{bmatrix} \Rightarrow a_0 = -\frac{1}{4}\text{Trace}[\mathbf{A}\mathbf{T}_0] = 2.$$

Therefore

$$\det[s\mathbf{I} - \mathbf{A}] = s^4 - 4s^3 + 2s^2 + 5s + 2, \tag{3.111}$$

$$\text{Adj}[s\mathbf{I} - \mathbf{A}] = \mathbf{I}s^3 + \begin{bmatrix} -2 & -1 & 1 & 2 \\ 0 & -3 & 1 & 0 \\ -1 & 1 & -3 & 1 \\ 1 & 1 & 1 & -4 \end{bmatrix} s^2 + \begin{bmatrix} -1 & 4 & 0 & -3 \\ -1 & 0 & -2 & 1 \\ 2 & 0 & 0 & -5 \\ -3 & -3 & -1 & 5 \end{bmatrix} s$$

$$+ \begin{bmatrix} 0 & 2 & 0 & -2 \\ 1 & 5 & -2 & -4 \\ -1 & -7 & 2 & 4 \\ 0 & 4 & -2 & -2 \end{bmatrix}, \tag{3.112}$$

and

$$(s\mathbf{I} - \mathbf{A})^{-1} = \frac{1}{s^4 - 4s^3 + 2s^2 + 5s + 2}$$

$$\begin{bmatrix} s^3 - 2s^2 - s & -s^2 + 4s + 2 & s^2 & 2s^2 - 3s - 2 \\ -s + 1 & s^3 - 3s^2 + 5 & s^2 - 2s - 2 & s - 4 \\ -s^2 + 2s - 1 & s^2 - 7 & s^3 - 3s^2 + 2 & s^2 - 5s + 4 \\ s^2 - 3s & s^2 - 3s + 4 & s^2 - s - 2 & s^3 - 4s^2 + 5s - 2 \end{bmatrix}. \tag{3.113}$$

MATLAB Solution

```
syms s
A=[2 -1 1 2; 0 1 1 0; -1 1 1 1; 1 1 1 0];
det(s*eye(4)-A)
```

MATLAB: From State-Space Description to Transfer Function

The following MATLAB script converts a given state-space description to the equivalent transfer function. It is noted that the conversion must be done one column at a time, corresponding to the transfer functions related to each input.

```
A=[0 1; -2 -2];
B=[1 0; 1 1];
C=eye(2);
D=zeros(2);
[num1,den]=ss2tf(A,B,C,D,1);
[num2,den]=ss2tf(A,B,C,D,2);
```

The resulting outputs are

```
num1=        0   1.0000    3.0000
             0   1.0000   -2.0000
num2=        0        0         1
             0        1         0
den=    1.0000   2.0000    2.0000
```

These represent

$$\begin{bmatrix} \dfrac{s+3}{s^2+2s+2} & \dfrac{1}{s^2+2s+2} \\ \dfrac{s-2}{s^2+2s+2} & \dfrac{s}{2+2s+2} \end{bmatrix}. \tag{3.114}$$

Conversely, the following script converts a transfer function to an equivalent state-space model. Similar to the above, a state-space model is returned in controllable canonical form one input at a time.

```
[A,B,C,D]=tf2ss(num1,den);
[A,B,C,D]=tf2ss(num2,den);
```

The transfer function expression may also be obtained by the following command:

```
syms s
C*inv(s*eye(2)-A)*B+D
```

3.7 Block Diagram Algebra

The state-space representation $[\mathbf{A}, \mathbf{B}, \mathbf{C}, \mathbf{D}]$ corresponds to the transfer function

$$\mathbf{G}(s) = \mathbf{C}(s\mathbf{I} - \mathbf{A})^{-1}\mathbf{B} + \mathbf{D}. \tag{3.115}$$

This formula is valid when the building blocks of the system are integrators, multipliers, or adders. In general, the building blocks may be arbitrary transfer functions $G_i(s)$.

In this section, we develop a general approach to writing equations for an arbitrary block diagram containing transfer functions $G_i(s)$, and show that the overall input–output transfer function $\mathbf{T}(s)$ can be obtained from state-space-like formulas, in closed form, with $G_i(s)$ replacing each integrator. This result is a simplification and we expect it to play a useful role in applications.

3.7.1   A Simplified Approach

Consider an arbitrary block diagram with m blocks each with transfer function $G_i(s)$ for $i \in \mathbf{m}$. Let $v_i(s)$ and $z_i(s)$ denote the input and output of $G_i(s)$ for $i \in \mathbf{m}$, respectively, and define (suppressing the dependence on s)

$$\mathbf{v} := \begin{bmatrix} v_1 \\ v_2 \\ \vdots \\ v_m \end{bmatrix}, \quad \mathbf{z} := \begin{bmatrix} z_1 \\ z_2 \\ \vdots \\ z_m \end{bmatrix}, \quad \mathbf{G} := \begin{bmatrix} G_1 & & & \\ & G_2 & & \\ & & \ddots & \\ & & & G_m \end{bmatrix}. \tag{3.116}$$

Let $\mathbf{r}$ denote the vector of external inputs and $\mathbf{y}$ the vector of external outputs. Then the system equations take the form:

$$\mathbf{v} = \mathbf{Az} + \mathbf{Br}, \tag{3.117a}$$

$$\mathbf{z} = \mathbf{Gv}, \tag{3.117b}$$

$$\mathbf{y} = \mathbf{Cz} + \mathbf{Dr}. \tag{3.117c}$$

In Equations (3.117a)–(3.117c), boldface vectors and matrices are functions of s, and $(\mathbf{A}, \mathbf{B}, \mathbf{C}, \mathbf{D})$ are numerical matrices. Solving for $\mathbf{v}$, $\mathbf{z}$, and $\mathbf{y}$ from Equations (3.117a) and (3.117b), we get $((\mathbf{I} - \mathbf{AG})^{-1}$ exists by a well-posedness assumption, namely that $\mathbf{r}$ determines $\mathbf{v}$ uniquely)

$$\mathbf{v} = (\mathbf{I} - \mathbf{AG})^{-1} \mathbf{Br}, \tag{3.118a}$$

$$\mathbf{z} = \mathbf{G}(\mathbf{I} - \mathbf{AG})^{-1} \mathbf{Br}, \tag{3.118b}$$

$$\mathbf{y} = \left[\mathbf{CG}(\mathbf{I} - \mathbf{AG})^{-1} \mathbf{B} + \mathbf{D} \right] \mathbf{r}. \tag{3.118c}$$

This can be represented in a canonical block diagram (Figure 3.17).

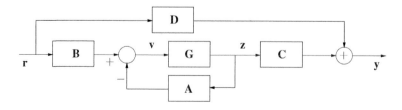

Figure 3.17 Canonical structure of any block diagram

3.7.2   Examples

EXAMPLE 3.13   Consider the block diagram in Figure 3.18.

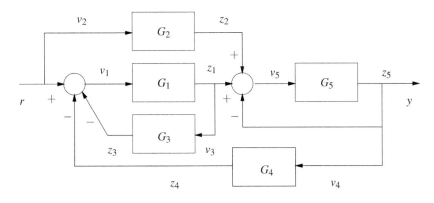

Figure 3.18 A block diagram for Example 3.13

The system equations are, in the Laplace domain:

$$v_1 = -z_3 - z_4 + r,$$
$$v_2 = r,$$
$$v_3 = z_1,$$
$$v_4 = z_5, \tag{3.119}$$
$$v_5 = z_1 + z_2 - z_5,$$
$$z_i = G_i v_i, \qquad \text{for } i = 1, 2, 3, 4, 5,$$
$$y = z_5,$$

which are of the form in Equations (3.117a)–(3.117c) with

$$\mathbf{A} = \begin{bmatrix} 0 & 0 & -1 & -1 & 0 \\ 0 & 0 & 0 & 0 & 0 \\ 1 & 0 & 0 & 0 & 0 \\ 0 & 0 & 0 & 0 & 1 \\ 1 & 1 & 0 & 0 & -1 \end{bmatrix}, \quad \mathbf{b} = \begin{bmatrix} 1 \\ 1 \\ 0 \\ 0 \\ 0 \end{bmatrix}, \quad \mathbf{c} = [\,0\,0\,0\,0\,1\,], \quad d = 0. \tag{3.120}$$

The formula in Equation (3.118c) can now be used to solve for $y(s)$.

Alternatively, we may proceed by defining

$$\mathbf{H}(s) := \mathbf{G}(s)^{-1} = \text{diag}\,[h_i(s)], \quad \text{for } i \in \mathbf{m} = \{1, 2, \ldots, m\}, \tag{3.121}$$

so that

$$\mathbf{v} = \mathbf{Hz} \tag{3.122}$$

and Equations (3.117a) and (3.117b) can be rewritten as

$$(\mathbf{H} - \mathbf{A})\mathbf{z} = \mathbf{Br}. \tag{3.123}$$

For the example at hand, Equation (3.123) becomes

$$\begin{bmatrix} h_1 & 0 & 1 & 1 & 0 \\ 0 & h_2 & 0 & 0 & 0 \\ -1 & 0 & h_3 & 0 & 0 \\ 0 & 0 & 0 & h_4 & -1 \\ -1 & 1 & 0 & 0 & h_5+1 \end{bmatrix} \mathbf{z} = \begin{bmatrix} 1 \\ 0 \\ 0 \\ 0 \\ 0 \end{bmatrix} r. \qquad (3.124)$$

Solving for $z_5 = y$ using Cramer's rule, we get

$$\begin{aligned} y &= \left(\frac{h_2 h_3 h_4}{h_3 + h_4 + h_4 h_5 + h_1 h_2 h_3 h_4 + h_1 h_2 h_3 h_4 h_5} \right) r \\ &= \left(\frac{G_1 G_5}{1 + G_5 + G_1 G_2 G_3 + G_1 G_2 G_3 G_5 + G_1 G_2 G_4 G_5} \right) r. \end{aligned} \qquad (3.125)$$

EXAMPLE 3.14   Consider the block diagram in Figure 3.19.

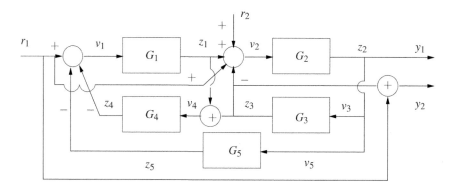

Figure 3.19 A block diagram for Example 3.14

There are two inputs r_1 and r_2, two outputs y_1 and y_2, five blocks with respective inputs v_i and outputs z_i with transfer functions $G_i(s)$. The equations of the system, in the Laplace domain, are

$$\begin{aligned} v_1 &= -z_4 - z_5 + r_1, \\ v_2 &= z_1 - z_3 + r_1 + r_2, \\ v_3 &= z_2, \\ v_4 &= z_1 + z_3, \\ v_5 &= z_2, \\ z_i &= G_i v_i, \quad i = 1, 2, 3, 4, 5, \\ y_1 &= z_2, \\ y_2 &= z_3 + r_1. \end{aligned} \qquad (3.126)$$

Using vector–matrix notation with

$$
\mathbf{v} = \begin{bmatrix} v_1 \\ v_2 \\ \vdots \\ v_5 \end{bmatrix}, \quad \mathbf{z} = \begin{bmatrix} z_1 \\ z_2 \\ \vdots \\ z_5 \end{bmatrix}, \quad \mathbf{r} = \begin{bmatrix} r_1 \\ r_2 \end{bmatrix}, \quad \mathbf{y} = \begin{bmatrix} y_1 \\ y_2 \end{bmatrix}, \tag{3.127}
$$

we have

$$
\begin{aligned}
\mathbf{v} &= A\mathbf{z} + B\mathbf{r}, \\
\mathbf{z} &= G\mathbf{v}, \\
\mathbf{y} &= C\mathbf{z} + D\mathbf{r},
\end{aligned}
\tag{3.128}
$$

where

$$
A = \begin{bmatrix} 0 & 0 & 0 & -1 & -1 \\ 1 & 0 & -1 & 0 & 0 \\ 0 & 1 & 0 & 0 & 0 \\ 1 & 0 & 1 & 0 & 0 \\ 0 & 1 & 0 & 0 & 0 \end{bmatrix}, \quad B = \begin{bmatrix} 1 & 0 \\ 1 & 1 \\ 0 & 0 \\ 0 & 0 \\ 0 & 0 \end{bmatrix}, \tag{3.129}
$$

$$
C = \begin{bmatrix} 0 & 1 & 0 & 0 & 0 \\ 0 & 0 & 1 & 0 & 0 \end{bmatrix}, \quad D = \begin{bmatrix} 0 & 0 \\ 1 & 0 \end{bmatrix},
$$

and

$$
G = \begin{bmatrix} G_1 & & & \\ & G_2 & & \\ & & \ddots & \\ & & & G_5 \end{bmatrix}. \tag{3.130}
$$

We now see that the transfer function matrix, denoted $\mathbf{T}$, is given by

$$
\mathbf{T} = CG(I - AG)^{-1}B + D = C\left(G^{-1} - A\right)^{-1}B + D. \tag{3.131}
$$

Remark 3.1 Equation (3.131) is a generalization of state-space equations where each $G_i = \frac{1}{s}$ and $G^{-1} = sI$. This method is a bridge between Mason's rule for solving block diagrams and state-space formulas.

3.8 State-Space Models for Interconnected Systems

The state-space model allows us to interconnect systems easily. We discuss series, parallel, and feedback connections of S_1 and S_2 described through their state-space models.

Let

$$S_1 : \begin{cases} \dot{\mathbf{x}}_1(t) &= \mathbf{A}_1\mathbf{x}_1(t) + \mathbf{B}_1\mathbf{u}_1(t), \\ \mathbf{y}_1(t) &= \mathbf{C}_1\mathbf{x}_1(t) + \mathbf{D}_1\mathbf{u}_1(t), \end{cases} \tag{3.132}$$

$$S_2 : \begin{cases} \dot{\mathbf{x}}_2(t) &= \mathbf{A}_2\mathbf{x}_2(t) + \mathbf{B}_2\mathbf{u}_2(t), \\ \mathbf{y}_2(t) &= \mathbf{C}_2\mathbf{x}_2(t) + \mathbf{D}_2\mathbf{u}_2(t). \end{cases} \tag{3.133}$$

3.8.1 Series Connection

This connection is specified by Figure 3.20.

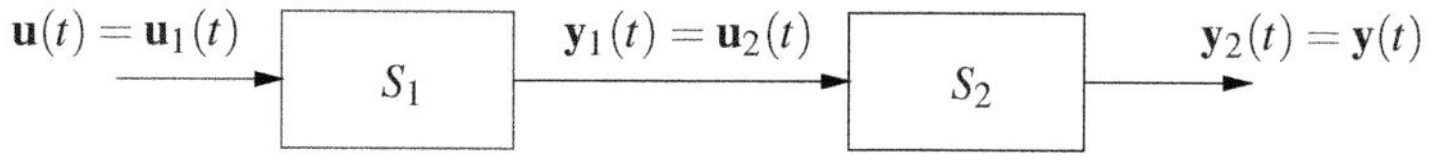

Figure 3.20 Series connection

Then

$$\dot{\mathbf{x}}_1(t) = \mathbf{A}_1\mathbf{x}_1(t) + \mathbf{B}_1\mathbf{u}(t),$$

$$\dot{\mathbf{x}}_2(t) = \mathbf{A}_2\mathbf{x}_2(t) + \mathbf{B}_2\mathbf{y}_1(t) = \mathbf{A}_2\mathbf{x}_2(t) + \mathbf{B}_2\mathbf{C}_1\mathbf{x}_1(t) + \mathbf{B}_2\mathbf{D}_1\mathbf{u}(t)$$

$$= \mathbf{B}_2\mathbf{C}_1\mathbf{x}_1(t) + \mathbf{A}_2\mathbf{x}_2(t) + \mathbf{B}_2\mathbf{D}_1\mathbf{u}(t), \tag{3.134}$$

$$\mathbf{y}(t) = \mathbf{C}_2\mathbf{x}_2(t) + \mathbf{D}_2\mathbf{y}_1(t) = \mathbf{C}_2\mathbf{x}_2(t) + \mathbf{D}_2\mathbf{C}_1\mathbf{x}_1(t) + \mathbf{D}_2\mathbf{D}_1\mathbf{u}(t)$$

$$= \mathbf{D}_2\mathbf{C}_1\mathbf{x}_1(t) + \mathbf{C}_2\mathbf{x}_2(t) + \mathbf{D}_2\mathbf{D}_1\mathbf{u}(t).$$

Therefore, the state-space model of the series connection is

$$\begin{bmatrix} \dot{\mathbf{x}}_1(t) \\ \dot{\mathbf{x}}_2(t) \end{bmatrix} = \begin{bmatrix} \mathbf{A}_1 & \mathbf{0} \\ \mathbf{B}_2\mathbf{C}_1 & \mathbf{A}_2 \end{bmatrix} \begin{bmatrix} \mathbf{x}_1(t) \\ \mathbf{x}_2(t) \end{bmatrix} + \begin{bmatrix} \mathbf{B}_1 \\ \mathbf{B}_2\mathbf{D}_1 \end{bmatrix} \mathbf{u}(t),$$

$$\mathbf{y}(t) = \begin{bmatrix} \mathbf{D}_2\mathbf{C}_1 & \mathbf{C}_2 \end{bmatrix} \begin{bmatrix} \mathbf{x}_1(t) \\ \mathbf{x}_2(t) \end{bmatrix} + \begin{bmatrix} \mathbf{D}_2\mathbf{D}_1 \end{bmatrix} \mathbf{u}(t). \tag{3.135}$$

MATLAB: Series Connection of Two Systems
After defining the state-space models of two systems, ($\mathbf{A}_1$, $\mathbf{B}_1$, $\mathbf{C}_1$, $\mathbf{D}_1$) and ($\mathbf{A}_2$, $\mathbf{B}_2$, $\mathbf{C}_2$, $\mathbf{D}_2$), the following MATLAB script may be used to find a state-space model of the series-connected system:

```
sys1=ss(A1,B1,C1,D1);
sys2=ss(A2,B2,C2,D2);
sys=series(sys1,sys2)
```

3.8.2 Parallel Connection

In the connection shown in Figure 3.21,

$$\mathbf{u}(t) = \mathbf{u}_1(t) = \mathbf{u}_2(t) \qquad \text{and} \qquad \mathbf{y}(t) = \mathbf{y}_1(t) + \mathbf{y}_2(t). \tag{3.136}$$

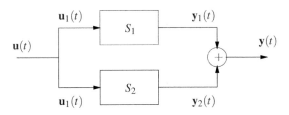

Figure 3.21 Parallel connection

The state-space model of the parallel connection is

$$
\left[\begin{array}{c} \dot{x}_1(t) \\ \hline \dot{x}_2(t) \end{array}\right] = \left[\begin{array}{c:c} A_1 & 0 \\ \hline 0 & A_2 \end{array}\right]\left[\begin{array}{c} x_1(t) \\ \hline x_2(t) \end{array}\right] + \left[\begin{array}{c} B_1 \\ \hline B_2 \end{array}\right]u(t),
$$

$$
y(t) = \left[\begin{array}{c:c} C_1 & C_2 \end{array}\right]\left[\begin{array}{c} x_1(t) \\ \hline x_2(t) \end{array}\right] + \left[\begin{array}{c} D_1 + D_2 \end{array}\right]u(t).
$$

(3.137)

MATLAB: Parallel Connection of Two Systems
Similar to the case of the series connection, the following MATLAB script may be used
to find a state-space model of the connected system:

```
sys1=ss(A1,B1,C1,D1);
sys2=ss(A2,B2,C2,D2);
sys=parallel(sys1,sys2)
```

Remark 3.2 In the series and parallel connections, the eigenvalues of the new "A"
matrix in each case equal the union of the eigenvalues of A_1 and those of A_2. Thus, in
particular, the interconnected system is stable if and only if the individual systems are
stable. This will *not* hold in the feedback connection discussed next.

3.8.3 Feedback Connection

In the connection shown in Figure 3.22 we have

$$
u_1(t) = u(t) - y_2(t) \qquad \text{and} \qquad y(t) = y_1(t) = u_2(t).
$$
(3.138)

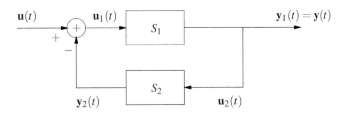

Figure 3.22 Feedback connection

Therefore

$$\begin{aligned} \mathbf{y}_1(t) &= \mathbf{C}_1\mathbf{x}_1(t) + \mathbf{D}_1\mathbf{u}(t) - \mathbf{D}_1\mathbf{y}_2(t) \\ &= \mathbf{C}_1\mathbf{x}_1(t) + \mathbf{D}_1\mathbf{u}(t) - \mathbf{D}_1\mathbf{C}_2\mathbf{x}_2(t) - \mathbf{D}_1\mathbf{D}_2\mathbf{y}_1(t) \end{aligned} \tag{3.139}$$

and

$$[\mathbf{I}+\mathbf{D}_1\mathbf{D}_2]\,\mathbf{y}_1(t) = \mathbf{C}_1\mathbf{x}_1(t) - \mathbf{D}_1\mathbf{C}_2\mathbf{x}_2(t) + \mathbf{D}_1\mathbf{u}(t). \tag{3.140}$$

Similarly

$$\begin{aligned} \mathbf{y}_2(t) &= \mathbf{C}_2\mathbf{x}_2(t) + \mathbf{D}_2\mathbf{y}_1(t) = \mathbf{C}_2\mathbf{x}_2(t) + \mathbf{D}_2\mathbf{C}_1\mathbf{x}_1(t) + \mathbf{D}_2\mathbf{D}_1\mathbf{u}_1(t) \\ &= \mathbf{C}_2\mathbf{x}_2(t) + \mathbf{D}_2\mathbf{C}_1\mathbf{x}_1(t) + \mathbf{D}_2\mathbf{D}_1\mathbf{u}(t) - \mathbf{D}_2\mathbf{D}_1\mathbf{y}_2(t) \end{aligned}$$

and

$$(\mathbf{I}+\mathbf{D}_2\mathbf{D}_1)\,\mathbf{y}_2(t) = \mathbf{D}_2\mathbf{C}_1\mathbf{x}_1(t) + \mathbf{C}_2\mathbf{x}_2(t) + \mathbf{D}_2\mathbf{D}_1\mathbf{u}(t). \tag{3.141}$$

To proceed, bring in the well-posedness assumption. This assumption states that $\mathbf{y}_1(t)$ and $\mathbf{y}_2(t)$ must uniquely be determined by $\mathbf{x}_1(t)$, $\mathbf{x}_2(t)$, and $\mathbf{u}(t)$. From Equations (3.140) and (3.141), the well-posedness assumption amounts to

(1) $\mathbf{I}+\mathbf{D}_1\mathbf{D}_2$ is invertible,
(2) $\mathbf{I}+\mathbf{D}_2\mathbf{D}_1$ is invertible.

Actually, conditions (1) and (2) are equivalent, that is, one holds if and only if the other does.

To proceed, let

$$\mathbf{E}_1 = (\mathbf{I}+\mathbf{D}_1\mathbf{D}_2)^{-1} \quad \text{and} \quad \mathbf{E}_2 = (\mathbf{I}+\mathbf{D}_2\mathbf{D}_1)^{-1} \tag{3.142}$$

so that from Equations (3.140) and (3.141):

$$\begin{aligned} \mathbf{y}_1(t) &= \mathbf{E}_1\mathbf{C}_1\mathbf{x}_1(t) - \mathbf{E}_1\mathbf{D}_1\mathbf{C}_2\mathbf{x}_2(t) + \mathbf{E}_1\mathbf{D}_1\mathbf{u}(t), \\ \mathbf{y}_2(t) &= \mathbf{E}_2\mathbf{D}_2\mathbf{C}_1\mathbf{x}_1(t) + \mathbf{E}_2\mathbf{C}_2\mathbf{x}_2(t) + \mathbf{E}_2\mathbf{D}_2\mathbf{D}_1\mathbf{u}(t). \end{aligned} \tag{3.143}$$

Similarly

$$\begin{aligned} \mathbf{u}_2(t) &= \mathbf{E}_1\mathbf{C}_1\mathbf{x}_1(t) - \mathbf{E}_1\mathbf{D}_1\mathbf{C}_2\mathbf{x}_2(t) + \mathbf{E}_1\mathbf{D}_1\mathbf{u}(t), \\ \mathbf{u}_2(t) &= -\mathbf{E}_2\mathbf{D}_2\mathbf{C}_1\mathbf{x}_1(t) - \mathbf{E}_2\mathbf{C}_2\mathbf{x}_2(t) + \mathbf{E}_2\mathbf{u}(t). \end{aligned} \tag{3.144}$$

Therefore, in the feedback system

$$\begin{aligned} \dot{\mathbf{x}}_1(t) &= \mathbf{A}_1\mathbf{x}_1(t) + \mathbf{B}_1\mathbf{u}(t) - \mathbf{B}_1\mathbf{y}_2(t) \\ &= \mathbf{A}_1\mathbf{x}_1(t) + \mathbf{B}_1\mathbf{u}(t) - \mathbf{B}_1\mathbf{E}_2\mathbf{D}_2\mathbf{C}_1\mathbf{x}_1(t) - \mathbf{B}_1\mathbf{E}_2\mathbf{C}_2\mathbf{x}_2(t) - \mathbf{B}_1\mathbf{E}_2\mathbf{D}_2\mathbf{D}_1\mathbf{u}(t) \\ &= (\mathbf{A}_1 - \mathbf{B}_1\mathbf{E}_2\mathbf{D}_2\mathbf{C}_1)\,\mathbf{x}_1(t) - \mathbf{B}_1\mathbf{E}_2\mathbf{C}_2\mathbf{x}_2(t) + (\mathbf{B}_1 - \mathbf{B}_1\mathbf{E}_2\mathbf{D}_2\mathbf{D}_1)\,\mathbf{u}(t) \end{aligned} \tag{3.145}$$

and

$$\begin{aligned} \dot{\mathbf{x}}_2(t) &= \mathbf{A}_2\mathbf{x}_2(t) + \mathbf{B}_2\mathbf{E}_1\mathbf{C}_1\mathbf{x}_1(t) - \mathbf{B}_2\mathbf{E}_1\mathbf{D}_1\mathbf{D}_2\mathbf{x}_2(t) + \mathbf{B}_2\mathbf{E}_1\mathbf{D}_1\mathbf{u}(t) \\ &= \mathbf{B}_2\mathbf{E}_1\mathbf{C}_1\mathbf{x}_1(t) + (\mathbf{A}_2 - \mathbf{B}_2\mathbf{E}_1\mathbf{D}_1\mathbf{C}_2)\,\mathbf{x}_2(t) + \mathbf{B}_2\mathbf{E}_1\mathbf{D}_1\mathbf{u}(t). \end{aligned} \tag{3.146}$$

The state-space representation of the feedback system is now

$$
\left[\begin{array}{c} \dot{\mathbf{x}}_1(t) \\ \hline \dot{\mathbf{x}}_2(t) \end{array} \right] = \left[\begin{array}{c:c} \mathbf{A}_1 - \mathbf{B}_1\mathbf{E}_2\mathbf{D}_2\mathbf{C}_1 & -\mathbf{B}_1\mathbf{E}_2\mathbf{C}_2 \\ \hdashline \mathbf{B}_2\mathbf{E}_1\mathbf{C}_1 & \mathbf{A}_2 - \mathbf{B}_2\mathbf{E}_1\mathbf{D}_1\mathbf{C}_2 \end{array} \right] \left[\begin{array}{c} \mathbf{x}_1(t) \\ \hline \mathbf{x}_2(t) \end{array} \right]
$$
$$
+ \left[\begin{array}{c} \mathbf{B}_1 - \mathbf{B}_1\mathbf{E}_2\mathbf{D}_2\mathbf{D}_1 \\ \hline \mathbf{B}_2\mathbf{E}_1\mathbf{D}_1 \end{array} \right] \mathbf{u}(t), \tag{3.147}
$$
$$
\mathbf{y}(t) = \left[\begin{array}{c:c} \mathbf{E}_1\mathbf{C}_1 & -\mathbf{E}_1\mathbf{D}_1\mathbf{C}_2 \end{array} \right] \left[\begin{array}{c} \mathbf{x}_1(t) \\ \hline \mathbf{x}_2(t) \end{array} \right] + \mathbf{E}_1\mathbf{D}_1\mathbf{u}(t).
$$

Remark 3.3 The dynamics, and in particular the stability of the feedback connection is determined by the new "**A**" matrix. Simple examples can be constructed to show that the feedback connection can be stable even when S_1 and S_2 are unstable. It is also possible to show that the feedback system may be unstable even though S_1 and S_2 are stable. Thus, feedback systems need to be designed with caution in order to take advantage of the potential benefits of feedback while avoiding its pitfalls.

Remark 3.4 The formulas given in Equation (3.147) simplify considerably if $\mathbf{D}_1$ or $\mathbf{D}_2$ is zero. In this case the well-posedness assumption holds automatically, $\mathbf{E}_1$ and $\mathbf{E}_2$ reduce to identity matrices, and Equation (3.147) simplifies to (assuming $\mathbf{D}_1 = 0$)

$$
\left[\begin{array}{c} \dot{\mathbf{x}}_1(t) \\ \dot{\mathbf{x}}_2(t) \end{array} \right] = \left[\begin{array}{cc} \mathbf{A}_1 - \mathbf{B}_1\mathbf{D}_2\mathbf{C}_1 & -\mathbf{B}_1\mathbf{C}_2 \\ \mathbf{B}_2\mathbf{C}_1 & \mathbf{A}_2 \end{array} \right] \left[\begin{array}{c} \mathbf{x}_1(t) \\ \mathbf{x}_2(t) \end{array} \right]. \tag{3.148}
$$

EXAMPLE 3.15 Consider the feedback connection shown in Figure 3.22 and let

$$
S_1 : \begin{cases} \dot{\mathbf{x}}_1(t) = \underbrace{\left[\begin{array}{cc} 0 & 1 \\ -2 & -1 \end{array} \right]}_{\mathbf{A}_1} \mathbf{x}_1(t) + \underbrace{\left[\begin{array}{c} 0 \\ 1 \end{array} \right]}_{\mathbf{B}_1} \mathbf{u}_1(t), \\[2em] \mathbf{y}_1(t) = \underbrace{\left[\begin{array}{cc} 1 & 1 \end{array} \right]}_{\mathbf{C}_1} \mathbf{x}_1(t) + \underbrace{1}_{\mathbf{D}_1} \mathbf{u}_1(t), \end{cases} \tag{3.149}
$$

$$
S_2 : \begin{cases} \dot{\mathbf{x}}_2 = \underbrace{\left[\begin{array}{cc} 0 & 1 \\ -3 & -2 \end{array} \right]}_{\mathbf{A}_2} \mathbf{x}_2(t) + \underbrace{\left[\begin{array}{c} 0 \\ 1 \end{array} \right]}_{\mathbf{B}_2} \mathbf{u}_2(t), \\[2em] \mathbf{y}_2(t) = \underbrace{\left[\begin{array}{cc} 1 & 2 \end{array} \right]}_{\mathbf{C}_2} \mathbf{x}_2(t) + \underbrace{1}_{\mathbf{D}_2} \mathbf{u}_2(t). \end{cases} \tag{3.150}
$$

Using the formulas in Equation (3.147), we have

$$
\mathbf{E}_1 = (1 + \mathbf{D}_1\mathbf{D}_2)^{-1} = (1 + 1 \cdot 1)^{-1} = \frac{1}{2}, \tag{3.151}
$$

$$
\mathbf{E}_2 = (1 + \mathbf{D}_2\mathbf{D}_1)^{-1} = (1 + 1 \cdot 1)^{-1} = \frac{1}{2}, \tag{3.152}
$$

and

$$\mathbf{A}_1 - \mathbf{B}_1\mathbf{E}_2\mathbf{D}_2\mathbf{C}_1 = \begin{bmatrix} 0 & 1 \\ -2 & -1 \end{bmatrix} - \begin{bmatrix} 0 \\ 1 \end{bmatrix} \cdot \frac{1}{2} \cdot 1 \begin{bmatrix} 1 & 1 \end{bmatrix} = \begin{bmatrix} 0 & 1 \\ -\frac{5}{2} & -\frac{3}{2} \end{bmatrix},$$

$$-\mathbf{B}_1\mathbf{E}_2\mathbf{C}_2 = -\begin{bmatrix} 0 \\ 1 \end{bmatrix} \cdot \frac{1}{2} \cdot \begin{bmatrix} 1 & 2 \end{bmatrix} = \begin{bmatrix} 0 & 0 \\ -\frac{1}{2} & -1 \end{bmatrix},$$

$$\mathbf{B}_2\mathbf{E}_1\mathbf{C}_1 = \begin{bmatrix} 0 \\ 1 \end{bmatrix} \cdot \frac{1}{2} \cdot \begin{bmatrix} 1 & 1 \end{bmatrix} = \begin{bmatrix} 0 & 0 \\ \frac{1}{2} & \frac{1}{2} \end{bmatrix},$$

$$\mathbf{A}_2 - \mathbf{B}_2\mathbf{E}_1\mathbf{D}_1\mathbf{C}_2 = \begin{bmatrix} 0 & 1 \\ -3 & -2 \end{bmatrix} - \begin{bmatrix} 0 \\ 1 \end{bmatrix} \cdot \frac{1}{2} \cdot 1 \begin{bmatrix} 1 & 2 \end{bmatrix} = \begin{bmatrix} 0 & 1 \\ -\frac{7}{2} & -3 \end{bmatrix},$$

$$\mathbf{B}_1 - \mathbf{B}_2\mathbf{E}_2\mathbf{D}_2\mathbf{D}_1 = \begin{bmatrix} 0 \\ 1 \end{bmatrix} - \begin{bmatrix} 0 \\ 1 \end{bmatrix} \cdot \frac{1}{2} \cdot 1 \cdot 1 = \begin{bmatrix} 0 \\ \frac{1}{2} \end{bmatrix},$$

$$\mathbf{B}_2\mathbf{E}_1\mathbf{D}_1 = \begin{bmatrix} 0 \\ 1 \end{bmatrix} \cdot \frac{1}{2} \cdot 1 = \begin{bmatrix} 0 \\ \frac{1}{2} \end{bmatrix},$$

$$\mathbf{E}_1\mathbf{C}_1 = \frac{1}{2} \cdot \begin{bmatrix} 1 & 1 \end{bmatrix} = \begin{bmatrix} \frac{1}{2} & \frac{1}{2} \end{bmatrix},$$

$$-\mathbf{E}_1\mathbf{D}_1\mathbf{C}_2 = -\frac{1}{2} \cdot 1 \cdot \begin{bmatrix} 1 & 2 \end{bmatrix} = \begin{bmatrix} -\frac{1}{2} & -1 \end{bmatrix},$$

$$\mathbf{E}_1\mathbf{D}_1 = \frac{1}{2} \cdot 1 = \frac{1}{2}.$$

Therefore

$$\begin{bmatrix} \dot{\mathbf{x}}_1(t) \\ \dot{\mathbf{x}}_2(t) \end{bmatrix} = \begin{bmatrix} 0 & 1 & \vdots & 0 & 0 \\ -\frac{5}{2} & -\frac{3}{2} & \vdots & -\frac{1}{2} & -1 \\ \cdots & \cdots & & \cdots & \cdots \\ 0 & 0 & \vdots & 0 & 1 \\ \frac{1}{2} & \frac{1}{2} & \vdots & -\frac{7}{2} & -3 \end{bmatrix} \begin{bmatrix} \mathbf{x}_1(t) \\ \mathbf{x}_2(t) \end{bmatrix} + \begin{bmatrix} 0 \\ \frac{1}{2} \\ \cdots \\ 0 \\ \frac{1}{2} \end{bmatrix} u(t),$$

$$y(t) = \begin{bmatrix} \frac{1}{2} & \frac{1}{2} & \vdots & -\frac{1}{2} & -1 \end{bmatrix} \begin{bmatrix} \mathbf{x}_1(t) \\ \mathbf{x}_2(t) \end{bmatrix} + \frac{1}{2}u(t). \tag{3.153}$$

MATLAB: Feedback Connection of Two Systems

Similar to the case of the series and parallel connections shown before, the following MATLAB script may be used to find a state-space model of the connected system. Note that the MATLAB command used assumes that the feedback connection is as shown in Figure 3.22.

```
sys1=ss(A1,B1,C1,D1);
sys2=ss(A2,B2,C2,D2);
sys=feedback(sys1,sys2)
```

3.9   Discrete-Time Systems

The basic building blocks of a discrete-time system are the unit delay, multiplier, and adder shown in Figure 3.23.

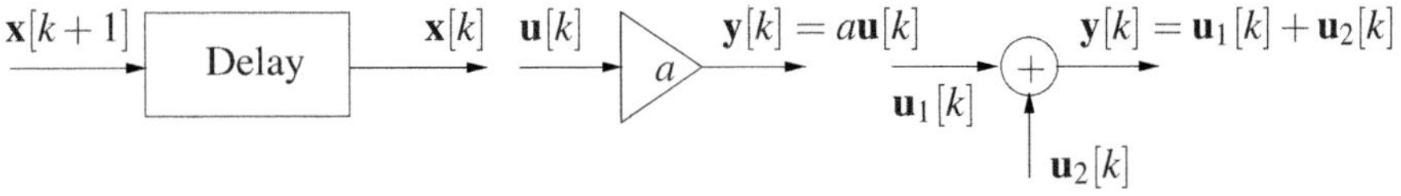

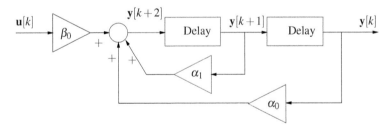

Figure 3.23 Basic building blocks of a discrete-time system

The notation $\mathbf{x}[k]$ denotes the sequence

$$\{x(0), x(1), \ldots, x(j), \ldots\}. \tag{3.154}$$

The basic building blocks can be interconnected to implement difference equations as we show below.

EXAMPLE 3.16   Consider the difference equation

$$\mathbf{y}[k+2] = \alpha_0 \mathbf{y}[k] + \alpha_1 \mathbf{y}[k+1] + \beta_0 \mathbf{u}[k]. \tag{3.155}$$

It may be realized by the circuit in Figure 3.24.

Figure 3.24   A realization (Example 3.16)

More generally, an $[\mathbf{A}, \mathbf{B}, \mathbf{C}, \mathbf{D}]$ state-variable model can be developed for an arbitrary connection of delays, multipliers, and adders.

EXAMPLE 3.17   Consider the realization shown in Figure 3.25.
This realization results in

$$\mathbf{x}[k+1] = \begin{bmatrix} a_{11} & a_{12} \\ a_{21} & a_{22} \end{bmatrix} \mathbf{x}[k] + \begin{bmatrix} b_1 \\ b_2 \end{bmatrix} \mathbf{u}[k],$$

$$\mathbf{y}[k] = \begin{bmatrix} c_1 & c_2 \end{bmatrix} \mathbf{x}[k] + d\mathbf{u}[k]. \tag{3.156}$$

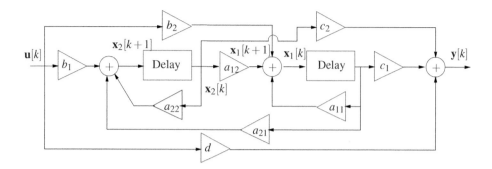

Figure 3.25 A realization (Example 3.17)

Generalizing the above example, we have the $[\mathbf{A}, \mathbf{B}, \mathbf{C}, \mathbf{D}]$ representation of a discrete-time system:

$$
\begin{aligned}
\mathbf{x}[k+1] &= \mathbf{A}\mathbf{x}[k] + \mathbf{B}\mathbf{u}[k], \\
\mathbf{y}[k] &= \mathbf{C}\mathbf{x}[k] + \mathbf{D}\mathbf{u}[k].
\end{aligned}
\tag{3.157}
$$

3.9.1   Solution of Discrete-Time Systems

Linear time-invariant discrete-time dynamic systems can be described by

$$
\mathbf{x}[k+1] = \mathbf{A}\mathbf{x}[k] + \mathbf{B}\mathbf{u}[k],
\tag{3.158a}
$$

$$
\mathbf{y}[k] = \mathbf{C}\mathbf{x}[k] + \mathbf{D}\mathbf{u}[k].
\tag{3.158b}
$$

The solution of the discrete-time state equation can be obtained from the recursion

$$
\begin{aligned}
\mathbf{x}[k] &= \mathbf{A}\mathbf{x}[k-1] + \mathbf{B}\mathbf{u}[k-1] \\
&= \mathbf{A}\left(\mathbf{A}\mathbf{x}[k-2] + \mathbf{B}\mathbf{u}[k-2]\right) + \mathbf{B}\mathbf{u}[k-1] \\
&= \mathbf{A}^2\mathbf{x}[k-2] + \mathbf{A}\mathbf{B}\mathbf{u}[k-2] + \mathbf{B}\mathbf{u}[k-1] \\
&\;\;\vdots \\
&= \mathbf{A}^k\mathbf{x}[0] + \sum_{j=0}^{k-1}\mathbf{A}^{k-1-j}\mathbf{B}\mathbf{u}[j]
\end{aligned}
\tag{3.159}
$$

and yields the output response

$$
\begin{bmatrix} \mathbf{y}[0] \\ \mathbf{y}[1] \\ \vdots \\ \mathbf{y}[k] \end{bmatrix} = \underbrace{\begin{bmatrix} \mathbf{C} \\ \mathbf{CA} \\ \vdots \\ \mathbf{CA}^k \end{bmatrix} \mathbf{x}[0]}_{\text{zero-input response}} + \underbrace{\begin{bmatrix} \mathbf{D} & 0 & \cdots & \ddots & 0 \\ \mathbf{CB} & \mathbf{D} & 0 & & \\ \mathbf{CAB} & & \mathbf{D} & & \\ \vdots & & & \ddots & 0 \\ \mathbf{CA}^{k-1}\mathbf{B} & & & & \mathbf{D} \end{bmatrix} \begin{bmatrix} \mathbf{u}[0] \\ \mathbf{u}[1] \\ \vdots \\ \mathbf{u}[k-1] \end{bmatrix}}_{\text{zero-state response}}.
$$

$$(3.160)$$

3.9.2 Transfer Function

Taking the $\mathscr{Z}$-transform of Equation (3.158b), we have

$$
\begin{aligned}
z\mathbf{X}(z) - z\mathbf{x}[0] &= \mathbf{AX}(z) + \mathbf{BU}(z), \\
\mathbf{Y}(z) &= \mathbf{CX}(z) + \mathbf{DU}(z),
\end{aligned}
$$
$$(3.161)$$

and

$$
\begin{aligned}
\mathbf{X}(z) &= (z\mathbf{I} - \mathbf{A})^{-1}\mathbf{Bu}(z) + (z\mathbf{I} - \mathbf{A})^{-1}z\mathbf{x}[0], \\
\mathbf{Y}(z) &= \mathbf{C}(z\mathbf{I} - \mathbf{A})^{-1}\mathbf{BU}(z) + \mathbf{DU}(z) + \mathbf{C}(z\mathbf{I} - \mathbf{A})^{-1}z\mathbf{x}[0].
\end{aligned}
$$
$$(3.162)$$

With zero initial conditions $\mathbf{x}[0] = 0$:

$$
\mathbf{Y}_u(z) = \underbrace{\left[\mathbf{C}(z\mathbf{I} - \mathbf{A})^{-1}\mathbf{B} + \mathbf{D} \right]}_{\mathbf{G}(z)} \mathbf{U}(z)
$$
$$(3.163)$$

and $\mathbf{G}(z)$ is the transfer function.

3.9.3 Discretization of Continuous-Time Systems

Suppose a continuous-time system as in Equation (3.58a) is sampled every T second. With the approximation

$$
\dot{\mathbf{x}}(t) \approx \frac{\mathbf{x}(t+T) - \mathbf{x}(t)}{T},
$$
$$(3.164)$$

we have

$$
\mathbf{x}(t+T) = \mathbf{x}(t) + \mathbf{Ax}(t)T + \mathbf{Bu}(t)T.
$$
$$(3.165)$$

For $t = kT$, $k = 0, 1, 2, \ldots$, we have

$$
\begin{aligned}
\mathbf{x}((k+1)T) &= \mathbf{x}(kT) + \mathbf{Ax}(kT)T + \mathbf{Bu}(kTt)T \\
&= (\mathbf{I} + T\mathbf{A})\mathbf{x}(kT) + T\mathbf{Bu}(kT).
\end{aligned}
$$
$$(3.166)$$

From

$$
\mathbf{x}(kT) = e^{\mathbf{A}kT}\mathbf{x}(0) + \int_0^{kT} e^{\mathbf{A}(kT-\tau)}\mathbf{Bu}(\tau)d\tau,
$$
$$(3.167)$$

$$\mathbf{x}((k+1)T) = e^{\mathbf{A}(k+1)T}\mathbf{x}(0) + \int_0^{(k+1)T} e^{\mathbf{A}((k+1)T-\tau)}\mathbf{B}\mathbf{u}(\tau)d\tau$$

$$= e^{\mathbf{A}T}\left(e^{\mathbf{A}kT}\mathbf{x}(0) + \int_0^{kT} e^{\mathbf{A}(kT-\tau)}\mathbf{B}\mathbf{u}(\tau)d\tau + \int_{kT}^{(k+1)T} e^{\mathbf{A}(kT-\tau)}\mathbf{B}\mathbf{u}(\tau)d\tau\right)$$

$$= e^{\mathbf{A}T}\mathbf{x}(kT) + \int_{kT}^{(k+1)T} e^{\mathbf{A}((k+1)T-\tau)}\mathbf{B}\mathbf{u}(\tau)d\tau. \tag{3.168}$$

With the so-called zero-order hold (ZOH):

$$\mathbf{u}(t) =: \mathbf{u}[kT] =: \mathbf{u}[k], \quad \mathbf{x}(t) =: \mathbf{x}[kT] =: \mathbf{x}[k], \quad \text{for } kT \le t < (k+1)T, \tag{3.169}$$

we have

$$\mathbf{x}[k+1] = e^{\mathbf{A}T}\mathbf{x}[k] + \left(\int_{kT}^{(k+1)T} e^{\mathbf{A}(k+1)T-\tau}d\tau\right)\mathbf{B}\mathbf{u}[k]. \tag{3.170}$$

Let $\eta := (k+1)T - \tau$, then

$$\int_{kT}^{(k+1)T} e^{\mathbf{A}(k+1)T-\tau}d\tau = \int_0^T e^{\mathbf{A}\eta}d\eta \tag{3.171}$$

and we have a discrete-time state-space system

$$\mathbf{x}[k+1] = e^{\mathbf{A}T}\mathbf{x}[k] + \left(\int_0^T e^{\mathbf{A}\eta}d\eta\right)\mathbf{B}\mathbf{u}[k],$$

$$\mathbf{y}[k] = \mathbf{C}\mathbf{x}[k] + \mathbf{D}\mathbf{u}[k]. \tag{3.172}$$

Moreover, write

$$\int_0^T e^{\mathbf{A}\eta}d\eta = \int_0^T \left(\mathbf{I} + \mathbf{A}\eta + \frac{\mathbf{A}^2\eta^2}{2!} + \cdots\right)d\eta$$

$$= \mathbf{I}T + \frac{\mathbf{A}T^2}{2!} + \frac{\mathbf{A}^2T^3}{3!} + \cdots. \tag{3.173}$$

If $\mathbf{A}$ is invertible, we have

$$\int_0^T e^{\mathbf{A}\eta}d\eta = \mathbf{A}^{-1}\left(e^{\mathbf{A}T} - \mathbf{I}\right). \tag{3.174}$$

Therefore, for an invertible $\mathbf{A}$, the discrete-time system in Equation (3.172) becomes

$$\mathbf{x}[k+1] = e^{\mathbf{A}T}\mathbf{x}[k] + \mathbf{A}^{-1}\left(e^{\mathbf{A}T} - \mathbf{I}\right)\mathbf{B}\mathbf{u}[k],$$

$$\mathbf{y}[k] = \mathbf{C}\mathbf{x}[k] + \mathbf{D}\mathbf{u}[k]. \tag{3.175}$$

EXAMPLE 3.18   Consider a continuous-time system

$$\dot{\mathbf{x}}(t) = \begin{bmatrix} 0 & 1 \\ -2 & -3 \end{bmatrix}\mathbf{x}(t) + \begin{bmatrix} 0 \\ 1 \end{bmatrix}\mathbf{u}(t),$$

$$\mathbf{y}(t) = [1 \quad 1]\mathbf{x}(t) + \mathbf{u}(t). \tag{3.176}$$

A discrete-time state-space system with ZOH and the sampling period $T = 0.01$ s can be conveniently calculated from the expression in Equation (3.175):

$$e^{AT} = \begin{bmatrix} 0.9999 & 0.0099 \\ -0.0197 & 0.9703 \end{bmatrix}, \quad A^{-1}\left(e^{AT} - I\right)B = \begin{bmatrix} 0 \\ 0.0099 \end{bmatrix}. \tag{3.177}$$

Thus, we obtain the following discrete-time state-space system:

$$\dot{x}(t) = \begin{bmatrix} 0.9999 & 0.0099 \\ -0.0197 & 0.9703 \end{bmatrix} x(t) + \begin{bmatrix} 0 \\ 0.0099 \end{bmatrix} u(t),$$
$$y(t) = \begin{bmatrix} 1 & 1 \end{bmatrix} x(t) + u(t). \tag{3.178}$$

MATLAB Solution

```
A=[0 1; -2 -3];
B=[0 1]';
C=[1 1];
D=1;
sysc=ss(A,B,C,D);
sysd=c2d(sysc,0.01,'zoh')
```

The output of this MATLAB script is the following discrete-time system:

$$A = \begin{bmatrix} 0.9999 & 0.0099 \\ -0.0197 & 0.9703 \end{bmatrix}, \quad B = \begin{bmatrix} 0 \\ 0.0099 \end{bmatrix}, \quad C = \begin{bmatrix} 1 & 1 \end{bmatrix}, \quad D = 1. \tag{3.179}$$

In addition to the ZOH there are several other discretization techniques also available in MATLAB.

EXAMPLE 3.19 A discrete-time system can also be simulated in Simulink. It is shown below. Let

$$x[k+1] = \begin{bmatrix} 0 & 1 \\ 0.25 & 0.5 \end{bmatrix} x[k] + \begin{bmatrix} 1 & 0 \\ 1 & 1 \end{bmatrix} u[k], \quad y[k] = I_2 x[k], \tag{3.180}$$

where

$$u[k] = \begin{bmatrix} u[k] \\ \sin(kT_s) \end{bmatrix} \quad \text{and the sampling time } T_s = 0.1 \text{ s}. \tag{3.181}$$

A Simulink model, the plant input data, and the response obtained by simulation are shown in Figures 3.26(a) and 3.27.

MATLAB Solution
The following MATLAB script also solves the problem and produces Figure 3.27.

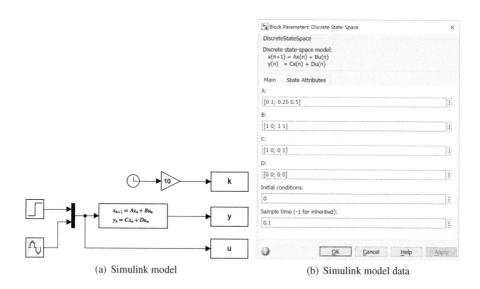

(a) Simulink model

(b) Simulink model data

Figure 3.26 Simulink simulation (Example 3.19)

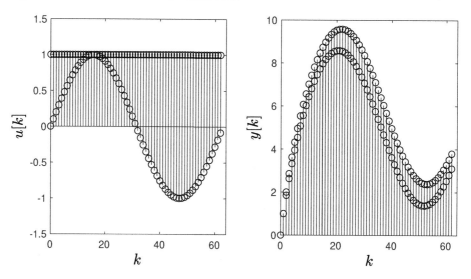

Figure 3.27 Simulink simulation data (Example 3.19)

```
A=[0 1; 0.25 0.5];
B=[1 0; 1 1];
C=eye(2);
D=zeros(2,2);
Ts=0.1;
T=2*pi;
sys=ss(A,B,C,D,Ts);
t=0:Ts:T;
u=[ones(size(t))'  sin(t)'];
[y,t]=lsim(sys,u,t);
k=t/Ts;
```

```
figure(1), stem(k,u), xlabel('k'), ylabel('u[k]')
axis([0 max(k) -1.5 1.5])
figure(2), stem(k,y), xlabel('k'), ylabel('y[k]')
axis([0 max(k) 0 10])
```

Conversion Between Continuous-Time Systems and Discrete-Time Systems via MATLAB

Consider the continuous-time system

$$\mathbf{A} = \begin{bmatrix} -1.5 & 1 \\ -1.25 & 0.7 \end{bmatrix}, \ \mathbf{B} = \mathbf{I}_2, \ \mathbf{C} = \mathbf{I}_2. \ \mathbf{D} = \mathbf{0}_2. \tag{3.182}$$

The following MATLAB script converts this system to an equivalent discrete-time system by assuming the sampling time of T s.

```
Ac=[-1.5 1; -1.25 0.7];
Bc=[1 0; 0 1];
Cc=[1 0; 0 1];
Dc=zeros(2,2);
sysc=ss(Ac,Bc,Cc,Dc);
T=0.01;
sysd=c2d(sysc,T);
```

The script above produces the following discrete-time system:

$$\mathbf{A} = \begin{bmatrix} 0.985 & 0.00996 \\ -0.01245 & 1.007 \end{bmatrix}, \ \mathbf{B} = \begin{bmatrix} 0.009925 & 4.987 \times 10^{-5} \\ -6.233 \times 10^{-5} & 0.01003 \end{bmatrix}, \tag{3.183}$$

$$\mathbf{C} = \mathbf{I}_2, \ \mathbf{D} = \mathbf{0}_2.$$

With the MATLAB command

```
d2c(sysd)   % converting discrete-time to continuous-time,
```

we recover the original continuous-time system in Equation (3.182).

3.10 Stability of State-Space Systems

The notion of stability refers to the ability of a system to "recover" from spurious or random inputs which may result in arbitrary changes in the states. For systems described by state variables which are physical quantities, the appropriate notion of stability is *internal stability*. A state-space model is said to be *internally stable* if the zero-input state response converges asymptotically to zero for every initial condition $\mathbf{x}(0)$. For a continuous-time system, the zero-input response is

$$\mathbf{x}_o(t) = e^{\mathbf{A}t}\mathbf{x}(0) \tag{3.184}$$

and therefore we must have, for internal stability

$$\lim_{t \to \infty} e^{\mathbf{A}t} = 0. \tag{3.185}$$

The condition in Equation (3.185) holds if and only if all eigenvalues of $\mathbf{A}$ have negative real parts. Write

$$\sigma(\mathbf{A}) = \{\lambda \mid \det(\lambda \mathbf{I} - \mathbf{A}) = 0\} \tag{3.186}$$

to denote the eigenvalue set or *spectrum* of $\mathbf{A}$. Then Equation (3.185) is equivalent to

$$\sigma(\mathbf{A}) \subset \mathbb{C}^-, \tag{3.187}$$

where $\mathbb{C}^-$ is the open left-half of the complex plane (LHP). This is evident from Equation (3.88), since the characteristic roots of $\mathbf{A}$ are just the eigenvalues of $\mathbf{A}$.

For a discrete-time system, the zero-input state response is

$$\mathbf{x}_o[k] = \mathbf{A}^k \mathbf{x}[0] \tag{3.188}$$

and internal stability requires that

$$\lim_{k \to \infty} \mathbf{A}^k = 0. \tag{3.189}$$

Clearly, Equation (3.189) is equivalent to the condition

$$\sigma(\mathbf{A}) \subset \mathbf{D}^1, \tag{3.190}$$

where $\mathbb{D}^1$ is the open unit disc in the complex plane. The restrictions in Equations (3.187) and (3.190) are called Hurwitz stability and Schur stability conditions, respectively, on $\mathbf{A}$.

EXAMPLE 3.20 (Internal vs. external stability)   Consider the following state-space system of a continuous-time system:

$$\mathbf{x}(t) = \begin{bmatrix} 0 & 1 \\ 1 & 0 \end{bmatrix} \mathbf{x}(t) + \begin{bmatrix} 0 \\ 1 \end{bmatrix} \mathbf{u}(t), \qquad y(t) = [-1 \quad 1]\mathbf{x}(t). \tag{3.191}$$

We say that this system is internally unstable because an eigenvalue of $\mathbf{A}$ is located in the right-half complex plane. On the other hand, the transfer function calculated from this state-space description seems to show that the system is stable.

$$\mathbf{G}(s) = \mathbf{C}(s\mathbf{I} - \mathbf{A})^{-1}\mathbf{B} + \mathbf{D} = \frac{1}{s+1}. \tag{3.192}$$

A system satisfying the stability condition $\sigma(\mathbf{A}) \subset \mathbf{C}^-$ is called *internally stable*. A system with all transfer function poles being located in the open left-half complex plane is called *externally stable*. Obviously internal stability implies external stability because transfer function poles are always a subset of the eigenvalues of $\mathbf{A}$, but the converse implication does not hold in general.

Internal Stability Check via MATLAB
The following two sets of MATLAB script will verify the internal stability of a given $\mathbf{A} \in \mathbb{R}^{n \times n}$:

```
%  for continuous-time systems
Ac=[-1.5 1 0; -1.25 0.7 0.4; -2.5 0.4 -6.2];
if  max(real(eig(Ac)))<0,
    disp('A is Hurwitz stable'),
else
    disp('A is Hurwitz unstable'),
end

%  for discrete-time systems
Ad=[-1.5 1 0; -1.25 0.7 0.4; -2.5 0.4 -6.2];
if  max(abs(eig(Ad)))<1,
    disp('Ad is Schur stable'),
else
    disp('Ad is Schur unstable'),
end
```

3.11   Exercises

3.1 Calculate (i) $(s\mathbf{I} - \mathbf{A})^{-1}$, (ii) $e^{\mathbf{A}t}$, and (iii) $\det[s\mathbf{I} - \mathbf{A}]$ for the following cases:

(a) $\mathbf{A} = \mathbf{0}_n$ ($n \times n$ zero matrix);

(b) $\mathbf{A} = \mathbf{I}_n$ ($n \times n$ identity matrix);

(c) $\mathbf{A} = \begin{bmatrix} \lambda_1 & & \\ & \ddots & \\ & & \lambda_n \end{bmatrix}$;

(d) $\mathbf{A} = \begin{bmatrix} \lambda_1 & 1 & 0 & \cdots & 0 \\ & \lambda_2 & 1 & & \vdots \\ & & \ddots & & 0 \\ & & & \ddots & 1 \\ & & & & \lambda_n \end{bmatrix}$;

(e) $\mathbf{A} = \begin{bmatrix} -\sigma & -\omega \\ \omega & -\sigma \end{bmatrix}$;

(f) $\mathbf{A} = \begin{bmatrix} 1 & 1 & 1 \\ 1 & -1 & 1 \\ -1 & 1 & 1 \end{bmatrix}$;

(g) $\mathbf{A} = \begin{bmatrix} 0 & 1 & 0 \\ 0 & 0 & 1 \\ 0 & -1 & 0 \end{bmatrix}.$

3.2   Give three different state-space realizations of the transfer functions:

(a) $g(s) = \dfrac{s-1}{s(s+1)(s+2)};$

(b) $g(s) = \dfrac{s-1}{s(s^2+1)}.$

3.3   Give a state-space realization of

$$\mathbf{G}(s) = \begin{bmatrix} \dfrac{1}{s+1} & \dfrac{s-1}{s(s+1)} \\ \dfrac{1}{s+2} & \dfrac{s}{s+1} \end{bmatrix}.$$ (3.193)

3.4   Show that if $[\mathbf{A}, \mathbf{b}, \mathbf{c}, \mathbf{d}]$ is a realization of $g(s)$, so is $[\mathbf{A}^T, \mathbf{c}^T, \mathbf{b}^T, \mathbf{d}]$.

3.5   Use the above result and the controllable companion realization in Equation (3.49a) to show that

$$\mathbf{A} = \begin{bmatrix} 0 & \cdots & \cdots & 0 & -a_0 \\ 1 & & & & -a_1 \\ & 1 & & & \vdots \\ & & \ddots & & \vdots \\ & & & 1 & -a_{n-1} \end{bmatrix}, \quad \mathbf{b} = \begin{bmatrix} c_0 \\ c_1 \\ \vdots \\ \\ c_{n-1} \end{bmatrix}, \quad \mathbf{c} = \begin{bmatrix} 0 & 0 & \cdots & \cdots & 1 \end{bmatrix}$$ (3.194)

is also a realization of

$$g(s) = \frac{c_0 + c_1 s + \cdots + c_{n-1} s^{n-1}}{s^n + a_{n-1} s^{n-1} + \cdots + a_1 s + a_0}.$$ (3.195)

3.6   Show that the impulse response of a state-space system $[\mathbf{A}, \mathbf{b}, \mathbf{c}, \mathbf{d}]$ is

$$\mathbf{c} e^{\mathbf{A}t} \mathbf{b} + \mathbf{d}\delta(t).$$ (3.196)

3.7   Show that the state-space system $[\mathbf{A}, \mathbf{b}, \mathbf{c}, \mathbf{d}]$ has a unit step response given by

$$\int_0^t \mathbf{c} e^{\mathbf{A}\tau} \mathbf{b}\, d\tau + \mathbf{d}u(t).$$ (3.197)

3.8   Find the state variable equations for the system in Figure 3.28 and determine the characteristic equation using the formula in Section 3.5.

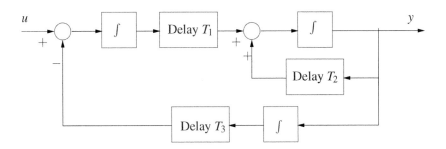

Figure 3.28 Exercise 3.8

3.9 Consider the multivariable servomechanism shown in Figure 3.29 with controllers

$$C_1(s) = K_p^1 + \frac{K_I^1}{s}, \qquad C_2(s) = K_p^2 + \frac{K_I^2}{s} \qquad (3.198)$$

and plant

$$\dot{\mathbf{x}}_p(t) = \mathbf{A}_p \mathbf{x}_p(t) + \mathbf{b}_1 u_1(t) + \mathbf{b}_2 u_2(t), \qquad (3.199)$$
$$y_1(t) = \mathbf{c}_1 \mathbf{x}_p(t), \qquad (3.200)$$
$$y_2(t) = \mathbf{c}_2 \mathbf{x}_p(t). \qquad (3.201)$$

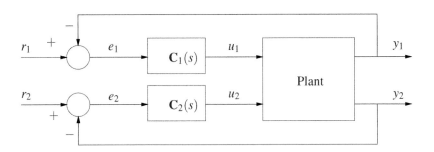

Figure 3.29 Exercise 3.9

Write the overall state-variable equations in the form

$$\dot{\mathbf{x}}(t) = \mathbf{A}\mathbf{x}(t) + \mathbf{B}\mathbf{r}(t), \qquad (3.202)$$
$$\mathbf{e}(t) = \mathbf{C}\mathbf{x}(t) + \mathbf{D}\mathbf{r}(t), \qquad (3.203)$$

where

$$\mathbf{e}(t) = \begin{bmatrix} e_1(t) \\ e_2(t) \end{bmatrix}, \quad \mathbf{r}(t) = \begin{bmatrix} r_1(t) \\ r_2(t) \end{bmatrix}. \qquad (3.204)$$

Choose your favorite plant model and the controller gains by trial and error to (a) internally stabilize (b) internally destabilize, the closed-loop system and verify in each case whether asymptotic tracking of step reference inputs does or does not occur by using Simulink.

3.10 Discretize the continuous-time system

$$\dot{\mathbf{x}}(t) = \begin{bmatrix} 0 & 1 \\ -1 & -1 \end{bmatrix} \mathbf{x}(t) + \begin{bmatrix} 0 \\ 1 \end{bmatrix} \mathbf{u}(t) \qquad (3.205)$$

with a zero-order hold and sampling period T to obtain the discretized system

$$\mathbf{z}[k+1] = \bar{\mathbf{A}}\mathbf{z}[k] + \bar{\mathbf{B}}\mathbf{u}[k]. \qquad (3.206)$$

Find $(\bar{\mathbf{A}}, \bar{\mathbf{B}})$ for $T = 0.1, 0.5$, and 1.

3.11 In the previous exercise, compare the unit step responses of the continuous and discrete-time systems for the three different sampling periods given, using MATLAB or Simulink.

3.12 Prove that when a continuous-time system

$$\dot{\mathbf{x}}(t) = \mathbf{A}\mathbf{x}(t) + \mathbf{B}\mathbf{u}(t) \qquad (3.207)$$

is discretized with a zero-order hold and sampling period T, the discrete-time system is Schur stable if and only if the continuous-time system is Hurwitz stable.

Hint: Prove that the eigenvalues $\bar{\lambda}_i$ of $e^{\mathbf{A}T}$ are $e^{\lambda_i T}$, where $\lambda_i \in \sigma(\mathbf{A})$.

3.13 Show that the unit impulse response of a discrete-time system is

$$\begin{bmatrix} y[0] \\ y[1] \\ y[2] \\ \vdots \\ y[k] \end{bmatrix} = \begin{bmatrix} \mathbf{d} \\ \mathbf{cb} \\ \mathbf{cAb} \\ \vdots \\ \mathbf{cA}^{k-1}\mathbf{b} \end{bmatrix}, \quad \text{for } k = 0, 1, 2, \dots.$$

Note that the sequence $\mathbf{b}, \mathbf{cb}, \mathbf{cAb}, \dots, \mathbf{cA}^{k-1}\mathbf{b}$ is called the *Markov parameters* of a system.

3.14 Consider the system in Figure 3.30

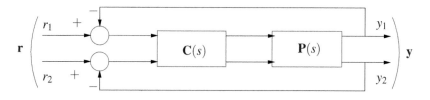

Figure 3.30 Exercise 3.14

where

$$C(s) = \begin{bmatrix} \dfrac{1}{s} & \dfrac{s+1}{s} \\ \dfrac{s+1}{s} & \dfrac{1}{s} \end{bmatrix} \quad \text{and} \quad P(s) = \begin{bmatrix} \dfrac{1}{s+1} & \dfrac{1}{s+2} \\ \dfrac{1}{s+1} & \dfrac{1}{s+1} \end{bmatrix}. \tag{3.208}$$

Find the state-space model

$$\dot{x}(t) = Ax(t) + Br(t), \tag{3.209}$$
$$y(t) = Cx(t) + Dr(t), \tag{3.210}$$

and determine its response to (a) unit steps and (b) unit ramps.

3.15 Show that a 2×2 matrix A is Hurwitz stable if and only if $\text{Trace}[A] < 0$ and $\det[A] > 0$.

3.16 Provide op-amp circuit realizations of the PI and PID controllers:

(a) $K_p + \dfrac{K_i}{s}$,

(b) $K_p + \dfrac{K_i}{s} + K_d s$,

(c) and the lead–lag controller $\dfrac{K(s+z)}{s+p}$.

3.17 Show an op-amp circuit realization of the state-space system:

$$\begin{aligned} \dot{x}_1(t) &= x_2(t), \\ \dot{x}_2(t) &= -a_0 x_1(t) - a_1 x_2(t) + u(t), \\ y(t) &= c_1 x_1(t) + c_2 x_2(t) + du(t). \end{aligned} \tag{3.211}$$

3.18 Given

$$S_1: \quad \dot{x}_1(t) = \begin{bmatrix} 0 & 1 & 0 \\ 0 & 0 & 1 \\ -1 & -1 & -1 \end{bmatrix} x_1(t) + \begin{bmatrix} 1 & 0 \\ 0 & 0 \\ 0 & 1 \end{bmatrix} u_1(t), \tag{3.212}$$

$$y_1(t) = \begin{bmatrix} 1 & 0 & 0 \\ 0 & 0 & 1 \end{bmatrix} x_1(t) + \begin{bmatrix} 0 & 1 \\ 0 & 0 \end{bmatrix} u_1(t),$$

$$S_2: \quad \dot{x}_2(t) = \begin{bmatrix} 0 & 0 & -1 \\ 1 & 0 & -1 \\ 0 & 1 & -1 \end{bmatrix} x_2(t) + \begin{bmatrix} 1 & 0 \\ 0 & 0 \\ 0 & 1 \end{bmatrix} u_2(t), \tag{3.213}$$

$$y_2(t) = \begin{bmatrix} 1 & 0 & 0 \\ 0 & 1 & 0 \end{bmatrix} x_2(t) + \begin{bmatrix} 0 & 0 \\ 0 & 0 \end{bmatrix} u_2(t).$$

(1) Find state-space representations of the systems in Figures 3.31 to 3.34.

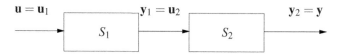

Figure 3.31 (a) Series connection

Figure 3.32 (b) Series connection

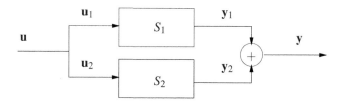

Figure 3.33 (c) Parallel connection

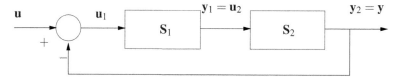

Figure 3.34 (d) Feedback connection

(2) In each case above, find

   (a) the eigenvalues of the **A** matrix of the interconnected system and

   (b) the overall transfer function matrix.

(3) Verify your calculations by using MATLAB.

3.19 Consider the feedback system in Figure 3.35, where the system S is

$$\dot{\mathbf{x}}(t) = \begin{bmatrix} 0 & 1 & 0 \\ 0 & 0 & 1 \\ 0 & -3 & -2 \end{bmatrix} \mathbf{x}(t) + \begin{bmatrix} 1 & 0 \\ 0 & 0 \\ 0 & 1 \end{bmatrix} \mathbf{u}(t),$$

$$\mathbf{y}(t) = \begin{bmatrix} 1 & 0 & 0 \\ 0 & 0 & 1 \end{bmatrix} \mathbf{x}(t) + \begin{bmatrix} 0 & 1 \\ 0 & 0 \end{bmatrix} \mathbf{u}(t),$$

(3.214)

with initial condition

$$\mathbf{x}(0) = \begin{bmatrix} 1 \\ 0 \\ -1 \end{bmatrix} \qquad (3.215)$$

and input

$$\mathbf{u}(t) = \begin{bmatrix} \delta(t) \\ \sin t \end{bmatrix}. \qquad (3.216)$$

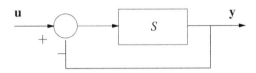

Figure 3.35 A feedback system (Exercise 3.19)

Determine the following for the closed-loop system:

(a) The zero-input state and output responses.

(b) The zero-state or forced response for state and output.

(c) The transfer function.

(d) Response to a unit step reference input.

(e) Verify your answer in (d) by comparing the plot of a mathematical expression obtained in (d) by using MATLAB and the output from a Simulink simulation.

3.20 A discrete-time system is shown below. Let

$$\mathbf{x}[k+1] = \begin{bmatrix} 0 & 1 \\ 0.25 & 1.5 \end{bmatrix} \mathbf{x}[k] + \begin{bmatrix} 1 & 0 \\ 1 & 1 \end{bmatrix} \mathbf{u}[k], \qquad (3.217a)$$

$$\mathbf{y}[k] = \mathbf{I}_2 \mathbf{x}[k], \qquad (3.217b)$$

where

$$\mathbf{u}[k] = \begin{bmatrix} u[k] \\ \sin(kT_s) \end{bmatrix} \qquad (3.218)$$

and the sampling time $T_s = 0.1$ s.

(a) Calculate the output response when the first input is a unit step and the second input is sinusoidal as shown.

(b) Verify your answer to (a) by plotting the solution using MATLAB, and comparing it with the output from a Simulink simulation.

3.12   Notes and References

The theory of state-space systems has an extensive literature. Excellent classic texts are those in Zadeh and Desoer (1963), Wolovich (1974), Wonham (1985), Callier and Desoer (1982), Kailath (1980), and Antsaklis and Michel (2005). Mason's rule is described in Mason (1953) and Mason (1956). The Faddeev–LeVerrier algorithm was reported in LeVerrier (1840) and Faddeev and Sominsky (1972). The stability of time-delay systems was investigated in Pontryagin (1955). The formula for the characteristic equation given in Section 3.5 was developed in Keel and Bhattacharyya (2017).

4 Modal Structure of State-Space Systems

The time-domain behavior of state-space systems can be explicitly characterized by its eigenvalues, eigenvectors, and more generally its modal structure. In this chapter we describe the modal coordinates and the interrelated topics of the Jordan form and its computation, the invariant polynomials and elementary divisors of the matrix $s\mathbf{I} - \mathbf{A}$, and the modal decomposition of the time-domain solutions for the state and output.

4.1 Similarity Transformation and the Jordan Form

The state variables describing a system are not unique and the state-space representation of a given system depends on the choice of state variables. By choosing different bases for the underlying n-dimensional state space we obtain different nth-order representations of the same system. Such systems are called *similar*. Similar systems, it turns out, have the same input–output transfer function, although their internal state-space representations are different. This fact can be exploited to develop equivalent state-space representations that display various structural properties or simplify computation. One useful form that can often be obtained in this way is a diagonal $\mathbf{A}$ matrix which simplifies the computation of solutions and allows for a decomposition or "visualization" of the motion of the state vector along eigenvectors. A generalization of this idea, when diagonalization is not possible, leads to a maximally diagonal form called the Jordan form. The derivation of the Jordan form is aided by the concepts of generalized eigenvector chains, invariant polynomials, Smith forms, and elementary divisors of polynomial matrices. These notions, discussed below, are also of independent interest and play an important role in the theory of minimal realizations described in Chapter 5.

4.1.1 Similarity Transformation

In the state-space representation

$$
\begin{aligned}
\dot{\mathbf{x}}(t) &= \mathbf{A}\mathbf{x}(t) + \mathbf{B}\mathbf{u}(t), \\
\mathbf{y}(t) &= \mathbf{C}\mathbf{x}(t) + \mathbf{D}\mathbf{u}(t),
\end{aligned} \tag{4.1}
$$

let $\mathbf{x}(t) = \mathbf{T}\mathbf{z}(t)$ where $\mathbf{T}$ is a square invertible real or complex matrix. Then

$$
\begin{aligned}
\mathbf{T}\dot{\mathbf{z}}(t) &= \mathbf{A}\mathbf{T}\mathbf{z}(t) + \mathbf{B}\mathbf{u}(t), \\
\mathbf{y}(t) &= \mathbf{C}\mathbf{T}\mathbf{z}(t) + \mathbf{D}\mathbf{u}(t),
\end{aligned} \tag{4.2}
$$

or

$$\dot{z}(t) = \underbrace{T^{-1}AT}_{A_n}z(t) + \underbrace{T^{-1}B}_{B_n}u(t),$$

$$y(t) = \underbrace{CT}_{C_n}z(t) + \underbrace{D}_{D_n}u(t).$$

(4.3)

The state-space model $\{A, B, C, D\}$ is the original representation with state x, whereas $\{A_n, B_n, C_n, D_n\}$ is the new representation with state z. The transfer function of the "new" system may be computed as:

$$
\begin{aligned}
C_n(sI - A_n)^{-1}B_n + D_n &= CT\left[T^{-1}(sI-A)T\right]^{-1}T^{-1}B + D \\
&= CT\left[T^{-1}(sI-A)^{-1}T\right]T^{-1}B + D \\
&= CTT^{-1}(sI-A)^{-1}TT^{-1}B + D \\
&= C(sI-A)^{-1}B + D,
\end{aligned}
$$

(4.4)

which shows that the representations in Equations (4.1) and (4.3) have the same transfer function.

EXAMPLE 4.1   Consider the following state-space representation $\{A, B, C, D\}$ where

$$A = \begin{bmatrix} 0 & 1 \\ -2 & -3 \end{bmatrix}, \quad B = \begin{bmatrix} 0 \\ 1 \end{bmatrix}, \quad C = [1 \ 3], \quad D = 1.$$

(4.5)

The transfer function of the system calculated from this state-space representation is

$$G(s) = C(sI-A)^{-1}B + D = \frac{s^2 + 4s + 1}{s^2 + 3s + 2}.$$

(4.6)

Now we select the non-singular matrix

$$T = \begin{bmatrix} 1 & 3 \\ 1 & 2 \end{bmatrix}$$

(4.7)

and calculate the "new" equivalent state-space system

$$A_n = T^{-1}AT = \begin{bmatrix} -17 & -40 \\ 6 & 14 \end{bmatrix}, \quad B_n = T^{-1}B = \begin{bmatrix} 3 \\ -1 \end{bmatrix},$$

$$C_n = CT = [0 \ -1], \quad D_n = D = 1.$$

(4.8)

The transfer function from this state-space representation is

$$G_n(s) = C_n(sI - A_n)^{-1}B_n + D_n = \frac{s^2 + 4s + 1}{s^2 + 3s + 2}.$$

(4.9)

4.1.2   Diagonalization: Distinct Eigenvalues

State-space calculations clearly become simpler when the matrix $\mathbf{A}$ is diagonal. In general, when the states represent physical variables such as currents and voltages in a circuit, the matrix $\mathbf{A}$ is *not* diagonal. However, by a coordinate transformation

$$\mathbf{x}(t) = \mathbf{T}\mathbf{z}(t), \tag{4.10}$$

with $\mathbf{T}$ nonsingular, the new $\mathbf{A}$,

$$\mathbf{A}_n = \mathbf{T}^{-1}\mathbf{A}\mathbf{T}, \tag{4.11}$$

may often be rendered diagonal. The new state variables $\mathbf{z}$ in general do not represent physical quantities and may not even have physical units, but they can facilitate the computation of $\mathbf{x}(t)$.

Distinct Eigenvalues
When $\mathbf{A}$ has distinct eigenvalues, that is, when

$$\det[s\mathbf{I} - \mathbf{A}] = (s - \lambda_1) \cdots (s - \lambda_n) \tag{4.12}$$

with

$$\lambda_i \neq \lambda_j, \quad \text{for } i \neq j, \tag{4.13}$$

then the n eigenvectors $\mathbf{v}_i$, satisfying,

$$\mathbf{A}\mathbf{v}_i = \lambda_i\mathbf{v}_i, \quad \text{for } i \in \mathbf{n}, \tag{4.14}$$

form a *linearly independent* set. Then choosing

$$\mathbf{T} = \begin{bmatrix} \mathbf{v}_1 & \mathbf{v}_2 & \cdots & \mathbf{v}_n \end{bmatrix} \tag{4.15}$$

we have $\det[\mathbf{T}] \neq 0$ and

$$\mathbf{A}_n = \begin{bmatrix} \lambda_1 & & & \\ & \lambda_2 & & \\ & & \ddots & \\ & & & \lambda_n \end{bmatrix}. \tag{4.16}$$

EXAMPLE 4.2 (Diagonalization)   Consider

$$\mathbf{A} = \begin{bmatrix} -5 & 3 & -2 \\ -6 & 4 & -4 \\ -4 & 4 & -5 \end{bmatrix} \tag{4.17}$$

with the eigenvalue set $\sigma(\mathbf{A}) = \{-1, -2, -3\}$. An eigenvector matrix $\mathbf{T}$ satisfying

$$\mathbf{A}\underbrace{\begin{bmatrix} \mathbf{v}_1 & \mathbf{v}_2 & \mathbf{v}_3 \end{bmatrix}}_{\mathbf{T}} = \underbrace{\begin{bmatrix} \mathbf{v}_1 & \mathbf{v}_2 & \mathbf{v}_3 \end{bmatrix}}_{\mathbf{T}} \begin{bmatrix} -1 & & \\ & -2 & \\ & & -3 \end{bmatrix} \tag{4.18}$$

is formed as

$$\mathbf{T} = \begin{bmatrix} 1 & 1 & 1 \\ 2 & 1 & 2 \\ 1 & 0 & 2 \end{bmatrix} \tag{4.19}$$

and

$$\mathbf{A}_n = \mathbf{T}^{-1}\mathbf{A}\mathbf{T} = \begin{bmatrix} -1 & & \\ & -2 & \\ & & -3 \end{bmatrix}. \tag{4.20}$$

MATLAB Solution

An eigenvector matrix $\mathbf{T}$ and the diagonalized matrix $\mathbf{A}_n$ are calculated from the following MATLAB script:

```
A=[-5 3 -2; -6 4 -4; -4 4 -5];
[T,An]=eig(A)
```

The following output is obtained from the script:

$$\mathbf{T} = \begin{bmatrix} 0.4082 & 0.7071 & 0.3333 \\ 0.8165 & 0.7071 & 0.6667 \\ 0.4082 & 0.0000 & 0.6667 \end{bmatrix}, \quad \mathbf{A}_n = \begin{bmatrix} -1.0000 & 0 & 0 \\ 0 & -2.0000 & 0 \\ 0 & 0 & -3.0000 \end{bmatrix}.$$

When $\mathbf{A}$ has complex eigenvalues, the Jordan form $\mathbf{J}$ and the transformation $\mathbf{T}$ are complex. It is sometimes useful to have an alternative approach avoiding complex numbers.

Suppose that $\mathbf{A}$ is a 2×2 real matrix with complex eigenvalues $\sigma \pm j\omega$ and corresponding eigenvectors $\mathbf{u} \pm j\mathbf{v}$ with $\mathbf{u}, \mathbf{v}$ being real vectors. Then

$$\mathbf{A} \underbrace{[\mathbf{u} \ \mathbf{v}]}_{\mathbf{T}} = \underbrace{[\mathbf{u} \ \mathbf{v}]}_{\mathbf{T}} \underbrace{\begin{bmatrix} \sigma & \omega \\ -\omega & \sigma \end{bmatrix}}_{\mathbf{A}_n}. \tag{4.21}$$

It is easily shown that $\mathbf{T}$ is nonsingular over the real field. Thus Equation (4.21) can be used to construct a real quasi-diagonal form with 1×1 and 2×2 blocks as the next example shows.

EXAMPLE 4.3 (Quasi-diagonalization of a matrix with complex eigenvalues)   Consider the matrix

$$\mathbf{A} = \begin{bmatrix} 0 & 1 & 0 & 0 \\ 0 & 0 & 1 & 0 \\ 0 & 0 & 0 & 1 \\ -10 & -14 & -11 & -4 \end{bmatrix} \tag{4.22}$$

with eigenvalue set $\sigma(\mathbf{A}) = \{-1 \pm j2, -1 \pm j\}$. A complex eigenvector matrix $\mathbf{T}$ satisfying

$$\mathbf{AT} = \mathbf{TA}_n \qquad (4.23)$$

is

$$\mathbf{T} = \begin{bmatrix} 0.0788 + j0.0143 & 0.0788 - j0.0143 & 0.1826 - j0.1826 & 0.1826 + j0.1826 \\ -0.1074 + j0.1432 & -0.1074 - j0.1432 & 0.0000 + j0.3651 & 0.0000 - j0.3651 \\ -0.1790 - j0.3581 & -0.1790 + j0.3581 & -0.3651 - j0.3651 & -0.3651 + j0.3651 \\ 0.8951 & 0.8951 & 0.7303 & 0.7303 \end{bmatrix}$$

and the diagonalized complex matrix is

$$\mathbf{A}_n = \begin{bmatrix} -1 + j2 & & & \\ & -1 - j2 & & \\ & & -1 + j & \\ & & & -1 - j \end{bmatrix}. \qquad (4.24)$$

The diagonal complex matrix could be replaced by a block-diagonal matrix with real entries by modifying the transformation matrix $\mathbf{T}$. From the real and imaginary parts of the columns in the complex eigenvector matrix $\mathbf{T}$, we construct a real transformation matrix

$$\mathbf{V} = \begin{bmatrix} 0.0788 & 0.0143 & 0.1826 & -0.1826 \\ -0.1074 & 0.1432 & 0.0000 & 0.3651 \\ -0.1790 & -0.3581 & -0.3651 & -0.3651 \\ 0.8951 & 0 & 0.7303 & 0 \end{bmatrix}. \qquad (4.25)$$

With this transformation matrix $\mathbf{V}$, we obtain the block-diagonal matrix with real entries

$$\mathbf{A}_n = \mathbf{V}^{-1}\mathbf{AV} = \begin{bmatrix} -1 & 2 & 0 & 0 \\ -2 & -1 & 0 & 0 \\ 0 & 0 & -1 & 1 \\ 0 & 0 & -1 & -1 \end{bmatrix}. \qquad (4.26)$$

MATLAB Solution

```
A=[0 1 0 0; 0 0 1 0; 0 0 0 1; -10 -14 -11 -4];
[T,D]=eig(A)        % complex matrices T and D
[V,An]=cdf2rdf(T,D)   % real matrices V and An
```

4.1.3   Jordan Form: Repeated Eigenvalues

We have seen in the previous section that an $n \times n$ matrix $\mathbf{A}$ can be converted through a similarity transformation into a diagonal matrix when the eigenvalues of $\mathbf{A}$ are distinct. When the eigenvalues of $\mathbf{A}$ are not distinct, that is, some are repeated, it is not always possible to "diagonalize" $\mathbf{A}$, as before. In this subsection, we deal with this case, and develop the so-called *Jordan form* of $\mathbf{A}$, which may be regarded as the "maximally" diagonal form that can be obtained.

We begin with a fundamental result known as the Cayley–Hamilton Theorem.

THEOREM 4.1 (Cayley–Hamilton Theorem)   Let $a(s) = \det[s\mathbf{I} - \mathbf{A}]$ denote the characteristic polynomial of $\mathbf{A}$. Then

$$a(\mathbf{A}) = 0. \tag{4.27}$$

Proof   The proof of the Cayley–Hamilton Theorem is given in Appendix A.4.   ∎

Now let the characteristic polynomial

$$a(s) = \det[s\mathbf{I} - \mathbf{A}] \tag{4.28}$$

be factored into monic coprime factors $a_i(s)$:

$$a(s) = a_1(s)a_2(s) \cdots a_t(s) \tag{4.29}$$

and introduce the subspaces $\mathscr{V}_i$, for $i = 1, 2, \ldots, t$:

$$\mathscr{V}_i := \mathrm{Ker}\,[a_i(\mathbf{A})] := \{\mathbf{x} \,:\, a_i(\mathbf{A})\mathbf{x} = 0\}. \tag{4.30}$$

LEMMA 4.1   The subspaces $\mathscr{V}_i$ are $\mathbf{A}$-invariant, disjoint, and span $\mathscr{X}$:

$$\mathscr{X} = \mathscr{V}_1 \oplus \mathscr{V}_2 \oplus \cdots \oplus \mathscr{V}_t. \tag{4.31}$$

Proof   If $\mathbf{x} \in \mathscr{V}_i$, then

$$a_i(\mathbf{A})\mathbf{A}\mathbf{x} = \mathbf{A}a_i(\mathbf{A})\mathbf{x} = 0 \tag{4.32}$$

and hence $\mathbf{A}\mathbf{x} \in \mathscr{V}_i$. Thus $\mathscr{V}_i$ is $\mathbf{A}$-invariant for $i = 1, 2, \ldots, t$.

To proceed, consider first the case $t = 2$. Since $a_1(s)$, $a_2(s)$ are coprime, it follows from a fundamental result of linear algebra, that there exist real polynomials $\beta_i(s), i = 1, 2$ such that

$$1 = \beta_1(s)a_1(s) + \beta_2(s)a_2(s). \tag{4.33}$$

Therefore,

$$\mathbf{I} = \beta_1(\mathbf{A})a_1(\mathbf{A}) + \beta_2(\mathbf{A})a_2(\mathbf{A}). \tag{4.34}$$

Thus, for arbitrary $\mathbf{x} \in \mathscr{X}$:

$$\mathbf{x} = \underbrace{\beta_1(\mathbf{A})a_1(\mathbf{A})\mathbf{x}}_{\mathbf{x}_2} + \underbrace{\beta_2(\mathbf{A})a_2(\mathbf{A})\mathbf{x}}_{\mathbf{x}_1}, \tag{4.35}$$

where

$$\mathbf{x}_2 \in \mathscr{V}_2 \quad \text{and} \quad \mathbf{x}_1 \in \mathscr{V}_1 \tag{4.36}$$

since by the Cayley–Hamilton Theorem,

$$a_2(\mathbf{A})\mathbf{x}_2 = 0 \quad \text{and} \quad a_1(\mathbf{A})\mathbf{x}_1 = 0. \tag{4.37}$$

Therefore,

$$\mathscr{X} = \mathscr{V}_1 + \mathscr{V}_2. \tag{4.38}$$

To prove that the right-hand side of Equation (4.38) is a direct sum, suppose that

$$\mathbf{x} \in \mathscr{V}_1 \cap \mathscr{V}_2. \tag{4.39}$$

Then

$$a_1(\mathbf{A})\mathbf{x} = 0 \quad \text{and} \quad a_2(\mathbf{A})\mathbf{x} = 0. \tag{4.40}$$

It follows from Equation (4.35) that $\mathbf{x} = 0$. This proves that

$$\mathscr{X} = \mathscr{V}_1 \oplus \mathscr{V}_2. \tag{4.41}$$

A similar proof can be constructed for $t > 2$ by repeatedly decomposing $\mathscr{V}_1$ and $\mathscr{V}_2$. $\blacksquare$

Let $\mathbf{T}_i$ be a matrix with columns spanning $\mathscr{V}_i$ for $i \in \mathbf{t}$. Then

$$\mathbf{T} := [\mathbf{T}_1 \ \mathbf{T}_2 \ \cdots \ \mathbf{T}_t] \tag{4.42}$$

is a coordinate transformation, since $\text{rank}[\mathbf{T}] = n$.

THEOREM 4.2

$$\mathbf{A}_J := \mathbf{T}^{-1}\mathbf{A}\mathbf{T} = \begin{bmatrix} \mathbf{A}_1 & & & \\ & \mathbf{A}_2 & & \\ & & \ddots & \\ & & & \mathbf{A}_t \end{bmatrix} \tag{4.43}$$

and

$$\det[s\mathbf{I} - \mathbf{A}_i] = a_i(s), \quad \text{for } i \in \mathbf{t}. \tag{4.44}$$

Proof The block-diagonal matrix in Equation (4.43) results from the $\mathbf{A}$-invariance of the $\mathscr{V}_i$ for $i \in \mathbf{t} := \{1, 2, \ldots, t\}$, so that

$$\mathbf{A}\mathbf{T}_i = \mathbf{T}_i\mathbf{A}_i, \quad \text{for } i \in \mathbf{t}, \tag{4.45}$$

proving Equation (4.43). From Equation (4.45) it follows that

$$a_i(\mathbf{A})\mathbf{T}_i = \mathbf{T}_i a_i(\mathbf{A}_i) \quad \text{for } i \in \mathbf{t}. \tag{4.46}$$

By definition of the $\mathbf{T}_i$:

$$a_i(\mathbf{A})\mathbf{T}_i = 0, \quad \text{for } i \in \mathbf{t} \tag{4.47}$$

and since $\mathbf{T}_i$ has full column rank, it follows that

$$a_i(\mathbf{A}_i) = 0, \quad \text{for } i \in \mathbf{t}. \tag{4.48}$$

Thus, Equation (4.44) is true. $\blacksquare$

In the rest of this section, we develop the Jordan decomposition of the matrices $\mathbf{A}_i$. Therefore, we can now consider, without loss of generality, an $n \times n$ matrix $\mathbf{A}$ with characteristic polynomial:

$$a(s) = \det(s\mathbf{I} - \mathbf{A}) = (s - \lambda)^n. \tag{4.49}$$

The Jordan structure of such an $\mathbf{A}$ matrix may be found by forming the matrices

$$(\mathbf{A} - \lambda\mathbf{I})^k, \quad \text{for } k = 0, 1, 2, \ldots, n. \tag{4.50}$$

Let $\mathcal{N}_k$ denote the null space of $(\mathbf{A} - \lambda\mathbf{I})^k$:

$$\mathcal{N}_k := \left\{ \mathbf{x} : (\mathbf{A} - \lambda\mathbf{I})^k\mathbf{x} = 0 \right\} \tag{4.51}$$

and denote the dimension of $\mathcal{N}_k$ by v_k:

$$\dim[\mathcal{N}_k] =: v_k, \quad \text{for } k = 0, 1, 2, \ldots, n+1. \tag{4.52}$$

We remark that

$$0 = v_0 \le v_1 \le \cdots \le v_{n-1} \le v_n = n = v_{n+1}. \tag{4.53}$$

The following theorem then gives the Jordan structure of $\mathbf{A}$.

THEOREM 4.3   Given the $n \times n$ matrix $\mathbf{A}$ with

$$\det(s\mathbf{I} - \mathbf{A}) = (s - \lambda)^n, \tag{4.54}$$

there exists an $n \times n$ matrix $\mathbf{T}$ with $\det(\mathbf{T}) \ne 0$, such that

$$\mathbf{A}_J = \mathbf{T}^{-1}\mathbf{A}\mathbf{T} = \begin{bmatrix} \mathbf{A}_1 & & & \\ & \mathbf{A}_2 & & \\ & & \ddots & \\ & & & \mathbf{A}_r \end{bmatrix}, \tag{4.55}$$

where

$$\mathbf{A}_j = \begin{bmatrix} \lambda & 1 & 0 & \cdots & 0 \\ 0 & \lambda & 1 & \cdots & 0 \\ \vdots & & \ddots & & \\ 0 & & & \lambda & 1 \\ 0 & 0 & \cdots & 0 & \lambda \end{bmatrix}, \quad \text{for } j = 1, 2, \ldots, r \tag{4.56}$$

is called a Jordan block of size $n_j \times n_j$.

A constructive proof of the above theorem will be given below. For those familiar with the notion of elementary divisors, to be developed below, it follows from the fact that $\mathbf{A}$ and $\mathbf{T}^{-1}\mathbf{A}\mathbf{T}$ have the same elementary divisors $(s - \lambda)^{n_j}, j = 1, 2, \ldots, r$

The number r and sizes n_j of the Jordan blocks $\mathbf{A}_i$ can be found from the formula given in the following lemma.

LEMMA 4.2   Under the assumptions of the previous theorem, the number of Jordan blocks of size $k \times k$ is given by

$$2v_k - v_{k-1} - v_{k+1}, \quad \text{for } k = 1, 2, \ldots, n. \tag{4.57}$$

Proof The number of blocks of size 1×1 or larger is clearly $v_1 - v_0$, of size 2×2 or larger is $v_2 - v_1$, and of size $k \times k$ or larger is $v_k - v_{k-1}$. Therefore, the exact number of blocks of size $k \times k$ is

$$v_k - v_{k-1} - (v_{k+1} - v_k) = 2v_k - v_{k-1} - v_{k+1}. \tag{4.58}$$

∎

EXAMPLE 4.4 Let **A** be an 11×11 matrix with

$$\det(s\mathbf{I} - \mathbf{A}) = (s - \lambda)^{11}.$$

Suppose that

$$v_1 = 6, \ v_2 = 9, \ v_3 = 10, \ v_4 = 11 = v_5 = v_6 = \cdots = v_{11} = v_{12}.$$

Then the number of Jordan blocks:

(a) of size $1 \times 1 = 2v_1 - v_0 - v_2 = 3$
(b) of size $2 \times 2 = 2v_2 - v_1 - v_3 = 2$
(c) of size $3 \times 3 = 2v_3 - v_2 - v_4 = 0$
(d) of size $4 \times 4 = 2v_4 - v_3 - v_5 = 1$
(e) of size $5 \times 5 = 2v_5 - v_4 - v_6 = 0$

$\vdots$

(k) of size $11 \times 11 = 2v_{11} - v_{12} - v_{10} = 0.$

Therefore, the Jordan form of **A** is as shown below:

$$\mathbf{A}_J = \begin{bmatrix} \lambda & 1 & 0 & 0 & & & & & & & \\ 0 & \lambda & 1 & 0 & & & & & & & \\ 0 & 0 & \lambda & 1 & & & & & & & \\ 0 & 0 & 0 & \lambda & & & & & & & \\ & & & & \lambda & 1 & & & & & \\ & & & & 0 & \lambda & & & & & \\ & & & & & & \lambda & 1 & & & \\ & & & & & & 0 & \lambda & & & \\ & & & & & & & & \lambda & & \\ & & & & & & & & & \lambda & \\ & & & & & & & & & & \lambda \end{bmatrix}.$$

This result can now be applied to each of the blocks $\mathbf{A}_i$ in the previous theorem to obtain the complete Jordan decomposition of **A** in the general case. It is best to illustrate the procedure with an example.

EXAMPLE 4.5   Let **A** be a 15×15 matrix with

$$\det(s\mathbf{I} - \mathbf{A}) = (s - \lambda_1)^8 (s - \lambda_2)^4 (s - \lambda_3)^3,$$

with λ_i being distinct. We compute

$$(\mathbf{A} - \lambda_i \mathbf{I})^k, \qquad \text{for } k = 0,1,2,\ldots,8,9$$

and set

$$h_k = \dim \mathcal{N} (\mathbf{A} - \lambda_1 \mathbf{I})^k, \qquad \text{for } k = 0,1,2,\ldots,9.$$

Similarly, let

$$i_j = \dim \mathcal{N} (\mathbf{A} - \lambda_2 \mathbf{I})^j, \qquad \text{for } j = 0,1,2,\ldots,5$$

and

$$l_s = \dim \mathcal{N} (\mathbf{A} - \lambda_3 \mathbf{I})^s, \qquad \text{for } s = 0,1,2,3,4.$$

Suppose, for example:

$$h_0 = 0, \ h_1 = 2, \ h_2 = 4, \ h_3 = 6, \ h_4 = 7, \ h_5 = h_6 = h_7 = h_8 = h_9 = 8,$$
$$i_0 = 0, \ i_1 = 3, \ i_2 = 4, \ i_3 = i_4 = i_5 = 4,$$
$$l_0 = 0, \ l_1 = 1, \ l_2 = 2, \ l_3 = l_4 = 3.$$

The Jordan form of **A** has the following "high-level" structure:

$$\mathbf{T}^{-1}\mathbf{A}\mathbf{T} = \begin{bmatrix} \mathbf{A}_1 & 0 & 0 \\ 0 & \mathbf{A}_2 & 0 \\ 0 & 0 & \mathbf{A}_3 \end{bmatrix},$$

with $\mathbf{A}_1 \in \mathbb{R}^{8\times8}$, $\mathbf{A}_2 \in \mathbb{R}^{4\times4}$, and $\mathbf{A}_3 \in \mathbb{R}^{3\times3}$. Furthermore, the detailed structure of $\mathbf{A}_1$, by Theorem 4.3, consists of the following numbers of Jordan blocks:

(a) of size $1 \times 1 = 2h_1 - h_0 - h_2 = 0$ blocks
(b) of size $2 \times 2 = 2h_2 - h_1 - h_3 = 0$ blocks
(c) of size $3 \times 3 = 2h_3 - h_2 - h_4 = 1$ block
(d) of size $4 \times 4 = 2h_4 - h_3 - h_5 = 0$ blocks
(e) of size $5 \times 5 = 2h_5 - h_4 - h_6 = 1$ block
(f) of size $6 \times 6 = 2h_6 - h_5 - h_7 = 0$ blocks
(g) of size $7 \times 7 = 2h_7 - h_6 - h_8 = 0$ blocks
(h) of size $8 \times 8 = 2h_8 - h_7 - h_9 = 0$ blocks.

Similarly, the numbers of Jordan blocks in $\mathbf{A}_2$ are:

(a) of size $1 \times 1 = 2i_1 - i_0 - i_2 = 2$ blocks
(b) of size $2 \times 2 = 2i_2 - i_1 - i_3 = 1$ block
(c) of size $3 \times 3 = 2i_3 - i_2 - i_4 = 0$ blocks
(d) of size $4 \times 4 = 2i_4 - i_3 - i_5 = 0$ blocks

and the numbers of Jordan blocks in $\mathbf{A}_3$ are:

(a) of size $1 \times 1 = 2l_1 - l_0 - l_2 = 0$ blocks
(b) of size $2 \times 2 = 2l_2 - l_1 - l_3 = 0$ blocks
(c) of size $3 \times 3 = 2l_3 - l_2 - l_4 = 1$ block.

With this information, we know that the Jordan form of $\mathbf{A}$ is given by

$$\mathbf{T}^{-1}\mathbf{A}\mathbf{T} = \begin{bmatrix} \lambda_1 & 1 & 0 & 0 & 0 & & & & & & & & & & \\ 0 & \lambda_1 & 1 & 0 & 0 & & & & & & & & & & \\ 0 & 0 & \lambda_1 & 1 & 0 & & & & & & & & & & \\ 0 & 0 & 0 & \lambda_1 & 1 & & & & & & & & & & \\ 0 & 0 & 0 & 0 & \lambda_1 & & & & & & & & & & \\ & & & & & \lambda_1 & 1 & 0 & & & & & & & \\ & & & & & 0 & \lambda_1 & 1 & & & & & & & \\ & & & & & 0 & 0 & \lambda_1 & & & & & & & \\ & & & & & & & & \lambda_2 & 1 & & & & & \\ & & & & & & & & 0 & \lambda_2 & & & & & \\ & & & & & & & & & & \lambda_2 & & & & \\ & & & & & & & & & & & \lambda_2 & & & \\ & & & & & & & & & & & & \lambda_3 & 1 & 0 \\ & & & & & & & & & & & & 0 & \lambda_3 & 1 \\ & & & & & & & & & & & & 0 & 0 & \lambda_3 \end{bmatrix}.$$

EXAMPLE 4.6 (Jordan form via MATLAB)   The following MATLAB script calculates the Jordan form of $\mathbf{A}$:

```
A=[5 -1 0 0; 9 -1 0 0; 0 0 7 -2; 0 0 12 -3];
jordan(A)
```

The output of the above MATLAB script is:

$$\mathbf{J} = \begin{bmatrix} 2 & 1 & 0 & 0 \\ 0 & 2 & 0 & 0 \\ 0 & 0 & 1 & 0 \\ 0 & 0 & 0 & 3 \end{bmatrix}. \tag{4.59}$$

4.1.4   Finding the Transformation Matrix $\mathbf{T}$: Generalized Eigenvector Chains

Once the Jordan form of $\mathbf{A}$ is known, it is relatively easy to find the transformation matrix. We illustrate using Example 4.5. Write $\mathbf{T}$ in terms of its columns:

$$\mathbf{T} = [\ \underbrace{\mathbf{t}_1\ \mathbf{t}_2\ \mathbf{t}_3\ \mathbf{t}_4\ \mathbf{t}_5}_{\text{block 1}}\ \underbrace{\mathbf{t}_6\ \mathbf{t}_7\ \mathbf{t}_8}_{\text{block 2}}\ \underbrace{\mathbf{t}_9\ \mathbf{t}_{10}}_{\text{block 3}}\ \underbrace{\mathbf{t}_{11}}_{\text{block 4}}\ \underbrace{\mathbf{t}_{12}}_{\text{block 5}}\ \underbrace{\mathbf{t}_{13}\ \mathbf{t}_{14}\ \mathbf{t}_{15}}_{\text{block 6}}\ ].$$

The set of vectors in each block above constitutes a *generalized eigenvector chain* and the transformation above is made up of six such chains. The first two sets of chains must satisfy

$$
\begin{aligned}
\mathbf{At}_1 &= \lambda_1 \mathbf{t}_1 &\quad \text{or} \quad& (\mathbf{A} - \lambda_1 \mathbf{I})\mathbf{t}_1 = 0, \\
\mathbf{At}_2 &= \lambda_1 \mathbf{t}_2 + \mathbf{t}_1 &\quad \text{or} \quad& (\mathbf{A} - \lambda_1 \mathbf{I})\mathbf{t}_2 = \mathbf{t}_1, \\
\mathbf{At}_3 &= \lambda_1 \mathbf{t}_3 + \mathbf{t}_2 &\quad \text{or} \quad& (\mathbf{A} - \lambda_1 \mathbf{I})\mathbf{t}_3 = \mathbf{t}_2, \\
\mathbf{At}_4 &= \lambda_1 \mathbf{t}_4 + \mathbf{t}_3 &\quad \text{or} \quad& (\mathbf{A} - \lambda_1 \mathbf{I})\mathbf{t}_4 = \mathbf{t}_3, \\
\mathbf{At}_5 &= \lambda_1 \mathbf{t}_5 + \mathbf{t}_4 &\quad \text{or} \quad& (\mathbf{A} - \lambda_1 \mathbf{I})\mathbf{t}_5 = \mathbf{t}_4,
\end{aligned}
\tag{4.60}
$$

as well as

$$
\begin{aligned}
\mathbf{At}_6 &= \lambda_1 \mathbf{t}_6 &\quad \text{or} \quad& (\mathbf{A} - \lambda_1 \mathbf{I})\mathbf{t}_6 = 0, \\
\mathbf{At}_7 &= \lambda_1 \mathbf{t}_7 + \mathbf{t}_6 &\quad \text{or} \quad& (\mathbf{A} - \lambda_1 \mathbf{I})\mathbf{t}_7 = \mathbf{t}_6, \\
\mathbf{At}_8 &= \lambda_1 \mathbf{t}_8 + \mathbf{t}_7 &\quad \text{or} \quad& (\mathbf{A} - \lambda_1 \mathbf{I})\mathbf{t}_8 = \mathbf{t}_7.
\end{aligned}
\tag{4.61}
$$

Therefore, we need to find two linearly independent vectors $\mathbf{t}_1$, $\mathbf{t}_6$ in $\mathcal{N}(\mathbf{A} - \lambda_1 \mathbf{I})$ such that the chains of vectors

$$
\mathbf{t}_1, \quad \mathbf{t}_2, \quad \mathbf{t}_3, \quad \mathbf{t}_4, \quad \mathbf{t}_5 \quad \text{and} \quad \mathbf{t}_6, \quad \mathbf{t}_7, \quad \mathbf{t}_8
\tag{4.62}
$$

are linearly independent. Alternatively, we can attempt to find $\mathbf{t}_5$ such that

$$
(\mathbf{A} - \lambda_i \mathbf{I})^5 \mathbf{t}_5 = 0
\tag{4.63}
$$

and find $\mathbf{t}_4, \mathbf{t}_3, \mathbf{t}_2, \mathbf{t}_1$ from the set in Equation (4.60). Similarly, we can find $\mathbf{t}_8$ such that

$$
(\mathbf{A} - \lambda_i \mathbf{I})^3 \mathbf{t}_8 = 0
\tag{4.64}
$$

and $\mathbf{t}_6, \mathbf{t}_7, \mathbf{t}_8$ found from Equation (4.61) are linearly independent.

Moving to the blocks associated with λ_2, we see that we need to find $\mathbf{t}_9, \mathbf{t}_{10}, \mathbf{t}_{11}, \mathbf{t}_{12}$ so that

$$
(\mathbf{A} - \lambda_2 \mathbf{I})\mathbf{t}_9 = 0, \qquad (\mathbf{A} - \lambda_2 \mathbf{I})\mathbf{t}_{11} = 0, \qquad (\mathbf{A} - \lambda_2 \mathbf{I})\mathbf{t}_{12} = 0
\tag{4.65}
$$

and $(\mathbf{t}_9, \quad \mathbf{t}_{10}, \quad \mathbf{t}_{11}, \quad \mathbf{t}_{12})$ are independent with

$$
\mathbf{At}_{10} = \lambda_2 \mathbf{t}_{10} + \mathbf{t}_9.
\tag{4.66}
$$

Likewise, we need to find $\mathbf{t}_{13}$ such that

$$
(\mathbf{A} - \lambda_3 \mathbf{I})\mathbf{t}_{13} = 0, \qquad (\mathbf{A} - \lambda_3 \mathbf{I})\mathbf{t}_{14} = \mathbf{t}_{13}, \qquad (\mathbf{A} - \lambda_3 \mathbf{I})\mathbf{t}_{15} = \mathbf{t}_{14}
\tag{4.67}
$$

with $(\mathbf{t}_{13}, \quad \mathbf{t}_{14}, \quad \mathbf{t}_{15})$ linearly independent. In each of the above cases, the existence of the vectors $\mathbf{t}_{ij}$ is guaranteed by the existence of the appropriate-sized Jordan blocks.

EXAMPLE 4.7   Consider the matrix with five repeated eigenvalues at $\lambda = 1$:

$$
\mathbf{A} = \begin{bmatrix}
0 & 1 & 0 & 0 & 0 \\
0 & 0 & 1 & 0 & 0 \\
0 & 0 & 0 & 1 & 0 \\
0 & 0 & 0 & 0 & 1 \\
1 & -5 & 10 & -10 & 5
\end{bmatrix}.
\tag{4.68}
$$

The dimensions of $\mathcal{N}_k$ in Equation (4.51) are

$$v_1 = 1, \; v_2 = 2, \; v_3 = 3, \; v_4 = 4, \; v_5 = 5 = v_6. \tag{4.69}$$

Then the number of Jordan blocks is

(a) of size $1 \times 1 = 2v_1 - v_0 - v_2 = 2 - 0 - 2 = 0$
(b) of size $2 \times 2 = 2v_2 - v_1 - v_3 = 4 - 1 - 3 = 0$
(c) of size $3 \times 3 = 2v_3 - v_2 - v_4 = 6 - 2 - 4 = 0$
(d) of size $4 \times 4 = 2v_4 - v_3 - v_5 = 8 - 3 - 5 = 0$
(e) of size $5 \times 5 = 2v_5 - v_4 - v_6 = 10 - 4 - 5 = 1.$

The transformation

$$\mathbf{T} = [\mathbf{t}_1 \; \mathbf{t}_2 \; \mathbf{t}_3 \; \mathbf{t}_4 \; \mathbf{t}_5] \tag{4.70}$$

is obtained from

$$\begin{aligned}(\mathbf{A} - \mathbf{I})\mathbf{t}_1 &= 0, \\ (\mathbf{A} - \mathbf{I})\mathbf{t}_2 &= \mathbf{t}_1, \\ (\mathbf{A} - \mathbf{I})\mathbf{t}_3 &= \mathbf{t}_2, \\ (\mathbf{A} - \mathbf{I})\mathbf{t}_4 &= \mathbf{t}_3, \\ (\mathbf{A} - \mathbf{I})\mathbf{t}_5 &= \mathbf{t}_4.\end{aligned} \tag{4.71}$$

Therefore, we have the transformation matrix $\mathbf{T}$ and the Jordan form $\mathbf{J}$ with one 5×5 block as shown below:

$$\mathbf{T} = \begin{bmatrix} 1 & -1 & 1 & -1 & 1 \\ 1 & 0 & 0 & 0 & 0 \\ 1 & 1 & 0 & 0 & 0 \\ 1 & 2 & 1 & 0 & 0 \\ 1 & 3 & 3 & 1 & 0 \end{bmatrix} \quad \text{and} \quad \mathbf{J} = \begin{bmatrix} 1 & 1 & 0 & 0 & 0 \\ 0 & 1 & 1 & 0 & 0 \\ 0 & 0 & 1 & 1 & 0 \\ 0 & 0 & 0 & 1 & 1 \\ 0 & 0 & 0 & 0 & 1 \end{bmatrix}. \tag{4.72}$$

MATLAB Solution
For the matrix $\mathbf{A}$ in this example, the following MATLAB script calculates the Jordan-form matrix $\mathbf{A}_J$ and the transformation matrix $\mathbf{T}$, where

$$\mathbf{A}_J = \mathbf{T}^{-1}\mathbf{A}\mathbf{T}. \tag{4.73}$$

```
A=[0 1 0 0 0; 0 0 1 0 0; 0 0 0 1 0 ...
0 0 0 0 1; 1 -5 10 -10 5];
[T,J]=jordan(A)
```

EXAMPLE 4.8 Consider the matrix below with three repeated eigenvalues at $\lambda = 2$:

$$\mathbf{A} = \begin{bmatrix} 2 & 4 & -8 \\ 0 & 0 & 4 \\ 0 & -1 & 4 \end{bmatrix}. \tag{4.74}$$

The dimensions of the $\mathcal{N}_k$ are

$$v_1 = 2, \ v_2 = v_3 = 3 = v_4. \tag{4.75}$$

Then the number of Jordan blocks is

(a) of size $1 \times 1 = 2v_1 - v_0 - v_2 = 4 - 0 - 3 = 1$
(b) of size $2 \times 2 = 2v_2 - v_1 - v_3 = 6 - 2 - 3 = 1$.

Therefore, the Jordan form has one 1×1 and one 2×2 block. The transformation

$$\mathbf{T} = [\, \mathbf{t}_1 \ \mathbf{t}_2 \ \mathbf{t}_3 \,] \tag{4.76}$$

can be obtained from

$$\begin{aligned} (\mathbf{A} - 2\mathbf{I})\mathbf{t}_1 &= 0, \\ (\mathbf{A} - 2\mathbf{I})\mathbf{t}_2 &= \mathbf{t}_1, \end{aligned} \tag{4.77}$$

and

$$(\mathbf{A} - 2\mathbf{I})\mathbf{t}_3 = 0. \tag{4.78}$$

Therefore we obtain,

$$\mathbf{T} = \begin{bmatrix} 4 & 0 & 0 \\ -2 & 1 & 2 \\ -1 & 0 & 1 \end{bmatrix} \quad \text{and} \quad \mathbf{J} = \begin{bmatrix} 2 & 1 & 0 \\ 0 & 2 & 0 \\ 0 & 0 & 2 \end{bmatrix}. \tag{4.79}$$

MATLAB Solution
The following MATLAB script calculates the Jordan form of $\mathbf{A}$:

```
A=[2 4 -8; 0 0 4; 0 -1 4];
[T,J]=jordan(A)
```

Remark 4.1 Note that the MATLAB command "jordan" results in a Jordan form with complex entries if $\sigma(\mathbf{A})$ has complex elements. The MATLAB help documentation also recommends forcing the matrix provided to the "jordan()" function be symbolic in order to force MATLAB to use its symbolic engine and avoid floating-point calculations. A sample is shown below:

```
[T,J]=jordan(sym(A));
T=double(T);
J=double(J);
```

4.1.5   Jordan Form of Companion Matrices

The companion matrix appears frequently in state-variable representations of differential equations and transfer functions. In this subsection, we discuss the Jordan form of a companion matrix and the similarity transformation taking a companion matrix into its

Jordan form. For this special case these can be determined in closed form once the eigenvalues are known. Since the Jordan form facilitates the computation of time-domain solutions, the results are useful in computing solutions of the state-space equations when the system matrix is in companion form.

We consider the companion matrix

$$\mathbf{A} = \begin{bmatrix} 0 & 1 & 0 & \cdots & 0 \\ 0 & 0 & 1 & & 0 \\ & & & \ddots & \vdots \\ & & & & 1 \\ -a_0 & -a_1 & \cdots & \cdots & -a_{n-1} \end{bmatrix} \tag{4.80}$$

with characteristic polynomial

$$a(s) := \det(s\mathbf{I} - \mathbf{A}) = s^n + a_{n-1}s^n + \cdots + a_1 s + a_0. \tag{4.81}$$

Let $\mathbf{J}$ denote the Jordan form of $\mathbf{A}$ and $\mathbf{T}$ a similarity transformation such that:

$$\mathbf{T}^{-1}\mathbf{A}\mathbf{T} = \mathbf{J}. \tag{4.82}$$

We give below, explicit formulas for $\mathbf{J}$ and $\mathbf{T}$, when $\mathbf{A}$ is as in Equation (4.80).

Distinct Eigenvalues

First, consider the special case where the eigenvalues of $\mathbf{A}$ in Equation (4.80) are distinct:

$$a(s) = (s - \lambda_1)(s - \lambda_2)\cdots(s - \lambda_n), \quad \text{for } \lambda_i \neq \lambda_j, \ i \neq j. \tag{4.83}$$

LEMMA 4.3 When the eigenvalues of $\mathbf{A}$ in Equation (4.80) are distinct, Equation (4.82) is satisfied by

$$\mathbf{J} = \begin{bmatrix} \lambda_1 & 0 & \cdots & 0 \\ 0 & \lambda_2 & & \vdots \\ \vdots & & \ddots & 0 \\ 0 & 0 & \cdots & \lambda_n \end{bmatrix} \tag{4.84}$$

and

$$\mathbf{T} = \begin{bmatrix} 1 & 1 & \cdots & 1 \\ \lambda_1 & \lambda_2 & & \lambda_n \\ \lambda_1^2 & \lambda_2^2 & & \lambda_n^2 \\ \vdots & \vdots & & \vdots \\ \lambda_1^{n-1} & \lambda_2^{n-1} & & \lambda_n^{n-1} \end{bmatrix}. \tag{4.85}$$

Proof The proof follows from the fact that

$$\det[\mathbf{T}] = \prod_{i=1}^{n} \prod_{j=1, i\neq j}^{n} (\lambda_i - \lambda_j) \tag{4.86}$$

and so **T** is nonsingular when the eigenvalues are distinct. That **J** satisfies Equation (4.82) follows from the fact that with

$$\mathbf{v}_i := \begin{bmatrix} 1 \\ \lambda_i \\ \lambda_i^2 \\ \vdots \\ \lambda_i^{n-1} \end{bmatrix} \tag{4.87}$$

we have

$$\mathbf{A}\mathbf{v}_i = \lambda_i \mathbf{v}_i, \quad \text{for } i \in \mathbf{n}. \tag{4.88}$$

∎

Repeated Eigenvalues

In this case, we have in general

$$a(s) = (s - \lambda_1)^{n_1} (s - \lambda_2)^{n_2} \cdots (s - \lambda_k)^{n_k}, \quad \text{for } \lambda_i \neq \lambda_j, \ i \neq j,$$
$$n_1 + n_2 + \cdots + n_k = n. \tag{4.89}$$

Introduce the Jordan block of size $n_i \times n_i$:

$$\mathbf{J}_i(\lambda_i) := \begin{bmatrix} \lambda_i & 1 & 0 & 0 & \cdots & 0 \\ 0 & \lambda_i & 1 & 0 & & \vdots \\ \vdots & & \ddots & \ddots & & \vdots \\ \vdots & & & \ddots & \ddots & 0 \\ \vdots & & & & \ddots & 1 \\ 0 & \cdots & \cdots & \cdots & \cdots & \lambda_i \end{bmatrix}, \quad \text{for } i \in \mathbf{k} \tag{4.90}$$

$$\underbrace{\qquad\qquad}_{n_i \times n_i}$$

and define the generalized eigenvector chains

$$[\mathbf{v}_1(\lambda_i) \quad \mathbf{v}_2(\lambda_i) \quad \cdots \quad \mathbf{v}_{n_i}(\lambda_i)] =: \mathbf{T}_i, \quad \text{for } i \in \mathbf{k}, \tag{4.91}$$

where

$$\mathbf{v}_1(\lambda_i) = \begin{bmatrix} 1 \\ \lambda_1 \\ \vdots \\ \lambda_i^{n-1} \end{bmatrix}, \quad \mathbf{v}_j(\lambda_i) = \frac{1}{j-1} \left[\frac{d}{d\lambda_i} \mathbf{v}_{j-1}(\lambda_i) \right] \tag{4.92}$$

for $j = 2, \ldots, n_i, \ i \in \mathbf{k}$.

LEMMA **4.4**

$$\mathbf{A}\mathbf{T}_i = \mathbf{T}_i \mathbf{J}_i(\lambda_i). \tag{4.93}$$

Proof The proof follows from the easily verified fact that

$$
\begin{aligned}
\mathbf{A}\mathbf{v}_1\left(\lambda_i\right) &= \lambda_i \mathbf{v}_1\left(\lambda_i\right), \\
\mathbf{A}\mathbf{v}_2\left(\lambda_i\right) &= \lambda_i \mathbf{v}_2\left(\lambda_i\right) + \mathbf{v}_1\left(\lambda_i\right), \\
\mathbf{A}\mathbf{v}_3\left(\lambda_i\right) &= \lambda_i \mathbf{v}_3\left(\lambda_i\right) + \mathbf{v}_3\left(\lambda_i\right),
\end{aligned}
\tag{4.94}
$$

$$
\vdots
$$

$$
\mathbf{A}\mathbf{v}_{n_i}\left(\lambda_i\right) = \lambda_i \mathbf{v}_{n_i}\left(\lambda_i\right) + \mathbf{v}_{n_i-1}\left(\lambda_i\right), \quad \text{for } i \in \mathbf{k},
$$

which is equivalent to Equation (4.93). ∎

This leads to the final result, which shows that in this case there is one and only one Jordan block associated with each distinct eigenvalue. The Jordan form of the companion matrix in Equation (4.80) is thus given by

$$
\mathbf{J} =
\begin{bmatrix}
\mathbf{J}(\lambda_1) & 0 & \cdots & 0 \\
0 & \mathbf{J}(\lambda_2) & & \vdots \\
\vdots & & \ddots & 0 \\
0 & \cdots & 0 & \mathbf{J}(\lambda_k)
\end{bmatrix},
\tag{4.95}
$$

and the similarity transformation $\mathbf{T}$ satisfying Equation (4.82) is

$$
\mathbf{T} = [\mathbf{T}_1 \ \mathbf{T}_2 \ \cdots \ \mathbf{T}_k],
\tag{4.96}
$$

where the columns of $\mathbf{T}_i$ are the generalized eigenvector chains defined in Equation (4.91).

EXAMPLE 4.9 Consider

$$
\mathbf{A} =
\begin{bmatrix}
0 & 1 & 0 \\
0 & 0 & 1 \\
-1 & -2 & -2
\end{bmatrix}
\tag{4.97}
$$

so that the characteristic polynomial is

$$
a(s) = s^3 + 2s^2 + 2s + 1 = (s+1)\left(\left(s+\frac{1}{2}\right)^2 + \frac{3}{4}\right).
\tag{4.98}
$$

The eigenvalues are

$$
\lambda_1 = -1, \quad \lambda_2 = -\frac{1}{2} + j\frac{\sqrt{3}}{2}, \quad \lambda_3 = -\frac{1}{2} - j\frac{\sqrt{3}}{2}.
\tag{4.99}
$$

By Lemma 4.3, the Jordan form of $\mathbf{A}$ is

$$
\mathbf{J} =
\begin{bmatrix}
\lambda_1 & 0 & 0 \\
0 & \lambda_2 & 0 \\
0 & 0 & \lambda_3
\end{bmatrix}
=
\begin{bmatrix}
-1 & 0 & 0 \\
0 & -\frac{1}{2} + j\frac{\sqrt{3}}{2} & 0 \\
0 & 0 & -\frac{1}{2} - j\frac{\sqrt{3}}{2}
\end{bmatrix}
\tag{4.100}
$$

and the similarity transformation $\mathbf{T}$ is given by

$$\mathbf{T} = \begin{bmatrix} 1 & 1 & 1 \\ \lambda_1 & \lambda_2 & \lambda_3 \\ \lambda_1^2 & \lambda_2^2 & \lambda_3^2 \end{bmatrix} = \begin{bmatrix} 1 & 1 & 1 \\ -1 & -\dfrac{1}{2}+j\dfrac{\sqrt{3}}{2} & -\dfrac{1}{2}-j\dfrac{\sqrt{3}}{2} \\ 1 & -\dfrac{1}{2}-j\dfrac{\sqrt{3}}{2} & -\dfrac{1}{2}+j\dfrac{\sqrt{3}}{2} \end{bmatrix}. \tag{4.101}$$

EXAMPLE 4.10   Let $\mathbf{A}$ be a 5×5 companion matrix in Equation (4.80) with characteristic polynomial

$$a(s) = (s-\lambda)^5. \tag{4.102}$$

Then

$$\mathbf{J} = \begin{bmatrix} \lambda & 1 & 0 & 0 & 0 \\ 0 & \lambda & 1 & 0 & 0 \\ 0 & 0 & \lambda & 1 & 0 \\ 0 & 0 & 0 & \lambda & 1 \\ 0 & 0 & 0 & 0 & \lambda \end{bmatrix} \tag{4.103}$$

and the similarity transformation $\mathbf{T}$ is

$$\mathbf{T} = \begin{bmatrix} 1 & 0 & 0 & 0 & 0 \\ \lambda & 1 & 0 & 0 & 0 \\ \lambda^2 & 2\lambda & 1 & 0 & 0 \\ \lambda^3 & 3\lambda^2 & 3\lambda & 1 & 0 \\ \lambda^4 & 4\lambda^3 & 6\lambda^2 & 4\lambda & 1 \end{bmatrix}. \tag{4.104}$$

EXAMPLE 4.11   Suppose $\mathbf{A}$ is a 6×6 companion matrix as in Equation (4.80) with characteristic polynomial

$$a(s) = (s-\lambda)^4(s-\mu)^2, \quad \text{for } \lambda \neq \mu. \tag{4.105}$$

Then the Jordan form of $\mathbf{A}$ is

$$\mathbf{J} = \left[\begin{array}{cccc:cc} \lambda & 1 & 0 & 0 & 0 & 0 \\ 0 & \lambda & 1 & 0 & 0 & 0 \\ 0 & 0 & \lambda & 1 & 0 & 0 \\ 0 & 0 & 0 & \lambda & 0 & 0 \\ \hdashline 0 & 0 & 0 & 0 & \mu & 1 \\ 0 & 0 & 0 & 0 & 0 & \mu \end{array}\right] \tag{4.106}$$

and the similarity transformation **T** is

$$
\mathbf{T} = \left[\ \mathbf{T}_1 \ \vdots \ \mathbf{T}_2 \ \right] = \begin{bmatrix}
1 & 0 & 0 & 0 & \vdots & 1 & 0 \\
\lambda & 1 & 0 & 0 & \vdots & \mu & 1 \\
\lambda^2 & 2\lambda & 1 & 0 & \vdots & \mu^2 & 2\mu \\
\lambda^3 & 3\lambda^2 & 3\lambda & 1 & \vdots & \mu^3 & 3\mu^2 \\
\lambda^4 & 4\lambda^3 & 6\lambda^2 & 4\lambda & \vdots & \mu^4 & 4\mu^3 \\
\lambda^5 & 5\lambda^4 & 10\lambda^3 & 10\lambda^2 & \vdots & \mu^5 & 5\mu^4 \\
\lambda^6 & 6\lambda^5 & 15\lambda^4 & 20\lambda^3 & \vdots & \mu^6 & 6\mu^5
\end{bmatrix} .
\tag{4.107}
$$

It is easy to verify that

$$
\operatorname{rank}[\mathbf{T}_1] = 4, \quad \operatorname{rank}[\mathbf{T}_2] = 2
\tag{4.108}
$$

and

$$
\operatorname{rank}[\mathbf{T}] = 6.
\tag{4.109}
$$

4.2 Jordan Form Using Polynomial Matrix Theory

The Jordan form can also be determined from polynomial matrix theory using the invariant polynomials and elementary divisors of the polynomial matrix $s\mathbf{I} - \mathbf{A}$. The details are given in Appendix A.8.4. Here we provide a couple of examples utilizing this approach.

4.3 Examples

EXAMPLE **4.12**   Suppose

$$
\mathbf{A} = \begin{bmatrix}
2 & 0 & 0 \\
1 & 2 & 0 \\
-1 & 0 & 1
\end{bmatrix} .
\tag{4.110}
$$

Then

$$
\mathbf{A}(s) = s\mathbf{I} - \mathbf{A} = \begin{bmatrix}
s-2 & 0 & 0 \\
-1 & s-2 & 0 \\
1 & 0 & s-1
\end{bmatrix} .
\tag{4.111}
$$

The greatest commom divisor (gcd) of miniors of various orders are

$$
m_0(s) := 1, \ m_1(s) = 1, \ m_2(s) = 1, \ m_3(s) = (s-1)(s-2)^2,
\tag{4.112}
$$

so that the invariant polynomials are

$$i_1(s) = 1 = i_2(s), \ i_3(s) = (s-1)(s-2)^2. \tag{4.113}$$

Therefore, the elementary divisors are

$$(s-1), (s-2)^2 \tag{4.114}$$

and thus the Jordan form of $\mathbf{A}$ is

$$\mathbf{J} = \begin{bmatrix} 1 & 0 & 0 \\ 0 & 2 & 1 \\ 0 & 0 & 2 \end{bmatrix}. \tag{4.115}$$

To determine $\mathbf{T}$, we note that it must be of the form

$$\mathbf{T} = [\underbrace{\mathbf{t}_1}_{\mathbf{T}_1} \ \underbrace{\mathbf{t}_2 \ \mathbf{t}_3}_{\mathbf{T}_2}], \tag{4.116}$$

where

$$(\mathbf{A} - \mathbf{I})\mathbf{t}_1 = 0, \qquad (\mathbf{A} - 2\mathbf{I})\mathbf{t}_2 = 0, \qquad (\mathbf{A} - 2\mathbf{I})\mathbf{t}_3 = \mathbf{t}_2. \tag{4.117}$$

Since

$$\mathbf{A} - \mathbf{I} = \begin{bmatrix} 1 & 0 & 0 \\ 1 & 1 & 0 \\ 0 & 0 & 0 \end{bmatrix} \quad \text{and} \quad \mathbf{A} - 2\mathbf{I} = \begin{bmatrix} 0 & 0 & 0 \\ 1 & 0 & 0 \\ -1 & 0 & -1 \end{bmatrix}, \tag{4.118}$$

we see that

$$\mathbf{t}_1 = \begin{bmatrix} 0 \\ 0 \\ 1 \end{bmatrix}, \quad \mathbf{t}_2 = \begin{bmatrix} 0 \\ 1 \\ 0 \end{bmatrix}, \quad \mathbf{t}_3 = \begin{bmatrix} 1 \\ 0 \\ -1 \end{bmatrix} \tag{4.119}$$

satisfies Equation (4.117) and $\mathbf{T}$ in Equation (4.116) is nonsingular.

MATLAB Solution

Unimodular matrices $\mathbf{U}(s)$, $\mathbf{V}(s)$, and the Smith form of $(s\mathbf{I} - \mathbf{A})$ can be calculated by the following MATLAB script:

```
syms s
A=[2 0 0;1 2 0;-1 0 1];
As=s*eye(size(A))-A;
[U,V,S]=smithForm(As)
```

The outputs are given below:

$$\mathbf{U}(s) = \begin{bmatrix} 0 & 0 & 1 \\ 0 & 1 & 1 \\ 1 & (s-1)(s-2) & (s-2)^2 \end{bmatrix}, \tag{4.120}$$

$$\mathbf{V}(s) = \begin{bmatrix} 1 & -s+1 & (s-1)(s-2) \\ 0 & -1 & s-1 \\ 0 & 1 & -s+2 \end{bmatrix}, \tag{4.121}$$

and the Smith form of $(s\mathbf{I} - \mathbf{A})$ is

$$
\begin{bmatrix}
1 & 0 & 0 \\
0 & 1 & 0 \\
0 & 0 & (s-1)(s-2)^2
\end{bmatrix}. \tag{4.122}
$$

The following MATLAB script finds the transformation matrix $\mathbf{T}$ carrying $\mathbf{A}$ into its Jordan form:

```
syms s
A=[2 0 0; 1 2 0; 1 0 1];
As=s*eye(size(A))-A;
[P1,Q1,S1]=smithForm(As);
%
J=jordan(A);
Js=s*eye(size(J))-J;
[P2,Q2,S2]=smithForm(Js);
Q=simplify(Q1*inv(Q2));
Qn=[];
for k=0:length(A),    % constructing Equation (A.86a)
  Qnk=subs(diff(Q,k)/factorial(k),s,0);
  Qn=[Qnk Qn];
end,
T=zeros(size(A));
for i=0:length(A),    % evaluating Equation (A.87a)
  Qv1=Qn(:,length(A)*i+1:length(A)*(i+1))*...
      (J^(length(A)-i));
  T=T+Qv1;
end,
simplify(inv(T)*A*T)    % verifying Equation (A.88)
```

The transformation matrix $\mathbf{T}$ is obtained below and it is verified that it transforms the given matrix to its Jordan form:

$$
\mathbf{T} =
\begin{bmatrix}
0 & 0 & -1 \\
0 & -1 & 0 \\
1 & 0 & 1
\end{bmatrix}, \quad
\mathbf{T}^{-1}\mathbf{A}\mathbf{T} =
\begin{bmatrix}
1 & 0 & 0 \\
1 & 2 & 1 \\
0 & 0 & 2
\end{bmatrix}. \tag{4.123}
$$

EXAMPLE 4.13   Let

$$
\mathbf{A} =
\begin{bmatrix}
1 & 0 & 0 & 0 \\
1 & 1 & -1 & 0 \\
0 & 0 & 1 & 0 \\
0 & 0 & 1 & 1
\end{bmatrix}. \tag{4.124}
$$

Find the invariant polynomials and elementary divisors of $(s\mathbf{I} - \mathbf{A})$ and the Jordan form $\mathbf{J}$ of $\mathbf{A}$. Find also the nonsingular transformation $\mathbf{T}$ so that $\mathbf{T}^{-1}\mathbf{A}\mathbf{T} = \mathbf{J}$.

Solution   We have

$$s\mathbf{I} - \mathbf{A} = \begin{bmatrix} s-1 & 0 & 0 & 0 \\ -1 & s-1 & 1 & 0 \\ 0 & 0 & s-1 & 0 \\ 0 & 0 & -1 & s-1 \end{bmatrix} \tag{4.125}$$

and so the gcd of minors of various orders

$$1 =: m_0(s) = m_1(s) = m_2(s), \quad m_3(s) = (s-1)^2, \quad m_4(s) = (s-1)^4. \tag{4.126}$$

Therefore, the invariant polynomials are

$$i_1(s) = \frac{m_1(s)}{m_0} = 1, \quad i_2(s) = \frac{m_2(s)}{m_1(s)} = 1, \quad i_3(s) = \frac{m_3(s)}{m_2(s)} = (s-1)^2,$$

$$i_4(s) = \frac{m_4(s)}{m_3(s)} = (s-1)^2. \tag{4.127}$$

The elementary divisors are then

$$(s-1)^2, \quad (s-1)^2 \tag{4.128}$$

and thus the Jordan form is

$$\mathbf{J} = \begin{bmatrix} 1 & 1 & 0 & 0 \\ 0 & 1 & 0 & 0 \\ 0 & 0 & 1 & 1 \\ 0 & 0 & 0 & 1 \end{bmatrix}. \tag{4.129}$$

The transformation $\mathbf{T}$ must be of the form

$$\mathbf{T} = [\underbrace{\mathbf{t}_1 \ \mathbf{t}_2}_{\mathbf{T}_1} \ \underbrace{\mathbf{t}_3 \ \mathbf{t}_4}_{\mathbf{T}_2}], \tag{4.130}$$

where

$$\mathbf{A}\mathbf{t}_1 = \mathbf{t}_1, \quad \mathbf{A}\mathbf{t}_2 = \mathbf{t}_1 + \mathbf{t}_2, \quad \mathbf{A}\mathbf{t}_3 = \mathbf{t}_3, \quad \mathbf{A}\mathbf{t}_4 = \mathbf{t}_4 + \mathbf{t}_3, \tag{4.131}$$

and $\mathbf{T}$ is nonsingular.

The above equations can be rewritten as

$$(\mathbf{A} - \mathbf{I})\mathbf{t}_1 = 0, \quad (\mathbf{A} - \mathbf{I})\mathbf{t}_2 = \mathbf{t}_1, \quad (\mathbf{A} - \mathbf{I})\mathbf{t}_3 = 0, \quad (\mathbf{A} - \mathbf{I})\mathbf{t}_4 = \mathbf{t}_3, \tag{4.132}$$

where

$$(\mathbf{A} - \mathbf{I}) = \begin{bmatrix} 0 & 0 & 0 & 0 \\ 1 & 0 & -1 & 0 \\ 0 & 0 & 0 & 0 \\ 0 & 0 & 1 & 0 \end{bmatrix}. \tag{4.133}$$

It is easy to see that a solution of Equation (4.132) is given by

$$
\mathbf{t}_1 = \begin{bmatrix} 0 \\ 1 \\ 0 \\ 0 \end{bmatrix}, \quad
\mathbf{t}_2 = \begin{bmatrix} 1 \\ 0 \\ 0 \\ 0 \end{bmatrix}, \quad
\mathbf{t}_3 = \begin{bmatrix} 0 \\ 0 \\ 0 \\ 1 \end{bmatrix}, \quad
\mathbf{t}_4 = \begin{bmatrix} 1 \\ 0 \\ 1 \\ 0 \end{bmatrix}.
\tag{4.134}
$$

MATLAB Solution
Unimodular matrices $\mathbf{U}(s)$, $\mathbf{V}(s)$ transforming $(s\mathbf{I} - \mathbf{A})$ to its Smith form can be calculated by the following MATLAB script:

```
syms s
A=[1 0 0 0; 1 1 -1 0; 0 0  1  0; 0 0 1 1];
As=s*eye(size(A))-A;
[U,V,S]=smithForm(As)
```

The MATLAB script results in the following:

$$
\mathbf{U}(s) = \begin{bmatrix} 0 & 0 & 0 & 1 \\ 0 & 1 & 0 & 1 \\ 0 & 0 & -1 & -s+1 \\ 1 & s-1 & 0 & s-1 \end{bmatrix}, \quad
\mathbf{V}(s) = \begin{bmatrix} 0 & -1 & 0 & s-1 \\ 0 & 0 & 1 & 1 \\ -1 & 0 & s+1 & 0 \\ 0 & 0 & -1 & 0 \end{bmatrix}
\tag{4.135}
$$

and the Smith form of $(s\mathbf{I} - \mathbf{A})$ is

$$
\begin{bmatrix} 1 & 0 & 0 & 0 \\ 0 & 1 & 0 & 0 \\ 0 & 0 & (s-1)^2 & 0 \\ 0 & 0 & 0 & (s-1)^2 \end{bmatrix}.
\tag{4.136}
$$

Using the same MATLAB script as in Example 4.12, we obtain

$$
\mathbf{T} = \begin{bmatrix} 0 & 1 & 0 & 0 \\ 1 & 0 & -1 & 0 \\ 0 & 0 & 0 & 1 \\ 0 & 0 & 1 & 0 \end{bmatrix} \quad \text{and} \quad
\mathbf{T}^{-1}\mathbf{AT} = \begin{bmatrix} 1 & 1 & 0 & 0 \\ 0 & 1 & 0 & 0 \\ 0 & 0 & 1 & 1 \\ 0 & 0 & 0 & 1 \end{bmatrix}.
\tag{4.137}
$$

4.4   Matrix Exponentials Evaluated Using Jordan Forms

Having the Jordan form $\mathbf{J}$ of $\mathbf{A}$ and the transformation $\mathbf{T}$ rendering $\mathbf{A}$ into $\mathbf{J}$ is useful in calculations involving state-space systems. If, for instance:

$$
\mathbf{J}_i = \begin{bmatrix} \lambda_i & 1 & 0 & 0 & \cdots & 0 \\ 0 & \lambda_i & 1 & 0 & \cdots & 0 \\ 0 & & \ddots & \ddots & & \vdots \\ \vdots & & & \ddots & \ddots & 0 \\ \vdots & & & & \ddots & 1 \\ 0 & \cdots & \cdots & \cdots & \ddots & \lambda_i \end{bmatrix}, \underbrace{\qquad}_{n_i \times n_i} \tag{4.138}
$$

then

$$
(s\mathbf{I} - \mathbf{J}_i)^{-1} = \begin{bmatrix} \dfrac{1}{(s-\lambda_i)} & \dfrac{1}{(s-\lambda_i)^2} & \cdots & \dfrac{1}{(s-\lambda_i)^{n_i}} \\ 0 & \dfrac{1}{(s-\lambda_i)} & & \vdots \\ \vdots & & \ddots & \vdots \\ \vdots & & & \dfrac{1}{(s-\lambda_i)^2} \\ 0 & \cdots & 0 & \dfrac{1}{(s-\lambda_i)} \end{bmatrix} \tag{4.139}
$$

so that

$$
e^{\mathbf{J}_i t} = \begin{bmatrix} e^{\lambda_i t} & te^{\lambda_i t} & \dfrac{t^2 e^{\lambda_i t}}{2} & \cdots & \cdots \\ & \ddots & \ddots & & \\ & & \ddots & \ddots & \dfrac{t^2 e^{\lambda_i t}}{2} \\ & & & \ddots & te^{\lambda_i t} \\ 0 & & & & e^{\lambda_i t} \end{bmatrix}. \tag{4.140}
$$

Then since

$$
\mathbf{T}^{-1}\mathbf{A}\mathbf{T} = \mathbf{J} = \begin{bmatrix} \mathbf{J}_1 & & 0 \\ & \ddots & \\ 0 & & \mathbf{J}_l \end{bmatrix}, \tag{4.141}
$$

we have

$$
(s\mathbf{I} - \mathbf{A})^{-1} = \mathbf{T}(s\mathbf{I} - \mathbf{J})^{-1}\mathbf{T}^{-1} = \mathbf{T} \begin{bmatrix} (s\mathbf{I} - \mathbf{J}_1)^{-1} & & 0 \\ & \ddots & \\ 0 & & (s\mathbf{I} - \mathbf{J}_l)^{-1} \end{bmatrix} \mathbf{T}^{-1} \tag{4.142}
$$

and

$$e^{\mathbf{A}t} = \mathbf{T} \begin{bmatrix} e^{\mathbf{J}_1 t} & & 0 \\ & \ddots & \\ 0 & & e^{\mathbf{J}_l t} \end{bmatrix} \mathbf{T}^{-1}. \tag{4.143}$$

4.5   Modal Decomposition: Zero-Input Response

The Jordan form of the matrix $\mathbf{A}$ indicates the form of the functions that make up the zero-input state response. When $\mathbf{A}$ can be diagonalized, these functions are exponentials of the form $e^{\lambda_i t}$, where λ_i are the eigenvalues of $\mathbf{A}$. These functions are called the modes of the system. In the general case when a Jordan block associated with an eigenvalue of $\mathbf{A}$ is of size $k \times k$, the modes are the functions

$$t^j e^{\lambda t}, \quad \text{for } j = 0, 1, \dots, k-1. \tag{4.144}$$

We consider now, for simplicity, a system where $\mathbf{A}$ can be diagonalized. Then there exists a set of n linearly independent eigenvectors $\mathbf{v}_i$, for $i = 1, 2, \dots, n$. The zero-input state response can be decomposed as a sum of motions along these eigenvectors. This decomposition is called *modal decomposition*.

The zero-input state response is given by

$$\mathbf{x}_o(t) = e^{\mathbf{A}t} \mathbf{x}(0), \tag{4.145}$$

where $\mathbf{x}(0)$ is the initial condition vector. Then we can write

$$\mathbf{x}(0) = \alpha_1 \mathbf{v}_1 + \alpha_2 \mathbf{v}_2 + \cdots + \alpha_n \mathbf{v}_n \tag{4.146}$$

for unique constants α_i for $i = 1, 2, \dots, n$.

Now since

$$\mathbf{A}\mathbf{v}_i = \lambda_i \mathbf{v}_i, \quad \text{for } i = 1, 2, \dots, n, \tag{4.147}$$

it follows that

$$e^{\mathbf{A}t} \mathbf{v}_i = e^{\lambda_i t} \mathbf{v}_i, \quad \text{for } i = 1, 2, \dots, n. \tag{4.148}$$

Therefore, from Equations (4.145)–(4.148):

$$\mathbf{x}_0(t) = \alpha_1 e^{\lambda_1 t} \mathbf{v}_1 + \alpha_2 e^{\lambda_2 t} \mathbf{v}_2 + \cdots + \alpha_n e^{\lambda_n t} \mathbf{v}_n. \tag{4.149}$$

Equation (4.149) shows that the motion along $\mathbf{v}_i$, consists of $\alpha_i e^{\alpha_i t}$ for $i = 1, 2, \dots, n$ and the sum of these modal components makes up the motion of $\mathbf{x}_0(t)$, the state vector. Also, the output response is

$$\mathbf{y}_o(t) = \alpha_1 \mathbf{C}\mathbf{v}_1 e^{\lambda_1 t} + \cdots + \alpha_n \mathbf{C}\mathbf{v}_n e^{\lambda_n t}. \tag{4.150}$$

EXAMPLE 4.14   Suppose $\mathbf{A}$ is 2×2 and

$$\dot{\mathbf{x}}(t) = \mathbf{A}\mathbf{x}(t) \tag{4.151}$$

with

$$\mathbf{A} = \begin{bmatrix} 0 & 1 \\ 1 & 0 \end{bmatrix}. \tag{4.152}$$

The eigenvalues are $\lambda_1 = 1$, $\lambda_2 = -1$ and the eigenvectors are

$$\mathbf{v}_1 = \begin{bmatrix} 1 \\ 1 \end{bmatrix}, \qquad \mathbf{v}_2 = \begin{bmatrix} 1 \\ -1 \end{bmatrix}. \tag{4.153}$$

Thus, the modal decomposition of the state response is, in this case (Figure 4.1):

$$\mathbf{x}_o(t) = \alpha_1 e^t \begin{bmatrix} 1 \\ 1 \end{bmatrix} + \alpha_2 e^{-t} \begin{bmatrix} 1 \\ -1 \end{bmatrix}. \tag{4.154}$$

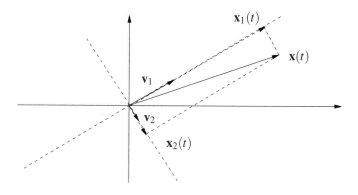

Figure 4.1 Modal decomposition (Example 4.14)

Note that the first component in the above decomposition is "unstable" while the second component is "stable."

EXAMPLE 4.15   In this example, we consider a matrix $\mathbf{A}$ with complex eigenvalues. Let

$$\mathbf{A} = \begin{bmatrix} 0 & 1 \\ -2 & -2 \end{bmatrix}. \tag{4.155}$$

The eigenvalues are

$$\lambda_1 = -1 + j, \quad \lambda_2 = -1 - j. \tag{4.156}$$

The eigenvectors are

$$\mathbf{v}_1 = \begin{bmatrix} 1 \\ -1+j \end{bmatrix}, \qquad \mathbf{v}_2 = \begin{bmatrix} 1 \\ -1-j \end{bmatrix} \tag{4.157}$$

and therefore the modal decomposition of the state response is

$$\mathbf{x}_o(t) = \alpha_1 e^{-t} e^{jt} \begin{bmatrix} 1 \\ -1+j \end{bmatrix} + \alpha_2 e^{-t} e^{-jt} \begin{bmatrix} 1 \\ -1-j \end{bmatrix}, \tag{4.158}$$

where α_1, α_2 are in general complex constants determined from

$$\mathbf{x}(0) = \alpha_1 \mathbf{v}_1 + \alpha_2 \mathbf{v}_2. \tag{4.159}$$

For example, if

$$\mathbf{x}(0) = \begin{bmatrix} 1 \\ 1 \end{bmatrix}, \qquad \alpha_1 = \frac{1}{2} - j, \quad \alpha_2 = \frac{1}{2} + j. \tag{4.160}$$

4.6   Modal Decomposition: Zero-State Response

Consider now the forced or zero-state response of a single-input system:

$$\mathbf{x}_u(t) = \int_0^t e^{\mathbf{A}(t-\tau)} \mathbf{b} u(\tau) d\tau. \tag{4.161}$$

Assume that $\mathbf{A}$ has n linearly independent eigenvectors $\mathbf{v}_i$, so that

$$\mathbf{b} = \beta_1 \mathbf{v}_1 + \beta_2 \mathbf{v}_2 + \cdots + \beta_n \mathbf{v}_n, \tag{4.162}$$

for unique β_i for $i \in \mathbf{n}$. Substituting Equation (4.162) in Equation (4.161) and using Equation (4.148), we have

$$\mathbf{x}_u(t) = \int_0^t \beta_1 e^{\lambda_i(t-\tau)} \mathbf{v}_1 u(\tau) d\tau + \cdots + \int_0^t \beta_n e^{\lambda_n(t-\tau)} \mathbf{v}_n u(\tau) d\tau$$
$$= \sum_{i=1}^n \beta_i \left(e^{\lambda_i t} * u(t) \right) \mathbf{v}_i, \tag{4.163}$$

where "*" denotes convolution.

Equation (4.163) shows that the zero-state or forced response is the sum of motions along the eigenvectors. In each such direction the motion consists of the convolution of the associated mode $e^{\lambda_i t}$ with the input $u(t)$ weighted by β_i, where the latter are determined from Equation (4.162).

The forced response of the output is

$$\mathbf{y}_u(t) = \sum_{i=1}^n \beta_i \left(e^{\lambda_i t} * u(t) \right) \mathbf{C} \mathbf{v}_i + \mathbf{D} u(t). \tag{4.164}$$

The following examples show how a given state-space system $(\mathbf{A}, \mathbf{B}, \mathbf{C}, \mathbf{D})$ may be converted into an equivalent modal form by using MATLAB.

EXAMPLE 4.16   Consider the system

$$
A = \begin{bmatrix} 0 & 1 & 0 & 0 \\ 0 & 0 & 1 & 0 \\ 0 & 0 & 0 & 1 \\ -24 & -50 & -35 & -10 \end{bmatrix}, \quad
B = \begin{bmatrix} 1 & -1 \\ 0 & 1 \\ 1 & 1 \\ 0 & 1 \end{bmatrix},
$$

$$
C = \begin{bmatrix} 1 & 1 & 1 & 0 \\ 1 & -1 & 1 & 1 \end{bmatrix}, \quad
D = \begin{bmatrix} 0 & 0 \\ 0 & 0 \end{bmatrix}.
$$
(4.165)

The following MATLAB script converts the given state-space system into modal form:

```
A=[0 1 0 0;0 0 1 0; 0 0 0 1; -24 -50 -35 -10];
B=[1 -1; 0 1; 1 1; 0 1];
C=[1 1 1 0; 1 -1 1 1];
D=zeros(2,2);
sys=ss(A,B,C,D);
sysC=canon(sys,'modal')
```

The output of the MATLAB script is

$$
A = \begin{bmatrix} -4 & 0 & 0 & 0 \\ 0 & -3 & 0 & 0 \\ 0 & 0 & -2 & 0 \\ 0 & 0 & 0 & -1 \end{bmatrix}, \quad
B = \begin{bmatrix} -39.4 & -39.4 \\ -86.3 & -80.54 \\ 60.83 & 48.66 \\ 16.51 & 6.005 \end{bmatrix},
$$

$$
C = \begin{bmatrix} 0.66 & -0.6084 & -0.4932 & 0.333 \\ -2.183 & 1.217 & 0.1644 & 0.6661 \end{bmatrix}, \quad
D = \begin{bmatrix} 0 & 0 \\ 0 & 0 \end{bmatrix}.
$$
(4.166)

EXAMPLE 4.17   We consider now the case where **A** has complex eigenvalues and do modal decomposition using MATLAB. Consider

$$
A = \begin{bmatrix} 1 & -2 & 0 & 1 \\ 1 & 0 & -1 & 1 \\ 2 & 0 & -3 & -1 \\ -1 & 0 & 1 & 1 \end{bmatrix}, \quad
B = \begin{bmatrix} 1 & 0 \\ 1 & 1 \\ 1 & -1 \\ 0 & 1 \end{bmatrix},
$$

$$
C = \begin{bmatrix} 1 & 1 & 0 & -1 \\ 1 & 0 & 1 & 1 \end{bmatrix}, \quad
D = \begin{bmatrix} 0 & 0 \\ 0 & 0 \end{bmatrix}.
$$
(4.167)

With the same MATLAB script used in Example 4.16 applied to $(\mathbf{A},\mathbf{B},\mathbf{C},\mathbf{D})$ in this example, we obtain the following real modal form:

$$\mathbf{A} = \begin{bmatrix} -2.117 & 0 & 0 & 0 \\ 0 & 1.242 & 0 & 0 \\ 0 & 0 & -0.06272 & 1.232 \\ 0 & 0 & -1.232 & -0.06272 \end{bmatrix}, \quad \mathbf{B} = \begin{bmatrix} -0.5607 & 1.469 \\ -0.2997 & -1.491 \\ -0.335 & -0.6066 \\ -0.7173 & -0.8731 \end{bmatrix},$$

$$\mathbf{C} = \begin{bmatrix} -0.8353 & 0.7987 & -1.318 & -1.853 \\ -0.9724 & 0.3496 & -0.2594 & -2.053 \end{bmatrix}, \quad \mathbf{D} = \begin{bmatrix} 0 & 0 \\ 0 & 0 \end{bmatrix}. \tag{4.168}$$

4.7   Modal Decomposition of the State Space

Let $\mathbb{C}$, $\mathbb{C}^+$, and $\mathbb{C}^-$ denote the complex plane, the closed right-half plane (RHP), and the open left-half plane (LHP), respectively. Write

$$\Pi(s) := \det[s\mathbf{I} - \mathbf{A}] = \Pi^+(s)\Pi^-(s) \tag{4.169}$$

with the zeros of Π^+ (Π^-) lying in $\mathbb{C}^+$ ($\mathbb{C}^-$), corresponding to the unstable (stable) eigenvalues of $\mathbf{A}$. Let also $\mathscr{V}^+$ ($\mathscr{V}^-$) denote the unstable (stable) eigenspaces, that is, with "Ker" denoting kernel or null space:

$$\mathscr{V}^+ = \text{Ker}\left[\Pi^+(\mathbf{A})\right], \tag{4.170}$$
$$\mathscr{V}^- = \text{Ker}\left[\Pi^-(\mathbf{A})\right]. \tag{4.171}$$

Now let V^+, V^- denote matrices whose columns span $\mathscr{V}^+, \mathscr{V}^-$, respectively. Then using

$$\mathbf{T} = \begin{bmatrix} V^+ & V^- \end{bmatrix} \tag{4.172}$$

as a similarity or coordinate transformation on $[\mathbf{A}, \mathbf{B}, \mathbf{C}, \mathbf{D}]$, the new representation takes the form

$$\begin{bmatrix} \mathbf{A}^+ & \mathbf{0} \\ \mathbf{0} & \mathbf{A}^- \end{bmatrix}, \quad \begin{bmatrix} \mathbf{B}^+ \\ \mathbf{B}^- \end{bmatrix}, \quad \begin{bmatrix} \mathbf{C}^+ & \mathbf{C}^- \end{bmatrix}, \quad \mathbf{D}, \tag{4.173}$$

where the eigenvalues of $\mathbf{A}^+$ ($\mathbf{A}^-$) are in $\mathbb{C}^+$ ($\mathbb{C}^-$). This constitutes a modal decomposition of the state-space system into stable and unstable components. In the new coordinates, the stable and unstable subspaces are orthogonal. The transfer function of the system admits the corresponding decomposition:

$$\mathbf{G}(s) = \mathbf{G}^+(s) + \mathbf{G}^-(s) + \mathbf{D}, \tag{4.174}$$

where

$$\mathbf{G}^+(s) = \mathbf{C}^+\left(s\mathbf{I} - \mathbf{A}^+\right)^{-1}\mathbf{B}^+, \tag{4.175}$$
$$\mathbf{G}^-(s) = \mathbf{C}^-\left(s\mathbf{I} - \mathbf{A}^-\right)^{-1}\mathbf{B}^-. \tag{4.176}$$

EXAMPLE 4.18   Consider a state-space system

$$
\mathbf{A} = \begin{bmatrix} 3 & -1 & -2 \\ 2 & 0 & -4 \\ -4 & 4 & -5 \end{bmatrix}, \quad \mathbf{B} = \begin{bmatrix} 1 \\ 1 \\ 1 \end{bmatrix}, \quad \mathbf{C} = [1 \ 0 \ -2], \quad \mathbf{D} = 0. \quad (4.177)
$$

The transfer function is

$$
\mathbf{G}(s) = \frac{-s^2 + 8s + 3}{(s+1)(s-2)(s+3)}. \quad (4.178)
$$

Note that we have

$$
\Pi(s) = \Pi^+(s)\Pi^-(s) = (s-2)(s+1)(s+3), \quad (4.179)
$$

$$
\mathbf{T} = \begin{bmatrix} V^+ & | & V^- \end{bmatrix} = \begin{bmatrix} 1 & | & 1 & -1 \\ 1 & | & 2 & -2 \\ 0 & | & 1 & 1 \end{bmatrix}. \quad (4.180)
$$

Applying this coordinate transformation to $\{\mathbf{A},\mathbf{B},\mathbf{C},\mathbf{D}\}$:

$$
\mathbf{T}^{-1}\mathbf{A}\mathbf{T} = \begin{bmatrix} \mathbf{A}^+ & 0 \\ 0 & \mathbf{A}^- \end{bmatrix} = \begin{bmatrix} 2 & | & 0 & 0 \\ 0 & | & -1 & -6 \\ 0 & | & 0 & -3 \end{bmatrix},
$$

$$
\mathbf{T}^{-1}\mathbf{B} = \begin{bmatrix} \mathbf{B}^+ \\ \mathbf{B}^- \end{bmatrix} = \begin{bmatrix} 1 \\ 0.5 \\ 0.5 \end{bmatrix}, \quad (4.181)
$$

$$
\mathbf{C}\mathbf{T} = \begin{bmatrix} \mathbf{C}^+ & \mathbf{C}^- \end{bmatrix} = \begin{bmatrix} 1 & | & -1 & -3 \end{bmatrix}, \quad \mathbf{D} = 0
$$

and

$$
\mathbf{G}^+(s) = \mathbf{C}^+ \left(s\mathbf{I} - \mathbf{A}^+\right)^{-1} \mathbf{B}^+ = \frac{1}{s-2},
$$

$$
\mathbf{G}^-(s) = \mathbf{C}^- \left(s\mathbf{I} - \mathbf{A}^-\right)^{-1} \mathbf{B}^- = \frac{-2s}{(s+1)(s+3)}, \quad (4.182)
$$

$$
\mathbf{G}(s) = \frac{1}{s-2} - \frac{2s}{(s+1)(s+3)} = \frac{-s^2 + 8s + 3}{(s-2)(s+1)(s+3)}.
$$

MATLAB Solution

The modal decomposition can be done by using the MATLAB script below:

```
A=[3 -1 -2; 2 0 -4; -4 4 -5];
B=[1 1 1]';
C=[1 0 -2];
D=0;
sys=ss(A,B,C,D);
[Gs,Gu]=stabsep(sys)
```

The output of the above MATLAB script is

$$\mathbf{G}_s = \left\{ \mathbf{A} = \begin{bmatrix} -1 & -7.071 \\ 0 & -3 \end{bmatrix}, \ \mathbf{B} = \begin{bmatrix} 0.8165 \\ 0.5774 \end{bmatrix}, \right. \tag{4.183}$$

$$\left. \mathbf{C} = [-0.8165 \ -2.309], \ \mathbf{D} = 0 \right\}$$

$$\mathbf{G}_u = \{\mathbf{A} = 2, \ \mathbf{B} = -1.414, \ \mathbf{C} = -0.7071, \ \mathbf{D} = 0\}. \tag{4.184}$$

The following additional MATLAB script gives $\mathbf{G}^+(s)$ and $\mathbf{G}^-(s)$:

```
Gplus=tf(Gu)
Gminus=tf(Gs)
G=Gplus+Gminus
```

This script results in

$$\mathbf{G}^+(s) = \frac{1}{s-2}, \quad \mathbf{G}^-(s) = \frac{-2s}{(s+1)(s+3)}, \tag{4.185}$$

and

$$G(s) = \mathbf{G}^+(s) + \mathbf{G}^-(s) = \frac{-s^2 + 8s + 3}{s^3 + 2s^2 - 5s - 6}. \tag{4.186}$$

4.8   Exercises

4.1   Find a similarity transformation to convert

$$\mathbf{A} = \begin{bmatrix} -1 & 0 & 0 \\ 0 & -1 & 1 \\ 0 & -1 & -1 \end{bmatrix} \tag{4.187}$$

into its upper companion form. Verify your answer by using MATLAB.

4.2   Find a similarity transformation to transform

$$\mathbf{A} = \begin{bmatrix} 0 & 1 & 0 \\ 0 & 0 & 1 \\ 1 & -1 & 1 \end{bmatrix} \tag{4.188}$$

into its Jordan form. Verify your answer by using MATLAB.

4.3   Find a similarity transformation to convert

$$\mathbf{A} = \begin{bmatrix} 0 & 1 & 0 \\ 0 & 0 & 1 \\ -6 & -8 & -5 \end{bmatrix} \tag{4.189}$$

into its Jordan form. Further transform this Jordan form to a quasi-diagonal form with real entries. Verify your answer by using MATLAB.

4.4 Find the invariant polynomials, elementary divisors, and Smith forms of $(s\mathbf{I} - \mathbf{A})$:

(a) $\mathbf{A} = \begin{bmatrix} \lambda & 0 \\ 0 & \lambda \end{bmatrix}$.

(b) $\mathbf{A} = \begin{bmatrix} \lambda & 1 \\ 0 & \lambda \end{bmatrix}$.

(c) $\mathbf{A} = \begin{bmatrix} \lambda & 0 & 0 \\ 0 & \mu & 1 \\ 0 & 0 & \mu \end{bmatrix}$, for $\lambda \neq \mu$.

(d) $\mathbf{A} = \begin{bmatrix} \lambda & 0 & 0 \\ 0 & \mu & 0 \\ 0 & 0 & \mu \end{bmatrix}$, for $\lambda \neq \mu$.

(e) $\mathbf{A} = \begin{bmatrix} \lambda & 1 & 0 & 0 \\ 0 & \lambda & 0 & 0 \\ 0 & 0 & \lambda & 1 \\ 0 & 0 & 0 & \lambda \end{bmatrix}$.

(f) $\mathbf{A} = \begin{bmatrix} \lambda & 0 & 0 & 0 \\ 0 & \mu & 0 & 0 \\ 0 & 0 & \mu & 1 \\ 0 & 0 & 0 & \mu \end{bmatrix}$, for $\lambda \neq \mu$.

4.5 The state-space equations of a system are given as

$$\dot{\mathbf{x}}(t) = \begin{bmatrix} 0 & 1 & 0 \\ 0 & 0 & 1 \\ -2 & -4 & -3 \end{bmatrix} \mathbf{x}(t) + \begin{bmatrix} 0 \\ 0 \\ 1 \end{bmatrix} u(t),$$

$$y(t) = [1\ 1\ 1]\mathbf{x}(t).$$

(4.190)

(a) Transform the system to Jordan canonical form and determine the new state-space model $(\mathbf{A}_n, \mathbf{B}_n, \mathbf{C}_n)$.

(b) Retransform $(\mathbf{A}_n, \mathbf{B}_n, \mathbf{C}_n)$ to a real form with the new $\mathbf{A}_n$ in real block-diagonal form, and the new $\mathbf{B}_n$, $\mathbf{C}_n$ real.

(c) For $u(t) = U(t)$ a unit step input, decompose the forced response of $\mathbf{x}_n(t)$ along eigenvectors. Also calculate $\mathbf{x}(t)$ from $\mathbf{x}_n(t)$.

4.6 The state-space model of a discrete-time system is given as

$$\mathbf{x}[k+1] = \begin{bmatrix} 0 & 1 & 0 \\ 0 & 0 & 1 \\ 0 & 1/4 & 0 \end{bmatrix} \mathbf{x}[k] + \begin{bmatrix} 0 \\ 0 \\ 1 \end{bmatrix} u[k].$$

(4.191)

(a) Determine a new representation with $\mathbf{A}_n$ diagonal and state $\mathbf{z}[k]$.

(b) With $u[k] = U[k]$ a discrete-time unit step, decompose the forced response of $\mathbf{z}[k]$ along eigenvectors.

(c) From the answer to (b), determine the forced response $\mathbf{x}[k]$.

4.7

(a) Transform the state-space model

$$\dot{\mathbf{x}}(t) = \begin{bmatrix} 0 & 1 & 0 \\ 0 & 0 & 1 \\ -1 & -3 & -3 \end{bmatrix} \mathbf{x}(t) + \begin{bmatrix} 0 \\ 1 \\ 1 \end{bmatrix} u(t) \qquad (4.192)$$

so that the new "**A**" matrix with state $\mathbf{z}(t)$ is in Jordan form.

(b) Find the forced response of $\mathbf{z}(t)$ in Jordan form.

(c) Find the forced response of $\mathbf{x}(t)$ using the answer to (b) above.

4.8 The state-space equations of a system are

$$\dot{\mathbf{x}}(t) = \begin{bmatrix} 0 & 1 & 0 \\ 0 & 0 & 1 \\ 0 & 1 & 0 \end{bmatrix} \mathbf{x}(t) + \begin{bmatrix} 1 & 0 \\ 0 & 0 \\ 0 & 1 \end{bmatrix} \mathbf{u}(t),$$

$$\mathbf{y}(t) = \begin{bmatrix} 1 & 0 & 0 \\ 0 & 0 & 1 \end{bmatrix} \mathbf{x}(t). \qquad (4.193)$$

(a) Transform the system equations to

$$\dot{\mathbf{z}}(t) = \mathbf{A}_n \mathbf{z}(t) + \mathbf{B}_n \mathbf{u}(t),$$
$$\mathbf{y}(t) = \mathbf{C}_n \mathbf{z}(t), \qquad (4.194)$$

so that $\mathbf{A}_n$ is diagonal.

(b) Determine the solution $\mathbf{z}(t)$ for $u_1(t) = \delta(t)$, $u_2(t) = U(t)$ unit step and initial condition $\mathbf{z}(0) = 0$.

(c) Find the solution $\mathbf{x}(t)$ for the inputs and initial condition given in (b) using the solution for $\mathbf{z}(t)$ obtained in (b).

(d) Determine the forced response of $\mathbf{y}(t)$ from the solution obtained in (b).

(e) Find the zero-input response for $\mathbf{x}(t)$ and $\mathbf{y}(t)$ for arbitrary initial conditions $\mathbf{x}(0)$.

4.9 The state-space model of a system is

$$\dot{\mathbf{x}}(t) = \begin{bmatrix} 0 & 1 & 0 \\ 0 & 0 & 1 \\ 2 & 0 & -1 \end{bmatrix} \mathbf{x}(t) + \begin{bmatrix} 1 & 0 \\ 1 & 1 \\ 0 & 1 \end{bmatrix} \mathbf{u}(t),$$

$$\mathbf{y}(t) = \begin{bmatrix} 1 & 0 & 0 \\ 0 & 0 & 1 \end{bmatrix} \mathbf{x}(t) + \begin{bmatrix} 1 & 0 \\ 0 & 0 \end{bmatrix} \mathbf{u}(t). \qquad (4.195)$$

(a) Determine a coordinate transformation so that

$$\mathbf{A}_n = \begin{bmatrix} \mathbf{A}^+ & 0 \\ 0 & \mathbf{A}^- \end{bmatrix}, \quad \mathbf{B}_n = \begin{bmatrix} \mathbf{B}^+ \\ \mathbf{B}^- \end{bmatrix}, \quad \mathbf{C}_n = \begin{bmatrix} \mathbf{C}^+ & \mathbf{C}^- \end{bmatrix}, \qquad (4.196)$$

where $\mathbf{A}^+$ is "unstable" and $\mathbf{A}^-$ is "stable" and all matrices are real.

(b) Find the unstable and stable components $\mathbf{G}^+(s)$, $\mathbf{G}^-(s)$ of the transfer function:

$$\mathbf{G}(s) = \mathbf{G}^+(s) + \mathbf{G}^-(s) + \mathbf{D}.$$

4.10 Repeat Exercise 4.9 for the discrete-time system

$$
\mathbf{x}[k+1] = \begin{bmatrix} 0 & 1 & 0 \\ 0 & 0 & 1 \\ 0 & -2 & -2 \end{bmatrix} \mathbf{x}[k] + \begin{bmatrix} 1 & 0 \\ 0 & 1 \\ 0 & 1 \end{bmatrix} \mathbf{u}[k],
$$

$$
\mathbf{y}[k] = \begin{bmatrix} 1 & 0 & 0 \\ 0 & 0 & 1 \end{bmatrix} \mathbf{x}[k] + \begin{bmatrix} 1 & 0 \\ 0 & 1 \end{bmatrix} \mathbf{u}[k].
$$

(4.197)

4.9   Notes and References

The Jordan form and modal structure are standard topics in state-space theory and Zadeh and Desoer (1963), Wolovich (1974), Wonham (1985), Callier and Desoer (1982), and Kailath (1980) are excellent references for this part of state-space theory. The connection between the Jordan form and its computation via polynomial matrix theory is made very nicely in Gantmacher's classic text (Gantmacher, 1959). See also the Appendix of this book. The proof of Lemma A.6 is given in Gantmacher (1959, p. 142) and Lemma A.9 is proved in Gantmacher (1959, p. 148). The computational procedure given in Equation (A.90) is also recommended in Gantmacher (1959, p. 162).

5 Controllability, Observability, and Realization Theory

In this chapter we introduce the fundamental state-space concepts of controllability and observability. Using them we develop the Kalman canonical decomposition, and the central result that a state-space realization is of minimal order if and only if it is controllable and observable. The importance of minimal realizations lies in the fact that feedback control design cannot be initiated unless a controllable and observable realization of the plant is first obtained. We also briefly describe the time-domain interpretations of controllability and observability. It is then shown that controllability and observability of a realization are equivalent to coprimeness of the matrix fraction description of the associated transfer function matrix. Some useful minimal realizations are described, including Gilbert's realization, companion form realizations, Jordan form realizations, realizations from Lyapunov equations, and finally Wolovich's realization from matrix fraction descriptions.

5.1   Introduction

Controllability and observability are properties of state-variable models that have come to play a central role in the description, understanding, and design of state-space systems. In fact, feedback control design in the state-space framework cannot be initiated without first obtaining a controllable and observable realization of the plant.

We first explore the role of controllability and observability in developing state-space realizations of a given transfer function matrix. The controllable and unobservable subspaces are **A**-invariant and can be combined to display the Kalman canonical decomposition of a system from which a minimal-order realization of the underlying system transfer matrix can be extracted. We prove that a realization is of minimal order if and only if it is controllable and observable and display some useful minimal realizations. These include Gilbert's realization, companion form realizations, realizations based on Lyapunov equations, and Wolovich's realization which is directly developed from matrix fraction descriptions.

5.2   Invariant Subspaces and the Kalman Decomposition

Let

$$\dot{\mathbf{x}}(t) = \mathbf{A}\mathbf{x}(t) + \mathbf{B}\mathbf{u}(t), \tag{5.1a}$$

$$\mathbf{y}(t) = \mathbf{C}\mathbf{x}(t) \tag{5.1b}$$

denote a dynamic system where $\mathscr{X}$, $\mathscr{U}$, and $\mathscr{Y}$ denote n, r, and m-dimensional vector spaces, and $\mathbf{A}$, $\mathbf{B}$, $\mathbf{C}$ are linear operators:

$$\mathbf{A} : \mathscr{X} \to \mathscr{X}, \quad \mathbf{B} : \mathscr{U} \to \mathscr{X}, \quad \mathbf{C} : \mathscr{X} \to \mathscr{Y}. \tag{5.2}$$

The structure of the system can be effectively explored and displayed in several different coordinate systems using various invariant subspaces.

5.2.1   Invariant Subspaces

A subspace $\mathscr{V} \subset \mathscr{X}$ is $\mathbf{A}$-invariant if $\mathbf{A}\mathscr{V} \subset \mathscr{V}$, that is, all elements of $\mathscr{V}$ are mapped by $\mathbf{A}$ back into $\mathscr{V}$. If $\mathscr{V}$ has dimension k and $\{v_1, v_2, \ldots, v_k\}$ is a basis for $\mathscr{V}$, one can choose any subspace $\mathscr{W} \subset \mathscr{X}$ such that $\mathscr{V} \oplus \mathscr{W} = \mathscr{X}$. A basis $\{w_{k+1}, \ldots, w_n\}$ for $\mathscr{W}$ is chosen, and used to form the coordinate transformation

$$\mathbf{T} = [v_1, \ldots, v_k, w_{k+1}, \ldots, w_n]. \tag{5.3}$$

Set

$$\mathbf{x}(t) = \mathbf{T}\mathbf{z}(t). \tag{5.4}$$

Equation (5.1) is then transformed into

$$\dot{\mathbf{z}}(t) = \mathbf{A}_n \mathbf{z}(t) + \mathbf{B}_n \mathbf{u}(t), \tag{5.5a}$$

$$\mathbf{y}(t) = \mathbf{C}_n \mathbf{z}(t), \tag{5.5b}$$

where

$$\mathbf{A}_n = \mathbf{T}^{-1}\mathbf{A}\mathbf{T}, \quad \mathbf{B}_n = \mathbf{T}^{-1}\mathbf{B}, \quad \mathbf{C}_n = \mathbf{C}\mathbf{T}. \tag{5.6}$$

Writing Equation (5.6) as

$$\mathbf{A}\mathbf{T} = \mathbf{T}\mathbf{A}_n, \quad \mathbf{T}\mathbf{B}_n = \mathbf{B}, \quad \mathbf{C}_n = \mathbf{C}\mathbf{T} \tag{5.7}$$

it is easy to see that the $\mathbf{A}$-invariance of $\mathscr{V}$ implies that

$$\mathbf{A}_n = \begin{bmatrix} \mathbf{A}_1 & \mathbf{A}_3 \\ \mathbf{0} & \mathbf{A}_2 \end{bmatrix}. \tag{5.8}$$

Thus, Equation (5.5) can be written as

$$\begin{bmatrix} \dot{\mathbf{z}}_1(t) \\ \dot{\mathbf{z}}_2(t) \end{bmatrix} = \begin{bmatrix} \mathbf{A}_1 & \mathbf{A}_3 \\ \mathbf{0} & \mathbf{A}_2 \end{bmatrix} \begin{bmatrix} \mathbf{z}_1(t) \\ \mathbf{z}_2(t) \end{bmatrix} + \begin{bmatrix} \mathbf{B}_1 \\ \mathbf{B}_2 \end{bmatrix} \mathbf{u}(t), \tag{5.9a}$$

$$\mathbf{y}(t) = \begin{bmatrix} \mathbf{C}_1 & \mathbf{C}_2 \end{bmatrix} \begin{bmatrix} \mathbf{z}_1(t) \\ \mathbf{z}_2(t) \end{bmatrix}. \tag{5.9b}$$

If $\mathscr{W}$ also happens to be $\mathbf{A}$-invariant, then $\mathbf{A}_3 = \mathbf{0}$ in Equation (5.9a).

Remark 5.1 $\mathbf{A}_1$ is the matrix representation of the map $\mathbf{A}|\mathscr{V}$, the restriction of $\mathbf{A}$ to $\mathscr{V}$. $\mathbf{A}_2$ is the matrix representation of the map $\bar{\mathbf{A}}$ induced by $\mathbf{A}$ in the factor space $\bar{\mathbf{X}} := \mathbf{X}/\mathscr{V}$, that is

$$\bar{\mathbf{A}} : \mathscr{X}/\mathscr{V} \to \mathscr{X}/\mathscr{V} \tag{5.10}$$

for $\bar{\mathbf{x}} \in \mathscr{X}/\mathscr{V}$:

$$\bar{\mathbf{A}}\bar{\mathbf{x}} = \overline{\mathbf{A}\bar{\mathbf{x}}}. \tag{5.11}$$

That $\mathbf{A}_1$ and $\mathbf{A}_2$ indeed represent maps as defined above can be seen from the fact that if a different basis is chosen for $\mathscr{V}$, and a different complement $\mathscr{W}$ of $\mathscr{V}$ is chosen, it is easy to show that the "new" $\mathbf{A}_1$ and $\mathbf{A}_2$ are related to the old $\mathbf{A}_1$ and $\mathbf{A}_2$, respectively, via similarity transformations.

Now we construct two special A-invariant subspaces for the dynamic system in Equation (5.1). These are the *controllable subspace* and the *unobservable subspace*. For this, let $\mathscr{B}$ denote the image of $\mathbf{B}$:

$$\mathscr{B} := \{\mathbf{Bu} \ : \ \mathbf{u} \in \mathscr{U}\} \tag{5.12}$$

and let

$$\mathscr{V}_u := \mathscr{B} + \mathbf{A}\mathscr{B} + \cdots + \mathbf{A}^{n-1}\mathscr{B}. \tag{5.13}$$

In the above equation $+$ denotes subspace addition, that is, set-theoretic union of subspaces.

Now let $\mathrm{Ker}[\mathbf{C}]$ denote the null space of $\mathbf{C}$:

$$\mathrm{Ker}[\mathbf{C}] := [\mathbf{x} \ : \ \mathbf{Cx} = 0] \tag{5.14}$$

and

$$\mathscr{V}_y := \bigcap_{i=0}^{n-1} \mathrm{Ker}\left[\mathbf{CA}^i\right]. \tag{5.15}$$

$\mathscr{V}_u$ is called the *controllable subspace* and $\mathscr{V}_y$ is called the *unobservable subspace* of the system in Equation (5.1).

LEMMA 5.1

$$\mathbf{A}\mathscr{V}_u \subset \mathscr{V}_u, \tag{5.16a}$$

$$\mathscr{B} \subset \mathscr{V}_u. \tag{5.16b}$$

Proof That $\mathscr{B} \subset \mathscr{V}_u$ follows from Equation (5.13). To prove Equation (5.16a), consider an arbitrary vector $\mathbf{v} \in \mathscr{V}_u$. Then

$$\mathbf{v} = \mathbf{Bu}_0 + \mathbf{ABu}_1 + \cdots + \mathbf{A}^{n-1}\mathbf{Bu}_{n-1} \tag{5.17}$$

for some $\mathbf{u}_i$, $i = 0, 1, \ldots, n-1$. Therefore

$$\mathbf{Av} = \mathbf{ABu}_0 + \mathbf{A}^2\mathbf{Bu}_1 + \cdots + \mathbf{A}^n\mathbf{Bu}_{n-1}. \tag{5.18}$$

Since $\mathbf{A}^n$ can be expressed as a linear combination of lower powers of $\mathbf{A}$ (Cayley–Hamilton Theorem), it follows that

$$\mathbf{A}\mathbf{v} = \mathbf{B}\bar{\mathbf{u}}_0 + \mathbf{A}\mathbf{B}\bar{\mathbf{u}}_1 + \mathbf{A}^2\mathbf{B}\bar{\mathbf{u}}_2 + \cdots + \mathbf{A}^{n-1}\mathbf{B}\bar{\mathbf{u}}_{n-1} \tag{5.19}$$

for some $\bar{\mathbf{u}}_i$, $i = 0, 1, 2, \ldots, n-1$. Therefore

$$\mathbf{A}\mathbf{v} \in \mathcal{V}_u. \tag{5.20}$$

Since $\mathbf{v}$ was an arbitrary vector in $\mathcal{V}_u$, it follows that

$$\mathbf{A}\mathcal{V}_u \subset \mathcal{V}_u. \tag{5.21}$$

∎

Remark 5.2   Consider the collection of subspaces

$$\underline{\mathcal{V}} = \{\mathcal{V} \;:\; \mathbf{A}\mathcal{V} \subset \mathcal{V}, \; \mathcal{B} \subset \mathcal{V}\}. \tag{5.22}$$

It is easy to show that $\mathcal{V}_u$ is the "smallest" element of $\underline{\mathcal{V}}$, that is, it is contained in every subspace in $\underline{\mathcal{V}}$.

Now consider the unobservable subspace.

LEMMA 5.2

$$\mathbf{A}\mathcal{V}_y \subset \mathcal{V}_y, \tag{5.23a}$$

$$\mathcal{V}_y \subset \mathrm{Ker}[\mathbf{C}]. \tag{5.23b}$$

Proof   That Equation (5.23b) is true follows from Equation (5.15). To prove Equation (5.23a), pick an arbitrary $\mathbf{v} \in \mathcal{V}_y$. Then

$$\mathbf{C}\mathbf{A}^i\mathbf{v} = 0, \quad \text{for } i = 0, 1, \ldots, n-1. \tag{5.24}$$

Now, it is easy to verify that

$$\mathbf{C}\mathbf{A}^i(\mathbf{A}\mathbf{v}) = 0, \quad \text{for } i = 0, 1, \ldots, n-1 \tag{5.25}$$

again using the Cayley–Hamilton Theorem. Thus, $\mathbf{A}\mathbf{v} \in \mathcal{V}_y$ and this proves Equation (5.23a). ∎

Remark 5.3   Consider the collection of all subspaces which are $\mathbf{A}$-invariant and contained in $\mathrm{Ker}[\mathbf{C}]$:

$$\underline{\mathcal{V}} = \{\mathcal{V} \;:\; \mathbf{A}\mathcal{V} \subset \mathcal{V}, \mathcal{V} \subset \mathrm{Ker}[\mathbf{C}]\}. \tag{5.26}$$

It is easy to see that $\mathcal{V}_y$ is the "largest" element of $\underline{\mathcal{V}}$, that is, every element $\mathcal{V} \in \underline{\mathcal{V}}$ satisfies

$$\mathcal{V} \subset \mathcal{V}_y. \tag{5.27}$$

5.2.2   Controllability and Observability Reductions

Suppose that $(\mathbf{A}, \mathbf{B}, \mathbf{C})$ is an nth-order realization with

$$\dim[\mathcal{V}_u] = \operatorname{rank} \begin{bmatrix} \mathbf{B} & \mathbf{AB} & \cdots & \mathbf{A}^{n-1}\mathbf{B} \end{bmatrix} = k. \tag{5.28}$$

DEFINITION 5.1 $(\mathbf{A}, \mathbf{B})$ is said to be *controllable* if

$$k = n. \tag{5.29}$$

Likewise if

$$\dim[\mathcal{V}_y] = n - \operatorname{rank} \begin{bmatrix} \mathbf{C} \\ \mathbf{CA} \\ \vdots \\ \mathbf{CA}^{n-1} \end{bmatrix} = \ell, \tag{5.30}$$

we have the following.

DEFINITION 5.2 $(\mathbf{C}, \mathbf{A})$ is said to be *observable* if

$$\ell = 0 \tag{5.31}$$

or equivalently

$$\operatorname{rank} \begin{bmatrix} \mathbf{C} \\ \mathbf{CA} \\ \vdots \\ \mathbf{CA}^{n-1} \end{bmatrix} = n. \tag{5.32}$$

If $(\mathbf{A}, \mathbf{B})$ is not controllable, it is possible to extract a controllable subsystem by constructing a coordinate transformation

$$\mathbf{T} = \begin{bmatrix} \mathbf{T}_u & \mathbf{T}_w \end{bmatrix} \in \mathbb{R}^{n \times n} \tag{5.33}$$

with the columns of $\mathbf{T}_u \in \mathbb{R}^{n \times n}$ spanning $\mathcal{V}_u$. Then since $\mathcal{B} \subset \mathcal{V}_u$ and $\mathbf{A}\mathcal{V}_u \subset \mathcal{V}_u$:

$$\mathbf{T}^{-1}\mathbf{AT} = \begin{bmatrix} \mathbf{A}_1 & \mathbf{A}_3 \\ 0 & \mathbf{A}_2 \end{bmatrix}, \quad \mathbf{T}^{-1}\mathbf{B} = \begin{bmatrix} \mathbf{B}_1 \\ 0 \end{bmatrix}, \quad \mathbf{CT} = \begin{bmatrix} \mathbf{C}_1 & \mathbf{C}_2 \end{bmatrix}. \tag{5.34}$$

It is easily verified that $[\mathbf{A}_1, \mathbf{B}_1, \mathbf{C}_1]$ is also a realization, of order k, with $(\mathbf{A}_1, \mathbf{B}_1)$ controllable, that is

$$\operatorname{rank} \begin{bmatrix} \mathbf{B}_1 & \mathbf{A}_1\mathbf{B}_1 & \cdots & \mathbf{A}_1^{k-1}\mathbf{B}_1 \end{bmatrix} = k. \tag{5.35}$$

The above procedure is called *controllability reduction*.

Dual Systems
DEFINITION 5.3 $(\mathbf{A}, \mathbf{B}, \mathbf{C})$ and $\left(\mathbf{A}^T, \mathbf{B}^T, \mathbf{C}^T\right)$ are said to be duals of each other.

It is easy to see that $(\mathbf{A}, \mathbf{B}, \mathbf{C})$ is controllable (observable) if and only if $(\mathbf{A}^T, \mathbf{B}^T, \mathbf{C}^T)$ is observable (controllable).

By dualizing the procedure for controllability reduction, we can extract an observable subsystem when $(\mathbf{C}, \mathbf{A})$ is unobservable. This leads to a coordinate transformation $\mathbf{Q}$ such that

$$\mathbf{Q}\mathbf{A}\mathbf{Q}^{-1} = \begin{bmatrix} \bar{\mathbf{A}}_1 & 0 \\ \bar{\mathbf{A}}_3 & \bar{\mathbf{A}}_2 \end{bmatrix}, \quad \mathbf{Q}\mathbf{B} = \begin{bmatrix} \bar{\mathbf{B}}_1 \\ \bar{\mathbf{B}}_2 \end{bmatrix}, \quad \mathbf{C}\mathbf{Q}^{-1} = \begin{bmatrix} \bar{\mathbf{C}}_1 & 0 \end{bmatrix}, \tag{5.36}$$

with $(\bar{\mathbf{A}}_1, \bar{\mathbf{B}}_1, \bar{\mathbf{C}}_1)$ being a realization of order $n - \ell =: h$, and $(\bar{\mathbf{C}}_1, \bar{\mathbf{A}}_1)$ observable, that is

$$\mathrm{rank} \begin{bmatrix} \bar{\mathbf{C}}_1^T \\ \bar{\mathbf{C}}_1^T \bar{\mathbf{A}}_1^T \\ \vdots \\ \bar{\mathbf{C}}_1^T \left(\bar{\mathbf{A}}_1^T\right)^{h-1} \end{bmatrix} = h. \tag{5.37}$$

This procedure is called *observability reduction*.

5.2.3   Kalman Canonical Decomposition

From the controllable and unobservable subspaces defined in Lemmas 5.1 and 5.2, we can construct two more $\mathbf{A}$-invariant subspaces.

LEMMA 5.3

$$\mathbf{A}\left(\mathcal{V}_u \cap \mathcal{V}_y\right) \subset \left(\mathcal{V}_u \cap \mathcal{V}_y\right), \tag{5.38a}$$

$$\mathbf{A}\left(\mathcal{V}_u + \mathcal{V}_y\right) \subset \left(\mathcal{V}_u + \mathcal{V}_y\right). \tag{5.38b}$$

Proof   If $\mathbf{v} \in \mathcal{V}_u \cap \mathcal{V}_y$, $\mathbf{A}\mathbf{v} \in \mathcal{V}_u$ and $\mathbf{A}\mathbf{v} \in \mathcal{V}_y$ and so $\mathbf{A}\mathbf{v} \in \mathcal{V}_u \cap \mathcal{V}_y$. If $\mathbf{v} \in \mathcal{V}_u + \mathcal{V}_y$, then $\mathbf{v} = \mathbf{v}_1 + \mathbf{v}_2$ with $\mathbf{v}_1 \in \mathcal{V}_u$ and $\mathbf{v}_2 \in \mathcal{V}_y$. Then $\mathbf{A}\mathbf{v} = \mathbf{A}\mathbf{v}_1 + \mathbf{A}\mathbf{v}_2 \in \mathcal{V}_u + \mathcal{V}_y$.   ∎

Now consider a direct decomposition of the state space as follows:

$$\mathcal{X} = \mathcal{V}_1 \oplus \mathcal{V}_2 \oplus \mathcal{V}_3 \oplus \mathcal{V}_4, \tag{5.39}$$

where

$$\mathcal{V}_1 := \mathcal{V}_u \cap \mathcal{V}_y \tag{5.40}$$

and $\mathcal{V}_2, \mathcal{V}_3, \mathcal{V}_4$ satisfy

$$\mathcal{V}_2 \oplus \mathcal{V}_1 = \mathcal{V}_u, \tag{5.41a}$$

$$\mathcal{V}_3 \oplus \mathcal{V}_1 = \mathcal{V}_y, \tag{5.41b}$$

and

$$\mathcal{V}_4 \oplus \mathcal{V}_3 \oplus \mathcal{V}_2 \oplus \mathcal{V}_1 = \mathcal{X} \tag{5.42}$$

but are otherwise arbitrary.

Let

$$\dim[\mathcal{V}_i] = k_i \tag{5.43}$$

and let $\mathbf{T}_i$ denote an $n \times k_i$ matrix whose columns form a basis for $\mathcal{V}_i$, $i = 1, 2, 3, 4$. Let

$$\mathbf{T} := \begin{bmatrix} \mathbf{T}_1 & \mathbf{T}_2 & \mathbf{T}_3 & \mathbf{T}_4 \end{bmatrix} \tag{5.44}$$

denote a coordinate transformation and set

$$\mathbf{x}(t) = \mathbf{T}\mathbf{z}(t). \tag{5.45}$$

In this coordinate

$$\dot{\mathbf{z}}(t) = \mathbf{A}_n \mathbf{z}(t) + \mathbf{B}_n \mathbf{u}(t), \tag{5.46a}$$
$$\mathbf{y}(t) = \mathbf{C}_n \mathbf{z}(t), \tag{5.46b}$$

and

$$\mathbf{A}_n = \mathbf{T}^{-1}\mathbf{A}\mathbf{T}, \quad \mathbf{B}_n = \mathbf{T}^{-1}\mathbf{B}, \quad \mathbf{C}_n = \mathbf{C}\mathbf{T}. \tag{5.47}$$

THEOREM 5.1 (Kalman canonical decomposition) In the coordinate system defined by Equations (5.40)–(5.47), $(\mathbf{A}_n, \mathbf{B}_n, \mathbf{C}_n)$ have the following structure:

$$\mathbf{A}_n = \begin{bmatrix} \mathbf{A}_1 & \mathbf{A}_3 & \mathbf{A}_5 & \mathbf{A}_7 \\ 0 & \mathbf{A}_2 & 0 & \mathbf{A}_8 \\ 0 & 0 & \mathbf{A}_4 & \mathbf{A}_9 \\ 0 & 0 & 0 & \mathbf{A}_6 \end{bmatrix}, \quad \mathbf{B}_n = \begin{bmatrix} \mathbf{B}_1 \\ \mathbf{B}_2 \\ 0 \\ 0 \end{bmatrix}, \quad \mathbf{C}_n = \begin{bmatrix} 0 & \mathbf{C}_2 & 0 & \mathbf{C}_6 \end{bmatrix}. \tag{5.48}$$

Proof The proof follows from the facts that

$$\mathbf{A}\mathcal{V}_1 \subset \mathcal{V}_1,$$
$$\mathbf{A}(\mathcal{V}_1 \oplus \mathcal{V}_2) \subset \mathcal{V}_1 \oplus \mathcal{V}_2, \tag{5.49}$$
$$\mathbf{A}(\mathcal{V}_1 \oplus \mathcal{V}_3) \subset \mathcal{V}_1 \oplus \mathcal{V}_3,$$

and

$$\mathbf{B} \subset \mathcal{V}_1 \oplus \mathcal{V}_2,$$
$$\mathcal{V}_1 \oplus \mathcal{V}_3 \subset \mathrm{Ker}[\mathbf{C}], \tag{5.50}$$

and the relations in Equation (5.47). ∎

COROLLARY 5.1

(a) $\mathbf{C}(s\mathbf{I} - \mathbf{A})^{-1}\mathbf{B} = \mathbf{C}_2(s\mathbf{I} - \mathbf{A}_2)^{-1}\mathbf{B}_2.$ $\quad$ (5.51)

(b) $(\mathbf{A}_2, \mathbf{B}_2)$ is controllable. $\quad$ (5.52)

(c) $(\mathbf{C}_2, \mathbf{A}_2)$ is observable. $\quad$ (5.53)

Proof The simple proof of Corollary 5.1 is left to the reader. ∎

Remark 5.4 The structure in Equation (5.48) specifies only the zero blocks in the matrices. The detailed structure of the non-zero blocks will depend on the actual subspaces $\mathcal{V}_1, \mathcal{V}_2, \mathcal{V}_3$, and $\mathcal{V}_4$ and the bases chosen for these subspaces.

EXAMPLE 5.1 (Kalman canonical decomposition)   Consider the system

$$\dot{\mathbf{x}}(t) = \mathbf{A}\mathbf{x}(t) + \mathbf{B}\mathbf{u}(t),$$
$$\mathbf{y}(t) = \mathbf{C}\mathbf{x}(t), \tag{5.54}$$

with

$$\mathbf{A} = \begin{bmatrix} 0 & 1 & 0 & 0 \\ 0 & 1 & 0 & 0 \\ 1 & 0 & 0 & 1 \\ 0 & 0 & 0 & 1 \end{bmatrix}, \quad \mathbf{B} = \begin{bmatrix} 0 & 1 \\ 0 & 0 \\ 1 & 0 \\ 0 & 0 \end{bmatrix}, \quad \mathbf{C} = \begin{bmatrix} 1 & 1 & 0 & 0 \end{bmatrix}. \tag{5.55}$$

The controllable subspace $\mathcal{V}_u$ is generated by

$$\left\{ \begin{matrix} 0 & 1 \\ 0 & 0 \\ 1 & 0 \\ 0 & 0 \end{matrix} \right\} = \mathcal{V}_u. \tag{5.56}$$

The unobservable subspace

$$\mathcal{V}_y = \mathrm{Ker}\begin{bmatrix} 1 & 1 & 0 & 0 \\ 0 & 2 & 0 & 0 \end{bmatrix} = \left\{ \begin{matrix} 0 & 0 \\ 0 & 0 \\ 1 & 0 \\ 0 & 1 \end{matrix} \right\}. \tag{5.57}$$

Then

$$\mathcal{V}_u \cap \mathcal{V}_y = \left\{ \begin{matrix} 0 \\ 0 \\ 1 \\ 0 \end{matrix} \right\} =: \mathcal{V}_1. \tag{5.58}$$

Define

$$\mathcal{V}_2 = \left\{ \begin{matrix} 1 \\ 0 \\ 0 \\ 0 \end{matrix} \right\}, \quad \mathcal{V}_3 = \left\{ \begin{matrix} 0 \\ 0 \\ 0 \\ 1 \end{matrix} \right\}, \quad \mathcal{V}_4 = \left\{ \begin{matrix} 0 \\ 1 \\ 0 \\ 0 \end{matrix} \right\} \tag{5.59}$$

to satisfy

$$\mathcal{X} = \mathcal{V}_1 \oplus \mathcal{V}_2 \oplus \mathcal{V}_3 \oplus \mathcal{V}_4 \tag{5.60}$$

and let

$$\mathbf{T} = \begin{bmatrix} 0 & 1 & 0 & 0 \\ 0 & 0 & 0 & 1 \\ 1 & 0 & 0 & 0 \\ 0 & 0 & 1 & 0 \end{bmatrix}. \tag{5.61}$$

It is easy to verify that

$$
\mathbf{T}^{-1}\mathbf{AT} = \mathbf{A}_n = \left[\begin{array}{cc|cc} 0 & 1 & 1 & 0 \\ 0 & \boxed{0} & 0 & 1 \\ \hline 0 & 0 & 1 & 0 \\ 0 & 0 & 0 & 1 \end{array}\right], \quad \mathbf{T}^{-1}\mathbf{B} = \mathbf{B}_n = \left[\begin{array}{cc} 1 & 0 \\ 0 & 1 \\ \hline 0 & 0 \\ 0 & 0 \end{array}\right], \tag{5.62}
$$

$$
\mathbf{CT} = \mathbf{C}_n = \left[\begin{array}{cc|cc} 0 & \boxed{1} & 0 & 1 \end{array}\right],
$$

which is the Kalman canonical decomposition of the system. The boxed elements represent $\mathbf{A}_2, \mathbf{B}_2, \mathbf{C}_2$, the reduced-order realization which will turn out to be of minimal order.

MATLAB Solution

The following two-step process consisting of observability reduction followed by controllability reduction, implemented below using MATLAB, will give the Kalman decomposition:

```
A=[0 1 0 0; 0 1 0 0; 1 0 0 1; 0 0 0 1];
B=[0 1; 0 0; 1 0; 0 0];
C=[1 1 0 0];
[Ao,Bo,Co,To]=obsvf(A,B,C);    % observability
[Ak,Bk,Ck,Tc]=ctrbf(Ao,Bo,Co);   % controllability
```

The resulting outputs of the MATLAB script above are

$$
\mathbf{A}_k = \left[\begin{array}{cccc} 1 & 0 & 0 & 0 \\ 0 & 1 & 0 & 0 \\ -0.7071 & -0.7071 & \boxed{0} & 0 \\ 0.7071 & -0.7071 & 1 & 0 \end{array}\right], \quad \mathbf{B}_k = \left[\begin{array}{cc} 0 & 0 \\ 0 & 0 \\ \hline 0 & -1 \\ -1 & 0 \end{array}\right], \tag{5.63}
$$

$$
\mathbf{C}_k = \left[\begin{array}{cccc} 0.7071 & 0.7071 & \boxed{-1} & 0 \end{array}\right].
$$

Remark 5.5   The MATLAB function "obsvf()" decomposes the state-space system (A, B, C) into the observability staircase form (Ao, Bo, Co). Similarly, "ctrbf()" decomposes it into the controllability staircase form (Ac, Bc, Cc). MATLAB users should keep in mind that numerical errors can affect the results of determining rank, null spaces, and other calculations using linear algebra.

In the next two sections we develop time-domain interpretations of controllability and observability.

5.3   Controllability: Steering to a Target State

Consider the problem of steering the state of a system to a prescribed target state $\mathbf{x}^*$ by applying inputs to it. The notion of controllability arises naturally in this context. Indeed, the concept of controllability first appeared in the 1950s in the work of Pontryagin when he solved the problem of steering the state of a system to the origin in minimum time using bounded inputs.

To fix our ideas, consider the system

$$\dot{\mathbf{x}}(t) = \mathbf{A}\mathbf{x}(t) + \mathbf{B}\mathbf{u}(t), \quad \text{with } \mathbf{x}(0) = 0 \tag{5.64}$$

and the problem of finding a piecewise continuous input defined over $[0, t^*]$:

$$\mathbf{u}^*(t), \quad \text{for } t \in [0, t^*], \ t^* < \infty, \tag{5.65}$$

so that the application of $\mathbf{u}^*(t)$ results in

$$\mathbf{x}(t^*) = \mathbf{x}^* \tag{5.66}$$

for a prescribed target state $\mathbf{x}^*$. When $t^* < \infty$ and $\mathbf{u}^*(t)$ satisfying Equation (5.66) exist, we say that $\mathbf{x}^*$ is *reachable* from the origin. The set of states reachable from the origin is denoted $\mathscr{R}_0$. The main result of this section states that $\mathscr{R}_0$ is precisely the controllable subspace $\mathscr{V}_u$.

THEOREM 5.2

$$\mathscr{R}_0 = \mathscr{V}_u := \mathrm{Im}\left[\mathbf{B} \ \mathbf{AB} \ \cdots \ \mathbf{A}^{n-1}\mathbf{B}\right]. \tag{5.67}$$

The proof of the above theorem depends on the concept of the *controllability Wronskian* defined as:

$$\mathbf{W}_{t^*} := \int_0^{t^*} e^{\mathbf{A}t}\mathbf{B}\mathbf{B}^T e^{\mathbf{A}^T t}dt, \quad \text{for } t^* > 0. \tag{5.68}$$

We first develop the auxiliary result below.

LEMMA 5.4

$$\mathrm{Im}\left[\mathbf{W}_{t^*}\right] = \mathscr{V}_u. \tag{5.69}$$

Proof   The matrix $\mathbf{W}_{t^*}$ is symmetric and therefore Equation (5.69) is equivalent to

$$\mathrm{Ker}\left[\mathbf{W}_{t^*}\right] = \mathscr{V}_u^{\perp}. \tag{5.70}$$

To prove Equation (5.70), consider $\mathbf{w} \in \mathrm{Ker}\left[\mathbf{W}_{t^*}\right]$ so that

$$\mathbf{w}^T\left[\int_0^{t^*} e^{\mathbf{A}t}\mathbf{B}\mathbf{B}^T e^{\mathbf{A}^T t}dt\right]\mathbf{w} = 0 \tag{5.71a}$$

$$\Leftrightarrow \quad \mathbf{w}^T e^{\mathbf{A}t}\mathbf{B} = 0, \quad \text{for } t \in [0, t^*] \tag{5.71b}$$

$$\Leftrightarrow \quad \frac{d^i}{dt^i}\left(\mathbf{w}^T e^{\mathbf{A}t}\mathbf{B}\right) = 0, \quad \text{for } t \in [0, t^*], \ i = 0, 1, 2, \dots \tag{5.71c}$$

$$\Leftrightarrow \quad \mathbf{w} \in \cap\mathrm{Ker}\left[\mathbf{B}^T\left(\mathbf{A}^T\right)^i\right], \quad \text{for } i = 0, 1, 2, \dots, n-1 \tag{5.71d}$$

$$\Leftrightarrow \quad \mathbf{w} \in \mathscr{V}_u^{\perp}.$$

∎

This result in fact also shows that $\mathrm{Im}\left[\mathbf{W}_{t^*}\right]$ is independent of t^*.

Proof of Theorem 5.2 We first prove that

$$\mathscr{V}_u \subset \mathscr{R}_0, \tag{5.72}$$

that is every target state $\mathbf{x}^*$ in $\mathscr{V}_u$ is reachable. If $\mathbf{x}^* \in \mathscr{V}_u$, then by Lemma 5.4

$$\mathbf{x}^* = \mathbf{W}_{t^*}\mathbf{w} \tag{5.73}$$

for some $\mathbf{w}$ and $t^* > 0$. Then with

$$\mathbf{u}^*(t) = \mathbf{B}^T e^{\mathbf{A}(t^*-\tau)}\mathbf{w}, \quad \text{for } \tau \in [0, t^*] \tag{5.74}$$

we have

$$\mathbf{x}(t^*) = \int_0^{t^*} e^{\mathbf{A}(t^*-\tau)}\mathbf{B}\mathbf{B}^T e^{\mathbf{A}^T(t^*-\tau)}\mathbf{w}d\tau = W_{t^*}\mathbf{w} = \mathbf{x}^* \tag{5.75}$$

so that $\mathbf{x}^* \in \mathscr{R}_0$. This proves Equation (5.72).

To prove the reverse inclusion, suppose that $\mathscr{V}_u \neq \mathscr{X}$. Then choose a coordinate system with basis vectors spanning $\mathscr{V}_u$ and a complement $\mathscr{W}$ so that

$$\mathbf{T} = [V_u \ W], \quad \text{for } \mathbf{x}(t) = \mathbf{T}\mathbf{z}(t), \tag{5.76}$$

is a similarity transformation. In this coordinate system

$$\begin{bmatrix} \dot{\mathbf{z}}_1(t) \\ \dot{\mathbf{z}}_2(t) \end{bmatrix} = \begin{bmatrix} \mathbf{A}_1 & \mathbf{A}_3 \\ 0 & \mathbf{A}_2 \end{bmatrix} \begin{bmatrix} \mathbf{z}_1(t) \\ \mathbf{z}_2(t) \end{bmatrix} + \begin{bmatrix} \mathbf{B}_1 \\ 0 \end{bmatrix} \mathbf{u}(t), \tag{5.77}$$

which is called the *controllability decomposition*. It is obvious that target states $\mathbf{z}^*$ with $\mathbf{z}_2^* \neq 0$ are not reachable from $\mathbf{z}(0) = 0$. But $\mathbf{z}_2^* \neq 0$ is precisely the condition that $\mathbf{x}^* \notin \mathscr{V}_u$. Therefore, $\mathbf{x}^* \notin \mathscr{V}_u$ implies $\mathbf{x}^* \notin \mathscr{R}_0$, so that $\mathscr{R}_0 \subset \mathscr{V}_u$. ∎

When the initial condition $\mathbf{x}(0) \neq 0$:

$$\bar{\mathbf{x}} := \mathbf{x}^* - e^{\mathbf{A}t^*}\mathbf{x}(0) \tag{5.78}$$

must lie in $\mathscr{R}_0$ for some $T > 0$ so that

$$\bar{\mathbf{x}} = \mathbf{W}_{t^*}\bar{\mathbf{w}} \tag{5.79}$$

and the control

$$\mathbf{u}^*(t) = \mathbf{B}^T e^{\mathbf{A}^T(t^*-t)}\bar{\mathbf{w}}, \quad \text{for } t \in [0, t^*] \tag{5.80}$$

accomplishes the transfer of $\mathbf{x}(0)$ to $\mathbf{x}^*$ in t^* s.

We say that the system is *controllable* when

$$\mathscr{V}_u = \mathscr{X} = \text{Im}[\mathbf{W}_{t^*}]. \tag{5.81}$$

In a controllable system any target state is reachable from any initial state and

$$\mathscr{R}_0 = \mathscr{X}. \tag{5.82}$$

In this case, $\mathbf{x}(0)$ can be transferred to $\mathbf{x}^*$ in t^* s by the control

$$\mathbf{u}^*(t) = \mathbf{B}^T e^{\mathbf{A}^T(t^*-\tau)}\mathbf{W}_{t^*}^{-1}\left(\mathbf{x}^* - e^{\mathbf{A}t^*}\mathbf{x}(0)\right), \quad \text{for } \tau \in [0, t^*], \ t^* > 0. \tag{5.83}$$

Since $t^* > 0$ is arbitrary, the transfer can be accomplished in as little time as desired, provided larger control signals can be applied. Indeed, instantaneous transfers can be achieved using impulsive inputs.

EXAMPLE 5.2   Consider

$$\begin{bmatrix} \dot{x}_1(t) \\ \dot{x}_2(t) \end{bmatrix} = \begin{bmatrix} -0.5 & 0 \\ 0 & -1 \end{bmatrix} \begin{bmatrix} x_1(t) \\ x_2(t) \end{bmatrix} + \begin{bmatrix} 0.5 \\ 1 \end{bmatrix} u(t). \tag{5.84}$$

Since

$$\text{rank}[\mathbf{B} \ \ \mathbf{AB}] = \text{rank} \begin{bmatrix} 0.5 & -0.25 \\ 1 & -1 \end{bmatrix} = 2, \tag{5.85}$$

the system is controllable. Let

$$\mathbf{x}(0) = \begin{bmatrix} x_1(0) \\ x_2(0) \end{bmatrix} = \begin{bmatrix} 10 \\ -1 \end{bmatrix}. \tag{5.86}$$

We want to find $\mathbf{u}(t)$ that moves the state $\mathbf{x}(0)$ to $\mathbf{x}(t) = [0 \ 0]^T$ in 2 s. Recall that

$$\mathbf{u}_{t_1}(t) = -\mathbf{B}^T e^{\mathbf{A}^T(t_1-t)} \mathbf{W}_c(t_1)^{-1} \left(e^{\mathbf{A}t_1} \mathbf{x}(0) - \mathbf{x}(t_1) \right), \tag{5.87}$$

where

$$\mathbf{W}_c(t) = \int_0^t e^{\mathbf{A}\tau} \mathbf{B} \mathbf{B}^T e^{\mathbf{A}^T \tau} d\tau. \tag{5.88}$$

We compute this input as follows:

$$\mathbf{W}_c(2) = \int_0^2 e^{\mathbf{A}\tau} \mathbf{B} \mathbf{B}^T e^{\mathbf{A}^T \tau} d\tau$$

$$= \int_0^2 \left(\begin{bmatrix} e^{-0.5\tau} & 0 \\ 0 & e^{-\tau} \end{bmatrix} \begin{bmatrix} 0.5 \\ 1 \end{bmatrix} [0.5 \ 1] \begin{bmatrix} e^{-0.5\tau} & 0 \\ 0 & e^{-\tau} \end{bmatrix} \right) d\tau \tag{5.89}$$

$$= \begin{bmatrix} 0.2162 & 0.3167 \\ 0.3167 & 0.4908 \end{bmatrix}.$$

$$\mathbf{u}_2(t) = -\mathbf{B}^T e^{\mathbf{A}^T(t_1-t)} \mathbf{W}_c(t_1)^{-1} \left(e^{\mathbf{A}t_1} \mathbf{x}(0) - \mathbf{x}(t_1) \right)$$

$$= -[0.5 \ 1] \begin{bmatrix} e^{-0.5(2-t)} & 0 \\ 0 & e^{-(2-t)} \end{bmatrix} \underbrace{\begin{bmatrix} 84.4450 & -54.4901 \\ -54.4901 & 37.1985 \end{bmatrix}}_{W_c(2)^{-1}} \tag{5.90}$$

$$\left(\begin{bmatrix} e^{-1} & 0 \\ 0 & e^{-2} \end{bmatrix} \begin{bmatrix} 10 \\ -1 \end{bmatrix} - \begin{bmatrix} 0 \\ 0 \end{bmatrix} \right)$$

$$= -58.50 e^{0.5t} + 27.81 e^t.$$

MATLAB Solution
The following MATLAB script calculates the states $\mathbf{x}(t)$ and the designed input $\mathbf{u}_2(t)$ for verification. The time-domain plots are shown in Figure 5.1.

```
x0=[10 -1];
A=[-0.5 0; 0 -1];
B=[0.5 1]';
C=[1 1];
D=0;
sys=ss(A,B,C,D);
t=0:0.01:2.5;
u=-58.50*exp(0.5*t)+27.81*exp(t);
[y,t,x]=lsim(sys,u,t,x0);
plot(t,x(:,1),'-b',t,x(:,2),'-.b',t,u,'-g'), grid
xlabel('t'), ylabel('x(t)')
```

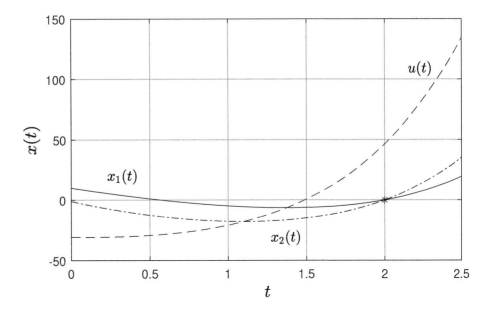

Figure 5.1 Controlled state trajectories and the designed input (Example 5.2)

Remark 5.6   In the above example, the states are transferred to the final states in 2 s. This was possible because the system was controllable and no restriction was imposed on the control effort $\mathbf{u}(t)$. However, the latter assumption often does not hold in practice. In the above example, if we restrict the control effort to say $|\mathbf{u}(t)| \leq M$ for all t, the transfer of the states to the final states may not be achieved in 2 s. Controllability only implies the existence of *some* control effort $\mathbf{u}(t)$ that transfers any state to any arbitrary state in finite time.

5.4   Observability: Extracting Internal States from Output Measurements

In general, the number n of states of a system exceeds the number m of measured outputs. The concept of observability arises in the problem of determining, if possible, the unmeasurable components of the internal state vector from output measurements

$$\dot{\mathbf{x}}(t) = \mathbf{A}\mathbf{x}(t) + \mathbf{B}\mathbf{u}(t),$$
$$\mathbf{y}(t) = \mathbf{C}\mathbf{x}(t) + \mathbf{D}\mathbf{u}(t), \qquad (5.91)$$
$$\mathbf{x}(0) = \mathbf{x}_0.$$

To proceed, first set $\mathbf{u}(t) = 0$ for all $t \geq 0$ and introduce the set of initial conditions resulting in zero output:

$$\mathcal{X}_0 := \{\mathbf{x}(0) \; : \; \mathbf{y}(t) = 0, \quad \text{for } t \geq 0 \text{ when } \mathbf{u}(t) = 0\}. \qquad (5.92)$$

Intuitively, the outputs resulting from initial conditions in $\mathcal{X}_0$ contain no information on the specific $\mathbf{x}(0) \in \mathcal{X}_0$ which results in that output. For this reason, $\mathcal{X}_0$ is defined as the set of *unobservable states*. The main result of this section is the determination of $\mathcal{X}_0$ below.

THEOREM 5.3

$$\mathcal{X}_0 = \cap_{i=0}^{n-1} \text{Ker}\left[\mathbf{C}\mathbf{A}^i\right] =: \mathcal{V}_y. \qquad (5.93)$$

To prove the theorem, it is convenient to introduce the *observability Wronskian*:

$$O_{t^*} := \int_0^{t^*} e^{\mathbf{A}^T t} \mathbf{C}^T \mathbf{C} e^{\mathbf{A}t} \, dt, \quad \text{for } t^* > 0. \qquad (5.94)$$

LEMMA 5.5

$$\mathcal{X}_0 = \text{Ker}\left[O_{t^*}\right]. \qquad (5.95)$$

Proof   With $\mathbf{u}(t) = 0$ for $t \geq 0$ in Equation (5.91), we have

$$\mathbf{y}(t) = \mathbf{C}e^{\mathbf{A}t}\mathbf{x}(0), \quad \text{for } t \geq 0. \qquad (5.96)$$

Therefore, $\mathbf{x}(0) \in \mathcal{X}_0$ implies that

$$\mathbf{y}(t) = \mathbf{C}e^{\mathbf{A}t}\mathbf{x}(0) = 0, \quad \text{for } t \in [0, t^*] \qquad (5.97\text{a})$$

$$\Leftrightarrow \quad \int_0^{t^*} \mathbf{y}^T(t)\mathbf{y}(t)dt = 0 \qquad (5.97\text{b})$$

$$\Leftrightarrow \quad \mathbf{x}^T(0)_{t^*} O_{t^*} \mathbf{x}(0) = 0 \qquad (5.97\text{c})$$

$$\Leftrightarrow \quad \mathbf{x}(0) \in \text{Ker}\left[O_{t^*}\right] \quad (O_{t^*} \text{ is symmetric}) \quad \Rightarrow \quad \mathcal{X}_0 \subset \text{Ker}\left[O_{t^*}\right]. \qquad (5.97\text{d})$$

The reverse inclusion $\text{Ker}\left[O_{t^*}\right] \subset \mathcal{X}_0$ is proved by using the fact that Equation (5.97b) implies Equation (5.97a). ∎

Proof of Theorem 5.3   The theorem is now proved by showing that

$$\mathrm{Ker}\left[O_{t^*}\right] = \mathscr{V}_y \tag{5.98}$$

using Lemma 5.5. The proof of Equation (5.98) is straightforward and follows from the fact that

$$\mathbf{y}(t) = 0, \quad \text{for } t \in [0,t^*] \quad \Leftrightarrow \quad \left.\frac{d^i\mathbf{y}(t)}{dt^i}\right|_{t=0} = 0, \quad \text{for } i = 0,1,\ldots,n-1. \tag{5.99}$$

■

To conclude this section, we show how $\mathbf{x}(0)$ can be determined uniquely from $\mathbf{y}(t)$, for $t \in [0,t^*]$ when $\mathscr{X}_0 = \mathscr{V}_y = 0$. For this, write

$$\mathbf{y}^T(t) = \mathbf{x}^T(0)e^{\mathbf{A}^T t}\mathbf{C}^T \tag{5.100a}$$

so that

$$\mathbf{y}^T(t)\mathbf{C}e^{\mathbf{A}t} = \mathbf{x}^T(0)e^{\mathbf{A}^T t}\mathbf{C}^T\mathbf{C}e^{\mathbf{A}t} \tag{5.100b}$$

and

$$\int_0^{t^*} \mathbf{y}^T(t)\mathbf{C}e^{\mathbf{A}t}dt = \mathbf{x}^T(0)\int_0^{t^*} e^{\mathbf{A}^T t}\mathbf{C}^T\mathbf{C}e^{\mathbf{A}t}dt \tag{5.100c}$$

$$= \mathbf{x}^T(0)O_{t^*} \tag{5.100d}$$

so that when $\mathscr{X}_0 = 0$:

$$\mathrm{Ker}\left[O_{t^*}\right] = 0 \tag{5.101}$$

and $O_{t^*}^{-1}$ exists, leading to

$$\mathbf{x}^T(0) = \left[\int_0^{t^*} \mathbf{y}^T(t)\mathbf{C}e^{\mathbf{A}t}dt\right]O_{t^*}^{-1} \tag{5.102}$$

as the determination of $\mathbf{x}(0)$. With $\mathbf{x}(0)$ computed, $\mathbf{x}(t)$ can be determined from

$$\mathbf{x}(t) = e^{\mathbf{A}t}\mathbf{x}(0). \tag{5.103}$$

If $\mathbf{u}(t) \neq 0$, the above calculations remain valid with $\mathbf{y}(t)$ in Equation (5.102) replaced by

$$\mathbf{y}_0(t) = \mathbf{y}(t) - \mathbf{y}_f(t) \tag{5.104}$$

where $\mathbf{y}_f(t)$ is the forced response:

$$\mathbf{y}_f(t) = \mathbf{D}\mathbf{u}(t) + \int_0^t \mathbf{C}e^{\mathbf{A}(t-\tau)}\mathbf{B}\mathbf{u}(\tau)d\tau. \tag{5.105}$$

5.5   Minimal Realizations

The *order* of a state-space system is defined to be the number of integrators contained in the realization and is clearly the same as the number of state variables. In general, a given transfer function matrix may be realized by state-space systems of various orders since the order of any realization can certainly be increased by adding integrators not connected to the inputs or to the outputs. An important question that arises therefore is: How can one test if a given realization is of minimal order and if not how can one construct a minimal-order realization? The answer to this question is intimately related to controllability and observability.

5.5.1   Controllability, Observability, and Minimality

Let $\mathbf{G}(s)$ be a proper rational matrix and let $\{\mathbf{A},\mathbf{B},\mathbf{C},\mathbf{D}\}$ be a realization. If $\mathbf{A}$ is $n \times n$, we say the order of the realization is n. An important question and one that will be addressed in this section is: Is it possible to realize $\mathbf{G}(s)$ with a lower-order dynamic system? If not, n is the *minimal-order* of the system. A related question is: What is the minimal order and how do we construct a *minimal-order realization*?

This problem was completely solved by Kalman in a classical paper in 1965. The solution involves the concepts of controllability and observability, which as we have seen are also important in other areas.

THEOREM 5.4 (Minimal realization)   A realization $\{\mathbf{A},\mathbf{B},\mathbf{C},\mathbf{D}\}$ of a proper rational matrix $\mathbf{G}(s)$ is of minimal order if and only if $(\mathbf{A},\mathbf{B})$ is controllable and $(\mathbf{C},\mathbf{A})$ is observable.

Proof   First, it is clear from the Kalman canonical decomposition that a realization of the original system is provided by $\mathbf{A}_2,\mathbf{B}_2,\mathbf{C}_2$. Thus a minimal-order realization must necessarily be controllable and observable, since its order could be reduced otherwise using the Kalman decomposition. The proof is now completed by showing the converse,namely that a controllable and observable realization must be of minimal order. This is done below by using contradiction.

Suppose that $[\mathbf{A}_1,\mathbf{B}_1,\mathbf{C}_1,\mathbf{D}_1]$ and $[\mathbf{A}_2,\mathbf{B}_2,\mathbf{C}_2,\mathbf{D}_2]$ are two controllable and observable realizations of orders n_1 and n_2, respectively, of a proper transfer function matrix

$$\mathbf{G}(s) = \mathbf{P}(s) + \mathbf{D} \tag{5.106}$$

where $\mathbf{P}(s)$ is strictly proper. Then we must have

$$\mathbf{D} = \mathbf{D}_1 = \mathbf{D}_2 \tag{5.107}$$

and

$$\mathbf{P}(s) = \mathbf{C}_1 \left(s\mathbf{I} - \mathbf{A}_1\right)^{-1} \mathbf{B}_1 = \mathbf{C}_2 \left(s\mathbf{I} - \mathbf{A}_2\right)^{-1} \mathbf{B}_2. \tag{5.108}$$

From Equation (5.108) it follows that

$$\frac{\mathbf{C}_1\mathbf{B}_1}{s} + \frac{\mathbf{C}_1\mathbf{A}_1\mathbf{B}_1}{s^2} + \cdots = \frac{\mathbf{C}_2\mathbf{B}_2}{s} + \frac{\mathbf{C}_2\mathbf{A}_2\mathbf{B}_2}{s^2} + \cdots, \tag{5.109}$$

so that

$$\mathbf{C}_1\mathbf{A}_1^j\mathbf{B}_1 = \mathbf{C}_2\mathbf{A}_2^j\mathbf{B}_2, \quad \text{for} \quad j = 0, 1, 2, \ldots. \tag{5.110}$$

Now suppose arbitrarily that

$$n_1 > n_2. \tag{5.111}$$

Introduce the *Hankel matrix* defined by

$$
H := \begin{bmatrix}
\mathbf{C}_1\mathbf{B}_1 & \mathbf{C}_1\mathbf{A}_1\mathbf{B}_1 & \cdots & \mathbf{C}_1\mathbf{A}_1^{n_1-1}\mathbf{B}_1 \\
\mathbf{C}_1\mathbf{A}_1\mathbf{B}_1 & \mathbf{C}_1\mathbf{A}_1^2\mathbf{B}_1 & \cdots & \mathbf{C}_1\mathbf{A}_1^{n_1}\mathbf{B}_1 \\
\vdots & & & \vdots \\
\mathbf{C}_1\mathbf{A}_1^{n_1-1}\mathbf{B}_1 & \mathbf{C}_1\mathbf{A}_1^{n_1}\mathbf{B}_1 & \cdots & \mathbf{C}_1\mathbf{A}_1^{2n_1-2}\mathbf{B}_1
\end{bmatrix}
$$

$$
= \underbrace{\begin{bmatrix}
\mathbf{C}_1 \\
\mathbf{C}_1\mathbf{A}_1 \\
\vdots \\
\mathbf{C}_1\mathbf{A}_1^{n_1-1}
\end{bmatrix}}_{:=O_1} \underbrace{\begin{bmatrix} \mathbf{B}_1 & \mathbf{A}_1\mathbf{B}_1 & \cdots & \mathbf{A}_1^{n_1-1}\mathbf{B}_1 \end{bmatrix}}_{:=C_1}. \tag{5.112}
$$

Since O_1 and C_1 have rank n_1, it follows that H has rank n_1 (use Sylvester's rank inequality):

$$\text{rank}[H] = n_1. \tag{5.113}$$

H can also be written as

$$
H = \underbrace{\begin{bmatrix}
\mathbf{C}_2 \\
\mathbf{C}_2\mathbf{A}_2 \\
\vdots \\
\mathbf{C}_2\mathbf{A}_2^{n_1-1}
\end{bmatrix}}_{:=\bar{O}_2} \underbrace{\begin{bmatrix} \mathbf{B}_2 & \mathbf{A}_2\mathbf{B}_2 & \cdots & \mathbf{A}_2^{n_1-1}\mathbf{B}_2 \end{bmatrix}}_{:=\bar{C}_2}. \tag{5.114}
$$

From Equation (5.111) it follows that

$$\text{rank}\left[\bar{O}_2\right] = \text{rank}\left[\bar{C}_2\right] = n_2 \tag{5.115}$$

and thus, again by Sylvester's rank inequality:

$$\text{rank}[H] = n_2. \tag{5.116}$$

The conditions in Equations (5.113) and (5.116) contradict Equation (5.111). Thus, all controllable and observable realizations must have the same order and this order is the minimal order. ∎

Remark 5.7 From the above result it follows that the minimal order of a system is a property of the transfer function matrix $\mathbf{G}(s)$. The minimal order of a transfer function matrix is called its *McMillan degree*.

5.5.2   Minimal Realization from the Kalman Canonical Decomposition

As shown earlier, any state-space system has the representation (Kalman canonical decomposition)

$$\mathbf{A} = \begin{bmatrix} \mathbf{A}_1 & \mathbf{A}_3 & \mathbf{A}_5 & \mathbf{A}_7 \\ 0 & \mathbf{A}_2 & 0 & \mathbf{A}_8 \\ 0 & 0 & \mathbf{A}_4 & \mathbf{A}_9 \\ 0 & 0 & 0 & \mathbf{A}_6 \end{bmatrix}, \quad \mathbf{B} = \begin{bmatrix} \mathbf{B}_1 \\ \mathbf{B}_2 \\ 0 \\ 0 \end{bmatrix}, \quad \mathbf{C} = [0 \ \mathbf{C}_2 \ 0 \ \mathbf{C}_6], \tag{5.117}$$

with

$$\begin{bmatrix} \mathbf{A}_1 & \mathbf{A}_3 \\ 0 & \mathbf{A}_2 \end{bmatrix}, \begin{bmatrix} \mathbf{B}_1 \\ \mathbf{B}_2 \end{bmatrix} \text{ controllable} \tag{5.118}$$

and

$$\begin{bmatrix} \mathbf{A}_2 & \mathbf{A}_8 \\ 0 & \mathbf{A}_6 \end{bmatrix}, [\mathbf{C}_2 \ \mathbf{C}_6] \text{ observable.} \tag{5.119}$$

It follows from Equation (5.117) that

$$\mathbf{C}(s\mathbf{I} - \mathbf{A})^{-1}\mathbf{B} = \mathbf{C}_2(s\mathbf{I} - \mathbf{A}_2)^{-1}\mathbf{B}_2 \tag{5.120}$$

and from Equations (5.118) and (5.119) that $(\mathbf{A}_2, \mathbf{B}_2, \mathbf{C}_2)$ is controllable and observable, and therefore of minimal order.

Suppose $\mathbf{G}(s)$ is a strictly proper transfer function of McMillan degree n and $(\mathbf{A}_1, \mathbf{B}_1, \mathbf{C}_1)$ and $(\mathbf{A}_2, \mathbf{B}_2, \mathbf{C}_2)$ are two minimal-order realizations, so that

$$\mathbf{C}_1(s\mathbf{I} - \mathbf{A}_1)^{-1}\mathbf{B}_1 = \mathbf{C}_2(s\mathbf{I} - \mathbf{A}_2)^{-1}\mathbf{B}_2. \tag{5.121}$$

The following result shows that $(\mathbf{A}_1, \mathbf{B}_1, \mathbf{C}_1)$ and $(\mathbf{A}_2, \mathbf{B}_2, \mathbf{C}_2)$ are related by a similarity transformation.

LEMMA 5.6   If $(\mathbf{A}_1, \mathbf{B}_1, \mathbf{C}_1)$ and $(\mathbf{A}_2, \mathbf{B}_2, \mathbf{C}_2)$ are two minimal realizations, then there exists a similarity transformation $\mathbf{T}$ such that

$$\mathbf{A}_2 = \mathbf{T}^{-1}\mathbf{A}_1\mathbf{T}, \quad \mathbf{B}_2 = \mathbf{T}^{-1}\mathbf{B}_1, \quad \mathbf{C}_2 = \mathbf{C}_1\mathbf{T}. \tag{5.122}$$

Proof   Let

$$O_1 = \begin{bmatrix} \mathbf{C}_1 \\ \mathbf{C}_1\mathbf{A}_1 \\ \vdots \\ \mathbf{C}_1\mathbf{A}_1^{n-1} \end{bmatrix}, \quad O_2 := \begin{bmatrix} \mathbf{C}_2 \\ \mathbf{C}_2\mathbf{A}_2 \\ \vdots \\ \mathbf{C}_2\mathbf{A}_2^{n-1} \end{bmatrix}, \tag{5.123}$$

$$\begin{aligned} \mathcal{C}_1 &:= [\ \mathbf{B}_1 \quad \mathbf{A}_1\mathbf{B}_1 \quad \cdots \quad \mathbf{A}_1^{n-1}\mathbf{B}_1 \ ], \\ \mathcal{C}_2 &:= [\ \mathbf{B}_2 \quad \mathbf{A}_2\mathbf{B}_2 \quad \cdots \quad \mathbf{A}_2^{n-1}\mathbf{B}_2 \ ]. \end{aligned} \tag{5.124}$$

From Equation (5.121) it follows that

$$\mathbf{C}_1\mathbf{A}_1^i\mathbf{B}_1 = \mathbf{C}_2\mathbf{A}_2^i\mathbf{B}_2, \quad \text{for } i = 1, 2, \dots \tag{5.125}$$

and therefore

$$O_1 C_1 = O_2 C_2, \tag{5.126a}$$
$$O_1 \mathbf{A}_1 C_1 = O_2 \mathbf{A}_2 C_2. \tag{5.126b}$$

Since both realizations, being minimal, are controllable and observable:

$$\text{rank}\,[O_1] = \text{rank}\,[O_2] = \text{rank}\,[C_1] = \text{rank}\,[C_2] = n. \tag{5.127}$$

Thus

$$\left(O_2^T O_2\right)^{-1} \quad \text{and} \quad \left(C_2 C_2^T\right)^{-1} \quad \text{exist.} \tag{5.128}$$

From Equations (5.126b) and (5.128) we have

$$\underbrace{\left(O_2^T O_2\right)^{-1} O_2^T O_1}_{\mathbf{P}} \mathbf{A}_1 \underbrace{C_1 C_2^T \left(C_2 C_2^T\right)^{-1}}_{\mathbf{Q}} = \mathbf{A}_2 \tag{5.129}$$

and using Equation (5.126a) we verify that

$$\mathbf{PQ} = \mathbf{I}. \tag{5.130}$$

Letting

$$\mathbf{Q} =: \mathbf{T}, \tag{5.131}$$

we have

$$\mathbf{P} = \mathbf{T}^{-1} \tag{5.132}$$

and

$$\mathbf{A}_2 = \mathbf{T}^{-1} \mathbf{A}_1 \mathbf{T}. \tag{5.133}$$

From

$$\mathcal{O}_1 \mathbf{B}_1 = \mathcal{O}_2 \mathbf{B}_2, \tag{5.134}$$

we also have

$$\mathbf{B}_2 = \mathbf{T}^{-1} \mathbf{B}_1 \tag{5.135}$$

and from

$$\mathbf{C}_1 \mathscr{C}_1 = \mathbf{C}_2 \mathscr{C}_2, \tag{5.136}$$

we have similarly

$$\mathbf{C}_2 = \mathbf{C}_1 \mathbf{T}. \tag{5.137}$$

∎

Hankel Matrix

To conclude this section, we show that the minimal order of a strictly proper transfer function $\mathbf{G}(s)$ can be found from its power-series expansion:

$$\mathbf{G}(s) = \frac{\mathbf{H}_0}{s} + \frac{\mathbf{H}_1}{s^2} + \frac{\mathbf{H}_2}{s^3} + \cdots + \frac{\mathbf{H}_i}{s^{i+1}} + \cdots . \tag{5.138}$$

The matrices $\mathbf{H}_i$ are called the Markov parameters of $\mathbf{G}(s)$, and we construct the doubly infinite matrix

$$\mathbf{H} := \begin{bmatrix} \mathbf{H}_0 & \mathbf{H}_1 & \mathbf{H}_2 & \cdots & \mathbf{H}_j & \cdots \\ \mathbf{H}_1 & \mathbf{H}_2 & \mathbf{H}_3 & \cdots & \mathbf{H}_{j+1} & \\ \mathbf{H}_2 & \mathbf{H}_3 & \vdots & & \vdots & \\ \vdots & \vdots & \vdots & & \vdots & \\ \mathbf{H}_j & \mathbf{H}_{j+1} & & & \mathbf{H}_{j+i} & \\ \vdots & \vdots & & & \vdots & \end{bmatrix} \tag{5.139}$$

called the *Hankel matrix* of the system.

THEOREM 5.5   The minimal-order realization of $\mathbf{G}(s)$ is given by rank$[\mathbf{H}]$.

Proof   If $(\mathbf{A}, \mathbf{B}, \mathbf{C})$ is a realization of $\mathbf{G}(s)$, we must have

$$\mathbf{H}_i = \mathbf{CA}^i\mathbf{B}. \tag{5.140}$$

Now $\mathbf{H}$ can be factored as

$$\mathbf{H} = OC, \tag{5.141}$$

where O and C are the semi-infinite matrices

$$O := \begin{bmatrix} \mathbf{C} \\ \mathbf{CA} \\ \mathbf{CA}^2 \\ \vdots \end{bmatrix}, \qquad C := \begin{bmatrix} \mathbf{B} & \mathbf{AB} & \mathbf{A}^2\mathbf{B} & \cdots \end{bmatrix}. \tag{5.142}$$

If $(\mathbf{A}, \mathbf{B}, \mathbf{C})$ is a minimal-order realization, then

$$\text{rank}[O] = n \quad \text{and} \quad \text{rank}[C] = n. \tag{5.143}$$

Therefore, Equation (5.141) implies that

$$\text{rank}[\mathbf{H}] = n. \tag{5.144}$$

∎

We illustrate next how a minimal realization can be obtained using MATLAB.

EXAMPLE 5.3 MATLAB provides a built-in function "minreal()" for minimal realization. Consider a state-space realization

$$
A = \begin{bmatrix} 0 & 0 & 0 & 0 \\ 0 & 0 & 1 & 0 \\ 0 & 0 & 0 & 1 \\ -2 & -3 & -4 & -5 \end{bmatrix}, \quad B = \begin{bmatrix} 0 & 0 \\ 0 & 0 \\ 1 & 0 \\ 0 & 1 \end{bmatrix},
$$

$$
C = \begin{bmatrix} 1 & 2 & 0 & 0 \\ 0 & 0 & 0 & 1 \end{bmatrix}, \quad D = \begin{bmatrix} 0 & 0 \\ 0 & 0 \end{bmatrix}.
$$
(5.145)

The following MATLAB script results in a minimal realization:

```
A=[0 0 0 0; 0 0 1 0; 0 0 0 1; -2 -3 -4 -5];
B=[0 0; 0 0; 1 0; 0 1];
C=[1 2 0 0; 0 0 0 1];
D-[0 0; 0 0];
sys=ss(A,B,C,D);
minreal(sys)
```

The outputs of the above script are

$$
A = \begin{bmatrix} 0 & 1 & 0 \\ 0 & 0 & 1 \\ -3 & -4 & -5 \end{bmatrix}, \quad B = \begin{bmatrix} 0 & 0 \\ 1 & 0 \\ 0 & 1 \end{bmatrix}, \quad C = \begin{bmatrix} 2 & 0 & 0 \\ 0 & 0 & 1 \end{bmatrix},
$$

$$
D = \begin{bmatrix} 0 & 0 \\ 0 & 0 \end{bmatrix}.
$$
(5.146)

Remark 5.8 The MATLAB function "minreal()" uses the Kalman decomposition described in the text.

5.5.3 Alternative Tests for Controllability and Observability

An alternative and sometimes convenient approach to verifying controllability and observability of a state-space system (A, B, C) is given in the following result.

THEOREM 5.6

A. (A, B) is controllable if and only if

$$
\text{rank}[sI - A, \ B] = n, \quad \text{for all } s \in \mathbb{C}.
$$
(5.147)

B. (C, A) is observable if and only if

$$
\text{rank} \begin{bmatrix} sI - A \\ C \end{bmatrix} = n, \quad \text{for all } s \in \mathbb{C}.
$$
(5.148)

Proof   To prove A, we show that Equation (5.147) is equivalent to

$$\text{rank} \begin{bmatrix} \mathbf{B} & \mathbf{AB} & \cdots & \mathbf{A}^{n-1}\mathbf{B} \end{bmatrix} = n. \tag{5.149}$$

If Equation (5.147) fails, there exist $\mathbf{x} \neq 0$ and $\lambda \in \mathbb{C}$ such that

$$\mathbf{x}^T(\lambda\mathbf{I} - \mathbf{A}) = 0, \quad \mathbf{x}^T\mathbf{B} = 0. \tag{5.150}$$

Then

$$\mathbf{x}^T \begin{bmatrix} \mathbf{B} & \mathbf{AB} & \cdots & \mathbf{A}^{n-1}\mathbf{B} \end{bmatrix} = 0 \tag{5.151}$$

and so Equation (5.149) fails. Therefore, Equation (5.149) implies Equation (5.147).

Now suppose Equation (5.149) fails. Then without loss of generality, we can assume that

$$\mathbf{A} = \begin{bmatrix} \mathbf{A}_1 & \mathbf{A}_3 \\ 0 & \mathbf{A}_2 \end{bmatrix}, \quad \mathbf{B} = \begin{bmatrix} \mathbf{B}_1 \\ 0 \end{bmatrix}. \tag{5.152}$$

Thus, there exist $\mathbf{x}_2 \neq 0$ and $\lambda_2 \in \mathbb{C}$ so that

$$\mathbf{x}_2^T[\lambda_2\mathbf{I} - \mathbf{A}_2] = 0. \tag{5.153}$$

Then with

$$\mathbf{x}^T := \begin{bmatrix} 0 & \mathbf{x}_2^T \end{bmatrix}, \tag{5.154}$$

we have

$$\mathbf{x}^T \begin{bmatrix} \lambda_2\mathbf{I} - \mathbf{A} & \mathbf{B} \end{bmatrix} = 0 \tag{5.155}$$

so that Equation (5.147) fails. Thus, Equation (5.147) implies Equation (5.149). The proof of B is similar and is left to the reader.   ∎

5.5.4   Stabilizability and Detectability

If $(\mathbf{A}, \mathbf{B})$ is not controllable, it may be desirable to know if the "uncontrollable subsystem" is stable. Using the controllability reduction, we may assume that

$$\mathbf{A} = \begin{bmatrix} \mathbf{A}_1 & \mathbf{A}_3 \\ 0 & \mathbf{A}_2 \end{bmatrix}, \quad \mathbf{B} = \begin{bmatrix} \mathbf{B}_1 \\ 0 \end{bmatrix} \tag{5.156}$$

with $(\mathbf{A}_1, \mathbf{B}_1)$ controllable.

DEFINITION 5.4 $(\mathbf{A}, \mathbf{B})$ in Equation (5.156) is said to be *stabilizable* if and only if $\mathbf{A}_2$ is stable.

Similarly, if $(\mathbf{C}, \mathbf{A})$ is not observable, it may be of interest to know if the "unobservable subsystem" is stable. Again, by the observability reduction, we may assume without loss of generality that

$$\mathbf{C} = \begin{bmatrix} \mathbf{C}_1 & 0 \end{bmatrix}, \quad \mathbf{A} = \begin{bmatrix} \mathbf{A}_1 & 0 \\ \mathbf{A}_3 & \mathbf{A}_2 \end{bmatrix} \tag{5.157}$$

with $(\mathbf{C}_1, \mathbf{A}_1)$ observable.

DEFINITION 5.5 (C, A) in Equation (5.157) is said to be *detectable* if and only if A_2 is stable.

These lead to the following result.

LEMMA 5.7

(a) (A, B) is stabilizable if and only if

$$\text{rank} \left[sI - A \;\; B \right] = n, \quad \text{for all } s \in \mathbb{C}^+. \tag{5.158}$$

(b) (C, A) is detectable if and only if

$$\text{rank} \left[\begin{array}{c} sI - A \\ C \end{array} \right] = n, \quad \text{for all } s \in \mathbb{C}^+. \tag{5.159}$$

5.6 Realizations via Lyapunov Equations

An alternative approach to realizations, using Lyapunov equations, is developed in this section. It is also useful for checking stability, controllability, observability, stabilizability, and detectability of a given state-space realization (A, B, C).

5.6.1 A Lyapunov Equation

The Lyapunov equation described below provides an important alternative test for controllability, observability, and stability. Besides being a characterization it is useful because reliable numerical algorithms are available for its solution.

THEOREM 5.7 (Lyapunov Theorem)   For given $A \in \mathbb{C}^{n \times n}$, suppose that $P \in \mathbb{C}^{n \times n}$ and $Q \in \mathbb{C}^{n \times n}$ satisfy the following matrix equation, called a Lyapunov equation:

$$A^* P + P A = -Q. \tag{5.160}$$

Then A is stable if and only if the solution P of Equation (5.160) is symmetric positive definite (denoted $P > 0$) for any symmetric positive definite Q ($Q > 0$). Moreover, the solution is unique in the class of symmetric positive definite matrices.

Proof   First observe that

$$-\underbrace{e^{A^* t} \left(A^* P + P A \right) e^{A t}}_{\frac{d}{dt} \left(e^{A^* t} P e^{A t} \right)} = e^{A^* t} Q e^{A t}, \quad \text{for all } t. \tag{5.161}$$

(Necessity) Now assume that A has all eigenvalues with negative real parts. Then by integrating both sides of the above equation, we have

$$-\int_0^\infty \frac{d}{dt} \left(e^{A^* t} P e^{A t} \right) dt = -\left[e^{A^* t} P e^{A t} \right]_0^\infty = P. \tag{5.162}$$

Note that since **A** is stable:

$$\lim_{t \to \infty} e^{\mathbf{A}^* t} \mathbf{P} e^{\mathbf{A} t} = 0 \qquad \text{and} \qquad \lim_{t \to 0} e^{\mathbf{A}^* t} \mathbf{P} e^{\mathbf{A} t} = \mathbf{P} \qquad (5.163)$$

and therefore,

$$\mathbf{P} = \int_0^\infty e^{\mathbf{A}^* t} \mathbf{Q} e^{\mathbf{A} t} dt \qquad (5.164)$$

is a solution of the Lyapunov equation. To establish the uniqueness of the solution, let $\mathbf{P}_1$ and $\mathbf{P}_2$ be two symmetric positive definite solutions. Then

$$\begin{aligned} \mathbf{A}^* \mathbf{P}_1 + \mathbf{P}_1 \mathbf{A} + \mathbf{Q} &= 0, \\ \mathbf{A}^* \mathbf{P}_2 + \mathbf{P}_2 \mathbf{A} + \mathbf{Q} &= 0. \end{aligned} \qquad (5.165)$$

From the above, we have

$$\mathbf{A}^* \left(\mathbf{P}_1 - \mathbf{P}_2 \right) + \left(\mathbf{P}_1 - \mathbf{P}_2 \right) \mathbf{A} = 0. \qquad (5.166)$$

Again, multiplying by $e^{\mathbf{A}^T t}$ on the left and $e^{\mathbf{A} t}$ on the right, we have

$$\begin{aligned} 0 &= e^{\mathbf{A}^* t} \left[\mathbf{A}^* \left(\mathbf{P}_1 - \mathbf{P}_2 \right) + \left(\mathbf{P}_1 - \mathbf{P}_2 \right) \mathbf{A} \right] e^{\mathbf{A} t} \\ &= e^{\mathbf{A}^* t} \mathbf{A}^* \left(\mathbf{P}_1 - \mathbf{P}_2 \right) e^{\mathbf{A} t} + e^{\mathbf{A}^* t} \left(\mathbf{P}_1 - \mathbf{P}_2 \right) \mathbf{A} e^{\mathbf{A} t} \\ &= \frac{d}{dt} \left[e^{\mathbf{A}^* t} \left(\mathbf{P}_1 - \mathbf{P}_2 \right) e^{\mathbf{A} t} \right]. \end{aligned} \qquad (5.167)$$

This implies that

$$e^{\mathbf{A}^* t} \left(\mathbf{P}_1 - \mathbf{P}_2 \right) e^{\mathbf{A} t} = \text{constant regardless of } t. \qquad (5.168)$$

So we pick an arbitrary t, say $t = 0$, then

$$e^{\mathbf{A}^* t} \left(\mathbf{P}_1 - \mathbf{P}_2 \right) e^{\mathbf{A} t} \Big|_{t=0} = \mathbf{P}_1 - \mathbf{P}_2. \qquad (5.169)$$

Now let $t \to \infty$. From the assumption that **A** has all the eigenvalues in LHP, the left-hand side of the above equation is zero:

$$0 = \mathbf{P}_1 - \mathbf{P}_2. \qquad (5.170)$$

This proves uniqueness of any positive definite solution if A is stable.

(Sufficiency) We need to show that if **P** and **Q** are positive definite, then all eigenvalues of A have negative real parts. Consider

$$\mathbf{A}^* \mathbf{P} + \mathbf{PA} = -\mathbf{Q}. \qquad (5.171)$$

Let λ be an eigenvalue of **A** and $\mathbf{x} \neq 0$ be an eigenvector associated with it. Pre-multiplying by $\mathbf{x}^*$ and post-multiplying by $\mathbf{x}$, we have

$$\mathbf{x}^* \mathbf{A}^* \mathbf{P} \mathbf{x} + \mathbf{x}^* \mathbf{PA} \mathbf{x} = -\mathbf{x}^* \mathbf{Q} \mathbf{x}. \qquad (5.172)$$

Note that

$$\mathbf{A}\mathbf{x} = \lambda \mathbf{x}, \quad \text{for } \mathbf{A} \in \mathbb{R}^{n \times n}, \ \mathbf{x} \in \mathbb{C}^n, \ \lambda \in \mathbb{C}. \qquad (5.173)$$

Taking the complex-conjugate transpose, we have

$$\mathbf{x}^*\mathbf{A}^* = \lambda^*\mathbf{x}^*. \tag{5.174}$$

Thus, we have

$$\begin{aligned}
\mathbf{x}^*\mathbf{A}^*\mathbf{Px} + \mathbf{x}^*\mathbf{PAx} &= \lambda^*\mathbf{x}^*\mathbf{Px} + \mathbf{x}^*\mathbf{P}(\lambda\mathbf{x}) \\
&= \lambda^*\mathbf{x}^*\mathbf{Px} + \lambda\mathbf{x}^*\mathbf{Px} \\
&= (\lambda^* + \lambda)\mathbf{x}^*\mathbf{Px} \\
&= 2\operatorname{Re}(\lambda)\mathbf{x}^*\mathbf{Px} = -\mathbf{x}^*\mathbf{Qx}.
\end{aligned} \tag{5.175}$$

Since $\mathbf{Q}$ and $\mathbf{P}$ are positive definite, $-\mathbf{x}^*\mathbf{Qx} < 0$ and $\mathbf{x}^*\mathbf{Px} > 0$. Thus, $\operatorname{Re}(\lambda)$ must be negative. ∎

Now we state a stronger version of the Lyapunov Theorem which will be needed to establish the results given in Section 5.6.3.

LEMMA 5.8 If $\mathbf{A}$ is stable, the Lyapunov equation

$$\mathbf{A}^T\mathbf{P} + \mathbf{PA} = -\mathbf{CC}^T \tag{5.176}$$

has the solution

$$\mathbf{P}_0 = \int_0^\infty e^{\mathbf{A}^T t}\mathbf{CC}^T e^{\mathbf{A}t}\,dt. \tag{5.177}$$

Proof From Equation (5.161), note that

$$\begin{aligned}
\mathbf{A}^T\mathbf{P}_0 + \mathbf{P}_0\mathbf{A} &= \int_0^\infty \frac{d}{dt}\left[e^{\mathbf{A}^T t}\mathbf{CC}^T e^{\mathbf{A}t}\right]dt \\
&= e^{\mathbf{A}^T t}\mathbf{CC}^T e^{\mathbf{A}t}\Big|_0^\infty = -\mathbf{CC}^T.
\end{aligned} \tag{5.178}$$

∎

THEOREM 5.8

(1) If $(\mathbf{C},\mathbf{A})$ is detectable and

$$\mathbf{A}^T\mathbf{P} + \mathbf{PA} = -\mathbf{CC}^T \tag{5.179}$$

has a solution $\mathbf{P}$ which is positive semi-definite, then $\mathbf{A}$ must be stable.
(2) If $(\mathbf{A},\mathbf{B})$ is stabilizable and

$$\mathbf{A}^T\mathbf{P} + \mathbf{PA} = -\mathbf{BB}^T \tag{5.180}$$

has a solution $\mathbf{P}$ which is symmetric positive semi-definite, then $\mathbf{A}$ must be stable.

Proof The statement (2) of the theorem is dual of the statement (1). Thus, we only prove the statement (1). Suppose that $\mathbf{A}$ is not stable and let

$$\lambda = \sigma + j\omega, \quad \text{for } \sigma \geq 0 \tag{5.181}$$

be an RHP eigenvalue of $\mathbf{A}$ with eigenvector $\mathbf{v}$. Then

$$\mathbf{Av} = \lambda\mathbf{v} \tag{5.182}$$

and

$$\mathbf{v}^* \mathbf{A}^T \mathbf{P} \mathbf{v} + \mathbf{v}^* \mathbf{P} \mathbf{A} \mathbf{v} = -\mathbf{v}^* \mathbf{C} \mathbf{C}^T \mathbf{v}, \tag{5.183a}$$

$$\lambda^* \mathbf{v}^* \mathbf{P} \mathbf{v} + \lambda \mathbf{v}^* \mathbf{P} \mathbf{v} = -\mathbf{v}^* \mathbf{C} \mathbf{C}^T \mathbf{v}, \tag{5.183b}$$

$$2\sigma \mathbf{v}^* \mathbf{P} \mathbf{v} = -\mathbf{v}^* \mathbf{C} \mathbf{C}^T \mathbf{v}. \tag{5.183c}$$

If $\sigma = 0$, the left-hand side of Equation (5.183c) is zero, whereas the right-hand side is negative since $\mathbf{C}^T \mathbf{v} \neq 0$ by detectability. On the other hand, if $\sigma > 0$, we have

$$\mathbf{v}^* \mathbf{P} \mathbf{v} < 0 \tag{5.184}$$

again by detectability. However, Equation (5.184) contradicts the positive semi-definiteness of $\mathbf{P}$. Thus, $\mathbf{A}$ must be stable. ∎

5.6.2   Singular Value Decomposition

The singular value decomposition (SVD) provides a numerically reliable way to compute the rank of a matrix in the presence of roundoff errors. In this section we provide an introduction to the SVD which, as we show, can be employed to check stabilizability, detectability, controllability, and observability via the solution of Lyapunov equations.

THEOREM 5.9 (Singular value decomposition)   For $\mathbf{A} \in \mathbb{C}^{n \times m}$, there exist unitary matrices $\mathbf{U} \in \mathbb{C}^{n \times n}$ and $\mathbf{V} \in \mathbb{C}^{m \times m}$ such that

$$\mathbf{U}^* \mathbf{A} \mathbf{V} = \begin{bmatrix} \sigma_1 & & & \vdots & \\ & \ddots & & 0 & \\ & & \sigma_p & \vdots & \\ \hdashline & 0 & & \vdots & 0 \end{bmatrix} \in \mathbb{R}^{n \times m}, \quad \text{for } p = \min\{n, m\}, \tag{5.185}$$

where $\sigma_1 \geq \sigma_2 \geq \cdots \geq \sigma_p \geq 0$.

The proof is given in Appendix A.5.

When the matrix $\mathbf{A}$ is real, the matrices $\mathbf{U}$ and $\mathbf{V}^T$ in the SVD are also real. In the special case that the matrix $\mathbf{A}$ is real, symmetric, and positive semi-definite, the SVD provides a "factorization" of $\mathbf{A}$.

LEMMA 5.9   If $\mathbf{A} \in \mathbb{R}^{n \times n}$ is symmetric and positive semi-definite with rank k, there exists $\mathbf{H} \in \mathbb{R}^{n \times k}$ such that

$$\mathbf{A} = \mathbf{H} \mathbf{H}^T. \tag{5.186}$$

Proof   The proof is by construction. Applying the SVD to $\mathbf{A}$, we have

$$\mathbf{A} = \mathbf{U} \Sigma \mathbf{U}^T \tag{5.187}$$

where $\mathbf{U}$ is unitary (i.e., $\mathbf{U}^T = \mathbf{U}^{-1}$) and

$$\Sigma = \begin{bmatrix} \Sigma_d & 0 \\ 0 & 0 \end{bmatrix} \tag{5.188}$$

with

$$\Sigma_d = \begin{bmatrix} \sigma_1^2 & & \\ & \ddots & \\ & & \sigma_k^2 \end{bmatrix}, \quad \text{for } \sigma_i \neq 0, \ i = 1, \ldots, k. \tag{5.189}$$

Partitioning $\mathbf{U}$ conformably with Σ, we have

$$\mathbf{U} = [\mathbf{U}_1 \ \mathbf{U}_2] \tag{5.190}$$

so that from Equation (5.187)

$$\mathbf{A} = \underbrace{\left(\mathbf{U}_1 \Sigma_d^{\frac{1}{2}}\right)}_{\mathbf{H}} \underbrace{\left(\mathbf{U}_1 \Sigma_d^{\frac{1}{2}}\right)^T}_{\mathbf{H}^T}, \tag{5.191}$$

proving Equation (5.186). ∎

5.6.3   Controllability and Observability Reductions via Lyapunov Equations

Suppose that $(\mathbf{A}, \mathbf{B}, \mathbf{C})$ is an nth-order realization of a system with $\mathbf{A}$ stable and $(\mathbf{A}, \mathbf{B})$ uncontrollable. Then the Lyapunov equation

$$\mathbf{AP} + \mathbf{PA}^T + \mathbf{BB}^T = 0 \tag{5.192}$$

is satisfied by

$$\mathbf{P} := \int_0^\infty e^{\mathbf{A}t} \mathbf{BB}^T e^{\mathbf{A}^T t} dt \tag{5.193}$$

and

$$\text{rank}[\mathbf{P}] = k < n. \tag{5.194}$$

By Lemma 5.9, $\mathbf{P}$ can be factored as

$$\mathbf{P} = \mathbf{U}\Sigma_P \mathbf{U}^{-1} \tag{5.195}$$

with $\mathbf{U}$ unitary and real:

$$\Sigma_P = \begin{bmatrix} \mathbf{P}_d & 0 \\ \hline 0 & 0 \end{bmatrix}, \tag{5.196}$$

where

$$\mathbf{P}_d = \begin{bmatrix} \sigma_1^2 & & \\ & \ddots & \\ & & \sigma_k^2 \end{bmatrix}, \quad \text{for } \sigma_i \neq 0, \ i = 1, \ldots, k. \tag{5.197}$$

Write

$$\bar{\mathbf{A}} := \mathbf{U}^{-1}\mathbf{A}\mathbf{U}, \quad \bar{\mathbf{B}} := \mathbf{U}^{-1}\mathbf{B}, \quad \bar{\mathbf{C}} := \mathbf{C}\mathbf{U} \tag{5.198}$$

and note that Equation (5.192) can be rewritten as:

$$\bar{\mathbf{A}}\Sigma + \Sigma\bar{\mathbf{A}}^T + \bar{\mathbf{B}}\bar{\mathbf{B}}^T = 0. \tag{5.199}$$

Partition $(\bar{\mathbf{A}}, \bar{\mathbf{B}})$ conformably with Equation (5.196):

$$\bar{\mathbf{A}} = \left[\begin{array}{c|c} \mathbf{A}_1 & \mathbf{A}_3 \\ \hline \mathbf{A}_2 & \mathbf{A}_4 \end{array}\right], \quad \mathbf{A}_1 \in \mathbb{R}^{k \times k}$$

$$\bar{\mathbf{B}} = \left[\begin{array}{c} \mathbf{B}_1 \\ \hline \mathbf{B}_2 \end{array}\right], \quad \bar{\mathbf{C}} = [\mathbf{C}_1 \ \mathbf{C}_2]. \tag{5.200}$$

Then Equation (5.199) is equivalent to

$$\mathbf{A}_1\mathbf{P}_d + \mathbf{P}_d\mathbf{A}_1^T + \mathbf{B}_1\mathbf{B}_1^T = 0, \tag{5.201a}$$
$$\mathbf{P}_d\mathbf{A}_2^T + \mathbf{B}_1\mathbf{B}_2^T = 0, \tag{5.201b}$$
$$\mathbf{B}_2\mathbf{B}_2^T = 0. \tag{5.201c}$$

From Equation (5.201c), $\mathbf{B}_2 = 0$ and since $\mathbf{P}_d$ is nonsingular (see Equation (5.197)) it follows from Equation (5.201b) that $\mathbf{A}_2 = 0$. Thus

$$\bar{\mathbf{A}} = \left[\begin{array}{cc} \mathbf{A}_1 & \mathbf{A}_3 \\ 0 & \mathbf{A}_4 \end{array}\right], \quad \bar{\mathbf{B}} = \left[\begin{array}{c} \mathbf{B}_1 \\ 0 \end{array}\right], \quad \bar{\mathbf{C}} = [\mathbf{C}_1 \ \mathbf{C}_2]. \tag{5.202}$$

Since $\mathbf{A}$ and $\bar{\mathbf{A}}$ are similar (see Equation (5.198)), it follows that $\mathbf{A}_1$ is stable and $(\mathbf{A}_1, \mathbf{B}_1)$ is controllable with controllability Wronskian $\mathbf{P}_d$. Thus, we have reduced the system order from n to k using the controllability Lyapunov equation in Equation (5.192).

The dual result is order reduction by using the observability Lyapunov equation:

$$\mathbf{A}^T\mathbf{Q} + \mathbf{Q}\mathbf{A} + \mathbf{C}^T\mathbf{C} = 0. \tag{5.203}$$

Assuming as before that $(\mathbf{A}, \mathbf{B}, \mathbf{C})$ is a realization of order n and $\mathbf{A}$ is stable, it follows that

$$\mathbf{Q} := \int_0^\infty e^{\mathbf{A}^T t}\mathbf{C}^T\mathbf{C}e^{\mathbf{A}t}\,dt \tag{5.204}$$

satisfies Equation (5.203). Suppose that

$$\text{rank}[\mathbf{Q}] = m, \tag{5.205}$$

then we have with $\mathbf{V}$ unitary:

$$\mathbf{Q} = \mathbf{V}\Sigma_Q\mathbf{V}^{-1}, \tag{5.206}$$

where

$$\Sigma_Q = \left[\begin{array}{cc} \mathbf{Q}_d & 0 \\ 0 & 0 \end{array}\right], \tag{5.207a}$$

$$\mathbf{Q}_d = \left[\begin{array}{ccc} \mu_1^2 & & \\ & \ddots & \\ & & \mu_m^2 \end{array}\right], \quad \text{for } \mu_i \neq 0, \ i = 1,\ldots,m. \tag{5.207b}$$

With

$$\bar{\mathbf{A}} := \mathbf{V}^{-1}\mathbf{A}\mathbf{V}, \quad \bar{\mathbf{C}} := \mathbf{C}\mathbf{V}, \quad \bar{\mathbf{B}} = \mathbf{V}^{-1}\mathbf{B}, \tag{5.208}$$

Equation (5.203) may be rewritten as

$$\bar{\mathbf{A}}^T \Sigma_Q + \Sigma_Q \bar{\mathbf{A}} + \bar{\mathbf{C}}^T \bar{\mathbf{C}} = 0. \tag{5.209}$$

Partitioning $(\bar{\mathbf{A}}, \bar{\mathbf{C}})$ conformably with Equation (5.207a):

$$\bar{\mathbf{A}} = \begin{bmatrix} \mathbf{A}_1 & \mathbf{A}_3 \\ \mathbf{A}_2 & \mathbf{A}_4 \end{bmatrix}, \quad \bar{\mathbf{B}} = \begin{bmatrix} \mathbf{B}_1 \\ \mathbf{B}_2 \end{bmatrix}, \quad \bar{\mathbf{C}} = [\mathbf{C}_1 \ \ \mathbf{C}_4], \tag{5.210}$$

Equation (5.209) can be rewritten as:

$$\mathbf{A}_1^T \mathbf{Q}_d + \mathbf{Q}_d \mathbf{A}_1 + \mathbf{C}_1^T \mathbf{C} = 0, \tag{5.211a}$$

$$\mathbf{Q}_d \mathbf{A}_3 + \mathbf{C}_1^T \mathbf{C}_4 = 0, \tag{5.211b}$$

$$\mathbf{C}_4^T \mathbf{C}_4 = 0. \tag{5.211c}$$

Thus, from Equation (5.211c), $\mathbf{C}_4 = 0$ and from Equation (5.211b) and the nonsingularity of $\mathbf{Q}_d$ (see Equation (5.207b)), $\mathbf{A}_3 = 0$. Therefore

$$\bar{\mathbf{A}} = \begin{bmatrix} \mathbf{A}_1 & 0 \\ \mathbf{A}_2 & \mathbf{A}_4 \end{bmatrix}, \quad \bar{\mathbf{C}} = [\mathbf{C}_1 \ \ 0]. \tag{5.212}$$

Since $\mathbf{A}$ and $\bar{\mathbf{A}}$ are similar, $\mathbf{A}_1$ is stable and $(\mathbf{A}_1, \mathbf{B}_1, \mathbf{C}_1)$ is a realization with $(\mathbf{C}_1, \mathbf{A}_1)$ observable and observability Wronskian $\mathbf{Q}_d$. Thus, we have reduced the system order from n to m using the observability Lyapunov equation in Equation (5.203).

5.6.4 Balanced Realizations

We now discuss another Lyapunov-based type of realization called a *balanced realization*. Consider a stable and minimal realization

$$\begin{aligned} \dot{\mathbf{x}}(t) &= \mathbf{A}\mathbf{x}(t) + \mathbf{B}\mathbf{u}(t), \\ \mathbf{y}(t) &= \mathbf{C}\mathbf{x}(t) + \mathbf{D}\mathbf{u}(t). \end{aligned} \tag{5.213}$$

Then the solutions of the following Lyapunov equations:

$$\mathbf{A}\mathbf{W}_c + \mathbf{W}_c \mathbf{A}^T = -\mathbf{B}\mathbf{B}^T, \tag{5.214}$$

$$\mathbf{A}^T \mathbf{W}_o + \mathbf{W}_o \mathbf{A} = -\mathbf{C}^T \mathbf{C}, \tag{5.215}$$

are, as shown before, the *controllability Gramian*

$$\mathbf{W}_c = \int_0^\infty e^{\mathbf{A}t} \mathbf{B}\mathbf{B}^T e^{\mathbf{A}^T t} \, dt \tag{5.216}$$

and the *observability Gramian*

$$\mathbf{W}_o = \int_0^\infty e^{\mathbf{A}^T t} \mathbf{C}^T \mathbf{C} e^{\mathbf{A}t} \, dt \tag{5.217}$$

and both are positive definite.

LEMMA 5.10 Consider two equivalent minimal realizations $(\mathbf{A}, \mathbf{B}, \mathbf{C})$ and $(\bar{\mathbf{A}}, \bar{\mathbf{B}}, \bar{\mathbf{C}})$. Let $\mathbf{W}_c$, $\bar{\mathbf{W}}_c$, $\mathbf{W}_o$, $\bar{\mathbf{W}}_o$ be the controllability and observability Gramians for the respective realizations. Then $\mathbf{W}_c\mathbf{W}_o$ and $\bar{\mathbf{W}}_c\bar{\mathbf{W}}_o$ are similar and positive definite.

Proof Write

$$\bar{\mathbf{A}} = \mathbf{T}^{-1}\mathbf{A}\mathbf{T}, \quad \bar{\mathbf{B}} = \mathbf{T}^{-1}\mathbf{B}, \quad \bar{\mathbf{C}} = \mathbf{C}\mathbf{T}, \tag{5.218}$$

then

$$\bar{\mathbf{A}}\bar{\mathbf{W}}_c + \bar{\mathbf{W}}_c\bar{\mathbf{A}}^T = -\bar{\mathbf{B}}\bar{\mathbf{B}}^T \tag{5.219}$$

yields

$$\mathbf{T}^{-1}\mathbf{A}\mathbf{T}\bar{\mathbf{W}}_c + \bar{\mathbf{W}}_c\mathbf{T}^T\mathbf{A}^T\mathbf{T}^{-T} = -\mathbf{T}^{-1}\mathbf{B}\mathbf{B}^T\mathbf{T}^{-T}$$
$$\mathbf{A}\mathbf{T}\bar{\mathbf{W}}_c + \mathbf{T}\bar{\mathbf{W}}_c\mathbf{T}^T\mathbf{A}^T\mathbf{T}^{-T} = -\mathbf{B}\mathbf{B}^T\mathbf{T}^{-T}$$
$$\mathbf{A}\underbrace{\mathbf{T}\bar{\mathbf{W}}_c\mathbf{T}^T}_{\mathbf{W}_c} + \underbrace{\mathbf{T}\bar{\mathbf{W}}_c\mathbf{T}^T}_{\mathbf{W}_c}\mathbf{A}^T = -\mathbf{B}\mathbf{B}^T, \tag{5.220}$$

where "$-T$" denotes transpose of the inverse of a matrix. Similarly, we can show that

$$\mathbf{W}_c = \mathbf{T}\bar{\mathbf{W}}_c\mathbf{T}^T, \mathbf{W}_o = \mathbf{T}^{-T}\bar{\mathbf{W}}_o\mathbf{T}^{-1}. \tag{5.221}$$

Therefore

$$\mathbf{W}_c\mathbf{W}_o = \mathbf{T}\bar{\mathbf{W}}_c\mathbf{T}^T\mathbf{T}^{-T}\bar{\mathbf{W}}_o\mathbf{T}^{-1} = \mathbf{T}\bar{\mathbf{W}}_c\bar{\mathbf{W}}_o\mathbf{T}^{-1}. \tag{5.222}$$

∎

THEOREM 5.10 (A balanced realization) For any stable minimal realization $(\mathbf{A}, \mathbf{B}, \mathbf{C})$ there exists a similarity transformation such that the controllability Gramian $\mathbf{W}_c$ and observability Gramian $\mathbf{W}_o$ of its equivalent state-space realization have the property

$$\bar{\mathbf{W}}_c = \bar{\mathbf{W}}_o. \tag{5.223}$$

Such an equivalent realization is called a balanced realization.

Proof Since $\mathbf{W}_c$ is symmetric positive definite, by using Lemma 5.9 we write

$$\mathbf{W}_c =: \mathbf{Q}^T\mathbf{D}^{\frac{1}{2}}\underbrace{\mathbf{D}^{\frac{1}{2}}\mathbf{Q}}_{\mathbf{R}} =: \mathbf{R}^T\mathbf{R} \tag{5.224}$$

where $\mathbf{Q}$ is orthonormal (i.e., $\mathbf{Q}^{-1} = \mathbf{Q}^T$). Since $\mathbf{R}\mathbf{W}_o\mathbf{R}^T$ is symmetric, we write

$$\mathbf{R}\mathbf{W}_o\mathbf{R}^T = \mathbf{U}\mathbf{\Sigma}^2\mathbf{U}^T = \mathbf{U}\mathbf{\Sigma}^{\frac{1}{2}}\mathbf{\Sigma}\mathbf{\Sigma}^{\frac{1}{2}}\mathbf{U}^T \tag{5.225}$$

where $\mathbf{U}$ is orthonormal. Then we write

$$\underbrace{\mathbf{\Sigma}^{-\frac{1}{2}}\mathbf{U}^T\mathbf{R}}_{\mathbf{T}^{-T}}\mathbf{W}_o\underbrace{\mathbf{R}^T\mathbf{U}\mathbf{\Sigma}^{-\frac{1}{2}}}_{\mathbf{T}^{-1}} = \mathbf{\Sigma} =: \bar{\mathbf{W}}_o. \tag{5.226}$$

Similarly, we can show

$$\underbrace{\mathbf{\Sigma}^{\frac{1}{2}}\mathbf{U}^T\mathbf{R}^{-T}}_{\mathbf{T}^{-1}}\mathbf{W}_c\underbrace{\mathbf{R}^{-1}\mathbf{U}\mathbf{\Sigma}^{\frac{1}{2}}}_{\mathbf{T}^{-T}} = \mathbf{\Sigma} =: \bar{\mathbf{W}}_c. \tag{5.227}$$

EXAMPLE 5.4 Using the same realization in Example 5.3, we use the following MAT-LAB script to obtain a balanced realization of the given state-space realization.

```
A=[0 0 0 0; 0 0 1 0; 0 0 0 1; -2 -3 -4 -5];
B=[0 0; 0 0; 1 0; 0 1];
C=[1 2 0 0; 0 0 0 1];
D-[0 0; 0 0];
sys=ss(A,B,C,D);
balreal(minreal(sys))
```

The outputs of the above script are

$$
\mathbf{A} = \begin{bmatrix} -0.2331 & -0.7545 & 0.1656 \\ 0.7813 & -0.5099 & -0.1307 \\ -0.2645 & 0.6471 & -4.257 \end{bmatrix}, \mathbf{B} = \begin{bmatrix} 1.177 & 0.1727 \\ -1.132 & -0.2536 \\ 0.518 & 0.9795 \end{bmatrix},
$$

$$
\mathbf{C} = \begin{bmatrix} 1.065 & 1.159 & 0.1122 \\ -0.5297 & -0.04622 & 1.102 \end{bmatrix}, \mathbf{D} = \begin{bmatrix} 0 & 0 \\ 0 & 0 \end{bmatrix}. \tag{5.228}
$$

Remark 5.9 The MATLAB function "balreal" does not necessarily provide a minimal realization. So it is recommended to use "minreal" prior to using "balreal" to obtain a balanced minimal realization.

A balanced realization is also an effective tool to obtain model order reduction by truncating the modes that are *almost uncontrollable* or *almost unobservable*. The following example illustrates this.

EXAMPLE 5.5 Consider the state-space description below:

$$
\mathbf{A} = \begin{bmatrix} -2.75 & -1.75 & 0.25 & 1.75 & 0 \\ -0.75 & -1.75 & 1.25 & -0.25 & 0 \\ 0.5 & 0.5 & -4.5 & -0.5 & 0 \\ 1 & 1 & 1 & -3 & 0 \\ 1.25 & -0.75 & -2.75 & -0.25 & -3 \end{bmatrix}, \mathbf{B} = \begin{bmatrix} 0.5 & -0.25 \\ 0 & 0.75 \\ 0 & -0.5 \\ 0.5 & 1 \\ 1 & -0.2499 \end{bmatrix},
$$

$$
\mathbf{C} = \begin{bmatrix} 1 & 0 & 1 & 3 & 1 \\ 2.0001 & 3.0001 & -0.9999 & -0.0001 & 1 \end{bmatrix}, \mathbf{D} = \mathbf{0}_2. \tag{5.229}
$$

As shown below, the smallest singular values of the controllability Gramian and the observability Gramian by using the following MATLAB script are close to zero, indicating that the given state-space representation has modes that are close to being uncontrollable and/or unobservable.

```
A=[-2.75 -1.75 0.25 1.75 0;
  -0.75 -1.75 1.25 -0.25 0;
  0.5 0.5 -4.5 -0.5 0;
  1 1 1 -3 0;
  1.25 -0.75 -2.75 -0.25 -3];
B=[0.5 -0.25; 0 0.75; 0 -0.5; 0.5 1; 1 -0.2499];
C=[1 0 1 3 1; 2.0001 3.0001 -0.9999 -0.0001 1];
D=zeros(2,2);
sys=ss(A,B,C,D);
Wc=gram(sys,'c');
Wo=gram(sys,'o'); eWc=svd(Wc);
eWo=svd(Wo);
```

$$\sigma\left(\mathbf{W}_c\right) = \left\{ \begin{array}{c} 0.4293 \\ 0.3439 \\ 0.0089 \\ 0.0020 \\ 0.0000 \end{array} \right\}, \quad \sigma\left(\mathbf{W}_o\right) = \left\{ \begin{array}{c} 3.9075 \\ 1.7533 \\ 0.1255 \\ 0.0139 \\ 4.571 \times 10^{-12} \end{array} \right\}. \tag{5.230}$$

Using the MATLAB command

```
sys=ss(A,B,C,D);
[sysb,g]=balreal(sys);
```

we obtain the balanced realization

$$\bar{\mathbf{A}} = \begin{bmatrix} -1.946 & 0.6732 & 0.356 & -0.000105 & -0.001994 \\ 0.6732 & -1.486 & -0.3542 & -0.01369 & -0.001463 \\ 0.356 & -0.3542 & -2.568 & 0.004212 & 0.02926 \\ 0.0001058 & 0.01369 & -0.0042 & -4 & -2.096 \\ -5.5 \times 10^{-14} & 3.4 \times 10^{-14} & 1.7 \times 10^{-14} & -6.8 \times 10^{-13} & -5 \end{bmatrix},$$

$$\bar{\mathbf{B}} = \begin{bmatrix} -1.724 & -1.16 \\ -0.02182 & 0.8019 \\ 0.1611 & 0.1032 \\ 0.00251 & -0.003629 \\ 9.02 \times 10^{-15} & -1.94 \times -14 \end{bmatrix}, \tag{5.231}$$

$$\bar{\mathbf{C}} = \begin{bmatrix} -1.724 & -0.02179 & 0.1611 & -0.00251 & -0.001521 \\ -1.16 & 0.8019 & 0.1032 & 0.003629 & 0.0003536 \end{bmatrix}, \quad \bar{\mathbf{D}} = 0_2$$

and the singular values of its controllability Gramian and observability Gramian are:

$$\sigma\left(\bar{\mathbf{W}}_c\right) = \left\{ \begin{array}{c} 1.1097 \\ 0.2165 \\ 0.0071 \\ 2.4338^{-6} \\ 0.0000 \end{array} \right\} \approx \sigma\left(\bar{\mathbf{W}}_o\right\} = \left\{ \begin{array}{c} 1.1097 \\ 0.2165 \\ 0.0071 \\ 2.4338^{-6} \\ 2.4382 \times 10^{-7} \end{array} \right\}. \tag{5.232}$$

The following additional MATALB script applied to this balanced realization results in a reduced-order model, that is minimal, considering the tolerance "g" for the states that can be removed:

```
[sysb,g]=balreal(sys);
elim=(g<1e-5);
sysm=modred(sysb,elim)
pole(sysm)
```

$$
\mathbf{A} = \begin{bmatrix} -1.946 & 0.6732 & 0.356 \\ 0.6732 & -1.486 & -0.3541 \\ 0.356 & -0.3542 & -2.568 \end{bmatrix}, \mathbf{B} = \begin{bmatrix} -1.724 & -1.16 \\ -0.02183 & 0.8019 \\ 0.1611 & 0.1032 \end{bmatrix}, \quad (5.233)
$$

$$
\mathbf{C} = \begin{bmatrix} -1.724 & -0.0218 & 0.1611 \\ -1.16 & 0.8019 & 0.1032 \end{bmatrix}, \mathbf{D} = \begin{bmatrix} -1.575 \times 10^{-6} & 2.277 \times 10^{-6} \\ 2.278 \times 10^{-6} & -3.293 \times 10^{-6} \end{bmatrix}.
$$

The eigenvalues of the reduced system $\mathbf{A}$ above are found as

$$
\lambda(\mathbf{A}) = \{-1.0000, -2.0000, -3.0000\}. \quad (5.234)
$$

Moreover, the reduced-order model realization above is balanced:

$$
\sigma(\mathbf{W}_c) = \sigma(\mathbf{W}_o) = \left\{ \begin{array}{c} 1.10969 \\ 0.21651 \\ 0.00713 \end{array} \right\}. \quad (5.235)
$$

Note that MATLAB provides more built-in functions for model reduction other than "modred()" used here. They are "balred()" and "balancmr()," but they require to provide the order of a reduced model.

5.7 Some Common Realizations

In this section, we present some common realizations of a transfer function matrix $\mathbf{G}(s)$. For convenience, we will assume that $\mathbf{G}(s)$ is strictly proper so that $\mathbf{D} = 0$.

5.7.1 Companion Form Realizations

In the last section, we established the important result that a state-space model is of minimal order if and only if it is controllable and observable. In this section, we establish the equivalence between minimality of the state-space realization and irreducibility of the associated transfer function or coprimeness of the numerator and denominator of the transfer function.

Single-Input Single-Output (SISO) Systems
Consider the strictly proper scalar transfer function

$$g(s) = \frac{c(s)}{a(s)} = \mathbf{c}^T(s\mathbf{I} - \mathbf{A})^{-1}\mathbf{b}, \tag{5.236}$$

where

$$a(s) = s^n + a_{n-1}s^{n-1} + \cdots + a_1 s + a_0, \tag{5.237a}$$

$$c(s) = c_0 + c_1 s + \cdots + c_{n-1}s^{n-1}. \tag{5.237b}$$

A controllable realization of $\mathbf{g}(s)$ is given by

$$\mathbf{A} = \begin{bmatrix} 0 & 1 & 0 & \cdots & 0 \\ 0 & 0 & 1 & & \\ & & & \ddots & \\ & & & & 1 \\ -a_0 & -a_1 & \cdots & \cdots & -a_{n-1} \end{bmatrix}, \mathbf{b} = \begin{bmatrix} 0 \\ 0 \\ \vdots \\ 0 \\ 1 \end{bmatrix}, \mathbf{c}^T = \begin{bmatrix} c_0 & c_1 & \cdots & c_{n-1} \end{bmatrix}. \tag{5.238}$$

The realization of Equation (5.238) is observable and therefore of minimal order if and only if

$$\mathrm{rank} \begin{bmatrix} s\mathbf{I} - \mathbf{A} \\ \mathbf{c}^T \end{bmatrix} = n, \quad \text{for all } s \in \mathbb{C}. \tag{5.239}$$

Equations (5.238) and (5.239) are equivalent to

$$\mathrm{rank} \begin{bmatrix} s & -1 & 0 & \cdots & & 0 \\ 0 & s & -1 & & & \\ \vdots & & & & & \\ 0 & & & & s & -1 \\ a_0 & a_1 & \cdots & \cdots & a_{n-2} & a_{n-1} \\ c_0 & c_1 & \cdots & \cdots & c_{n-2} & c_{n-1} \end{bmatrix} = n, \quad \text{for all } s \in \mathbb{C}. \tag{5.240}$$

Let

$$\psi(s) := \begin{bmatrix} 1 & 0 & \cdots & 0 \\ s & 1 & & 0 \\ s^2 & s & & \vdots \\ \vdots & \vdots & & 0 \\ s^{n-1} & s^{n-2} & & 1 \end{bmatrix} \tag{5.241}$$

and note that

$$\mathrm{rank} \begin{bmatrix} s\mathbf{I} - \mathbf{A} \\ \mathbf{c}^T \end{bmatrix} = \mathrm{rank} \left(\begin{bmatrix} s\mathbf{I} - \mathbf{A} \\ \mathbf{c}^T \end{bmatrix} \psi(s) \right) = \mathrm{rank} \begin{bmatrix} 0 & \vdots & \\ \vdots & & \mathbf{I}_{n-1} \\ 0 & \vdots & \\ \hline a(s) & \vdots & 0 \\ c(s) & \vdots & 0 \end{bmatrix}. \tag{5.242}$$

Therefore, Equation (5.239) is satisfied if and only if

$$\text{rank}[a(s) \quad c(s)] = 1, \quad \text{for all } s \in \mathbb{C} \tag{5.243}$$

or equivalently if $a(s)$ and $c(s)$ are coprime.

EXAMPLE 5.6 Consider the circuit with variables as shown in Figure 5.2.

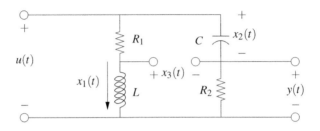

Figure 5.2 An electrical circuit (Example 5.6)

We write *state equations* as follows:

$$u(t) = R_1 x_1(t) + L\dot{x}_1(t), \tag{5.244a}$$
$$u(t) = R_2 C\dot{x}_2(t) + x_2(t), \tag{5.244b}$$
$$x_2(t) = R_1 x_1(t) + x_3(t), \tag{5.244c}$$

so that with

$$\tau_1 = \frac{R_1}{L} \quad \text{and} \quad \tau_2 = \frac{1}{R_2 C}$$

$$\dot{x}_1(t) = -\tau_1 x_1(t) + \frac{1}{L}u(t),$$
$$\dot{x}_2(t) = -\tau_2 x_2(t) + \tau_2 u(t),$$
$$\dot{x}_3(t) = -R_1\dot{x}_1(t) + \dot{x}_2(t) = R_1\tau_1 x_1(t) - \tau_2 x_2(t) + (\tau_2 - \tau_1)u(t),$$

and

$$y(t) = -x_2(t) + u(t).$$

In standard notation

$$\begin{bmatrix} \dot{x}_1(t) \\ \dot{x}_2(t) \\ \dot{x}_3(t) \end{bmatrix} = \begin{bmatrix} -\tau_1 & 0 & 0 \\ 0 & -\tau_2 & 0 \\ R_1\tau_1 & -\tau_2 & 0 \end{bmatrix} \begin{bmatrix} x_1(t) \\ x_2(t) \\ x_3(t) \end{bmatrix} + \begin{bmatrix} \frac{1}{L} \\ \tau_2 \\ \tau_2 - \tau_1 \end{bmatrix} u(t), \tag{5.245}$$

$$y(t) = [0 \quad -1 \quad 0]x(t) + [1]u(t).$$

The controllability matrix

$$
\mathscr{C}_1 := \begin{bmatrix} \mathbf{B} & \mathbf{AB} & \mathbf{A}^2\mathbf{B} \end{bmatrix} = \begin{bmatrix} \dfrac{1}{L} & -\dfrac{\tau_1}{L} & \dfrac{\tau_1^2}{L} \\[2ex] \tau_2 & -\tau_2^2 & \tau_2^3 \\[2ex] \tau_2 - \tau_1 & \tau_1^2 - \tau_2^2 & -\tau_1^3 + \tau_2^3 \end{bmatrix} \tag{5.246}
$$

and it is easy to verify that

$$
\operatorname{rank}[\mathscr{C}_1] \le 2, \quad \text{for all } \tau_1, \tau_2 \tag{5.247}
$$

and

$$
\operatorname{rank}[\mathscr{C}_1] = 1, \quad \text{for } \tau_1 = \tau_2. \tag{5.248}
$$

This means that the system in Equation (5.245) is generically uncontrollable.

This is expected since Equation (5.244c) shows that *independent* control over $x_1(t)$, $x_2(t)$, and $x_3(t)$ is impossible. This means that Equations (5.244a)–(5.244c) have redundant variables and a more meaningful model could be the second-order model

$$
\begin{bmatrix} \dot{x}_1(t) \\ \dot{x}_2(t) \end{bmatrix} = \begin{bmatrix} -\tau_1 & 0 \\ 0 & -\tau_2 \end{bmatrix} \begin{bmatrix} x_1(t) \\ x_2(t) \end{bmatrix} + \begin{bmatrix} \dfrac{1}{L} \\ \tau_2 \end{bmatrix} u(t),
$$
$$
y(t) = \begin{bmatrix} 0 & -1 \end{bmatrix} \begin{bmatrix} x_1(t) \\ x_2(t) \end{bmatrix} + [1]u(t). \tag{5.249}
$$

The controllability matrix of Equation (5.249) is

$$
\mathscr{C}_2 := \begin{bmatrix} \dfrac{1}{L} & -\dfrac{\tau_1}{L} \\[2ex] \tau_2 & -\tau_2^2 \end{bmatrix} \tag{5.250}
$$

and it is easy to see that

$$
\operatorname{rank}[\mathscr{C}_2] = 2, \quad \text{for } \tau_1 \ne \tau_2 \tag{5.251}
$$

and

$$
\operatorname{rank}[\mathscr{C}_2] = 1, \quad \text{for } \tau_1 = \tau_2. \tag{5.252}
$$

Thus, Equation (5.249) is generically controllable and loses controllability only when Equation (5.252) holds. In fact, it can easily be seen that when Equation (5.252) holds two inputs are necessary to render Equation (5.249) controllable.

It is easy to check that both the third and second-order models are unobservable independent of the circuit parameter values. Indeed, it is possible to check that the system transfer function is

$$
\frac{Y(s)}{U(s)} = \frac{-\tau_2 s (s + \tau_1)}{s (s + \tau_2)(s + \tau_1)} + 1 = -\frac{\tau_2}{s + \tau_2} + 1, \tag{5.253}
$$

showing that it is possible to realize it as a first-order system. This is due to the fact that $y(t)$ is determined independent of $x_1(t)$.

Applying the duality reasoning to the transfer function $\mathbf{g}(s)$, we have

$$
\mathbf{A}_o = \begin{bmatrix} 0 & \cdots & \cdots & 0 & -a_0 \\ 1 & & & & -a_1 \\ 0 & \ddots & & & \vdots \\ \vdots & \ddots & \ddots & & \vdots \\ 0 & \cdots & \cdots & 1 & -a_{n-1} \end{bmatrix}, \quad \mathbf{b}_o = \begin{bmatrix} c_0 \\ c_1 \\ \vdots \\ \vdots \\ c_{n-1} \end{bmatrix}, \quad \mathbf{C}_o = \begin{bmatrix} 0 & 0 & \cdots & 0 & 1 \end{bmatrix} \quad (5.254)
$$

as another realization of $\mathbf{g}(s)$. This realization is always observable and is called an *observable companion* realization. The realization will be controllable, and therefore of minimal order, if and only if $(c(s), a(s))$ is a coprime pair.

Single-Input Multi-Output (SIMO) Systems

Now consider the strictly proper single-input m-output system with transfer function

$$
\mathbf{G}(s) = \begin{bmatrix} \dfrac{c_1(s)}{a(s)} \\ \vdots \\ \dfrac{c_m(s)}{a(s)} \end{bmatrix} = \mathbf{C}^T (s\mathbf{I} - \mathbf{A})^{-1} \mathbf{b}, \quad (5.255)
$$

with

$$
a(s) = s^n + a_{n-1} s^{n-1} + \cdots + a_1 s + a_0, \quad (5.256a)
$$
$$
c_i(s) = c_0^i + c_1^i s + \cdots + c_{n-1}^i s^{n-1}, \quad \text{for } i \in \mathbf{m}. \quad (5.256b)
$$

A controllable realization is given by

$$
\mathbf{A} = \begin{bmatrix} 0 & 1 & 0 & \cdots & 0 \\ 0 & 0 & 1 & & 0 \\ & & & \ddots & \\ & & & & 1 \\ -a_0 & -a_1 & -a_2 & \cdots & -a_{n-1} \end{bmatrix}, \quad \mathbf{b} = \begin{bmatrix} 0 \\ 0 \\ \vdots \\ 0 \\ 1 \end{bmatrix},
$$

$$
\mathbf{C}^T = \begin{bmatrix} c_0^1 & \cdots & c_{n-1}^1 \\ \vdots & & \vdots \\ c_0^m & & c_{n-1}^m \end{bmatrix} := \begin{bmatrix} \mathbf{c}_1^T \\ \mathbf{c}_2^T \\ \vdots \\ \mathbf{c}_m^T \end{bmatrix}. \quad (5.257)
$$

The realization in Equation (5.257) is observable if and only if

$$
\operatorname{rank} \underbrace{\begin{bmatrix} s\mathbf{I} - \mathbf{A} \\ \mathbf{C}^T \end{bmatrix}}_{\mathcal{O}(s)} = n, \quad \text{for all } s \in \mathbb{C}. \quad (5.258)
$$

Note that with $\psi(s)$ as in Equation (5.241):

$$\text{rank}[\mathscr{O}(s)] = \text{rank}[\mathscr{O}(s)\psi(s)] = \text{rank} \begin{bmatrix} 0 & \\ 0 & \mathbf{I}_{n-1} \\ 0 & \\ \hline a(s) & 0 \\ c_1(s) & 0 \\ \vdots & \vdots \\ c_m(s) & 0 \end{bmatrix} \tag{5.259}$$

so that Equation (5.258) is satisfied if and only if

$$\text{rank}\,[a(s)\ c_1(s)\ \cdots\ c_m(s)] = 1, \quad \text{for all } s \in \mathbb{C}. \tag{5.260}$$

That is, if and only if the polynomials $\{a(s), c_1(s), \ldots, c_m(s)\}$ have no common factor.

A similar proof can be given for multi-input single-output systems by dualizing the above proof. To summarize, we have the following result.

THEOREM 5.11   A strictly proper vector transfer function

$$\mathbf{G}(s) = \frac{1}{a(s)} \begin{bmatrix} c_1(s) \\ c_2(s) \\ \vdots \\ c_m(s) \end{bmatrix} \tag{5.261}$$

has a minimal realization of order equal to the degree of $a(s)$ if and only if no zero of $a(s)$ is a common zero of $c_i(s)$ for all $i \in \mathbf{m}$, or equivalently if Equation (5.260) holds.

Dual Case

The dual case is where $\mathbf{G}(s)$ is the transfer function of a single-output system:

$$\mathbf{G}(s) = \begin{bmatrix} \dfrac{c_1(s)}{a(s)} & \dfrac{c_2(s)}{a(s)} & \cdots & \dfrac{c_m(s)}{a(s)} \end{bmatrix} \tag{5.262}$$

with $a(s)$ as in Equation (5.256a) and $c_i(s)$ as in Equation (5.256b). The realization, in this case, is

$$\mathbf{A}_o = \begin{bmatrix} 0 & \cdots & \cdots & 0 & -a_0 \\ 1 & & & & -a_1 \\ 0 & \ddots & & & \vdots \\ \vdots & \ddots & \ddots & & \vdots \\ 0 & \cdots & \cdots & 1 & -a_{n-1} \end{bmatrix}, \quad \mathbf{B}_o = \begin{bmatrix} c_0^1 & c_0^2 & \cdots & c_0^m \\ c_1^1 & c_1^2 & & c_1^m \\ \vdots & & & \\ c_{n-1}^1 & c_{n-1}^2 & \cdots & c_{n-1}^m \end{bmatrix}, \tag{5.263}$$

$$\mathbf{c}_o = [0\ 0\ \cdots\ 0\ 1].$$

The realization of Equation (5.263) is always observable and will be controllable, and therefore of minimal order, if and only if $[a(s), c_1(s), \ldots, c_m(s)]$ have no common factor.

When multiple inputs and multiple outputs are present, it is more difficult to find a minimal-order realization and this requires the matrix fraction description (MFD) of

$\mathbf{G}(s)$, developed later in this chapter. However, the procedures described above can be extended in an obvious way as shown below.

MIMO Systems

Consider the $m \times r$ strictly proper transfer function matrix $\mathbf{G}(s)$. Rewrite $\mathbf{G}(s)$ so that all elements of each column have the same denominator by introducing additional common factors if necessary. Then write

$$\mathbf{G}(s) = \begin{bmatrix} \dfrac{n_{11}(s)}{a_1(s)} & \dfrac{n_{12}(s)}{a_2(s)} & \cdots & \dfrac{n_{1r}(s)}{a_r(s)} \\ \dfrac{n_{21}(s)}{a_1(s)} & \dfrac{n_{22}(s)}{a_2(s)} & & \dfrac{n_{2r}(s)}{a_r(s)} \\ \vdots & & & \\ \dfrac{n_{m1}(s)}{a_1(s)} & \dfrac{n_{m2}(s)}{a_2(s)} & \cdots & \dfrac{n_{mr}(s)}{a_r(s)} \end{bmatrix} \tag{5.264}$$

with

$$a_i(s) = s^{d_i} + a_i^{d_i-1} s^{d_i-1} + \cdots + a_i^1 s + a_i^0, \quad i = 1, 2, \ldots, r, \tag{5.265}$$

$$n_{ij}(s) = n_{ij}^0 + n_{ij}^1 s + \cdots + n_{ij}^{d_i-1} s^{d_i-1}, \quad i = 1, 2, \ldots, m; j = 1, 2, \ldots, r. \tag{5.266}$$

Define

$$\bar{n}_{ij} := \begin{bmatrix} n_{ij}^0 & n_{ij}^1 & \cdots & n_{ij}^{d_i-1} \end{bmatrix}, \quad i = 1, 2, \ldots, m; j = 1, 2, \ldots, r. \tag{5.267}$$

Now it is readily verified that a realization of $\mathbf{G}(s)$ is

$$\mathbf{A}_c = \begin{bmatrix} \mathbf{A}_c^1 & 0 & \cdots & 0 \\ 0 & \mathbf{A}_c^2 & & \\ \vdots & & \ddots & \\ 0 & \cdots & & \mathbf{A}_c^r \end{bmatrix}, \quad \mathbf{B}_c = \begin{bmatrix} \mathbf{b}_c^1 & 0 & \cdots & 0 \\ 0 & \mathbf{b}_c^2 & & \\ \vdots & & \ddots & \\ 0 & \cdots & & \mathbf{b}_c^r \end{bmatrix},$$

$$\mathbf{C}_c = \begin{bmatrix} \bar{n}_{11} & \bar{n}_{12} & \cdots & \bar{n}_{1r} \\ \vdots & & & \\ \bar{n}_{m1} & \bar{n}_{m2} & \cdots & \bar{n}_{mr} \end{bmatrix}, \tag{5.268}$$

where

$$\mathbf{A}_c^i = \begin{bmatrix} 0 & 1 & 0 & \cdots & 0 \\ & & 1 & & 0 \\ & & & \ddots & \vdots \\ & & & & 1 \\ -a_i^0 & -a_i^1 & \cdots & \cdots & -a_i^{d_i-1} \end{bmatrix}, \quad \mathbf{b}_c^i = \begin{bmatrix} 0 \\ 0 \\ \vdots \\ 0 \\ 1 \end{bmatrix}. \tag{5.269}$$

Clearly, $(\mathbf{A}_c^i, \mathbf{b}_c^i)$ is controllable for every $i \in \mathbf{r}$. In general, $(\mathbf{C}_c, \mathbf{A}_c)$ may or may not be observable. If it is, Equation (5.268) will represent a minimal-order realization.

The above procedure can be "dualized." In that case, we rewrite $\mathbf{G}(s)$ with each row having the same denominator and develop a realization $(\mathbf{A}_o, \mathbf{B}_o, \mathbf{C}_o)$ with $(\mathbf{C}_o, \mathbf{A}_o)$ observable and $\mathbf{B}_o$ containing the numerator coefficients. The details are explained through an example.

EXAMPLE 5.7   Suppose

$$
\mathbf{G}(s) = \begin{bmatrix} \dfrac{s+2}{(s+1)^2} & \dfrac{s+4}{s^2+3s+4} \\[3mm] \dfrac{s+3}{(s+1)^2} & \dfrac{s+5}{s^2+3s+4} \end{bmatrix}.
\tag{5.270}
$$

Then from Equation (5.268) we have

$$
\mathbf{A}_c = \left[\begin{array}{cc|cc} 0 & 1 & & \\ -1 & -2 & & \\ \hline & & 0 & 1 \\ & & -4 & -3 \end{array}\right], \quad
\mathbf{B}_c = \left[\begin{array}{c|c} 0 & \\ 1 & \\ \hline & 0 \\ & 1 \end{array}\right],
\tag{5.271}
$$

$$
\mathbf{C}_c = \left[\begin{array}{cc|cc} 2 & 1 & 4 & 1 \\ 3 & 1 & 5 & 1 \end{array}\right].
$$

EXAMPLE 5.8   Consider

$$
\mathbf{G}(s) = \begin{bmatrix} \dfrac{s+2}{(s+1)^2} & \dfrac{s+3}{(s+1)^2} \\[3mm] \dfrac{s+4}{s^2+3s+4} & \dfrac{s+5}{s^2+3s+4} \end{bmatrix}.
\tag{5.272}
$$

Then we have

$$
\mathbf{A}_o = \left[\begin{array}{cc|cc} 0 & -1 & & \\ 1 & -2 & & \\ \hline & & 0 & -4 \\ & & 1 & -3 \end{array}\right], \quad
\mathbf{B}_o = \left[\begin{array}{c|c} 2 & 3 \\ 1 & 1 \\ \hline 4 & 5 \\ 1 & 1 \end{array}\right], \quad
\mathbf{C}_o = \left[\begin{array}{cc|cc} 0 & 1 & & \\ \hline & & 0 & 1 \end{array}\right].
\tag{5.273}
$$

5.7.2   Gilbert's Realization

Gilbert's realization is a particular minimal realization which can be obtained directly from a transfer function matrix $\mathbf{G}(s)$ when each entry of $\mathbf{G}(s)$ has distinct poles. When each entry $g_{ij}(s)$ of $\mathbf{G}(s)$ has no repeated poles, we may write

$$
\mathbf{G}(s) = \frac{[R_1]}{s - \lambda_1} + \frac{[R_2]}{s - \lambda_2} + \cdots + \frac{[R_p]}{s - \lambda_p}.
\tag{5.274}
$$

Suppose the R_i, which in general are complex, have ranks:

$$\text{rank}\,[R_i] = r_i, \quad \text{for } i = 1, 2, \ldots, p. \tag{5.275}$$

We may do a full rank factorization

$$R_i = C_i B_i, i = 1, 2, \ldots, p \quad \text{where } C_i \in \mathbb{C}^{m \times r_i}, \; B_i \in \mathbb{C}^{r_i \times r}. \tag{5.276}$$

Then

$$\mathbf{A} = \begin{bmatrix} \mathbf{A}_1 & & & \\ & \mathbf{A}_2 & & \\ & & \ddots & \\ & & & \mathbf{A}_p \end{bmatrix}, \quad \mathbf{B} = \begin{bmatrix} \mathbf{B}_1 \\ \mathbf{B}_2 \\ \vdots \\ \mathbf{B}_p \end{bmatrix}, \quad \mathbf{C} = [\mathbf{C}_1 \; \mathbf{C}_2 \; \cdots \; \mathbf{C}_p] \tag{5.277}$$

is a minimal-order realization, with

$$\mathbf{A}_i = \begin{bmatrix} \lambda_i & & \\ & \ddots & \\ & & \lambda_i \end{bmatrix} \in \mathbb{R}^{r_i \times r_i}, \quad \text{for } i \in \mathbf{p}. \tag{5.278}$$

The proof of minimality follows from the fact that

$$\text{rank}\,[\mathbf{A}_i - \lambda_i \mathbf{I} \; \mathbf{B}_i] = r_i, \quad \text{for } i \in \mathbf{p} \tag{5.279}$$

and

$$\text{rank}\begin{bmatrix} \mathbf{A}_i - \lambda_i \mathbf{I} \\ \mathbf{C}_i \end{bmatrix} = r_i, \quad \text{for } i \in \mathbf{p}, \tag{5.280}$$

which prove controllability and observability of $(\mathbf{A}_i, \mathbf{B}_i, \mathbf{C}_i)$ and therefore of $(\mathbf{A}, \mathbf{B}, \mathbf{C})$.

EXAMPLE 5.9 Find a minimal realization of the following transfer function:

$$\mathbf{G}(s) = \begin{bmatrix} \dfrac{1}{(s-1)(s-2)} & \dfrac{1}{(s-2)(s-3)} \\[3mm] \dfrac{1}{(s-2)(s-3)} & \dfrac{1}{(s-1)(s-2)} \end{bmatrix}. \tag{5.281}$$

Since all entries of $G(s)$ have simple poles, we can use the Gilbert realization

$$G(s) = \begin{bmatrix} \dfrac{-1}{s-1} + \dfrac{1}{s-2} & \dfrac{-1}{s-2} + \dfrac{1}{s-3} \\ \dfrac{-1}{s-2} + \dfrac{1}{s-3} & \dfrac{-1}{s-1} + \dfrac{1}{s-2} \end{bmatrix}$$

$$= \dfrac{1}{s-1}\begin{bmatrix} -1 & 0 \\ 0 & -1 \end{bmatrix} + \dfrac{1}{s-2}\begin{bmatrix} 1 & -1 \\ -1 & 1 \end{bmatrix}$$

$$+ \dfrac{1}{s-3}\begin{bmatrix} 0 & 1 \\ 1 & 0 \end{bmatrix} \tag{5.282}$$

$$= \dfrac{1}{s-1}\underbrace{\begin{bmatrix} -1 & 0 \\ 0 & -1 \end{bmatrix}}_{C_1}\underbrace{\begin{bmatrix} 1 & 0 \\ 0 & 1 \end{bmatrix}}_{B_1} + \dfrac{1}{s-2}\underbrace{\begin{bmatrix} 1 \\ -1 \end{bmatrix}}_{C_2}\underbrace{\begin{bmatrix} 1 & -1 \end{bmatrix}}_{B_2}$$

$$+ \dfrac{1}{s-3}\underbrace{\begin{bmatrix} 0 & 1 \\ 1 & 0 \end{bmatrix}}_{C_3}\underbrace{\begin{bmatrix} 1 & 0 \\ 0 & 1 \end{bmatrix}}_{B_3}.$$

Therefore, a minimal realization is given by

$$A = \left[\begin{array}{cc:c:cc} 1 & 0 & 0 & 0 & 0 \\ 0 & 1 & 0 & 0 & 0 \\ \hdashline 0 & 0 & 2 & 0 & 0 \\ \hdashline 0 & 0 & 0 & 3 & 0 \\ 0 & 0 & 0 & 0 & 3 \end{array}\right],$$

$$B = \left[\begin{array}{cc} 1 & 0 \\ 0 & 1 \\ \hdashline 1 & -1 \\ \hdashline 1 & 0 \\ 0 & 1 \end{array}\right], \tag{5.283}$$

$$C = \left[\begin{array}{cc:c:cc} -1 & 0 & 1 & 0 & 1 \\ 0 & -1 & -1 & 1 & 0 \end{array}\right].$$

Gilbert's realization may also be used for model reduction by considering the size of residues. We illustrate this by considering the following example, which is the same as in Example 5.5.

EXAMPLE 5.10   Consider the transfer function calculated from the state-space description of the system in Example 5.5:

$$\mathbf{G}(s) = \begin{bmatrix} \dfrac{3s^2 + 12s + 11}{(s+1)(s+2)(s+3)} & \dfrac{20,001s^2 + 150,005s + 250,006}{10,000(s+2)(s+3)(s+5)} \\[3mm] \dfrac{2s+5}{(s+2)(s+3)} & \dfrac{19,998s^3 + 299,981s^2 + 848,843s + 999,946}{10,000(s+2)(s+3)(s+4)(s+5)} \end{bmatrix}$$

$$= \begin{bmatrix} \dfrac{1}{s+1} + \dfrac{1}{s+2} + \dfrac{1}{s+3} & -\dfrac{1}{s+2} + \dfrac{1}{s+3} + \dfrac{\frac{1}{10,000}}{s+5} \\[3mm] \dfrac{1}{s+2} + \dfrac{1}{s+3} & \dfrac{1}{s+2} + \dfrac{1}{s+3} - \dfrac{\frac{1}{10,000}}{s+4} - \dfrac{\frac{1}{10,000}}{s+5} \end{bmatrix}$$

$$= \frac{1}{s+1}\begin{bmatrix} 1 & 0 \\ 0 & 0 \end{bmatrix} + \frac{1}{s+2}\begin{bmatrix} 1 & 1 \\ 1 & 1 \end{bmatrix} + \frac{1}{s+3}\begin{bmatrix} 1 & 1 \\ 1 & 1 \end{bmatrix} \qquad (5.284)$$

$$+ \frac{1}{s+4}\begin{bmatrix} 0 & 0 \\ 0 & -\frac{1}{10,000} \end{bmatrix} + \frac{1}{s+5}\begin{bmatrix} 0 & \frac{1}{10,000} \\ 0 & -\frac{1}{10,000} \end{bmatrix}$$

$$= \frac{1}{s+1}\begin{bmatrix} 1 \\ 0 \end{bmatrix}\begin{bmatrix} 1 & 0 \end{bmatrix} + \frac{1}{s+2}\begin{bmatrix} 1 \\ 1 \end{bmatrix}\begin{bmatrix} 1 & 1 \end{bmatrix} + \frac{1}{s+3}\begin{bmatrix} 1 \\ 1 \end{bmatrix}\begin{bmatrix} 1 & 1 \end{bmatrix}$$

$$+ \frac{1}{s+4}\begin{bmatrix} 0 \\ 1 \end{bmatrix}\begin{bmatrix} 0 & -\frac{1}{10,000} \end{bmatrix} + \frac{1}{s+5}\begin{bmatrix} \frac{1}{10,000} \\ -\frac{1}{10,000} \end{bmatrix}\begin{bmatrix} 0 & 1 \end{bmatrix}.$$

Therefore, we obtain the following state-space representation:

$$\mathbf{A} = \begin{bmatrix} -1 & 0 & 0 & 0 & 0 \\ 0 & -2 & 0 & 0 & 0 \\ 0 & 0 & -3 & 0 & 0 \\ 0 & 0 & 0 & -4 & 0 \\ 0 & 0 & 0 & 0 & -5 \end{bmatrix}, \; \mathbf{B} = \begin{bmatrix} 1 & 0 \\ 1 & 1 \\ 1 & 1 \\ 0 & -\frac{1}{10,000} \\ 0 & 1 \end{bmatrix},$$

$$\mathbf{C} = \begin{bmatrix} 1 & 1 & 1 & 0 & \frac{1}{10,000} \\ 0 & 1 & 1 & 1 & -\frac{1}{10,000} \end{bmatrix}. \qquad (5.285)$$

Clearly, the mode "−4" is nearly uncontrollable and the mode "−5" is nearly unobservable. Using this information, the third-order reduced model may be obtained.

5.8   Minimal Realizations from MFDs: Wolovich's Realization

In 1974, Wolovich developed a useful procedure to construct a minimal-order state-space realization from a polynomial matrix fraction description (MFD) of the transfer function matrix. In this section, we describe this realization. First, we need some elementary concepts from the theory of polynomial matrices.

5.8.1   Matrix Fraction Description

Let $\mathbf{G}(s)$ denote an $m \times r$ strictly proper real rational matrix. Any such matrix can be represented as:

$$\mathbf{G}(s) = \underbrace{N_R(s)}_{m \times r} \underbrace{D_R(s)^{-1}}_{r \times r} \tag{5.286a}$$

$$= \underbrace{D_L(s)^{-1}}_{m \times m} \underbrace{N_L(s)}_{m \times r}, \tag{5.286b}$$

where $N_R(s), D_R(s), N_L(s), D_L(s)$ are polynomial matrices and $\{N_R(s), D_R(s)\}$ ($\{N_L(s), D_L(s)\}$) is called a right (left) MFD. The MFDs in Equations (5.286a) and (5.286b) always exist, as we show below.

EXAMPLE 5.11

Right MFD

$$\mathbf{G}(s) = \begin{bmatrix} \dfrac{1}{s-1} & \dfrac{1}{s+1} & \dfrac{1}{s} \\[2ex] \dfrac{1}{s+1} & \dfrac{1}{s} & \dfrac{1}{s-1} \end{bmatrix} = \begin{bmatrix} \dfrac{1+s}{s^2-1} & \dfrac{s}{s^2+s} & \dfrac{-1+s}{s^2-s} \\[2ex] \dfrac{-1+s}{s^2-1} & \dfrac{1+s}{s^2+s} & \dfrac{s}{s^2-s} \end{bmatrix}$$

$$= \underbrace{\begin{bmatrix} 1+s & s & -1+s \\ -1+s & 1+s & s \end{bmatrix}}_{N_R(s)} \underbrace{\begin{bmatrix} s^2-1 & & \\ & s^2+s & \\ & & s^2-s \end{bmatrix}^{-1}}_{D_R(s)^{-1}}. \tag{5.287}$$

Left MFD

$$\mathbf{G}(s) = \begin{bmatrix} \dfrac{1}{s-1} & \dfrac{1}{s+1} & \dfrac{1}{s} \\[2ex] \dfrac{1}{s+1} & \dfrac{1}{s} & \dfrac{1}{s-1} \end{bmatrix} = \begin{bmatrix} \dfrac{s+s^2}{s^3-s} & \dfrac{-s+s^2}{s^3-s} & \dfrac{s^2-1}{s^3-s} \\[2ex] \dfrac{-s+s^2}{s^3-s} & \dfrac{-1+s^2}{s^3-s} & \dfrac{s+s^2}{s^3-s} \end{bmatrix}$$

$$= \underbrace{\begin{bmatrix} s^3-s & 0 \\ 0 & s^3-s \end{bmatrix}^{-1}}_{D_L(s)^{-1}} \underbrace{\begin{bmatrix} s+s^2 & -s+s^2 & -1+s^2 \\ -s+s^2 & -1+s^2 & s+s^2 \end{bmatrix}}_{N_L(s)}. \tag{5.288}$$

5.8.2 Matrix Divisors

If polynomial matrices $(\mathbf{A}(s), \mathbf{B}(s), \mathbf{C}(s))$ satisfy

$$\mathbf{A}(s) = \mathbf{B}(s)\mathbf{C}(s), \tag{5.289}$$

then $\mathbf{B}(s)$ ($\mathbf{C}(s)$) is a *left (right) divisor* of $\mathbf{A}(s)$. If

$$\mathbf{A}_1(s) = \mathbf{B}_1(s)\mathbf{C}(s), \quad \mathbf{A}_2(s) = \mathbf{B}_2(s)\mathbf{C}(s), \tag{5.290}$$

then $\mathbf{C}(s)$ is a *common right divisor* of $(\mathbf{A}_1(s), \mathbf{A}_2(s))$. A similar definition applies to *common left divisor*. $\mathbf{C}(s)$ is a *greatest common right (left) divisor (GCRD (GCLD))* of $(\mathbf{A}_1(s), \mathbf{A}_2(s))$ if every common right (left) divisor of $(\mathbf{A}_1(s), \mathbf{A}_2(s))$ is also a common right (left) divisor of $\mathbf{C}(s)$. Two polynomial matrices are right (left) *coprime* if every GCRD (GCLD) is unimodular. In the following we show how GCRDs or GCLDs may be computed and extracted via row or column compression using unimodular transformations.

5.8.3 Unimodular Matrices

A square polynomial matrix $U(s)$ is *unimodular* if $U(s)^{-1}$ exists and is polynomial. It is easy to see that $U(s)$ is unimodular if and only if

$$\det[U(s)] = \text{constant} \neq 0. \tag{5.291}$$

5.8.4 Computation of a GCRD by Row Compression

Suppose $N(s)D(s)^{-1}$ is a right MFD, not necessarily coprime. A GCRD of $(N(s), D(s))$ may be found by *row compressing* $N(s)$, stacked over $D(s)$:

$$\underbrace{\begin{bmatrix} U_{11}(s) & U_{12}(s) \\ U_{21}(s) & U_{22}(s) \end{bmatrix}}_{\text{unimodular}} \begin{bmatrix} N(s) \\ D(s) \end{bmatrix} = \begin{bmatrix} R(s) \\ \hline 0 \end{bmatrix}, \tag{5.292}$$

where $R(s)$ is a GCRD. Write

$$\begin{bmatrix} N(s) \\ D(s) \end{bmatrix} = U(s)^{-1} \begin{bmatrix} R(s) \\ 0 \end{bmatrix} = \begin{bmatrix} V_{11}(s) & V_{12}(s) \\ V_{21}(s) & V_{22}(s) \end{bmatrix} \begin{bmatrix} R(s) \\ 0 \end{bmatrix} \tag{5.293}$$

so that

$$\begin{bmatrix} N(s) \\ D(s) \end{bmatrix} = \begin{bmatrix} V_{11}(s)R(s) \\ V_{21}(s)R(s) \end{bmatrix}. \tag{5.294}$$

Therefore, $R(s)$ is a common right divisor of $(N(s), D(s))$.

From Equation (5.292):

$$U_{11}(s)N(s) + U_{12}(s)D(s) = R(s), \tag{5.295}$$

so that if $\bar{R}(s)$ is another common right divisor:

$$N(s) = \bar{N}(s)\bar{R}(s), \quad D(s) = \bar{D}(s)\bar{R}(s) \tag{5.296}$$

for some polynomial matrices $(\bar{N}(s), \bar{D}(s))$. Therefore, substituting Equation (5.296) in Equation (5.295):

$$[U_{11}(s)\bar{N}(s) + U_{12}(s)\bar{D}(s)]\bar{R}(s) = R(s), \tag{5.297}$$

proving that $\bar{R}(s)$ is a right divisor of $R(s)$. Since $\bar{R}(s)$ was an arbitrary common right divisor, $R(s)$ is a *greatest common right divisor*.

EXAMPLE 5.12 (Reduction to Hermite form)

$$\begin{bmatrix} s^2 & 0 \\ 0 & s^2 \\ 1 & s+1 \end{bmatrix} \rightarrow \begin{bmatrix} 1 & s+1 \\ 0 & s^2 \\ s^2 & 0 \end{bmatrix} \rightarrow \begin{bmatrix} 1 & s+1 \\ 0 & s^2 \\ 0 & -s^2(s+1) \end{bmatrix} \rightarrow \begin{bmatrix} 1 & s+1 \\ 0 & s^2 \\ 0 & 0 \end{bmatrix}.$$

$$\tag{5.298}$$

The corresponding unimodular matrix is

$$\begin{bmatrix} 1 & 0 & 0 \\ 0 & 1 & 0 \\ 0 & s+1 & 1 \end{bmatrix} \begin{bmatrix} 1 & 0 & 0 \\ 0 & 1 & 0 \\ -s^2 & 0 & 1 \end{bmatrix} \begin{bmatrix} 0 & 0 & 1 \\ 0 & 1 & 0 \\ 1 & 0 & 0 \end{bmatrix} = \begin{bmatrix} 0 & 0 & 1 \\ 0 & 1 & 0 \\ 1 & s+1 & s^2 \end{bmatrix}. \tag{5.299}$$

Remark 5.10 (Non-uniqueness of GCRDs)   Notice that the elementary operations to reduce to Hermite form are not unique and thus the unimodular matrices $U(s)$ and GCRDs are not unique. However, any two GCRDs, $R_1(s)$ and $R_2(s)$, say, must be related (by definition) as

$$R_1(s) = W_2(s)R_2(s), \quad R_2(s) = W_1(s)R_1(s), \quad W_i(s) \text{ polynomial.} \tag{5.300}$$

Since

$$R_1(s) = W_2(s)W_1(s)R_1(s) \tag{5.301}$$

it follows that

(1) If $R_1(s)$ is nonsingular, then the $W_i(s)$, $i = 1,2$, must be unimodular, and hence the GCRD $R_2(s)$ is also nonsingular. That is, if one GCRD is nonsingular, then all GCRDs must be so, and they can only differ by a unimodular (left) factor.
(2) If a GCRD is unimodular, then all GCRDs must be unimodular.

Remark 5.11 (Nonsingular GCRDs)   If

$$\begin{bmatrix} D(s) \\ N(s) \end{bmatrix} \text{ has full column rank,} \tag{5.302}$$

then *all* GCRDs of $(N(s), D(s))$ must be nonsingular and unimodular, and can differ only by unimodular (left) factors.

DEFINITION 5.6 $(N_R(s), D_R(s))$ is *right coprime* if and only if all its GCRDs are unimodular. Similarly, $(N_L(s), D_L(s))$ is *left coprime* if and only if all its GCLDs are unimodular.

5.8.5   Coprimeness of MFDs

The MFD $\{N_R(s), D_R(s)\}$ ($\{N_L(s), D_L(s)\}$) is right coprime (left coprime) if

$$\text{rank} \begin{bmatrix} D_R(s) \\ N_R(s) \end{bmatrix} = r, \quad \text{for all } s \in \mathbb{C} \tag{5.303a}$$

$$(\text{rank}\,[D_L(s) \ N_L(s)] = m, \quad \text{for all } s \in \mathbb{C}). \tag{5.303b}$$

EXAMPLE 5.13   Taking the right and left factorization in Example 5.11, we see that.

$$\text{rank} \begin{bmatrix} D_R(s) \\ N_R(s) \end{bmatrix} = \text{rank} \left[\begin{array}{ccc} s^2-1 & 0 & 0 \\ 0 & s^2+s & 0 \\ 0 & 0 & s^2-s \\ \hdashline s+s^2 & -s+s^2 & -1+s^2 \\ -s+s^2 & -1+s^2 & s+s^2 \end{array} \right] = 3, \quad \text{for all } s \tag{5.304}$$

and therefore, the pair $(D_R(s), N_R(s))$ is right coprime. Also

$$\text{rank}\,[D_L(s) \ N_L(s)] = \text{rank} \left[\begin{array}{cc:ccc} s^3-s & 0 & s+s^2 & -s+s^2 & -1+s^2 \\ 0 & s^3-s & -s+s^2 & -1+s^2 & s+s^2 \end{array} \right]$$

$$= 2, \tag{5.305}$$

and therefore the pair $(D_L(s), N_L(s))$ is left coprime.

5.8.6   Extraction of a GCRD (GCLD) from an MFD

If $(N_R(s), D_R(s))$ is not right coprime, it follows that it does not satisfy Equation (5.303a). This is equivalent to the existence of a GCRD, $R(s)$, which is *not* unimodular:

$$N_R(s) = \bar{N}_R(s) R(s),$$
$$D_R(s) = \bar{D}_R(s) R(s), \tag{5.306}$$

and

$$\text{rank} \begin{bmatrix} \bar{D}_R(s) \\ \bar{N}_R(s) \end{bmatrix} = r, \quad \text{for all } s \in \mathbb{C}. \tag{5.307}$$

Therefore, $\{\bar{N}_R(s), \bar{D}_R(s)\}$ is right coprime. This shows that, by extracting a GCRD if necessary, a right MFD that is right coprime can always be produced.

5.8.7   Column (Row) Properness

Now consider a square polynomial matrix $D(s)$ and let k_i denote the degree of the highest degree term in column i for $i \in \mathbf{r}$. Then $D(s)$ has the unique representation

$$D(s) = D_{\text{hc}} S(s) + D_{\text{lc}} \psi(s), \tag{5.308}$$

where

$$
S(s) := \begin{bmatrix} s^{k_1} & & & \\ & s^{k_2} & & \\ & & \ddots & \\ & & & s^{k_r} \end{bmatrix}, \quad \psi(s) := \begin{bmatrix} 1 \\ s \\ \vdots \\ s^{k_1-1} \\ \hline & 1 \\ & s \\ & \vdots \\ & s^{k_2-1} \\ \hline & & \ddots \\ \hline & & & 1 \\ & & & s \\ & & & \vdots \\ & & & s^{k_r-1} \end{bmatrix} \tag{5.309}
$$

and the square real matrix D_{hc} is called the *high-order column coefficient* matrix. It is easy to see that

$$
\det[D(s)] = \det[D_{\text{hc}}]\, s^{k_1+k_2+\cdots+k_r} + (\text{lower-order terms}) \tag{5.310}
$$

and therefore

$$
\deg\left(\det[D(s)]\right) = k_1 + k_2 + \cdots + k_r =: \nu \tag{5.311}
$$

if and only if

$$
\det[D_{\text{hc}}] \neq 0. \tag{5.312}
$$

$D(s)$ is said to be *column proper* if Equation (5.312) holds.

In case Equation (5.312) fails to hold, unimodular column operations can be performed on $D(s)$ to render it column proper. These operations essentially reduce the sum of the column degrees until Equation (5.312) holds.

Remark 5.12   If $D(s)$ is replaced by $D(s)U(s)$, we must replace $N(s)$ by $N(s)U(s)$ to preserve the validity of the right MFD.

Remark 5.13   It is noted that ν in Equation (5.311) will turn out to be the minimal-order of the system, which is invariant regardless of particular left or right factorizations used. It is the property of $\mathbf{G}(s)$.

5.8.8   Wolovich's Realization

With the above preliminary results in hand, we are ready to present Wolovich's realization. We present the results for a right MFD, leaving the dual results using left MFDs to the reader.

Assume that $\mathbf{G}(s)$ is a strictly proper $m \times r$ real rational matrix with a right MFD

$$\mathbf{G}(s) = \underbrace{N(s)}_{m \times r} \underbrace{D(s)^{-1}}_{r \times r} \tag{5.313}$$

satisfying the conditions

(i) $[N(s), D(s)]$ is right coprime. $\hspace{2cm}$ (5.314a)

(ii) $D(s)$ is column proper. $\hspace{2cm}$ (5.314b)

(iii) The degree of column $N_i(s)$ is less than the degree of column $D_i(s)$ for each i.
$\hspace{10cm}$ (5.314c)

The condition in Equation (5.314c) follows from the strict properness of $\mathbf{G}(s)$. We can write

$$N(s) = N_{\mathrm{lc}} \, \boldsymbol{\psi}(s), \tag{5.315}$$

since Equation (5.314c) holds.

From Equations (5.308) and (5.315), we have k_i, for $i \in \mathbf{r}$ and $D_{\mathrm{hc}}, D_{\mathrm{lc}}, N_{\mathrm{lc}}$ which are used to construct Wolovich's realization in two steps.

1. Define

$$\mathbf{A}_c^o := \begin{bmatrix} \mathbf{A}_{c_1}^o & & & \\ & \mathbf{A}_{c_2}^o & & \\ & & \ddots & \\ & & & \mathbf{A}_{c_r}^o \end{bmatrix}, \quad \mathbf{B}_c^o := \begin{bmatrix} \mathbf{b}_{c_1}^o & & & \\ & \mathbf{b}_{c_2}^o & & \\ & & \ddots & \\ & & & \mathbf{b}_{c_r}^o \end{bmatrix}, \tag{5.316}$$

where

$$\mathbf{A}_{c_i}^o := \underbrace{\begin{bmatrix} 0 & 1 & 0 & \cdots & 0 \\ & & \ddots & & \\ & & & \ddots & \\ & & & & 1 \\ 0 & \cdots & \cdots & \cdots & 0 \end{bmatrix}}_{k_i \times k_i}, \quad \mathbf{b}_{c_i}^o := \underbrace{\begin{bmatrix} 0 \\ 0 \\ \vdots \\ 0 \\ 1 \end{bmatrix}}_{k_i \times 1}, \quad \text{for } i \in \mathbf{r}. \tag{5.317}$$

2. Construct the minimal-order realization

$$\begin{aligned}
\mathbf{A}_c &:= \mathbf{A}_c^o - \mathbf{B}_c^o D_{\mathrm{hc}}^{-1} D_{\mathrm{lc}}, \\
\mathbf{B}_c &:= \mathbf{B}_c^o D_{\mathrm{hc}}^{-1}, \\
\mathbf{C}_c &:= N_{\mathrm{lc}}.
\end{aligned} \tag{5.318}$$

In other words:

$$\mathbf{C}_c \left(s\mathbf{I} - \mathbf{A}_c \right)^{-1} \mathbf{B}_c = \mathbf{G}(s) \tag{5.319}$$

and the minimal order is $k_1 + k_2 + \cdots + k_r$.

To prove that Equation (5.318) is a minimal realization of $\mathbf{G}(s)$, we verify first that it is a realization and then that it is controllable and observable.

From Equation (5.316) it follows that

$$(s\mathbf{I} - \mathbf{A}_c^o)^{-1}\mathbf{B}_c^o = \psi(s)S(s)^{-1} \tag{5.320}$$

and from Equations (5.318) and (5.308) it follows that

$$(s\mathbf{I} - \mathbf{A}_c)^{-1}\mathbf{B}_c = \psi(s)D(s)^{-1}. \tag{5.321}$$

Left multiplying Equation (5.321) by $\mathbf{C}_c$ and using Equations (5.318) and (5.315), we finally get Equation (5.319), proving that Equation (5.318) is a realization.

To prove minimality, note that from Equations (5.316) and (5.317):

$$(\mathbf{A}_c^o, \mathbf{B}_c^o) \quad \text{is controllable} \tag{5.322}$$

and so from Equation (5.318)

$$(\mathbf{A}_c, \mathbf{B}_c) \quad \text{is controllable.} \tag{5.323}$$

Suppose now, by contradiction, that $(\mathbf{C}_c, \mathbf{A}_c)$ is not observable. Then for some $\lambda \in \mathbb{C}$ and vector $\mathbf{p} \neq 0$:

$$\begin{bmatrix} \lambda\mathbf{I} - \mathbf{A}_c \\ \mathbf{C}_c \end{bmatrix}\mathbf{p} = \begin{bmatrix} 0 \\ 0 \end{bmatrix}. \tag{5.324}$$

Therefore

$$[\lambda\mathbf{I} - \mathbf{A}_c \quad -\mathbf{B}_c]\begin{bmatrix} \mathbf{p} \\ 0 \end{bmatrix} = 0. \tag{5.325}$$

By controllability of $(\mathbf{A}_c, \mathbf{B}_c)$, the matrix on the left in Equation (5.325) has rank n, and its null space has dimension r. From Equation (5.318), this null space is spanned by the columns of the matrix

$$\begin{bmatrix} \psi(\lambda) \\ D(\lambda) \end{bmatrix} \tag{5.326}$$

which has rank r, because of the structure of $\psi(s)$. Therefore, from Equation (5.321), it follows that there exists $\mathbf{q} \neq 0$ such that

$$\begin{bmatrix} \mathbf{p} \\ 0 \end{bmatrix} = \begin{bmatrix} \psi(\lambda) \\ D(\lambda) \end{bmatrix}\mathbf{q}. \tag{5.327}$$

Using Equation (5.318), we now have

$$\begin{bmatrix} N(\lambda) \\ D(\lambda) \end{bmatrix}\mathbf{q} = \begin{bmatrix} 0 \\ 0 \end{bmatrix}, \tag{5.328}$$

which contradicts the right coprimeness of $(N(s), D(s))$. Therefore, $(\mathbf{A}_c, \mathbf{B}_c, \mathbf{C}_c)$ is a controllable and observable, and therefore a minimal-order realization of $\mathbf{G}(s)$.

EXAMPLE 5.14 The following example illustrates the steps involved in constructing a minimal realization using Wolovich's method. Let

$$\mathbf{G}(s) = \begin{bmatrix} \dfrac{s}{(s-1)^4} & \dfrac{1}{(s-1)^3} \\ \dfrac{1}{(s-1)^4} & \dfrac{1}{(s-1)^3} \end{bmatrix}. \tag{5.329}$$

In the following, we describe Wolovich's minimal realization procedure in detail.

1. Obtain a right MFD:

$$\mathbf{G}(s) = \underbrace{\begin{bmatrix} s & 1 \\ 1 & 1 \end{bmatrix}}_{N_R(s)} \underbrace{\begin{bmatrix} (s-1)^4 & 0 \\ 0 & (s-1)^3 \end{bmatrix}^{-1}}_{D_R(s)^{-1}}. \tag{5.330}$$

2. Check right coprimeness of $(N_R(s), D_R(s))$:

$$\operatorname{rank} \begin{bmatrix} N_R(s) \\ D_R(s) \end{bmatrix}_{s=1} = \operatorname{rank} \begin{bmatrix} s & 1 \\ 1 & 1 \\ (s-1)^4 & 0 \\ 0 & (s-1)^3 \end{bmatrix}_{s=1} < 2. \tag{5.331}$$

Therefore, $(N_R(s), D_R(s))$ is *not* right coprime.

3. Obtain a GCRD using row compression. Using elementary row operations on

$$\begin{bmatrix} N_R(s) \\ D_R(s) \end{bmatrix} = \begin{bmatrix} s & 1 \\ 1 & 1 \\ (s-1)^4 & 0 \\ 0 & (s-1)^3 \end{bmatrix} \to \begin{bmatrix} 1 & 1 \\ s & 1 \\ (s-1)^4 & 0 \\ 0 & (s-1)^3 \end{bmatrix}$$

$$\to \begin{bmatrix} 1 & 1 \\ 0 & 1-s \\ 0 & -(s-1)^4 \\ 0 & (s-1)^3 \end{bmatrix} \to \begin{bmatrix} 1 & 1 \\ 0 & 1-s \\ \hline 0 & 0 \\ 0 & 0 \end{bmatrix}, \tag{5.332}$$

we obtain

$$R(s) = \begin{bmatrix} 1 & 1 \\ 0 & 1-s \end{bmatrix}. \tag{5.333}$$

4. Extract GCRD and obtain a right coprime MFD:

$$\bar{N}_R(s) := N_R(s)R(s)^{-1} = \begin{bmatrix} s & 1 \\ 1 & 1 \end{bmatrix} \begin{bmatrix} 1 & \dfrac{1}{s-1} \\ 0 & -\dfrac{1}{s-1} \end{bmatrix} = \begin{bmatrix} s & 1 \\ 1 & 0 \end{bmatrix}, \tag{5.334a}$$

$$\bar{D}_R(s) := D_R(s)R(s)^{-1} = \begin{bmatrix} (s-1)^4 & 0 \\ 0 & (s-1)^3 \end{bmatrix} \begin{bmatrix} 1 & \dfrac{1}{s-1} \\ 0 & -\dfrac{1}{s-1} \end{bmatrix}$$

$$= \begin{bmatrix} (s-1)^4 & (s-1)^3 \\ 0 & -(s-1)^2 \end{bmatrix}. \tag{5.334b}$$

5. Verify right coprimeness of $(\bar{N}_R(s), \bar{D}_R(s))$:

$$\text{rank} \begin{bmatrix} \bar{N}_R(s) \\ \bar{D}_R(s) \end{bmatrix} = \text{rank} \begin{bmatrix} s & 1 \\ 1 & 0 \\ (s-1)^4 & (s-1)^3 \\ 0 & -(s-1)^2 \end{bmatrix} = 2, \qquad \text{for all } s \in \mathbb{C}. \tag{5.335}$$

6. Check column properness of $\bar{D}_R(s)$:

$$\bar{D}_R(s) = \underbrace{\begin{bmatrix} 1 & 1 \\ 0 & 0 \end{bmatrix}}_{D_{R_{hc}}} \begin{bmatrix} s^4 & 0 \\ 0 & s^3 \end{bmatrix} + (\text{terms of lower degree}). \tag{5.336}$$

Since $D_{R_{hc}}$ is singular, $\bar{D}_R(s)$ is *not* column proper.

7. Render $\bar{D}_R(s)$ column proper by unimodular column operations (right multiplication by $U(s)$):

$$D(s) := \bar{D}_R(s)U(s) = \begin{bmatrix} (s-1)^4 & (s-1)^3 \\ 0 & -(s-1)^2 \end{bmatrix} \underbrace{\begin{bmatrix} 1 & 0 \\ -(s-1) & 1 \end{bmatrix}}_{U(s)}$$

$$= \begin{bmatrix} 0 & (s-1)^3 \\ \hline (s-1)^3 & -(s-1)^2 \end{bmatrix}. \tag{5.337}$$

8. Check if $D(s)$ is column proper:

$$D(s) = \underbrace{\begin{bmatrix} 0 & 1 \\ 1 & 0 \end{bmatrix}}_{D_{hc}} \underbrace{\begin{bmatrix} s^3 & 0 \\ 0 & s^3 \end{bmatrix}}_{S(s)}$$

$$+ \underbrace{\begin{bmatrix} 0 & 0 & 0 & -1 & 3 & -3 \\ -1 & 3 & -3 & -1 & 2 & -1 \end{bmatrix}}_{D_{lc}} \underbrace{\begin{bmatrix} 1 & 0 \\ s & 0 \\ s^2 & 0 \\ 0 & 1 \\ 0 & s \\ 0 & s^2 \end{bmatrix}}_{\psi(s)}. \qquad (5.338)$$

Since D_{hc} is nonsingular, $D(s)$ is column proper.

9. Construct

$$N(s) := \bar{N}_R(s)U(s) = \begin{bmatrix} s & 1 \\ 1 & 0 \end{bmatrix} \begin{bmatrix} 1 & 0 \\ -(s-1) & 1 \end{bmatrix} = \begin{bmatrix} 1 & 1 \\ 1 & 0 \end{bmatrix}$$

$$= \underbrace{\begin{bmatrix} 1 & 0 & 0 & 1 & 0 & 0 \\ 1 & 0 & 0 & 0 & 0 & 0 \end{bmatrix}}_{C_c} \psi(s). \qquad (5.339)$$

10. Construct the minimal realization

$$\mathbf{A}_c^o = \begin{bmatrix} 0 & 1 & 0 & 0 & 0 & 0 \\ 0 & 0 & 1 & 0 & 0 & 0 \\ 0 & 0 & 0 & 0 & 0 & 0 \\ \hline 0 & 0 & 0 & 0 & 1 & 0 \\ 0 & 0 & 0 & 0 & 0 & 1 \\ 0 & 0 & 0 & 0 & 0 & 0 \end{bmatrix}, \qquad \mathbf{B}_c^o = \begin{bmatrix} 0 & 0 \\ 0 & 0 \\ 1 & 0 \\ \hline 0 & 0 \\ 0 & 0 \\ 0 & 1 \end{bmatrix} \qquad (5.340)$$

and

$$\mathbf{A}_c = \mathbf{A}_c^o - \mathbf{B}_c^o D_{hc}^{-1} D_{lc} = \begin{bmatrix} 0 & 1 & 0 & 0 & 0 & 0 \\ 0 & 0 & 1 & 0 & 0 & 0 \\ 1 & -3 & 3 & 1 & -2 & 1 \\ 0 & 0 & 0 & 0 & 1 & 0 \\ 0 & 0 & 0 & 0 & 0 & 1 \\ 0 & 0 & 0 & 1 & -3 & 3 \end{bmatrix},$$

$$\qquad (5.341)$$

$$\mathbf{B}_c = \mathbf{B}_c^o D_{hc}^{-1} = \begin{bmatrix} 0 & 0 \\ 0 & 0 \\ 0 & 1 \\ 0 & 0 \\ 0 & 0 \\ 1 & 0 \end{bmatrix}.$$

11. (1) Check that $(\mathbf{C}_c, \mathbf{B}_c, \mathbf{C}_c)$ is a realization:

$$\mathbf{C}_c (s\mathbf{I} - \mathbf{A}_c)^{-1} \mathbf{B}_c = \mathbf{G}(s). \tag{5.342}$$

(2) Check that $(\mathbf{A}_c, \mathbf{B}_c)$ is controllable and $(\mathbf{C}_c, \mathbf{A}_c)$ is observable.

More detailed information on the structure of the minimal realization constructed here can be obtained. We start with the following result.

LEMMA 5.11   Let $(\mathbf{A}_c, \mathbf{B}_c)$ be a controllable realization of $\Psi(s)\mathbf{D}(s)^{-1}$. Then there exist polynomial matrices $\mathbf{P}(s), \mathbf{Q}(s), \mathbf{U}(s), \mathbf{V}(s)$ such that

$$\begin{bmatrix} s\mathbf{I} - \mathbf{A}_c & \mathbf{B}_c \\ \mathbf{P}(s) & \mathbf{Q}(s) \end{bmatrix} \begin{bmatrix} \mathbf{U}(s) & -\Psi(s) \\ \mathbf{V}(s) & \mathbf{D}(s) \end{bmatrix} = \begin{bmatrix} \mathbf{I} & 0 \\ 0 & \mathbf{I} \end{bmatrix}. \tag{5.343}$$

Proof   Since $(\mathbf{A}_c, \mathbf{B}_c)$ is controllable:

$$\operatorname{rank} \begin{bmatrix} s\mathbf{I} - \mathbf{A}_c & \mathbf{B}_c \end{bmatrix} = n, \quad \text{for all } s \in \mathbb{C}. \tag{5.344}$$

Therefore $((s\mathbf{I} - \mathbf{A}_c), \mathbf{B}_c)$ are left coprime and there exist $\mathbf{U}(s), \mathbf{V}(s)$ such that

$$\begin{bmatrix} s\mathbf{I} - \mathbf{A}_c & \mathbf{B}_c \end{bmatrix} \begin{bmatrix} \mathbf{U}(s) \\ \mathbf{V}(s) \end{bmatrix} = \mathbf{I}. \tag{5.345}$$

Similarly, $(\Psi(s), \mathbf{D}(s))$ are right coprime and there exist $\bar{\mathbf{P}}(s), \bar{\mathbf{Q}}(s)$ polynomial satisfying

$$\bar{\mathbf{P}}(s)\Psi(s) + \bar{\mathbf{Q}}(s)\mathbf{D}(s) = \mathbf{I}. \tag{5.346}$$

Now write

$$\bar{\mathbf{P}}(s)\mathbf{U}(s) + \bar{\mathbf{Q}}(s)\mathbf{V}(s) = \mathbf{W}(s) \tag{5.347}$$

and note that $\mathbf{W}(s)$ is polynomial and

$$\begin{bmatrix} s\mathbf{I} - \mathbf{A}_c & \mathbf{B}_c \\ -\bar{\mathbf{P}}(s) & \bar{\mathbf{Q}}(s) \end{bmatrix} \begin{bmatrix} \mathbf{U}(s) & -\Psi(s) \\ \mathbf{V}(s) & \bar{\mathbf{Q}}(s) \end{bmatrix} = \begin{bmatrix} \mathbf{I} & 0 \\ \mathbf{W}(s) & \mathbf{I} \end{bmatrix}. \tag{5.348}$$

Multiplying Equation (5.348) by

$$\begin{bmatrix} \mathbf{I} & 0 \\ -\mathbf{W}(s) & \mathbf{I} \end{bmatrix} \tag{5.349}$$

on the left we have Equation (5.343) with

$$\mathbf{P}(s) = -\mathbf{W}(s)(s\mathbf{I} - \mathbf{A}_c) - \bar{\mathbf{P}}(s), \tag{5.350a}$$

$$\mathbf{Q}(s) = -\mathbf{W}(s)\mathbf{B}_c + \bar{\mathbf{Q}}(s). \tag{5.350b}$$

The matrix

$$\begin{bmatrix} \mathbf{U}(s) & -\Psi(s) \\ \mathbf{V}(s) & \mathbf{D}(s) \end{bmatrix} \tag{5.351}$$

must be unimodular since the right-hand side is the identity matrix and is a product of the polynomial matrices on the left. From Equation (5.343) we have

$$\begin{bmatrix} s\mathbf{I} - \mathbf{A}_c & \mathbf{B}_c \\ 0 & \mathbf{I} \end{bmatrix} \underbrace{\begin{bmatrix} \mathbf{U}(s) & -\mathbf{\Psi}(s) \\ \mathbf{V}(s) & \mathbf{D}(s) \end{bmatrix}}_{\text{unimodular}} = \begin{bmatrix} \mathbf{I} & 0 \\ \mathbf{V}(s) & \mathbf{D}(s) \end{bmatrix} \qquad (5.352)$$

and therefore,

$$\begin{bmatrix} s\mathbf{I} - \mathbf{A}_c & \mathbf{B}_c \\ 0 & \mathbf{I} \end{bmatrix} \quad \text{and} \quad \begin{bmatrix} \mathbf{I} & 0 \\ \mathbf{V}(s) & \mathbf{D}(s) \end{bmatrix} \qquad (5.353)$$

are equivalent, that is they have the same invariant polynomials and elementary divisors. By elementary row operations on the matrices in Equation (5.353), it follows that the elementary divisors of $(s\mathbf{I} - \mathbf{A}_c)$ are identical to those of $\mathbf{D}(s)$. Thus, the elementary divisors of $\mathbf{D}(s)$ determine the Jordan structure of $\mathbf{A}_c$. ∎

5.9 Transfer Function Conditions for Stabilizability and Detectability

Controllability, observability, stabilizability, and detectability are properties of a state-space model. In this section, we show that they can also be expressed as properties of appropriate transfer functions.

Let $\mathbf{G}(s)$ denote a real, rational, proper transfer function matrix with a realization $[\mathbf{A}, \mathbf{B}, \mathbf{C}, \mathbf{D}]$, not necessarily minimal, of order n:

$$\mathbf{G}(s) = \mathbf{C}(s\mathbf{I} - \mathbf{A})^{-1}\mathbf{B} + \mathbf{D} \qquad (5.354a)$$
$$= \mathbf{G}^+(s) + \mathbf{G}^-(s) + \mathbf{D}, \qquad (5.354b)$$

with all poles of the strictly proper matrices $\mathbf{G}^+(s)$ $(\mathbf{G}^-(s)) \in \mathbb{C}^+$ $(\mathbb{C}^-)$. Let $\nu[\mathbf{H}(s)]$ denote the McMillan degree of a real, rational, proper matrix $\mathbf{H}(s)$. $\nu[\mathbf{H}(s)]$ is also the order of a minimal realization of $\mathbf{H}(s)$.

THEOREM 5.12

(a) $\nu\left[(s\mathbf{I} - \mathbf{A})^{-1}\mathbf{B}\right] = \nu\left[(s\mathbf{I} - \mathbf{A})^{-1}\right] = n$ (5.355a)
    if and only if $(\mathbf{A}, \mathbf{B})$ is controllable.

(b) $\nu\left[\mathbf{C}(s\mathbf{I} - \mathbf{A})^{-1}\right] = \nu\left[(s\mathbf{I} - \mathbf{A})^{-1}\right] = n$ (5.355b)
    if and only if $(\mathbf{C}, \mathbf{A})$ is controllable.

(c) $\nu\left[\mathbf{C}(s\mathbf{I} - \mathbf{A})^{-1}\mathbf{B}\right] = \nu\left[(s\mathbf{I} - \mathbf{A})^{-1}\right] = n$ (5.355c)
    if and only if $(\mathbf{A}, \mathbf{B}, \mathbf{C})$ is controllable and observable.

Proof Part (c) is obvious from minimal realization theory. Part (a) follows from Equation (5.355c) with $\mathbf{C} = \mathbf{I}$. Part (b) follows from Equation (5.355c) with $\mathbf{B} = \mathbf{I}$. ∎

To proceed, represent $(\mathbf{A}, \mathbf{B}, \mathbf{C}, \mathbf{D})$ as

$$\mathbf{A} = \begin{bmatrix} \mathbf{A}^+ & 0 \\ 0 & \mathbf{A}^- \end{bmatrix}, \quad \mathbf{B} = \begin{bmatrix} \mathbf{B}^+ \\ \mathbf{B}^- \end{bmatrix}, \quad \mathbf{C} = \begin{bmatrix} \mathbf{C}^+ & \mathbf{C}^- \end{bmatrix}, \quad \mathbf{D} = \mathbf{D}, \qquad (5.356)$$

where the eigenvalues of $\mathbf{A}^+(\mathbf{A}^-)$ lie in $\mathbb{C}^+(\mathbb{C}^-)$ and $\mathbf{A}^+$ is $n^+ \times n^+$. Then

$$\mathbf{G}^+(s) = \mathbf{C}^+ \left(s\mathbf{I} - \mathbf{A}^+\right)^{-1} \mathbf{B}^+, \qquad (5.357a)$$

$$\mathbf{G}^-(s) = \mathbf{C}^- \left(s\mathbf{I} - \mathbf{A}^-\right)^{-1} \mathbf{B}^-. \qquad (5.357b)$$

Using the notation

$$v\left[\mathbf{G}^+(s)\right] =: v^+[\mathbf{G}(s)], \qquad (5.358)$$

an immediate corollary of Theorem 5.12 is the following.

COROLLARY 5.2 (of Theorem 5.12)

(a) $(\mathbf{A}, \mathbf{B})$ is stabilizable if and only if

$$v^+\left[(s\mathbf{I} - \mathbf{A})^{-1}\right] = v^+\left[(s\mathbf{I} - \mathbf{A})^{-1}\mathbf{B}\right]. \qquad (5.359)$$

(b) $(\mathbf{C}, \mathbf{A})$ is detectable if and only if

$$v^+\left[(s\mathbf{I} - \mathbf{A})^{-1}\right] = v^+\left[\mathbf{C}(s\mathbf{I} - \mathbf{A})^{-1}\right]. \qquad (5.360)$$

(c) $(\mathbf{A}, \mathbf{B}, \mathbf{C})$ is stabilizable and detectable if and only if

$$v^+\left[(s\mathbf{I} - \mathbf{A})^{-1}\right] = v^+\left[\mathbf{C}(s\mathbf{I} - \mathbf{A})^{-1}\mathbf{B}\right]. \qquad (5.361)$$

Proof   Follows from Theorem 5.12 applied to Equation (5.357a) and noting that stabilizability is equivalent to controllability of $(\mathbf{A}^+, \mathbf{B}^+)$ and detectability to observability of $(\mathbf{C}^+, \mathbf{A}^+)$.   ∎

A generalized plant framework is sometimes used to formulate and solve control problems (see Chapter 8). This can be described by Figure 5.3.

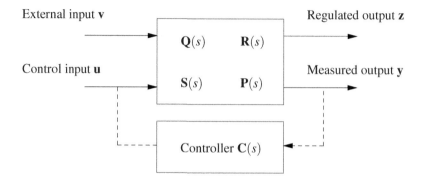

Figure 5.3 Generalized plant and controller

Let

$$\mathbf{G}(s) =: \begin{bmatrix} \mathbf{Q}(s) & \mathbf{R}(s) \\ \mathbf{S}(s) & \mathbf{P}(s) \end{bmatrix} \tag{5.362}$$

denote the real, rational, proper generalized plant $\mathbf{G}(s)$ and $\mathbf{P}(s)$ the physical plant. A question of immediate interest is: Does there exist a controller $\mathbf{C}(s)$ stabilizing the feedback system? The answer is given by the following result, a second corollary of Theorem 5.12.

COROLLARY 5.3 (of Theorem 5.12)   There exists a stabilizing controller in Figure 5.3 if and only if

$$v^+[\mathbf{P}(s)] = v^+[\mathbf{G}(s)]. \tag{5.363}$$

Proof   A stabilizing controller exists if and only if the generalized plant $\mathbf{G}(s)$ is stabilizable from $\mathbf{u}$ and detectable from $\mathbf{y}$.   ∎

5.10   Exercises

5.1   For the circuit in Figure 5.4, write state equations and show that the system is controllable as long as $R_1 C_1 \neq R_2 C_2$.

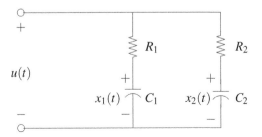

Figure 5.4   Exercise 5.1

(a)   With $R_1 C_1 = 1$, $R_2 C_2 = 2$, find the input that can charge C_1 to 100 V and C_2 to -100 V starting from 0 V each in 10 s. Verify by plotting $x_1(t)$, $x_2(t)$, and $u(t)$ for $t \in [0, 10]$.

(b)   Explore the behavior of $u(t)$ in (a) as $R_2 C_2 \to 1$.

(c)   Explore the behavior $u(t)$, $x_1(t)$, $x_2(t)$ in (a) as the time of transfer goes from 10 s to 0 s.

5.2   Suppose

$$\dot{\mathbf{x}}(t) = \mathbf{A}\mathbf{x}(t) + \mathbf{B}\mathbf{u}(t) \tag{5.364}$$

and let $\mathcal{V}_u$ denote the controllable subspace:

$$\mathcal{V}_u := \mathcal{B} + \mathbf{A}\mathcal{B} + \cdots + \mathbf{A}^{n-1}\mathcal{B}. \tag{5.365}$$

The controllability Wronskian is

$$\mathbf{W}_{t^*} := \int_0^{t^*} e^{\mathbf{A}t} \mathbf{B} \mathbf{B}' e^{\mathbf{A}'t} dt, \quad \text{for } t^* > 0. \tag{5.366}$$

Prove that

$$\mathscr{V}_u = \text{Im}\,[\mathbf{W}_{t^*}] \tag{5.367}$$

where Im[·] denotes image.

5.3 For a state-space pair $(\mathbf{A}, \mathbf{B})$, let

$$\mathscr{V}_k := \mathscr{B} + \mathbf{A}\mathscr{B} + \cdots + \mathbf{A}^k \mathscr{B}, \quad \text{for } k = 0, 1, 2, \ldots \tag{5.368}$$

and define the controllability index

$$k_u := \min\{k \mid \mathscr{V}_k = \mathscr{V}_{k+1}\}. \tag{5.369}$$

Prove that in the discrete-time system

$$\mathbf{x}[k+1] = \mathbf{A}\mathbf{x}[k] + \mathbf{B}\mathbf{u}[k], \tag{5.370}$$

there exists a control sequence

$$\mathbf{u}[0], \mathbf{u}[1], \mathbf{u}[2], \ldots, \mathbf{u}[k_u] \tag{5.371}$$

that transfers $\mathbf{x}[0] = 0$ to an arbitrary prescribed point $\mathbf{x}^*$ in the controllable subspace $\mathscr{V}_u = \mathscr{V}_{k_u}$.

5.4 Let $\mathbf{A}$ be $n \times n$ and consider the linear equation

$$\mathbf{A}\mathbf{x} = \mathbf{b}. \tag{5.372}$$

Prove the following:
(a) If Equation (5.372) has a unique solution, it lies in the controllable subspace of $(\mathbf{A}, \mathbf{b})$.
(b) There cannot be more than one solution of Equation (5.372) in the controllable subspace of $(\mathbf{A}, \mathbf{b})$.

5.5 For

$$\dot{\mathbf{x}}(t) = \mathbf{A}\mathbf{x}(t) + \mathbf{B}\mathbf{u}(t) \tag{5.373}$$

with $\mathbf{x}(0^-) = 0$, consider the "impulsive" input

$$\mathbf{u}(t) = \alpha_0 \delta(t) + \alpha_1 \frac{d\delta(t)}{dt} + \cdots + \alpha_{n-1} \frac{d^{n-1}\delta(t)}{dt^{n-1}} \tag{5.374}$$

and prove that there exist $[\alpha_0, \alpha_1, \ldots \alpha_{n-1}]$ such that $\mathbf{x}(0^-)$ is transferred instantaneously to any

$$\mathbf{x}^* \in \mathscr{B} + \mathbf{A}\mathscr{B} + \cdots + \mathbf{A}^{n-1}\mathscr{B}; \tag{5.375}$$

that is $\mathbf{x}(0^+) = \mathbf{x}^*$.

5.6 Prove that the minimum number of inputs (outputs) necessary for controllability (observability) of a state-space system is the largest of the numbers n_i of Jordan blocks associated with each distinct eigenvalue λ_i for $i = 1, 2, \ldots$.

5.7 Give three different minimal realizations of

$$\mathbf{G}(s) = \frac{s-1}{s(s+1)(s+2)}. \tag{5.376}$$

5.8 For

$$\mathbf{G}(s) = \frac{1}{s+2}, \tag{5.377}$$

provide:

(a) a controllable and unstable realization of order 100;

(b) an observable and unstable realization of order 100.

5.9 Give a minimal realization of

$$\mathbf{G}(s) = \begin{bmatrix} \dfrac{1}{s-1} & \dfrac{2}{(s-1)(s+1)} \\ \dfrac{1}{s-1} & \dfrac{1}{s-1} \end{bmatrix}. \tag{5.378}$$

(a) Find the order of the minimal realization of the $(1,2)$ element of $\mathbf{G}(s)$ perturbed to

$$\frac{2+\delta}{(s-1)(s+1)}. \tag{5.379}$$

(b) Find the order of the minimal realization of $\mathbf{G}(s)$ when it is subject to perturbations or uncertainties as shown below:

$$\begin{bmatrix} \dfrac{1}{s-1+\delta_1} & \dfrac{2}{(s-1+\delta_3)(s+1)} \\ \dfrac{1}{s-1+\delta_2} & \dfrac{1}{s-1+\delta_4} \end{bmatrix}. \tag{5.380}$$

5.10 For

$$\mathbf{G}(s) = \begin{bmatrix} \dfrac{s-1}{s(s+1)} & \dfrac{1}{(s+1)(s+2)} \\ \dfrac{1}{s(s+20)} & \dfrac{s+1}{s(s+2)} \end{bmatrix}, \tag{5.381}$$

find minimal realizations using

(a) Gilbert's method;

(b) Wolovich's method.

5.11   Find minimal realizations for

$$\mathbf{G}(s) = \begin{bmatrix} \dfrac{1}{(s+1)^2} & \dfrac{s}{(s+1)(s+2)} \\[3mm] \dfrac{s}{(s-1)(s+2)} & \dfrac{1}{(s-1)^2} \end{bmatrix}, \qquad (5.382)$$

using a left and a right MFD, respectively.

5.12   Prove that if $(\mathbf{A}, \mathbf{B}, \mathbf{C})$ is controllable and observable, so is $(\mathbf{A} + \mathbf{BKC}, \mathbf{BG}, \mathbf{LC})$ for all square nonsingular $\mathbf{G}$ and $\mathbf{L}$, and arbitrary $\mathbf{K}$.

5.13 $\mathbf{U}(s)$ is unimodular with

$$\mathbf{U}(s) = U_0 + U_1 s + \cdots + U_n s^n. \qquad (5.383)$$

Find $U^{-1}(s)$ in terms of U_i for $i = 0, 1, 2, \ldots, n$.

5.14 $\mathbf{G}(s)$ is square with state-space representation $[\mathbf{A}, \mathbf{B}, \mathbf{C}, \mathbf{D}]$ and $\mathbf{D}$ is invertible. Prove that $\mathbf{G}^{-1}(s)$ exists, is proper, and has a state-space representation $[\mathbf{A}_i, \mathbf{B}_i, \mathbf{C}_i, \mathbf{D}_i]$ with

$$\begin{aligned} \mathbf{A}_i &= \mathbf{A} - \mathbf{BD}^{-1}\mathbf{C}, \\ \mathbf{B}_i &= \mathbf{BD}^{-1}, \\ \mathbf{C}_i &= -\mathbf{D}^{-1}\mathbf{C}, \\ \mathbf{D}_i &= \mathbf{D}^{-1}. \end{aligned} \qquad (5.384)$$

Hint: $(\mathbf{u}(t), \mathbf{y}(t))$ is the (output, input) of $\mathbf{G}^{-1}(s)$.

5.15   Consider a state-space representation $[\mathbf{A}, \mathbf{B}, \mathbf{C}, \mathbf{D}]$ with $\mathbf{A}$ in Jordan form. Determine conditions on the rows of $\mathbf{B}$ (columns of $\mathbf{C}$) for controllability (observability).

5.16   Consider the system in Figure 5.5.

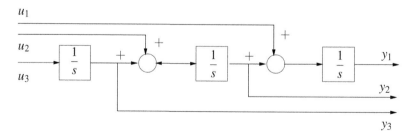

Figure 5.5 A control system (Exercise 5.16)

(a)   Prove that the system is observable from y_1, but not from y_2 or $\begin{bmatrix} y_2 \\ y_3 \end{bmatrix}$.

(b)   Prove that the system is controllable from u_3, but not from u_1 or $\begin{bmatrix} u_1 \\ u_2 \end{bmatrix}$.

5.17   Consider the system in Figure 5.6.

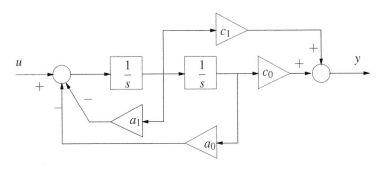

Figure 5.6 A control system (Exercise 5.17)

Prove that the system is always controllable, but is observable if and only if

$$c_0^2 - a_1 c_1 c_0 + a_0 c_1^2 \neq 0. \tag{5.385}$$

5.11   Notes and References

Kalman's seminal work on minimal realizations was published in Kalman (1965). The singular value decomposition is described in Golub and Van Loan (1996). The staircase algorithm used in the MATLAB functions discussed in Remark 5.5 can be found in Rosenbrock (1970). Wolovich's realization was developed in Wolovich (1974) and Gilbert's realization was presented in Gilbert (1963). The results on transfer function conditions for stabilizability and detectability were derived in Bhattacharyya and Howze (1984).

6 State Feedback, Observers, and Regulators

In this chapter we deal with the control of a state-space system via feedback of the states and outputs, respectively. The first main result that is proved is that *all* eigenvalues of a controllable state-space system can be reassigned to arbitrary new locations by a static state feedback control law.

Next we turn to the output feedback problem. In most real systems the states are unavailable for measurement and state feedback cannot be implemented directly. In these cases the states can be reconstructed asymptotically by a dynamic system, driven only by plant inputs and outputs. Such a system is known as an observer. The observer eigenvalues can also be assigned arbitrarily provided the system is observable. The computed state feedback control law can then be replaced by the "observer-reconstructed" state feedback control law.

We develop the theory of the full-order observer as well as the reduced-order observer. In each case the closed-loop eigenvalues under observer-reconstructed state feedback, consist of the union of the eigenvalues assigned by computed state feedback and those of the observer. The final result of this exercise is an equivalent output feedback dynamic controller, which assigns a preassigned set of complex numbers as the closed-loop eigenvalues of the feedback system. In particular, a controllable, observable plant can always be feedback stabilized by such a controller. This problem of feedback stabilization is called a regulator problem and the controller constructed thus via observer theory is called a regulator. We also discuss the issue of proper implementation of observers and show that a feedback structure as in the Kalman filter is more robust. Such an observer is called a robust observer.

Next we discuss the question of what realizations of the plant should be used for reliable stabilization. We show that realizations in which the unstable part of the plant model has ımaximal order under the perturbations expected lead to nonfragile stabilization that holds up when the plant parameters are subject to small uncertainties, whereas minimal realizations may, in general, not withstand such perturbations.

6.1   Introduction

A fundamental function of feedback control is the stabilization of an unstable plant or improving the stability margin of an already stable plant. For example, an inverted pendulum on a cart is unstable in the upright vertical position in the absence of feedback control. A carefully designed feedback signal applied to the cart and driven by deviations of the pendulum from the vertical can be used to stabilize the vertical position in the controlled system. In systems which are stable, feedback may be necessary to increase the damping of lightly damped or oscillatory modes. In the design of servomechanisms one needs to adjoin the dynamics of the external reference and disturbance signals, which are often unstable, to the plant and stabilize the resulting system. In each of these cases we are dealing with a feedback stabilization or regulator problem.

In 1960 Kalman introduced the concept, quite new at the time, that feedback of *all* the states of a system, as opposed to only the output, could deliver much higher levels of performance. This ushered in the era of modern control, with its emphasis on optimal state feedback, designed to minimize a quadratic cost index as proposed by Kalman. Later it was shown that with state feedback one could reassign *all* the closed-loop eigenvalues arbitrarily. This contrasted sharply with static output feedback, where the closed-loop eigenvalues could only migrate along the branches of the "root loci."

We first develop below, this eigenvalue assignability property of state feedback. Next we deal with the constraint that the states are most often not available for measurement in real systems. This leads naturally to the notion of a system that dynamically reconstructs the states from available input and output signals. Such a device is known as an *observer*. We discuss the design of full and reduced-order observers. Finally we show that by closing the feedback loop through observer-reconstructed feedback we can obtain a predesigned set of eigenvalues for the closed-loop system.

We also discuss robust and nonrobust implementation of observers and show that robust implementation corresponds to the Kalman filter architecture. Finally we discuss the issue of what plant order should be used for reliable feedback stabilization. The important result is established that the correct order realization that must be used is one that has *maximal* order for the unstable part of the plant, since otherwise miniscule perturbations can destabilize the closed loop.

6.2 State Feedback versus Output Feedback

In this section we briefly discuss, compare, and contrast state feedback, static output feedback, and dynamic output feedback. We begin our discussion with a simple illustrative example.

EXAMPLE 6.1   The discrete-time system

$$\begin{bmatrix} x_1[k+1] \\ x_2[k+1] \end{bmatrix} = \begin{bmatrix} 0 & 1 \\ -a_0 & -a_1 \end{bmatrix} \begin{bmatrix} x_1[k] \\ x_2[k] \end{bmatrix} + \begin{bmatrix} 0 \\ 1 \end{bmatrix} u[k], \tag{6.1a}$$

$$y[k] = \begin{bmatrix} 1 & 0 \end{bmatrix} \begin{bmatrix} x_1[k] \\ x_2[k] \end{bmatrix}, \tag{6.1b}$$

with state $\mathbf{x}[k] = [x_1[k] \ x_2[k]]^T$ and measured output $y[k]$, is to be controlled by state feedback

$$u[k] = f_1 x_1[k] + f_2 x_2[k] + v[k]. \tag{6.2}$$

This transforms Equation (6.1a) into

$$\begin{bmatrix} x_1[k+1] \\ x_2[k+1] \end{bmatrix} = \begin{bmatrix} 0 & 1 \\ -a_0 + f_1 & -a_1 + f_2 \end{bmatrix} \begin{bmatrix} x_1[k] \\ x_2[k] \end{bmatrix} + \begin{bmatrix} 0 \\ 1 \end{bmatrix} v[k]. \tag{6.3}$$

The characteristic equation of Equation (6.3) is

$$z^2 + (a_1 - f_2)z + (a_0 - f_1) = 0 \tag{6.4}$$

and thus the feedback gains (f_1, f_2) can be used to freely assign the two eigenvalues as long as complex eigenvalues occur in conjugate pairs.

The control law in Equation (6.2) can be implemented only if $x_1[k]$ and $x_2[k]$ can be measured. If we impose the restriction that only $y[k]$ can be measured, we need to express the control law in Equation (6.2) in terms of the output signal $y[\cdot]$. This results in

$$u[k] = f_1 y[k] + f_2 y[k+1] + v[k]. \tag{6.5}$$

To implement Equation (6.5) one must know $y[k+1]$ at time k. In other words, the control law is *noncausal* and therefore *unrealizable* in "real time" as written.

There are two approaches to dealing with the noncausality inherent in Equation (6.5). The first approach is to build a device that can construct an approximation $\hat{\mathbf{x}}[k]$ of the true state $\mathbf{x}[k]$ from measurements of $y[k]$ and $u[k]$. Then Equation (6.2) is replaced by

$$\hat{u}[k] = f_1 \hat{x}_1[k] + f_2 \hat{x}_2[k] + v[k]. \tag{6.6}$$

The device that produces $\hat{\mathbf{x}}[k]$ is known as a Kalman filter in the stochastic setting wherein Equations (6.1a) and (6.1b) are subject to random inputs. In the deterministic setting, the device producing $\hat{\mathbf{x}}[k]$ is known as an *observer*. Note that Equation (6.6) can be rewritten as

$$\hat{u}[k] = f_1 y[k] + f_2 \hat{y}[k+1] + v[k], \tag{6.7}$$

which makes explicit the predictive nature of observer-based state feedback.

A second approach to dealing with the noncausality in Equation (6.5) is to introduce a "delay" in Equation (6.1a) by adjoining the equation

$$u[k+1] = v[k]. \tag{6.8}$$

Now consider the augmented state-space system

$$\begin{bmatrix} x_1[k+1] \\ x_2[k+1] \\ u[k+1] \end{bmatrix} = \underbrace{\begin{bmatrix} 0 & 1 & 0 \\ -a_0 & -a_1 & 1 \\ 0 & 0 & 0 \end{bmatrix}}_{\bar{A}} \underbrace{\begin{bmatrix} x_1[k] \\ x_2[k] \\ u[k] \end{bmatrix}}_{\bar{x}[k]} + \underbrace{\begin{bmatrix} 0 \\ 0 \\ 1 \end{bmatrix}}_{\bar{b}} v[k]. \tag{6.9}$$

Consider state feedback applied to Equation (6.9):

$$v[k] = g_0 x_1[k] + g_1 x_2[k] + g_2 u[k] = \underbrace{[g_0 \quad g_1 \quad g_2]}_{\bar{\mathbf{g}}} \begin{bmatrix} x_1[k] \\ x_2[k] \\ u[k] \end{bmatrix}, \tag{6.10}$$

resulting in

$$\bar{\mathbf{x}}[k+1] = [\bar{\mathbf{A}} + \bar{\mathbf{b}}\bar{\mathbf{g}}] \, \bar{\mathbf{x}}[k]. \tag{6.11}$$

It is easily verified that $(\bar{\mathbf{A}}, \bar{\mathbf{b}})$ is controllable and the vector of gains $\bar{\mathbf{g}}$ can be chosen to freely assign the eigenvalues of $\bar{\mathbf{A}} + \bar{\mathbf{b}}\bar{\mathbf{g}}$. Moreover, rewriting Equation (6.10) in terms of inputs and outputs, we get:

$$u[k+1] = g_0 y[k] + g_1 y[k+1] + g_2 u[k]. \tag{6.12}$$

Equation (6.12) is causal and can be implemented if $u[0]$ is chosen or known. It also represents the proper discrete-time transfer function

$$\frac{U(z)}{Y(z)} = \frac{g_0 + g_1 z}{z - g_2} =: \mathbf{C}(z), \tag{6.13}$$

which is a feedback *compensator* driven by the plant or system output as shown in Figure 6.1.

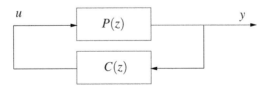

Figure 6.1 Output feedback

This illustrates the notion that dynamic compensation is equivalent to state feedback in the augmented system in Equation (6.9).

6.3 Eigenvalue Assignment by State Feedback

A fundamental result in state-space-based control theory is the ability of state feedback to freely reassign the eigenvalues of a controllable system. The direct implementation of state feedback requires that the state variables be accessible for measurement, a condition that is rarely satisfied in real systems. Nevertheless, this result on eigenvalue assignment plays a key intermediate role in the design of observers and dynamic compensators which address the realistic problem of output feedback controller synthesis. This will be shown in subsequent sections of this chapter.

Consider the state-space system

$$\dot{\mathbf{x}}(t) = \mathbf{A}\mathbf{x}(t) + \mathbf{B}u(t), \tag{6.14}$$

where $\mathbf{A} \in \mathbb{R}^{n \times n}$, $\mathbf{B} \in \mathbb{R}^{n \times r}$, and $(\mathbf{A}, \mathbf{B})$ is controllable, that is,

$$\operatorname{rank} \begin{bmatrix} \mathbf{B} & \mathbf{AB} & \cdots & \mathbf{A}^{n-1}\mathbf{B} \end{bmatrix} = n. \tag{6.15}$$

Applying state feedback control

$$\mathbf{u}(t) = -\mathbf{Kx}(t) + \mathbf{v}(t), \tag{6.16}$$

where $\mathbf{K} \in \mathbb{R}^{r \times n}$, results in the new system:

$$\dot{\mathbf{x}}(t) = (\mathbf{A} - \mathbf{BK})\mathbf{x}(t) + \mathbf{Bv}(t). \tag{6.17}$$

The control law in Equation (6.16) transforms Equation (6.14) into Equation (6.17), as depicted in Figure 6.2.

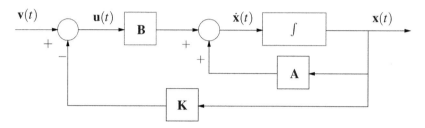

Figure 6.2 State feedback

Effectively then, state feedback transforms $\mathbf{A}$ into $\mathbf{A} - \mathbf{BK}$. Since the eigenvalues of a state-space matrix strongly influence the system dynamic response, it is natural to inquire what effect $\mathbf{K}$ has on the set of eigenvalues of $\mathbf{A} - \mathbf{BK}$, denoted by $\sigma(\mathbf{A} - \mathbf{BK})$. This question is answered in the following theorem, often called the *Pole Assignment Theorem*.

Let $\chi_{\mathbf{A}}(s)$ denote the characteristic polynomial of $\mathbf{A}$:

$$\chi_{\mathbf{A}}(s) := \det[s\mathbf{I} - \mathbf{A}] = s^n + a_{n-1}s^{n-1} + \cdots + a_1 s + a_0 \tag{6.18a}$$

$$= s^n + \underbrace{[a_0 \ a_1 \ \cdots \ a_{n-1}]}_{:=\mathbf{a}^T} \underbrace{\begin{bmatrix} 1 \\ s \\ s^2 \\ \vdots \\ s^{n-1} \end{bmatrix}}_{:=\theta(s)} = s^n + \mathbf{a}^T \theta(s). \tag{6.18b}$$

Clearly, using the above notation:

$$\chi_{\mathbf{A}-\mathbf{BK}}(s) = s^n + \mathbf{a}^T(\mathbf{K})\theta(s), \tag{6.19}$$

where $\mathbf{a}^T(\mathbf{K})$ denotes the coefficients of the characteristic polynomial of $\mathbf{A} - \mathbf{BK}$.

THEOREM 6.1 (Pole Assignment Theorem)   If $(\mathbf{A}, \mathbf{B})$ is real and controllable, the coefficients of the characteristic polynomial of $\mathbf{A} - \mathbf{BK}$, denoted by $\mathbf{a}^T(\mathbf{K})$, can be made to equal any prescribed vector $\underline{\mu}^T \in \mathbb{R}^{1 \times n}$ by choice of $\mathbf{K} \in \mathbb{R}^{r \times n}$.

Proof
Case 1: ($r = 1$) It is instructive to first prove the theorem for the case of a single-input system, that is, when $r = 1$ and hence

$$\mathbf{B} = \mathbf{b}. \tag{6.20}$$

Let

$$\mathbf{L} := \begin{bmatrix} \mathbf{b} & \mathbf{Ab} & \cdots & \mathbf{A}^{n-1}\mathbf{b} \end{bmatrix} \tag{6.21}$$

and $\mathbf{q}^T$ be the last row of $\mathbf{L}^{-1}$, which exists by the controllability hypothesis. Introduce

$$\mathbf{Q} := \begin{bmatrix} \mathbf{q}^T \\ \mathbf{q}^T \mathbf{A} \\ \vdots \\ \mathbf{q}^T \mathbf{A}^{n-1} \end{bmatrix} \tag{6.22}$$

and note that $\mathbf{Q}$ is nonsingular, since

$$\mathbf{QL} = \begin{bmatrix} 0 & \cdots & \cdots & 0 & 1 \\ 0 & & & 1 & * \\ \vdots & & \cdot^{\cdot^{\cdot}} & & \vdots \\ \vdots & \cdot^{\cdot^{\cdot}} & & & \vdots \\ 1 & * & \cdots & \cdots & * \end{bmatrix}. \tag{6.23}$$

Writing

$$\mathbf{z}(t) := \mathbf{Qx}(t), \tag{6.24}$$

we have

$$\dot{\mathbf{z}}(t) = \mathbf{A}_c \mathbf{z}(t) + \mathbf{b}_c \mathbf{u}(t), \tag{6.25}$$

where

$$\mathbf{A}_c = \mathbf{QAQ}^{-1}, \tag{6.26a}$$

$$\mathbf{b}_c = \mathbf{Qb}, \tag{6.26b}$$

and, as is easily verified, $(\mathbf{A}_c, \mathbf{b}_c)$ are in the controllable companion form:

$$\mathbf{A}_c = \begin{bmatrix} 0 & 1 & 0 & \cdots & 0 \\ 0 & 0 & 1 & & 0 \\ & & & \ddots & \\ & & & & 1 \\ -a_0 & -a_1 & \cdots & \cdots & -a_{n-1} \end{bmatrix}, \quad \mathbf{b}_c = \begin{bmatrix} 0 \\ 0 \\ \vdots \\ 0 \\ 1 \end{bmatrix}. \tag{6.27}$$

From Equation (6.27):

$$\chi_{\mathbf{A}_c}(s) = s^n + \underbrace{\begin{bmatrix} a_0 & a_1 & \cdots & a_{n-1} \end{bmatrix}}_{\mathbf{a}^T} \boldsymbol{\theta}(s). \tag{6.28}$$

Now introduce

$$\mathbf{u}(t) = -\underbrace{\begin{bmatrix} k_0 & k_1 & \cdots & k_{n-1} \end{bmatrix}}_{\mathbf{k}_c^T} \begin{bmatrix} z_1(t) \\ \vdots \\ z_n(t) \end{bmatrix} + \mathbf{v}(t) \tag{6.29}$$

and note that

$$\chi_{\mathbf{A}_c - \mathbf{b}_c \mathbf{k}_c^T}(s) = s^n + \left(\mathbf{a}^T + \mathbf{k}_c^T \right) \theta(s). \tag{6.30}$$

We set

$$\mathbf{k}_c^T = \underline{\mu}^T - \mathbf{a}^T \tag{6.31}$$

so that

$$\chi_{\mathbf{A}_c - \mathbf{b}_c \mathbf{k}_c^T}(s) = s^n + \underline{\mu}^T \theta(s), \tag{6.32}$$

and let

$$\mathbf{k}^T = \mathbf{k}_c^T \mathbf{Q}. \tag{6.33}$$

Then

$$\mathbf{A} - \mathbf{b}\mathbf{k}^T = \mathbf{Q}^{-1} \left(\mathbf{A}_c - \mathbf{b}_c \mathbf{k}_c^T \right) \mathbf{Q} \tag{6.34}$$

so that

$$\chi_{\mathbf{A} - \mathbf{b}\mathbf{k}^T}(s) = s^n + \underline{\mu}^T \theta(s), \tag{6.35}$$

completing the proof for the single-input case.

Case 2: $(r > 1)$ In this case, the controllability matrix

$$\mathbf{C} = \begin{bmatrix} \mathbf{B} & \mathbf{A}\mathbf{B} & \cdots & \mathbf{A}^{n-1}\mathbf{B} \end{bmatrix} \tag{6.36}$$

has n linearly independent columns and these may always be chosen and arranged as the columns of $\mathbf{L}$ below:

$$\mathbf{L} := \begin{bmatrix} \mathbf{b}_1 & \mathbf{A}\mathbf{b}_1 & \cdots & \mathbf{A}^{v_1-1}\mathbf{b}_1 & \vdots & \cdots & \vdots & \mathbf{b}_r & \mathbf{A}\mathbf{b}_r & \cdots & \mathbf{A}^{v_r-1}\mathbf{b}_r \end{bmatrix}. \tag{6.37}$$

The matrix $\mathbf{L}$ can be constructed by scanning $\mathscr{C}$ from left to right and collecting "chains" of linearly independent vectors.

Let $\mathbf{q}_i^T$ be the ith row of $\mathbf{L}^{-1}$ for $i = v_1, v_1 + v_2, \ldots, v_1 + v_2 + \cdots + v_r$, forming the matrix

$$
\mathbf{Q} := \begin{bmatrix}
\mathbf{q}_1^T \\
\mathbf{q}_1^T \mathbf{A} \\
\vdots \\
\mathbf{q}_1^T \mathbf{A}^{\nu_1 - 1} \\
\mathbf{q}_2^T \\
\mathbf{q}_2^T \mathbf{A} \\
\vdots \\
\mathbf{q}_2^T \mathbf{A}^{\nu_2 - 1} \\
\vdots \\
\mathbf{q}_r^T \\
\mathbf{q}_r^T \mathbf{A} \\
\vdots \\
\mathbf{q}_r^T \mathbf{A}^{\nu_r - 1}
\end{bmatrix}
\tag{6.38}
$$

and note that $\mathbf{QL}$ can be verified to be nonsingular, which proves that $\mathbf{Q}$ is nonsingular. Now set

$$
\mathbf{z}(t) = \mathbf{Q}\mathbf{x}(t) \tag{6.39}
$$

and consider the pair $(\mathbf{A}_h, \mathbf{B}_h)$:

$$
\mathbf{A}_h := \mathbf{Q}\mathbf{A}\mathbf{Q}^{-1}, \quad \mathbf{B}_h := \mathbf{Q}\mathbf{B}. \tag{6.40}
$$

It is easily verified that $(\mathbf{A}_h, \mathbf{B}_h)$ is of the form

$$
\mathbf{A}_h = \begin{bmatrix}
\mathbf{A}_{11} & \mathbf{A}_{12} & \cdots & \mathbf{A}_{1r} \\
\mathbf{A}_{21} & \mathbf{A}_{22} & & \vdots \\
\vdots & & \ddots & \vdots \\
\mathbf{A}_{r1} & & & \mathbf{A}_{rr}
\end{bmatrix}, \quad
\mathbf{B}_h = \begin{bmatrix}
\mathbf{b}_{11} & 0 & \cdots & 0 \\
0 & \mathbf{b}_{22} & & \vdots \\
\vdots & & \ddots & 0 \\
0 & & & \mathbf{b}_{rr}
\end{bmatrix} \mathbf{G}
\tag{6.41}
$$

with $\mathbf{G} \in \mathbb{R}^{r \times r}$ nonsingular:

$$
\mathbf{A}_{ii} = \begin{bmatrix}
0 & 1 & 0 & \cdots & 0 \\
\vdots & & 1 & & \vdots \\
\vdots & & & \ddots & \vdots \\
0 & & & & 1 \\
& & -\mathbf{a}_{ii}^T & &
\end{bmatrix}, \quad
\mathbf{b}_{ii} = \begin{bmatrix}
0 \\
\vdots \\
\vdots \\
0 \\
1
\end{bmatrix},
\tag{6.42}
$$

and

$$
\mathbf{A}_{ij} = \begin{bmatrix}
0 & \cdots & \cdots & \cdots & 0 \\
\vdots & & & & \vdots \\
0 & \cdots & & \cdots & 0 \\
& & -\mathbf{a}_{ij}^T & &
\end{bmatrix} \quad \text{for } i \in \mathbf{r}, \ j \in \mathbf{r}, \ i \neq j.
\tag{6.43}
$$

Introduce

$$\mathbf{a}_i^T := \begin{bmatrix} a_{i1}^T & a_{i2}^T & \cdots & a_{ir}^T \end{bmatrix} \quad \text{for } i \in \mathbf{r}. \tag{6.44}$$

Let $\mathbf{e}_i$ denote the ith standard basis vector (with zeros everywhere except the ith element which is 1), and define

$$\mathbf{M} := \begin{bmatrix} \mathbf{a}_1^T \\ \mathbf{a}_2^T \\ \vdots \\ \mathbf{a}_r^T \end{bmatrix}, \quad \mathbf{N} := \begin{bmatrix} \mathbf{e}_{v_1+1}^T \\ \mathbf{e}_{v_1+v_2+1}^T \\ \vdots \\ \mathbf{e}_{v_1+v_2+\cdots+v_{r-1}+1}^T \\ -\underline{\mu}^T \end{bmatrix}. \tag{6.45}$$

Then

$$\mathbf{K}_h := -\mathbf{G}^{-1}[\mathbf{M} + \mathbf{N}] \tag{6.46}$$

renders the matrix

$$\mathbf{A}_h - \mathbf{B}_h \mathbf{K}_h \tag{6.47}$$

into the form

$$\begin{bmatrix} 0 & 1 & 0 & \cdots & 0 \\ 0 & 0 & 1 & & 0 \\ & & & \ddots & \\ & & & & 1 \\ & & -\underline{\mu}^T & & \end{bmatrix} \tag{6.48}$$

so that

$$\chi_{\mathbf{A}_h - \mathbf{B}_h \mathbf{K}_h}(s) = s^n + \underline{\mu}^T \theta(s). \tag{6.49}$$

Now, set

$$\mathbf{K} = \mathbf{K}_h \mathbf{Q} \tag{6.50}$$

so that

$$\begin{aligned} \mathbf{A} - \mathbf{BK} &= \mathbf{Q}^{-1} \mathbf{A}_h \mathbf{Q} - \mathbf{Q}^{-1} \mathbf{B}_h \mathbf{K}_h \mathbf{Q} \\ &= \mathbf{Q}^{-1} (\mathbf{A}_h - \mathbf{B}_h \mathbf{K}_h) \mathbf{Q} \end{aligned} \tag{6.51}$$

and thus

$$\chi_{\mathbf{A}-\mathbf{BK}}(s) = s^n + \underline{\mu}^T \theta(s) \tag{6.52}$$

as required, thus proving the theorem. ∎

EXAMPLE 6.2 Let

$$\mathbf{A} = \begin{bmatrix} 0 & 1 & 0 & 0 \\ 1 & 0 & 1 & 0 \\ 0 & 1 & 0 & 1 \\ 0 & 0 & 1 & 0 \end{bmatrix}, \quad \mathbf{B} = \begin{bmatrix} 1 & 0 \\ 0 & 0 \\ 0 & 1 \\ 0 & 0 \end{bmatrix}. \tag{6.53}$$

Clearly,

$$\mathbf{L} = [\mathbf{b}_1 \ \mathbf{A}\mathbf{b}_1 \ \mathbf{b}_2 \ \mathbf{A}\mathbf{b}_2] = \begin{bmatrix} 1 & 0 & 0 & 0 \\ 0 & 1 & 0 & 1 \\ 0 & 0 & 1 & 0 \\ 0 & 0 & 0 & 1 \end{bmatrix} \tag{6.54}$$

has rank 4. It is easy to determine $\mathbf{q}_1^T$ and $\mathbf{q}_2^T$, the second and fourth rows of $\mathbf{L}^{-1}$ from

$$\mathbf{q}_1^T \mathbf{L} = [0 \ 1 \ 0 \ 0], \qquad \mathbf{q}_2^T \mathbf{L} = [0 \ 0 \ 0 \ 1], \tag{6.55}$$

and we get

$$\mathbf{q}_1^T = [0 \ 1 \ 0 \ -1], \qquad \mathbf{q}_2^T = [0 \ 0 \ 0 \ 1]. \tag{6.56}$$

Then

$$\mathbf{Q} = \begin{bmatrix} \mathbf{q}_1^T \\ \mathbf{q}_1^T \mathbf{A} \\ \mathbf{q}_2^T \\ \mathbf{q}_2^T \mathbf{A} \end{bmatrix} = \begin{bmatrix} 0 & 1 & 0 & -1 \\ 1 & 0 & 0 & 0 \\ 0 & 0 & 0 & 1 \\ 0 & 0 & 1 & 0 \end{bmatrix} \tag{6.57}$$

and $(\mathbf{A}_h, \mathbf{B}_h)$ can be found from

$$\mathbf{Q}\mathbf{A} = \mathbf{A}_h\mathbf{Q}, \quad \mathbf{Q}\mathbf{B} = \mathbf{B}_h. \tag{6.58}$$

We find that

$$\mathbf{A}_h = \begin{bmatrix} 0 & 1 & 0 & 0 \\ 1 & 0 & 1 & 0 \\ 0 & 0 & 0 & 1 \\ 1 & 0 & 2 & 0 \end{bmatrix}, \quad \mathbf{B}_h = \begin{bmatrix} 0 & 0 \\ 1 & 0 \\ 0 & 0 \\ 0 & 1 \end{bmatrix} \tag{6.59}$$

so that

$$\mathbf{B}_h = \begin{bmatrix} 0 & 0 \\ 1 & 0 \\ 0 & 0 \\ 0 & 1 \end{bmatrix} \underbrace{\begin{bmatrix} 1 & 0 \\ 0 & 1 \end{bmatrix}}_{\mathbf{G}}. \tag{6.60}$$

Now

$$\mathbf{M} = \begin{bmatrix} -1 & 0 & -1 & 0 \\ -1 & 0 & -2 & 0 \end{bmatrix}, \quad \mathbf{N} = \begin{bmatrix} 0 & 0 & 1 & 0 \\ -\mu_0 & -\mu_1 & -\mu_2 & -\mu_3 \end{bmatrix} \tag{6.61}$$

and

$$\mathbf{G}^{-1} = \begin{bmatrix} 1 & 0 \\ 0 & 1 \end{bmatrix} \tag{6.62}$$

so that

$$\begin{aligned} \mathbf{K}_h &= - \begin{bmatrix} 1 & 0 \\ 0 & 1 \end{bmatrix} \begin{bmatrix} -1 & 0 & 0 & 0 \\ -1-\mu_0 & -\mu_1 & -2-\mu_2 & -\mu_3 \end{bmatrix} \\ &= \begin{bmatrix} 1 & 0 & 0 & 0 \\ 1+\mu_0 & \mu_1 & 2+\mu_2 & \mu_3 \end{bmatrix} \end{aligned} \tag{6.63}$$

and

$$\mathbf{K} = \mathbf{K}_h \mathbf{Q} = - \begin{bmatrix} 0 & 1 & 0 & -1 \\ \mu_1 & 1+\mu_0 & \mu_3 & 1+\mu_2-\mu_0 \end{bmatrix}. \tag{6.64}$$

Now

$$\mathbf{A} - \mathbf{BK} = \begin{bmatrix} 0 & 0 & 0 & 1 \\ 1 & 0 & 1 & 0 \\ -\mu_1 & -\mu_0 & -\mu_3 & \mu_0-\mu_2 \\ 0 & 0 & 1 & 0 \end{bmatrix}. \tag{6.65}$$

We leave it to the reader to verify that

$$\chi_{\mathbf{A}-\mathbf{BK}}(s) = s^4 + \mu_3 s^3 + \mu_2 s^2 + \mu_1 s + \mu_0, \tag{6.66}$$

which validates the procedure of the theorem.

EXAMPLE 6.3 (A MATLAB example)   Consider a controllable pair $(\mathbf{A}, \mathbf{B})$ below:

$$\mathbf{A} = \begin{bmatrix} -1 & 0 & 1 & 1 \\ -2 & 2 & 1 & 0 \\ 0 & 0 & 1 & -4 \\ 2 & 1 & -1 & 0 \end{bmatrix}, \quad \mathbf{B} = \begin{bmatrix} 1 & 0 \\ 1 & -1 \\ 1 & 1 \\ 0 & 1 \end{bmatrix}. \tag{6.67}$$

The following MATLAB script calculates a state feedback matrix $\mathbf{K}$ such that the eigenvalues of $\mathbf{A} - \mathbf{BK}$ are assigned to

$$\Lambda = \{-1, -2, -1 \pm 2j\}. \tag{6.68}$$

```
clear
A=[-1 0 1 1; -2 2 1 0; 0 0 1 -4; 2 1 -1 0];
B=[1 0; 1 -1; 1 1; 0 1];
p=[-1 -2 -1-j*2 -1+j*2];
K=place(A,B,p)
eig(A-B*K)
```

The resulting state feedback is:

$$\mathbf{K} = \begin{bmatrix} 2.6061 & 5.9165 & -3.0140 & 8.7992 \\ 1.0753 & -2.3533 & 0.8937 & -1.7556 \end{bmatrix}. \tag{6.69}$$

6.4   Pole Assignment via Sylvester's Equation

The procedure developed in the last section to prove the theorem on pole assignment by state feedback essentially involves transformation of $\mathbf{A}$ and $\mathbf{B}$ into companion or block-companion controllable forms. In general, such transformations may be numerically ill-conditioned and can lead to large errors in the computed feedback matrix when there are round-off errors, especially in high-order systems, say with $n \geq 10$.

An alternative computational approach that avoids such transformations is described in this section. Given real matrices $\mathbf{A}$ and $\mathbf{B}$, form the matrix equations

$$\mathbf{AX} - \mathbf{X}\tilde{\mathbf{A}} = \mathbf{BG}, \tag{6.70a}$$

$$\mathbf{KX} = \mathbf{G}, \tag{6.70b}$$

for a real matrix pair $(\tilde{\mathbf{A}}, \mathbf{G})$ with $\mathbf{X}$ and $\tilde{\mathbf{A}}$ being $n \times n$ and $\tilde{\mathbf{A}}$ being in companion form with

$$\chi_{\tilde{\mathbf{A}}}(s) = s^n + \underline{\mu}^T \theta(s) \tag{6.71}$$

for a prescribed vector $\underline{\mu} \in \mathbb{R}^{1 \times n}$. If $\mathbf{X}$ were nonsingular, it would follow that

$$\mathbf{A} - \mathbf{BK} = \mathbf{X}\tilde{\mathbf{A}}\mathbf{X}^{-1} \tag{6.72}$$

and this would render

$$\chi_{\mathbf{A}-\mathbf{BK}}(s) = s^n + \underline{\mu}^T \theta(s). \tag{6.73}$$

The following theorem characterizes the solution $\mathbf{X}$ of Equation (6.70a).

THEOREM 6.2

(1) The solution of Equation (6.70a) exists and has the form

$$\mathbf{X} = \begin{bmatrix} \mathbf{B} & \mathbf{AB} & \mathbf{A}^{n-1}\mathbf{B} \end{bmatrix} \Gamma \begin{bmatrix} \mathbf{G} \\ \mathbf{G}\tilde{\mathbf{A}} \\ \vdots \\ \mathbf{G}\tilde{\mathbf{A}}^{n-1} \end{bmatrix} \tag{6.74}$$

where Γ has full rank if

$$\sigma(\mathbf{A}) \cap \sigma(\tilde{\mathbf{A}}) = \emptyset. \tag{6.75}$$

(2) If $(\mathbf{A}, \mathbf{B})$ is controllable, $(\mathbf{G}, \tilde{\mathbf{A}})$ is observable, and Equation (6.75) holds, then $\mathbf{X}$ is nonsingular for almost every $\mathbf{G}$.

The proof of this result is omitted and the interested reader is referred to the Notes and References section for the sources of the proof.

The above theorem suggests the following algorithm for calculating $\mathbf{K}$, the state feedback pole assignment matrix.

(a) Given $\mathbf{A}, \mathbf{B}$, and $\underline{\mu}$, choose $\tilde{\mathbf{A}}$ such that

$$\chi_{\tilde{\mathbf{A}}}(s) = s^n + \underline{\mu}^T \theta(s). \tag{6.76}$$

(b) Choose $\mathbf{G}$ so that $(\mathbf{G}, \tilde{\mathbf{A}})$ is observable.
(c) Solve Equation (6.70a) for $\mathbf{X}$ and if $\mathbf{X}$ is nonsingular, solve for $\mathbf{K}$ from Equation (6.70b). If $\mathbf{X}$ is singular, perturb $\mathbf{G}$ and repeat steps (a) and (b).

An additional advantage of this algorithm is that there exist effective and reliable numerical algorithms for solving Equation (6.70a), which is known as Sylvester's equation.

EXAMPLE 6.4 (A MATLAB example)   Consider a controllable pair $(\mathbf{A}, \mathbf{B})$ below:

$$\mathbf{A} = \begin{bmatrix} -1 & 0 & 1 & 1 \\ -2 & 2 & 1 & 0 \\ 0 & 0 & 1 & -4 \\ 2 & 1 & -1 & 0 \end{bmatrix}, \quad \mathbf{B} = \begin{bmatrix} 1 & 0 \\ 1 & -1 \\ 1 & 1 \\ 0 & 1 \end{bmatrix}. \tag{6.77}$$

The following MATLAB script calculates a state feedback matrix $\mathbf{K}$ such that the eigenvalues of $\mathbf{A} - \mathbf{BK}$ are assigned to

$$\Lambda = \{-1, -2, -1 \pm j2\}. \tag{6.78}$$

```
clear
A=[-1 0 1 1; -2 2 1 0; 0 0 1 -4; 2 1 -1 0];
B=[1 0; 1 -1; 1 1;0 1];
Ah=[-1 0 0 0; 0 -2 0 0; 0 0 -1 2; 0 0 2 -1];
G=rand(2,4);
X=sylvester(A,At,B*G);
K=G*inv(X)
eig(A-B*K)
```

The resulting state feedback is:

$$\mathbf{K} = \begin{bmatrix} 1.7057 & -20.8249 & 5.2200 & -16.6800 \\ 3.5149 & -43.8336 & 9.7314 & -32.6658 \end{bmatrix}. \tag{6.79}$$

6.5 Partial Pole Assignment by State Feedback

In the previous sections, we considered the problem of assigning all the eigenvalues of a system by state feedback. In this section, we extend this result to the problem of assigning $k(< n)$ eigenvalues leaving the remaining $n - k$ eigenvalues unchanged. This could be useful when there are unstable (RHP) or lightly damped modes and the remaining modes are stable and well-damped.

Consider again

$$\dot{\mathbf{x}}(t) = \mathbf{A}\mathbf{x}(t) + \mathbf{B}\mathbf{u}(t) \tag{6.80}$$

with $\mathbf{A}, \mathbf{B}$ real and controllable and let

$$\sigma(\mathbf{A}) = \Lambda_1 \cup \Lambda_2 \tag{6.81}$$

where Λ_1 has k elements that need to be reassigned by state feedback, leaving the elements of Λ_2, assumed to be disjoint from Λ_1, unchanged. Let $\mathcal{V}_i$ denote the eigenspaces associated with Λ_i for $i = 1, 2$ and let $\mathbf{T}_i$ denote a matrix whose columns form a basis for $\mathcal{V}_i$ for $i = 1, 2$. Then with

$$\mathbf{T} = [\mathbf{T}_1 \ \mathbf{T}_2], \quad \text{where} \quad \mathbf{T}_1 \in \mathbb{R}^{n \times k}, \ \mathbf{T}_2 \in \mathbb{R}^{n \times (n-k)} \tag{6.82}$$

set

$$\mathbf{x}(t) = \mathbf{T}\mathbf{z}(t) = \mathbf{T}_1 \mathbf{z}_1(t) + \mathbf{T}_2 \mathbf{z}_2(t). \tag{6.83}$$

With this coordinate transformation, Equation (6.80) is transformed into

$$\begin{bmatrix} \dot{\mathbf{z}}_1(t) \\ \dot{\mathbf{z}}_2(t) \end{bmatrix} = \begin{bmatrix} \mathbf{A}_1 & 0 \\ 0 & \mathbf{A}_2 \end{bmatrix} \begin{bmatrix} \mathbf{z}_1(t) \\ \mathbf{z}_2(t) \end{bmatrix} + \begin{bmatrix} \mathbf{B}_1 \\ \mathbf{B}_2 \end{bmatrix} \mathbf{u}(t). \tag{6.84}$$

Since $(\mathbf{A}, \mathbf{B})$ is controllable, so are $(\mathbf{A}_1, \mathbf{B}_1)$ and $(\mathbf{A}_2, \mathbf{B}_2)$. Now apply the state feedback control

$$\mathbf{u}(t) = -\mathbf{K}_1 \mathbf{z}_1(t) + \mathbf{v}(t) \tag{6.85}$$

which changes Equation (6.84) into

$$\begin{bmatrix} \dot{\mathbf{z}}_1(t) \\ \dot{\mathbf{z}}_2(t) \end{bmatrix} = \begin{bmatrix} \mathbf{A}_1 - \mathbf{B}_1 \mathbf{K}_1 & 0 \\ -\mathbf{B}_2 \mathbf{K}_1 & \mathbf{A}_2 \end{bmatrix} \begin{bmatrix} \mathbf{z}_1(t) \\ \mathbf{z}_2(t) \end{bmatrix} + \begin{bmatrix} \mathbf{B}_1 \\ \mathbf{B}_2 \end{bmatrix} \mathbf{u}(t). \tag{6.86}$$

In the original coordinates

$$\mathbf{u}(t) = -\underbrace{\begin{bmatrix} \mathbf{K}_1 & 0 \end{bmatrix} \mathbf{T}^{-1}}_{\mathbf{K}} \mathbf{x}(t) + \mathbf{v}(t) \tag{6.87}$$

and

$$\mathbf{A} - \mathbf{B}\mathbf{K} = \mathbf{T} \begin{bmatrix} \mathbf{A}_1 - \mathbf{B}_1 \mathbf{K}_1 & 0 \\ -\mathbf{B}_2 \mathbf{K}_1 & \mathbf{A}_2 \end{bmatrix} \mathbf{T}^{-1} \tag{6.88}$$

so that

$$\chi_{\mathbf{A} - \mathbf{B}\mathbf{K}}(s) = \chi_{\mathbf{A}_1 - \mathbf{B}_1 \mathbf{K}_1}(s) \cdot \chi_{\mathbf{A}_2}(s). \tag{6.89}$$

Since $\mathbf{A}_1, \mathbf{B}_1$ is controllable, the matrix $\mathbf{K}_1$ can be chosen to assign $\chi_{\mathbf{A}_1 - \mathbf{B}_1 \mathbf{K}_1}(s)$ to be any monic polynomial of degree k. Thus the control law in Equation (6.88) reassigns k eigenvalues, and leaves the $n - k$ eigenvalues of Λ_2 unchanged.

EXAMPLE 6.5   Consider a state-space description of the system

$$\dot{\mathbf{x}}(t) = \begin{bmatrix} 0 & 1 & 0 & 0 \\ 0 & 0 & 1 & 0 \\ 0 & 0 & 0 & 1 \\ -4 & 0 & 5 & 0 \end{bmatrix} \mathbf{x}(t) + \begin{bmatrix} 1 & 0 \\ 1 & 1 \\ 1 & -1 \\ 1 & 1 \end{bmatrix} \mathbf{u}(t) =: \mathbf{A}\mathbf{x}(t) + \mathbf{B}\mathbf{u}(t), \qquad (6.90)$$

$$\mathbf{y}(t) = \begin{bmatrix} 1 & 0 & 0 & 0 \\ 1 & 1 & -1 & 1 \end{bmatrix} \mathbf{u}(t) =: \mathbf{C}\mathbf{x}(t) + \mathbf{D}\mathbf{u}(t). \qquad (6.91)$$

The following MATLAB script calculates a state feedback $\mathbf{K}$ using feedback of only the states corresponding to the unstable part. The reassigned eigenvalues are at $\{-3, -4\}$.

```
clear
A=[0 1 0 0; 0 0 1 0; 0 0 0 1; -4 0 5 0];
B=[1 0; 1 1; 1 -1; 1 1];
C=[1 0 0 0; 1 1 -1 1];
D=[0 0; 0 0];
[n,m]=size(B); [r,n]=size(C);
[V,D]=eig(A);
ku=0; ks=0;
for i=1:n,
    if real(D(i,i))>0,
        ku=ku+1;
        T1(:,ku)=V(:,i);
    else
        ks=ks+1;
        T2(:,ks)=V(:,i);
    end,
end,
T=[T1   T2];
Ao=inv(T)*A*T;
Bo=inv(T)*B;
Co=C*T; Do=D;
p=[-3 -4];    % desired poles of the unstable part
K1=place(Ao(1:ku,1:ku),Bo(1:ku,:),p);
K=[K1 zeros(m,n-ku)]*inv(T);
eig(A-B*K)
```

The feedback gain obtained is

$$\mathbf{K} = \begin{bmatrix} \mathbf{K}_1 & 0 \end{bmatrix} \mathbf{T}^{-1}$$

$$= \begin{bmatrix} -2.1693 & 2.5 & 0 & 0 \\ 3.2540 & 0 & 0 & 0 \end{bmatrix} \begin{bmatrix} -0.1085 & 0.5 & 0.1085 & 0.5 \\ -0.2169 & 0.5 & -0.2169 & -0.5 \\ -0.4339 & 0.5 & 0.4339 & 0.5 \\ -0.8677 & 0.5 & -0.8677 & -0.5 \end{bmatrix}^{-1} \quad (6.92)$$

$$= \begin{bmatrix} 0 & 1.6667 & 2.5000 & 0.8333 \\ 5 & 2.5000 & -5.0000 & -2.5000 \end{bmatrix}.$$

The closed-loop eigenvalues are:

$$\sigma(\mathbf{A} - \mathbf{BK}) = \Lambda_1 \cup \Lambda_2, \quad (6.93)$$

where $\Lambda_1 = \{-3, -4\}$ are the reassigned eigenvalues and $\Lambda_2 = \{-1, -2\}$ are the original eigenvalues of the system.

6.6 Output Feedback: Dynamic Compensators

As shown above, state feedback offers the possibility of reassigning all closed-loop eigenvalues arbitrarily. In the next chapter it will be shown that state feedback can also optimize the dynamics of the controlled system to minimize a quadratic cost functional. It will also be shown in the next chapter that optimal state feedback can provide excellent gain and phase margins.

In practice, however, it is impossible to measure all the states of a system and feedback control must be implemented based on the available measurements which, in general, are much fewer in number than the states. In such systems in general the outputs have to be processed dynamically to generate the feedback control signal. This processor is what we commonly refer to in control systems as the dynamic controller or compensator. In this section, we discuss the design of such controllers using the machinery of state-variable models.

Consider the plant described by

$$\dot{\mathbf{x}}_p(t) = \mathbf{A}_p \mathbf{x}_p(t) + \mathbf{B}_p \mathbf{u}_p(t), \quad (6.94a)$$

$$\mathbf{y}_p(t) = \mathbf{C}_p \mathbf{x}_p(t), \quad (6.94b)$$

where $\mathbf{A}_p \in \mathbb{R}^{n_p \times n_p}$, $\mathbf{B}_p \in \mathbb{R}^{n_p \times r}$, $\mathbf{C}_p \in \mathbb{R}^{m \times n_p}$, and $\mathbf{x}_p(t)$, $\mathbf{u}_p(t)$, and $\mathbf{y}_p(t)$ represent the state, input, and output vectors, respectively. Typically, in large-scale systems, $n_p >> m$, that is there are far fewer outputs that can be measured than the number of plant states.

The dynamic feedback controller is described by

$$\dot{\mathbf{x}}_c(t) = \mathbf{A}_c \mathbf{x}_c(t) + \mathbf{B}_c \mathbf{y}_p(t), \quad (6.95a)$$

$$\mathbf{u}_p(t) = \mathbf{C}_c \mathbf{x}_c(t) + \mathbf{D}_c \mathbf{y}_p(t), \quad (6.95b)$$

with $\mathbf{A}_c \in \mathbb{R}^{n_c \times n_c}$, $\mathbf{B}_c \in \mathbb{R}^{n_c \times m}$, $\mathbf{C}_c \in \mathbb{R}^{r \times n_c}$, $\mathbf{D}_c \in \mathbb{R}^{r \times m}$, and has $\mathbf{y}_p(t)$ as input and $\mathbf{u}_p(t)$ as output. Thus, Equations (6.94) and (6.95) form a closed-loop system described by

$$\begin{bmatrix} \dot{\mathbf{x}}_p(t) \\ \dot{\mathbf{x}}_c(t) \end{bmatrix} = \underbrace{\begin{bmatrix} \mathbf{A}_p + \mathbf{B}_p \mathbf{D}_c \mathbf{C}_p & \mathbf{B}_p \mathbf{C}_c \\ \mathbf{B}_c \mathbf{C}_p & \mathbf{A}_c \end{bmatrix}}_{\mathbf{A}_{cl}} \begin{bmatrix} \mathbf{x}_p(t) \\ \mathbf{x}_c(t) \end{bmatrix}. \tag{6.96}$$

It is now clear that the closed-loop system is internally stable if and only if

$$\sigma(\mathbf{A}_{cl}) \subset \mathbb{C}^-, \tag{6.97}$$

that is, all eigenvalues of $\mathbf{A}_{cl}$ have negative real parts.

The stabilizing compensator design problem can thus be stated as follows. Given the plant description $(\mathbf{A}_p, \mathbf{B}_p, \mathbf{C}_p)$, find n_c the controller order and $(\mathbf{A}_c, \mathbf{B}_c, \mathbf{C}_c, \mathbf{D}_c)$ so that $\mathbf{A}_{cl}$ in Equation (6.96) satisfies Equation (6.97), that is, has all $n_p + n_c$ eigenvalues in the left-half plane. In the following sections, we develop solutions to this problem. We begin with the basic result below.

THEOREM 6.3   There exists a solution to the stabilizing compensator design problem if and only if the plant is stabilizable and detectable.

The "only if " part of the above theorem can easily be established by considering the Kalman canonical decomposition of the plant:

$$\mathbf{A}_p = \begin{bmatrix} \mathbf{A}_1 & \mathbf{A}_3 & \mathbf{A}_5 & \mathbf{A}_7 \\ 0 & \mathbf{A}_2 & 0 & \mathbf{A}_8 \\ 0 & 0 & \mathbf{A}_4 & \mathbf{A}_9 \\ 0 & 0 & 0 & \mathbf{A}_6 \end{bmatrix}, \quad \mathbf{B}_p = \begin{bmatrix} \mathbf{B}_1 \\ \mathbf{B}_2 \\ 0 \\ 0 \end{bmatrix}, \tag{6.98}$$

$$\mathbf{C}_p = \begin{bmatrix} 0 & \mathbf{C}_2 & 0 & \mathbf{C}_4 \end{bmatrix}.$$

Substituting Equation (6.98) into Equation (6.96), we get

$$\mathbf{A}_{cl} = \begin{bmatrix} \mathbf{A}_1 & \mathbf{A}_3 + \mathbf{B}_1 \mathbf{D}_c \mathbf{C}_2 & \mathbf{A}_5 & \mathbf{A}_7 + \mathbf{B}_1 \mathbf{D}_c \mathbf{C}_4 & \mathbf{B}_1 \mathbf{C}_c \\ 0 & \mathbf{A}_2 + \mathbf{B}_2 \mathbf{D}_c \mathbf{C}_2 & 0 & \mathbf{A}_8 + \mathbf{B}_2 \mathbf{D}_c \mathbf{C}_4 & \mathbf{B}_2 \mathbf{C}_c \\ 0 & 0 & \mathbf{A}_4 & \mathbf{A}_0 & 0 \\ 0 & 0 & 0 & \mathbf{A}_6 & 0 \\ 0 & \mathbf{B}_c \mathbf{C}_2 & 0 & \mathbf{B}_c \mathbf{C}_4 & \mathbf{A}_c \end{bmatrix} \tag{6.99}$$

which shows that, regardless of the controller, or for every controller, it holds true that,

$$\sigma(\mathbf{A}_1) \cup (\mathbf{A}_4) \cup (\mathbf{A}_6) \subset \sigma(\mathbf{A}_{cl}). \tag{6.100}$$

Thus, $\mathbf{A}_1$, $\mathbf{A}_4$, and $\mathbf{A}_6$ must have eigenvalues in $\mathbb{C}^-$ and this is equivalent to stabilizability ($\mathbf{A}_4, \mathbf{A}_6$ stable) and detectability ($\mathbf{A}_1, \mathbf{A}_4$ stable) of the plant.

To complete the proof of the theorem above, we will need to show that

$$\bar{\mathbf{A}}_{cl} = \begin{bmatrix} \mathbf{A}_2 + \mathbf{B}_2 \mathbf{D}_c \mathbf{C}_2 & \mathbf{B}_2 \mathbf{C}_c \\ \mathbf{B}_c \mathbf{C}_2 & \mathbf{A}_c \end{bmatrix} \tag{6.101}$$

can be made to be stable by choosing $(\mathbf{A}_c, \mathbf{B}_c, \mathbf{C}_c, \mathbf{D}_c)$. Because $(\mathbf{A}_2, \mathbf{B}_2, \mathbf{C}_2)$ is controllable and observable, we see that we can construct the controller for a controllable, observable, and therefore minimal-order plant. This is done constructively below using first an observer in this chapter and then a dynamic compensator in the next chapter.

6.7   State Observers

An observer is a dynamic system that has been designed to asymptotically produce an approximation of the unmeasurable states of a system, based on measurements of its outputs and inputs. In the following subsections we develop the theory of design of full as well as reduced-order observers.

6.7.1   Full-Order Observers

The Observer Design Problem
The observer design problem may be stated as: Design a device which will dynamically produce an approximation of the state $\mathbf{x}(t)$ (n-dimensional state vector) from external measurements $\mathbf{y}(t)$ (m vector) and $\mathbf{u}(t)$ (r vector) (see Figure 6.3).

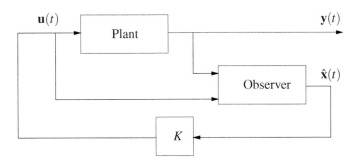

Figure 6.3 A feedback system with plant and observer

Consider the plant

$$\dot{\mathbf{x}}(t) = \mathbf{A}\mathbf{x}(t) + \mathbf{B}\mathbf{u}(t),$$
$$\mathbf{y}(t) = \mathbf{C}\mathbf{x}(t),$$
(6.102)

and a dynamic system, driven by plant inputs and outputs, called an *observer*:

$$\dot{\mathbf{z}}(t) = \mathbf{M}\mathbf{z}(t) + \mathbf{L}\mathbf{y}(t) + \mathbf{G}\mathbf{u}(t),$$
$$\hat{\mathbf{x}}(t) = \mathbf{P}\mathbf{z}(t) + \mathbf{Q}\mathbf{y}(t) + \mathbf{R}\mathbf{u}(t).$$
(6.103)

The observer design problem then reduces to determining $(\mathbf{M}, \mathbf{L}, \mathbf{G}, \mathbf{P}, \mathbf{Q}, \mathbf{R})$ so that

$$\lim_{t \to \infty} (\hat{\mathbf{x}}(t) - \mathbf{x}(t)) = 0 \qquad \text{for all} \quad \mathbf{x}(0), \mathbf{z}(0), \mathbf{u}(t).$$
(6.104)

It will turn out that this will be possible if $(\mathbf{C}, \mathbf{A})$ is observable or detectable.

Consider the special case $\mathbf{P} = \mathbf{I}_n$, $\mathbf{Q} = \mathbf{R} = \mathbf{0}$. Then we must have

$$\hat{\mathbf{x}}(t) = \mathbf{z}(t). \tag{6.105}$$

Let

$$\mathbf{e}(t) = \mathbf{z}(t) - \mathbf{x}(t). \tag{6.106}$$

Then

$$\begin{aligned}
\dot{\mathbf{e}}(t) &= \dot{\mathbf{z}}(t) - \dot{\mathbf{x}}(t) \\
&= \underbrace{\mathbf{Mz}(t) + \mathbf{Ly}(t) + \mathbf{Gu}(t)}_{\dot{\mathbf{z}}(t)} - \underbrace{(\mathbf{Ax}(t) + \mathbf{Bu}(t))}_{\dot{\mathbf{x}}(t)} \\
&= \mathbf{Mz}(t) + \mathbf{LCx}(t) + \mathbf{Gu}(t) - \mathbf{Ax}(t) - \mathbf{Bu}(t) \\
&= \mathbf{Me}(t) + (\mathbf{M} - \mathbf{A} + \mathbf{LC})\mathbf{x}(t) + (\mathbf{G} - \mathbf{B})\mathbf{u}(t).
\end{aligned} \tag{6.107}$$

Therefore, if we set

$$\mathbf{G} = \mathbf{B}, \qquad \mathbf{M} = \mathbf{A} - \mathbf{LC} \tag{6.108}$$

we will have

$$\dot{\mathbf{e}}(t) = \mathbf{Me}(t) = (\mathbf{A} - \mathbf{LC})\mathbf{e}(t) \tag{6.109}$$

and the influence of $\mathbf{x}(t)$ and $\mathbf{u}(t)$ on $\mathbf{e}(t)$ is canceled. For convergence of $\mathbf{e}(t) \to 0$, we need that

$$\sigma(\mathbf{A} - \mathbf{LC}) \subset \mathbb{C}^-. \tag{6.110}$$

To proceed it is useful to state a result which is the dual of Theorem 6.1, the Pole Assignment Theorem.

THEOREM 6.4 (Dual of Pole Assignment Theorem (Theorem 6.1)) If $(\mathbf{C}, \mathbf{A})$ is real and observable, the coefficients of the characteristic polynomial of $\mathbf{A} - \mathbf{LC}$, denoted $a(\mathbf{L})$, can be made to equal any prescribed vector $\underline{v}^T \in \mathbb{R}^{1 \times n}$ by choice of $\mathbf{L} \in \mathbb{R}^{n \times m}$.

Proof The proof follows from Theorem 6.1 upon noting that the eigenvalues of $\mathbf{A} - \mathbf{LC}$ are identical to those of $\mathbf{A}^T - \mathbf{C}^T\mathbf{L}^T$ and that $(\mathbf{C}, \mathbf{A})$ is observable if and only if $(\mathbf{A}^T, \mathbf{C}^T)$ is controllable. ∎

The observer designed above is sometimes called an identity observer because each component of $\mathbf{z}(t)$ estimates the corresponding component of $\mathbf{x}(t)$.

EXAMPLE 6.6 Consider the following single-input two-output system:

$$\dot{\mathbf{x}}(t) = \begin{bmatrix} 0 & 1 & 0 \\ 0 & 0 & 0 \\ 1 & 0 & 0 \end{bmatrix} \mathbf{x}(t) + \begin{bmatrix} 1 \\ 0 \\ 1 \end{bmatrix} u(t), \tag{6.111a}$$

$$\mathbf{y}(t) = \begin{bmatrix} 1 & 0 & 0 \\ 0 & 0 & 1 \end{bmatrix} \mathbf{x}(t). \tag{6.111b}$$

We want to design a full-order observer with eigenvalues at $-1, -2, -3$ to estimate the state $\mathbf{x}(t)$:

$$\mathbf{A} - \mathbf{LC} = \begin{bmatrix} 0 & 1 & 0 \\ 0 & 0 & 0 \\ 1 & 0 & 0 \end{bmatrix} - \begin{bmatrix} l_{11} & l_{12} \\ l_{21} & l_{22} \\ l_{31} & l_{32} \end{bmatrix} \begin{bmatrix} 1 & 0 & 0 \\ 0 & 0 & 1 \end{bmatrix}$$

$$= \begin{bmatrix} -l_{11} & 1 & -l_{12} \\ -l_{21} & 0 & -l_{22} \\ 1-l_{31} & 0 & -l_{32} \end{bmatrix}.$$

(6.112)

One easy choice is

$$l_{31} = 1, \qquad l_{32} = 3, \qquad s^2 + l_{11}s + l_{21} = s^2 + 3s + 2. \tag{6.113}$$

Thus, we have

$$l_{11} = 3, \qquad l_{21} = 2, \qquad l_{12}, l_{22} \text{ (arbitrary, say 0)}. \tag{6.114}$$

Then

$$\mathbf{L} = \begin{bmatrix} 3 & 0 \\ 2 & 0 \\ 1 & 3 \end{bmatrix}, \qquad \mathbf{A} - \mathbf{LC} = \begin{bmatrix} -3 & 1 & 0 \\ -2 & 0 & 0 \\ 0 & 0 & -3 \end{bmatrix}, \tag{6.115}$$

and therefore the observer is the dynamic system

$$\dot{\mathbf{z}}(t) = \begin{bmatrix} -3 & 1 & 0 \\ -2 & 0 & 0 \\ 0 & 0 & -3 \end{bmatrix} \mathbf{z}(t) + \begin{bmatrix} 3 & 0 \\ 2 & 0 \\ 1 & 3 \end{bmatrix} \mathbf{y}(t) + \begin{bmatrix} 1 \\ 0 \\ 1 \end{bmatrix} \mathbf{u}(t) \tag{6.116}$$

and $\mathbf{z}(t) = \hat{\mathbf{x}}(t)$ is the estimate of $\mathbf{x}(t)$.

EXAMPLE 6.7   Let the given stable plant be

$$\mathbf{A} = \begin{bmatrix} 0 & 1 & 0 \\ 0 & 0 & 1 \\ -1 & -2 & -2 \end{bmatrix}, \quad \mathbf{B} = \begin{bmatrix} 1 \\ 0 \\ 1 \end{bmatrix}, \quad \mathbf{C} = \begin{bmatrix} 1 & 0 & 0 \\ 0 & 0 & 1 \end{bmatrix}. \tag{6.117}$$

The following MATLAB script designs a full state observer. The system and the observer are shown in the Simulink model in Figure 6.4. Note that the observer poles are chosen to be $\{-1, -2, -3\}$ and

$$\mathbf{x}(0) = \begin{bmatrix} 2 & -2 & -3 \end{bmatrix}^T \quad \text{and} \quad \mathbf{z}(0) = \begin{bmatrix} 0 & 0 & 0 \end{bmatrix}^T. \tag{6.118}$$

```
clear
A=[0 1 0; 0 0 1; -1 -2 -2];    % original plant (A,B,C,D)
B=[1 0 1]';
C=[1 0 0; 0 0 1];
D=zeros(2,1);
C1=eye(3);     % modified plant (A,B,C1,D1) for SIMULINK
D1=zeros(3,1);    % model to plot states x
p=[-1 -2 -3];
L=place(A',C',p);
Ao=A-L'*C;    % full order observer (Ao,Bo,Co,Do)
Bo=[L' B];
Co=eye(3);
Do=zeros(3,3);
```

The full-order observer found is

$$\dot{z}(t) = \begin{bmatrix} -2.3425 & 1 & 0.8086 \\ -0.5073 & 0 & 1.2473 \\ 0.7043 & -2 & -3.6575 \end{bmatrix} z(t) + \begin{bmatrix} 2.3425 & -0.8086 & 1 \\ 0.5073 & -0.2473 & 0 \\ -1.7043 & 1.6575 & 1 \end{bmatrix} \begin{bmatrix} y(t) \\ u(t) \end{bmatrix},$$

$$z(t) = I_3 z(t).$$

$$(6.119)$$

The Simulink model in Figure 6.4 illustrates state "estimation" by the observer designed.

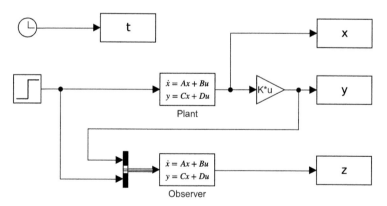

Figure 6.4 Simulink model for testing the full-order observer (Example 6.7)

The simulation was carried out with the initial conditions of the plant states $x(0)$ and the observer states $z(0)$ given in Equation (6.118). The reference input given is the unit step. Finally, Figure 6.5 shows that the estimated states $z(t) = \hat{x}(t)$ converge to the true states $x(t)$:

```
plot(t,x,'-',t,'z','-')
xlabel('t'), grid,
```

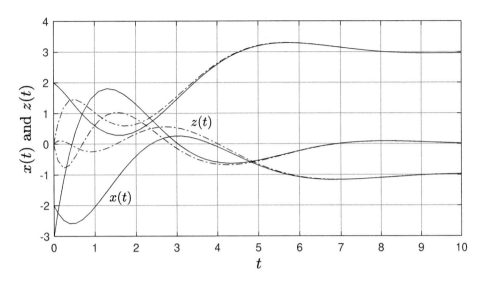

Figure 6.5 Estimate states $\mathbf{z}(t)$ (dashed lines) track the true states $\mathbf{x}(t)$ (solid lines) (Example 6.7)

6.7.2 Reduced-Order Observers

Since some of the n system states are measurable in the form of $\mathbf{y}(t)$, it should be possible to estimate only the remaining $(n-m)$ "unmeasurable"states. Without loss of generality (change coordinates if necessary), let

$$\mathbf{y}(t) = \mathbf{x}_1(t) \tag{6.120}$$

denote the first m components of the state vector which are measurable and let $\mathbf{x}_2(t)$ be the remaining $n-m$ components which are not measurable. The reduced-order observer will simply be an identity observer for $\mathbf{x}_2(t)$.

The system equations now become

$$\left[\begin{array}{c} \dot{\mathbf{x}}_1(t) \\ \dot{\mathbf{x}}_2(t) \end{array} \right] = \left[\begin{array}{cc} \mathbf{A}_{11} & \mathbf{A}_{12} \\ \mathbf{A}_{21} & \mathbf{A}_{22} \end{array} \right] \left[\begin{array}{c} \mathbf{x}_1(t) \\ \mathbf{x}_2(t) \end{array} \right] + \left[\begin{array}{c} \mathbf{B}_1 \\ \mathbf{B}_2 \end{array} \right] \mathbf{u}(t),$$

$$\mathbf{y}(t) = \mathbf{x}_1(t) = \left[\begin{array}{cc} \mathbf{I} & 0 \end{array} \right] \left[\begin{array}{c} \mathbf{x}_1(t) \\ \mathbf{x}_2(t) \end{array} \right]. \tag{6.121}$$

Since $\mathbf{x}_2(t)$ is the state to be estimated, write

$$\dot{\mathbf{x}}_2(t) = \mathbf{A}_{22}\mathbf{x}_2(t) + (\mathbf{B}_2\mathbf{u}(t) + \mathbf{A}_{21}\mathbf{y}(t)), \tag{6.122a}$$

$$\mathbf{A}_{12}\mathbf{x}_2(t) = \dot{\mathbf{y}}(t) - \mathbf{A}_{11}\mathbf{y}(t) - \mathbf{B}_1\mathbf{u}(t). \tag{6.122b}$$

Regard Equation (6.122a) as the "dynamic equation" for $\mathbf{x}_2(t)$ and Equation (6.122b) as the "measurement equation" for $\mathbf{x}_2(t)$.

LEMMA 6.1 If $(\mathbf{C}, \mathbf{A})$ is observable (detectable) then so is $(\mathbf{A}_{12}, \mathbf{A}_{22})$.

Proof   If $(\mathbf{C}, \mathbf{A})$ is observable:

$$\text{rank} \left[\begin{array}{c:c} s\mathbf{I} - \mathbf{A}_{11} & -\mathbf{A}_{12} \\ -\mathbf{A}_{21} & s\mathbf{I} - \mathbf{A}_{22} \\ \hdashline \mathbf{I} & 0 \end{array} \right] = n, \qquad \text{for all } s \in \mathbb{C}. \tag{6.123}$$

Therefore,

$$\text{rank} \left[\begin{array}{c} s\mathbf{I} - \mathbf{A}_{22} \\ \mathbf{A}_{12} \end{array} \right] = n - m, \qquad \text{for all } s \in \mathbb{C}. \tag{6.124}$$

Hence, $(\mathbf{A}_{12}, \mathbf{A}_{22})$ is observable. The proof for detectability is similar. ∎

Now we "apply" the full-order observer formulas, namely

$$\dot{\mathbf{z}}(t) = (\mathbf{A} - \mathbf{L}\mathbf{C})\mathbf{z}(t) + \mathbf{L}\mathbf{y}(t) + \mathbf{B}\mathbf{u}(t), \tag{6.125}$$

to the state $\mathbf{x}_2(t)$ and obtain the dynamic equation:

$$\begin{aligned} \dot{\mathbf{z}}_2(t) = (\mathbf{A}_{22} - \mathbf{L}_2\mathbf{A}_{12})\, \mathbf{z}_2(t) + \mathbf{L}_2\, (\dot{\mathbf{y}}(t) - \mathbf{A}_{11}\mathbf{y}(t) - \mathbf{B}_1\mathbf{u}(t)) \\ + (\mathbf{B}_2\mathbf{u}(t) + \mathbf{A}_{21}\mathbf{y}(t)). \end{aligned} \tag{6.126}$$

It is easily verified that the error

$$\mathbf{e}_2(t) := \mathbf{z}_2(t) - \mathbf{x}_2(t) \tag{6.127}$$

will satisfy

$$\dot{\mathbf{e}}_2(t) = (\mathbf{A}_{22} - \mathbf{L}_2\mathbf{A}_{12})\, \mathbf{e}_2(t). \tag{6.128}$$

As in the case of a full-order observer, for $\mathbf{e}_2(t) \to 0$, it is necessary to ensure that

$$\sigma \left(\mathbf{A}_{22} - \mathbf{L}_2\mathbf{A}_{12} \right) \subset \mathbb{C}^- \tag{6.129}$$

by some choice of $\mathbf{L}_2$. This is possible by observability (detectability) of $\mathbf{A}_{22}, \mathbf{A}_{12}$. All that remains now is to eliminate $\dot{\mathbf{y}}(t)$ from the reduced-order observer equation. Write

$$\begin{aligned} \dot{\mathbf{z}}_2(t) - \mathbf{L}_2\dot{\mathbf{y}}(t) &= (\mathbf{A}_{22} - \mathbf{L}_2\mathbf{A}_{12})\, \mathbf{z}_2(t) + \mathbf{L}_2\, (-\mathbf{A}_{11}\mathbf{y}(t) - \mathbf{B}_1\mathbf{u}(t)) \\ &\quad + (\mathbf{B}_2\mathbf{u}(t) + \mathbf{B}_{21}\mathbf{y}(t)) \\ &= (\mathbf{A}_{22} - \mathbf{L}_2\mathbf{A}_{12})\, (\mathbf{z}_2(t) - \mathbf{L}_2\mathbf{y}(t)) \\ &\quad + (\mathbf{A}_{22} - \mathbf{L}_2\mathbf{A}_{12})\, \mathbf{L}_2\mathbf{y}(t) - \mathbf{L}_2\mathbf{A}_{11}\mathbf{y}(t) - \mathbf{L}_2\mathbf{B}_2\mathbf{u}(t) + \mathbf{A}_{21}\mathbf{y}(t) \\ &= (\mathbf{A}_{22} - \mathbf{L}_2\mathbf{A}_{12})\, (\mathbf{z}_2(t) - \mathbf{L}_2\mathbf{y}(t)) \\ &\quad + [(\mathbf{A}_{22} - \mathbf{L}_2\mathbf{A}_{12})\, \mathbf{L}_2 - \mathbf{L}_2\mathbf{A}_{11} + \mathbf{A}_{21}]\, \mathbf{y}(t) + (\mathbf{B}_2 - \mathbf{L}_2\mathbf{B}_1)\, \mathbf{u}(t). \end{aligned} \tag{6.130}$$

Let

$$\mathbf{w}(t) = \mathbf{z}_2(t) - \mathbf{L}_2\mathbf{y}(t). \tag{6.131}$$

Then Equation (6.126) can be rewritten as follows:

$$\begin{aligned} \dot{\mathbf{w}}(t) &= (\mathbf{A}_{22} - \mathbf{L}_2\mathbf{A}_{12})\, \mathbf{w}(t) + [(\mathbf{A}_{22} - \mathbf{L}_2\mathbf{A}_{12})\, \mathbf{L}_2 - \mathbf{L}_2\mathbf{A}_{11} + \mathbf{A}_{21}]\, \mathbf{y}(t) \\ &\quad + (\mathbf{B}_2 - \mathbf{L}_2\mathbf{B}_1)\, \mathbf{u}(t), \end{aligned} \tag{6.132a}$$

$$\mathbf{z}_2(t) = \mathbf{w}(t) + \mathbf{L}_2\mathbf{y}(t), \tag{6.132b}$$

thus eliminating the derivative of $\mathbf{y}(t)$ as an input.

EXAMPLE 6.8   For the system given in Example 6.6, we want to design a minimal-order observer with eigenvalue at -3:

$$\mathbf{y}(t) = \begin{bmatrix} y_1(t) \\ y_2(t) \end{bmatrix} = \begin{bmatrix} x_1(t) \\ x_3(t) \end{bmatrix} := \begin{bmatrix} \bar{x}_1(t) \\ \bar{x}_2(t) \end{bmatrix}. \tag{6.133}$$

Now, with $\bar{x}_3(t) = x_2(t)$:

$$\dot{\bar{x}}_1(t) = \dot{x}_1(t) = x_2(t) + u(t) = \bar{x}_3 + u(t), \tag{6.134a}$$

$$\dot{\bar{x}}_2(t) = \dot{x}_3(t) = x_1(t) + u(t) = \bar{x}_1(t) + u(t), \tag{6.134b}$$

$$\dot{\bar{x}}_3(t) = \dot{x}_2(t) = 0, \tag{6.134c}$$

$$\begin{bmatrix} \dot{\bar{x}}_1(t) \\ \dot{\bar{x}}_2(t) \\ \dot{\bar{x}}_3(t) \end{bmatrix} = \begin{bmatrix} 0 & 0 & 1 \\ 1 & 0 & 0 \\ 0 & 0 & 0 \end{bmatrix} \begin{bmatrix} \bar{x}_1(t) \\ \bar{x}_2(t) \\ \bar{x}_3(t) \end{bmatrix} + \begin{bmatrix} 1 \\ 1 \\ 0 \end{bmatrix} u(t), \tag{6.135}$$

$$\begin{bmatrix} y_1(t) \\ y_2(t) \end{bmatrix} = \begin{bmatrix} \bar{x}_1(t) \\ \bar{x}_2(t) \end{bmatrix} = \begin{bmatrix} 1 & 0 & 0 \\ 0 & 1 & 0 \end{bmatrix} \begin{bmatrix} \bar{x}_1(t) \\ \bar{x}_2(t) \\ \bar{x}_3(t) \end{bmatrix}. \tag{6.136}$$

Thus

$$\mathbf{A}_{11} = \begin{bmatrix} 0 & 0 \\ 1 & 0 \end{bmatrix}, \qquad \mathbf{A}_{12} = \begin{bmatrix} 1 \\ 0 \end{bmatrix}, \qquad \mathbf{B}_1 = \begin{bmatrix} 1 \\ 1 \end{bmatrix}, \tag{6.137}$$

$$\mathbf{A}_{21} = \begin{bmatrix} 0 & 0 \end{bmatrix}, \qquad \mathbf{A}_{22} = 0, \qquad \mathbf{B}_2 = 0. \tag{6.138}$$

The reduced-order observer is

$$\dot{w}_2(t) = (\mathbf{A}_{22} - \mathbf{L}_2\mathbf{A}_{12})w_2 + [(\mathbf{A}_{22} - \mathbf{L}_2\mathbf{A}_{12})\mathbf{L}_2 - \mathbf{L}_2\mathbf{A}_{11} + \mathbf{A}_{21}]\mathbf{y}(t)$$

$$+ (\mathbf{B}_2 - \mathbf{L}_2\mathbf{B}_1)u(t), \tag{6.139}$$

$$\hat{\bar{x}}_3(t) = w_2(t) - \mathbf{L}_2\mathbf{y}(t). \tag{6.140}$$

Since the eigenvalue of $\mathbf{A}_{22} - \mathbf{L}_2\mathbf{A}_{12}$ should be -3, we have

$$0 - \begin{bmatrix} l_{21} & l_{22} \end{bmatrix} \begin{bmatrix} 1 \\ 0 \end{bmatrix} := -3. \tag{6.141}$$

Here, we set $l_{21} = 3$ and $l_{22} = 0$. Therefore, the reduced-order observer is:

$$\dot{w}_2(t) = -3w_2(t) - 9y_1(t) - 3u(t), \tag{6.142a}$$

$$\hat{\bar{x}}_3(t) = w_2(t) + 3y_1(t) = \hat{x}_2(t). \tag{6.142b}$$

EXAMPLE 6.9   Consider the system

$$
\mathbf{A} = \begin{bmatrix} 0 & 1 & 0 \\ 0 & 0 & 1 \\ -1 & -2 & -2 \end{bmatrix}, \quad \mathbf{B} = \begin{bmatrix} 1 \\ 0 \\ 0 \end{bmatrix}, \quad \mathbf{C} = \begin{bmatrix} 1 & 0 & 0 \\ 0 & 0 & 1 \end{bmatrix}. \tag{6.143}
$$

The following MATLAB script designs a reduced-order observer.

```
clear
A=[0 1 0; 0 0 1; -1 -2 -2];    % plant (A,B,C,D)
B=[1 0 0]';
C=[1 0 0; 0 0 1];
[r,n]=size(C);
[n,m]=size(B);
D=zeros(r,m);
p=[-1];    % observer pole
V=inv([C; null(C)']);
Cv=C*V;
Av=V*A*inv(V);    % coordinate transformation
Bv=V*B;
A11=Av(1:r,1:r);
A12=Av(1:r,r+1:n);
A21=Av(r+1:n,1:r);
A22=Av(r+1:n,r+1:n);
L2=place(A22',A12',p);
B1=Bv(1:r,:);
B2=Bv(r+1:n,:);
Ao=A22-L2'*A12;    % reduced order observer (Ao,Bo,Co,Do)
Bo=[(A22-L2'*A12)*L2'-L2'*A11+A21 (B2-L2'*B1)];
                % [y(t) u(t)]
Co=eye(n-r);
Do=[L2' zeros*B2];
```

The reduced-observer obtained is

$$
\dot{w}_3(t) = -w_3(t) + \begin{bmatrix} -0.6 & -0.6 & -0.2 \end{bmatrix} \begin{bmatrix} \mathbf{y}(t) \\ u(t) \end{bmatrix},
$$

$$
\hat{x}_3(t) = w_3(t) + \begin{bmatrix} 0.2 & 0.4 & 0 \end{bmatrix} \begin{bmatrix} \mathbf{y}(t) \\ u(t) \end{bmatrix}. \tag{6.144}
$$

The Simulink model in Figure 6.6 illustrates the fact that the reduced-order observer correctly estimates the states asymptotically as shown in Figure 6.7.

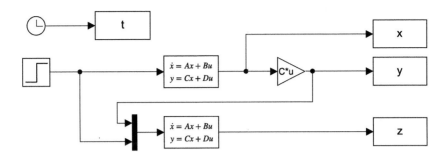

Figure 6.6 Simulink model for testing an observer (Example 6.9)

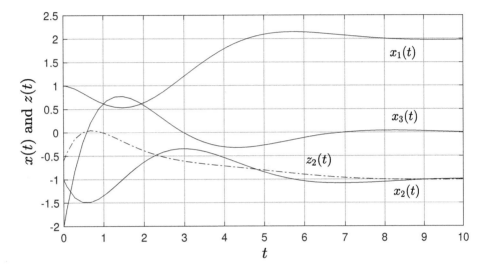

Figure 6.7 The estimated state $\hat{x}_2(t)$ (dashed lines) converges to the true state $x_2(t)$ (solid lines) (Example 6.9)

6.7.3 Closed-Loop Eigenvalues under Observer-Based Feedback

Suppose now that we have calculated a state feedback control law $\mathbf{u}(t) = -\mathbf{Kx}(t)$ to place the eigenvalues of $\mathbf{A} - \mathbf{BK}$. Now we design an observer with eigenvalues of $\mathbf{M}$ chosen by us to implement this control law. From the observer we obtain $\hat{\mathbf{x}}(t)$. The question of interest now is: What if we close the loop with the observer reconstructed control law

$$\mathbf{u}(t) = -\mathbf{K}\hat{\mathbf{x}}(t) \tag{6.145}$$

instead of $\mathbf{u}(t) = -\mathbf{Kx}(t)$? What are the closed-loop eigenvalues? The answer is given below.

THEOREM 6.5   The closed-loop system under observer-reconstructed state feedback
has eigenvalues

$$\sigma(\mathbf{A} - \mathbf{B}\mathbf{K}) \cup \sigma(\mathbf{M}). \tag{6.146}$$

Proof   Consider the general observer

$$\dot{\mathbf{z}}(t) = \mathbf{M}\mathbf{z}(t) + \mathbf{L}\mathbf{y}(t) + \mathbf{G}\mathbf{u}(t). \tag{6.147}$$

Let

$$\mathbf{z}(t) - \mathbf{T}\mathbf{x}(t) = \mathbf{e}(t). \tag{6.148}$$

Then

$$\dot{\mathbf{e}}(t) = \mathbf{M}\mathbf{e} + (\mathbf{M}\mathbf{T} - \mathbf{T}\mathbf{A} + \mathbf{L}\mathbf{C})\mathbf{x} + (\mathbf{G} - \mathbf{T}\mathbf{B})\mathbf{u}(t). \tag{6.149}$$

Therefore to make $\mathbf{z}(t)$ estimate $\mathbf{T}\mathbf{x}(t)$, we want $\mathbf{e}(t) \to 0$. Thus, set

$$\mathbf{M}\mathbf{T} - \mathbf{T}\mathbf{A} + \mathbf{L}\mathbf{C} = 0, \tag{6.150}$$
$$\mathbf{G} - \mathbf{T}\mathbf{B} = 0, \tag{6.151}$$

and

$$\sigma(\mathbf{M}) \subset \mathbb{C}^-. \tag{6.152}$$

To "estimate" $\mathbf{K}\mathbf{x}(t)$ we use the observer output

$$\begin{aligned}\mathbf{h}(t) := &= \mathbf{S}_1\mathbf{z}(t) + \mathbf{S}_2\mathbf{y}(t) \\ &= (\mathbf{S}_1\mathbf{T} + \mathbf{S}_2\mathbf{C})\mathbf{x}(t) + \mathbf{S}_1\mathbf{e}(t)\end{aligned} \tag{6.153}$$

and set

$$\mathbf{S}_1\mathbf{T} + \mathbf{S}_2\mathbf{C} = -\mathbf{K}. \tag{6.154}$$

The loop is closed by setting $\mathbf{u}(t) = \mathbf{h}(t)$ (instead of $\mathbf{u}(t) = -\mathbf{K}\mathbf{x}(t)$). Now the closed-loop equations are

$$\dot{\mathbf{x}}(t) = \mathbf{A}\mathbf{x}(t) + \mathbf{B}\mathbf{u}(t), \tag{6.155a}$$
$$\dot{\mathbf{z}}(t) = \mathbf{M}\mathbf{z}(t) + \mathbf{L}\mathbf{y}(t) + \mathbf{G}\mathbf{u}(t), \tag{6.155b}$$
$$\mathbf{u}(t) = \mathbf{S}_1\mathbf{z}(t) + \mathbf{S}_2\mathbf{y}(t). \tag{6.155c}$$

We have

$$\begin{bmatrix} \dot{\mathbf{x}}(t) \\ \dot{\mathbf{z}}(t) \end{bmatrix} = \underbrace{\begin{bmatrix} \mathbf{A} + \mathbf{B}\mathbf{S}_2\mathbf{C} & \mathbf{B}\mathbf{S}_1 \\ \mathbf{L}\mathbf{C} + \mathbf{G}\mathbf{S}_2\mathbf{C} & \mathbf{M} + \mathbf{G}\mathbf{S}_1 \end{bmatrix}}_{\mathbf{A}_{cl}} \begin{bmatrix} \mathbf{x}(t) \\ \mathbf{z}(t) \end{bmatrix}. \tag{6.156}$$

Make the coordinate transformation

$$\begin{bmatrix} \mathbf{x}(t) \\ \mathbf{z}(t) \end{bmatrix} = \begin{bmatrix} \mathbf{I} & 0 \\ \mathbf{T} & \mathbf{I} \end{bmatrix} \begin{bmatrix} \mathbf{x}(t) \\ \mathbf{e}(t) \end{bmatrix}, \tag{6.157}$$

$$\begin{bmatrix} \mathbf{x}(t) \\ \mathbf{e}(t) \end{bmatrix} = \begin{bmatrix} \mathbf{I} & 0 \\ -\mathbf{T} & \mathbf{I} \end{bmatrix} \begin{bmatrix} \mathbf{x}(t) \\ \mathbf{z}(t) \end{bmatrix}. \tag{6.158}$$

Then using Equations (6.150)–(6.156), we have

$$\begin{bmatrix} \mathbf{I} & 0 \\ -\mathbf{T} & \mathbf{I} \end{bmatrix} \mathbf{A}_{cl} \begin{bmatrix} \mathbf{I} & 0 \\ \mathbf{T} & \mathbf{I} \end{bmatrix} = \begin{bmatrix} \mathbf{A} - \mathbf{BK} & \mathbf{BS}_1 \\ 0 & \mathbf{M} \end{bmatrix}. \tag{6.159}$$

Therefore

$$\sigma(\mathbf{A}_{cl}) = \sigma(\mathbf{A} - \mathbf{BK}) \cup \sigma(\mathbf{M}). \tag{6.160}$$

The transfer function of the equivalent observer-based controller is derived as follows:

$$\dot{\mathbf{z}}(t) = \mathbf{Mz}(t) + \mathbf{Ly}(t) + \mathbf{Gu}(t), \tag{6.161}$$

$$\mathbf{h}(t) = \mathbf{S}_1 \mathbf{z}(t) + \mathbf{S}_2 \mathbf{y}(t), \tag{6.162}$$

$$\dot{\mathbf{z}}(t) = (\mathbf{M} + \mathbf{GS}_1)\,\mathbf{z}(t) + (\mathbf{L} + \mathbf{GS}_2)\,\mathbf{y}(t), \tag{6.163}$$

Therefore

$$U(s) = \underbrace{\left[\mathbf{S}_1 (s\mathbf{I} - \mathbf{M} - \mathbf{GS}_1)^{-1} (\mathbf{L} + \mathbf{GS}_2) + \mathbf{S}_2 \right]}_{\mathbf{C}(s)} Y(s). \tag{6.164}$$

∎

EXAMPLE 6.10

$$\mathbf{A} = \begin{bmatrix} 0 & 1 \\ 0 & 0 \end{bmatrix}, \quad \mathbf{b} = \begin{bmatrix} 0 \\ 1 \end{bmatrix}, \quad \mathbf{c} = \begin{bmatrix} 1 & 1 \end{bmatrix}. \tag{6.165}$$

(1) Design a reduced-order observer with its pole at -1:

$$\dot{\mathbf{x}}(t) = \mathbf{Ax}(t) + \mathbf{bu}(t), \qquad y(t) = \mathbf{cx}(t). \tag{6.166}$$

We need to choose an invertible matrix $\mathbf{V}$ so that

$$\mathbf{cV} = \begin{bmatrix} 1 & 0 \end{bmatrix}. \tag{6.167}$$

We have

$$\mathbf{V} = \begin{bmatrix} 1 & 1 \\ 0 & -1 \end{bmatrix} \quad \text{so that} \quad \mathbf{V}^{-1} = \begin{bmatrix} 1 & 1 \\ 0 & -1 \end{bmatrix}. \tag{6.168}$$

In the new coordinate system

$$\dot{\bar{\mathbf{x}}}(t) = \bar{\mathbf{A}}\bar{\mathbf{x}}(t) + \bar{\mathbf{b}}u(t), \qquad y(t) = \bar{\mathbf{c}}\bar{\mathbf{x}}(t), \tag{6.169}$$

where

$$\bar{\mathbf{A}} = \mathbf{V}^{-1}\mathbf{AV} = \left[\begin{array}{c|c} 0 & -1 \\ \hline 0 & 0 \end{array} \right], \quad \bar{\mathbf{b}} = \mathbf{V}^{-1}\mathbf{b} = \left[\begin{array}{c} 1 \\ \hline -1 \end{array} \right], \tag{6.170}$$

$$\bar{\mathbf{c}} = \mathbf{cV} = \begin{bmatrix} 1 & 0 \end{bmatrix}.$$

The reduced-order observer is

$$\begin{aligned} \dot{w}(t) &= (\bar{\mathbf{A}}_{22} - \mathbf{L}_2 \bar{\mathbf{A}}_{12})\,w(t) + \left[(\bar{\mathbf{A}}_{22} - \mathbf{L}_2 \bar{\mathbf{A}}_{12})\,\mathbf{L}_2 + \mathbf{L}_2 \bar{\mathbf{A}}_{11} + \bar{\mathbf{A}}_{21} \right] y(t) \\ &\quad + (\bar{\mathbf{b}}_2 - \mathbf{L}_2 \bar{\mathbf{b}}_1)\,u(t). \end{aligned} \tag{6.171}$$

Let us select $\mathbf{L}_2$ such that

$$\sigma\left(\bar{\mathbf{A}}_{22} - \mathbf{L}_2\bar{\mathbf{A}}_{12}\right) = -1. \tag{6.172}$$

Then we have $\mathbf{L}_2 = -1$. herefore, the reduced-order observer becomes

$$\dot{w}(t) = -w(t) + y(t), \tag{6.173a}$$
$$\hat{x}_2(t) = w(t) + \mathbf{L}_2 y(t). \tag{6.173b}$$

(2) Combine the state feedback and the observer into an output feedback controller, and give the state equations as well as the transfer function of this equivalent controller. Choose a state feedback to assign poles of the original system all at -1.

Now we need to consider the coordinate transformation

$$\hat{\mathbf{x}}(t) = \mathbf{V}\hat{\bar{\mathbf{x}}}, \quad y(t) = \bar{\mathbf{c}}\bar{\mathbf{x}}(t), \quad w(t) = \hat{\bar{x}}_2(t) - \mathbf{L}_2 y(t), \tag{6.174}$$

$$\begin{bmatrix} y(t) \\ w(t) \end{bmatrix} = \begin{bmatrix} 1 & 0 \\ -\mathbf{L}_2 & 1 \end{bmatrix} \begin{bmatrix} \bar{x}_1(t) \\ \hat{\bar{x}}_2(t) \end{bmatrix}. \tag{6.175}$$

Thus

$$\hat{\mathbf{x}}(t) = \mathbf{V}\hat{\bar{\mathbf{x}}}(t) = \mathbf{V}\begin{bmatrix} 1 & 0 \\ \mathbf{L}_2 & 1 \end{bmatrix}\begin{bmatrix} y(t) \\ w(t) \end{bmatrix} = \begin{bmatrix} 1 & 0 \\ 1 & -1 \end{bmatrix}\begin{bmatrix} y(t) \\ w(t) \end{bmatrix}. \tag{6.176}$$

From the control law, $-\mathbf{k}^T = [k_1 \; k_2] = [-1 \; -2]$, which assigns the eigenvalues at -1, we have

$$u(t) = -\mathbf{k}^T\hat{\mathbf{x}}(t) = \begin{bmatrix} -1 & -2 \end{bmatrix}\begin{bmatrix} 1 & 0 \\ 1 & -1 \end{bmatrix}\begin{bmatrix} y(t) \\ w(t) \end{bmatrix} \tag{6.177}$$
$$= -3y(t) + 2w(t).$$

Therefore, the controller state equations are

$$\dot{w}(t) = -w(t) + y(t),$$
$$u(t) = -\mathbf{k}^T\hat{\mathbf{x}}(t) = 2w(t) - 3y(t), \tag{6.178}$$

and the equivalent controller transfer function is

$$C(s) = 2(s+1)^{-1}1 + (-3) = \frac{-3s - 1}{s + 1}. \tag{6.179}$$

EXAMPLE 6.11 (Closing the loop through a full-order observer)   Recall the observer design problem described in Example 6.7. For this example, we now close the loop by observer-reconstructed state feedback $\mathbf{K}$ calculated from the following MATLAB script in addition to the MATLAB script in Example 6.7. The poles of the closed-loop system are chosen to be $\{-1, -1 \pm j\}$.

```
P=[-1 -1+j -1-j];
K=place(A,B,P);
```

The state feedback matrix obtained is

$$\mathbf{K} = \begin{bmatrix} 0.6667 & 1 & 0.3333 \end{bmatrix}. \tag{6.180}$$

The Simulink model in Figure 6.8 was created to compare the sinusoidal responses of the observed feedback system and the system with the state feedback.

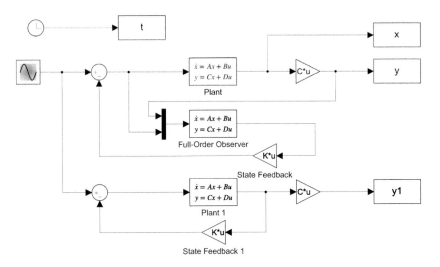

Figure 6.8 Output response (Example 6.11)

The initial state conditions chosen for the states of the plant $\mathbf{x}(0)$ and the states of the observer $\hat{\mathbf{x}}(0)$ are:

$$\mathbf{x}(0) = \begin{bmatrix} 1 & 2 & -1 \end{bmatrix}^T \quad \text{and} \quad \hat{\mathbf{x}}(0) = \begin{bmatrix} 0 & 0 & 0 \end{bmatrix}. \tag{6.181}$$

Figure 6.9 shows that the response of the observed state feedback system works as expected.

EXAMPLE 6.12 (Closing the loop through a reduced-order observer) Using Example 6.9 and its solution, we complete the design by closing the loop. The state feedback $\mathbf{K}$ is designed so that

$$\sigma(\mathbf{A} - \mathbf{BK}) = \{-1, -2, -3\}. \tag{6.182}$$

```
A=[0 1 0; 0 0 1; -1 -2 -2];
B=[1 0 0]';
P=[-1, -2 -3];
K=place(A,B,P)
```

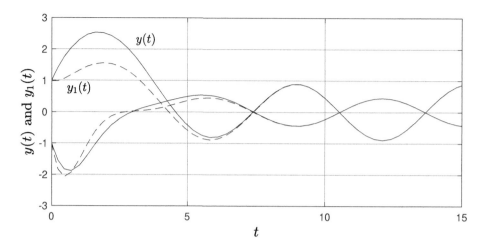

Figure 6.9 Outputs of the observer-based closed-loop system (dashed lines) converge to outputs of the closed-loop system with true state feedback (solid lines) (Example 6.11)

The state feedback is

$$\mathbf{KV}^{-1} = \begin{bmatrix} 4 & 3 & -1 \end{bmatrix} \begin{bmatrix} 1 & 0 & 0 \\ 0 & 0 & 1 \\ 0 & 1 & 0 \end{bmatrix} = \begin{bmatrix} 4 & -1 & 3 \end{bmatrix}. \tag{6.183}$$

We now close the loop as shown in Figure 6.10. Finally, the output of the overall system to the sinusoidal reference input is depicted in Figure 6.11 with the unit step reference input..

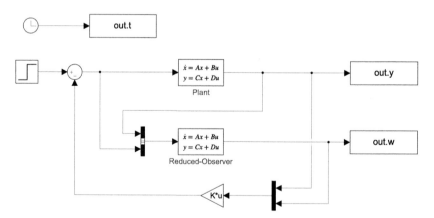

Figure 6.10 Simulink model for testing an observer (Example 6.12)

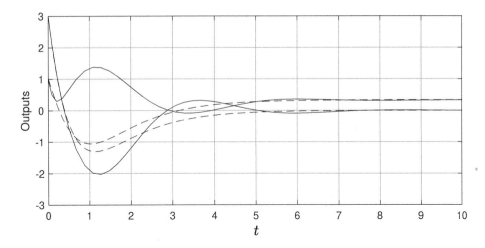

Figure 6.11 Outputs of the reduced-observer-based closed-loop system (solid lines) converge to outputs of the closed-loop system with true state feedback (dashed lines) (Example 6.12)

6.7.4 Dual Observers and Controllers

Let $(\mathbf{A}_p, \mathbf{B}_p, \mathbf{C}_p)$ denote the state-space model of a plant of order n with r inputs and m outputs. The dual plant is $(\mathbf{A}_p^T, \mathbf{C}_p^T, \mathbf{B}_p^T)$. By applying the previous results on reduced-order observer-reconstructed state feedback, we obtain a "controller" called a dual controller of order $n - r$. Denote this dual controller as $(\mathbf{A}_c^T, \mathbf{C}_c^T, \mathbf{B}_c^T, \mathbf{D}_c^T)$ and let the $2n - r$ eigenvalues of

$$\mathbf{A}_{cl}^T := \begin{bmatrix} \mathbf{A}_p^T + \mathbf{C}_p^T \mathbf{D}_c^T \mathbf{B}_p^T & \mathbf{C}_p^T \mathbf{B}_c^T \\ \mathbf{C}_c^T \mathbf{B}_p^T & \mathbf{A}_c^T \end{bmatrix} \tag{6.184}$$

assigned by the dual controller be denoted as Λ. Then the feedback controller $(\mathbf{A}_c, \mathbf{B}_c, \mathbf{C}_c, \mathbf{D}_c)$ applied to the original plant $(\mathbf{A}_p, \mathbf{B}_p, \mathbf{C}_p)$ assigns Λ as the closed-loop eigenvalues. This result shows that the order of a controller that can assign closed-loop eigenvalues arbitrarily need be no higher than

$$n_c = \min\{(n - m), (n - r)\}. \tag{6.185}$$

6.7.5 Structure of Robust Observers

The following example illustrates a certain fragiliy of the standard observer architecture.

EXAMPLE 6.13   Consider a system

$$A = \begin{bmatrix} 1 & 0 & 0 \\ 0 & 0 & 0 \\ 0 & 1 & 0 \end{bmatrix}, \quad B = \begin{bmatrix} 1 & 0 \\ 0 & -1 \\ 1 & 1 \end{bmatrix}, \quad C = \begin{bmatrix} 1 & 0 & 0 \\ 0 & 0 & 1 \end{bmatrix}. \tag{6.186}$$

A full-order observer is:

$$\dot{z}(t) = Mz(t) + Ly(t) + Gu(t), \tag{6.187a}$$

$$\hat{x}(t) = Pz(t) + Qy(t). \tag{6.187b}$$

We choose the observer poles as:

$$\Lambda_o = \{-0.1, \ -0.2, \ -0.3\}, \tag{6.188}$$

then from $M = A - LC$ we have

$$L = \begin{bmatrix} 1.1998 & -0.0008 \\ -0.0148 & 0.0301 \\ -0.0644 & 0.4002 \end{bmatrix}, \tag{6.189}$$

$$M = A - LC = \begin{bmatrix} -0.1998 & 0 & 0.0008 \\ 0.0148 & 0 & -0.0301 \\ 0.0644 & 1 & -0.4002 \end{bmatrix}. \tag{6.190}$$

Thus, the observer equations are (setting $G = B$, $P = I$, $Q = R = 0$)

$$\dot{z}(t) = \begin{bmatrix} -0.1998 & 0 & 0.0008 \\ 0.0148 & 0 & -0.0301 \\ 0.0644 & 1 & -0.4002 \end{bmatrix} z(t) + \begin{bmatrix} 1.1998 & -0.0008 \\ -0.0148 & 0.0301 \\ -0.0644 & 0.4002 \end{bmatrix} y(t)$$

$$+ \begin{bmatrix} 1 & 0 \\ 0 & -1 \\ 1 & 1 \end{bmatrix} u(t), \tag{6.191}$$

$$\hat{x}(t) = \begin{bmatrix} 1 & 0 & 0 \\ 0 & 1 & 0 \\ 0 & 0 & 1 \end{bmatrix} z(t).$$

Now we choose the initial states of the plant and the observer to be

$$x(0) = [1 \ 2 \ 3]^T \quad \text{and} \quad z(0) = [-5 \ 0 \ 5]^T. \tag{6.192}$$

By using the Simulink model in Figure 6.4, the errors $e(t)$ between the true states $x(t)$ and the estimated states $z(t)$ are computed and displayed in Figure 6.12. This shows that the estimated states $z(t)$ do not converge to the true states $x(t)$. Obviously the full-order observer we designed does not function properly. This interesting fact can be explained in terms of the architecture of the observer (Figure 6.12).

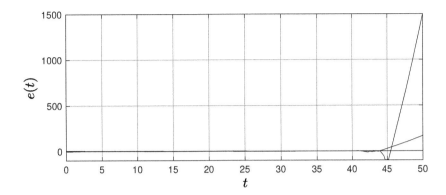

Figure 6.12 The state estimation error $\mathbf{e}(t) = \mathbf{x}(t) - \mathbf{z}(t)$ grows due to non-robustness of the observer ($\mathbf{x}(0) = [1\ 2\ 3]$, $\mathbf{z}(0) = [-5\ 0\ 5]$) (Example 6.13)

Let us recall the error equation

$$\dot{\mathbf{e}}(t) = \mathbf{M}\mathbf{e}(t) + (\mathbf{M} - \mathbf{A} + \mathbf{L}\mathbf{C})\mathbf{x}(t) + (\mathbf{G} - \mathbf{B})\mathbf{u}(t). \tag{6.193}$$

For $\mathbf{e}(t) \longrightarrow 0$, it is required that

$$\mathbf{M} = \mathbf{A} - \mathbf{L}\mathbf{C}, \quad \mathbf{G} = \mathbf{B}. \tag{6.194}$$

However, we note that it is unlikely that $\mathbf{L}$ and $\mathbf{M}$, being independent physical blocks, could be computed or implemented such that $\mathbf{M} = \mathbf{A} - \mathbf{L}\mathbf{C}$ exactly. This means that small mismatch error between $\mathbf{M}$ and $\mathbf{A} - \mathbf{L}\mathbf{C}$ will cause the term $(\mathbf{M} - \mathbf{A} + \mathbf{L}\mathbf{C}) =: \mathbf{E}$ to be nonzero. Furthermore, under such perturbations

$$\dot{\mathbf{e}}(t) = \mathbf{M}\mathbf{e}(t) + \mathbf{E}\mathbf{x}(t) \tag{6.195}$$

and if the state $\mathbf{x}(t)$ is unstable, $\mathbf{e}(t)$, which is now driven by $\mathbf{E}\mathbf{x}(t)$, will certainly "blow up" for large values of t. This can be prevented if we design an observer of the following form:

$$\dot{\mathbf{z}}(t) = \mathbf{A}\mathbf{z}(t) + \mathbf{B}\mathbf{u}(t) + \mathbf{L}(\mathbf{y}(t) - \mathbf{C}\mathbf{z}(t)), \tag{6.196a}$$

$$\hat{\mathbf{x}}(t) = \mathbf{z}(t). \tag{6.196b}$$

Let us write the error equation again:

$$
\begin{aligned}
\dot{\mathbf{e}}(t) &= \dot{\mathbf{x}}(t) - \dot{\mathbf{z}}(t) \\
&= \mathbf{A}\mathbf{x}(t) + \mathbf{B}\mathbf{u}(t) - \mathbf{A}\mathbf{z}(t) - \mathbf{B}\mathbf{u}(t) - \mathbf{L}\mathbf{C}(\mathbf{x}(t) - \mathbf{z}(t)) \\
&= (\mathbf{A} - \mathbf{L}\mathbf{C} + \mathbf{E})(\mathbf{x}(t) - \mathbf{z}(t)) \\
&= (\mathbf{A} - \mathbf{L}\mathbf{C} + \mathbf{E})\mathbf{e}(t).
\end{aligned}
\tag{6.197}
$$

As seen, the error signal $\mathbf{e}(t)$ will converge even if $\mathbf{L}$ deviates from its calculated value, as long as it maintains the stability of $\mathbf{A} - \mathbf{L}\mathbf{C} + \mathbf{E}$. However, this mechanism requires exact duplication of the system model $(\mathbf{A}, \mathbf{B}, \mathbf{C})$ into the observer. This is called a *robust observer* (Figure 6.14). We remark that the robust observer architecture is the same as that of the Kalman filter.

Figure 6.13 An observer

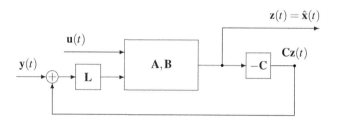

Figure 6.14 A robust observer

Remark 6.1   The case we treated in the previous example would not present a problem if we use an observer for stabilization. Once the loop is closed by using the estimated states, $\mathbf{x}(t)$ will no longer "blow up" since the closed-loop system is stable. In other words, $\mathbf{e}(t)$ will converge.

Of course, by the same argument, the state observer will work without loop closure if the plant is stable. Therefore, if one wants to use an observer as a state "estimator" alone (without loop closure) for an unstable system, one will encounter a convergence problem. In this case, one may want to consider using a *robust observer* as discussed above.

Case 1: Designing a Robust Observer
Consider again the last example of the previous section. A robust observer for this system is

$$\dot{\mathbf{z}}(t) = \mathbf{A}\mathbf{z}(t) + \mathbf{B}\mathbf{u}(t) + \mathbf{L}(\mathbf{y}(t) - \mathbf{C}\mathbf{z}(t)),$$
$$\hat{\mathbf{x}}(t) = \mathbf{z}(t), \tag{6.198}$$

which becomes:

$$\dot{\mathbf{z}}(t) = \begin{bmatrix} 1 & 0 & 0 \\ 0 & 0 & 0 \\ 0 & 1 & 0 \end{bmatrix} \mathbf{z}(t) + \begin{bmatrix} 1 & 0 \\ 0 & -1 \\ 1 & 1 \end{bmatrix} \mathbf{u}(t)$$

$$+ \begin{bmatrix} 1.1998 & -0.0008 \\ -0.0148 & 0.0301 \\ -0.0644 & 0.4002 \end{bmatrix} \left(\mathbf{y}(t) - \begin{bmatrix} 1 & 0 & 0 \\ 0 & 0 & 1 \end{bmatrix} \mathbf{z}(t) \right), \tag{6.199}$$

$$\hat{\mathbf{x}}(t) = \mathbf{z}(t).$$

We modified the structure of the observer and the robust observer is now shown in the Simulink model in Figure 6.15. The error $\mathbf{e}(t)$ is depicted in Figure 6.16.

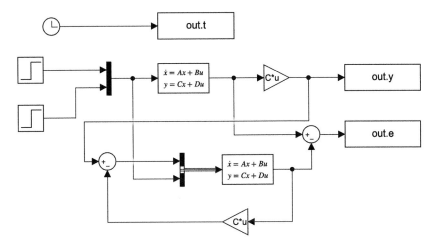

Figure 6.15 Simulink model of a robust observer

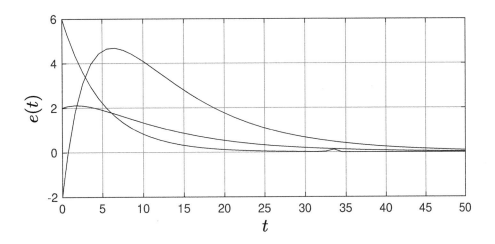

Figure 6.16 Error $\mathbf{e}(t) = \mathbf{x}(t) - \mathbf{z}(t)$ converges to zero with a robust observer

Remark 6.2 To determine the systems in the Simulink model in Figure 6.15, the same MATLAB script as in Example 6.7 may be used, except that the state matrix of the observer must be set to $\mathbf{A}$ instead of $\mathbf{A}_o$ following Equation (6.199).

Case 2: Closing the Loop

Consider first the observer in the last example of the previous section and let the desired poles of the closed-loop system be

$$\Lambda_p = \{-1, \ -2, \ -3\}. \tag{6.200}$$

Then we have a state feedback

$$-\mathbf{K} = \begin{bmatrix} -4.5636 & 1.5970 & 0.9457 \\ -3.3733 & 5.4395 & 2.0574 \end{bmatrix} \tag{6.201}$$

which places the poles of $\mathbf{A} - \mathbf{BK}$ at Λ_p.

6.8   Minimal or Maximal Realizations?

The dynamic models that arise in control engineering depend on various system pa-rameters which are imprecisely known at best and are subject at least to small pertur-bations. Nevertheless, we proceed with controller designs by setting these parameters to nominal values, designing a controller for the nominal system, and validating its performance over the expected uncertainty set. In the context of linear multivariable control theory, this often amounts to starting with a parametrized plant transfer matrix $\mathbf{P}(s,\mathbf{p})$, setting $\mathbf{p} = \mathbf{p}_0$ the nominal value, constructing a minimal state-space realization of $\mathbf{P}_0(s) = \mathbf{P}(s,\mathbf{p}_0)$, and designing a feedback controller that stabilizes this minimal realization.

The purpose of this section is to point out a potential pitfall in this procedure. Specif-ically, we show that there are situations in which stabilization of any minimal-order state-space realization of $\mathbf{P}(s,\mathbf{p}_0)$ will lead to a closed-loop system which becomes un-stable for arbitrarily small perturbations of $\mathbf{p}_0$. This occurs because the McMillan degree of $\mathbf{P}(s,\mathbf{p})$ is, in general, discontinuous with respect to $\mathbf{p}$ at $\mathbf{p}_0$.

We show that such a catastrophic failure can be avoided if the correct class of paramet-rizations to which $\mathbf{P}_0(s)$ belongs is known. Let v_{max}^+ denote the maximal McMillan degree of the antistable component of $\mathbf{P}(s,\mathbf{p})$ under the given parametrization. It can easily be found by arbitrarily perturbing $\mathbf{p}_0$ by a small amount within its perturbation class and recalculating the McMillan degree, since the McMillan degree drops only on an algebraic variety.

By stabilizing a perturbed nominal plant whose minimal-order realization has v_{max}^+ unstable poles, the structural instability discussed above can be overcome as the con-troller also stabilizes a "ball" of plants centered at the new nominal. Such a stability ball around the perturbed nominal plant, however, cannot include the original system..

6.8.1   Motivation

By way of motivation, consider the transfer function

$$\mathbf{P}_0(s) = \begin{bmatrix} \dfrac{2}{(s+1)(s-1)} & \dfrac{1}{s-1} \\ \dfrac{1}{s-1} & \dfrac{1}{s-1} \end{bmatrix} \tag{6.202}$$

which represents an unstable plant to be stabilized by feedback. The order of a minimal realization of $\mathbf{P}_0(s)$ is 2, and it can be stabilized by a compensator $\mathbf{C}_0$, say of order q.

Now suppose that $\mathbf{P}_0(s)$ perturbs to

$$\mathbf{P}_1(s) = \mathbf{P}(s,\delta) = \begin{bmatrix} \dfrac{2}{(s+1)(s-1)} & \dfrac{1+\delta}{s-1} \\[2ex] \dfrac{1}{s-1} & \dfrac{1}{s-1} \end{bmatrix}, \qquad (6.203)$$

where δ is a real parameter perturbation. It is easily seen that the closed-loop system with compensator $\mathbf{C}_0$ and plant $\mathbf{P}_1(s)$ is unstable with a closed-loop pole near $s = 1$. Moreover, this occurs for every nominally stabilizing controller $\mathbf{C}_0$ of $\mathbf{P}_0(s)$, and for infinitesimally small perturbations δ.

To avoid the undesirable situation discussed above, it is necessary to know the class of uncertain systems to which $\mathbf{P}_0(s)$ belongs, and to stabilize a perturbed version of $\mathbf{P}_0(s)$ which has the generic maximal order of unstable poles in the perturbation class.

In the example above, the nominal plant should have been chosen after perturbation ($\mathbf{P}_1(s)$ with $\delta \neq 0$) and realized minimally to be of order 3, and a stabilizing controller $\mathbf{C}_1$ designed for it. The controller $\mathbf{C}_1$ remains stabilizing under small perturbations. It is important to note, however, that such a stability "ball" around the perturbation cannot include the nominal system $\mathbf{P}_0(s)$!

6.8.2 Nominal and Parametrized Models

Consider a plant parametrized by a family of rational proper transfer function matrices, $\mathbf{P}(s,\mathbf{p})$, where $\mathbf{p}$ is an ℓ-dimensional real parameter vector which ranges over an uncertainty set $\Omega \subset \mathbb{R}^\ell$.

We assume that the coefficients of the transfer functions in $\mathbf{P}(s,\mathbf{p})$ are continuous functions of $\mathbf{p}$ and that $\mathbf{P}(s,\mathbf{p})$ has a state-space realization $[\mathbf{A}(\mathbf{p}),\mathbf{B}(\mathbf{p}),\mathbf{C}(\mathbf{p}),\mathbf{D}(\mathbf{p})]$ with matrix entries being continuous functions of $\mathbf{p}$. Let $\mathbf{p} = \mathbf{p}_0$ be the nominal parameter and denote

$$\mathbf{P}(s,\mathbf{p}_0) = \mathbf{P}_0(s). \qquad (6.204)$$

Now let $\nu[\mathbf{P}(s,\mathbf{p})]$ denote the McMillan degree of $\mathbf{P}(s,\mathbf{p})$. Decompose $\mathbf{P}(s,\mathbf{p})$ into its stable and antistable (all poles unstable) components

$$\mathbf{P}(s,\mathbf{p}) = \mathbf{P}^-(s,\mathbf{p}) + \mathbf{P}^+(s,\mathbf{p}) \qquad (6.205)$$

and let

$$\nu\left[\mathbf{P}^+(s,\mathbf{p})\right] =: \nu^+[\mathbf{P}(s,\mathbf{p})], \qquad (6.206)$$

$$\nu\left[\mathbf{P}^-(s,\mathbf{p})\right] =: \nu^-[\mathbf{P}(s,\mathbf{p})]. \qquad (6.207)$$

When the context is clear, we write $\nu(\mathbf{p})$, $\nu^+(\mathbf{p})$, etc. instead of $\nu[\mathbf{P}(s,\mathbf{p}), \nu^+[\mathbf{P}(s,\mathbf{p})]$.

In general, the functions $\nu(\mathbf{p})$, $\nu^+(\mathbf{p})$, $\nu^-(\mathbf{p})$ are discontinuous functions of $\mathbf{p}$. Moreover, the generic, maximal McMillan degree depends on the specific structure of the parametrization, as we show next.

EXAMPLE 6.14   Continuing with our previous plant

$$
\mathbf{P}_0(s) =
\begin{bmatrix}
\dfrac{2}{(s+1)(s-1)} & \dfrac{1}{s-1} \\[2ex]
\dfrac{1}{s-1} & \dfrac{1}{s-1}
\end{bmatrix}
\tag{6.208}
$$

consider, for example, the four parametrized families to which $\mathbf{P}_0(s)$ might belong:

$$
\mathbf{P}_1(s,\mathbf{a}) =
\begin{bmatrix}
\dfrac{2+a_1}{(s-1+a_2)(s+1)} & \dfrac{1+a_1}{s-1+a_2} \\[2ex]
\dfrac{1+a_1}{s-1+a_2} & \dfrac{1+a_1}{s-1+a_2}
\end{bmatrix},
\text{ for }
\begin{cases}
\mathbf{a} = [a_1\ a_2] \\
\mathbf{a}_0 = [0\ 0]
\end{cases};
\tag{6.209a}
$$

$$
\mathbf{P}_2(s,\mathbf{b}) =
\begin{bmatrix}
\dfrac{2+b_1}{(s+1)(s-1+b_5)} & \dfrac{1+b_2}{s-1+b_5} \\[2ex]
\dfrac{1+b_3}{s-1+b_6} & \dfrac{1+b_4}{s-1+b_6}
\end{bmatrix},
\text{ for }
\begin{cases}
\mathbf{b} = [b_1\ b_2\ \cdots\ b_6] \\
\mathbf{b}_0 = [0\ 0\ \cdots\ 0]
\end{cases};
\tag{6.209b}
$$

$$
\mathbf{P}_3(s,\mathbf{c}) =
\begin{bmatrix}
\dfrac{2+c_1}{(s+1)(s-1+c_5)} & \dfrac{1+c_2}{s-1+c_5} \\[2ex]
\dfrac{1+c_3}{s-1+c_5} & \dfrac{1+c_4}{s-1+c_6}
\end{bmatrix},
\text{ for }
\begin{cases}
\mathbf{c} = [c_1\ c_2\ \cdots\ c_6] \\
\mathbf{c}_0 = [0\ 0\ \cdots\ 0]
\end{cases};
\tag{6.209c}
$$

$$
\mathbf{P}_4(s,\mathbf{d}) =
\begin{bmatrix}
\dfrac{2+d_1}{(s+1)(s-1+d_5)} & \dfrac{1+d_2}{s-1+d_6} \\[2ex]
\dfrac{1+d_3}{s-1+d_7} & \dfrac{1+d_4}{s-1+d_8}
\end{bmatrix},
\text{ for }
\begin{cases}
\mathbf{d} = [d_1\ d_2\ \cdots\ d_8] \\
\mathbf{d}_0 = [0\ 0\ \cdots\ 0]
\end{cases}.
\tag{6.209d}
$$

Write

$$
v\left[\mathbf{P}_i^+(s,\mathbf{a})\right] = v_i^+(\mathbf{a}), \quad \text{for } i = 1,\ldots,4
\tag{6.210}
$$

and note that the nominal transfer functions are identical:

$$
\mathbf{P}_0(s) = \mathbf{P}_1(s,\mathbf{a}_0) = \mathbf{P}_2(s,\mathbf{b}_0) = \mathbf{P}_3(s,\mathbf{c}_0) = \mathbf{P}_4(s,\mathbf{d}_0),
\tag{6.211}
$$

with McMillan degrees all equal to 2 and antistable McMillan degrees equal to 1:

$$
1 = v^+[\mathbf{P}_0(s)] = v_1^+(\mathbf{a}_0) = v_2^+(\mathbf{b}_0) = v_3^+(\mathbf{c}_0) = v_4^+(\mathbf{d}_0).
\tag{6.212}
$$

Under arbitrary but infinitesimal perturbations, however, we have generically

$$
v_1^+(\mathbf{a}) = 2, \quad v_2^+(\mathbf{b}) = 2, \quad v_3^+(\mathbf{c}) = 3, \quad v_4^+(\mathbf{d}) = 4.
\tag{6.213}
$$

These values are the values of v_{max}^+ for each perturbation class, respectively, and in each case there is a drop in the value of v^+ at the nominal.

Based on the above discussion we see that the McMillan degree

$$\nu(\mathbf{p}) = \nu[\mathbf{P}(s,\mathbf{p})], \tag{6.214}$$

as well as $\nu^+(\mathbf{p})$, is a discontinuous function of $\mathbf{p}$ and in general its value drops on an algebraic variety. Indeed, if $\mathscr{B}$ denotes an arbitrarily small ball in $\mathbb{R}^\ell$, centered at the origin, we have

$$\nu_{max} = \max_{\delta\mathbf{p}\in\mathscr{B}} \nu(\mathbf{p}+\delta\mathbf{p}). \tag{6.215}$$

Then the algebraic variety $\mathscr{V}$ is

$$\mathscr{V} = \{\mathbf{p} : \nu(\mathbf{p}) \neq \nu_{max}\}. \tag{6.216}$$

In an exactly similar manner, we can write

$$\mathscr{V}^+ := \{\mathbf{p} : \nu^+(\mathbf{p}) \neq \nu^+{}_{max}\} \tag{6.217}$$

to denote the algebraic variety where the McMillan degree of the antistable component drops. Referring to the previous example we see, for example, that for $P_1(s,\mathbf{a})$:

$$\mathscr{V}_1^+ = \{\mathbf{a} : a_1 + a_2 + a_1 a_2 = 0\}. \tag{6.218}$$

Next, we discuss the effect of the discontinuity of $\nu^+(\mathbf{p})$ on feedback stabilization.

6.8.3   Structurally Stable Stabilization

A parametrized system is *structurally stable* if it remains stable under small but arbitrary perturbations of its parameters. Consider, in the previous example, a feedback controller $\mathbf{C}_0$, of order q_0, that stabilizes the plant $\mathbf{P}_0$ represented by the transfer function $\mathbf{P}_0(s) = \mathbf{P}(s,\mathbf{p}_0)$ in Equation (6.208).

Equivalently, $\mathbf{C}_0$ internally stabilizes a second-order realization of $\mathbf{P}_0$. Now let us introduce small but arbitrary perturbations of $\mathbf{p}_0$. It is easy to see that:

(a) If $\mathbf{P}(s,\mathbf{p})$ is parametrized as in Equation (6.209a) such perturbations will introduce one unstable closed-loop pole close to $s = 1$.

(b) If $\mathbf{P}(s,\mathbf{p})$ is parametrized as in Equation (6.209b) such perturbations will introduce one unstable closed-loop pole close to $s = 1$.

(c) If $\mathbf{P}(s,\mathbf{p})$ is parametrized as in Equation (6.209c) such perturbations will introduce two unstable closed-loop poles close to $s = 1$.

(d) If $\mathbf{P}(s,\mathbf{p})$ is parametrized as in Equation (6.209d) such perturbations will introduce three unstable closed-loop poles close to $s = 1$.

Therefore, in each case, the closed loop is rendered unstable by infinitesimal arbitrary perturbations. Such a system is called *structurally* unstable.

To remedy the situation it is obviously necessary to know the perturbation class to which $\mathbf{P}_0(s)$ belongs. This requires at least partial knowledge of internal structure and parametrization beyond the nominal transfer function. Once the correct parametrization is known, we can perturb the parameter $\mathbf{p}_0$ to $\mathbf{p}_1$ in this class, so that

$$v^+[\mathbf{P}(s,\mathbf{p}_1)] = v_{max}^+ \qquad (6.219)$$

constructs a minimal realization of order v_{max} and stabilizes it with a compensator $\mathbf{C}_1$. Note that such a compensator generically stabilizes $\mathbf{P}(s,\mathbf{p}_1)$ for small perturbations about $\mathbf{p}_1$. Thus, if $\mathbf{P}(s,\mathbf{p})$ belongs to the parametrization $\mathbf{P}_1(s,\mathbf{b})$, the controller must stabilize a second-order plant, if it belongs to $\mathbf{P}_2(s,\mathbf{c})$ it must stabilize a third-order plant, and if it belongs to $\mathbf{P}_3(s,\mathbf{d})$ it must stabilize a fourth-order plant, respectively.

In each case, structurally stable stabilization will be achieved by nominally stabilizing a system of order v_{max} for that class, rather than the minimal-order realization of $\mathbf{P}_0(s)$. In each case, however, the minimal realization of order v_{max} of an arbitrary neighborhood of $\mathbf{P}(s,\mathbf{p}_0)$ is destabilized by the compensator. These observations are summarized in the following.

THEOREM 6.6   Assume that small perturbations in $\mathbf{p}_0$ introduces some additional open RHP poles in $\mathbf{P}(s,\mathbf{p})$. A plant with transfer function $\mathbf{P}(s,\mathbf{p}_0)$ can be stabilized by a linear time-invariant feedback controller in a structurally stable manner if and only if

$$v^+(\mathbf{p}_0) = v_{max}^+. \qquad (6.220)$$

If $v^+(\mathbf{p}_0) < v_{max}^+$, then

(a) any stabilizing controller for $\mathbf{P}(s,\mathbf{p}_0)$ renders the closed loop is not structurally stable, that is, the closed loop is destabilized by arbitrarily small perturbations of the parameter $\mathbf{p}_0$;

(b) any controller that stabilizes $\mathbf{P}(s,\mathbf{p}_1)$ with $v^+(\mathbf{p}_1) = v_{max}^+$ fails to stabilize a "ball" around $\mathbf{p}_1$ that includes plant $\mathbf{P}(s,\mathbf{p}_0)$.

In other words, failure of the condition stated in the theorem implies that one must either give up structural stability or nominal stability as no controller can simultaneously achieve both.

Remark 6.3   It is important to note that it is possible to have a controller that simultaneously stabilizes the perturbed plant $\mathbf{P}(\mathbf{p}_1)$ of higher order and the nominal plant $\mathbf{P}(s,\mathbf{p}_0)$ of lower order. However, there is no controller that stabilizes the entire ball around $\mathbf{p}_1$ that includes the nominal $\mathbf{p}_0$. This means that some plant of higher order along any path connecting $(\mathbf{p}_1,\mathbf{p}_0)$ will be destabilized even though the controller simultaneously stabilizes $\mathbf{P}(\mathbf{p}_1)$ and $\mathbf{P}(\mathbf{p}_0)$.

Remark 6.4   In this theorem, we assume that the balls of plants being considered are all of degree v_{max}.

Remark 6.5   On any continuous path $\mathbf{p}_\lambda$ connecting $\mathbf{p}_1$ and $\mathbf{p}_0$, there will be a point λ^* such that a system of order $v_{max} +$ (order of controller) with realization of the transfer function $\mathbf{P}(s,\lambda^*)$ of order v_{max} will have $j\omega$ eigenvalues. These may or may not correspond to uncontrollable/unobservable eigenvalues. This is determined by whether the path intersects the algebraic variety $\mathscr{V}^+$ or not.

The proof is a formalization of the above remarks and is best stated utilizing the following lemma. We say an $n \times n$ matrix is stable (unstable) if it contains all (some) eigenvalues in the open LHP (closed RHP).

LEMMA 6.2 Let $\mathbf{A}(\mathbf{p})$ denote an $n \times n$ real or complex matrix whose entries are continuous functions of the real parameter vector $\mathbf{p} \in \mathbb{R}^l$. If $\mathbf{A}(\mathbf{p}_0)$ is stable, then $\mathbf{A}(\mathbf{p}_0 + \varepsilon)$ is stable, for sufficiently small ε, in a ball $\mathscr{B} \in \mathbb{R}^l$. Similarly, If $\mathbf{A}(\mathbf{p}_0)$ is unstable with at least one pole at the open RHP, then $\mathbf{A}(\mathbf{p}_0 + \varepsilon)$ is unstable with at least one pole at the open RHP, for sufficiently small ε, in a ball $\mathscr{B} \in \mathbb{R}^l$.

The proof of the lemma is a straightforward consequence of the facts that the characteristic polynomial of $\mathbf{A}$ is degree invariant and the eigenvalues of A are continuous functions of its entries. The proof of the theorem can now be stated.

Proof of Theorem 6.6 Let $\mathbf{C}_0$ be a controller, say of order q_0, internally stabilizing a stabilizable and detectable realization of $\mathbf{P}(s, \mathbf{p}_0)$, say of order n_0. Denote the $n_0 + q_0$ closed-loop eigenvalues by Λ_0. To discuss structural stability, we fix $\mathbf{C}_0$, perturb $\mathbf{p}_0$ to $\mathbf{p}_1 = \mathbf{p}_0 + \varepsilon$, and replace the plant by a stabilizable and detectable realization of $\mathbf{P}(s, \mathbf{p}_1)$, say of order n_1. Now suppose $v^+(\mathbf{p}_0) = v_{max}^+$. Then for ε sufficiently small but arbitrary, $v^+(\mathbf{p}_1) = v^+(\mathbf{p}_0)$ and we may take $n_1 = n_0$. By Lemma 6.2 above the closed-loop eigenvalues with plant $\mathbf{P}(s, \mathbf{p}_1)$ and controller $\mathbf{C}_0$ remain close to Λ_0 and therefore in the LHP.

On the other hand, suppose that $v^+(\mathbf{p}_0) < v_{max}^+$. Thus, $v^+(\mathbf{p}_1) = v_{max}^+ > v^+(\mathbf{p}_0)$. Now let n_1 denote any stabilizable and detectable realization of $\mathbf{P}(s, \mathbf{p}_1)$ and consider the $n_1 + q_0$ closed-loop eigenvalues, denoted $\Lambda_1(\mathbf{p}_0 + \varepsilon)$, of this plant with the fixed controller $\mathbf{C}_0$. In this case it is impossible to take $n_1 = n_0$.

Moreover, as $\varepsilon \to 0$, any state-space realization of $\mathbf{P}(s, \mathbf{p}_1)$ contains $v^+(\mathbf{p}_1) - v^+(\mathbf{p}_0)$ uncontrollable and/or unobservable RHP eigenvalues, therefore contained in $\Lambda_1(\mathbf{p}_0)$. By Lemma 6.2 above, we conclude that the closed loop is unstable for sufficiently small, but arbitrary ε. ∎

EXAMPLE 6.15 Consider the plant with transfer function parametrization:

$$\mathbf{Y}(s) = \mathbf{P}(s, \delta)\mathbf{U}(s), \tag{6.221}$$

where

$$\mathbf{P}(s, \delta) = \begin{bmatrix} \dfrac{2}{(s+1)(s-1)} & \dfrac{1+\delta}{s-1} \\[3mm] \dfrac{1}{s-1} & \dfrac{1}{s-1} \end{bmatrix}. \tag{6.222}$$

With $\delta = 0$:

$$\mathbf{P}(s, 0) =: \mathbf{P}_0(s) = \begin{bmatrix} \dfrac{2}{(s+1)(s-1)} & \dfrac{1}{s-1} \\[3mm] \dfrac{1}{s-1} & \dfrac{1}{s-1} \end{bmatrix}. \tag{6.223}$$

We consider the stabilizing controller

$$\mathbf{u}(t) = -\mathbf{K}\mathbf{y}(t) + \mathbf{v}(t) \tag{6.224}$$

with

$$K = \begin{bmatrix} 1 & 2 \\ 3 & 4 \end{bmatrix}. \tag{6.225}$$

A minimal realization of $\mathbf{P}_0(s)$ is

$$\dot{\mathbf{x}}(t) = \begin{bmatrix} -1 & 0 \\ 0 & 1 \end{bmatrix} \mathbf{x}(t) + \begin{bmatrix} 1 & 0 \\ 1 & 1 \end{bmatrix} \mathbf{u}(t),$$

$$\mathbf{y}(t) = \begin{bmatrix} -1 & 1 \\ 0 & 1 \end{bmatrix} \mathbf{x}(t) \tag{6.226}$$

and the closed-loop system

$$\dot{\mathbf{x}}(t) = (\mathbf{A} - \mathbf{BKC})\mathbf{x}(t) + \mathbf{Bv}(t) \tag{6.227}$$

is internally stable with the controller in Equation (6.225), with characteristic polynomial

$$s^2 + 9s + 12. \tag{6.228}$$

Now consider a "small" perturbation of $\mathbf{P}_0(s)$ obtained by letting δ be nonzero. A minimal realization of Equation (6.222) with $\delta \neq 0$ is

$$\dot{\mathbf{x}}(t) = \begin{bmatrix} -1 & 0 & 0 \\ 0 & 1 & 0 \\ 0 & 0 & 1 \end{bmatrix} \mathbf{x}(t) + \begin{bmatrix} 1 & 0 \\ 1 & 1+\delta \\ 1 & 1 \end{bmatrix} \mathbf{u}(t),$$

$$\mathbf{y}(t) = \begin{bmatrix} -1 & 1 & 0 \\ 0 & 0 & 1 \end{bmatrix} \mathbf{x}(t), \tag{6.229}$$

and the closed-loop system with the previous controller is

$$\dot{\mathbf{x}}(t) = \begin{bmatrix} 0 & -1 & -2 \\ 4+3\delta & -3-3\delta & -6-4\delta \\ 4 & -4 & -5 \end{bmatrix} \mathbf{x}(t) + \begin{bmatrix} 1 & 0 \\ 1 & 1+\delta \\ 1 & 1 \end{bmatrix} \mathbf{v}(t). \tag{6.230}$$

The characteristic polynomial of Equation (6.230) is

$$s^3 + (8+3\delta)s^2 + (3+2\delta)s - \delta - 12 \tag{6.231}$$

and is seen to be unstable for "small" values of δ, and in this particular case for all values of δ. Moreover, as $\delta \to 0$, a root of Equation (6.231) tends to $s = 1$.

A remedy to the structural instability above would be to design a stabilizing controller for the third-order model in Equation (6.229), say for some $\delta = \delta^0$. Such a controller would stabilize a ball of plants around δ^0 but cannot obviously stabilize the plant with $\delta = 0$.

EXAMPLE 6.16 Consider the plant with transfer function

$$\mathbf{P}(s) = \begin{bmatrix} \dfrac{1}{s-1} & \dfrac{1}{s-1} \\[3mm] \dfrac{1}{s-1} & \dfrac{1}{s-1} \end{bmatrix}.$$ (6.232)

Its minimal realization is

$$\dot{x}(t) = [1]x(t) + \begin{bmatrix} 1 & 1 \end{bmatrix} \mathbf{u}(t) =: ax(t) + \mathbf{b}^T \mathbf{u}(t),$$

$$\mathbf{y}(t) = \begin{bmatrix} 1 \\ 1 \end{bmatrix} x(t) =: \mathbf{c}x(t).$$ (6.233)

The controller below is used to stabilize the nominal system in Equation (6.233):

$$\mathbf{K}_o = \begin{bmatrix} k_0 & 0 \\ 0 & k_0 \end{bmatrix}.$$ (6.234)

The closed-loop characteristic polynomial of the system is

$$\Pi_o(s) = \det\left[s - \left(a - \mathbf{b}^T \mathbf{K}_o \mathbf{c}\right)\right] = s + (2k - 1).$$ (6.235)

It shows that the closed-loop system is stable for $k > \frac{1}{2}$.

Now consider a perturbed plant

$$\mathbf{P}(s,\delta) = \begin{bmatrix} \dfrac{1}{s-1} & \dfrac{1+\delta}{s-1} \\[3mm] \dfrac{1}{s-1} & \dfrac{1}{s-1} \end{bmatrix}.$$ (6.236)

Its minimal realization is

$$\dot{\mathbf{x}}(t) = \begin{bmatrix} 1 & 0 \\ 0 & 1 \end{bmatrix} \mathbf{x}(t) + \begin{bmatrix} 1 & 0 \\ 0 & 1 \end{bmatrix} \mathbf{u}(t) =: \mathbf{A}\mathbf{x}(t) + \mathbf{B}\mathbf{u}(t),$$

$$\mathbf{y}(t) = \begin{bmatrix} 1 & 1+\delta \\ 1 & 1 \end{bmatrix} \mathbf{x}(t) = \mathbf{C}\mathbf{x}(t).$$ (6.237)

For convenience, let $\mathbf{p} = \delta > 0$ (real). Then the following controller stabilizes the system:

$$\mathbf{K}_1 = \begin{bmatrix} -1 & 3 \\ k_1 & -k_1 \end{bmatrix}.$$ (6.238)

The closed-loop characteristic polynomial of this system is

$$\Pi_1(s) = \det\left[s\mathbf{I} - (\mathbf{A} - \mathbf{B}\mathbf{K}\mathbf{C})\right] = (s+1)(s+k_1\delta - 1).$$ (6.239)

It shows that for any given perturbation δ, there is a controller value that robustly stabilizes the entire family in $\mathbf{p} \in (0, \delta]$. Furthermore, the order of a minimal realization of every plant in the family is 2. However, when $\delta = 0$, the controller $\mathbf{K}_0$ in Equation (6.234) should be applied to its minimal realization, which is of order 1. In summary, for any given perturbation δ^*, no controller can stabilize the entire family of plants for all $\delta \in (0, \delta^*]$ (i.e., including the nominal plant) if and only if

$$v^+(\mathbf{p}_0) \neq v^+(\mathbf{p}). \tag{6.240}$$

6.9   Exercises

6.1   Consider the plants

$$P_1(s) = \frac{s-1}{s^2}, \qquad P_2(s) = \frac{1}{s(s-1)}, \qquad P_3(s) = \frac{1}{(s+1)(s-1)}. \tag{6.241}$$

(a)   In each case, show that static-output feedback cannot stabilize the system but state feedback can.

(b)   In each case, design a dynamic feedback controller to place closed-loop characteristic roots at $-2, -1 \pm j$.

(c)   In each case, design an observer-based state feedback controller to place closed-loop eigenvalues at $-1, -1 \pm j$.

6.2   Consider the state-space model of a vehicle:

$$\dot{x}_1(t) = x_2(t),$$
$$\dot{x}_2(t) = u(t), \tag{6.242}$$
$$y(t) = x_1(t),$$

where $x_1(t)$ is position and $x_2(t)$ is velocity. Design a first-order observer to "estimate" the velocity $x_2(t)$, from measurements of position $x_1(t)$ and $u(t)$. Use simulations to check the closeness of $\dot{y}(t)$ to $\hat{x}_2(t)$ the observer output, for several inputs $u(t)$ and several observer eigenvalues.

6.3   Consider an extension of the model in Example 6.2 where we now have $x_1(t)$ (position), $x_2(t)$ (velocity), and $x_3(t)$ (acceleration):

$$\dot{x}_1(t) = x_2(t),$$
$$\dot{x}_2(t) = x_3(t),$$
$$\dot{x}_3(t) = u(t), \tag{6.243}$$
$$y(t) = x_1(t).$$

Design a second-order observer to estimate velocity $x_2(t)$ and acceleration $x_3(t)$ from measurements of $x_1(t) = y(t)$ and $u(t)$. Using simulations, check the closeness of $\hat{x}_2(t)$, $\hat{x}_3(t)$ to $\dot{y}$ and $\ddot{y}(t)$, respectively, for various inputs $u(t)$ and several sets of observer eigenvalues.

6.4 (Reliable stabilization) Consider the plant **P** with transfer function

$$\mathbf{P}(s) = \begin{bmatrix} \dfrac{2s}{s^2 - 1} & \dfrac{3s + 1}{s^2 - 1} \\ \dfrac{3s + 1}{s^2 - 1} & \dfrac{5s + 3}{s^2 - 1} \end{bmatrix}. \tag{6.244}$$

(a) Find a minimal-order realization $\mathscr{R}_{\min}$.

(b) Design a stabilizing controller for **P** based on $\mathscr{R}_{\min}$.

(c) Suppose **P**(s) is subject to parameter perturbations as shown:

$$\mathbf{P}(s, \underline{\delta}) = \begin{bmatrix} \dfrac{(2 + \delta_1)s}{s^2 - 1} & \dfrac{(3 + \delta_2)s + 1}{s^2 - 1} \\ \dfrac{(3 + \delta_3)s + 1}{s^2 - 1} & \dfrac{(5 + \delta_4)s + 3}{s^2 - 1} \end{bmatrix}. \tag{6.245}$$

Design a stabilizing controller for **P**$(s, \underline{\delta})$ based on a realization of maximal-order $\mathscr{R}_{\max}(\underline{\delta})$ to reliably stabilize **P** over the ball of perturbations

$$\{ \underline{\delta} : \|\underline{\delta}\| < \varepsilon \}. \tag{6.246}$$

(d) Consider another class of perturbations as shown:

$$\mathbf{P}(s, \underline{\alpha}) = \begin{bmatrix} \dfrac{2s}{s^2 - 1 + \alpha_1} & \dfrac{3s + 1}{s^2 - 1 + \alpha_2} \\ \dfrac{3s + 1}{s^2 - 1 + \alpha_3} & \dfrac{5s + 3}{s^2 - 1 + \alpha_4} \end{bmatrix}. \tag{6.247}$$

Design a stabilizing controller for **P**$(s, \underline{\alpha})$ based on a realization $\mathscr{R}_{\max}(\underline{\alpha})$ of maximal order to reliably stabilize **P** over the ball of perturbations

$$\{ \underline{\alpha} : \|\underline{\alpha}\| < \varepsilon, \underline{\alpha} \neq 0 \}. \tag{6.248}$$

(e) In (c) and (d), check closed-loop stability when $\underline{\delta} = 0$, $\underline{\alpha} = 0$, respectively.

(f) Discuss the results obtained in (a)–(e) above.

6.5 (Reliable stabilization) Design an observer-based state feedback controller to stabilize the plant below using:

(a) a full-order observer;

(b) a reduced-order observer.

$$\mathbf{P}(s) = \begin{bmatrix} \dfrac{1}{s - 1} & \dfrac{1}{(s - 1)^2} \\ \dfrac{1}{(s - 1)^2} & \dfrac{1}{s - 1} \end{bmatrix}. \tag{6.249}$$

Check whether the closed-loop stability is preserved if **P**(s) is subject to small parameter perturbations by simulation.

6.6   Consider the plant

$$
\mathbf{P}(s) = \begin{bmatrix} \dfrac{1}{(s+1)(s-1)} & \dfrac{1}{s+2} \\[3mm] \dfrac{1}{(s+1)(s-1)} & \dfrac{1}{s-2} \end{bmatrix}.
\tag{6.250}
$$

(a)   Find a minimal realization of the plant $\mathbf{P}(s)$.
(b)   Find a state feedback control law to stabilize $\mathbf{P}(s)$.
(c)   Implement the control law determined in (b) by a reduced-order observer.
(d)   Find the equivalent feedback controller.
(e)   Check if closed-loop stability is preserved if all poles of $\mathbf{P}(s)$ are subject to perturbations.

6.7   In

$$
\dot{\mathbf{x}}(t) = \mathbf{A}\mathbf{x}(t) + \mathbf{B}\mathbf{u}(t),
\tag{6.251}
$$

where

$$
\mathbf{A} = \begin{bmatrix} 0 & 1 & 0 & 1 \\ 1 & 0 & 1 & 1 \\ 0 & 0 & 0 & 1 \\ 0 & 0 & 0 & 0 \end{bmatrix}, \quad \mathbf{B} = \begin{bmatrix} 1 & 0 \\ 0 & 0 \\ 0 & 0 \\ 0 & 1 \end{bmatrix},
\tag{6.252}
$$

find $\mathbf{K}$ so that

$$
\sigma(\mathbf{A} - \mathbf{B}\mathbf{K}) = \{-1, -1, -1 \pm j\}
\tag{6.253}
$$

using the procedure in the proof of the Pole Assignment Theorem.

6.8   Solve Exercise 6.7 using the Sylvester equation method.

6.9   Design an observer of reduced order to implement the control law obtained in Exercise 6.7. Find the equivalent feedback controller.

6.10   Notes and References

The Pole Assignment Theorem was proved by Wonham in 1968 and is described in Wonham (1985). The use of Sylvester's equation for pole assignment was proposed in Bhattacharyya and de Souza (1982). It was based on properties of the solution of Sylvester's equation developed in de Souza and Bhattacharyya (1981). Observer theory was developed by Luenberger and the exposition given here follows that in Luenberger (1979). The structure of robust observers was developed in Bhattacharyya (1976). That maximal realizations, rather than minimal realizations, should be used for reliable stabilization was pointed out in Keel and Bhattacharyya (2010b).

7 The Optimal Linear Quadratic Regulator

This chapter develops the theory of the optimal linear quadratic regulator (LQR) problem as formulated and developed by Kalman in 1960. This theory is now a central result in the optimal control of state-space systems. The general approach to optimal control problems is developed using Bellman's invariant embedding and the Principle of Optimality. For the special case of continuous-time linear systems and quadratic cost functionals, we obtain a solution to the finite-time LQR problem in the form of a state feedback solution computable by solving the Ricatti matrix differential equation.

Next, the infinite-horizon time-invariant LQR problem is considered and once again Bellman's approach is invoked, leading this time to a time-invariant state feedback optimal control law determined by solving the algebraic Ricatti equation (ARE). A geometric approach to the infinite-time LQR is also described briefly, leading to a subspace inclusion condition by considering the LQR as a problem of zeroing the output with bounded control energy. We then discuss solvability of the ARE and the existence of stabilizing solutions. It is shown that the closed-loop eigenvalues of the optimal LQR can be determined directly from the performance index using the Hamiltonian matrix. Next we derive the universal guaranteed gain and phase margins delivered by the optimal LQR whenever state feedback can be implemented.

The guaranteed excellent stability margins obtainable with state feedback can disappear in an output feedback implementation using for example "observer-reconstructed" optimal state feedback. Besides loss of optimality, a more serious problem with such an implementation is the potential loss of robustness. This observation, due to Doyle, was responsible for the renewed interest in the subject of robust control in the 1980s.

We then describe an approach to optimal output feedback dynamic compensation proposed by Pearson in the late 1960s. In this approach the plant is augmented by a suitable number of integrators so that state feedback in the augmented system can be implemented by output feedback in the actual system. The LQR design approach is then applied to the augmented system, thus leading to an optimal dynamic compensator.

Finally, a brief derivation of the results for the LQR problem for discrete-time systems is included.

7.1   An Optimal Control Problem

A dynamic system described by the equations

$$\dot{\mathbf{x}}(t) = f(\mathbf{x}(t), \mathbf{u}(t), t) \tag{7.1}$$

and operating over the time interval $[t_0, T]$ with T, t_0, $x(t_0)$ fixed is given. Consider the problem of determining the control function $\mathbf{u}[t_0, T]$ that minimizes the "cost" functional defined by

$$I(t_0, \mathbf{x}(t_0), \mathbf{u}[t_0, T]) := \int_{t_0}^{T} \phi(\mathbf{x}(t), \mathbf{u}(t), t) dt + \theta(\mathbf{x}(T), T). \tag{7.2}$$

In Equation (7.1), $\mathbf{x}(t)$ represents the n-dimensional vector of state variables $x_i(t)$, $i = 1, 2, \ldots, n$ at time t, $\mathbf{u}(t)$ represents the r-dimensional vector of controls $u_i(t)$, $i = 1, 2, \ldots, r$ at time t, f is an n-vector of functions, and ϕ and θ are scalar functions of their respective arguments. The notation $\mathbf{u}[t_0, T]$ represents the time function $\mathbf{u}(t)$, $t \in [t_0, T]$. We assume that $\mathbf{u}(t)$ is allowed to be piecewise continuous, that Equation (7.1) has a unique solution $\mathbf{x}[t_0, T]$, and ϕ and θ are continuous functions of their arguments. The term $\theta(\mathbf{x}(T), T)$ is called a *terminal penalty term*.

Following Bellman, we introduce the value function defined as

$$V(\mathbf{x}(t), t) := \inf_{\mathbf{u}[t, T]} \int_t^T \phi(\mathbf{x}, \mathbf{u}, \tau) d\tau + \theta(\mathbf{x}(T), T) \tag{7.3}$$

for arbitrary $t \in [t_0, T]$ and arbitrary starting state $\mathbf{x}(t)$. In Equation (7.3), we have dropped the explicit dependence of $\mathbf{x}$ and $\mathbf{u}$ on t. The problem of minimizing Equation (7.2), assuming $V(\mathbf{x}(t), t)$ can be found for all $\mathbf{x}(t)$ and t, is simply the special case of finding $V(\mathbf{x}(t_0), t_0)$. This process of introducing the general class of problems in Equation (7.3) of which the given problem is a special case, is known as "invariant embedding" and was used by Bellman to derive a mathematical statement of the so-called Principle of Optimality.

7.1.1   Principle of Optimality

Observe that for $t_1, \in [t, T]$:

$$V(\mathbf{x}(t), t) = \inf_{\mathbf{u}[t, T]} \left[\int_t^{t_1} \phi(\mathbf{x}, \mathbf{u}, \tau) d\tau + \int_{t_1}^{T} \phi(\mathbf{x}, \mathbf{u}, \tau) d\tau + \theta(\mathbf{x}(T), T) \right]$$

$$= \inf_{\mathbf{u}[t, t_1) \cup \mathbf{u}[t_1, T]} \left[\int_t^{t_1} \phi(\mathbf{x}, \mathbf{u}, \tau) d\tau + \int_{t_1}^{T} \phi(\mathbf{x}, \mathbf{u}, \tau) d\tau + \theta(\mathbf{x}(T), T) \right] \tag{7.4}$$

$$= \inf_{\mathbf{u}[t, t_1)} \left[\int_t^{t_1} \phi(\mathbf{x}, \mathbf{u}, \tau) d\tau + \inf_{\mathbf{u}[t_1, T]} \int_{t_1}^{T} \phi(\mathbf{x}, \mathbf{u}, \tau) d\tau + \theta(\mathbf{x}(T), T) \right],$$

so that

$$V(\mathbf{x}(t), t) = \inf_{\mathbf{u}[t, t_1)} \left[\int_t^{t_1} \phi(\mathbf{x}, \mathbf{u}, \tau) d\tau + V(\mathbf{x}(t_1), t_1) \right], \tag{7.5}$$

where $\mathbf{x}(t_1)$ is the state at t_1, resulting from application of the control $\mathbf{u}[t, t_1)$.

Equation (7.5) is the mathematical statement of Bellman's famous *Principle of Optimality*. It states that the optimal control $\mathbf{u}[t,T]$ can be found by concatenating the optimal $\mathbf{u}[t,t_1]$ for Equation (7.5) to the optimal $\mathbf{u}[t_1,T]$ for the system starting at $\mathbf{x}(t_1)$ at time t_1, that is the solution of Equation (7.3) with $t = t_1$.

7.1.2   Hamilton–Jacobi–Bellman Equation

The Hamilton–Jacobi–Bellman (HJB) equations are a differential form of Equation (7.5) obtained by letting $t_1 \to t$ and making appropriate assumptions regarding the smoothness of $V(\mathbf{x}(t),t)$. Letting $t_1 = t + \Delta t$ in Equation (7.5):

$$V(\mathbf{x}(t),t) = \inf_{\mathbf{u}[t,t+\Delta t]} \left[\int_t^{t+\Delta t} \phi(\mathbf{x},\mathbf{u},\tau)d\tau + V(\mathbf{x}(t+\Delta t),t+\Delta t) \right]. \quad (7.6)$$

Applying the Mean Value Theorem to the integral in Equation (7.6) and assuming that $V(\mathbf{x}(t+\Delta t),t+\Delta t)$ can be Taylor series expanded about $V(\mathbf{x}(t),t)$ we have, for some $\alpha \in [0,1]$:

$$
\begin{aligned}
V(\mathbf{x}(t),t) = \inf_{\mathbf{u}[t,t+\Delta t]} & \left[\phi(\mathbf{x}(t+\alpha\Delta t),\mathbf{u}(t+\alpha\Delta t),t+\alpha\Delta t)\Delta t + V(\mathbf{x}(t),t) \right. \\
& \left. + \frac{\partial V^T}{\partial \mathbf{x}}\dot{\mathbf{x}}\Delta t + \frac{\partial V}{\partial t}\Delta t + O(\Delta t)^2 \right] \\
= V(\mathbf{x}(t),t) & + \frac{\partial V}{\partial t}\Delta t \qquad\qquad\qquad\qquad\qquad (7.7) \\
+ \inf_{\mathbf{u}[t,t+\Delta t]} & \left[\phi(\mathbf{x}(t+\alpha\Delta t),\mathbf{u}(t+\alpha\Delta t),t+\alpha\Delta t)\Delta t \right. \\
& \left. + \frac{\partial V^T}{\partial \mathbf{x}}f(\mathbf{x},\mathbf{u},t)\Delta t + O(\Delta t)^2 \right],
\end{aligned}
$$

where $O(\Delta t^2)$ represents terms of second and higher order in Δt. Now, Equation (7.7) can be rewritten as

$$
\begin{aligned}
-\frac{\partial V}{\partial t} = \inf_{\mathbf{u}[t,t+\Delta t]} & \left[\phi(\mathbf{x}(t+\alpha\Delta t),\mathbf{u}(t+\alpha\Delta t),t+\alpha\Delta t) \right. \\
& \left. + \frac{\partial V^T}{\partial \mathbf{x}}f(\mathbf{x},\mathbf{u},t) + O(\Delta t) \right].
\end{aligned} \quad (7.8)
$$

Letting $\Delta t \to 0$, we obtain

$$-\frac{\partial V}{\partial t} = \inf_{\mathbf{u}(t)} \left[\phi(\mathbf{x},\mathbf{u},t) + \frac{\partial V^T}{\partial \mathbf{x}}f(\mathbf{x},\mathbf{u},t) \right], \quad (7.9)$$

which is known as the HJB equation for optimality.

Define

$$J\left(\mathbf{x},\mathbf{u},\frac{\partial V}{\partial \mathbf{x}},t\right) := \phi(\mathbf{x},\mathbf{u},t) + \frac{\partial V^T}{\partial \mathbf{x}}f(\mathbf{x},\mathbf{u},t) \quad (7.10)$$

and

$$\inf_{\mathbf{u}(t)} J\left(\mathbf{x}, \mathbf{u}, \frac{\partial V}{\partial \mathbf{x}}, t\right) := J^*. \tag{7.11}$$

Then the HJB equation can be rewritten as:

$$-\frac{\partial V}{\partial t} = J^*, \tag{7.12}$$

which represents a partial differential equation for $V(\mathbf{x}, t)$ with the boundary condition

$$V(\mathbf{x}(T), T) = \theta(\mathbf{x}(T), T). \tag{7.13}$$

It is emphasized that the minimization in Equation (7.11) yields $\mathbf{u}(t)$, implicitly or explicitly as a function of $\mathbf{x}(t)$ and $\frac{\partial V}{\partial \mathbf{x}}$. If a solution $V(\mathbf{x}, t)$ to Equations (7.12) and (7.13) can be found, this finally gives $\mathbf{u}(t)$ as a function of $\mathbf{x}(t)$, that is, a state feedback control.

Note that so far, we have only established that Equations (7.12) and (7.13) represent necessary conditions on the optimal control and the value function if they exist. The following theorem provides the desired sufficient conditions. To state these let

$$\mathbf{u}^*(t) = \mathrm{Arg}\left[\inf_{\mathbf{u}(t)} J\left(\mathbf{x}, \mathbf{u}, \frac{\partial V}{\partial \mathbf{x}}, t\right)\right], \quad \text{for } t \in [t_0, T]. \tag{7.14}$$

THEOREM 7.1   Assume that a continuously differentiable function $V(\mathbf{x}, t)$ satisfying Equations (7.12) and (7.13), and a piecewise continuous function $\mathbf{u}^*[t_0, T]$ satisfying Equations (7.11) and (7.14) exist. Then $\mathbf{u}^*[t_0, T]$ is the optimal control minimizing the cost function in Equation (7.2) and the minimum value of the cost function is $V(\mathbf{x}(t_0), t_0)$.

Proof   The cost function in Equation (7.2) is unchanged if we add

$$\int_{t_0}^T \frac{dV(\mathbf{x}(t), t)}{dt} dt - [V(\mathbf{x}(T), T) - V(\mathbf{x}(t_0), t_0)] = 0 \tag{7.15}$$

to it. Since

$$\frac{dV(\mathbf{x}(t), t)}{dt} = \frac{\partial V(\mathbf{x}(t), t)}{\partial t} + \frac{\partial V(t, \mathbf{x}(t))^T}{\partial \mathbf{x}} f(\mathbf{x}(t), \mathbf{u}(t), t), \tag{7.16}$$

it follows that

$$I(t_0, \mathbf{x}(t_0), \mathbf{u}[t_0, T])$$

$$= \int_{t_0}^T \left(\phi(\mathbf{x}, \mathbf{u}, t) + \frac{\partial V(t, \mathbf{x}(t))^T}{\partial \mathbf{x}} f(\mathbf{x}(t), \mathbf{u}(t), t) + \frac{\partial V(\mathbf{x}(t), t)}{\partial t}\right) dt$$

$$+ \theta(\mathbf{x}(T), T) - V(\mathbf{x}(T), T) + V(\mathbf{x}(t_0), t_0) \tag{7.17}$$

$$= V(\mathbf{x}(t_0), t_0) + \int_{t_0}^T \left[J\left(\mathbf{x}(t), \mathbf{u}(t), \frac{\partial V}{\partial \mathbf{x}}, t\right) - J^*\left(\mathbf{x}(t), \mathbf{u}^*(t), \frac{\partial V}{\partial \mathbf{x}}, t\right)\right] dt$$

using Equations (7.10), (7.11), (7.12), and (7.13). Since

$$J^*\left(\mathbf{x}(t), \mathbf{u}^*(t), \frac{\partial V}{\partial \mathbf{x}}, t\right) \leq J\left(\mathbf{x}(t), \mathbf{u}(t), \frac{\partial V}{\partial \mathbf{x}}, t\right), \quad \text{for all } \mathbf{u}(t) \tag{7.18}$$

it follows that the integral in Equation (7.17) has a minimum value of zero, which is attained for

$$\mathbf{u}[t_0,T] = \mathbf{u}^*[t_0,T]. \tag{7.19}$$

Thus, the minimum cost is $V(\mathbf{x}(t_0),t_0)$, that is

$$I(t_0,\mathbf{x}(t_0),\mathbf{u}[t_0,T]) \geq I(t_0,\mathbf{x}(t_0),\mathbf{u}^*[t_0,T]) = V(\mathbf{x}(t_0),t_0), \tag{7.20}$$

for all $\mathbf{u}[t_0,T]$.   ∎

7.2   The Finite-Time LQR Problem

The results of the previous section can be used to derive a solution to the optimal LQR problem. We first consider the finite-horizon problem. Here the system equations are:

$$\dot{\mathbf{x}}(t) = \mathbf{A}(t)\mathbf{x}(t) + \mathbf{B}(t)\mathbf{u}(t) \tag{7.21}$$

where t_0 and $\mathbf{x}(t_0)$ are given, and the cost functional is:

$$I = \int_{t_0}^{T} \left[\mathbf{x}(t)^T\mathbf{Q}(t)\mathbf{x}(t) + \mathbf{u}(t)^T\mathbf{R}(t)\mathbf{u}(t)\right] dt + \mathbf{x}(T)^T\mathbf{M}\mathbf{x}(T). \tag{7.22}$$

In Equations (7.21) and (7.22), $\mathbf{A}(t)$, $\mathbf{B}(t)$, $\mathbf{Q}(t)$, $\mathbf{R}(t)$ are matrices of appropriate size with possibly time-varying entries and $\mathbf{Q}(t)$ and $\mathbf{R}(t)$ are both assumed to be symmetric and positive semi-definite and positive definite, respectively (denoted $\mathbf{Q}(t) \geq 0$, $\mathbf{R}(t) > 0$). The matrix $\mathbf{M}$ is symmetric positive semi-definite, $\mathbf{M} \geq 0$.

Let $V(\mathbf{x}(t),t)$ denote the value function. Then Equations (7.10), (7.11), (7.12), and (7.13) imply that

$$-\frac{\partial V(\mathbf{x}(t),t)}{\partial t}$$
$$= \inf_{\mathbf{u}(t)}\left[\mathbf{x}(t)^T\mathbf{Q}(t)\mathbf{x}(t) + \mathbf{u}(t)^T\mathbf{R}(t)\mathbf{u}(t) + \frac{\partial V^T}{\partial \mathbf{x}}(\mathbf{A}(t)\mathbf{x}(t) + \mathbf{B}(t)\mathbf{u}(t))\right] \tag{7.23}$$

$$V(\mathbf{x}(T),T) = \mathbf{x}(T)^T\mathbf{M}\mathbf{x}(T). \tag{7.24}$$

The optimal control $\mathbf{u}^*(t)$ must satisfy Equation (7.14) and therefore

$$2\mathbf{R}(t)\mathbf{u}^*(t) + \mathbf{B}(t)^T\frac{\partial V}{\partial \mathbf{x}} = 0 \tag{7.25}$$

so that

$$\mathbf{u}^*(t) = -\frac{1}{2}\mathbf{R}(t)^{-1}\mathbf{B}(t)^T\frac{\partial V}{\partial \mathbf{x}}. \tag{7.26}$$

Substituting Equation (7.26) back into Equation (7.23), we obtain

$$-\frac{\partial V}{\partial t} = \mathbf{x}(t)^T\mathbf{Q}(t)\mathbf{x}(t) + \frac{\partial V^T}{\partial \mathbf{x}}\mathbf{A}(t)\mathbf{x}(t) - \frac{1}{4}\frac{\partial V^T}{\partial \mathbf{x}}\mathbf{B}(t)\mathbf{R}(t)^{-1}\mathbf{B}(t)^T\frac{\partial V}{\partial \mathbf{x}} \tag{7.27}$$

as the partial differential equation to be satisfied by $V(\mathbf{x},t)$ subject to the boundary condition

$$V(\mathbf{x}(T),T) = \mathbf{x}(T)^T \mathbf{M} \mathbf{x}(T). \tag{7.28}$$

To solve Equations (7.27) and (7.28), we make the reasonable guess that a V-function that is quadratic in $\mathbf{x}$ might work, and propose

$$V(\mathbf{x}(t),t) = \mathbf{x}(t)^T \mathbf{P}(t) \mathbf{x}(t), \tag{7.29}$$

where $P(t)$ is a symmetric matrix, as a candidate solution. Then Equation (7.27) becomes

$$-\mathbf{x}(t)^T \dot{\mathbf{P}}(t) \mathbf{x}(t) = \left[\mathbf{x}(t)^T \mathbf{Q}(t) \mathbf{x}(t) + 2\mathbf{x}(t)^T \mathbf{P}(t) \mathbf{A}(t) \mathbf{x}(t) \\ -\mathbf{x}(t)^T \mathbf{P}(t) \mathbf{B}(t) \mathbf{R}(t)^{-1} \mathbf{B}(t)^T \mathbf{P}(t) \mathbf{x}(t) \right] \tag{7.30}$$

and Equation (7.28) becomes

$$\mathbf{x}(T)^T \mathbf{P}(T) \mathbf{x}(T) = \mathbf{x}(T)^T \mathbf{M} \mathbf{x}(T). \tag{7.31}$$

Since

$$2\mathbf{P}(t)\mathbf{A}(t) = \underbrace{\mathbf{P}(t)\mathbf{A}(t) + \mathbf{A}(t)^T \mathbf{P}(t)}_{\text{symmetric}} + \underbrace{\mathbf{P}(t)\mathbf{A}(t) - \mathbf{A}(t)^T \mathbf{P}(t)}_{\text{antisymmetric}} \tag{7.32}$$

and

$$\mathbf{x}(t)^T \mathbf{S}(t) \mathbf{x}(t) = -\mathbf{x}(t)^T \mathbf{S}(t) \mathbf{x}(t) \tag{7.33}$$

for $\mathbf{S}(t)$ antisymmetric, we can rewrite Equation (7.30) as

$$-\mathbf{x}(t)^T \dot{\mathbf{P}}(t) \mathbf{x}(t) = \mathbf{x}(t)^T \left[\mathbf{Q}(t) + \mathbf{P}(t)\mathbf{A}(t) + \mathbf{A}(t)^T \mathbf{P}(t) \\ -\mathbf{P}(t)\mathbf{B}(t)\mathbf{R}(t)^{-1}\mathbf{B}(t)^T \mathbf{P}(t) \right] \mathbf{x}(t). \tag{7.34}$$

It is now clear that we have obtained a solution of Equations (7.23) and (7.24), the sufficient conditions for optimality, if $\mathbf{P}(t)$ can be chosen to satisfy:

$$-\dot{\mathbf{P}}(t) = \mathbf{Q}(t) + \mathbf{P}(t)\mathbf{A}(t) + \mathbf{A}(t)^T \mathbf{P}(t) - \mathbf{P}(t)\mathbf{B}(t)\mathbf{R}(t)^{-1}\mathbf{B}(t)^T \mathbf{P}(t) \tag{7.35}$$

for $t \in [t_0, T]$ with

$$\mathbf{P}(T) = \mathbf{M}. \tag{7.36}$$

If a solution $\mathbf{P}(t)$ to Equations (7.35) and (7.36) can be found, the optimal control is given, from Equation (7.26), by

$$\mathbf{u}^*(t) = - \underbrace{\mathbf{R}(t)^{-1}\mathbf{B}(t)^T \mathbf{P}(t)}_{\mathbf{K}(t)} \mathbf{x}(t), \tag{7.37}$$

which represents a time-varying state feedback control law. To implement this control law, $P(t)$ must be precomputed and stored starting from the boundary condition in Equation (7.36), and satisfying Equation (7.35) for $t \in [t_0, T]$.

7.2.1 Solution of the Matrix Ricatti Differential Equation

The solution of the nonlinear matrix differential equation in Equation (7.35), known as the Ricatti differential equation, can be obtained by considering the associated linear matrix differential equation:

$$\begin{bmatrix} \dot{\mathbf{X}}(t) \\ \dot{\mathbf{Y}}(t) \end{bmatrix} = \begin{bmatrix} \mathbf{A}(t) & -\mathbf{B}(t)\mathbf{R}(t)^{-1}\mathbf{B}(t)^T \\ -\mathbf{Q}(t) & -\mathbf{A}(t)^T \end{bmatrix} \begin{bmatrix} \mathbf{X}(t) \\ \mathbf{Y}(t) \end{bmatrix}, \tag{7.38a}$$

$$\begin{bmatrix} \mathbf{X}(T) \\ \mathbf{Y}(T) \end{bmatrix} = \begin{bmatrix} \mathbf{I} \\ \mathbf{M} \end{bmatrix}, \tag{7.38b}$$

and setting

$$\mathbf{P}(t) = \mathbf{Y}(t)\mathbf{X}(t)^{-1}. \tag{7.39}$$

The verification that Equation (7.39) satisfies Equation (7.35) is straightforward and is left to the reader.

The solution of Equation (7.38a) can be represented in terms of its state transition matrix $\Phi(t,T)$, appropriately partitioned:

$$\begin{bmatrix} \mathbf{X}(t)] \\ \mathbf{Y}(t) \end{bmatrix} = \underbrace{\begin{bmatrix} \Phi_{11}(t,T) & \Phi_{12}(t,T) \\ \Phi_{21}(t,T) & \Phi_{22}(t,T) \end{bmatrix}}_{\Phi(t,T)} \begin{bmatrix} \mathbf{I} \\ \mathbf{M} \end{bmatrix} \tag{7.40}$$

so that

$$\mathbf{P}(t) = [\Phi_{21}(t,T) + \Phi_{22}(t,T)\mathbf{M}] [\Phi_{11}(t,T) + \Phi_{12}(t,T)\mathbf{M}]^{-1}. \tag{7.41}$$

7.2.2 Accommodating Cross-Product Terms

A slightly more general LQR problem can be formulated by considering the performance index

$$I = \int_{t_0}^{T} \begin{bmatrix} \mathbf{x}^T & \mathbf{u}^T \end{bmatrix} \begin{bmatrix} \mathbf{Q} & \mathbf{S} \\ \mathbf{S}^T & \mathbf{R} \end{bmatrix} \begin{bmatrix} \mathbf{x} \\ \mathbf{u} \end{bmatrix} dt + \theta(\mathbf{x}(T), T)$$

$$= \int_{t_0}^{T} \left(\mathbf{x}^T \mathbf{Q} \mathbf{x} + 2\mathbf{x}^T \mathbf{S} \mathbf{u} + \mathbf{u}^T \mathbf{R} \mathbf{u} \right) dt + \theta(\mathbf{x}(T), T). \tag{7.42}$$

To reduce this to the standard problem, note that

$$\mathbf{x}^T \mathbf{Q} \mathbf{x} + \mathbf{u}^T \mathbf{R} \mathbf{u} + 2\mathbf{x}^T \mathbf{S} \mathbf{u}$$

$$= \left(\mathbf{u} + \mathbf{R}^{-1} \mathbf{S}^T \mathbf{x} \right)^T \mathbf{R} \left(\mathbf{u} + \mathbf{R}^{-1} \mathbf{S}^T \mathbf{x} \right) + \mathbf{x}^T \left(\mathbf{Q} - \mathbf{S} \mathbf{R}^{-1} \mathbf{S}^T \right) \mathbf{x}. \tag{7.43}$$

Assume that

$$\mathbf{Q} - \mathbf{S} \mathbf{R}^{-1} \mathbf{S}^T \geq 0 \tag{7.44}$$

and define

$$\bar{\mathbf{u}} := \mathbf{u} + \mathbf{R}^{-1} \mathbf{S}^T \mathbf{x}. \tag{7.45}$$

Then

$$\dot{\mathbf{x}}(t) = \left(\mathbf{A} - \mathbf{B}\mathbf{R}^{-1}\mathbf{S}^T\right)\mathbf{x}(t) + \mathbf{B}\bar{\mathbf{u}}, \tag{7.46}$$

$$I = \int_{t_0}^{T} \left(\mathbf{x}^T \left(\mathbf{Q} - \mathbf{S}\mathbf{R}^{-1}\mathbf{S}^T\right)\mathbf{x} + \bar{\mathbf{u}}^T\mathbf{R}\bar{\mathbf{u}}\right)dt + \theta(\mathbf{x}(T), T), \tag{7.47}$$

so that the optimal $\bar{\mathbf{u}}(t)$ is

$$\bar{\mathbf{u}}(t) = -\mathbf{R}^{-1}\mathbf{B}^T\mathbf{P}(t)\mathbf{x}(t) = -\mathbf{K}(t)\mathbf{x}(t), \tag{7.48}$$

$$-\dot{\mathbf{P}} = \mathbf{P}\left(\mathbf{A} - \mathbf{B}\mathbf{R}^{-1}\mathbf{S}^T\right) + \left(\mathbf{A} - \mathbf{B}\mathbf{R}^{-1}\mathbf{S}^T\right)^T\mathbf{P} + \left(\mathbf{Q} - \mathbf{S}\mathbf{R}^{-1}\mathbf{S}^T\right) - \mathbf{P}\mathbf{B}\mathbf{R}^{-1}\mathbf{B}^T\mathbf{P}. \tag{7.49}$$

Therefore, the optimal control $\mathbf{u}(t)$ is

$$\mathbf{u}(t) = -\mathbf{R}(t)^{-1}\left(\mathbf{B}(t)^T\mathbf{P}(t) + \mathbf{S}(t)^T\right)\mathbf{x}(t) \tag{7.50}$$

and the optimal performance is

$$\mathbf{x}(t_0)^T\mathbf{P}(t_0)\mathbf{x}(t_0). \tag{7.51}$$

7.3   The Infinite-Horizon LQR Problem

We first consider a general infinite-horizon optimal control problem before specializing to the linear quadratic case.

7.3.1   General Conditions for Optimality

For the infinite-horizon problem, the cost functional assumes the form

$$I(\mathbf{x}(0), \mathbf{u}[0, \infty)) = \int_0^\infty \psi(\mathbf{x}(t), \mathbf{u}(t))dt. \tag{7.52}$$

The system dynamics are described by

$$\dot{\mathbf{x}}(t) = f(\mathbf{x}(t), \mathbf{u}(t)), \quad \text{with } \mathbf{x}(0) \text{ given} \tag{7.53}$$

and we seek feedback controls of the form

$$\mathbf{u}(t) = \mu(\mathbf{x}(t)) \tag{7.54}$$

which minimize the cost I in Equation (7.52), which can be rewritten as

$$I(\mathbf{x}(0), \mu) = \int_0^\infty \psi(\mathbf{x}(t), \mu(\mathbf{x}(t)))dt. \tag{7.55}$$

The closed-loop optimal system is also required to be asymptotically stable. This means that we require

$$\lim_{t \to \infty} \mathbf{x}(t) = 0, \quad \text{for all } \mathbf{x}(0). \tag{7.56}$$

Thus, we introduce the family Ω of admissible control laws $\mu(\cdot)$ which are continuous functions of $\mathbf{x}(t)$, for which Equation (7.53) has a unique and continuously differentiable solution $\mathbf{x}(t)$ and for which Equation (7.56) holds. The optimal control problem is to find a control law $\mu^*(\mathbf{x}(t)) \in \Omega$ that minimizes the integral in Equation (7.52):

$$\mu^* = \text{Arg} \inf_{\mu \in \Omega} \int_0^\infty \psi(\mathbf{x}(t), \mu(\mathbf{x}(t)))dt. \tag{7.57}$$

Let us now consider the value function approach for this problem and define:

$$V(\mathbf{x}(0)) := \inf_{\mu \in \Omega} I(\mathbf{x}(0), \mu). \tag{7.58}$$

This suggests that we search for the general value function $V(\mathbf{x}(t))$ with the given problem being the evaluation of V at $\mathbf{x} = \mathbf{x}(0)$. The following result provides the basis for this approach.

THEOREM 7.2 If there exists a control law $\mathbf{u} = \mu^*(\mathbf{x}(t))$ and a continuously differentiable $V(\mathbf{x})$ such that

$$0 \le V(\mathbf{x}) \le \mathbf{x}^T \mathbf{T} \mathbf{x} \quad \text{for some } \mathbf{T} = \mathbf{T}^T > 0 \quad \text{for all } \mathbf{x}; \tag{7.59}$$

(a)

$$\frac{\partial V^T}{\partial \mathbf{x}} f(\mathbf{x}(t), \mu^*(\mathbf{x}(t))) + \psi(\mathbf{x}(t), \mu^*(\mathbf{x}(t))) = 0, \quad \text{for all } \mathbf{x}; \tag{7.60}$$

(b)

$$\frac{\partial V^T}{\partial \mathbf{x}} f(\mathbf{x}(t), \mathbf{u}) + \psi(\mathbf{x}(t), \mathbf{u}(t)) \ge 0 \quad \text{for all } \mathbf{x}, \mathbf{u}; \tag{7.61}$$

that is

$$J\left(\mathbf{x}, \frac{\partial V}{\partial \mathbf{x}}, \mathbf{u}\right) \ge J\left(\mathbf{x}, \frac{\partial V}{\partial \mathbf{x}}, \mu^*(\mathbf{x})\right) = 0, \tag{7.62}$$

then $\mu^*(\mathbf{x}(t))$ is the optimal control minimizing Equation (7.52).

Proof Let $\mathbf{u}(t) = \mu^*(\mathbf{x}(t))$. Then

$$\frac{dV}{dt} \mathbf{x}(t, \mathbf{x}(0), \mu^*) = \frac{\partial V^T}{\partial \mathbf{x}} f(\mathbf{x}, \mu^*) + \underbrace{\frac{\partial V}{\partial t}}_{=0}$$

$$= \frac{\partial V^T}{\partial \mathbf{x}} f(\mathbf{x}, \mu^*) \tag{7.63}$$

$$= -\psi(\mathbf{x}, \mu^*(\mathbf{x})) \quad \text{(by Equation (7.60))}.$$

Integrating from 0 to τ:

$$V(\mathbf{x}(\tau)) - V(\mathbf{x}(0)) = -\int_0^\tau \psi(\mathbf{x}, \mu^*(\mathbf{x}))dt. \tag{7.64}$$

Since $V(\mathbf{x}(\tau)) \le \mathbf{x}(\tau)^T \mathbf{T} \mathbf{x}(\tau)$ and $\mathbf{x}(\tau) \to 0$, it follows that

$$\lim_{\tau \to \infty} V(\mathbf{x}(\tau)) = 0 \tag{7.65}$$

and therefore

$$V(\mathbf{x}(0)) = I(\mathbf{x}(0), \boldsymbol{\mu}^*). \tag{7.66}$$

Now let $\mathbf{u}(t) = \boldsymbol{\mu}(\mathbf{x}(t))$ be an arbitrary admissible control law and let $\mathbf{x}(t, \mathbf{x}(0), \boldsymbol{\mu})$ denote the corresponding solution of Equation (7.53). Integrating Equation (7.61) with $\mathbf{u}(t) = \boldsymbol{\mu}(\mathbf{x}(t))$, we obtain

$$V(\mathbf{x}(\tau)) - V(\mathbf{x}(0)) = \int_0^\tau \frac{\partial V^T}{\partial \mathbf{x}} f(\mathbf{x}, \boldsymbol{\mu}(\mathbf{x})) dt \geq - \int_0^\tau \psi(\mathbf{x}(t), \boldsymbol{\mu}(\mathbf{x}(t))) dt \tag{7.67}$$

or

$$V(\mathbf{x}(0)) \leq V(\mathbf{x}(\tau)) + \int_0^\tau \psi(\mathbf{x}(t), \boldsymbol{\mu}(\mathbf{x}(t))) dt. \tag{7.68}$$

Letting $\tau \to \infty$ and using Equations (7.65) and (7.66), we obtain

$$I(\mathbf{x}(0), \boldsymbol{\mu}^*) \leq I(\mathbf{x}(0), \boldsymbol{\mu}) \tag{7.69}$$

so that

$$V(\mathbf{x}(0)) = \inf_{\boldsymbol{\mu} \in \Omega} I(\mathbf{x}(0), \boldsymbol{\mu}). \tag{7.70}$$

∎

In the next subsection we apply these results to the infinite-horizon LQR.

7.3.2 The Infinite-Horizon LQR Problem

The infinite-horizon LQR problem considers the linear *time-invariant* plant

$$\dot{\mathbf{x}}(t) = \mathbf{A}\mathbf{x}(t) + \mathbf{B}\mathbf{u}(t) \tag{7.71}$$

and the time-invariant quadratic cost functional

$$I(\mathbf{x}(0), \mathbf{u}[0, \infty)) = \int_0^\infty \left[\mathbf{x}(t)^T \mathbf{Q}\mathbf{x}(t) + \mathbf{u}(t)^T \mathbf{R}\mathbf{u}(t) \right] dt \tag{7.72}$$

with

$$\mathbf{Q} = \mathbf{Q}^T \geq 0 \quad \text{and} \quad \mathbf{R} = \mathbf{R}^T > 0. \tag{7.73}$$

Following the approach of the previous theorem, we search for functions $V(\mathbf{x})$ and $\boldsymbol{\mu}^*(\mathbf{x})$ satisfying Equations (7.60) and (7.61), the sufficient conditions for optimality:

$$\frac{\partial V^T}{\partial \mathbf{x}} (\mathbf{A}\mathbf{x} + \mathbf{B}\boldsymbol{\mu}^*(\mathbf{x})) + \mathbf{x}^T \mathbf{Q}\mathbf{x} + (\boldsymbol{\mu}^*(\mathbf{x}))^T \mathbf{R}\boldsymbol{\mu}^*(\mathbf{x}) = 0 \tag{7.74}$$

and for arbitrary admissible $\boldsymbol{\mu}(\mathbf{x}(t))$:

$$\frac{\partial V^T}{\partial \mathbf{x}} (\mathbf{A}\mathbf{x} + \mathbf{B}\boldsymbol{\mu}) + \mathbf{x}^T \mathbf{Q}\mathbf{x} + \boldsymbol{\mu}^T \mathbf{R}\boldsymbol{\mu} \geq 0. \tag{7.75}$$

It is easy to see that Equations (7.74) and (7.75) are satisfied by

$$V(\mathbf{x}) = \mathbf{x}^T \mathbf{P}\mathbf{x} \tag{7.76}$$

and

$$\mu^*(\mathbf{x}) = -\mathbf{R}^{-1}\mathbf{B}^T\mathbf{P}\mathbf{x} \qquad (7.77)$$

provided

$$\mathbf{A}^T\mathbf{P} + \mathbf{P}\mathbf{A} + \mathbf{Q} - \mathbf{P}\mathbf{B}\mathbf{R}^{-1}\mathbf{B}^T\mathbf{P} = 0. \qquad (7.78)$$

The above equation is the ARE. In the next section we discuss its solution and thus the solution of the optimal LQR.

7.4 Solution of the Algebraic Riccati Equation

The main result of this section gives conditions for the existence of stabilizing solutions of the ARE:

$$\mathbf{A}^T\mathbf{P} + \mathbf{P}\mathbf{A} - \mathbf{P}\mathbf{B}\mathbf{R}^{-1}\mathbf{B}^T\mathbf{P} + \underbrace{\mathbf{C}^T\mathbf{C}}_{\mathbf{Q}} = 0 \qquad (7.79)$$

and the computation of these solutions.

THEOREM 7.3 If $(\mathbf{A}, \mathbf{B})$ is stabilizable and $(\mathbf{C}, \mathbf{A})$ is detectable, then the ARE has a unique positive semi-definite solution $\mathbf{P}$, the optimal control for the corresponding LQR problem is

$$\mathbf{u}^*(t) = -\mathbf{R}^{-1}\mathbf{B}^T\mathbf{P}\mathbf{x}(t), \qquad (7.80)$$

and $(\mathbf{A} - \mathbf{B}\mathbf{R}^{-1}\mathbf{B}^T\mathbf{P})$ is stable.

To prove this result, we develop some auxiliary machinery. First, the ARE in Equation (7.79) can be written in the following three equivalent ways:

$$\begin{bmatrix} -\mathbf{P} & \mathbf{I} \end{bmatrix} \begin{bmatrix} \mathbf{A} & -\mathbf{B}\mathbf{R}^{-1}\mathbf{B}^T \\ -\mathbf{Q} & -\mathbf{A}^T \end{bmatrix} \begin{bmatrix} \mathbf{I} \\ \mathbf{P} \end{bmatrix} = 0, \qquad (7.81)$$

$$\begin{aligned} (\mathbf{A} - \mathbf{B}\mathbf{K})^T\mathbf{P} + \mathbf{P}(\mathbf{A} - \mathbf{B}\mathbf{K}) &= -\mathbf{Q} - \mathbf{P}\mathbf{B}\mathbf{R}^{-1}\mathbf{B}^T\mathbf{P}, \\ \mathbf{K} &= \mathbf{R}^{-1}\mathbf{B}^T\mathbf{P}, \end{aligned} \qquad (7.82)$$

$$\underbrace{\begin{bmatrix} \mathbf{A} & -\mathbf{B}\mathbf{R}^{-1}\mathbf{B}^T \\ -\mathbf{Q} & -\mathbf{A}^T \end{bmatrix}}_{\mathbf{H}} \begin{bmatrix} \mathbf{I} \\ \mathbf{P} \end{bmatrix} = \begin{bmatrix} \mathbf{I} \\ \mathbf{P} \end{bmatrix} (\mathbf{A} - \mathbf{B}\mathbf{K}). \qquad (7.83)$$

The matrix $\mathbf{H}$ is called the *associated Hamiltonian*. Let $\sigma(\mathbf{H})$ denote the spectrum or eigenvalue set of $\mathbf{H}$.

LEMMA 7.1 $\sigma(\mathbf{H})$ is symmetric about the imaginary axis.

Proof **H** is similar to $-\mathbf{H}^T$ since with

$$\mathbf{J} = \begin{bmatrix} 0 & -\mathbf{I} \\ \mathbf{I} & 0 \end{bmatrix}, \tag{7.84}$$

we have

$$\mathbf{J}^{-1}\mathbf{H}\mathbf{J} = -\mathbf{H}^T. \tag{7.85}$$

Therefore, if $\lambda \in \sigma(\mathbf{H})$, $-\lambda^*$ is also in $\sigma(\mathbf{H})$. ∎

Therefore, if **H** has no $j\omega$ eigenvalues, it has n eigenvalues each in the open left-half plane and open right-half plane, respectively. Suppose **H** has no $j\omega$ eigenvalues. Let the n-dimensional stable eigenspace of **H** be denoted

$$\mathbf{V} = \mathrm{Im} \underbrace{\begin{bmatrix} \mathbf{X}_1 \\ \mathbf{X}_2 \end{bmatrix}}_{2n \times n} \tag{7.86}$$

where $\mathbf{X}_1$, $\mathbf{X}_2$ are in general complex. Since **H** is real, it is possible to choose a basis so that $\mathbf{X}_1$, $\mathbf{X}_2$ are rendered real.

LEMMA 7.2 Suppose that **H**

(a) has no $j\omega$ eigenvalues, and
(b) has a stable eigenspace with $\mathbf{X}_1$ real and invertible and $\mathbf{X}_2$ real,

then

$$\mathbf{P} = \mathbf{X}_2\mathbf{X}_1^{-1} \tag{7.87}$$

is a solution of the ARE and $(\mathbf{A} - \mathbf{B}\mathbf{R}^{-1}\mathbf{B}^T\mathbf{P})$ is stable.

Proof We have, for some matrix $\mathbf{H}_-$:

$$\begin{bmatrix} \mathbf{A} & -\mathbf{B}\mathbf{R}^{-1}\mathbf{B}^T \\ -\mathbf{Q} & -\mathbf{A}^T \end{bmatrix} \begin{bmatrix} \mathbf{X}_1 \\ \mathbf{X}_2 \end{bmatrix} = \begin{bmatrix} \mathbf{X}_1 \\ \mathbf{X}_2 \end{bmatrix} \mathbf{H}_-, \tag{7.88}$$

where $\sigma(\mathbf{H}_-)$ is in the open left-half plane. Then

$$\begin{bmatrix} \mathbf{A} & -\mathbf{B}\mathbf{R}^{-1}\mathbf{B}^T \\ -\mathbf{Q} & \mathbf{A}^T \end{bmatrix} \begin{bmatrix} \mathbf{I} \\ \mathbf{X}_2\mathbf{X}_1^{-1} \end{bmatrix} = \begin{bmatrix} \mathbf{I} \\ \mathbf{X}_2\mathbf{X}_1^{-1} \end{bmatrix} \mathbf{X}_1\mathbf{H}_-\mathbf{X}_1^{-1} \tag{7.89}$$

and multiplying Equation (7.89) on the left by

$$\begin{bmatrix} -\mathbf{X}_2\mathbf{X}_1^{-1} & I \end{bmatrix}, \tag{7.90}$$

we obtain

$$\begin{bmatrix} -\mathbf{X}_2\mathbf{X}_1^{-1} & \mathbf{I} \end{bmatrix} \begin{bmatrix} \mathbf{A} & -\mathbf{B}\mathbf{R}^{-1}\mathbf{B}^T \\ -\mathbf{Q} & -\mathbf{A}^T \end{bmatrix} \begin{bmatrix} I \\ \mathbf{X}_2\mathbf{X}_1^{-1} \end{bmatrix} = 0 \tag{7.91}$$

so that

$$\mathbf{P} = \mathbf{X}_2\mathbf{X}_1^{-1} \tag{7.92}$$

solves the ARE from Equation (7.81). Also

$$\mathbf{A} - \mathbf{BR}^{-1}\mathbf{B}^T \underbrace{\left(\mathbf{X}_2\mathbf{X}_1^{-1}\right)}_{\mathbf{P}} = \mathbf{X}_1\mathbf{H}_-\mathbf{X}_1^{-1} \tag{7.93}$$

so that

$$\sigma\left(\mathbf{A} - \mathbf{BR}^{-1}\mathbf{B}^T\mathbf{P}\right) = \sigma(\mathbf{H}_-). \tag{7.94}$$

∎

Under an arbitrary change of basis:

$$\mathscr{V} = \text{Im}\begin{bmatrix} \mathbf{X}_1 \\ \mathbf{X}_2 \end{bmatrix} \mathbf{T} = \text{Im}\begin{bmatrix} \mathbf{X}_1\mathbf{T} \\ \mathbf{X}_2\mathbf{T} \end{bmatrix}, \tag{7.95}$$

where $\mathbf{T}$ is invertible, and so the solution P is unchanged, since

$$\mathbf{P} = (\mathbf{X}_2\mathbf{T})(\mathbf{X}_1\mathbf{T})^{-1} = \mathbf{X}_2\mathbf{X}_1^{-1}. \tag{7.96}$$

Thus, under the assumptions of Lemma 7.2, the solution

$$\mathbf{P} = \mathbf{X}_2\mathbf{X}_1^{-1} \tag{7.97}$$

is unique and is *stabilizing*, that is, $\sigma(\mathbf{A} - \mathbf{BR}^{-1}\mathbf{B}^T\mathbf{P})$ is in the open left-half plane.

To proceed let $\mathbf{U}^*$ denote the complex conjugate transpose of the matrix $\mathbf{U}$.

LEMMA 7.3   Under the assumptions of Lemma 7.2, except that $\mathbf{X}_1$, $\mathbf{X}_2$ may now be complex, we have:

(a) $\mathbf{X}_1^*\mathbf{X}_2$ is Hermitian, that is $(\mathbf{X}_1^*\mathbf{X}_2)^* = \mathbf{X}_1^*\mathbf{X}_2$;
(b) $\mathbf{P} = \mathbf{X}_2\mathbf{X}_1^{-1}$ is Hermitian, that is $\mathbf{P} = \mathbf{P}^*$.

Proof   We know that

$$\mathbf{H}\begin{bmatrix} \mathbf{X}_1 \\ \mathbf{X}_2 \end{bmatrix} = \begin{bmatrix} \mathbf{X}_1 \\ \mathbf{X}_2 \end{bmatrix}\mathbf{H}_-. \tag{7.98}$$

Pre-multiply by $\begin{bmatrix} \mathbf{X}_1^* & \mathbf{X}_2^* \end{bmatrix}\mathbf{J}$ to get

$$\begin{bmatrix} \mathbf{X}_1^* & \mathbf{X}_2^* \end{bmatrix}\mathbf{JH}\begin{bmatrix} \mathbf{X}_1 \\ \mathbf{X}_2 \end{bmatrix} = \begin{bmatrix} \mathbf{X}_1^* & \mathbf{X}_2^* \end{bmatrix}\mathbf{J}\begin{bmatrix} \mathbf{X}_1 \\ \mathbf{X}_2 \end{bmatrix}\mathbf{H}_-. \tag{7.99}$$

Now

$$\mathbf{JH} = \begin{bmatrix} 0 & -\mathbf{I} \\ \mathbf{I} & 0 \end{bmatrix}\begin{bmatrix} \mathbf{A} & -\mathbf{BR}^{-1}\mathbf{B}^T \\ -\mathbf{Q} & -\mathbf{A}^T \end{bmatrix} = \begin{bmatrix} \mathbf{Q} & \mathbf{A}^T \\ \mathbf{A} & -\mathbf{BR}^{-1}\mathbf{B}^T \end{bmatrix} \tag{7.100}$$

is symmetric. Therefore, both sides of Equation (7.99) are Hermitian so that

$$\underbrace{\left(-\mathbf{X}_1^*\mathbf{X}_2 + \mathbf{X}_2^*\mathbf{X}_1\right)}_{\mathbf{X}_3}\underbrace{\mathbf{H}_-}_{\mathbf{A}_3} = \mathbf{H}_-^*\begin{bmatrix} \mathbf{X}_1^* & \mathbf{X}_2^* \end{bmatrix}\begin{bmatrix} \mathbf{X}_2 \\ -\mathbf{X}_1 \end{bmatrix} \tag{7.101}$$

$$= -\mathbf{H}_-^*\left(-\mathbf{X}_1^*\mathbf{X}_2 + \mathbf{X}_2^*\mathbf{X}_1\right).$$

Equation (7.101) is of the type

$$\mathbf{X}_3 \mathbf{A}_3 + \mathbf{A}_3^* \mathbf{X}_3 = 0 \tag{7.102}$$

and since $\sigma(\mathbf{A}_3) \subset \mathbb{C}^-$, the unique solution is $\mathbf{X}_3 = 0$. Therefore

$$\mathbf{X}_1^* \mathbf{X}_2 = \mathbf{X}_2^* \mathbf{X}_1 \tag{7.103}$$

and

$$\mathbf{P} = \mathbf{X}_2 \mathbf{X}_1^{-1} \tag{7.104}$$

is Hermitian, since

$$\mathbf{P}^* = \left(\mathbf{X}_1^{-1}\right)^* \left(\mathbf{X}_1^* \mathbf{X}_2\right) \mathbf{X}_1^{-1} \tag{7.105}$$

is Hermitian again by Equation (7.103). ∎

LEMMA 7.4   If $\mathbf{H}$ has no $j\omega$ eigenvalues, then $\mathbf{X}_1$ is invertible if and only if $(\mathbf{A}, \mathbf{B})$ is stabilizable.

Proof   If $\mathbf{X}_1$ is invertible, $\mathbf{P} = \mathbf{X}_2 \mathbf{X}_1^{-1}$ is a stabilizing solution so $\mathbf{A} - \mathbf{B} \mathbf{R}^{-1} \mathbf{B}^T \mathbf{P}$ is stable, implying that $(\mathbf{A}, \mathbf{B})$ must be stabilizable. Conversely, let

$$\mathbf{H} \begin{bmatrix} \mathbf{X}_1 \\ \mathbf{X}_2 \end{bmatrix} = \begin{bmatrix} \mathbf{X}_1 \\ \mathbf{X}_2 \end{bmatrix} \mathbf{H}_-. \tag{7.106}$$

We now need to show that $\mathbf{X}_1$ must be invertible. Arguing by contradiction, suppose that there exists $\mathbf{x} \neq 0$ such that

$$\mathbf{X}_1 \mathbf{x} = 0 \quad \text{or} \quad \mathbf{x}^* \mathbf{X}_1^* = 0 \quad \text{or} \quad \mathbf{x} \in \text{Ker}\,[\mathbf{X}_1]. \tag{7.107}$$

Since

$$\mathbf{A} \mathbf{X}_1 - \mathbf{B} \mathbf{R}^{-1} \mathbf{B}^T \mathbf{X}_2 = \mathbf{X}_1 \mathbf{H}_-, \tag{7.108}$$

by multiplying on the left by $\mathbf{x}^* \mathbf{X}_2^*$ and on the right by $\mathbf{x}$, we get

$$-\mathbf{x}^* \mathbf{X}_2^* \mathbf{B} \mathbf{R}^{-1} \mathbf{B}^T \mathbf{X}_2 \mathbf{x} = \mathbf{x}^* \mathbf{X}_2^* \mathbf{X}_1 \mathbf{H}_- \mathbf{x}$$
$$= 0 \tag{7.109}$$

using Equation (7.103). Therefore

$$\mathbf{x}^* \mathbf{X}_2^* \mathbf{B} \mathbf{R}^{-1} \mathbf{B}^T \mathbf{X}_2 \mathbf{x} = 0 \tag{7.110}$$

or

$$\mathbf{B}^T \mathbf{X}_2 \mathbf{x} = 0. \tag{7.111}$$

From Equation (7.108), we then have

$$\mathbf{X}_1 \mathbf{H}_- \mathbf{x} = 0 \tag{7.112}$$

proving that

$$\mathbf{H}_-\left(\text{Ker}\,[\mathbf{X}_1]\right) \subset \text{Ker}\,[\mathbf{X}_1]. \tag{7.113}$$

If $\text{Ker}\,[\mathbf{X}_1] \neq 0$, there exists $\mathbf{x} \neq 0$ such that $\mathbf{x} \in \text{Ker}\,[\mathbf{X}_1]$ and

$$\mathbf{H}_-\mathbf{x} = \lambda\mathbf{x} \tag{7.114}$$

with $\lambda \in \mathbb{C}^-$. Using

$$-\mathbf{Q}\mathbf{X}_1 - \mathbf{A}^T\mathbf{X}_2 = \mathbf{X}_2\mathbf{H}_-, \tag{7.115}$$

post-multiplying by $\mathbf{x}$ above, and using Equation (7.114), we obtain

$$\left(\mathbf{A}^T + \lambda\mathbf{I}\right)\mathbf{X}_2\mathbf{x} = 0 \tag{7.116}$$

or

$$\mathbf{x}^*\mathbf{X}_2^*\left(\mathbf{A} + \lambda^*\mathbf{I}\right) = 0. \tag{7.117}$$

Combining Equations (7.111) and (7.117), we obtain

$$\mathbf{x}^*\mathbf{X}_2^*\begin{bmatrix} \mathbf{A} - (-\lambda^*)\mathbf{I} & \mathbf{B} \end{bmatrix} = 0. \tag{7.118}$$

Since $-\lambda^* \in \mathbb{C}^+$, it follows from the stabilizability of $(\mathbf{A}, \mathbf{B})$ that $\mathbf{x}^*\mathbf{X}_2^* = 0$, and hence

$$\begin{bmatrix} \mathbf{X}_1 \\ \mathbf{X}_2 \end{bmatrix}\mathbf{x} = 0. \tag{7.119}$$

However, $\begin{bmatrix} \mathbf{X}_1 \\ \mathbf{X}_2 \end{bmatrix}$ has rank n and so $\mathbf{x} = 0$. Thus, $\text{Ker}\,[\mathbf{X}_1] = 0$, that is $\mathbf{X}_1$ is invertible.

■

LEMMA 7.5 $\mathbf{H}$ has no $j\omega$ axis eigenvalues and $\mathbf{X}_1$ is nonsingular if and only if $(\mathbf{A}, \mathbf{B})$ is stabilizable and $(\mathbf{C}, \mathbf{A})$ has no $j\omega$ axis unobservable eigenvalues. Also the solution

$$\mathbf{P} = \mathbf{X}_2\mathbf{X}_1^{-1} \tag{7.120}$$

is positive semi-definite.

Proof   It has already been shown that if $\mathbf{H}$ has no $j\omega$ axis eigenvalues, then $\mathbf{X}_1$ is nonsingular if and only if $(\mathbf{A}, \mathbf{B})$ is stabilizable. It remains to prove that $\mathbf{H}$ has no $j\omega$ eigenvalues if and only if $(\mathbf{C}, \mathbf{A})$ has no $j\omega$ unobservable eigenvalues. Suppose, by contradiction, that $j\omega_0$ is an eigenvalue of $\mathbf{H}$, with eigenvector

$$\begin{bmatrix} \mathbf{x} \\ \mathbf{z} \end{bmatrix} \neq 0. \tag{7.121}$$

Then

$$\begin{bmatrix} \mathbf{A} & -\mathbf{B}\mathbf{R}^{-1}\mathbf{B}^T \\ -\mathbf{C}^T\mathbf{C} & -\mathbf{A}^T \end{bmatrix}\begin{bmatrix} \mathbf{x} \\ \mathbf{z} \end{bmatrix} = j\omega_0\begin{bmatrix} \mathbf{x} \\ \mathbf{z} \end{bmatrix}, \tag{7.122}$$

and

$$\begin{aligned} (\mathbf{A} - j\omega_0\mathbf{I})\mathbf{x} &= \mathbf{B}\mathbf{R}^{-1}\mathbf{B}^T\mathbf{z}, \\ -(\mathbf{A} + j\omega_0\mathbf{I})^T\mathbf{z} &= \mathbf{C}^T\mathbf{C}\mathbf{x}. \end{aligned} \tag{7.123}$$

Then

$$
\mathbf{z}^{*}\left(\mathbf{A}-j\omega_{0}\mathbf{I}\right)\mathbf{x}=\mathbf{z}^{*}\mathbf{B}\mathbf{R}^{-1}\mathbf{B}^{T}\mathbf{z}\geq 0\ \ (\text{real}),
$$
$$
\mathbf{x}^{*}\left(\mathbf{A}+j\omega_{0}\mathbf{I}\right)^{T}\mathbf{z}=-\mathbf{x}^{*}\mathbf{C}^{T}\mathbf{C}\mathbf{x}\leq 0\ \ (\text{real}).
\tag{7.124}
$$

Therefore, since the left-hand sides are equal, we must have

$$
\mathbf{B}^{T}\mathbf{z}=0,
$$
$$
\mathbf{C}\mathbf{x}=0,
\tag{7.125}
$$

and so

$$
\mathbf{z}^{T}\left[\ \mathbf{A}-(-j\omega_{0})\mathbf{I}\quad \mathbf{B}\ \right]=0,
$$
$$
\left[\begin{array}{c}\mathbf{A}-j\omega_{0}\mathbf{I}\\ \mathbf{C}\end{array}\right]\mathbf{x}=0.
\tag{7.126}
$$

Since $(\mathbf{A},\mathbf{B})$ is stabilizable, $\mathbf{z}=0$ and since $(\mathbf{C},\mathbf{A})$ has no $j\omega$ unobservable modes, $\mathbf{x}=0$. Thus, $j\omega_{0}$ cannot be an eigenvalue of $\mathbf{H}$. To prove that $\mathbf{P}=\mathbf{X}_{2}\mathbf{X}_{1}^{-1}$ is positive semi-definite, note that $\mathbf{P}$ satisfies the Lyapunov equation

$$
(\mathbf{A}-\mathbf{B}\mathbf{K})^{T}\mathbf{P}+\mathbf{P}(\mathbf{A}-\mathbf{B}\mathbf{K})=-\mathbf{Q}-\mathbf{P}\mathbf{B}\mathbf{R}^{-1}\mathbf{B}^{T}\mathbf{P},
\tag{7.127}
$$

with $\mathbf{A}-\mathbf{B}\mathbf{K}$ stable and thus the standard solution of the Lyapunov equation yields

$$
\mathbf{P}=\int_{0}^{\infty}e^{(\mathbf{A}-\mathbf{B}\mathbf{K})^{T}t}\left(\mathbf{Q}+\mathbf{P}\mathbf{B}\mathbf{R}^{-1}\mathbf{B}^{T}\mathbf{P}\right)e^{(\mathbf{A}-\mathbf{B}\mathbf{K})t}dt,
\tag{7.128}
$$

which is positive semi-definite since $\mathbf{Q}+\mathbf{P}\mathbf{B}\mathbf{R}^{-1}\mathbf{B}^{T}\mathbf{P}$ is positive semi-definite.   ∎

Proof of Theorem 7.3   Since $(\mathbf{A},\mathbf{B})$ is stabilizable and $(\mathbf{C},\mathbf{A})$ is detectable, $\mathbf{H}$ clearly has no $j\omega$ eigenvalues, and $\mathbf{X}_{1}$ is invertible. Suppose now that, $\mathbf{X}_{2}\mathbf{X}_{1}^{-1}=\mathbf{P}$ is not stabilizing, then

$$
\left(\mathbf{A}-\mathbf{B}\mathbf{R}^{-1}\mathbf{B}^{T}\mathbf{P}\right)\mathbf{x}=\lambda\mathbf{x},\qquad \text{for}\ \ \text{Re}(\lambda)\geq 0,\ \ \mathbf{x}\neq 0.
\tag{7.129}
$$

Since

$$
(\mathbf{A}-\mathbf{B}\mathbf{K})^{T}\mathbf{P}+\mathbf{P}(\mathbf{A}-\mathbf{B}\mathbf{K})+\mathbf{P}\mathbf{B}\mathbf{R}^{-1}\mathbf{B}^{T}\mathbf{P}+\mathbf{Q}=0,
\tag{7.130}
$$

pre-multiplying by $\mathbf{x}^{*}$ and post-multiplying by $\mathbf{x}$, we get

$$
(\lambda+\lambda^{*})\mathbf{x}^{*}\mathbf{P}\mathbf{x}+\mathbf{x}^{*}\mathbf{P}\mathbf{B}\mathbf{R}^{-1}\mathbf{B}^{T}\mathbf{P}\mathbf{x}+\mathbf{x}^{*}\mathbf{C}^{T}\mathbf{C}\mathbf{x}=0
\tag{7.131}
$$

so that

$$
\mathbf{B}^{T}\mathbf{P}\mathbf{x}=0,\qquad \mathbf{C}\mathbf{x}=0
\tag{7.132}
$$

and therefore

$$
\mathbf{A}\mathbf{x}=\lambda\mathbf{x},\qquad \mathbf{C}\mathbf{x}=0.
\tag{7.133}
$$

By detectability of $(\mathbf{C},\mathbf{A})$, we must have $\mathbf{x}=0$, so that λ with $\text{Re}(\lambda)\geq 0$ cannot be an eigenvalue of $\mathbf{A}-\mathbf{B}\mathbf{K}$.

It is now easy to prove optimality, that is the claim that $\mathbf{u}^* = -\mathbf{Kx}$ minimizes I. Note that

$$\int_0^\infty \frac{d}{dt} \left(\mathbf{x}(t)^T \mathbf{Px}(t) \right) dt = \mathbf{x}(t)^T \mathbf{Px}(t) \big|_{t\to\infty} - \mathbf{x}(t)^T \mathbf{Px}(t) \big|_{t=0}. \tag{7.134}$$

Let

$$\mathbf{u}(t) = -\mathbf{Kx}(t) + \mathbf{v}(t) \tag{7.135}$$

so that

$$\dot{\mathbf{x}}(t) = (\mathbf{A} - \mathbf{BK})\mathbf{x}(t) + \mathbf{Bv}(t) \tag{7.136}$$

and assume that $\mathbf{v}(t)$ is such that

$$\int_0^\infty \mathbf{v}(t)^T \mathbf{v}(t) dt < \infty, \tag{7.137}$$

and $\mathbf{x}(t) \to 0$ as $t \to \infty$. Now

$$I - \mathbf{x}(0)^T \mathbf{Px}(0) = \int_0^\infty \left(\mathbf{x}(t)^T \left[(\mathbf{A} - \mathbf{BK})^T \mathbf{P} + \mathbf{P}(\mathbf{A} - \mathbf{BK}) \right] \mathbf{x} \right.$$
$$\left. + \mathbf{x}(t)^T \left[\mathbf{Q} + \mathbf{PBR}^{-1}\mathbf{B}^T\mathbf{P} \right] \mathbf{x} + \mathbf{v}^T \mathbf{Rv} \right) dt. \tag{7.138}$$

Since

$$(\mathbf{A} - \mathbf{BK})^T \mathbf{P} + \mathbf{P}(\mathbf{A} - \mathbf{BK}) = -\mathbf{Q} - \mathbf{PBR}^{-1}\mathbf{B}^T\mathbf{P} \tag{7.139}$$

we have

$$I = \mathbf{x}(0)^T \mathbf{Px}(0) + \int_0^\infty \mathbf{v}^T \mathbf{Rv}\, dt, \tag{7.140}$$

which is minimized by setting $\mathbf{v} = 0$. ∎

7.5 The LQR as an Output Zeroing Problem

In this section we develop some geometric solvability conditions for the general LQR problem by regarding it as a problem of zeroing the output of a linear system. This existence condition will show that the unstable eigenspace of the open-loop system must be spanned or covered by the sum of the controllable subspace and the subspace unobservable from the performance index.

Consider the linear time-invariant (LTI) plant of order n:

$$\dot{\mathbf{x}}(t) = \mathbf{Ax}(t) + \mathbf{Bu}(t) \tag{7.141}$$

with the performance index

$$I = \int_0^\infty \left[\mathbf{x}(t)^T \mathbf{D}^T \mathbf{Dx}(t) + \mathbf{u}(t)^T \mathbf{Ru}(t) \right] dt \tag{7.142}$$

and introduce the following subspaces of the state space $\mathscr{X}$:

(a) $\mathscr{X}_u(\mathbf{A})$ the unstable eigenspace of $\mathbf{A}$

(b) θ the unobservable subspace corresponding to the pair $(\mathbf{D}, \mathbf{A})$

(c) $\mathscr{C}$ the controllable subspace of the pair $(\mathbf{A}, \mathbf{B})$.

If $\alpha(\lambda)$, the minimal polynomial of $\mathbf{A}$ is factored as

$$\alpha(\lambda) = \alpha_s(\lambda)\alpha_u(\lambda) \tag{7.143}$$

with $\alpha_s(\lambda)$ having all its zeros in the open LHP and $\alpha_u(\lambda)$ having all its zeros in the closed RHP, we have

$$\mathscr{X}_u(\mathbf{A}) = \mathrm{Ker}\left[\alpha_u(\mathbf{A})\right], \tag{7.144}$$

$$\theta = \cap_{i=0}^{n-1} \mathrm{Ker}\left[\mathbf{DA}^i\right], \tag{7.145}$$

and with $\mathrm{Im}[\mathbf{B}] = \mathscr{B}$:

$$\mathscr{C} = \mathscr{B} + \mathbf{A}\mathscr{B} + \cdots + \mathbf{A}^{n-1}\mathscr{B}. \tag{7.146}$$

THEOREM 7.4   There exists a state feedback control

$$\mathbf{u}(t) = -\mathbf{K}\mathbf{x}(t) \tag{7.147}$$

that minimizes

$$I = \int_0^\infty \left[\mathbf{x}(t)^T\mathbf{D}^T\mathbf{D}\mathbf{x}(t) + \mathbf{u}(t)^T R\mathbf{u}(t)\right] dt \tag{7.148}$$

if and only if

$$\mathscr{X}_u(\mathbf{A}) \subset \theta + \mathscr{C}. \tag{7.149}$$

Proof   Note that the subspaces defined by Equations (7.144), (7.145), and (7.146) are A-invariant subspaces. We introduce the factor space

$$\bar{\mathscr{X}} := \mathscr{X}/\theta \tag{7.150}$$

and the canonical projection

$$\mathbf{T} : \mathscr{X} \to \bar{\mathscr{X}} \tag{7.151}$$

given by

$$\mathbf{T}\mathbf{x} = \bar{\mathbf{x}}. \tag{7.152}$$

Then there exist maps $\bar{\mathbf{D}}$, $\bar{\mathbf{A}}$, and $\bar{\mathbf{B}}$ defined by

$$\mathbf{T}\mathbf{A} = \bar{\mathbf{A}}\mathbf{T}, \quad \bar{\mathbf{D}}\mathbf{T} = \mathbf{D}, \quad \mathbf{T}\mathbf{B} =: \bar{\mathbf{B}}, \tag{7.153}$$

where $\bar{\mathbf{A}}$ is the induced map in the factor space as shown in the commutative diagram of Figure 7.1.

Now Equation (7.141) implies

$$\dot{\bar{\mathbf{x}}}(t) = \bar{\mathbf{A}}\bar{\mathbf{x}}(t) + \bar{\mathbf{B}}\mathbf{u}(t),$$
$$\mathbf{y}(t) = \bar{\mathbf{D}}\bar{\mathbf{x}}(t), \tag{7.154}$$

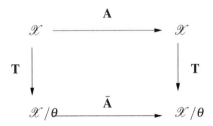

Figure 7.1 A commutative diagram

and Equation (7.142) can be rewritten

$$I = \int_0^\infty \left[\bar{\mathbf{x}}(t)^T \bar{\mathbf{D}}^T \bar{\mathbf{D}} \bar{\mathbf{x}}(t) + \mathbf{u}(t)^T \mathbf{R} \mathbf{u}(t) \right] dt. \tag{7.155}$$

It is easy to verify that $(\bar{D}, \bar{\mathbf{A}})$ is observable and thus an optimal control

$$\mathbf{u}(t) = -\bar{\mathbf{K}} \bar{\mathbf{x}}(t), \tag{7.156}$$

minimizing Equation (7.142) exists if and only if $(\bar{\mathbf{A}}, \bar{\mathbf{B}})$ is stabilizable, that is,

$$\bar{\mathscr{X}}_u(\bar{\mathbf{A}}) \subset \bar{\mathscr{C}}, \tag{7.157}$$

where $\bar{\mathscr{X}}_u(\bar{\mathbf{A}})$ is the unstable eigenspace of $\bar{\mathbf{A}}$ and $\bar{\mathscr{C}}$ is the controllable subspace of $(\bar{\mathbf{A}}, \bar{\mathbf{B}})$. It is now straightforward to prove that Equation (7.157) is equivalent to Equation (7.149). ∎

Remark 7.1   It is emphasized that the optimal control zeros the output **y** asymptotically, with $\mathbf{u}(t)$ tending asymptotically to zero. However, the optimal control can only stabilize the dynamics in the factor space and does not necessarily stabilize the original system.

7.6   Return Difference Relations

In this section we derive some loop transfer function properties for LQR systems known as return difference relations, which are helpful in subsequent sections to establish stability and robustness properties. We use the following standard LQR relations, restated below for convenience:

$$I = \int_0^\infty \left[\mathbf{x}(t)^T \mathbf{Q} \mathbf{x}(t) + \mathbf{u}(t)^T \mathbf{R} \mathbf{u}(t) \right] dt, \tag{7.158a}$$

$$\dot{\mathbf{x}}(t) = \mathbf{A} \mathbf{x}(t) + \mathbf{B} \mathbf{u}(t), \tag{7.158b}$$

$$0 = \mathbf{A}^T \mathbf{P} + \mathbf{P} \mathbf{A} + \mathbf{Q} - \mathbf{P} \mathbf{B} \mathbf{R}^{-1} \mathbf{B}^T \mathbf{P}, \tag{7.158c}$$

$$\mathbf{u}(t) = -\mathbf{R}^{-1} \mathbf{B}^T \mathbf{P} \mathbf{x}(t) = -\mathbf{K} \mathbf{x}(t), \tag{7.158d}$$

where

$$\mathbf{K} = \mathbf{R}^{-1} \mathbf{B}^T \mathbf{P}. \tag{7.159}$$

From the ARE we have

$$\mathbf{P}(s\mathbf{I} - \mathbf{A}) + (-s\mathbf{I} - \mathbf{A}^T)\mathbf{P} - \mathbf{Q} = -\mathbf{PBR}^{-1}\mathbf{RR}^{-1}\mathbf{B}^T\mathbf{P}$$
$$= -\mathbf{K}^T\mathbf{RK}. \tag{7.160}$$

Multiplying the above on the left by $\mathbf{B}^T(-s\mathbf{I} - \mathbf{A}^T)^{-1}$ and on the right by $(s\mathbf{I} - \mathbf{A})^{-1}\mathbf{B}$, we get using $(\mathbf{PB} = \mathbf{K}^T\mathbf{R})$

$$\mathbf{B}^T\left(-s\mathbf{I} - \mathbf{A}^T\right)^{-1}\mathbf{K}^T\mathbf{R} + \mathbf{RK}^T(s\mathbf{I} - \mathbf{A})^{-1}\mathbf{B}$$
$$+ \mathbf{B}^T\left(-s\mathbf{I} - \mathbf{A}^T\right)^{-1}\mathbf{K}^T\mathbf{RK}(s\mathbf{I} - \mathbf{A})^{-1}\mathbf{B} \tag{7.161}$$
$$= \mathbf{B}^T\left(-s\mathbf{I} - \mathbf{A}^T\right)^{-1}\mathbf{Q}(s\mathbf{I} - \mathbf{A})^{-1}\mathbf{B}.$$

Adding $\mathbf{R}$ to both sides, we obtain

$$\mathbf{R} + \mathbf{B}^T\left(-s\mathbf{I} - \mathbf{A}^T\right)^{-1}\mathbf{KRK}^T(s\mathbf{I} - \mathbf{A})^{-1}\mathbf{B}$$
$$+ \mathbf{B}^T\left(-s\mathbf{I} - \mathbf{A}^T\right)^{-1}\mathbf{KR} + \mathbf{RK}^T(s\mathbf{I} - \mathbf{A})^{-1}\mathbf{B} \tag{7.162}$$
$$= \mathbf{R} + \mathbf{B}^T\left(-s\mathbf{I} - \mathbf{A}^T\right)^{-1}\mathbf{Q}(s\mathbf{I} - \mathbf{A})^{-1}\mathbf{B}$$

and finally

$$\left[\mathbf{I} + \mathbf{B}^T\left(-s\mathbf{I} - \mathbf{A}^T\right)^{-1}\mathbf{K}^T\right]\mathbf{R}\left[\mathbf{I} + \mathbf{K}(s\mathbf{I} - \mathbf{A})^{-1}\mathbf{B}\right]$$
$$= \mathbf{R} + \mathbf{B}^T\left(-s\mathbf{I} - \mathbf{A}^T\right)^{-1}\mathbf{Q}(s\mathbf{I} - \mathbf{A})^{-1}\mathbf{B}, \tag{7.163}$$

which is known as Kalman's return difference identity.

7.7   Guaranteed Stability Margins for the LQR

The state feedback LQR enjoys some exceptional guaranteed or universal stability margins. We derive these here using Kalman's return difference relations established in the last section. Consider the multivariable system

$$\dot{\mathbf{x}}(t) = \mathbf{A}\mathbf{x}(t) + \mathbf{B}\mathbf{u}(t),$$
$$\mathbf{u}(t) = -\mathbf{K}\mathbf{x}(t) + \mathbf{v}(t), \tag{7.164}$$

where $\mathbf{K}$ has been determined using LQR theory so that $\mathbf{A} - \mathbf{BK}$ is stable.

This corresponds to the block diagram (see Figure 7.2) with the loop transfer function

$$\mathbf{L}(s) = \mathbf{K}(s\mathbf{I} - \mathbf{A})^{-1}\mathbf{B} \tag{7.165}$$

and

$$\mathbf{u}(s) = [\mathbf{I} + \mathbf{L}(s)]^{-1}\mathbf{v}(s). \tag{7.166}$$

The stability margins of the above system can be determined by inserting a perturbation matrix Δ at the loop breaking point ℓ shown in Figure 7.2, and determining the maximal "size" of Δ that preserves closed-loop stability.

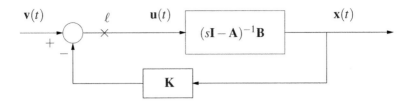

Figure 7.2 A state feedback loop

We assume for simplicity that Δ is a complex constant invertible matrix. The loop gain of the perturbed system is $\mathbf{L}(s)\Delta$. It is easily verified that

$$\mathbf{I} + \mathbf{L}(s)\Delta = \left[\left(\Delta^{-1} - \mathbf{I} \right) \left(\mathbf{I} + \mathbf{L}(s) \right)^{-1} + \mathbf{I} \right] \left(\mathbf{I} + \mathbf{L}(s) \right) \Delta. \qquad (7.167)$$

Now let $\mathscr{N}(\mathbf{F})$ denote the net change in the argument of the scalar function $\mathbf{F}(s)$ as s traverses the Nyquist D-contour, consisting of the imaginary axis along with a half circle of infinite radius enclosing the entire right half of the complex plane. From Equation (7.167):

$$\mathscr{N}\left(\det\left[\mathbf{I} + \mathbf{L}(s)\Delta\right]\right) = \mathscr{N}\left(\det\left[\left(\Delta^{-1} - \mathbf{I}\right)\left(\mathbf{I} + \mathbf{L}(s)\right)^{-1} + \mathbf{I}\right]\right)$$
$$+ \mathscr{N}\left(\det\left[\mathbf{I} + \mathbf{L}(s)\right]\right), \qquad (7.168)$$

since $\mathscr{N}(\det[\Delta]) = 0$. We will therefore have

$$\mathscr{N}\left(\det\left[\mathbf{I} + \mathbf{L}(s)\Delta\right]\right) = \mathscr{N}\left(\det\left[\mathbf{I} + \mathbf{L}(s)\right]\right) \qquad (7.169)$$

if

$$\mathscr{N}\left(\det\left[\left(\Delta^{-1} - \mathbf{I}\right)\left(\mathbf{I} + \mathbf{L}(s)\right)^{-1} + \mathbf{I}\right)\right]\right) = 0. \qquad (7.170)$$

Note that Equation (7.169) is a necessary and sufficient condition for stability of the perturbed closed loop since the nominal loop transfer function $\mathbf{L}(s)$ corresponds to a stable closed-loop system.

Let $\bar{\sigma}[\mathbf{U}]$ and $\underline{\sigma}[\mathbf{U}]$ denote the maximum and minimum singular values of the matrix $\mathbf{U}$. A sufficient condition for Equation (7.170) to hold, and therefore for stability of the perturbed system, is given below in terms of singular values:

$$\bar{\sigma}\left[\left(\Delta^{-1} - \mathbf{I}\right)\left(\mathbf{I} + \mathbf{L}(j\omega)\right)^{-1}\right] < 1, \quad \text{for all } \omega \qquad (7.171)$$

or

$$\bar{\sigma}\left[\Delta^{-1} - \mathbf{I}\right] \cdot \bar{\sigma}\left[\left(\mathbf{I} + \mathbf{L}(j\omega)\right)^{-1}\right] < 1, \quad \text{for all } \omega \qquad (7.172)$$

or

$$\bar{\sigma}\left[\Delta^{-1} - \mathbf{I}\right] < \underline{\sigma}[\mathbf{I} + \mathbf{L}(j\omega)], \quad \text{for all } \omega. \qquad (7.173)$$

From Kalman's return difference relation in Equation (7.163), we already know that

$$[\mathbf{I} + \mathbf{L}(j\omega)]^* \mathbf{R} [\mathbf{I} + \mathbf{L}(j\omega)] \geq \mathbf{R}. \qquad (7.174)$$

With $\mathbf{R} = \rho\mathbf{I}$, we obtain

$$[\mathbf{I}+\mathbf{L}(j\omega)]^*[\mathbf{I}+\mathbf{L}(j\omega)] \geq \mathbf{I}, \tag{7.175}$$

which can be expressed in terms of singular values as

$$\underline{\sigma}[\mathbf{I}+\mathbf{L}(j\omega)] \geq 1. \tag{7.176}$$

From Equations (7.173) and (7.176), it follows that

$$\bar{\sigma}\left[\Delta^{-1}-\mathbf{I}\right] < 1 \tag{7.177}$$

is a sufficient condition for closed-loop stability of the perturbed system.

7.7.1 Gain Margin

Let

$$\Delta = \begin{bmatrix} \mathbf{K}_1 & & \\ & \ddots & \\ & & \mathbf{K}_r \end{bmatrix}, \tag{7.178}$$

where the $\mathbf{K}_i$ are real. Then Equation (7.177) becomes

$$\left|\frac{1}{\mathbf{K}_i} - 1\right| < 1, \quad \text{for } i = 1, 2, \ldots, r \tag{7.179}$$

and this is equivalent to

$$\frac{1}{2} < \mathbf{K}_i < \infty, \quad \text{for } i = 1, 2, \ldots, r. \tag{7.180}$$

Thus, the LQR has a gain margin in each input channel that ranges from $\frac{1}{2}$ to ∞.

7.7.2 Phase Margin

In this case, let

$$\Delta = \begin{bmatrix} e^{-j\theta_1} & & \\ & \ddots & \\ & & e^{-j\theta_r} \end{bmatrix}, \tag{7.181}$$

so that Equation (7.177) is equivalent to

$$\left|e^{j\theta_i} - 1\right| < 1, \quad \text{for } i = 1, 2, \ldots, r \tag{7.182}$$

or

$$(\cos\theta_i - 1)^2 + \sin^2\theta_i < 1, \quad \text{for } i = 1, 2, \ldots, r \tag{7.183}$$

or

$$\cos\theta_i > \frac{1}{2}, \quad \text{for } i = 1, 2, \ldots, r. \tag{7.184}$$

The condition in Equation (7.184) is equivalent to a phase margin of 60°, that is

$$-60° < \theta_i < 60°$$ (7.185)

in each input channel of the LQR.

7.7.3 Single-Input Case

In the single-input case (see Figure 7.3), $\mathbf{R} = r$, $\mathbf{K} = \mathbf{k}$ a row vector, $\mathbf{B} = \mathbf{b}$ a column vector, and the return difference identity reduces to

$$\left|1 + \mathbf{k}(s\mathbf{I} - \mathbf{A})^{-1}\mathbf{b}\right|^2 r = \left|r + \mathbf{b}^T\left(-s\mathbf{I} - \mathbf{A}^T\right)^{-1}\mathbf{Q}(s\mathbf{I} - \mathbf{A})\mathbf{b}\right|$$ (7.186)

so that

$$\left|1 + \mathbf{k}(j\omega\mathbf{I} - \mathbf{A})^{-1}\mathbf{b}\right| \geq 1.$$ (7.187)

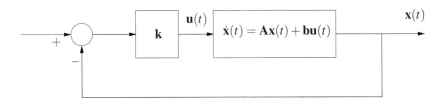

Figure 7.3 The optimal feedback system

The relation in Equation (7.187) means that the Nyquist plot

$$g(j\omega) = \mathbf{k}(j\omega\mathbf{I} - \mathbf{A})^{-1}\mathbf{b}$$ (7.188)

stays outside a circle of radius 1 centered at $-1 + j0$, as depicted in Figure 7.4. This clearly results in the gain and phase margins established above.

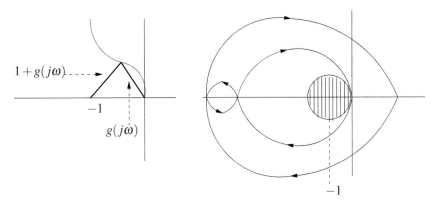

Figure 7.4 Nyquist plot and the unit circle centered at -1 © 2009 Taylor & Francis Group, LLC. Reproduced from Bhattacharyya, Datta, and Keel (2009) with permission.

7.8   Eigenvalues of the Optimal Closed-Loop System

It is possible to determine the eigenvalues of the closed-loop optimal LQR from the matrices defining the performance index, without computing the feedback control as we show below.

7.8.1   Closed-Loop Spectrum

Recall the definition of the Hamiltonian:

$$\mathbf{H} = \begin{bmatrix} \mathbf{A} & -\mathbf{B}\mathbf{R}^{-1}\mathbf{B}^T \\ -\mathbf{Q} & -\mathbf{A}^T \end{bmatrix}. \tag{7.189}$$

Since

$$\dot{\mathbf{x}}(t) = (\mathbf{A} - \mathbf{B}\mathbf{K})\mathbf{x}(t), \tag{7.190}$$

the closed-loop eigenvalues are the eigenvalues of $\mathbf{A} - \mathbf{B}\mathbf{K}$. Using the ARE, it is easily verified that

$$s\mathbf{I} - \mathbf{H} = \begin{bmatrix} \mathbf{I} & 0 \\ \mathbf{P} & -\mathbf{I} \end{bmatrix} \begin{bmatrix} s\mathbf{I} - (\mathbf{A} - \mathbf{B}\mathbf{K}) & \mathbf{B}\mathbf{R}^{-1}\mathbf{B}^T \\ 0 & -s\mathbf{I} - (\mathbf{A} - \mathbf{B}\mathbf{K})^T \end{bmatrix} \begin{bmatrix} \mathbf{I} & 0 \\ -\mathbf{P} & \mathbf{I} \end{bmatrix}, \tag{7.191}$$

so that

$$\det[s\mathbf{I} - \mathbf{H}] = (-1)^n \det[s\mathbf{I} - (\mathbf{A} - \mathbf{B}\mathbf{K})] \det\left[-s\mathbf{I} - (\mathbf{A} - \mathbf{B}\mathbf{K})^T\right]$$
$$= \det[s\mathbf{I} - (\mathbf{A} - \mathbf{B}\mathbf{K})] \det[s\mathbf{I} + (\mathbf{A} - \mathbf{B}\mathbf{K})]. \tag{7.192}$$

The LHP eigenvalues of $\mathbf{H}$ therefore consist of the closed-loop eigenvalues of the optimal system.

It is now quite natural to ask: How do the closed-loop eigenvalues of the optimal system behave with respect to, say, ρ where

$$\mathbf{R} = \rho\mathbf{R}_0 \tag{7.193}$$

for fixed $\mathbf{R}_0$ as ρ increases or decreases? Let

$$\alpha_k(s) := \det[s\mathbf{I} - (\mathbf{A} - \mathbf{B}\mathbf{K})],$$
$$\alpha_0(k) := \det[s\mathbf{I} - \mathbf{A}], \tag{7.194}$$

and with $\mathbf{Q} = \mathbf{C}^T\mathbf{C}$ let

$$\mathbf{P}(s) := \mathbf{C}(s\mathbf{I} - \mathbf{A})^{-1}\mathbf{B}. \tag{7.195}$$

From Equation (7.192), we have

$$\alpha_k(s)\alpha_k(-s) = (-1)^n \det[s\mathbf{I} - \mathbf{H}]. \tag{7.196}$$

Now using the identities

$$\det\begin{bmatrix} \mathbf{X} & \mathbf{Y} \\ \mathbf{Z} & \mathbf{W} \end{bmatrix} = \det[\mathbf{X}]\det\left[\mathbf{W} - \mathbf{Z}\mathbf{X}^{-1}Y\right], \qquad \text{for } \det[\mathbf{X}] \neq 0 \tag{7.197}$$

and

$$\det\left[\mathbf{I}_n + \mathbf{U}\mathbf{V}\right] = \det\left[\mathbf{I}_m + \mathbf{V}\mathbf{U}\right], \tag{7.198}$$

we have

$$
\begin{aligned}
\det[s\mathbf{I} - \mathbf{H}] &= \det[s\mathbf{I} - \mathbf{A}]\det\left[s\mathbf{I} + \mathbf{A}^T - \mathbf{C}^T\mathbf{C}(s\mathbf{I} - \mathbf{A})^{-1}\mathbf{B}\mathbf{R}^{-1}\mathbf{B}^T\right] \\
&= (-1)^n\det[s\mathbf{I} - \mathbf{A}]\det\left[-s\mathbf{I} - \mathbf{A}^T + \mathbf{C}^T\mathbf{C}(s\mathbf{I} - \mathbf{A})^{-1}\mathbf{B}\mathbf{R}^{-1}\mathbf{B}^T\right] \\
&= (-1)^n\det[s\mathbf{I} - \mathbf{A}]\det\left[-s\mathbf{I} - \mathbf{A}^T\right] \\
&\qquad \det\left[\mathbf{I} + \mathbf{C}^T\mathbf{C}(s\mathbf{I} - \mathbf{A})^{-1}\mathbf{B}\mathbf{R}^{-1}\mathbf{B}^T(-s\mathbf{I} - \mathbf{A}^T)^{-1}\right] \\
&= (-1)^n\det[s\mathbf{I} - \mathbf{A}]\det\left[-s\mathbf{I} - \mathbf{A}^T\right] \\
&\qquad \det\left[\mathbf{I} + \mathbf{C}(s\mathbf{I} - \mathbf{A})^{-1}\mathbf{B}\mathbf{R}^{-1}\mathbf{B}^T(-s\mathbf{I} - \mathbf{A}^T)^{-1}\mathbf{C}^T\right].
\end{aligned}
\tag{7.199}
$$

Therefore, using Equations (7.194)–(7.196), we have

$$\alpha_k(s)\alpha_k(-s) = \alpha_0(s)\alpha_0(-s)\det\left[\mathbf{I} + \mathbf{P}(s)\mathbf{R}^{-1}\mathbf{P}^T(-s)\right]. \tag{7.200}$$

When $\mathbf{R} = \rho I$ and ρ is large, it is seen from Equation (7.200) that the closed-loop eigenvalues of the optimal system approach the LHP open-loop eigenvalues and the reflection of the open-loop RHP eigenvalues about the imaginary axis. On the other hand, as $\rho \downarrow 0$ the closed-loop eigenvalues approach the LHP zeros of $\mathbf{P}(s)\mathbf{P}^T(-s)$ or approach ∞ in the LHP.

7.9 A Summary of LQR Theory

In this section we summarize the results on the LQR obtained above for the convenience of the reader.

When state feedback can be implemented, it allows for the arbitrary reassignment of all controllable eigenvalues as we have shown earlier. However, this leaves open the question: How does one choose the locations of these closed-loop eigenvalues? More generally, is there a rational way to design the state feedback gain matrix?

In 1960, Kalman gave an answer to the above question by proposing that linear control systems should be designed to be optimal with respect to a quadratic performance index which penalizes deviations of states and inputs from zero. Specifically, Kalman introduced the following cost functional, penalizing deviation of $\mathbf{D}^T\mathbf{x}(t)$ and $\mathbf{u}(t)$ from zero:

$$I = \int_0^\infty \left(\mathbf{x}(t)^T\mathbf{D}\mathbf{D}^T\mathbf{x}(t) + \mathbf{u}(t)^T\mathbf{R}\mathbf{u}(t)\right)dt, \quad \mathbf{R} = \mathbf{R}^T > 0. \tag{7.201}$$

This cost function is to be minimized over all control inputs $\mathbf{u}(t)$ for $t \in [0,\infty)$ subject to the constraint (system equation)

$$\dot{\mathbf{x}}(t) = \mathbf{A}\mathbf{x}(t) + \mathbf{B}\mathbf{u}(t). \tag{7.202}$$

This optimization problem came to be known as the linear quadratic regulator (LQR) problem.

As it turns out, the solution of the LQR problem is state feedback. We describe the solution here in this summary. The proof and detailed derivation are given earlier in this chapter.

To describe the optimal solution, introduce the ARE

$$\mathbf{A}^T\mathbf{P} + \mathbf{P}\mathbf{A} + \mathbf{D}\mathbf{D}^T - \mathbf{P}\mathbf{B}\mathbf{R}^{-1}\mathbf{B}^T\mathbf{P} = 0. \tag{7.203}$$

The following theorem summarizes the results on the LQR problem. The proof has been given earlier in this chapter.

THEOREM 7.5   If $(\mathbf{A},\mathbf{B})$ is stabilizable and $(\mathbf{D}^T,\mathbf{A})$ is detectable, the ARE in Equation (7.203) has a unique positive semi-definite solution $\mathbf{P}$, the optimal control is

$$\mathbf{u}^*(t) = -\underbrace{\mathbf{R}^{-1}\mathbf{B}^T\mathbf{P}}_{\mathbf{K}}\mathbf{x}(t), \tag{7.204}$$

and $\mathbf{A} - \mathbf{B}\mathbf{K}$ is stable with eigenvalues which consist of the n LHP eigenvalues of

$$\mathbf{H} = \begin{bmatrix} \mathbf{A} & -\mathbf{B}\mathbf{R}^{-1}\mathbf{B} \\ -\mathbf{D}\mathbf{D}^T & -\mathbf{A}^T \end{bmatrix}. \tag{7.205}$$

Moreover, the optimal control law in Equation (7.204) has the following robustness properties. If $\mathbf{K}$ in Equation (7.204) is replaced by $\Delta\mathbf{K}$ where Δ belongs to one of the following classes:

$$\Delta = \begin{bmatrix} k_1 & & \\ & \ddots & \\ & & k_r \end{bmatrix}, \quad \text{for } \frac{1}{2} < k_i < \infty, \quad \text{for } i \in \mathbf{r} \tag{7.206}$$

or

$$\Delta = \begin{bmatrix} e^{-j\theta_1} & & \\ & \ddots & \\ & & e^{-j\theta_r} \end{bmatrix}, \quad \text{for } 0 < \theta_i < \frac{\pi}{3}, \quad \text{for } i \in \mathbf{r} \tag{7.207}$$

then $(\mathbf{A} - \mathbf{B}\Delta\mathbf{K})$ continues to be stable.

7.10   LQR Computations Using MATLAB

In the following we illustrate the use of MATLAB to compute the optimal LQR state feedback.

EXAMPLE 7.1   Consider the plant described by

$$\dot{\mathbf{x}}(t) = \underbrace{\begin{bmatrix} -1 & 0 & 1 & 1 \\ -2 & 2 & 1 & 0 \\ 0 & 0 & 1 & -4 \\ 2 & 1 & -1 & 0 \end{bmatrix}}_{\mathbf{A}}\mathbf{x}(t) + \underbrace{\begin{bmatrix} 1 & 0 \\ 1 & -1 \\ 1 & 1 \\ 0 & 1 \end{bmatrix}}_{\mathbf{B}}\mathbf{u}(t). \tag{7.208}$$

Note that this plant has three RHP poles with the eigenvalues of $\mathbf{A}$ being

$$\{-3.2202, \ 2.1101 \pm j1.0664, \ 1\}. \tag{7.209}$$

The following MATLAB script calculates the optimal state feedback. The choices of weight matrices selected are

$$\mathbf{Q} := \mathbf{DD}^T = \mathbf{I}_4, \quad \mathbf{R} = \mathbf{I}_2. \tag{7.210}$$

MATLAB Solution

```
clear
A=[-1 0 1 1; -2 2 1 0; 0 0 1 -4; 2 1 -1 0];
B=[1 0; 1 -1; 1 1;0 1];
Q=eye(4);
R=eye(2);
K=lqr(A,B,Q,R);
eig(A-B*K)
```

The state feedback obtained is

$$\mathbf{K} = \begin{bmatrix} 4.1631 & 9.1194 & -5.3390 & 13.4096 \\ 2.7925 & -3.3205 & -0.5208 & 1.3675 \end{bmatrix} \tag{7.211}$$

and the eigenvalues of $\mathbf{A} - \mathbf{BK}$ are

$$\sigma(\mathbf{A} - \mathbf{BK}) = \{-3.6591, \ -2.3832 \pm j1.3394 \ -1.6853\}, \tag{7.212}$$

verifying stability of the optimal system.

Verification of the Guaranteed Stability Margins

Figure 7.5 plots the root locus of the eigenvalues of the closed-loop system

$$\lambda \left(\mathbf{A} - k\mathbf{BK}\right), \quad \text{for } 0.44 \leq k \leq 5 \ (\approx \infty). \tag{7.213}$$

It shows that the lower gain margin can go below $\frac{1}{2}$, verifying what the theory says. A similar verification can be done for the phase margin by plotting the eigenvalues of

$$\lambda \left(\mathbf{A} - e^{-j\theta}\mathbf{BK}\right), \quad \text{for } 0 \leq \theta. \tag{7.214}$$

The phase margin turns out to be $61.5°$ at which value an eigenvalue intersects the imaginary axis, once again verifying the theoretical guarantee of $60°$ (see Figure 7.5).

EXAMPLE 7.2 Consider the system with $(\mathbf{A}, \mathbf{B})$ below:

$$\mathbf{A} = \begin{bmatrix} -1 & 0 & 1 & 1 \\ -2 & 2 & 1 & 0 \\ 0 & 0 & 1 & -4 \\ 2 & 1 & -1 & 0 \end{bmatrix}, \quad \mathbf{B} = \begin{bmatrix} 1 & 0 \\ 1 & -1 \\ 1 & 1 \\ 0 & 1 \end{bmatrix}. \tag{7.215}$$

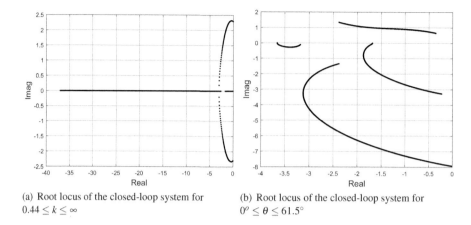

(a) Root locus of the closed-loop system for $0.44 \le k \le \infty$

(b) Root locus of the closed-loop system for $0^o \le \theta \le 61.5^\circ$

Figure 7.5 Verifying the margins (Example 7.1)

We choose $\mathbf{Q} = \rho \mathbf{I}_4$ and $\mathbf{R} = \mathbf{I}_2$. We now observe the change of the closed-loop eigenvalues as $\rho \in (2, 0.1)$. The following MATLAB script and Figure 7.6 shows that a designer may appropriately choose ρ such that the closed-loop eigenvalues are located in a certain region.

MATLAB Solution

```
A=[-1 0 1 1; -2 2 1 0; 0 0 1 -4; 2 1 -1 0];
B=[1 0; 1 -1; 1 1; 0 1];
rho=2:-0.05:0.1;
for i=1:length(rho),
    Q=rho(i)*eye(4);
    R=eye(2);
    K=lqr(A,B,Q,R);
    r(:,i)=eig(A-B*K);
end,
plot(real(r),imag(r),'.k'), grid,
axis([-4.5 0.5 -2.5 2.5]), hold on
plot(real(r(:,1)),imag(r(:,1)),'*k')
```

7.11   A Design Example: Control of a Rotary Double-Inverted Pendulum Experiment

In this example, we design a controller using LQR theory for the rotary double-inverted pendulum developed by Quanser, Inc. and shown in Figure 7.7(a). The pendulum system has one input $u(t)$, which is the generalized force acting on the rotary arm, and three outputs, which are angle measurements shown in Figure 7.7(b).

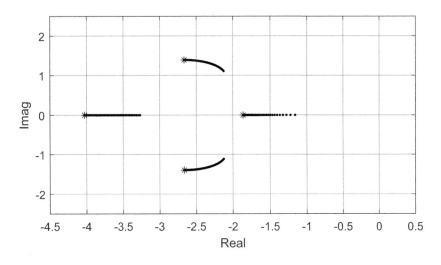

Figure 7.6 Closed-loop root locus as $\rho \in (2, 0.1)$ (Example 7.2)

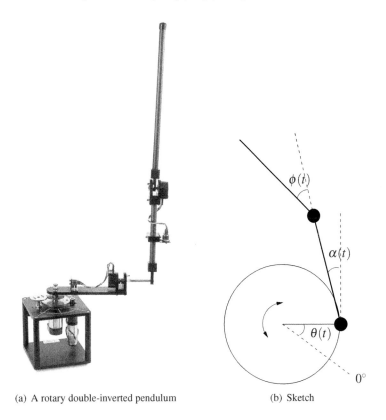

(a) A rotary double-inverted pendulum   (b) Sketch

Figure 7.7 A Quanser rotary double-inverted pendulum (Photo courtesy of Quanser, Inc. © Quanser, Inc.)

The linearized state-space model of the pendulum is given by

$$
\underbrace{\begin{bmatrix} \dot{\theta}(t) \\ \dot{\alpha}(t) \\ \dot{\phi}(t) \\ \ddot{\theta}(t) \\ \ddot{\alpha}(t) \\ \ddot{\phi}(t) \end{bmatrix}}_{\dot{x}(t)} = \underbrace{\begin{bmatrix} 0 & 0 & 0 & 1 & 0 & 0 \\ 0 & 0 & 0 & 0 & 1 & 0 \\ 0 & 0 & 0 & 0 & 0 & 1 \\ 0 & 290.83 & -15.84 & -18.55 & -0.75 & 0.99 \\ 0 & 256.22 & -28.89 & -14.13 & -0.69 & 1.14 \\ 0 & -339.59 & 127.42 & 18.73 & 1.14 & -3.10 \end{bmatrix}}_{\mathbf{A}} \underbrace{\begin{bmatrix} \theta(t) \\ \alpha(t) \\ \phi(t) \\ \dot{\theta}(t) \\ \dot{\alpha}(t) \\ \dot{\phi}(t) \end{bmatrix}}_{x(t)}
$$

$$
+ \underbrace{\begin{bmatrix} 0 \\ 0 \\ 0 \\ 84.99 \\ 64.75 \\ -85.82 \end{bmatrix}}_{\mathbf{B}} u(t), \tag{7.216}
$$

$$
\underbrace{\begin{bmatrix} \theta(t) \\ \alpha(t) \\ \phi(t) \end{bmatrix}}_{y(t)} = \underbrace{\begin{bmatrix} 1 & 0 & 0 & 0 & 0 & 0 \\ 0 & 1 & 0 & 0 & 0 & 0 \\ 0 & 0 & 1 & 0 & 0 & 0 \end{bmatrix}}_{\mathbf{C}} \underbrace{\begin{bmatrix} \theta(t) \\ \alpha(t) \\ \phi(t) \\ \dot{\theta}(t) \\ \dot{\alpha}(t) \\ \dot{\phi}(t) \end{bmatrix}}_{x(t)}.
$$

Note that θ has been introduced as a state variable, since we want to regulate it. Indeed, the first equation in the state-space model is just an identity. The objective of control is to maintain the pendulum in the upright vertical position and the rotary arm to track the position of $0°$ angle. In other words:

$$
\theta(t) = \alpha(t) = \phi(t) = 0 \tag{7.217}
$$

are the reference values to be regulated to. To design an LQR-based controller we assume that all states are available for measurement. This will mean that the corresponding feedback controller will be driven by the output $y(t)$ and its derivative $\dot{y}(t)$. Such a controller can be implemented as a proportional derivative (PD) controller driven only by the output $y(t)$.

To proceed, we design a state feedback $\mathbf{K}$ for the controllable pair $(\mathbf{A}, \mathbf{B})$ by using the "lqr()" command in MATLAB. The weight matrices are chosen to be

$$
\mathbf{Q} = \mathbf{I}_6 \quad \text{and} \quad \mathbf{R} = 30. \tag{7.218}
$$

The following MATLAB script is used to determine a state feedback gain by using LQR theory.

MATLAB Solution

```
clear
A=zeros(6,6);          % pendulum system (A,B,C,D)
A(1:3,4:6)=eye(3);
A(4:6,2:6)=[290.83 -15.84 -18.55 -0.75 0.99;
       256.22 -28.89 -14.13 -0.69 1.14;
       -339.59 127.42 18.73 1.14 -3.10];
B=[0; 0; 0; 84.99; 64.75; -85.82];
C=[eye(3) zeros(3,3)];
D=zeros(3,1);
Q=eye(7);
R=30;
K=lqr(A,B,Q,R);
eig(A-B*K)
```

Using the MATLAB script above, we obtain the state feedback for the extended system:

$$\mathbf{K} = \begin{bmatrix} k_1 & k_2 & k_3 & k_4 & k_5 & k_6 \end{bmatrix}$$
$$= \begin{bmatrix} 0.1826 & -7.9348 & -32.1164 & 0.2004 & -3.7425 & -3.1766 \end{bmatrix} \quad (7.219)$$

and the closed-loop poles are:

$$\sigma(\mathbf{A}-\mathbf{BK}) = \left\{ \begin{array}{c} -39.5968 \\ -10.1363 \pm j2.7837 \\ -6.6058 \\ -2.5324 \\ -0.6548 \end{array} \right\}. \quad (7.220)$$

Therefore, the PD output feedback controller constructed from $\mathbf{K}$ will have transfer function:

$$\mathbf{C}(s) = \begin{bmatrix} k_1 + k_4 s & k_2 + k_5 s & k_3 + k_6 s \end{bmatrix}. \quad (7.221)$$

The Simulink model in Figure 7.8 is used to verify the design. A simulation was done with the initial condition

$$\mathbf{x}(0) = \begin{bmatrix} 1 & 1 & 1 & 1 & 1 & 1 \end{bmatrix} \quad (7.222)$$

and the outputs are shown in Figure 7.9.

7.12   Observer-Based Optimal State Feedback: Loss of Guaranteed Margins

In general, all the states of a system cannot be directly measured. In this case, their "estimates" $\hat{x}(t)$ can be generated by an observer:

$$\dot{\hat{\mathbf{x}}}(t) = \mathbf{A}\hat{\mathbf{x}}(t) + \mathbf{B}\mathbf{u}(t) + \mathbf{L}(\mathbf{y}(t) - \mathbf{C}\hat{\mathbf{x}}(t)), \quad (7.223a)$$

$$\mathbf{u}(t) = -\mathbf{K}\hat{\mathbf{x}}(t). \quad (7.223b)$$

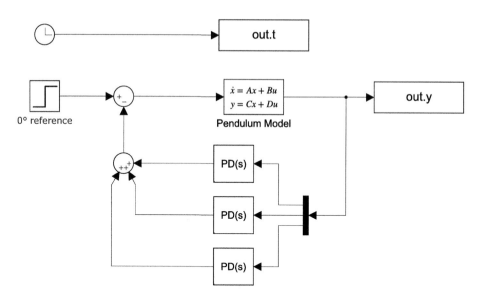

Figure 7.8 Simulink model for pendulum control

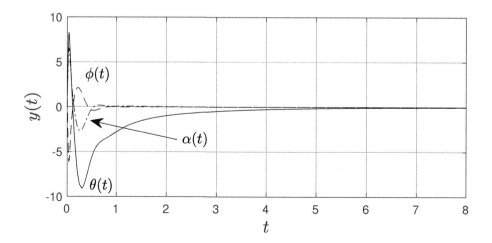

Figure 7.9 Simulation outputs of pendulum control

The observer-reconstructed states can be used to implement the state feedback control law as shown in the block diagram in Figure 7.10.

Observer-based optimal feedback clearly results in an increase in the value of the quadratic cost function and therefore loss of optimality. However, the convergence rate of the observer estimates can be controlled and this loss of optimality may not be serious. A more important issue, however, is the potential loss of the guaranteed stability margins provided by optimal state feedback. This issue is discussed next.

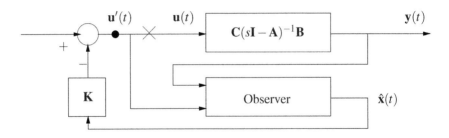

Figure 7.10 Observer-based output feedback

First note that the loop gain of the optimal system under state feedback is given by:

$$\mathbf{L}_0(s) = \mathbf{K}(s\mathbf{I} - \mathbf{A})^{-1}\mathbf{B}. \tag{7.224}$$

From the block diagram corresponding to the observer shown in Figure 7.10, the loop transfer function at the loop breaking point "×" can be calculated from the following relationships:

$$\mathbf{y}(s) = \mathbf{C}(s\mathbf{I} - \mathbf{A})^{-1}\mathbf{B}\mathbf{u}(s), \tag{7.225a}$$

$$\hat{\mathbf{x}}(s) = (s\mathbf{I} - \mathbf{A} + \mathbf{L}\mathbf{C})^{-1}\mathbf{B}\mathbf{u}'(s) + (s\mathbf{I} - \mathbf{A} + \mathbf{L}\mathbf{C})^{-1}\mathbf{L}\mathbf{y}(s), \tag{7.225b}$$

$$\mathbf{u}'(s) = -\mathbf{K}\hat{\mathbf{x}}(s). \tag{7.225c}$$

Substituting Equation (7.225b) into Equation (7.225c):

$$\mathbf{u}'(s) = -\mathbf{K}\left[(s\mathbf{I} - \mathbf{A} + \mathbf{L}\mathbf{C})^{-1}\mathbf{B}\mathbf{u}'(s) + (s\mathbf{I} - \mathbf{A} + \mathbf{L}\mathbf{C})^{-1}\mathbf{L}\mathbf{y}(s)\right], \tag{7.226a}$$

$$\mathbf{u}'(s) + \mathbf{K}(s\mathbf{I} - \mathbf{A} + \mathbf{L}\mathbf{C})^{-1}\mathbf{B}\mathbf{u}'(s) = -\mathbf{K}(s\mathbf{I} - \mathbf{A} + \mathbf{L}\mathbf{C})^{-1}\mathbf{L}\mathbf{y}(s), \tag{7.226b}$$

$$\left[\mathbf{I} + \mathbf{K}(s\mathbf{I} - \mathbf{A} + \mathbf{L}\mathbf{C})^{-1}\mathbf{B}\right]\mathbf{u}'(s) = -\mathbf{K}(s\mathbf{I} - \mathbf{A} + \mathbf{L}\mathbf{C})^{-1}\mathbf{L}\mathbf{y}(s). \tag{7.226c}$$

Substituting Equation (7.225a) into Equation (7.226c):

$$\left[\mathbf{I} + \mathbf{K}(s\mathbf{I} - \mathbf{A} + \mathbf{L}\mathbf{C})^{-1}\mathbf{B}\right]\mathbf{u}'(s) = -\mathbf{K}(s\mathbf{I} - \mathbf{A} + \mathbf{L}\mathbf{C})^{-1}\mathbf{L}\mathbf{C}(s\mathbf{I} - \mathbf{A})^{-1}\mathbf{B}\mathbf{u}(s). \tag{7.227}$$

Let

$$\mathbf{R}_1(s) := \mathbf{K}(s\mathbf{I} - \mathbf{A} + \mathbf{L}\mathbf{C})^{-1}\mathbf{B}, \tag{7.228a}$$

$$\mathbf{R}_2(s) := \mathbf{K}(s\mathbf{I} - \mathbf{A} + \mathbf{L}\mathbf{C})^{-1}\mathbf{L}\mathbf{C}(s\mathbf{I} - \mathbf{A})^{-1}\mathbf{B}. \tag{7.228b}$$

Equation (7.227) can now be rewritten as

$$[\mathbf{I} + \mathbf{R}_1(s)]\mathbf{u}'(s) = -\mathbf{R}_2(s)\mathbf{u}(s), \tag{7.229}$$

$$\mathbf{u}'(s) = -[\mathbf{I} + \mathbf{R}_1(s)]^{-1}\mathbf{R}_2(s)\mathbf{u}(s). \tag{7.230}$$

The loop transfer function $\mathbf{L}_1(s)$ is

$$\mathbf{L}_1(s) = [\mathbf{I} + \mathbf{R}_1(s)]^{-1}\mathbf{R}_2(s). \tag{7.231}$$

Therefore, the loop transfer function under observer-based optimal feedback $\mathbf{L}_1(s)$ is different from $\mathbf{L}_0(s)$, the loop transfer function of the optimal state feedback system. Thus, the stability margins corresponding to $\mathbf{L}_0(s)$, the optimal system, will differ from those attained by using observers. These stability margins correspond respectively to the loop breaking points "•" and "×" in the block diagram above. The loop breaking point "×" is physically meaningful, since it captures the realistic situation that the plant perturbs and the observer is "unaware" of this. This was shown in an example by Doyle and Stein, which is described below.

EXAMPLE 7.3   Consider the plant

$$\dot{\mathbf{x}}(t) = \begin{bmatrix} 0 & 1 \\ -3 & -4 \end{bmatrix} \mathbf{x}(t) + \begin{bmatrix} 0 \\ 1 \end{bmatrix} \mathbf{u}(t), \qquad \mathbf{y}(t) = \begin{bmatrix} 2 & 1 \end{bmatrix} \mathbf{x}(t) \tag{7.232}$$

and controller

$$\mathbf{u}(t) = \begin{bmatrix} -50 & -10 \end{bmatrix} \hat{\mathbf{x}}(t) + 50\mathbf{r}(t), \tag{7.233}$$

where $\hat{\mathbf{x}}(t)$ is the "estimate" of $\mathbf{x}(t)$. The controller is linear quadratic optimal, corresponding to the performance index

$$\mathscr{I} = \int_0^\infty \left(\mathbf{x}(t)^T \mathbf{H}^T \mathbf{H} \mathbf{x}(t) + \mathbf{u}(t)^2 \right) dt, \tag{7.234}$$

with

$$\mathbf{H} = 4\sqrt{5} \begin{bmatrix} \sqrt{35} & 1 \end{bmatrix}. \tag{7.235}$$

It places the closed-loop poles at $\{-7.0 \pm j2.0\}$.

Although Kalman's LQR provides the phase margin of $86°$ under optimal state feedback, the implemented system with an observer-based control law achieves less then $15°$. This is clearly seen in the Nyquist plots shown in Figure 7.11.

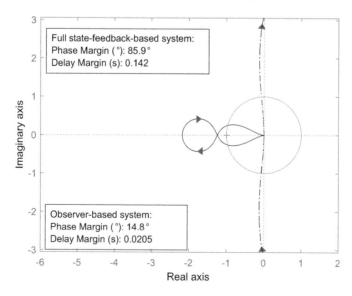

Figure 7.11 Nyquist plots of system with full state feedback, $\mathbf{L}_0$ (dash-dot line) and system with observer-based feedback, $\mathbf{L}_1$ (solid line) (Example 7.3)

Several observations are in order. First, the observer-based controller in this example certainly does not inherit Kalman's excellent guaranteed stability margins. This is because the loop gains $\mathbf{L}_1(s)$ and $\mathbf{L}_0(s)$ are different. The LQR problem is solved for an nth-order plant but the overall closed-loop system with controller is of order $2n$. The loop breaking point for the state feedback case does not possess the same physical meaning as in the case of observer-based output feedback. The phase margin of $15°$ is somewhat surprising because the plant is stable and minimum phase.

In the following section, we describe an approach to optimal output feedback controller design proposed by Pearson in 1969 using the LQR machinery, where an observer is not used and the system implemented is optimal. In the design method developed here, we also show how to employ a design parameter ρ and use it to "recover" stability margins or robustify the system.

7.13   Optimal Dynamic Compensators

Consider the controllable and observable system

$$\dot{\mathbf{x}}(t) = \mathbf{A}\mathbf{x}(t) + \mathbf{B}\mathbf{u}(t) \tag{7.236}$$

with output measurements

$$\mathbf{y}(t) = \mathbf{C}\mathbf{x}(t) \tag{7.237}$$

where $\mathbf{x}(t)$ is an n-vector, $\mathbf{u}(t)$ is an r-vector, and $\mathbf{y}(t)$ is an m-vector, with $m < n$. It is easily seen that implementation of state feedback control using the measurements $\mathbf{y}(t)$ would require that we obtain derivatives of $\mathbf{y}(t)$. In most practical systems, it is impossible to carry out pure differentiation and one must employ dynamic compensators as approximate differentiators. If optimal state feedback is computed for Equation (7.236) using LQR theory and dynamic compensation is introduced as an afterthought, the optimality of the closed-loop system would inevitably be deteriorated. This is also obviously true when an observer is used to obtain the compensator.

An intelligent approach to avoid this pitfall is to anticipate the use of dynamic compensation of a suitable order and introduce an equal number of integrators to augment the plant state, so that state feedback control in the augmented system can be exactly implemented by a proper dynamic compensator. In this way, the system that is implemented is optimal and there is no uncontrollable loss of performance from that designed. We describe the details of this approach below.

Since the system is observable, there exists q, an integer such that

$$\mathrm{Rank} \begin{bmatrix} \mathbf{C} \\ \mathbf{CA} \\ \mathbf{CA}^2 \\ \vdots \\ \mathbf{CA}^q \end{bmatrix} = n. \tag{7.238}$$

The minimum value of q satisfying the above equation is called the *observability index* of the system. Write

$$\dot{\mathbf{u}}(t) = \mathbf{u}_1(t),$$
$$\dot{\mathbf{u}}_1(t) = \mathbf{u}_2(t),$$
$$\dot{\mathbf{u}}_2(t) = \mathbf{u}_3(t),$$
$$\dot{\mathbf{u}}_3(t) = \mathbf{u}_4(t),$$

$$\vdots$$

$$\dot{\mathbf{u}}_{q-1}(t) = \mathbf{v}(t),$$

(7.239)

to denote a string of rq integrators. The augmented system consisting of Equations (7.236) and (7.239) has state vector

$$\mathbf{x}_a(t) := \begin{bmatrix} \mathbf{x}(t) \\ \mathbf{u}(t) \\ \mathbf{u}_1(t) \\ \vdots \\ \mathbf{u}_{q-1}(t) \end{bmatrix},$$

(7.240)

input $\mathbf{v}(t)$, and state equation

$$\dot{\mathbf{x}}_a(t) = \mathbf{A}_a \mathbf{x}_a(t) + \mathbf{B}_a \mathbf{v}(t),$$

(7.241)

where

$$\mathbf{A}_a = \begin{bmatrix} \mathbf{A} & \mathbf{B} & 0 & 0 & \cdots & 0 \\ 0 & 0 & \mathbf{I} & 0 & & \vdots \\ 0 & 0 & 0 & \mathbf{I} & & \vdots \\ \vdots & & & & \ddots & 0 \\ \vdots & & & & & \mathbf{I} \\ 0 & 0 & 0 & 0 & \cdots & 0 \end{bmatrix}, \qquad \mathbf{B}_a = \begin{bmatrix} 0 \\ 0 \\ 0 \\ \vdots \\ 0 \\ \mathbf{I} \end{bmatrix}.$$

(7.242)

Optimal control of Equation (7.241) using LQR theory requires that we specify a performance index

$$I_a = \int_0^\infty \left[\mathbf{x}_a(t)^T \mathbf{Q}_a \mathbf{x}_a(t) + \mathbf{v}(t)^T \mathbf{R}_a \mathbf{v}(t) \right] dt,$$

(7.243)

with $\mathbf{R}_a$ symmetric positive definite and $\mathbf{Q}_a$ symmetric positive semi-definite, with $(\mathbf{Q}_a, \mathbf{A}_a)$ observable. The optimal state feedback control can then be solved from the corresponding ARE, and can be written in the form

$$\mathbf{v}(t) = -\mathbf{K}_a^T \mathbf{x}_a(t)$$

(7.244)

or in the expanded form

$$\mathbf{v}(t) = -\left(\mathbf{K}_0 \mathbf{x}(t) + \mathbf{L}_0 \mathbf{u}(t) + \mathbf{L}_1 \mathbf{u}_1(t) + \cdots + \mathbf{L}_{q-1} \mathbf{u}_{q-1}(t) \right).$$

(7.245)

It is easy to see that $(\mathbf{A}_a, \mathbf{B}_a)$ is controllable and this along with the observability of $(Q_a, \mathbf{A}_a)$ implies, as we have seen before, that $(\mathbf{A}_a - \mathbf{B}_a \mathbf{K}_a^T)$ is stable.

From Equation (7.238), it follows that there exists

$$\mathbf{P} = [\mathbf{P}_0 \ \ \mathbf{P}_1 \ \cdots \ \mathbf{P}_q] \tag{7.246}$$

such that

$$[\mathbf{P}_0 \ \ \mathbf{P}_1 \ \cdots \ \mathbf{P}_q] \begin{bmatrix} \mathbf{C} \\ \mathbf{CA} \\ \mathbf{CA}^2 \\ \vdots \\ \mathbf{CA}^q \end{bmatrix} = \mathbf{K}_0. \tag{7.247}$$

Now observe that

$$\begin{aligned}
\mathbf{y}(t) &= \mathbf{C}\mathbf{x}(t), \\
\dot{\mathbf{y}}(t) &= \mathbf{CA}\mathbf{x}(t) + \mathbf{CB}\mathbf{u}(t), \\
\ddot{\mathbf{y}}(t) &= \mathbf{CA}^2\mathbf{x}(t) + \mathbf{CAB}\mathbf{u}(t) + \mathbf{CB}\mathbf{u}_1(t), \\
&\vdots \\
\mathbf{y}^{(q)}(t) &= \mathbf{CA}^q\mathbf{x}(t) + \mathbf{CA}^{q-1}\mathbf{B}\mathbf{u}(t) + \mathbf{CA}^{q-2}\mathbf{B}\mathbf{u}_1(t) + \cdots + \mathbf{CB}\mathbf{u}_{q-1}(t),
\end{aligned} \tag{7.248}$$

so that the optimal control in Equation (7.245) may be rewritten as

$$\begin{aligned}
\frac{d^q\mathbf{u}(t)}{dt^q} = -\bigg(&\mathbf{P}_0\mathbf{y}(t) + \mathbf{P}_1\frac{d\mathbf{y}(t)}{dt} + \cdots + \mathbf{P}_q\frac{d^q\mathbf{y}(t)}{dt^q} \\
&+ \mathbf{Q}_0\mathbf{u}(t) + \mathbf{Q}_1\frac{d\mathbf{u}(t)}{dt} + \cdots + \mathbf{Q}_{q-1}\frac{d^{q-1}\mathbf{u}(t)}{dt^{q-1}} \bigg),
\end{aligned} \tag{7.249}$$

where

$$\begin{aligned}
\mathbf{Q}_{q-1} &= \mathbf{L}_{q-1} - \mathbf{P}_q\mathbf{CB}, \\
\mathbf{Q}_{q-2} &= \mathbf{L}_{q-2} - \mathbf{P}_q\mathbf{CAB} - \mathbf{P}_{q-1}\mathbf{CB}, \\
&\vdots \\
\mathbf{Q}_1 &= \mathbf{L}_1 - \mathbf{P}_q\mathbf{CA}^{q-2}\mathbf{B} - \cdots - \mathbf{P}_2\mathbf{CB}, \\
\mathbf{Q}_0 &= \mathbf{L}_0 - \mathbf{P}_q\mathbf{CA}^{q-1}\mathbf{B} - \cdots - \mathbf{P}_1\mathbf{CB}.
\end{aligned} \tag{7.250}$$

Thus, Equation (7.249) corresponds to the optimal compensator transfer function matrix given by

$$\mathbf{u}(s) = \mathbf{Q}^{-1}(s)\mathbf{P}(s)(-\mathbf{y}(s)), \tag{7.251}$$

where

$$\mathbf{Q}(s) = \mathbf{I}s^q + \mathbf{Q}_{q-1}s^{q-1} + \mathbf{Q}_{q-2}s^{q-2} + \cdots + \mathbf{Q}_0, \tag{7.252}$$

$$\mathbf{P}(s) = \mathbf{P}_q s^q + \mathbf{P}_{q-1}s^{q-1} + \cdots + \mathbf{P}_0. \tag{7.253}$$

A state-space realization of Equations (7.251) and (7.253) of dynamic order rq is

$$
\begin{bmatrix} \dot{\mathbf{z}}_1(t) \\ \dot{\mathbf{z}}_2(t) \\ \vdots \\ \dot{\mathbf{z}}_q(t) \end{bmatrix} = \underbrace{\begin{bmatrix} 0 & \cdots & 0 & -\mathbf{Q}_0 \\ \mathbf{I} & & 0 & -\mathbf{Q}_1 \\ \vdots & \ddots & \vdots & \vdots \\ 0 & \cdots & \mathbf{I} & -\mathbf{Q}_{q-1} \end{bmatrix}}_{\mathbf{A}_c} \underbrace{\begin{bmatrix} \mathbf{z}_1(t) \\ \vdots \\ \mathbf{z}_q(t) \end{bmatrix}}_{\mathbf{z}(t)} + \underbrace{\begin{bmatrix} \mathbf{P}_0 - \mathbf{Q}_0 \mathbf{P}_q \\ \mathbf{P}_1 - \mathbf{Q}_1 \mathbf{P}_q \\ \vdots \\ \mathbf{P}_{q-1} - \mathbf{Q}_{q-1}\mathbf{P}_q \end{bmatrix}}_{\mathbf{B}_c} (-\mathbf{y}(t)),
$$

$$
\mathbf{u}(t) = \underbrace{[0 \ 0 \ \cdots \ 0 \ \mathbf{I}]}_{\mathbf{C}_c} \begin{bmatrix} \mathbf{z}_1(t) \\ \mathbf{z}_2(t) \\ \mathbf{z}_3(t) \\ \vdots \\ \mathbf{z}_q(t) \end{bmatrix} + \underbrace{\mathbf{P}_q}_{\mathbf{D}_c} (-\mathbf{y}(t)), \tag{7.254}
$$

or more compactly

$$
\begin{aligned}
\dot{\mathbf{z}}(t) &= \mathbf{A}_c \mathbf{z}(t) + \mathbf{B}_c(-\mathbf{y}(t)), \\
\mathbf{u}(t) &= \mathbf{C}_c \mathbf{z}(t) + \mathbf{D}_c(-\mathbf{y}(t)).
\end{aligned} \tag{7.255}
$$

The closed-loop dynamics are described by

$$
\begin{aligned}
\dot{\mathbf{x}}(t) &= \mathbf{A}\mathbf{x}(t) + \mathbf{B}\mathbf{u}(t) = \mathbf{A}\mathbf{x}(t) - \mathbf{B}(\mathbf{C}_c\mathbf{z}(t) - \mathbf{D}_c\mathbf{y}(t)) \\
&= (\mathbf{A} - \mathbf{B}\mathbf{D}_c\mathbf{C})\mathbf{x}(t) + \mathbf{B}\mathbf{C}_c\mathbf{z}(t), \\
\dot{\mathbf{z}}(t) &= \mathbf{A}_c\mathbf{z}(t) - \mathbf{B}_c\mathbf{y}(t) = -\mathbf{B}_c\mathbf{C}\mathbf{x}(t) + \mathbf{A}_c\mathbf{z}(t),
\end{aligned} \tag{7.256}
$$

so that

$$
\mathbf{A}_{cl} = \begin{bmatrix} \mathbf{A} - \mathbf{B}\mathbf{D}_c\mathbf{C} & \mathbf{B}\mathbf{C}_c \\ -\mathbf{B}_c\mathbf{C} & \mathbf{A}_c \end{bmatrix}. \tag{7.257}
$$

The next result shows that the closed-loop system has the same eigenvalues as the optimal system.

LEMMA 7.6 $(\mathbf{A}_a - \mathbf{B}_a\mathbf{K}_a)$ and $\mathbf{A}_{cl}$ are similar.

Proof Let

$$
\mathbf{T} := \begin{bmatrix} \mathbf{A}^q & \mathbf{A}^{q-1}\mathbf{B} & \cdots & \mathbf{A}\mathbf{B} & \mathbf{B} \\ \mathbf{P}_0\mathbf{C}\mathbf{A}^{q-1} & \mathbf{P}_0\mathbf{C}\mathbf{A}^{q-2}\mathbf{B} & \cdots & \mathbf{P}_0\mathbf{C}\mathbf{B} & \mathbf{Q}_0 \\ \vdots & \vdots & \ddots & \vdots & \vdots \\ \sum_{i=0}^{q-2}\mathbf{P}_i\mathbf{C}\mathbf{A}^{i+1} & \sum_{i=0}^{q-2}\mathbf{P}_i\mathbf{C}\mathbf{A}^i\mathbf{B} & \cdots & \mathbf{P}_{q-2}\mathbf{C}\mathbf{B} + \mathbf{Q}_{q-3} & \mathbf{Q}_{q-2} \\ \sum_{i=0}^{q-1}\mathbf{P}_i\mathbf{C}\mathbf{A}^i & \sum_{i=1}^{q-1}\mathbf{P}_i\mathbf{C}\mathbf{A}^{i-1}\mathbf{B} + \mathbf{Q}_0 & \cdots & \mathbf{P}_{q-1}\mathbf{C}\mathbf{B} + \mathbf{Q}_{q-2} & \mathbf{Q}_{q-1} \end{bmatrix}. \tag{7.258}
$$

Then it is not difficult to verify that

$$
\mathbf{T}\left(\mathbf{A}_a - \mathbf{B}_a\mathbf{K}_a^T\right) = \mathbf{A}_{cl}\mathbf{T}. \tag{7.259}
$$

and $\mathbf{T}$ is nonsingular. ∎

EXAMPLE 7.4   Consider the plant

$$\dot{\mathbf{x}}_p(t) = \mathbf{A}_p \mathbf{x}_p(t) + \mathbf{B}_p \mathbf{u}_p(t),$$
$$\mathbf{y}_p(t) = \mathbf{C}_p \mathbf{x}_p(t),$$

(7.260)

where

$$\mathbf{A}_p = \begin{bmatrix} 1 & 0 & 1 & 0 \\ 0 & 1 & 0 & 1 \\ 1 & 0 & 0 & 0 \\ 0 & 1 & 0 & 0 \end{bmatrix}, \quad \mathbf{B}_p = \begin{bmatrix} 0 & 1 \\ 1 & 1 \\ 1 & 0 \\ 0 & 0 \end{bmatrix}, \quad \mathbf{C}_p = \begin{bmatrix} 0 & 0 & 1 & 0 \\ 0 & 1 & 0 & 0 \end{bmatrix}.$$

(7.261)

Since

$$\text{Rank} \begin{bmatrix} \mathbf{C}_p \\ \mathbf{C}_p \mathbf{A}_p \end{bmatrix} = 4,$$

(7.262)

$q = 1$, we let ($\dot{\mathbf{u}}_p(t) = \mathbf{v}(t)$) and write

$$\mathbf{x}_a(t) = \begin{bmatrix} \mathbf{x}_p(t) \\ \mathbf{u}_p(t) \end{bmatrix}, \quad \mathbf{A}_a = \begin{bmatrix} \mathbf{A}_p & \mathbf{B}_p \\ 0 & 0 \end{bmatrix}, \quad \mathbf{B}_a = \begin{bmatrix} 0 \\ \mathbf{I} \end{bmatrix}.$$

(7.263)

Let

$$\mathbf{C}_a = \begin{bmatrix} \mathbf{C}_p & 0 \\ 0 & \mathbf{R} \end{bmatrix} = \begin{bmatrix} 0 & 0 & 1 & 0 & 0 & 0 \\ 0 & 1 & 0 & 0 & 0 & 0 \\ 0 & 0 & 0 & 0 & 1 & 0 \\ 0 & 0 & 0 & 0 & 0 & 1 \end{bmatrix},$$

(7.264)

so that $(\mathbf{C}_a, \mathbf{A}_a)$ is observable. Choose an LQR performance index with $(\mathbf{Q}_a = \mathbf{C}_a' \mathbf{C}_a)$ and $(\mathbf{R}_a = \mathbf{I})$.

Using MATLAB, we obtain the optimal feedback matrix

$\mathbf{K}_a =$

$$\begin{bmatrix} -25.0984 & 24.8637 & -15.5596 & 15.2837 & 4.4277 & -0.0631 \\ 25.4715 & -16.2511 & 15.9281 & -10.1213 & -0.0631 & 4.4087 \end{bmatrix},$$

(7.265)

and therefore

$$\mathbf{K}_0 := \begin{bmatrix} -25.0984 & 24.8637 & -15.5596 & 15.2837 \\ 25.4715 & -16.2511 & 15.9281 & -10.1213 \end{bmatrix},$$

$$\mathbf{L}_0 := \begin{bmatrix} 4.4277 & -0.0631 \\ -0.0631 & 4.4087 \end{bmatrix}.$$

(7.266)

From

$$\begin{bmatrix} \mathbf{P}_0 & \mathbf{P}_1 \end{bmatrix} = \mathbf{K}_0 \begin{bmatrix} \mathbf{C}_p \\ \mathbf{C}_p \mathbf{A}_p \end{bmatrix}^{-1},$$

(7.267)

we obtain

$$\mathbf{P}_0 = \begin{bmatrix} -15.5596 & 9.5801 \\ 15.9281 & -6.1298 \end{bmatrix}, \quad \mathbf{P}_1 = \begin{bmatrix} -25.0984 & 15.2837 \\ 25.4715 & -10.1213 \end{bmatrix},$$

(7.268)

and thus

$$\mathbf{Q}_0 = \mathbf{L}_0 - \mathbf{P}_1 \mathbf{C}_p \mathbf{B}_p = \begin{bmatrix} 14.2424 & -15.3468 \\ -15.4133 & 14.5300 \end{bmatrix}. \tag{7.269}$$

A state-space realization of the dynamic compensator is given by

$$\begin{aligned} \dot{\mathbf{z}}(t) &= -\mathbf{Q}_0 \mathbf{z}(t) - (\mathbf{P}_0 - \mathbf{Q}_0 \mathbf{P}_1) \mathbf{y}_p(t), \\ \mathbf{u}_p(t) &= \mathbf{z}(t) - \mathbf{P}_1 \mathbf{y}_p(t), \end{aligned} \tag{7.270}$$

and so the closed-loop system becomes

$$\begin{bmatrix} \dot{\mathbf{x}}_p(t) \\ \dot{\mathbf{z}}(t) \end{bmatrix} = \underbrace{\begin{bmatrix} \mathbf{A}_p - \mathbf{B}_p \mathbf{P}_1 \mathbf{C}_p & \mathbf{B}_p \\ -(\mathbf{P}_0 - \mathbf{Q}_0 \mathbf{P}_1) \mathbf{C}_p & -\mathbf{Q}_0 \end{bmatrix}}_{=:\mathbf{A}_{cl}} \begin{bmatrix} \mathbf{x}_p(t) \\ \mathbf{z}(t) \end{bmatrix} + \begin{bmatrix} \mathbf{B}_p \mathbf{P}_1 \\ \mathbf{P}_0 - \mathbf{Q}_0 \mathbf{P}_1 \end{bmatrix} \mathbf{r}(t),$$

$$\mathbf{y}_p(t) = \begin{bmatrix} \mathbf{C}_p & 0 \end{bmatrix} \begin{bmatrix} \mathbf{x}_p(t) \\ \mathbf{z}(t) \end{bmatrix}, \tag{7.271}$$

where $\mathbf{r}(t)$ is an external reference input.

The eigenvalues of $\mathbf{A}_{cl}$ are

$$\{-0.4313 + j0, -0.9146 \pm j0.4260, -1.4599 \pm j0.5006, -1.6560 + j0\}. \tag{7.272}$$

Thus, the dynamic compensator method utilizes the LQR theory to design a compensator driven by measurable outputs that minimizes the following cost function:

$$\begin{aligned} J_a &= \int_0^\infty \left[\mathbf{x}_a(t)^T \mathbf{Q}_a \mathbf{x}_a(t) + \mathbf{v}(t)^T \mathbf{R}_a \mathbf{v}(t) \right] dt \\ &= \int_0^\infty \left[\mathbf{x}_p(t)^T \mathbf{C}_p^t \mathbf{C}_p \mathbf{x}_p(t) + \mathbf{u}_p(t)^T \mathbf{R}^T R \mathbf{u}_p(t) + \dot{\mathbf{u}}(t)^T \mathbf{R}_a \dot{\mathbf{u}}(t) \right] dt, \end{aligned} \tag{7.273}$$

whereas the standard LQR minimizes

$$J = \int_0^\infty \left[\mathbf{x}_p(t)^T \mathbf{C}_p^t \mathbf{C}_p \mathbf{x}_p(t) + \mathbf{u}_p(t)^T \mathbf{R} \mathbf{u}_p(t) \right] dt. \tag{7.274}$$

EXAMPLE 7.5   In this example we show how a tuning parameter can be introduced into the approach outlined above to robustify the closed-loop optimal system. Recalling Example 7.3, we set $\mathbf{Q}_a$ using $\mathbf{H}$ in Equation (7.235):

$$\mathbf{Q}_a = \begin{bmatrix} \mathbf{H}^T \mathbf{H} & 0 \\ 0 & 1 \end{bmatrix} = \begin{bmatrix} 2800 & 80\sqrt{35} & 0 \\ 80\sqrt{35} & 80 & 0 \\ 0 & 0 & 1 \end{bmatrix} \tag{7.275}$$

and

$$\mathbf{R}_a = \rho \mathbf{I}, \tag{7.276}$$

where ρ is our design parameter. Since

$$\text{Rank} \begin{bmatrix} C \\ CA \end{bmatrix} = 2, \tag{7.277}$$

we have $q = 1$. Hence we have the following augmented matrices:

$$\mathbf{A}_a = \begin{bmatrix} \mathbf{A} & \mathbf{B} \\ 0 & 0 \end{bmatrix} = \begin{bmatrix} 0 & 1 & 0 \\ -3 & -4 & 1 \\ 0 & 0 & 0 \end{bmatrix}, \qquad \mathbf{B}_a = \begin{bmatrix} 0 \\ 0 \\ 1 \end{bmatrix}. \tag{7.278}$$

According to the performance index in Equation (7.243), the gain matrix is obtained as

$$\mathbf{K}_a = \begin{bmatrix} 39.0663 & 10.2859 & 4.6446 \end{bmatrix}. \tag{7.279}$$

Now, solve Equation (7.247) for $\mathbf{P}_0$ and $\mathbf{P}_1$ to get

$$\begin{bmatrix} \mathbf{P}_0 & \mathbf{P}_1 \end{bmatrix} = \begin{bmatrix} 47.2749 & 18.4945 \end{bmatrix} \tag{7.280}$$

and

$$\mathbf{Q}_0 = \mathbf{L}_0 - \mathbf{P}_1 \mathbf{CB} = 4.6446 - 18.4945 \begin{bmatrix} 2 & 1 \end{bmatrix} \begin{bmatrix} 0 \\ 1 \end{bmatrix} = -13.8500. \tag{7.281}$$

Following Equation (7.254), the state-variable parameters for the controller are

$$\mathbf{A}_c = 13.8500, \quad \mathbf{B}_c = 303.4230, \quad \mathbf{C}_c = 1, \quad \mathbf{D}_c = 18.4945 \tag{7.282}$$

and thus the transfer function is

$$\mathbf{C}_c (s\mathbf{I} - \mathbf{A}_c)^{-1} \mathbf{B}_c + \mathbf{D}_c = \frac{18.49s + 47.27}{s - 13.85}. \tag{7.283}$$

Figure 7.12 shows the comparison of the Nyquist plots for the state feedback, observer-based control, and dynamic compensator cases, respectively. Clearly the compensator is more robust than the observer, in this example, because its Nyquist plot is further from $-1 + j$ than the observer-based controller.

We may employ ρ as a design parameter. By choosing various values of ρ, say $\rho \in (0.01, 100)$, we have corresponding optimal controllers. For each controller we plot step responses and stability margins with respect to ρ. Since this is a regulator design problem, the step responses have nonzero steady-state errors. Figures 7.13(a) and (b) show the variation in the robustness as we choose different values for ρ.

Figure 7.14 shows the impulse responses. The final design can be obtained by selecting a ρ value that results in satisfactory robustness as well as transient behavior, referring to the plots in Figures 7.13 and 7.14.

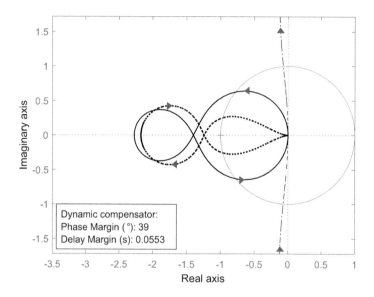

Figure 7.12 Nyquist plots of system with state feedback (dash-dot line), system with observer-based feedback (dotted line), and system with dynamic compensator (solid line) ($\rho = 1$) (Example 7.5)

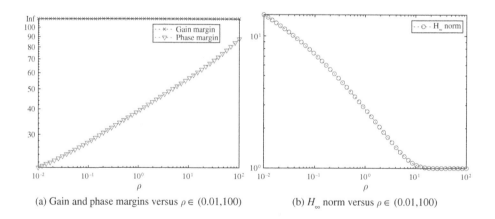

(a) Gain and phase margins versus $\rho \in (0.01, 100)$ (b) H_∞ norm versus $\rho \in (0.01, 100)$

Figure 7.13 Example 7.5

7.14   Dual Compensators

A dual procedure for optimal dynamic compensation that is sometimes useful for reducing the controller order is described next. For this, we consider the "dual plant"

$$\dot{\tilde{\mathbf{x}}}(t) = \mathbf{A}^T \tilde{\mathbf{x}}(t) + \mathbf{C}^T \tilde{\mathbf{u}}(t),$$
$$\tilde{\mathbf{y}}(t) = \mathbf{B}^T \tilde{\mathbf{x}}(t). \tag{7.284}$$

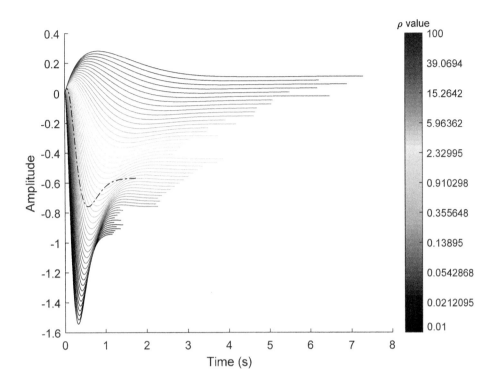

Figure 7.14 Step responses for $\rho \in (0.01, 100)$ (Example 7.5)

Let

$$\text{Rank}\begin{bmatrix} \mathbf{B} & \mathbf{AB} & \cdots & \mathbf{A}^p\mathbf{B} \end{bmatrix} = n. \tag{7.285}$$

The minimum value of p satisfying the above rank condition is called the controllability index. Introduce

$$\begin{aligned}
\mathring{\mathbf{u}}(t) &:= \tilde{\mathbf{u}}_1(t), \\
\mathring{\mathbf{u}}_1(t) &:= \tilde{\mathbf{u}}_2(t), \\
&\vdots \\
\mathring{\mathbf{u}}_{p-1}(t) &:= \tilde{\mathbf{v}}(t),
\end{aligned} \tag{7.286}$$

and consider the augmented system

$$\dot{\tilde{\mathbf{x}}}_a(t) = \tilde{\mathbf{A}}_a\tilde{\mathbf{x}}_a(t) + \tilde{\mathbf{B}}_a\tilde{\mathbf{v}}(t) \tag{7.287}$$

with quadratic performance index

$$I = \int_0^\infty \left[\tilde{\mathbf{x}}_a(t)^T \tilde{\mathbf{Q}}_a \tilde{\mathbf{x}}_a(t) + \tilde{\mathbf{v}}(t)^T \tilde{\mathbf{R}}\tilde{\mathbf{v}}(t) \right] dt, \tag{7.288}$$

where $(\tilde{\mathbf{Q}}_a, \tilde{\mathbf{A}}_a)$ is observable, $\tilde{\mathbf{R}} = \tilde{\mathbf{R}}^T$ is positive definite, and

$$
\tilde{\mathbf{A}}_a = \begin{bmatrix} \mathbf{A}^T & \mathbf{C}^T & \mathbf{0} & \cdots & \mathbf{0} \\ \mathbf{0} & \mathbf{0} & \mathbf{I} & & \mathbf{0} \\ \vdots & & & \ddots & \vdots \\ \vdots & & & & \mathbf{I} \\ \mathbf{0} & \mathbf{0} & \mathbf{0} & \cdots & \mathbf{0} \end{bmatrix}, \quad \tilde{\mathbf{B}}_a = \begin{bmatrix} \mathbf{0} \\ \mathbf{0} \\ \vdots \\ \mathbf{0} \\ \mathbf{I} \end{bmatrix}. \tag{7.289}
$$

The dual plant is observable and controllable, since the original plant in Equations (7.236) and (7.237) is controllable and observable. Thus, $(\tilde{\mathbf{A}}_a, \tilde{\mathbf{B}}_a)$ is controllable and therefore the optimal control

$$
\tilde{\mathbf{v}}(t) = -\tilde{\mathbf{K}}_a \tilde{\mathbf{x}}_a(t), \tag{7.290}
$$

written in expanded form as

$$
\tilde{\mathbf{v}}(t) = -\left(\tilde{\mathbf{K}}_0 \tilde{\mathbf{x}}(t) + \mathbf{L}_0 \tilde{\mathbf{u}}(t) + \mathbf{L}_1 \tilde{\mathbf{u}}_1(t) + \cdots + \mathbf{L}_{p-1} \tilde{\mathbf{u}}_{p-1}(t) \right), \tag{7.291}
$$

is stabilizing, that is

$$
\tilde{\mathbf{A}}_a - \tilde{\mathbf{B}}_a \tilde{\mathbf{K}}_a \tag{7.292}
$$

is stable.

Now introduce the "dual" compensator

$$
\begin{aligned}
\frac{d^p \tilde{\mathbf{u}}(t)}{dt^p} = -&\left(\mathbf{P}_0^T \tilde{\mathbf{u}}(t) + \mathbf{P}_1^T \frac{d\tilde{\mathbf{u}}(t)}{dt} + \cdots + \mathbf{P}_{p-1}^T \frac{d^{p-1}\tilde{\mathbf{u}}(t)}{dt^{p-1}} \right. \\
&\left. + \mathbf{Q}_0^T \tilde{\mathbf{y}}(t) + \mathbf{Q}_1^T \frac{d\tilde{\mathbf{y}}(t)}{dt} + \cdots + \mathbf{Q}_q^T \frac{d^p \tilde{\mathbf{y}}(t)}{dt^p} \right)
\end{aligned} \tag{7.293}
$$

with transfer function

$$
\tilde{\mathbf{u}}(s) = \left(\tilde{\mathbf{P}}(s)^T \right)^{-1} \tilde{\mathbf{Q}}(s)^T \left(-\tilde{\mathbf{y}}(s) \right), \tag{7.294}
$$

where

$$
\begin{aligned}
\tilde{\mathbf{P}}(s)^T &= \left[s^p \mathbf{I} + s^{p-1} \mathbf{P}_{p-1}^T + \cdots + \mathbf{P}_0^T \right]^{-1}, \\
\tilde{\mathbf{Q}}(s)^T &= \mathbf{Q}_0^T + \mathbf{Q}_1^T s + \cdots + \mathbf{Q}_q^T s^q.
\end{aligned} \tag{7.295}
$$

We take the "dual" of Equations (7.294) and (7.295) that is transpose them to finally obtain the transfer function matrix of the dynamic compensator for the original plant:

$$
\mathbf{u}(s) = \tilde{\mathbf{Q}}(s) \tilde{\mathbf{P}}^{-1}(s) \left(-\mathbf{y}(s) \right). \tag{7.296}
$$

A state-space realization of Equation (7.296) is

$$
\begin{bmatrix} \dot{\mathbf{w}}_1(t) \\ \dot{\mathbf{w}}_2(t) \\ \dot{\mathbf{w}}_3(t) \\ \vdots \\ \dot{\mathbf{w}}_p(t) \end{bmatrix} = \begin{bmatrix} 0 & \mathbf{I} & 0 & \cdots & 0 \\ 0 & 0 & \mathbf{I} & & \vdots \\ \vdots & & & \ddots & 0 \\ 0 & & & & \mathbf{I} \\ -\mathbf{P}_0 & -\mathbf{P}_1 & \cdots & \cdots & -\mathbf{P}_{p-1} \end{bmatrix} \begin{bmatrix} \mathbf{w}_1(t) \\ \mathbf{w}_2(t) \\ \mathbf{w}_3(t) \\ \vdots \\ \mathbf{w}_p(t) \end{bmatrix} + \begin{bmatrix} 0 \\ 0 \\ \vdots \\ 0 \\ \mathbf{I} \end{bmatrix} (-\mathbf{y}(t)),
$$

(7.297)

$$
\mathbf{u}(t) = [\mathbf{Q}_0 - \mathbf{Q}_p\mathbf{P}_0, \quad \cdots \quad, \mathbf{Q}_{p-1} - \mathbf{Q}_p\mathbf{P}_p] \begin{bmatrix} \mathbf{w}_1(t) \\ \mathbf{w}_2(t) \\ \mathbf{w}_3(t) \\ \vdots \\ \mathbf{w}_p(t) \end{bmatrix} + \mathbf{Q}_p(-\mathbf{y}(t)),
$$

where the Q_i, P_j are calculated from

$$
[\mathbf{B} \ \ \mathbf{AB} \ \ \cdots \ \ \mathbf{A}^p\mathbf{B}] \begin{bmatrix} \mathbf{Q}_0 \\ \mathbf{Q}_1 \\ \vdots \\ \mathbf{Q}_p \end{bmatrix} = \tilde{\mathbf{K}}_0
$$

(7.298)

and

$$
\begin{aligned}
\mathbf{P}_{p-1} &= \mathbf{L}_{p-1} - \mathbf{CBQ}_p, \\
\mathbf{P}_{p-2} &= \mathbf{L}_{p-2} - \mathbf{CABQ}_p - \mathbf{CBQ}_{p-1}, \\
&\vdots \\
\mathbf{P}_0 &= \mathbf{L}_0 - \mathbf{CA}^{p-1}\mathbf{BQ}_p - \mathbf{CA}^{p-2}\mathbf{BQ}_{p-1} - \cdots - \mathbf{CBQ}_1.
\end{aligned}
$$

(7.299)

Remark 7.2 It is straightforward to prove by dualizing the result for the previous case that the plant in Equations (7.236) and (7.237) with the compensator in Equation (7.297) has a closed-loop spectrum that is identical to the eigenvalues of $\tilde{\mathbf{A}}_a - \tilde{\mathbf{B}}_a\tilde{\mathbf{K}}_a$.

Remark 7.3 Finally, it should be pointed out that a controllable, observable system can be rendered controllable from each single input and observable from each single output by "almost any" output feedback. The proof of this result depends on the fact that the output feedback matrices rendering the closed-loop system not observable from a given output or not controllable from a given single-input lie on a proper algebraic variety. The details of the proof are omitted and the reader is referred to the literature in the Notes and References section. This result can be used to prove the important result that a pole placement controller of order no higher that the minimum of the observability and controllability indices exists for any plant.

7.15   Quadratic Optimal Discrete-Time Control

In this section, we briefly describe the discrete-time counterpart of the continuous-time linear quadratic optimal regulator problem. We begin with the finite-horizon problem.

7.15.1   Finite-Horizon Optimization Problem

Consider the discrete-time linear time-varying plant:

$$\mathbf{x}_{j+1} = \mathbf{A}_j \mathbf{x}_j + \mathbf{B}_j \mathbf{u}_j, \quad \text{for } j \in [k_0, N-1] \tag{7.300}$$

and the quadratic cost functional

$$J = \sum_{j=k_0}^{N-1} \left(\mathbf{x}_j^T \mathbf{Q}_j \mathbf{x}_j + \mathbf{u}_j^T \mathbf{R}_j \mathbf{u}_j \right) + \mathbf{x}_N^T \mathbf{M} \mathbf{x}_N \tag{7.301}$$

with

$$\mathbf{Q}_j = \mathbf{Q}_j^T \geq 0, \quad \mathbf{R}_j = \mathbf{R}_j^T > 0, \quad \mathbf{M} = \mathbf{M}^T \geq 0. \tag{7.302}$$

We seek to find the optimal input

$$\left(\mathbf{u}_{k_0}, \mathbf{u}_{k_0+1}, \ldots, \mathbf{u}_{N-1} \right) \tag{7.303}$$

that minimizes J for a prescribed initial condition $\mathbf{x}_{k_0}$. The cist functional J reflects the control objective of maintaining a "small" value of $\sum_{k=k_0}^{N-1} \mathbf{x}_k^T \mathbf{Q}_k \mathbf{x}_k$. Using a "small" amount of control, $\sum_{k=k_0}^{N-1} \mathbf{u}_k^T \mathbf{R}_k \mathbf{u}_k$, $\mathbf{x}_N^T \mathbf{M} \mathbf{x}_N$ represents the terminal cost or penalty.

The solution will be developed using Bellman's Principle of Optimality or dynamic programming. To that end, we introduce the value function

$$V_k(\mathbf{x}_k) := \min_{\mathbf{u}_k, \mathbf{u}_{k+1}, \cdots, \mathbf{u}_{N-1}} \left(\sum_{j=k}^{N-1} \mathbf{x}_j^T \mathbf{Q}_j \mathbf{x}_j + \mathbf{u}_j^T \mathbf{R}_j \mathbf{u}_j \right) + \mathbf{x}_N^T \mathbf{M} \mathbf{x}_N, \tag{7.304}$$

for $k \in [k_0, N-1]$. By the Principle of Optimality:

$$V_k[\mathbf{x}_k] = \min_{\mathbf{u}_k, \mathbf{u}_{k+1}, \cdots, \mathbf{u}_{N-1}} \left[\mathbf{x}_k^T \mathbf{Q}_k \mathbf{x}_k + \mathbf{u}_k^T \mathbf{R}_k \mathbf{u}_k + \sum_{j=k+1}^{N-1} \left(\mathbf{x}_j^T \mathbf{Q}_j \mathbf{x}_j + \mathbf{u}_j^T \mathbf{R}_j \mathbf{u}_j \right) + \mathbf{x}_N^T \mathbf{M} \mathbf{x}_N \right]$$

$$= \min_{\mathbf{u}_k} \left[\mathbf{x}_k^T \mathbf{Q}_k \mathbf{x}_k + \mathbf{u}_k^T \mathbf{R}_k \mathbf{u}_k + \min_{\mathbf{u}_{k+1}, \cdots, \mathbf{u}_{N-1}} V_{k+1}(\mathbf{x}_{k+1}) \right]. \tag{7.305}$$

Assume now that the value function is quadratic in $\mathbf{x}_K$, that is:

$$V_k(\mathbf{x}_k) = \mathbf{x}_k^T \mathbf{P}_k \mathbf{x}_k, \quad \text{for } k \in [k_0, N], \tag{7.306a}$$

$$\mathbf{P}_k = \mathbf{P}_k^T \geq 0, \quad \mathbf{P}_N = \mathbf{M}, \tag{7.306b}$$

for some $\mathbf{P}_k, k = k_0, \ldots, N$.

Substituting Equation (7.306a) in Equation (7.305), we obtain

$$V_k(\mathbf{x}_k) =$$

$$\min_{\mathbf{u}_k} \left[\mathbf{x}_k^T \mathbf{Q}_k \mathbf{x}_k + \mathbf{u}_k^T \mathbf{R}_k \mathbf{u}_k + (\mathbf{A}_k \mathbf{x}_k + \mathbf{B}_k \mathbf{u}_k)^T \mathbf{P}_{k+1} (\mathbf{A}_k \mathbf{x}_k + \mathbf{B}_k \mathbf{u}_k) \right]. \tag{7.307}$$

Carrying out the minimization in Equation (7.307), we obtain the optimal $\mathbf{u}_k = \mathbf{u}_k^*$:

$$2\mathbf{u}_k^{*T}\mathbf{R}_k + 2\mathbf{B}_k^T\mathbf{P}_{k+1}\mathbf{P}_k + 2\mathbf{x}_k^T\mathbf{A}_k^T\mathbf{P}_{k+1}\mathbf{B}_k = 0, \tag{7.308a}$$

$$\mathbf{u}_k^* = -\underbrace{\left(\mathbf{R}_k + \mathbf{B}_k^T\mathbf{P}_{k+1}\mathbf{B}_k\right)^{-1}\mathbf{B}_k^T\mathbf{P}_{k+1}\mathbf{A}_k}_{\mathbf{F}_k}\mathbf{x}_k = -\mathbf{F}_k\mathbf{x}_k. \tag{7.308b}$$

Substituting Equation (7.308a) in Equation (7.307), we have

$$\mathbf{x}_k^T\mathbf{P}_k\mathbf{x}_k =$$
$$\mathbf{x}_k\left[\mathbf{Q}_k + \mathbf{A}_k^T\mathbf{P}_{k+1}\mathbf{A}_k - \mathbf{A}_k^T\mathbf{P}_{k+1}\mathbf{B}_k\left(\mathbf{R}_k + \mathbf{B}_k^T\mathbf{P}_{k+1}\mathbf{B}_k\right)^{-1}\mathbf{B}_k^T\mathbf{P}_{k+1}\mathbf{A}_k\right]\mathbf{x}_k, \tag{7.309}$$

which is satisfied if the sequence $\mathbf{P}_k$ satisfies

$$\mathbf{P}_k = \mathbf{Q}_k + \mathbf{A}_k^T\mathbf{P}_{k+1}\mathbf{A}_k - \mathbf{A}_k^T\mathbf{P}_{k+1}\mathbf{B}_k\left(\mathbf{R}_k + \mathbf{B}_k^T\mathbf{P}_{k+1}\mathbf{B}_k\right)^{-1}\mathbf{B}_k^T\mathbf{P}_{k+1}\mathbf{A}_k, \tag{7.310}$$

$$\mathbf{P}_N = \mathbf{M}. \tag{7.311}$$

Summarizing these developments, we see that the optimal control is given by Equation (7.308b) where $\mathbf{P}_K$ satisfies Equation (7.310) with the boundary condition $\mathbf{P}_N = \mathbf{M}$. Clearly, $\mathbf{P}_K$ has to be solved offline, backwards in time following the sequence

$$\mathbf{P}_N \to \mathbf{P}_{N-1} \to \mathbf{P}_{N-2} \to \cdots \to \mathbf{P}_{K_0} \tag{7.312}$$

and stored to apply the optimal control

$$\mathbf{u}_k^*, \quad \text{for } k = k_0, k_0 + 1, \ldots, N - 1 \tag{7.313}$$

in real time. The minimum cost is given by $\mathbf{x}_{k_0}^T\mathbf{P}_{k_0}\mathbf{x}_{k_0}$.

7.15.2 Infinite-Horizon Optimal Control

For the infinite-horizon optimal control problem, we assume the plant is a linear time-invariant system

$$\mathbf{x}_{k+1} = \mathbf{A}\mathbf{x}_k + \mathbf{B}\mathbf{u}_k, \tag{7.314a}$$

$$\mathbf{y}_k = \mathbf{C}\mathbf{x}_k, \tag{7.314b}$$

and the cost function

$$J = \sum_{j=0}^{\infty}\left(\mathbf{y}_j^T\mathbf{y}_j + \mathbf{u}_j\mathbf{R}\mathbf{u}_j\right) = \sum_{j=0}^{\infty}\left(\mathbf{x}_j^T\underbrace{\mathbf{C}^T\mathbf{C}}_{\mathbf{Q}}\mathbf{x}_j + \mathbf{u}_j^T\mathbf{R}\mathbf{u}_j\right) \tag{7.315a}$$

$$= \sum_{j=0}^{\infty}\left(\mathbf{x}_j^T\mathbf{Q}\mathbf{x}_j + \mathbf{u}_j^T\mathbf{R}\mathbf{u}_j\right), \tag{7.315b}$$

with $\mathbf{Q} = \mathbf{Q}^T \geq 0$ and $\mathbf{R} = \mathbf{R}^T > 0$. We seek the state feedback optimal control

$$\mathbf{u}_k = -\mathbf{K}\mathbf{x}_k \tag{7.316}$$

that minimizes J and renders $\mathbf{A} - \mathbf{B}\mathbf{K}$ Schur stable.

Proceeding as before by invariant embedding, introduce the value function

$$V(\mathbf{x}_k) := \inf_{\mathbf{u}_k, \cdots, \mathbf{u}_\infty} \left[\sum_{j=k}^{\infty} \left(\mathbf{x}_j^T \mathbf{Q} \mathbf{x}_j + \mathbf{u}_j^T \mathbf{R} \mathbf{u}_j \right) \right], \quad \text{for } k \in [0, \infty). \tag{7.317}$$

By the Principle of Optimality or dynamic programming:

$$V(\mathbf{x}_k) = \inf_{\mathbf{u}_k} \left[\mathbf{x}_k^T \mathbf{Q} \mathbf{x}_k + \mathbf{u}_k^T \mathbf{R} \mathbf{u}_k + V(\mathbf{x}_{k+1}) \right]. \tag{7.318}$$

Assuming that

$$V(\mathbf{x}_k) = \mathbf{x}_k^T \mathbf{P} \mathbf{x}_k, \quad \mathbf{P} = \mathbf{P}^T \geq 0, \tag{7.319}$$

we have

$$\mathbf{x}_k^T \mathbf{P} \mathbf{x}_k = \inf_{\mathbf{u}_k} \left[\mathbf{x}_k^T \mathbf{Q} \mathbf{x}_k + \mathbf{u}_k^T \mathbf{R} \mathbf{u}_k + \mathbf{x}_{k+1}^T \mathbf{P} \mathbf{x}_{k+1} \right] \tag{7.320a}$$

$$= \inf_{\mathbf{u}_k} \underbrace{\left[\mathbf{x}_k^T \mathbf{Q} \mathbf{x}_k + \mathbf{u}_k^T \mathbf{R} \mathbf{u}_k + (\mathbf{A} \mathbf{x}_k + \mathbf{B} \mathbf{u}_k)^T \mathbf{P} (\mathbf{A} \mathbf{x}_k + \mathbf{B} \mathbf{u}_k) \right]}_{\mathbf{H}}. \tag{7.320b}$$

The optimal control $\mathbf{u}_K^*$ is obtained by differentiating the right-hand side of Equation (7.320b) and setting it to zero:

$$2\mathbf{u}_k^{*T} \left(\mathbf{R} + \mathbf{B}^T \mathbf{P} \mathbf{B} \right) + 2\mathbf{x}_k^T \mathbf{A}^T \mathbf{P} \mathbf{B} = 0 \tag{7.321}$$

or

$$\mathbf{u}_k^* = - \underbrace{\left(\mathbf{R} + \mathbf{B}^T \mathbf{P} \mathbf{B} \right)^{-1} \mathbf{B}^T \mathbf{P} \mathbf{A}}_{\mathbf{K}} \mathbf{x}_k. \tag{7.322}$$

Substituting Equation (7.322) into Equation (7.320b), we obtain

$$\mathbf{x}_k^T \mathbf{P} \mathbf{x}_k = \mathbf{x}_k^T \left[\mathbf{Q} + \mathbf{K}^T \mathbf{R} \mathbf{K} + (\mathbf{A} - \mathbf{B} \mathbf{K})^{-1} \mathbf{P} (\mathbf{A} - \mathbf{B} \mathbf{K}) \right] \mathbf{x}_k. \tag{7.323}$$

Equation (7.323) is satisfied if $\mathbf{P}$ is chosen to satisfy

$$(\mathbf{A} - \mathbf{B} \mathbf{K})^T \mathbf{P} (\mathbf{A} - \mathbf{B} \mathbf{K}) - \mathbf{P} = -\mathbf{Q} - \mathbf{K}^T \mathbf{R} \mathbf{K}, \tag{7.324}$$

which can be shown to be equivalent to

$$\mathbf{A}^T \mathbf{P} \mathbf{A} - \mathbf{P} = -\mathbf{Q} + \mathbf{A}^T \mathbf{P} \mathbf{B} \left(\mathbf{R} + \mathbf{B}^T \mathbf{P} \mathbf{B} \right)^{-1} \mathbf{B}^T \mathbf{P} \mathbf{A} \tag{7.325}$$

using Equation (7.322).

It is possible to show that Equation (7.325) has a unique positive semi-definite solution $\mathbf{P}$ and $(\mathbf{A} - \mathbf{B} \mathbf{K})$ is stable if $(\mathbf{C}, \mathbf{A})$ is detectable. The proof is similar to that for the continuous-time case and is omitted.

To summarize the above results, we have obtained the solution of the discrete-time linear quadratic regulator (DTLQR) under the assumption that $(\mathbf{A}, \mathbf{B})$ is stabilizable and $(\mathbf{C}, \mathbf{A})$ is detectable. The solution consists of the state feedback in Equation (7.322) where $\mathbf{P}$ is the positive semi-definite solution of Equation (7.325), which is known as the discrete-time algebraic Riccati equation (DTARE).

The MATLAB program "dtlqr" solves the discrete-time LQR problem.

7.15.3   Solution of the Discrete-Time ARE

We have shown that the DTARE

$$\mathbf{A}^T \mathbf{PA} - \mathbf{P} = -\mathbf{Q} + \mathbf{A}^T \mathbf{PB} \left(\mathbf{R} + \mathbf{B}^T \mathbf{PB}\right)^{-1} \mathbf{B}^T \mathbf{PA} \qquad (7.326)$$

is equivalent to the two equations

$$(\mathbf{A} - \mathbf{BK})^T \mathbf{P}(\mathbf{A} - \mathbf{BK}) - \mathbf{P} + \mathbf{Q} + \mathbf{K}^T \mathbf{RK} = 0, \qquad (7.327a)$$

$$\mathbf{K} = \left(\mathbf{R} + \mathbf{B}^T \mathbf{PB}\right)^{-1} \mathbf{B}^T \mathbf{PA}. \qquad (7.327b)$$

An iterative algorithm can be set up to solve for $\mathbf{P}$ and $\mathbf{K}$ as follows:

$$\mathbf{P}_k = (\mathbf{A} - \mathbf{BK}_k)^T \mathbf{P}_k (\mathbf{A} - \mathbf{BK}_k) + \mathbf{Q} + \mathbf{K}_k^T \mathbf{RK}_k, \quad \text{for } k = 0, 1, 2, \qquad (7.328a)$$

$$\mathbf{K}_k = \left(\mathbf{R} + \mathbf{B}^T \mathbf{P}_{k-1} \mathbf{B}\right)^{-1} \mathbf{B}^T \mathbf{P}_{k-1} \mathbf{A}, \quad \text{for } k = 1, 2, 3, \qquad (7.328b)$$

and can be initiated with $\mathbf{K}_0$ chosen arbitrarily to Schur stabilize $\mathbf{A} - \mathbf{BK}_0$. With $\mathbf{Q} > 0$ and $(\mathbf{A} - \mathbf{BK}_k)$ stable, it follows that $\mathbf{P}_k > 0$ and it is possible to show that

$$\lim_{k \to \infty} \mathbf{P}_k = \mathbf{P} \qquad (7.329)$$

exists, satisfies Equation (7.326) and is unique. The condition $\mathbf{Q} > 0$ can be relaxed to $(\mathbf{Q}, \mathbf{A})$ detectable, and in this case we have $\mathbf{P} \geq 0$ instead of $\mathbf{P} > 0$. The detailed proofs, which are omitted, follow from properties of the discrete-time Lyapunov equation in Equation (7.328a).

Remark 7.4   It turns out that there are no guaranteed stability margins for the discrete-time LQR, unlike what was true in the continuous-time case.

EXAMPLE 7.6 (Discrete-time LQR)   Consider the following continuous-time system with a controllable state-space realization:

$$\mathbf{x}(t) = \begin{bmatrix} 0 & 1 & 0 \\ 0 & 0 & 1 \\ -1 & -2 & -3 \end{bmatrix} \mathbf{x}(t) + \begin{bmatrix} 1 & 0 \\ 0 & 1 \\ 1 & 1 \end{bmatrix} \mathbf{u}(t). \qquad (7.330)$$

The system is discretized with sampling time $T_s = 0.01$. This example shows how to use MATLAB to design a discrete-time LQR controller and verify the results by plotting the states of the closed-loop system.

MATLAB Solution

```
clear
A1=[0 1 0; 0 0 1; -1 -2 -3];   % continuous-time system model
B1=[1 0; 0 1; 1 1];
C1=eye(3);
D1=zeros(3,2);
sys_ss=ss(A1,B1,C1,D1);
Ts=1/100;
sys_d=c2d(sys_ss,Ts,'zoh');
```

```
% checking controllability
controllability=rank(ctrb(sys_d));
A=sys_d.a;  B=sys_d.b;  % discrete-time system model
C=sys_d.c;  D=sys_d.d;
Q=C'*C;
R=eye(2);
[K]=dlqr(A,B,Q,R);
sys_cl=ss(A-B*K,B,C,D,Ts);
t=0:Ts:5;
[y,t,x]=step(sys_cl,t);
stairs(t,y(:,1),'b-'), hold on
stairs(t,y(:,2),'b-.'),
stairs(t,y(:,3),'-b--'), hold off
```

Figure 7.15 (upper) shows the unit step response with the LQR control obtained by setting $\mathbf{Q} = \mathbf{C}^T\mathbf{C}$ as above. The closed-loop eigenvalues and the spectral radius $\rho(\cdot)$ (largest value over the magnitudes of all eigenvalues) are

$$\sigma(\mathbf{A} - \mathbf{BF}) = \{0.9845 \pm j0.0082, 0.9768\} \text{ and } \rho(\mathbf{A} - \mathbf{BF}) = 0.9845, \quad (7.331)$$

which verifies the closed-loop stability. Figure 7.15 (lower) shows the case when $\mathbf{Q} = 50\mathbf{C}^T\mathbf{C}$ and it shows faster convergence. The closed-loop eigenvalues are

$$\sigma(\mathbf{A} - \mathbf{BF}) = \{0.8856, 0.9698, 0.9332\}. \quad (7.332)$$

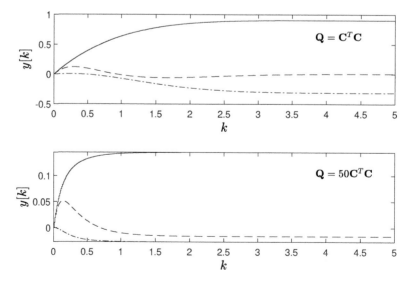

Figure 7.15 Step responses with digital LQR control with $\mathbf{Q} = \mathbf{C}^T\mathbf{C}$ and $\mathbf{Q} = 50\mathbf{C}^T\mathbf{C}$ (Example 7.6)

7.16   Exercises

7.1   Consider the system

$$\dot{x}(t) = u(t), \qquad x(0) = x_0 \tag{7.333}$$

and the performance index

$$I = \int_0^T \left(x^2(t) + u^2(t) + \frac{1}{2}x^4(t) \right) dt. \tag{7.334}$$

Write down the Hamilton–Jacobi–Bellman equation for the optimal system.

Answer:

$$H = x^2(t) + u^2(t) + \frac{1}{2}x^4(t) + \frac{\partial V}{\partial x(t)}u(t), \tag{7.335}$$

$$\left(H = \phi + \frac{\partial V}{\partial x(t)}f \right) 2u^*(t) + \frac{\partial V}{\partial x(t)} = 0, \tag{7.336}$$

$$H^* = x^2(t) + \frac{1}{4}\left(\frac{\partial V}{\partial x(t)} \right)^2 + \frac{1}{2}x^4(t) - \frac{1}{2}\left(\frac{\partial V}{\partial x(t)} \right)^2, \tag{7.337}$$

$$\frac{\partial V}{\partial t} + x^2(t) + \frac{1}{2}x^4(t) - \frac{1}{4}\left(\frac{\partial V}{\partial x(t)} \right)^2 = 0 \quad \text{(HJB)}, \tag{7.338}$$

$$V(x(T), T) = 0, \qquad \left(H^* + \frac{\partial V}{\partial t} = 0 \right), \tag{7.339}$$

$$\dot{x}(t) = -\frac{1}{2}\frac{\partial V}{\partial x(t)} \qquad \text{(optimal system).} \tag{7.340}$$

7.2   Determine the optimal control for

$$\dot{x}(t) = u(t) \tag{7.341}$$

and

$$I = \int_0^T \left(x^2(t) + u^2(t) \right) dt. \tag{7.342}$$

Discuss what happens when T tends to ∞.

Answer:

$$u^*(t) = -P(t)x(t), \tag{7.343}$$

where

$$\dot{P}(t) = P^2(t) - 1, \qquad P(T) = 0, \qquad P(t) = \frac{e^{2(T-t)} - 1}{e^{2(T-t)} + 1}. \tag{7.344}$$

When $T \to \infty$, $P(t) \to 1$ and $u^*(t) = -x(t)$.

7.3   Determine the optimal control for

$$\dot{x}(t) = u(t) \tag{7.345}$$

and

$$I = \frac{1}{2}sx^2(2) + \frac{1}{2}\int_0^2 u^2(t)dt, \quad s > 0. \tag{7.346}$$

Discuss the behavior of the solution as $s \to 0$.

7.4   Find the optimal control for the double integrator

$$\ddot{y}(t) = u(t) \tag{7.347}$$

with

$$I = \int_0^\infty \left(y^2(t) + u^2(t)\right) dt. \tag{7.348}$$

7.5   Design an LQ optimal first-order controller for the double-integrator plant in Exercise 7.4 using the performance index

$$I = \int_0^\infty \left[y^2(t) + \dot{u}(t)^2\right] dt. \tag{7.349}$$

7.6   Given the state-space system for the double integrator

$$\dot{\mathbf{x}}(t) = \begin{bmatrix} 0 & 1 \\ 0 & 0 \end{bmatrix} \mathbf{x}(t) + \begin{bmatrix} 0 \\ 1 \end{bmatrix} \mathbf{u}(t), \tag{7.350}$$

choose a quadratic cost functional with $R = 1$ and $Q = I$ and show that for finite T:

$$\mathbf{K} = \begin{bmatrix} K_1 & K_2 \end{bmatrix}, \tag{7.351}$$

where

$$K_1 = \frac{\cosh\left[\sqrt{3}(t-T)\right] - 1}{2 + \cosh\left[\sqrt{3}(t-T)\right]}, \quad K_2 = \frac{-\sqrt{3}\sinh\left[\sqrt{3}(t-T)\right]}{2 + \cosh\left[\sqrt{3}(t-T)\right]}. \tag{7.352}$$

Show that when T tends to ∞, the gain tends to $\mathbf{K} = [1 \ \sqrt{3}]$.

7.7   The inverted-pendulum cart shown in Figure 7.16 is linearized about the straight-up position and has the following equations when the disturbance d is perpendicular to the pendulum:

$$\dot{\mathbf{x}}(t) = \begin{bmatrix} 0 & 1 & 0 & 0 \\ 0 & 0 & -\dfrac{m}{Mg} & 0 \\ 0 & 0 & 0 & 1 \\ 0 & 0 & \dfrac{(M+m)g}{ML} & 0 \end{bmatrix} \mathbf{x}(t) + \begin{bmatrix} 0 \\ \dfrac{1}{M} \\ 0 \\ -\dfrac{1}{ML} \end{bmatrix} u(t) + \begin{bmatrix} 0 \\ -\dfrac{1}{M} \\ 0 \\ \dfrac{M+m}{mML} \end{bmatrix} d(t). \tag{7.353}$$

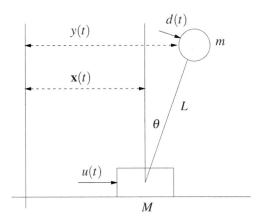

Figure 7.16 The inverted pendulum (Exercise 7.7)

Assume that

$$d(t) = 0, \quad M = 2 \text{ kg}, \quad m = 1 \text{ kg}, \quad L = 0.5 \text{ m}, \quad g = 9.18 \text{ m/s}^2. \qquad (7.354)$$

Find the infinite-horizon LQR controller when $Q = \text{diag}[1 \ 1 \ 10 \ 10]$ and $R = [1]$ and $R = [10]$. Plot θ and $u(t)$ for both cases when $\mathbf{x}(0) = [0 \ 0 \ 1 \ 0]^T$.

7.8 The system shown in Figure 7.17 is described by the equations

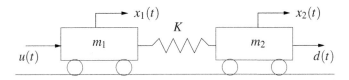

Figure 7.17 Two-mass spring system (Exercise 7.8)

$$\begin{bmatrix} \dot{x}_1(t) \\ \dot{x}_2(t) \\ \dot{x}_3(t) \\ \dot{x}_4(t) \end{bmatrix} = \begin{bmatrix} 0 & 0 & 1 & 0 \\ 0 & 0 & 0 & 1 \\ -\dfrac{k}{m_1} & \dfrac{k}{m_1} & 0 & 0 \\ \dfrac{k}{m_2} & -\dfrac{k}{m_2} & 0 & 0 \end{bmatrix} \begin{bmatrix} x_1(t) \\ x_2(t) \\ x_3(t) \\ x_4(t) \end{bmatrix} + \begin{bmatrix} 0 \\ 0 \\ \dfrac{1}{m_1} \\ 0 \end{bmatrix} u(t) + \begin{bmatrix} 0 \\ 0 \\ 0 \\ \dfrac{1}{m_2} \end{bmatrix} d(t),$$

$$(7.355)$$

where
$x_1(t)$ is the position of body 1 (m)
$x_2(t)$ is the position of body 2 (m)
$x_3(t)$ is the velocity of body 1 (m/s)
$x_4(t)$ is the velocity of body 2 (m/s)
$u(t)$ is the control-force input (N)

$d(t)$ is a disturbance-force input (N).

Assume that $m_1 = m_2 = k = 1$ and $d(t) = 0$.

(1) Design a time-invariant LQR controller with

$$\mathbf{Q}_1 = \text{diag}\begin{bmatrix} 1 & 0 & 0 & 0 \end{bmatrix} \quad \text{and} \quad R = [1]. \tag{7.356}$$

(2) Compare the transient behavior of the state components $x_1(t)$ and $x_2(t)$ and the control effort $u(t)$, given the initial condition

$$\mathbf{x}(0) = \begin{bmatrix} 1 & 0 & 0 & 0 \end{bmatrix}^T \tag{7.357}$$

and the control law above, with that obtained if $\mathbf{Q}_1$ were changed to $\mathbf{Q}_2 = \text{diag}[10 \ 0 \ 0 \ 0]$.

7.9 Find the optimal control for the double integrator

$$\ddot{y}(t) = u(t) \tag{7.358}$$

and

$$I = \int_0^\infty \left[y^2(t) + \rho u^2(t) \right] dt, \qquad \rho > 0. \tag{7.359}$$

Determine the closed-loop eigenvalues as a function of ρ. Discuss the behavior of the eigenvalues as $\rho \to 0$ (cheap control) and $\rho \to \infty$ (expensive control).

7.10 Design an LQ optimal first-order compensator for the double integrator

$$\ddot{y}(t) = u(t) \tag{7.360}$$

using the cost functional

$$I = \int_0^\infty \left[y^2(t) + \rho \left(\dot{u}(t) \right)^2 \right] dt. \tag{7.361}$$

Determine the compensator and closed-loop spectrum as a function of ρ and their behavior as $\rho \to 0$ and $\rho \to \infty$. Also determine the gain and phase margins of the closed-loop system at the input to the plant and study their behavior as $\rho \to 0$ and $\rho \to \infty$.

7.11 Consider the scalar plant

$$\dot{x}(t) = ax(t) + bu(t) \tag{7.362}$$

with

$$I = \int_0^\infty \left[qx^2(t) + \rho u^2(t) \right] dt. \tag{7.363}$$

Show that the optimal control is

$$u^*(t) = -Kx(t), \tag{7.364}$$

where

$$K = \frac{\sqrt{a^2 + \frac{q}{\rho b^2}} + a}{b} \tag{7.365}$$

and the closed-loop eigenvalue is

$$\lambda = -\sqrt{a^2 + \frac{qb^2}{\rho}}. \qquad (7.366)$$

Discuss the behavior of the system as $\rho \to 0$ and $\rho \to \infty$ for fixed q.

7.12 Show that the optimal control problem for

$$\dot{x}_1(t) = x_2(t), \qquad \dot{x}_2(t) = u(t) \qquad (7.367)$$

with

$$I = \int_0^\infty \left[(x_1^2(t) + q x_2^2(t)) + \rho u^2(t) \right] dt \qquad (7.368)$$

results in

$$u^*(t) = -K\mathbf{x}(t), \qquad (7.369)$$

with

$$K = \left(\frac{1}{\sqrt{\rho}}, \sqrt{\frac{q + 2\sqrt{\rho}}{\rho}} \right) = \frac{1}{\sqrt{\rho}} \left(1, \sqrt{q + 2\sqrt{\rho}} \right) \qquad (7.370)$$

and the closed-loop eigenvalues are

$$\lambda_1, \lambda_2 = \frac{1}{2\sqrt{\rho}} \left(-\sqrt{q + 2\sqrt{\rho}} \pm \sqrt{q - 2\sqrt{\rho}} \right). \qquad (7.371)$$

Discuss the behavior of the closed-loop system as $\rho \to 0$ and $\rho \to \infty$ for fixed q.

7.13 Prove that if $(\mathbf{A}, \mathbf{B})$ is controllable and $(\mathbf{C}, \mathbf{A})$ is observable, then the ARE has a unique positive-definite solution P.

Hint: Argue by contradiction and suppose there exists $\mathbf{x}(0) \neq 0$ such that

$$0 = \mathbf{x}(0)^T P \mathbf{x}(0) = \int_0^\infty \left[\mathbf{x}(t)^T \mathbf{C} \mathbf{C}^T \mathbf{x}(t) + \mathbf{u}(t)^T \mathbf{R} \mathbf{u}(t) \right] dt. \qquad (7.372)$$

7.17 Notes and References

The LQR problem was introduced by Kalman in his now historic paper Kalman (1960). The return difference relations were developed in Kalman (1964). Our treatment of LQR theory, which is now a standard topic, has closely followed and liberally borrowed from the excellent treatments in Wonham (1985), Dorato, Abdallah, and Cerone (1994), and Green and Limebeer (1995).

The existence condition for the LQR problem, obtained by treating it as a problem of zeroing the output, was obtained in Bhattacharyya (1973). The rotary double-inverted pendulum has been developed by Quanser, Inc. The details of the hardware and the derivation of the linearized model are described in Quanser (2012). The photo in Figure 7.7(a) is credited to Quanser, Inc.

Example 7.3 is taken from Doyle and Stein (1979). Optimal dynamic compensators were introduced by Pearson (1968). The result that a pole placement controller of order no higher than the minima of the controllability and observability indices was proved in Brasch and Pearson (1970).

Sections 7.1–7.8, 7.13, and 7.14 are adapted from Bhattacharyya, Datta, and Keel (2009), with permission. © 2008 Taylor & Francis LLC Books.

8 H_∞ and H_2 Optimal Control

The control design problem can sometimes be regarded as a disturbance-reduction problem. The controller must then be designed to stabilize the system and attenuate the transmission of the disturbances to outputs of particular interest, such as errors. This can be accomplished by using feedback control to reduce or minimize the "gain" or norm of the transfer function from disturbances to outputs. The H_∞ and H_2 norms can be used to quantify this gain, when the input and output signal magnitudes are measured using Euclidean or 2-norms. Minimizing the H_∞ norm of the disturbance transfer function can also enhance the stability margin of the closed-loop system. The H_2 norm can also be identified with a quadratic cost function as in the LQR problem.

In this chapter we develop the basic solutions of the H_∞ and H_2 optimal control problems using the state-space framework. As will be seen, the solutions have characteristics similar to the LQR problem and use Ricatti equations extensively to solve the optimization problems that arise. However, there are important differences, namely that an output feedback solution is obtained which in addition can have prescribed levels of tolerance or robustness to plant uncertainty.

We begin with a discussion of norms for linear systems and introduce the H_∞ and H_2 norms, their significance, and their computation. Then the generalized plant (GP) approach is described as the standard framework in which to formulate norm-optimization problems. This framework is then used to develop the H_∞ and H_2 optimal controllers. An illustrative example is used to demonstrate how a familiar control problem can be cast in the GP framework and MATLAB is used to compute the H_∞ and H_2 solutions.

8.1 Norms for Signals and Systems

In this section, we briefly introduce, mostly without proofs, some common notions of norms applicable to linear time-invariant dynamic systems.

Consider first a linear map

$$\mathbf{A} : \mathcal{X} \to \mathcal{Z}, \tag{8.1}$$

where $\mathcal{X}$ and $\mathcal{Z}$ are finite-dimensional vector spaces over the real or complex fields. Let $\|\mathbf{x}\|_X$ denote a norm of the vector $\mathbf{x}$, $\|\mathbf{z}\|_Z$ a norm of $\mathbf{z}$, and let sup denote supremum.

Then the quantity

$$\sup_{\mathbf{x} \in \mathscr{X}} \frac{\|\mathbf{z}\|_Z}{\|\mathbf{x}\|_X} = \sup_{\mathbf{x} \in \mathscr{X}} \frac{\|\mathbf{Ax}\|_Z}{\|\mathbf{x}\|_X} =: \|\mathbf{A}\| \tag{8.2}$$

is the *operator norm of* $\mathbf{A}$ *induced* by the vector norms.

Suppose now that $\mathscr{X}$ and $\mathscr{Z}$ have dimensions n and m, respectively, and we choose the vector norms, denoted by $\|\mathbf{x}\|_2$ and $\|\mathbf{z}\|_2$, to be the Euclidean or 2-norms as defined by

$$\|\mathbf{x}\|_2^2 := \sum_{i=1}^{n} |x_i|^2, \tag{8.3a}$$

$$\|\mathbf{z}\|_2^2 := \sum_{i=1}^{m} |z_i|^2. \tag{8.3b}$$

By definition, the operator norm induced by this vector norm and denoted by $|\mathbf{A}|_2$ is

$$\|\mathbf{A}\|_2 = \sup_{\mathbf{x} \in \mathscr{X}} \frac{\|\mathbf{z}\|_2}{\|\mathbf{x}\|_2} = \sup_{\mathbf{x} \in \mathscr{X}} \frac{\|\mathbf{Ax}\|_2}{\|\mathbf{x}\|_2}. \tag{8.4}$$

It is possible to compute $\|\mathbf{A}\|_2$ without executing the sup operation, as we show below. Let $\lambda_{\max}(\mathbf{M})$ denote the magnitude of the the largest-magnitude eigenvalue of a Hermitian matrix $\mathbf{M}$. A matrix $\mathbf{M}$ is Hermitian if and only if $\mathbf{M} = \mathbf{M}^*$.

LEMMA 8.1

$$\|\mathbf{A}\|_2 = \sqrt{\lambda_{\max}(\mathbf{A}^*\mathbf{A})} \tag{8.5a}$$

$$= \sigma_{\max}(\mathbf{A}) \quad \text{(largest singular value of } \mathbf{A}\text{).} \tag{8.5b}$$

Proof First $\mathbf{A}^*\mathbf{A}$ is a Hermitian matrix, so that the eigenvalues of $\mathbf{A}^*\mathbf{A}$ are real and non-negative. Furthermore

$$\mathbf{x}^*\mathbf{A}^*\mathbf{Ax} \le \max_i \lambda_i(\mathbf{A}^*\mathbf{A})\mathbf{x}^*\mathbf{x}$$

$$\Rightarrow \|\mathbf{Ax}\|_2^2 \le \max_i \lambda_i(\mathbf{A}^*\mathbf{A})\|\mathbf{x}\|_2^2 \tag{8.6}$$

$$\Rightarrow \|\mathbf{Ax}\|_2 \le \left[\max_i \lambda_i(\mathbf{A}^*\mathbf{A})\right]^{\frac{1}{2}} |\mathbf{x}|_2.$$

Thus, for $\mathbf{x} \ne 0$ we have

$$\frac{\|\mathbf{Ax}\|_2}{\|\mathbf{x}\|_2} \le \left[\max_i \lambda_i(\mathbf{A}^*\mathbf{A})\right]^{\frac{1}{2}} \quad \Rightarrow \quad \|\mathbf{A}\|_2 \le \left[\max_i \lambda_i(\mathbf{A}^*\mathbf{A})\right]^{\frac{1}{2}}. \tag{8.7}$$

To show equality, let $\mathbf{x}_m$ be the eigenvector of $\mathbf{A}^*\mathbf{A}$ corresponding to $\max_i \lambda_i$. Furthermore, without any loss of generality, we assume that $\|\mathbf{x}_m\|_2 = 1$. Also let

$$\lambda_{\max} \overset{\Delta}{=} \max_i \lambda_i(\mathbf{A}^*\mathbf{A}). \tag{8.8}$$

Then

$$\mathbf{A}^*\mathbf{A}\mathbf{x}_m = \lambda_{\max}\mathbf{x}_m \quad \Rightarrow \quad \mathbf{x}_m^*\mathbf{A}^*\mathbf{A}\mathbf{x}_m = \lambda_{\max}\mathbf{x}_m^*\mathbf{x}_m$$

$$\Rightarrow \quad \|\mathbf{A}\mathbf{x}_m\|_2^2 = \lambda_{\max}\|\mathbf{x}_m\|_2^2$$

$$\Rightarrow \quad \|\mathbf{A}\mathbf{x}_m\|_2 = \lambda_{\max}^{\frac{1}{2}}\|\mathbf{x}_m\|_2 \tag{8.9}$$

$$\Rightarrow \quad \|\mathbf{A}\|_2 \geq \left[\max_i \lambda_i\left(\mathbf{A}^*\mathbf{A}\right)\right]^{\frac{1}{2}}.$$

Combining Equations (8.7) and (8.9), we obtain

$$\|\mathbf{A}\|_2 = \max_i \left[\lambda_i\left(\mathbf{A}^*\mathbf{A}\right)\right]^{\frac{1}{2}}. \tag{8.10}$$

The quantity on the right-hand side of Equation (8.10) is called the *largest singular value* of the matrix $\mathbf{A}$ and is denoted by $\sigma_{\max}(\mathbf{A})$. Thus, this lemma shows that the matrix-induced norm corresponding to the vector 2-norm is the largest-singular value of the matrix. ∎

8.1.1   Norms of Signals

Consider an n-dimensional time-domain signal $\mathbf{x}(t)$ for $t \geq 0$. Let $\mathscr{L}_2[0,\infty]$ denote the space of square-integrable signals and define the 2-norm of the signal $\mathbf{x}$ by

$$\|\mathbf{x}(t)\|_2^2 := \int_0^\infty \sum_{i=1}^n |x_i(t)|^2\, dt. \tag{8.11}$$

When the time interval is finite, say $[0,T]$, the notation $\mathscr{L}_2[0,\infty]$ is replaced by $\mathscr{L}_2[0,T]$.

8.1.2   Norms for Linear Systems

The H_∞ Norm

Consider a linear time-invariant system, viewed as an operator denoted $\mathbf{H}$, mapping input signals into output signals. Suppose that $\mathbf{H}$ is described by a stable proper transfer function matrix $\mathbf{H}(s)$ with input $\mathbf{u}(t)$ and output $\mathbf{y}(t)$ defined for $t \geq 0$. Clearly, if $\mathbf{u}(\cdot) \in \mathscr{L}_2[0,\infty]$, so does the corresponding $\mathbf{y}(\cdot)$. We may now define the induced norm of $\mathbf{H}(s)$, also called its H_∞ norm, by

$$\|\mathbf{H}(s)\|_\infty := \sup_{\mathbf{u}(\cdot)\in\mathscr{L}_2[0,\infty]} \frac{\|\mathbf{y}(\cdot)\|_2}{\|\mathbf{u}(\cdot)\|_2}. \tag{8.12}$$

The definition in Equation (8.12) is in the time domain. It is possible to show that

$$\|\mathbf{H}(s)\|_\infty := \sup_{\omega\in[0,\infty]} \sigma_{\max}(\mathbf{H}(j\omega)), \tag{8.13}$$

where $\sigma_{\max}(\cdot)$ denotes the largest singular value. This is a frequency-domain characterization of the H_∞ norm of a multivariable system. The definitions in Equations (8.12) and (8.13) are consistent, in other words, equivalent. This can be proved using Parseval's Theorem. The proof is omitted here and the reader is referred to the literature in the Notes and References section.

As seen from the above analysis, computation of the H_∞ norm from Equation (8.13) requires a singular value decomposition (SVD) at each frequency and sweeping over frequency $\omega \in [0, \infty]$. The following algorithm greatly simplifies the process by eliminating the frequency sweep and reducing this calculation to a bisection algorithm over a single variable.

A Bisection Method for Calculating the H_∞ Norm

Consider the linear dynamical system

$$\dot{\mathbf{x}}(t) = \mathbf{A}\mathbf{x}(t) + \mathbf{B}\mathbf{u}(t),$$
$$\mathbf{y}(t) = \mathbf{C}\mathbf{x}(t) + \mathbf{D}\mathbf{u}(t),$$

(8.14)

and its transfer function

$$\mathbf{H}(s) = \mathbf{C}(s\mathbf{I} - \mathbf{A})^{-1}\mathbf{B} + \mathbf{D}.$$

(8.15)

If $\mathbf{A}$ is stable, we define the H_∞ norm of the transfer function matrix $\mathbf{H}(s)$ to be as in Equation (8.13). An important interpretation of $\|\mathbf{H}(s)\|_\infty$ is as the $\mathbf{L}_2$ gain of the system in Equation (8.14). Let $\mathbf{x}(t), \mathbf{u}(t), \mathbf{y}(t)$ satisfy Equation (8.14), $\mathbf{x}(0) = 0$, and let $T_f > 0$, then we have

$$\int_0^{T_f} \mathbf{y}(t)^T \mathbf{y}(t)dt \leq \|\mathbf{H}(s)\|_\infty^2 \int_0^{T_f} \mathbf{u}(t)^T \mathbf{u}(t)dt.$$

(8.16)

To describe the bisection algorithm we start by establishing a connection between the singular values of the transfer function matrix and the imaginary-axis eigenvalues of an associated *Hamiltonian matrix*. Let $\gamma > 0$, and not be a singular value of $\mathbf{D}$. Define

$$\mathbf{M}_\gamma := \begin{bmatrix} \mathbf{A} & 0 \\ 0 & -\mathbf{A}^T \end{bmatrix} + \begin{bmatrix} \mathbf{B} & 0 \\ 0 & -\mathbf{C}^T \end{bmatrix} \begin{bmatrix} -\mathbf{D} & \gamma\mathbf{I} \\ \gamma\mathbf{I} & -\mathbf{D}^T \end{bmatrix}^{-1} \begin{bmatrix} \mathbf{C} & 0 \\ 0 & \mathbf{B}^T \end{bmatrix}$$
$$= \begin{bmatrix} \mathbf{A} - \mathbf{B}\mathbf{D}^T\mathbf{S}^{-1}\mathbf{C} & -\gamma\mathbf{B}\mathbf{R}^{-1}\mathbf{B}^T \\ \gamma\mathbf{C}^T\mathbf{S}^{-1}\mathbf{C} & -\mathbf{A}^T + \mathbf{C}^T\mathbf{D}\mathbf{R}^{-1}\mathbf{B}^T \end{bmatrix},$$

(8.17)

where

$$\mathbf{R} = \mathbf{D}^T\mathbf{D} - \gamma^2\mathbf{I} \quad \text{and} \quad \mathbf{S} = \mathbf{D}\mathbf{D}^T - \gamma^2\mathbf{I}.$$

(8.18)

$\mathbf{M}_\gamma$ is a Hamiltonian matrix, meaning thereby that

$$\mathbf{J}^{-1}\mathbf{M}_\gamma\mathbf{J} = -\mathbf{M}_\gamma^T, \quad \text{where} \quad \mathbf{J} = \begin{bmatrix} 0 & \mathbf{I} \\ -\mathbf{I} & 0 \end{bmatrix}.$$

(8.19)

The following theorem relates the singular values of $\mathbf{H}(j\omega)$ and the imaginary-axis eigenvalues of $\mathbf{M}_\gamma$.

THEOREM 8.1   Assume $\mathbf{A}$ has no imaginary-axis eigenvalues, $\gamma > 0$ is not a singular value of $\mathbf{D}$, and $\omega_0 \in \mathbb{R}$. Then γ is a singular value of $\mathbf{H}(j\omega_0)$ if and only if $(\mathbf{M}_\gamma - j\omega_0\mathbf{I})$ is singular.

Proof Let γ be a singular value of $\mathbf{H}(j\omega_0)$. Then there exists a nonzero $\mathbf{u}$ such that

$$\begin{aligned} \mathbf{H}(j\omega_0)\mathbf{u} &= \gamma\mathbf{v}, \\ \mathbf{H}(j\omega_0)^*\mathbf{v} &= \gamma\mathbf{u}, \end{aligned} \tag{8.20}$$

so that

$$\begin{aligned} \left[\mathbf{C}(j\omega_0\mathbf{I}-\mathbf{A})^{-1}\mathbf{B}+\mathbf{D}\right]\mathbf{u} &= \gamma\mathbf{v}, \\ \left[\mathbf{B}^T(-j\omega_0\mathbf{I}-\mathbf{A}^T)^{-1}\mathbf{C}^T+\mathbf{D}^T\right]\mathbf{v} &= \gamma\mathbf{u}. \end{aligned} \tag{8.21}$$

Define

$$\begin{aligned} \mathbf{r} &:= (j\omega_0\mathbf{I}-\mathbf{A})^{-1}\mathbf{B}\mathbf{u}, \\ \mathbf{s} &:= (-j\omega_0\mathbf{I}-\mathbf{A}^T)^{-1}\mathbf{C}^T\mathbf{v}. \end{aligned} \tag{8.22}$$

Solving for $\mathbf{u}$ and $\mathbf{v}$ in terms of $\mathbf{r}$ and $\mathbf{s}$, we have

$$\begin{bmatrix} \mathbf{u} \\ \mathbf{v} \end{bmatrix} = \begin{bmatrix} -\mathbf{D} & \gamma\mathbf{I} \\ \gamma\mathbf{I} & -\mathbf{D}^T \end{bmatrix}^{-1} \begin{bmatrix} \mathbf{C} & 0 \\ 0 & \mathbf{B}^T \end{bmatrix} \begin{bmatrix} \mathbf{r} \\ \mathbf{s} \end{bmatrix}. \tag{8.23}$$

Equation (8.23) guarantees that

$$\begin{bmatrix} \mathbf{r} \\ \mathbf{s} \end{bmatrix} \neq 0. \tag{8.24}$$

From Equation (8.22):

$$\begin{aligned} (j\omega_0\mathbf{I}-\mathbf{A})\mathbf{r} &= \mathbf{B}\mathbf{u}, \\ (-j\omega_0\mathbf{I}-\mathbf{A}^T)\mathbf{s} &= \mathbf{C}^T\mathbf{v}, \end{aligned} \tag{8.25}$$

and so from Equations (8.23) and (8.25), we obtain

$$\left(\begin{bmatrix} \mathbf{A} & 0 \\ 0 & -\mathbf{A}^T \end{bmatrix} + \begin{bmatrix} \mathbf{B} & 0 \\ 0 & -\mathbf{C}^T \end{bmatrix} \begin{bmatrix} -\mathbf{D} & \gamma\mathbf{I} \\ \gamma\mathbf{I} & -\mathbf{D}^T \end{bmatrix}^{-1} \begin{bmatrix} \mathbf{C} & 0 \\ 0 & \mathbf{B}^T \end{bmatrix}\right) \begin{bmatrix} \mathbf{r} \\ \mathbf{s} \end{bmatrix}$$
$$= j\omega_0 \begin{bmatrix} \mathbf{r} \\ \mathbf{s} \end{bmatrix}. \tag{8.26}$$

Thus

$$\mathbf{M}_\gamma \begin{bmatrix} \mathbf{r} \\ \mathbf{s} \end{bmatrix} = j\omega_0 \begin{bmatrix} \mathbf{r} \\ \mathbf{s} \end{bmatrix}. \tag{8.27}$$

To prove the converse, suppose that $\mathbf{M}_\gamma$ has eigenvalue $j\omega_0$, that is, Equation (8.26) holds for some $\begin{bmatrix} \mathbf{r} \\ \mathbf{s} \end{bmatrix} \neq 0$. Define $\mathbf{u}$ and $\mathbf{v}$ via Equation (8.23), so that, $[\mathbf{u}^T \ \mathbf{v}^T] \neq 0$. Then from Equations (8.23) and (8.26) we conclude Equation (8.21), which establishes that γ is a singular value of $\mathbf{H}(j\omega_0)$. ∎

Remark 8.1 When $\mathbf{D} = 0$, $\mathbf{M}_\gamma$ reduces to the form

$$\mathbf{M}_\gamma = \begin{bmatrix} \mathbf{A} & \gamma^{-1}\mathbf{B}\mathbf{B}^T \\ -\gamma^{-1}\mathbf{C}^T\mathbf{C} & -\mathbf{A}^T \end{bmatrix}. \tag{8.28}$$

A useful collorary of Theorem 8.1 is the following.

COROLLARY 8.1   Let **A** be stable and $\gamma > \sigma_{\max}(\mathbf{D})$. Then $\|\mathbf{H}(s)\|_\infty \geq \gamma$ if and only if $\mathbf{M}_\gamma$ has at least one imaginary-axis eigenvalue.

Proof   Since

$$\gamma > \sigma_{\max}(\mathbf{D}) = \lim_{\omega \to \infty} \sigma_{\max}(\mathbf{H}(j\omega)) \tag{8.29}$$

and $\sigma_{\max}(\mathbf{H}(j\omega))$ is a continuous function of ω, we have $\|\mathbf{H}(s)\|_\infty \geq \gamma$ if and only if there exists ω_0 such that $\sigma_{\max}(\mathbf{H}(j\omega_0)) = \gamma$. Thus the corollary follows immediately from Theorem 8.1.   ∎

Remark 8.2   If **A** is stable, then

$$\beta(\mathbf{A}) = \inf\left\{\sigma_{\max}(\mathbf{E}) \ : \ \mathbf{E} \in \mathbb{C}^{n \times n} \text{ and } \mathbf{A}+\mathbf{E} \text{ has imaginary eigenvalues}\right\} \tag{8.30}$$

is referred to as the *distance to the nearest unstable matrix* in the numerical analysis literature. Note that

$$\begin{aligned}
\beta(\mathbf{A}) &= \inf_{\omega \in \mathbb{R}} \inf\{\sigma_{\max}(\mathbf{E}) \ : \ \mathbf{A}+\mathbf{E} - j\omega\mathbf{I} \text{ is singular}\} \\
&= \inf_{\omega \in \mathbb{R}} \sigma_{\max}(\mathbf{A} - j\omega\mathbf{I}) \\
&= \left[\sup_{\omega \in \mathbb{R}} \sigma_{\max}\left((j\omega\mathbf{I} - \mathbf{A})^{-1}\right)\right]^{-1} \\
&= \left\|(s\mathbf{I} - \mathbf{A})^{-1}\right\|_\infty^{-1}.
\end{aligned} \tag{8.31}$$

Thus $\beta(\mathbf{A})^{-1}$ can be determined by an H_∞-norm calculation of the resolvent of **A**, and the Corollary 8.1:

$$\beta(\mathbf{A}) = \inf\left\{\alpha \ : \ \begin{bmatrix} \mathbf{A} & \alpha\mathbf{I} \\ -\alpha\mathbf{I} & -\mathbf{A}^T \end{bmatrix} \text{ has imaginary-axis eigenvalues}\right\}. \tag{8.32}$$

A Bisection Algorithm

Corollary 8.1 suggests a bisection algorithm for computing $\|\mathbf{H}(s)\|_\infty$. Let γ_l and γ_u be some lower and upper bound, respectively, on $\|\mathbf{H}(s)\|_\infty$. The bisection algorithm is as follows:

Set $\gamma_L := \gamma_l$; $\gamma_H := \gamma_u$
While $|\gamma_H - \gamma_L| \geq 2\varepsilon\gamma_L$
  $\gamma := (\gamma_L + \gamma_H)/2$;
  if $\mathbf{M}_\gamma$ has no imaginary eigenvalues
    Set $\gamma_H := \gamma$
  else Set $\gamma_L := \gamma$
end

Note that we always have $\gamma_L \leq \|\mathbf{H}(s)\|_\infty \leq \gamma_H$. Moreover, after M iterations:

$$\gamma_H - \gamma_L = 2^{-M}(\gamma_u - \gamma_l). \tag{8.33}$$

On exit, $(\gamma_L + \gamma_H)/2$ is guaranteed to approximate $\|\mathbf{H}(s)\|_\infty$ within a relative accuracy of ε, that is,

$$\left| \frac{\gamma_L + \gamma_H}{2} - \|\mathbf{H}(s)\|_\infty \right| \leq \varepsilon \|\mathbf{H}(s)\|_\infty. \tag{8.34}$$

The computation of $\mathbf{R}^{-1}$ and $\mathbf{S}^{-1}$, while forming $\mathbf{M}_\gamma$ in the $\mathbf{D} \neq 0$ case, is reduced by the following initial transformation on $(\mathbf{A}, \mathbf{B}, \mathbf{C}, \mathbf{D})$. Let $\mathbf{D} = \mathbf{U}\Sigma\mathbf{V}^T$ be an SVD of $\mathbf{D}$, where $\mathbf{U} \in \mathbb{R}^{p \times p}$, $\mathbf{V} \in \mathbb{R}^{m \times m}$ and orthogonal. Of course,

$$\|\mathbf{H}(s)\|_\infty = \|\mathbf{U}^T \mathbf{H}(s)\mathbf{V}\|_\infty = \|\mathbf{U}^T \mathbf{C}(s\mathbf{I} - \mathbf{A})^{-1}\mathbf{B}\mathbf{V} + \Sigma\|_\infty. \tag{8.35}$$

The bisection algorithm is then applied to the transformed system $(\mathbf{A}, \mathbf{B}\mathbf{V}, \mathbf{U}^T\mathbf{C}, \Sigma)$. For this system, $\mathbf{R}$ and $\mathbf{S}$ are diagonal, so computing $\mathbf{R}^{-1}$ and $\mathbf{S}^{-1}$ is fast. The H_∞ norm of a dynamic system may be calculated conveniently by using MATLAB, as shown below.

EXAMPLE 8.1   Consider the system

$$\mathbf{A} = \begin{bmatrix} 0 & 1 & 0 \\ 0 & 0 & 1 \\ -1 & -2 & -3 \end{bmatrix}, \quad \mathbf{B} = \begin{bmatrix} 0 & 1 \\ 1 & 1 \\ 1 & 0 \end{bmatrix},$$

$$\mathbf{C} = \begin{bmatrix} 1 & 1 & 0 \\ 0 & 0 & 1 \end{bmatrix}, \quad \mathbf{D} = \begin{bmatrix} 1 & 0 \\ 0 & 1 \end{bmatrix}. \tag{8.36}$$

The following MATLAB script calculates the H_∞ norm of the above dynamic system. The calculation is verified from the singular value plot shown in Figure 8.1. The H_∞ norm of the given system is

$$\|\mathbf{G}(\mathbf{A}, \mathbf{B}, \mathbf{C}, \mathbf{D})\|_\infty = 8.1769 = 18.2518 \text{ dB}. \tag{8.37}$$

```
clear
A=[0 1 0; 0 0 1; -1 -2 -3];
B=[0 1; 1 1; 1 0];
C=[1 1 0; 0 0 1];
D=[1 0; 0 1];
sys=ss(A,B,C,D);
ninf=hinfnorm(sys)
ninf_db=20*log10(ninf)
```

The H_2 Norm

Let us now define the space H_2 and the H_2 norm.

DEFINITION 8.1 H_2 is the set of Laplace transforms of $L_2[0, \infty]$, or equivalently H_2 is the set of rational, stable, strictly proper transfer function matrices $G(s)$. The norm on H_2 is defined by

$$\|\mathbf{G}(s)\|_{H_2} = \left(\frac{1}{2\pi} \int_{-\infty}^{\infty} \text{Trace}\, [\mathbf{G}^*(j\omega)\mathbf{G}(j\omega)]\, d\omega \right)^{\frac{1}{2}}. \tag{8.38}$$

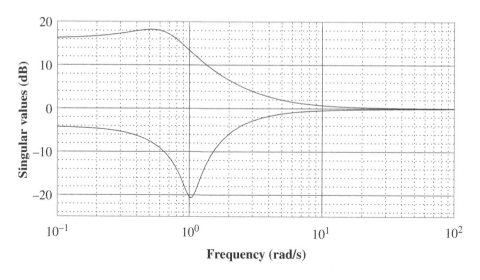

Figure 8.1 Singular values of dynamic system (Example 8.1)

The H_2 norm of a strictly proper stable transfer function matrix $\mathbf{G}(s)$ can also be defined via its impulse response matrix denoted by $\hat{\mathbf{G}}(t)$:

$$\|\mathbf{G}(s)\|_2^2 := \int_0^\infty \text{Trace}\left[\hat{\mathbf{G}}(t)^T\hat{\mathbf{G}}(t)\right] dt. \tag{8.39}$$

It can be shown using Parseval's Theorem that the above time-domain definition is equivalent to the previous frequency-domain definition. It is easy to verify that the time-domain definition states that the entries of the impulse-response matrix must be square integrable.

Robustness via the Small Gain Theorem

The Nyquist criterion shows that a unity feedback system which is stable with forward transfer function $g(s)$ remains stable when $g(s)$ is replaced by $\tilde{g}(s)$ provided $\tilde{g}(j\omega) - g(j\omega)$ has magnitude less than 1, for all $\omega \in [0, \infty]$. The Small Gain Theorem generalizes this result to multivariable systems as shown in Figure 8.2.

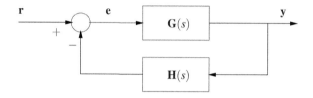

Figure 8.2 A feedback system ($\mathbf{G}(s)$, $\mathbf{H}(s)$ are open-loop stable transfer functions)

It gives us an intuitive feel for the stability of a feedback system since it gives us sufficient conditions under which the "boundedness" of the exogenous signals guarantees the "boundedness" of all the closed-loop signals. To formalize this notion, we introduce the definition[1] of $\mathscr{L}$-stability:

$$\max_{\omega} \|\mathbf{GH}(j\omega)\| < 1. \tag{8.40}$$

An important application of the Small Gain Theorem is in the derivation of robustness conditions for a nominally stable closed-loop system that is being perturbed by norm-bounded uncertainties. To illustrate the steps involved, we derive the robustness condition for one particular type of uncertainty called *additive uncertainty*. Towards this end, consider the feedback control system shown in Figure 8.3.

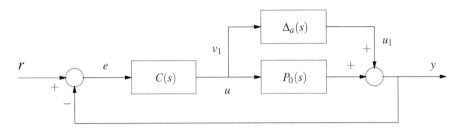

Figure 8.3 A feedback system with additive plant uncertainty

Here $P_0(s)$ represents the nominal plant and $C(s)$ the feedback controller. Due to modeling errors, the actual plant is the sum of the nominal plant $P_0(s)$ and an additive perturbation[1] $\Delta_a(s)$.

Let us assume that $\Delta_a(s)$ is stable. Now the controller $C(s)$ is usually designed based on $P_0(s)$ so that $\frac{P_0(s)C(s)}{1+P_0(s)C(s)}$ is a stable transfer function, by design. The problem of interest now is to find conditions on $\Delta_a(s)$ to guarantee that the closed-loop system remains stable. In what follows, we show that such a condition can be developed quite easily by using small gain-type arguments.

From Figure 8.3, we have

$$\begin{aligned} y(s) &= P_0(s)u(s) + \Delta_a(s)u(s), \\ u(s) &= C(s)[r(s) - y(s)] \quad \Rightarrow \quad u(s) = C(s)\left(r(s) - P_0(s)u(s) - \Delta_a(s)u(s)\right) \end{aligned} \tag{8.41}$$

or

$$(1 + P_0(s)C(s))u(s) = C(s)r(s) - C(s)\Delta_a(s)u(s) \tag{8.42}$$

or

$$u(s) = \frac{C(s)}{1+P_0(s)C(s)}r(s) - \frac{C(s)}{1+P_0(s)C(s)}\Delta_a(s)u(s). \tag{8.43}$$

Thus

$$\|u_t\|_2 \le c\|r_t\|_2 + \left\| \frac{C(s)}{1+P_0(s)C(s)}\Delta_a(s) \right\|_{H_\infty} \cdot \|u_t\|_2. \tag{8.44}$$

---

[1] The choice of an additive perturbation is without any loss of generality; indeed, similar conditions can be developed for other types of perturbations, such as multiplicative perturbations.

So the given system is L_2-stable provided

$$\left\| \frac{C(s)}{1+P_0(s)C(s)} \Delta_a(s) \right\|_{H_\infty} < 1, \tag{8.45}$$

which is the required condition on $\Delta_a(s)$. A similar result can be derived for the multi-variable case. Note that the transfer function seen by the uncertainty $\Delta_a(s)$ is:

$$T_{v_1 u_1}(s) = -\frac{C(s)}{1+P_0(s)C(s)}. \tag{8.46}$$

Thus, the condition in Equation (8.45) states that for the stability of the actual closed-loop system, the H_∞ norm of the product of (i) the transfer function of the uncertainty and (ii) the transfer function seen by it must be less than 1. This is the celebrated "loop gain less than one" stability result in norm-bounded robust control.

Computation of the H_2 Norm

Suppose that $\mathbf{G}(s)$ is strictly proper and has a minimal, stable, state-space realization $(\mathbf{A}, \mathbf{B}, \mathbf{C})$:

$$\mathbf{G}(s) = \mathbf{C}(s\mathbf{I} - \mathbf{A})^{-1}\mathbf{B}. \tag{8.47}$$

Then the impulse-response matrix with some abuse of notation is denoted by

$$\mathbf{G}(t) := \mathbf{C}e^{\mathbf{A}t}\mathbf{B}, \quad \mathbf{G}(t)^T := \mathbf{B}^T e^{\mathbf{A}^T t}\mathbf{C}^T. \tag{8.48}$$

Substituting the above in Equation (8.39):

$$\|\mathbf{G}(s)\|_2^2 = \int_0^\infty \mathrm{Trace}\left[\mathbf{B}^T e^{\mathbf{A}^T t}\mathbf{C}^T \mathbf{C}e^{\mathbf{A}t}\mathbf{B}\right] dt \tag{8.49a}$$

$$= \mathrm{Trace}\left[\mathbf{B}^T \underbrace{\left(\int_0^\infty e^{\mathbf{A}t}\mathbf{C}^T \mathbf{C}e^{\mathbf{A}t} dt\right)}_{\mathbf{L}_0} \mathbf{B}\right] \tag{8.49b}$$

$$= \mathrm{Trace}\left[\mathbf{B}^T \mathbf{L}_0 \mathbf{B}\right], \tag{8.49c}$$

where $\mathbf{L}_0$ is the observability Wronskian and is the unique solution of the observability Lyapunov equation:

$$\mathbf{A}^T \mathbf{L}_0 + \mathbf{L}_0 \mathbf{A} = -\mathbf{C}\mathbf{C}^T. \tag{8.50}$$

Since

$$\left[\mathbf{G}(t)^T \mathbf{G}(t)\right]^T = \mathbf{G}(t)^T \mathbf{G}(t), \tag{8.51}$$

it also follows that

$$\|\mathbf{G}(s)\|_2^2 = \mathrm{Trace}\left[\mathbf{C}^T \underbrace{\left(\int_0^\infty e^{\mathbf{A}t}\mathbf{B}\mathbf{B}^T e^{\mathbf{A}t} dt\right)}_{\mathbf{L}_c} \mathbf{C}\right] \tag{8.52}$$

$$= \mathrm{Trace}\left[\mathbf{C}^T \mathbf{L}_c \mathbf{C}\right], \tag{8.53}$$

where $\mathbf{L}_c$ is the controllability Wronskian and is the unique solution of the controllability Lyapunov equation

$$\mathbf{AL}_c + \mathbf{L}_c \mathbf{A}^T = -\mathbf{BB}^T. \tag{8.54}$$

In the multivariable case, consider a system described by a rational, stable, strictly proper transfer function matrix $\mathbf{G}(s)$ with r inputs and m outputs. A generalization of the definition for the scalar case can be obtained by considering the vector impulse responses $\mathbf{y}_i(t)$, for $i = 1, 2, \ldots, r$ of r experiments, where r unit impulse inputs $\mathbf{u}^i(t) = \delta(t)\mathbf{e}_i$, for $i = 1, 2, \ldots, r$ are applied to the system (e_i denotes the ith standard basis vector in n-space). Then the H_2 norm of $\mathbf{G}(s)$ is

$$\|\mathbf{G}(s)\|_{H_2}^2 = \sum_{i=1}^{r} \|\mathbf{y}_i\|_2^2. \tag{8.55}$$

8.2 Control Problems in a Generalized Plant Framework

Consider the general multi-input multi-output feedback control configuration shown in Figure 8.4.

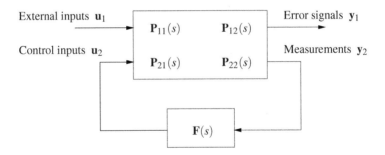

Figure 8.4 A general feedback control configuration

Here

$$\mathbf{P}(s) = [\mathbf{P}_{ij}(s)], \quad \text{for } i, j = 1, 2 \tag{8.56}$$

represents the *generalized plant*, u_1 is the vector of exogenous inputs such as references and disturbances, y_1 is the vector made up of signals that we are trying to keep small, such as errors, y_2 is the vector of available measurement signals, and u_2 is the vector of control inputs.

The control design problem is to choose the controller $\mathbf{F}(s)$ so that the closed-loop system is stable and the closed-loop transfer function map from $\mathbf{u}_1$ to $\mathbf{y}_1$, denoted $\mathbf{T}_{\mathbf{y}_1\mathbf{u}_1}$, is small in some sense. Clearly, the weighted sensitivity minimization problem from Figure 8.4 can be cast within this framework.

We next derive the expression for $\mathbf{T}_{\mathbf{y}_1\mathbf{u}_1}$. From Figure 8.4, we have

$$\mathbf{y}_1 = \mathbf{P}_{11}\mathbf{u}_1 + \mathbf{P}_{12}\mathbf{u}_2, \tag{8.57a}$$

$$\mathbf{y}_2 = \mathbf{P}_{21}\mathbf{u}_1 + \mathbf{P}_{22}\mathbf{u}_2, \tag{8.57b}$$

$$\mathbf{u}_2 = \mathbf{F}\mathbf{y}_2. \tag{8.57c}$$

Substituting Equation (8.57c) into Equation (8.57b), we have

$$\mathbf{y}_2 = \mathbf{P}_{21}\mathbf{u}_1 + \mathbf{P}_{22}\mathbf{F}\mathbf{y}_2 \quad \text{or} \quad (\mathbf{I} - \mathbf{P}_{22}\mathbf{F})\mathbf{y}_2 = \mathbf{P}_{21}\mathbf{u}_1. \tag{8.58}$$

Thus

$$\mathbf{y}_2 = (\mathbf{I} - \mathbf{P}_{22}F)^{-1}\mathbf{P}_{21}\mathbf{u}_1. \tag{8.59}$$

Substituting this expression into Equation (8.57c), we obtain

$$\mathbf{u}_2 = \mathbf{F}(\mathbf{I} - \mathbf{P}_{22}\mathbf{F})^{-1}\mathbf{P}_{21}\mathbf{u}_1. \tag{8.60}$$

Thus, from Equation (8.57a), we get

$$\mathbf{y}_1 = \left[\mathbf{P}_{11} + \mathbf{P}_{12}\mathbf{F}(\mathbf{I} - \mathbf{P}_{22}\mathbf{F})^{-1}\mathbf{P}_{21}\right]\mathbf{u}_1. \tag{8.61}$$

Hence

$$\mathbf{T}_{\mathbf{y}_1\mathbf{u}_1} = \mathbf{P}_{11} + \mathbf{P}_{12}\mathbf{F}(\mathbf{I} - \mathbf{P}_{22}\mathbf{F})^{-1}\mathbf{P}_{21}. \tag{8.62}$$

Now introduce the *standard H_∞ control problem* and the *H_∞ optimal control problem*. In the standard H_∞ control problem, our objective is to find a stabilizing $F(s)$ such that

$$\|\mathbf{T}_{\mathbf{y}_1\mathbf{u}_1}\|_\infty < 1. \tag{8.63}$$

In the H_∞ optimal control problem, the goal is to minimize $\|\mathbf{T}_{\mathbf{y}_1\mathbf{u}_1}\|_\infty$ subject to $\mathbf{F}(s)$ stabilizing. In practice, the H_∞ optimal control problem can be solved by an iterative technique. Consider the problem of making $\|\mathbf{T}_{\mathbf{y}_1\mathbf{u}_1}\|_\infty < \gamma$, where $\gamma > 0$ is some prescribed constant. This problem can be converted into a standard H_∞ control problem by absorbing the factor $\frac{1}{\gamma}$ in $\mathbf{T}_{\mathbf{y}_1\mathbf{u}_1}$. Solving this standard problem, and progressively reducing γ at each stage, we will ultimately reach a stage when the standard H_∞ control problem can no longer be solved. At that stage, we can say that the terminal value of γ is very close to the minimal H_∞ norm. This technique is called the *γ-iteration* technique.

We next present two examples to show how the closed-loop disturbance rejection problem and the robust stability problem can be accommodated within the H_∞ framework.

EXAMPLE 8.2 (Closed-loop disturbance response for the SISO plant case)   Consider the feedback control configuration shown in Figure 8.5.

Here d is a disturbance that enters the system at the plant output. Let y_{CL} and y_{OL} denote the disturbance responses at the plant output in the closed-loop and open-loop configurations. Then clearly

$$y_{CL} = \frac{1}{1+GF}d \quad \text{and} \quad y_{OL} = d. \tag{8.64}$$

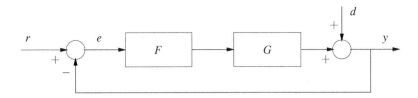

Figure 8.5 Disturbance attenuation using feedback control

Thus

$$y_{CL} = \frac{1}{1+GF}y_{OL},$$ (8.65)

so that

$$|y_{CL}|_2 \le \left|\frac{1}{1+GF}\right|_\infty \cdot |y_{OL}|_2.$$ (8.66)

Thus, by using feedback control and making

$$\left|\frac{1}{1+GF}\right|_\infty < 1,$$ (8.67)

we can attenuate the effect of disturbances (plant uncertainty) on the plant output. Here, the attenuation is measured in terms of the energy in the signal before and after feedback. Note that

$$T_{er} = T_{yd} = \frac{1}{1+GF} = S,$$ (8.68)

the sensitivity function. Thus, good tracking goes hand in hand with good disturbance rejection and both require that the sensitivity function be small.

EXAMPLE 8.3 (Robust stability for the SISO plant case)   Consider the feedback control system shown in Figure 8.6.

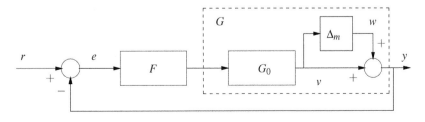

Figure 8.6 Stability robustness of feedback control

Here the nominal plant $G_o(s)$ is perturbed by a multiplicative perturbation $\Delta_m(s)$.

Note that in the absence of $\Delta_m(s)$, the complementary sensitivity function is

$$T_{yr} = \frac{G_oF}{1+G_oF} = T. \tag{8.69}$$

Furthermore, the transfer function seen by $\Delta_m(s)$, T_{vw}, can be computed as follows:

$$v = -G_oF(v+w) \quad \text{and} \quad (1+G_oF)v = -G_oFw, \tag{8.70}$$

so that

$$T_{vw} = -\frac{G_oF}{1+G_oF} = -T. \tag{8.71}$$

Assuming that $\Delta_m(s)$ is stable, it follows from the Small Gain Theorem that if the controller $F(s)$ stabilizes the nominal plant $G_o(s)$, then the closed-loop system with the multiplicative perturbation will continue to be stable provided

$$|\Delta_m(s)T(s)|_\infty < 1. \tag{8.72}$$

Thus, choosing $F(s)$ to make $T(s)$ small will enhance the robustness of the design.

Since

$$T+S = 1, \tag{8.73}$$

both T and S cannot be simultaneously made small. This brings us to a *fundamental trade-off* in feedback design. However, fortunately, good tracking and good disturbance rejection are required in the low-frequency range, while good stability margin is required in the high-frequency range where there is maximum model uncertainty. Thus, the trade-off between performance and robustness can be achieved by assigning frequency-dependent weights to the sensitivity and complementary sensitivity functions. Suppose it is desired that

$$\sigma_{\max}(S(j\omega)) \le |W_1(j\omega)|, \qquad \text{for all } \omega, \tag{8.74}$$

where $W_1(s)$ is a stable and minimum phase weighting function chosen by the designer to have a small-magnitude response at low frequencies. Since the H_∞ norm is the frequency domain supremum of the largest singular value of a transfer function matrix, it follows that the requirement in Equation (8.74) can be satisfied by imposing the H_∞-norm constraint

$$\left\| \frac{1}{W_1}S \right\|_\infty \le 1. \tag{8.75}$$

Similarly, if it is desired that

$$\sigma_{\max}(T(j\omega)) \le |W_2(j\omega)|, \qquad \text{for all } \omega, \tag{8.76}$$

where $W_2(s)$ is an appropriately chosen weight, then we arrive at the H_∞-norm constraint

$$\left\| \frac{1}{W_2}T \right\|_\infty \le 1. \tag{8.77}$$

It can be shown that for any two matrices **A** and **B** having the same number of columns:

$$\max\left\{\sigma_{\max}(\mathbf{A}), \sigma_{\max}(\mathbf{B})\right\} \leq \sigma_{\max}\left(\begin{bmatrix} \mathbf{A} \\ \mathbf{B} \end{bmatrix}\right) \leq \max\left\{\sqrt{2}\sigma_{\max}(\mathbf{A}), \sqrt{2}\sigma_{\max}(\mathbf{B})\right\}.$$

(8.78)

Thus, to within a factor of 3 dB, the requirements in Equations (8.75) and (8.77) can be combined into the single requirement

$$\left\|\begin{array}{c} \frac{1}{W_1}S \\ \frac{1}{W_2}T \end{array}\right\|_\infty \leq 1.$$

(8.79)

Figure 8.7 shows how the sensitivity and complementary sensitivity requirements in Equations (8.74) and (8.76) can be converted into a standard H_∞ problem involving an augmented plant

$$\mathbf{P}(s) = [P_{ij}(s)], \quad \text{for } i, j = 1, 2.$$

(8.80)

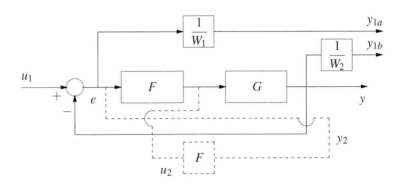

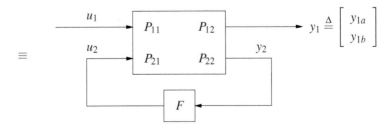

Figure 8.7 Converting sensitivity and complementary sensitivity requirements into a standard H_∞ problem

8.3   State-Space Solution of the Multivariable H_∞ Optimal Control Problem

In H_∞ optimal control, the design objective is to find a controller to minimize the worst-case energy amplification of certain signals of interest. Thus, the H_∞ control problem can be viewed as a min-max problem where the controller objective is to minimize the cost function (or the performance index) while the uncertainties such as external disturbances are assumed to be the worst-case inputs. Such min-max problems can be studied within the framework of game theory and the state-space solution does precisely that. As we will see in this section, by using a game-theoretic approach, the state-space solution bypasses the need for the YJBK parametrization. The result is obtained by a simple completion of squares and involves the solution of two Ricatti equations. To keep the presentation simple, we will impose certain restrictions to be satisfied by the augmented plant. These restrictions can be relaxed but at the expense of making the formulas more complicated. It turns out that the state-space approach to H_∞ presented here can be extended to the H_2 optimal control problem in a fairly transparent, albeit not so rigorous fashion and leads to the recovery of well-known results from the H_2 optimal control literature, provided one sets all initial conditions to zero throughout. We will also be presenting this extension applicable to the H_2 case. For the entire subsequent development, we use the following notation:

$$\mathbf{P} = Ric \begin{bmatrix} \mathbf{A} & -\mathbf{R} \\ -\mathbf{Q} & -\mathbf{A}^T \end{bmatrix} \tag{8.81}$$

means that $\mathbf{P}$ satisfies the algebraic Ricatti equation

$$0 = \begin{bmatrix} -\mathbf{P} & \mathbf{I} \end{bmatrix} \begin{bmatrix} \mathbf{A} & -\mathbf{R} \\ -\mathbf{Q} & -\mathbf{A}^T \end{bmatrix} \begin{bmatrix} \mathbf{I} \\ \mathbf{P} \end{bmatrix} \tag{8.82}$$

with $\mathbf{A} - \mathbf{R}\mathbf{P}$ stable. In general, the algebraic Ricatti equation has several solutions. However, under appropriate assumptions, it has a unique stabilizing solution $\mathbf{P}$, that is $\mathbf{P}$ makes the matrix $\mathbf{A} - \mathbf{R}\mathbf{P}$ stable.

8.3.1 H_∞ Solution

We will develop the state-space solution to H_∞ optimal control in two parts. First, we consider the simpler case where all states are measured. Thereafter, we extend the results to observer-reconstructed-state H_∞.

Full State Feedback H_∞

Suppose that the full state $\mathbf{x}$ is available as $\mathbf{y}_2$, as shown in Figure 8.8.

Here the packed $(\mathbf{A}, \mathbf{B}, \mathbf{C}, \mathbf{D})$ matrix representation describes the augmented plant while $F(s)$ represents the feedback controller to be designed. The equations describing

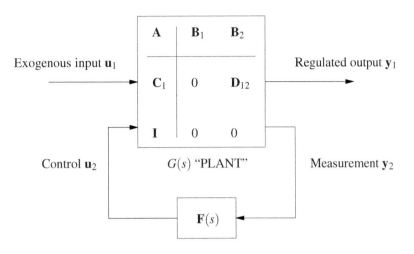

Figure 8.8 The standard plant controller setup

the augmented plant are

$$\dot{\mathbf{x}}(t) = \mathbf{A}\mathbf{x}(t) + \mathbf{B}_1\mathbf{u}_1(t) + \mathbf{B}_2\mathbf{u}_2(t), \tag{8.83a}$$

$$\mathbf{y}_1(t) = \mathbf{C}_1\mathbf{x}(t) + \mathbf{D}_{12}\mathbf{u}_2(t), \tag{8.83b}$$

$$\mathbf{y}_2(t) = \mathbf{x}(t). \tag{8.83c}$$

For simplicity, we make the following assumptions:

$$\mathbf{D}_{12}^T\mathbf{C}_1 = 0, \quad \mathbf{D}_{11} = 0, \quad \mathbf{D}_{12}^T\mathbf{D}_{12} = \mathbf{I}. \tag{8.84}$$

In this section, we focus on the general problem

$$\sup_{\mathbf{u}_1 \in L_2[0,\infty)} \frac{\|\mathbf{y}_1\|_{L_2[0,\infty)}}{\|\mathbf{u}_1\|_{L_2[0,\infty)}} < \gamma, \tag{8.85}$$

where $\gamma > 0$ is some given constant, and refer to it as the standard H_∞ control problem. Note that the H_∞ norm of a transfer function matrix captures the input–output behavior of the system and the initial condition has no role in its definition. Clearly, this problem can be equivalently expressed as

$$\|\mathbf{y}_1\|_{L_2}^2 - \gamma^2\|\mathbf{u}_1\|_{L_2}^2 < 0 \quad \text{for all } \mathbf{u}_1. \tag{8.86}$$

Let

$$J(\mathbf{x},\mathbf{u}_1,\mathbf{u}_2) \overset{\Delta}{=} \|\mathbf{y}_1\|_{L_2[0,\infty)}^2 - \gamma^2\|\mathbf{u}_1\|_{L_2[0,\infty)}^2$$
$$= \int_0^\infty \left(\mathbf{y}_1^T\mathbf{y}_1 - \gamma^2\mathbf{u}_1^T\mathbf{u}_1\right) dt. \tag{8.87}$$

Here J is the infinite-horizon cost and the goal is to make J negative. In a typical game-theoretic problem, there are two adversarial players, one trying to maximize the cost while the other tries to minimize it. The situation here is similar to the exogenous input $\mathbf{u}_1$ trying to maximize J while the control input $\mathbf{u}_2$ tries to minimize J so that the optimization problem may be expressed as

$$\min_{\mathbf{u}_2} \max_{\mathbf{u}_1} J(\mathbf{x}, \mathbf{u}_1, \mathbf{u}_2). \tag{8.88}$$

We will start with the cost in Equation (8.87) and employ a simple completion of squares to arrive at a solution to the min-max problem. From Equation (8.87):

$$\begin{aligned}
J &= \int_0^\infty \left(\mathbf{y}_1^T(t)\mathbf{y}_1(t) - \gamma^2 \mathbf{u}_1^T(t)\mathbf{u}_1(t) \right) dt \\
&= \int_0^\infty \left((\mathbf{C}_1\mathbf{x}(t) + \mathbf{D}_{12}\mathbf{u}_2(t))^T (\mathbf{C}_1\mathbf{x}(t) + \mathbf{D}_{12}\mathbf{u}_2(t)) - \gamma^2 \mathbf{u}_1^T(t)\mathbf{u}_1(t) \right) dt
\end{aligned}$$

(using Equation (8.83b)) $\tag{8.89}$

$$= \int_0^\infty \left(\mathbf{x}^T(t)\mathbf{C}_1^T\mathbf{C}_1\mathbf{x}(t) + \mathbf{u}_2^T(t)\mathbf{u}_2(t) - \gamma^2 \mathbf{u}_1^T(t)\mathbf{u}_1(t) \right) dt$$

(using Equation (8.84)).

Now for J to be well-defined, we must have $J = \lim_{T \to \infty} J_T$, where

$$J_T \triangleq \int_0^T \left(\mathbf{x}^T(t)\mathbf{C}_1^T\mathbf{C}_1\mathbf{x}(t) + \mathbf{u}_2^T(t)\mathbf{u}_2(t) - \gamma^2 \mathbf{u}_1^T(t)\mathbf{u}_1(t) \right) dt. \tag{8.90}$$

Let $\mathbf{P}(t) = \mathbf{P}^T(t)$ be an arbitrary symmetric matrix and consider

$$\begin{aligned}
&\mathbf{x}^T(T)\mathbf{P}(T)\mathbf{x}(T) - \mathbf{x}^T(0)\mathbf{P}(0)\mathbf{x}(0) + J_T \\
&= \int_0^T \frac{d}{dt}\left(\mathbf{x}^T(t)\mathbf{P}\mathbf{x}(t) \right) dt + \int_0^T \left(\mathbf{x}^T(t)\mathbf{C}_1^T\mathbf{C}_1\mathbf{x}(t) + \mathbf{u}_2^T(t)\mathbf{u}_2(t) - \gamma^2 \mathbf{u}_1^T(t)\mathbf{u}_1(t) \right) dt \\
&= \int_0^T \left[\dot{\mathbf{x}}^T(t)\mathbf{P}\mathbf{x}(t) + \mathbf{x}^T(t)\dot{\mathbf{P}}\mathbf{x}(t) + \mathbf{x}^T(t)\mathbf{P}\dot{\mathbf{x}}(t) + \mathbf{x}^T(t)\mathbf{C}_1^T\mathbf{C}_1\mathbf{x}(t) \right. \\
&\qquad\qquad \left. + \mathbf{u}_2^T(t)\mathbf{u}_2(t) - \gamma^2 \mathbf{u}_1^T(t)\mathbf{u}_1(t) \right] dt.
\end{aligned} \tag{8.91}$$

By using Equation (8.83a)

$$\begin{aligned}
&= \int_0^T \left[(\mathbf{A}\mathbf{x}(t) + \mathbf{B}_1\mathbf{u}_1(t) + \mathbf{B}_2\mathbf{u}_2(t))^T \mathbf{P}\mathbf{x}(t) + \mathbf{x}^T(t)\dot{\mathbf{P}}\mathbf{x}(t) \right. \\
&\qquad\qquad + \mathbf{x}^T(t)\mathbf{P}(\mathbf{A}\mathbf{x}(t) + \mathbf{B}_1\mathbf{u}_1(t) + \mathbf{B}_2\mathbf{u}_2(t)) \\
&\qquad\qquad \left. + \mathbf{x}^T(t)\mathbf{C}_1^T\mathbf{C}_1\mathbf{x}(t) + \mathbf{u}_2^T(t)\mathbf{u}_2(t) - \gamma^2 \mathbf{u}_1^T(t)\mathbf{u}_1(t) \right] dt
\end{aligned} \tag{8.92}$$

and completing the squares to eliminate cross terms, we have

$$\begin{aligned}
&= \int_0^T \left(\mathbf{x}^T(t) \left[\dot{\mathbf{P}} + \mathbf{A}^T\mathbf{P} + \mathbf{P}\mathbf{A} + \mathbf{P}\left(\frac{\mathbf{B}_1\mathbf{B}_1^T}{\gamma^2} - \mathbf{B}_2\mathbf{B}_2^T \right)\mathbf{P} + \mathbf{C}_1^T\mathbf{C}_1 \right] \mathbf{x}(t) \right. \\
&\qquad\qquad + \left(\mathbf{u}_2(t) + \mathbf{B}_2^T\mathbf{P}\mathbf{x}(t) \right)^T \left(\mathbf{u}_2(t) + \mathbf{B}_2^T\mathbf{P}\mathbf{x}(t) \right) \\
&\qquad\qquad \left. - \left(\gamma\mathbf{u}_1(t) - \frac{\mathbf{B}_1^T\mathbf{P}}{\gamma}\mathbf{x}(t) \right)^T \left(\gamma\mathbf{u}_1(t) - \frac{\mathbf{B}_1^T\mathbf{P}}{\gamma}\mathbf{x}(t) \right) \right) dt.
\end{aligned} \tag{8.93}$$

Since $\mathbf{P}(t)$ is arbitrary, we choose $\mathbf{P}(t)$ to simplify the above expression, to be the solution of the Ricatti differential equation

$$-\dot{\mathbf{P}} = \mathbf{A}^T\mathbf{P} + \mathbf{P}\mathbf{A} + \mathbf{P}\left(\frac{\mathbf{B}_1\mathbf{B}_1^T}{\gamma^2} - \mathbf{B}_2\mathbf{B}_2^T\right)\mathbf{P} + \mathbf{C}_1^T\mathbf{C}_1, \tag{8.94}$$

$$\mathbf{P}(T) = 0. \tag{8.95}$$

Also let us define

$$\mathbf{u}_1^*(\mathbf{x}(t)) \triangleq \frac{\mathbf{B}_1^T\mathbf{P}}{\gamma^2}\mathbf{x}(t), \tag{8.96}$$

$$\mathbf{u}_2^*(\mathbf{x}(t)) \triangleq -\mathbf{B}_2^T\mathbf{P}\mathbf{x}(t). \tag{8.97}$$

Then

$$J_T = \mathbf{x}(0)^T\mathbf{P}(0)\mathbf{x}(0) + \left(\|\mathbf{u}_2(t) - \mathbf{u}_2^*(\mathbf{x}(t))\|_{L_2[0,T]}\right)^2$$
$$- \gamma^2\left(\|\mathbf{u}_1(t) - \mathbf{u}_1^*(\mathbf{x}(t))\|_{L_2[0,T]}\right)^2. \tag{8.98}$$

Thus, if the state $\mathbf{x}(t)$ is available for measurement, the cost minimizing $\mathbf{u}_2(t)$ is simply

$$\mathbf{u}_2(t) = \mathbf{u}_2^*(\mathbf{x}(t)) = -\mathbf{B}_2^T\mathbf{P}\mathbf{x}(t), \tag{8.99}$$

the cost-maximizing disturbance $\mathbf{u}_1(t)$ is

$$\mathbf{u}_1(t) = \mathbf{u}_1^*(\mathbf{x}(t)) = \frac{\mathbf{B}_1^T\mathbf{P}}{\gamma^2}\mathbf{x}(t), \tag{8.100}$$

and the optimal cost is

$$J_T = \mathbf{x}(0)^T\mathbf{P}(0)\mathbf{x}(0). \tag{8.101}$$

Furthermore, if

$$\bar{\mathbf{P}} = \lim_{T \to \infty} \mathbf{P}(t) \tag{8.102}$$

exists, then using arguments similar to those used for the infinite horizon LQR problem in Section 7.3, it can be shown that $\bar{\mathbf{P}}$ satisfies the algebraic Ricatti equation

$$0 = \mathbf{A}^T\mathbf{P} + \mathbf{P}\mathbf{A} + \mathbf{P}\left(\frac{\mathbf{B}_1\mathbf{B}_1^T}{\gamma^2} - \mathbf{B}_2\mathbf{B}_2^T\right)\mathbf{P} + \mathbf{C}_1^T\mathbf{C}_1 \tag{8.103}$$

or equivalently

$$\mathbf{P} = Ric\left[\begin{array}{c|c} \mathbf{A} & \dfrac{\mathbf{B}_1\mathbf{B}_1^T}{\gamma^2} - \mathbf{B}_2\mathbf{B}_2^T \\ \hline -\mathbf{C}_1^T\mathbf{C}_1 & -\mathbf{A}^T \end{array}\right]. \tag{8.104}$$

Summarizing, we now have the following theorem.

THEOREM 8.2 (Full state feedback H_∞)   Let

$$G(s) = \left[\begin{array}{c|cc} \mathbf{A} & \mathbf{B}_1 & \mathbf{B}_2 \\ \hline \mathbf{C}_1 & 0 & \mathbf{D}_{12} \\ \mathbf{I} & 0 & 0 \end{array} \right], \tag{8.105a}$$

$$\mathbf{D}_{12}^T \mathbf{C}_1 = 0, \tag{8.105b}$$

$$\mathbf{D}_{12}^T \mathbf{D}_{12} = \mathbf{I}. \tag{8.105c}$$

Then a solution to the standard H_∞ control problem $\left\| T_{\mathbf{y}_1(t)\mathbf{u}_1(t)} \right\|_\infty \leq \gamma$ is

$$\mathbf{F}(s) = \mathbf{F} = -\mathbf{B}_2^T \mathbf{P}, \tag{8.106}$$

where $\mathbf{P}$ is a solution to the algebraic Ricatti equation

$$\mathbf{P} = Ric \left[\begin{array}{c|c} \mathbf{A} & \dfrac{\mathbf{B}_1 \mathbf{B}_1^T}{\gamma^2} - \mathbf{B}_2 \mathbf{B}_2^T \\ \hline -\mathbf{C}_1^T \mathbf{C}_1 & -\mathbf{A}^T \end{array} \right]. \tag{8.107}$$

Some of the assumptions in the above theorem can be relaxed. However, in the process, the formulas become more complicated. The following corollary demonstrates this for the assumption in Equation (8.105b).

COROLLARY 8.2 (Full state feedback without assumption in Equation (8.105b))   Let everything be as in Theorem 8.2, except that the condition in Equation (8.105b) does not necessarily hold, that is

$$\mathbf{D}_{12}^T \mathbf{C}_1 \neq 0. \tag{8.108}$$

Then a solution to the standard H_∞ problem

$$\left\| \mathbf{T}_{\mathbf{y}_1(t)\mathbf{u}_1(t)} \right\|_\infty \leq \gamma \tag{8.109}$$

is

$$\mathbf{u}_2(t) = \mathbf{F}\mathbf{x}(t), \quad \mathbf{F} = -\left(\mathbf{D}_{12}^T \mathbf{C}_1 + \mathbf{B}_2^T \mathbf{P} \right), \tag{8.110}$$

where $\mathbf{P}$ is a solution to the algebraic Ricatti equation

$$\mathbf{P} = Ric \left[\begin{array}{c|c} \mathbf{A} - \mathbf{B}_2 \mathbf{D}_{12}^T \mathbf{C}_1 & \dfrac{\mathbf{B}_1 \mathbf{B}_1^T}{\gamma^2} - \mathbf{B}_2 \mathbf{B}_2^T \\ \hline -\mathbf{C}_1^T \left(\mathbf{I} - \mathbf{D}_{12} \mathbf{D}_{12}^T \right) \mathbf{C}_1 & -\left(\mathbf{A} - \mathbf{B}_2 \mathbf{D}_{12}^T \mathbf{C}_1 \right)^T \end{array} \right]. \tag{8.111}$$

Proof   Note that the change of variables

$$\mathbf{u}_2(t) = \tilde{\mathbf{u}}_2(t) - \mathbf{D}_{12}^T \mathbf{C}_1 \mathbf{x}(t) \tag{8.112}$$

does not in any way affect the map from $\mathbf{u}_1(t)$ to $\mathbf{y}_1(t)$. Moreover, the new control input $\tilde{\mathbf{u}}_2(t)$ results in

$$\tilde{\mathbf{G}}(s) = \left[\begin{array}{c|cc} \mathbf{A} - \mathbf{B}_2 \mathbf{D}_{12}^T \mathbf{C}_1 & \mathbf{B}_1 & \mathbf{B}_2 \\ \hline \left(\mathbf{I} - \mathbf{D}_{12} \mathbf{D}_{12}^T \right) \mathbf{C}_1 & 0 & \mathbf{D}_{12} \\ \mathbf{I} & 0 & 0 \end{array} \right] \tag{8.113}$$

and this $\tilde{\mathbf{G}}(s)$ satisfies the conditions of Theorem 8.2. Hence, it follows that the optimal

$$\tilde{\mathbf{u}}_2(t) = -\mathbf{B}_2^T \mathbf{P} \mathbf{x}(t) \tag{8.114}$$

where $\mathbf{P}$ satisfies Equation (8.111). Thus, the optimal $\mathbf{u}_2(t)$ is given by

$$\mathbf{u}_2(t) = -\left(\mathbf{D}_{12}^T \mathbf{C}_1 + \mathbf{B}_2^T \mathbf{P}\right) \mathbf{x}(t). \tag{8.115}$$

∎

Remark 8.3   Note that in both the above theorems, the introduction of a nonzero $\mathbf{D}_{21}$ does not affect the solution since the state $\mathbf{x}(t)$ and not the output $\mathbf{y}_2(t)$ is used to generate the control input $\mathbf{u}_2(t)$.

Remark 8.4   The existence of a stabilizing solution $\mathbf{P}$ to the Ricatti equations in the above theorems depends on the value of γ. Typically, one could start with a large value of γ (for which a stabilizing solution exists as long as the subplant $\mathbf{G}_{22}$ is stabilizable) and step down γ using a bisection algorithm until finally a stabilizing solution to the appropriate Ricatti equation ceases to exist. It is clear that using such an approach, one could get very close to the minimal value of $\left\|\mathbf{T}_{\mathbf{y}_1(t)\mathbf{u}_1(t)}\right\|_\infty$.

8.3.2   Observer-Reconstructed State Feedback H_∞ Optimal Control

Here we solve the H_∞ problem when the state x is not directly measurable. The net result of our endeavors will be that we will arrive at a two-Ricatti solution to the general H_∞ control problem. This procedure involves a number of steps. The first step is to show that an augmented plant $\mathbf{G}(s)$ for the standard H_∞ problem can be replaced by an equivalent one in which $\tilde{\mathbf{G}}(s)$ has $\mathbf{D}_{12} = \mathbf{I}$. This is done in the following lemma.

LEMMA 8.2   Consider the following two plants:

$$\mathbf{G}(s) \triangleq \left[\begin{array}{c|cc} \mathbf{A} & \mathbf{B}_1 & \mathbf{B}_2 \\ \hline \mathbf{C}_1 & 0 & \mathbf{D}_{12} \\ \mathbf{C}_2 & \mathbf{D}_{21} & 0 \end{array} \right], \tag{8.116}$$

$$\tilde{\mathbf{G}}(s) \triangleq \left[\begin{array}{c|cc} \mathbf{A} + \dfrac{\mathbf{B}_1 \mathbf{B}_1^T}{\gamma^2}\mathbf{P} & \mathbf{B}_1 & \mathbf{B}_2 \\ \hline \mathbf{B}_2^T \mathbf{P} & 0 & \mathbf{I} \\ \mathbf{C}_2 & \mathbf{D}_{21} & 0 \end{array} \right]. \tag{8.117}$$

Assume that the conditions of Theorem 8.2 hold, that is $\mathbf{D}_{12}^T \mathbf{C}_1 = 0$ and $\mathbf{D}_{12}^T \mathbf{D}_{12} = \mathbf{I}$. In addition, suppose $\mathbf{B}_1 \mathbf{D}_{21}^T = 0$ and

$$\mathbf{P} = Ric \left[\begin{array}{c|c} \mathbf{A} & \dfrac{\mathbf{B}_1 \mathbf{B}_1^T}{\gamma^2} - \mathbf{B}_2 \mathbf{B}_2^T \\ \hline -\mathbf{C}_1^T \mathbf{C}_1 & -\mathbf{A}^T \end{array} \right]. \tag{8.118}$$

Then a control law

$$\mathbf{u}_2(t) = \mathbf{F}(s)\mathbf{y}_2(t) \tag{8.119}$$

solves the standard H_∞ problem for $\mathbf{G}(s)$ if and only if it solves it for the plant $\tilde{\mathbf{G}}(s)$.

Proof   Define

$$\tilde{\mathbf{u}}_1(t) = \mathbf{u}_1(t) - \frac{\mathbf{B}_1^T \mathbf{P}}{\gamma^2}\mathbf{x}(t), \tag{8.120a}$$

$$\tilde{\mathbf{y}}_1(t) = \mathbf{u}_2(t) + \mathbf{B}_2^T \mathbf{P}\mathbf{x}(t). \tag{8.120b}$$

Then, from Equation (8.98), we have

$$
\begin{aligned}
J_T &= \|\mathbf{y}_1(t)\|_{L_2[0,T]}^2 - \gamma^2 \|\mathbf{u}_1(t)\|_{L_2[0,T]}^2 \\
&= \mathbf{x}(0)^T \mathbf{P}(0)\mathbf{x}(0) + \|\tilde{\mathbf{y}}_1(t)\|_{L_2[0,T]}^2 - \gamma^2 \|\tilde{\mathbf{u}}_1(t)\|_{L_2[0,T]}^2 .
\end{aligned}
\tag{8.121}
$$

Since the initial condition $\mathbf{x}(0)$ does not affect the H_∞ norm (which is an input–output property), by setting $\mathbf{x}(0) = 0$, we conclude that

$$\left\|\mathbf{T}_{\mathbf{y}_1(t)\mathbf{u}_1(t)}\right\|_\infty \leq \gamma \tag{8.122}$$

if and only if

$$\left\|\mathbf{T}_{\tilde{\mathbf{y}}_1(t)\tilde{\mathbf{u}}_1(t)}\right\|_\infty \leq \gamma. \tag{8.123}$$

So, solving the H_∞ problem for

$$
\mathbf{G}(s) \overset{\Delta}{=}
\left[
\begin{array}{c|cc}
\mathbf{A} & \mathbf{B}_1 & \mathbf{B}_2 \\
\hline
\mathbf{C}_1 & 0 & \mathbf{D}_{12} \\
\mathbf{C}_2 & \mathbf{D}_{21} & 0
\end{array}
\right]
\tag{8.124}
$$

is equivalent to solving the H_∞ problem for the plant $\tilde{\mathbf{G}}(s)$ corresponding to $\tilde{\mathbf{y}}_1(t), \tilde{\mathbf{u}}_1(t)$. Now let us derive the packed matrix representation for $\tilde{\mathbf{G}}(s)$. Substituting

$$\mathbf{u}_1(t) = \tilde{\mathbf{u}}_1(t) + \frac{\mathbf{B}_1^T \mathbf{P}}{\gamma^2}\mathbf{x}(t) \tag{8.125}$$

into the state and output equations

$$\dot{\mathbf{x}}(t) = \mathbf{A}\mathbf{x}(t) + \mathbf{B}_1\mathbf{u}_1(t) + \mathbf{B}_2\mathbf{u}_2(t), \tag{8.126a}$$

$$\mathbf{y}_1(t) = \mathbf{C}_1\mathbf{x}(t) + \mathbf{D}_{12}\mathbf{u}_2(t), \tag{8.126b}$$

$$\mathbf{y}_2(t) = \mathbf{C}_2\mathbf{x}(t) + \mathbf{D}_{21}\mathbf{u}_1(t), \tag{8.126c}$$

we obtain the new state and output equations

$$\dot{\mathbf{x}}(t) = \mathbf{A}\mathbf{x}(t) + \mathbf{B}_1\left(\tilde{\mathbf{u}}_1(t) + \frac{\mathbf{B}_1^T \mathbf{P}}{\gamma^2}\mathbf{x}(t)\right) + \mathbf{B}_2\mathbf{u}(t)_2, \tag{8.127a}$$

$$\tilde{\mathbf{y}}_1(t) = \mathbf{B}_2^T P\mathbf{x}(t) + \mathbf{u}_2(t), \tag{8.127b}$$

$$\mathbf{y}_2(t) = \mathbf{C}_2\mathbf{x}(t) + \mathbf{D}_{21}\left(\tilde{\mathbf{u}}_1(t) + \frac{\mathbf{B}_1^T \mathbf{P}}{\gamma^2}\mathbf{x}(t)\right) = \mathbf{C}_2\mathbf{x}(t) + \mathbf{D}_{21}\tilde{\mathbf{u}}_1(t), \tag{8.127c}$$

from which it follows that

$$
\tilde{\mathbf{G}}(s) = \left[\begin{array}{c:cc} \mathbf{A} + \dfrac{\mathbf{B}_1 \mathbf{B}_1^T}{\gamma^2}\mathbf{P} & \mathbf{B}_1 & \mathbf{B}_2 \\ \hdashline \mathbf{B}_2^T \mathbf{P} & 0 & \mathbf{I} \\ \mathbf{C}_2 & \mathbf{D}_{21} & 0 \end{array} \right] \tag{8.128}
$$

is the plant corresponding to $\tilde{\mathbf{y}}_1(t)$, $\tilde{\mathbf{u}}_1(t)$. ∎

Lemma 8.2 has a dual result, which is the following.

LEMMA 8.3 Consider

$$
\mathbf{G}(s) = \left[\begin{array}{c:cc} \mathbf{A} & \mathbf{B}_1 & \mathbf{B}_2 \\ \hdashline \mathbf{C}_1 & 0 & \mathbf{D}_{12} \\ \mathbf{C}_2 & \mathbf{D}_{21} & 0 \end{array} \right] \tag{8.129}
$$

and

$$
\tilde{\mathbf{G}}(s) = \left[\begin{array}{c:cc} \mathbf{A} + \mathbf{Q}\dfrac{\mathbf{C}_1^T \mathbf{C}_1}{\gamma^2} & \mathbf{Q}\mathbf{C}_2^T & \mathbf{B}_2 \\ \hdashline \mathbf{C}_1 & 0 & \mathbf{D}_{12} \\ \mathbf{C}_2 & \mathbf{I} & 0 \end{array} \right], \tag{8.130}
$$

where $\mathbf{B}_1 \mathbf{D}_{21}^T = 0$, $\mathbf{D}_{21}\mathbf{D}_{21}^T = \mathbf{I}$, $\mathbf{D}_{12}^T \mathbf{C}_1 = 0$, and

$$
\mathbf{Q} = Ric \left[\begin{array}{c:c} \mathbf{A}^T & \dfrac{\mathbf{C}_1^T \mathbf{C}_1}{\gamma^2} - \mathbf{C}_2^T \mathbf{C}_2 \\ \hdashline -\mathbf{B}_1 \mathbf{B}_1^T & -\mathbf{A} \end{array} \right]. \tag{8.131}
$$

Then a control law $\mathbf{u}_2(t) = \mathbf{F}(s)\mathbf{y}_2(t)$ solves the standard H_∞ control problem for $\mathbf{G}(s)$ if and only if it solves it for the plant $\tilde{\mathbf{G}}(s)$.

Proof Noting that

$$
\|\mathbf{G}(s)\|_\infty = \|\mathbf{G}^T(s)\|_\infty, \tag{8.132}
$$

the result follows directly from Lemma 8.2 by taking the transpose of everything, applying Lemma 8.2, and then transposing all of the resultant equations. ∎

LEMMA 8.4 (Observer-based H_∞ feedback) Consider the plant

$$
\mathbf{G}(s) = \left[\begin{array}{c:cc} \mathbf{A} & \mathbf{H} & \mathbf{B}_2 \\ \hdashline \mathbf{C}_1 & 0 & \mathbf{D}_{12} \\ \mathbf{C}_2 & \mathbf{I} & 0 \end{array} \right] \tag{8.133}
$$

and suppose that $\mathbf{A} - \mathbf{H}\mathbf{C}_2$ is stable, and that $\mathbf{D}_{12}^T \mathbf{C}_1 = 0$ and $\mathbf{D}_{12}^T \mathbf{D}_{12} = \mathbf{I}$. Let

$$
\mathbf{F} \triangleq -\mathbf{B}_2^T \mathbf{P} \tag{8.134}
$$

be the optimal full state feedback

$$\mathbf{u}_2(t) = \mathbf{F}\mathbf{x}(t) \tag{8.135}$$

given by Theorem 8.2. Then the observer-based feedback

$$\mathbf{u}_2(t) = \mathbf{K}(s)\mathbf{y}_2(t) \tag{8.136}$$

given below solves the standard H_∞ problem for $\mathbf{G}(s)$:

$$\mathbf{u}_2(t) = -\mathbf{B}_2^T\mathbf{P}\hat{\mathbf{x}}(t),$$
$$\dot{\hat{\mathbf{x}}}(t) = \mathbf{A}\hat{\mathbf{x}}(t) + \mathbf{B}_2\mathbf{u}_2(t) + \mathbf{H}\left(\mathbf{y}_2(t) - \mathbf{C}_2\hat{\mathbf{x}}(t)\right),$$

$$\mathbf{P} = Ric\left[\begin{array}{c|c} \mathbf{A} & \dfrac{\mathbf{H}\mathbf{H}^T}{\gamma^2} - \mathbf{B}_2\mathbf{B}_2^T \\ \hline -\mathbf{C}_1^T\mathbf{C}_1 & -\mathbf{A}^T \end{array}\right], \tag{8.137}$$

that is $\mathbf{K}(s) = -\mathbf{B}_2^T\mathbf{P}\left(s\mathbf{I} - \mathbf{A} + \mathbf{B}_2\mathbf{B}_2^T\mathbf{P} + \mathbf{H}\mathbf{C}_2\right)^{-1}\mathbf{H}.$

Proof   Define the state observation error

$$\mathbf{e}(t) = \mathbf{x}(t) - \hat{\mathbf{x}}(t). \tag{8.138}$$

Now the plant state equation is

$$\dot{\mathbf{x}}(t) = \mathbf{A}\mathbf{x}(t) + \mathbf{H}\mathbf{u}_1(t) + \mathbf{B}_2\mathbf{u}_2(t) \tag{8.139}$$

while the observed state equation is

$$\dot{\hat{\mathbf{x}}}(t) = (\mathbf{A} - \mathbf{H}\mathbf{C}_2)\hat{\mathbf{x}}(t) + \mathbf{B}_2\mathbf{u}_2(t) + \mathbf{H}\mathbf{y}_2(t),$$
$$\text{or } \dot{\hat{\mathbf{x}}}(t) = (\mathbf{A} - \mathbf{H}\mathbf{C}_2)\hat{\mathbf{x}}(t) + \mathbf{B}_2\mathbf{u}_2(t) + \mathbf{H}(\mathbf{C}_2\mathbf{x}(t) + \mathbf{u}_1(t)), \tag{8.140}$$
$$\text{or } \dot{\hat{\mathbf{x}}}(t) = (\mathbf{A} - \mathbf{H}\mathbf{C}_2)\hat{\mathbf{x}}(t) + \mathbf{B}_2\mathbf{u}_2(t) + \mathbf{H}\mathbf{C}_2\mathbf{x}(t) + \mathbf{H}\mathbf{u}_1(t).$$

Subtracting Equation (8.140) from Equation (8.139), we obtain

$$\dot{\mathbf{e}}(t) = (\mathbf{A} - \mathbf{H}\mathbf{C}_2)\mathbf{e}(t). \tag{8.141}$$

Noting that $(\mathbf{A} - \mathbf{H}\mathbf{C}_2)$ is stable and $\mathbf{e}(t)$ is not excited by $\mathbf{u}_1(t)$, it follows that if $\mathbf{x}(0) = \hat{\mathbf{x}}(0)$, then

$$\mathbf{e}(t) \equiv 0, \qquad \text{for all } t \tag{8.142}$$

and hence

$$\mathbf{x}(t) = \hat{\mathbf{x}}(t), \qquad \text{for all } t. \tag{8.143}$$

The H_∞ norm being an input–output property, nonzero initial conditions will not affect its value. Therefore, one may substitute $\hat{\mathbf{x}}(t)$ for $\mathbf{x}(t)$ in the control law

$$\mathbf{u}_2(t) = -\mathbf{B}_2^T\mathbf{P}\mathbf{x}(t) \tag{8.144}$$

without affecting $\mathbf{T}_{\mathbf{y}_1(t)\mathbf{u}_1(t)}$ or its H_∞ norm.

   Combining Lemmas 8.3 and 8.4 one has a solution to the standard H_∞ control problem which is stated formally in the following lemma.   ∎

LEMMA 8.5 (A 2-Ricatti solution to the standard H_∞ control problem)   Let

$$\mathbf{G}(s) = \left[\begin{array}{c|cc} \mathbf{A} & \mathbf{B}_1 & \mathbf{B}_2 \\ \hline \mathbf{C}_1 & 0 & \mathbf{D}_{12} \\ \mathbf{C}_2 & \mathbf{D}_{21} & 0 \end{array}\right]. \tag{8.145}$$

with $\mathbf{D}_{12}^T\mathbf{D}_{12} = \mathbf{I}$, $\mathbf{D}_{21}\mathbf{D}_{21}^T = \mathbf{I}$, $\mathbf{D}_{12}^T\mathbf{C}_1 = 0$, and $\mathbf{B}_1\mathbf{D}_{21}^T = 0$. Define

$$\mathbf{Q} \triangleq Ric\left[\begin{array}{c|c} \mathbf{A}^T & \dfrac{\mathbf{C}_1^T\mathbf{C}_1}{\gamma^2} - \mathbf{C}_2^T\mathbf{C}_2 \\ \hline -\mathbf{B}_1\mathbf{B}_1^T & -\mathbf{A} \end{array}\right]. \tag{8.146}$$

Let

$$\tilde{\mathbf{G}}(s) \triangleq \left[\begin{array}{c|cc} \tilde{\mathbf{A}} & \mathbf{H} & \mathbf{B}_2 \\ \hline \mathbf{C}_1 & 0 & \mathbf{D}_{12} \\ \mathbf{C}_2 & \mathbf{I} & 0 \end{array}\right] \tag{8.147}$$

where

$$\tilde{\mathbf{A}} \triangleq \mathbf{A} + \mathbf{Q}\dfrac{\mathbf{C}_1^T\mathbf{C}_1}{\gamma^2}, \quad \mathbf{H} \triangleq \mathbf{Q}\mathbf{C}_2^T \tag{8.148}$$

and let $\mathbf{F} = \mathbf{B}_2^T\tilde{\mathbf{P}}$ where

$$\begin{aligned} \tilde{\mathbf{P}} &= Ric\left[\begin{array}{c|c} \tilde{\mathbf{A}} & \dfrac{\mathbf{H}\mathbf{H}^T}{\gamma^2} - \mathbf{B}_2\mathbf{B}_2^T \\ \hline -\mathbf{C}_1^T\mathbf{C}_1 & -\tilde{\mathbf{A}}^T \end{array}\right] \\[2em] &\triangleq Ric\left[\begin{array}{c|c} \mathbf{A} + \mathbf{Q}\dfrac{\mathbf{C}_1^T\mathbf{C}_1}{\gamma^2} & \mathbf{Q}\dfrac{\mathbf{C}_2^T\mathbf{C}_2}{\gamma^2}\mathbf{Q} - \mathbf{B}_2\mathbf{B}_2^T \\ \hline -\mathbf{C}_1^T\mathbf{C}_1 & -\left(\mathbf{A} + \mathbf{Q}\dfrac{\mathbf{C}_1^T\mathbf{C}_1}{\gamma^2}\right)^T \end{array}\right]. \end{aligned} \tag{8.149}$$

Then the observer-based control law $\mathbf{u}_2(t) = K(s)[\mathbf{y}_2(t)]$ solves the standard H_∞ control problem with

$$\begin{aligned} K(s) &= -\mathbf{B}_2^T\tilde{\mathbf{P}}\left(s\mathbf{I} - \tilde{\mathbf{A}} + \mathbf{B}_2\mathbf{B}_2^T\tilde{\mathbf{P}} + \mathbf{H}\mathbf{C}_2\right)^{-1}\mathbf{H} \\ &= -\mathbf{B}_2^T\tilde{\mathbf{P}}\left(s\mathbf{I} - \mathbf{A} - \mathbf{Q}\dfrac{\mathbf{C}_1^T\mathbf{C}_1}{\gamma^2} + \mathbf{B}_2\mathbf{B}_2^T\tilde{\mathbf{P}} + \mathbf{Q}\mathbf{C}_2^T\mathbf{C}_2\right)^{-1}\mathbf{Q}\mathbf{C}_2^T \end{aligned} \tag{8.150}$$

or equivalently

$$\begin{aligned} \mathbf{u}_2(t) &= -\mathbf{B}_2^T\tilde{\mathbf{P}}\hat{\mathbf{x}}(t), \\ \dot{\hat{\mathbf{x}}}(t) &= \tilde{\mathbf{A}}\hat{\mathbf{x}}(t) + \mathbf{B}_2\mathbf{u}_2(t) + \mathbf{H}\left(\mathbf{y}_2(t) - \mathbf{C}_2\hat{\mathbf{x}}(t)\right) \\ &= \left(\mathbf{A} + \mathbf{Q}\dfrac{\mathbf{C}_1^T\mathbf{C}_1}{\gamma^2}\right)\hat{\mathbf{x}}(t) + \mathbf{B}_2\mathbf{u}_2(t) + \mathbf{Q}\mathbf{C}_2^T\left(\mathbf{y}_2(t) - \mathbf{C}_2\hat{\mathbf{x}}(t)\right). \end{aligned} \tag{8.151}$$

Proof   The proof is completed by using Lemma 8.3 followed by Lemma 8.4.   ■

The solution in Lemma 8.5 is stated in terms of the Ricatti solution $\tilde{\mathbf{P}}$, which is different from the Ricatti solution $\mathbf{P}$ used for the full state feedback case in Theorem 8.2. We next show that $\tilde{\mathbf{P}}$ is related to $\mathbf{P}$. We first note that the $\tilde{\mathbf{P}}$ Ricatti Hamiltonian is similar to the $\mathbf{P}$ Ricatti Hamiltonian, that is

$$
\mathbf{T}^{-1}
\begin{bmatrix}
\mathbf{A} + \mathbf{Q}\dfrac{\mathbf{C}_1^T\mathbf{C}_1}{\gamma^2} & \mathbf{Q}\dfrac{\mathbf{C}_2^T\mathbf{C}_2}{\gamma^2}\mathbf{Q} - \mathbf{B}_2\mathbf{B}_2^T \\
-\mathbf{C}_1^T\mathbf{C}_1 & -\left(\mathbf{A} + \mathbf{Q}\dfrac{\mathbf{C}_1^T\mathbf{C}_1}{\gamma^2}\right)^T
\end{bmatrix}
\mathbf{T}
$$
$$
=
\begin{bmatrix}
\mathbf{A} & \dfrac{\mathbf{B}_1\mathbf{B}_1^T}{\gamma^2} - \mathbf{B}_2\mathbf{B}_2^T \\
-\mathbf{C}_1^T\mathbf{C}_1 & -\mathbf{A}^T
\end{bmatrix}
\tag{8.152}
$$

where

$$
\mathbf{T} =
\begin{bmatrix}
\mathbf{I} & -\dfrac{\mathbf{Q}}{\gamma^2} \\
\mathbf{0} & \mathbf{I}
\end{bmatrix},
\quad
\mathbf{T}^{-1} =
\begin{bmatrix}
\mathbf{I} & \dfrac{\mathbf{Q}}{\gamma^2} \\
\mathbf{0} & \mathbf{I}
\end{bmatrix}
\tag{8.153}
$$

and the fact that

$$
\mathbf{Q}\mathbf{A}^T + \mathbf{A}\mathbf{Q} + \mathbf{Q}\left(\frac{\mathbf{C}_1^T\mathbf{C}_1}{\gamma^2} - \mathbf{C}_2^T\mathbf{C}_2\right)\mathbf{Q} + \mathbf{B}_1\mathbf{B}_1^T = 0
\tag{8.154}
$$

is used to simplify the upper right entry to

$$
\frac{\mathbf{B}_1\mathbf{B}_1^T}{\gamma^2} - \mathbf{B}_2\mathbf{B}_2^T.
\tag{8.155}
$$

Thus the two Hamiltonians will have the same eigenvalues and their corresponding eigenvectors will also be related through the nonsingular matrix T.

Recall from Lemma 7.2 in Chapter 7 that if

$$
\mathbf{P} = Ric
\begin{bmatrix}
\mathbf{A} & -\mathbf{R} \\
-\mathbf{Q} & -\mathbf{A}^T
\end{bmatrix}
\tag{8.156}
$$

then $\mathbf{P}$ may be computed as

$$
\mathbf{P} = \mathbf{P}_2\mathbf{P}_1^{-1},
\tag{8.157}
$$

where $\begin{bmatrix} \mathbf{P}_1 \\ \mathbf{P}_2 \end{bmatrix}$ is any matrix whose columns form a basis for the stable eigenspace of

$$
\begin{bmatrix}
\mathbf{A} & -\mathbf{R} \\
-\mathbf{Q} & -\mathbf{A}^T
\end{bmatrix}.
\tag{8.158}
$$

Furthermore, since the two Hamiltonian matrices are similar, we have

$$
\begin{bmatrix} \tilde{\mathbf{P}}_1 \\ \tilde{\mathbf{P}}_2 \end{bmatrix} = T \begin{bmatrix} \mathbf{P}_1 \\ \mathbf{P}_2 \end{bmatrix} = \begin{bmatrix} \mathbf{I} & -\dfrac{\mathbf{Q}}{\gamma^2} \\ 0 & \mathbf{I} \end{bmatrix} \begin{bmatrix} \mathbf{P}_1 \\ \mathbf{P}_2 \end{bmatrix} = \begin{bmatrix} \mathbf{P}_1 - \dfrac{\mathbf{Q}}{\gamma^2} \mathbf{P}_2 \\ \mathbf{P}_2 \end{bmatrix}.
\tag{8.159}
$$

Thus

$$
\begin{aligned}
\tilde{\mathbf{P}} &= \tilde{\mathbf{P}}_2 \tilde{\mathbf{P}}_1^{-1} \\
&= \mathbf{P}_2 \left(\mathbf{P}_1 - \frac{\mathbf{Q}}{\gamma^2} \mathbf{P}_2 \right)^{-1} \\
&= \mathbf{P}_2 \left[\left(\mathbf{I} - \frac{\mathbf{Q}}{\gamma^2} \mathbf{P}_2 \mathbf{P}_1^{-1} \right) \mathbf{P}_1 \right]^{-1} \\
&= \mathbf{P}_2 \mathbf{P}_1^{-1} \left(\mathbf{I} - \frac{\mathbf{Q}}{\gamma^2} \mathbf{P}_2 \mathbf{P}_1^{-1} \right)^{-1} \\
&= \mathbf{P} \left(\mathbf{I} - \frac{\mathbf{QP}}{\gamma^2} \right)^{-1}.
\end{aligned}
\tag{8.160}
$$

For the above expression to be well-defined, $\rho(\mathbf{QP})$, the spectral radius of $\mathbf{QP}$ must be less than γ^2. Under this condition, $\tilde{\mathbf{P}}$ can be replaced by

$$
\mathbf{P} \left(\mathbf{I} - \frac{\mathbf{QP}}{\gamma^2} \right)^{-1}
\tag{8.161}
$$

in Lemma 8.5 to obtain

$$
\begin{aligned}
\mathbf{K}(s) = -\mathbf{B}_2^T \mathbf{P} \left(\mathbf{I} - \frac{\mathbf{QP}}{\gamma^2} \right)^{-1} \\
\left(s\mathbf{I} - \mathbf{A} - \mathbf{Q}\frac{\mathbf{C}_1^T \mathbf{C}_1}{\gamma^2} + \mathbf{B}_2 \mathbf{B}_2^T \mathbf{P} \left[\mathbf{I} - \frac{\mathbf{QP}}{\gamma^2} \right]^{-1} + \mathbf{Q}\mathbf{C}_2^T \mathbf{C}_2 \right)^{-1} \mathbf{Q}\mathbf{C}_2^T
\end{aligned}
\tag{8.162}
$$

or equivalently

$$
\mathbf{u}_2(t) = -\mathbf{B}_2^T \mathbf{P} \left(\mathbf{I} - \frac{\mathbf{QP}}{\gamma^2} \right)^{-1} \hat{\mathbf{x}}(t),
\tag{8.163}
$$

$$
\dot{\hat{\mathbf{x}}}(t) = \left(\mathbf{A} + \mathbf{Q}\frac{\mathbf{C}_1^T \mathbf{C}_1}{\gamma^2} \right) \hat{\mathbf{x}}(t) + \mathbf{B}_2 \mathbf{u}_2(t) + \mathbf{Q}\mathbf{C}_2^T \left(\mathbf{y}_2(t) - \mathbf{C}_2 \hat{\mathbf{x}}(t) \right).
\tag{8.164}
$$

Restating Lemma 8.5 in terms of $\mathbf{P}$, we have the following main result.

THEOREM 8.3 (2-Ricatti solution to the standard H_∞ problem) Let

$$
\mathbf{G}(s) = \left[\begin{array}{c|cc} \mathbf{A} & \mathbf{B}_1 & \mathbf{B}_2 \\ \hline \mathbf{C}_1 & 0 & \mathbf{D}_{12} \\ \mathbf{C}_2 & \mathbf{D}_{21} & 0 \end{array} \right]
\tag{8.165}
$$

with

$$\mathbf{D}_{12}^T \mathbf{D}_{12} = \mathbf{I},$$
$$\mathbf{D}_{21} \mathbf{D}_{21}^T = \mathbf{I},$$
$$\mathbf{D}_{12}^T \mathbf{C}_1 = 0, \qquad\qquad (8.166)$$
$$\mathbf{B}_1 \mathbf{D}_{21}^T = 0.$$

Let

$$
\left.
\begin{aligned}
\mathbf{P} &= Ric
\begin{bmatrix}
\mathbf{A} & \vdots & \dfrac{\mathbf{B}_1 \mathbf{B}_1^T}{\gamma^2} - \mathbf{B}_2 \mathbf{B}_2^T \\
\hdashline
-\mathbf{C}_1^T \mathbf{C}_1 & \vdots & -\mathbf{A}^T
\end{bmatrix} \\[2em]
\mathbf{Q} &= Ric
\begin{bmatrix}
\mathbf{A}^T & \vdots & \dfrac{\mathbf{C}_1^T \mathbf{C}_1}{\gamma^2} - \mathbf{C}_2^T \mathbf{C}_2 \\
\hdashline
-\mathbf{B}_1 \mathbf{B}_1^T & \vdots & -\mathbf{A}
\end{bmatrix}
\end{aligned}
\right\}
\text{ two Ricattis,} \qquad (8.167)
$$

$$
\left.
\begin{aligned}
\mathbf{F} &= \mathbf{B}_2^T \mathbf{P} \\
\mathbf{H} &= \mathbf{Q} \mathbf{C}_2^T
\end{aligned}
\right\}
\text{ (optimal H_∞ state feedback and its dual),} \qquad (8.168)
$$

and

$$\rho \mathbf{Q} \mathbf{P} < \gamma^2. \qquad\qquad (8.169)$$

Then an H_∞ control law for which

$$\left\| \mathbf{T}_{\mathbf{y}_1(t)\mathbf{u}_1(t)} \right\|_\infty \leq \gamma \qquad\qquad (8.170)$$

is given (in observer-reconstructed state feedback form) by

$$\mathbf{u}_2(t) = -\mathbf{F} \left(\mathbf{I} - \dfrac{\mathbf{QP}}{\gamma^2} \right)^{-1} \hat{\mathbf{x}}(t), \qquad\qquad (8.171)$$

$$\dot{\hat{\mathbf{x}}}(t) = \left(\mathbf{A} + \mathbf{Q} \dfrac{\mathbf{C}_1^T \mathbf{C}_1}{\gamma^2} \right) \hat{\mathbf{x}}(t) + \mathbf{B}_2 \mathbf{u}_2(t) + \mathbf{Q} \mathbf{C}_2^T \left(\mathbf{y}_2(t) - \mathbf{C}_2 \hat{\mathbf{x}}(t) \right). \qquad (8.172)$$

Remark 8.5   From the above theorem, it is clear that for the existence of a stabilizing controller such that

$$\left\| \mathbf{T}_{\mathbf{y}_1(t)\mathbf{u}_1(t)} \right\|_\infty < \gamma, \qquad\qquad (8.173)$$

we must have (i) **P** stabilizing, (ii) **Q** stabilizing, and (iii) $\rho(\mathbf{QP}) < \gamma^2$. Thus, one can progressively step down γ using a bisection algorithm until one of these three conditions breaks down. At that stage, one would be very close to having a controller that is H_∞ optimal.

8.4 H_2 **Optimal Solution**

In this section, we present a solution to the H_2 optimal control problem using an approach similar to the one that was used in the last section for the standard H_∞ optimal control problem. Towards this end, let let

$$G(s) = \left[\begin{array}{c|cc} \mathbf{A} & \mathbf{B}_1 & \mathbf{B}_2 \\ \hline \mathbf{C}_1 & 0 & \mathbf{D}_{12} \\ \mathbf{C}_2 & \mathbf{D}_{21} & 0 \end{array} \right] \tag{8.174}$$

with $\mathbf{D}_{12}^T\mathbf{D}_{12} = \mathbf{I}$, $\mathbf{D}_{21}\mathbf{D}_{21}^T = \mathbf{I}$, $\mathbf{D}_{12}^T\mathbf{C}_1 = 0$, and $\mathbf{B}_1\mathbf{D}_{21}^T = 0$. Furthermore, let us define

$$\mathbf{Q} \stackrel{\Delta}{=} Ric \left[\begin{array}{c|c} \mathbf{A}^T & -\mathbf{C}_2^T\mathbf{C}_2 \\ \hline -\mathbf{B}_1\mathbf{B}_1^T & -\mathbf{A} \end{array} \right], \tag{8.175}$$

$$\mathbf{P} \stackrel{\Delta}{=} Ric \left[\begin{array}{c|c} \mathbf{A} & -\mathbf{B}_2\mathbf{B}_2^T \\ \hline -\mathbf{C}_1^T\mathbf{C}_1 & -\mathbf{A}^T \end{array} \right]. \tag{8.176}$$

We will show that the optimal H_2 solution is given by (dropping the dependency on t, henceforth)

$$\mathbf{u}_2 = -\mathbf{B}_2^T\mathbf{P}\hat{\mathbf{x}}, \tag{8.177}$$

where

$$\dot{\hat{\mathbf{x}}} = \mathbf{A}\hat{\mathbf{x}} + \mathbf{B}_2\mathbf{u}_2 + \mathbf{Q}\mathbf{C}_2^T\left(\mathbf{y}_2 - \mathbf{C}_2\hat{\mathbf{x}}\right). \tag{8.178}$$

This will be carried out in two stages. First, we will solve the H_2 optimal control problem using full state feedback. Thereafter, we will extend the results to observer-reconstructed state feedback.

8.4.1 Full State Feedback H_2 Optimal Control

Suppose that the full state x is available as y_2, as shown in Figure 8.9.

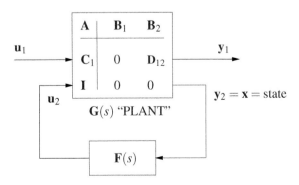

Figure 8.9 The standard plant controller setup

Here, the packed $(\mathbf{A}, \mathbf{B}, \mathbf{C}, \mathbf{D})$ matrix representation describes the augmented plant while $F(s)$ represents the feedback controller to be designed. The equations describing the augmented plant are

$$\dot{\mathbf{x}} = \mathbf{A}\mathbf{x} + \mathbf{B}_1\mathbf{u}_1 + \mathbf{B}_2\mathbf{u}_2,$$
$$\mathbf{y}_1 = \mathbf{C}_1\mathbf{x} + \mathbf{D}_{12}\mathbf{u}_2, \tag{8.179}$$
$$\mathbf{y}_2 = \mathbf{x}.$$

For simplicity, we make the following assumptions, which are a subset of the assumptions introduced at the beginning of this section:

$$\mathbf{D}_{12}^T\mathbf{C}_1 = 0, \tag{8.180a}$$
$$\mathbf{D}_{11} = 0, \tag{8.180b}$$
$$\mathbf{D}_{12}^T\mathbf{D}_{12} = \mathbf{I}, \tag{8.180c}$$
$$\mathbf{D}_{22} = 0. \tag{8.180d}$$

Now we want to minimize $\|\mathbf{T}_{\mathbf{y}_1\mathbf{u}_1}\|_2$. To carry out this minimization, we first consider the case when $\mathbf{T}_{\mathbf{y}_1\mathbf{u}_1}$ is *scalar*. Then

$$\mathbf{y}_1 = \mathbf{T}_{\mathbf{y}_1\mathbf{u}_1}(s)[\mathbf{u}_1] \tag{8.181}$$

$$\Rightarrow \int_0^\infty \mathbf{y}_1^T(t)\mathbf{y}_1(t)dt = \int_{-\infty}^\infty |\mathbf{Y}_1(j\omega)|^2 \frac{d\omega}{2\pi} \quad \text{(by Parseval's Theorem)}$$

$$= \int_{-\infty}^\infty |\mathbf{T}_{\mathbf{y}_1\mathbf{u}_1}(j\omega)|^2 \frac{d\omega}{2\pi}$$

$$\text{(assuming } \mathbf{U}_1(j\omega) = 1, \, \forall\, \omega, \text{ that is } \mathbf{u}_1(t) = \delta(t))$$

$$= \|\mathbf{T}_{\mathbf{y}_1\mathbf{u}_1}\|_2.$$

Thus, to minimize $\|\mathbf{T}_{\mathbf{y}_1\mathbf{u}_1}\|_2$, one could define the cost

$$J = \|\mathbf{y}_1\|_{L_2[0,\infty)}^2 \tag{8.182}$$

and minimize J over u_2, that is solve the problem

$$\min_{\mathbf{u}_2} J(\mathbf{x}_0, \mathbf{u}_2). \tag{8.183}$$

We note that in contrast to the H_∞ case, $\mathbf{u}_1$ here is a *fixed* signal, namely an impulse function.

An important point to note here is that the H_2 norm is defined for a transfer function matrix and thus does not depend on initial conditions which can be set to zero. We will demonstrate that the well-known H_2 solution can be derived along the lines of the state-space H_∞ solution presented earlier. To do so, we have converted a transfer-function-domain performance index into a time-domain performance index. As we will see, as a consequence initial condition terms appear in the time-domain performance index possibly affecting the rigor of the optimality arguments. Recognizing that such initial condition terms have only appeared as an artifact of our simplistic approach and have nothing to do with the frequency domain H_2 norm, we will set them equal to zero as we go along.

Now

$$J = \|\mathbf{y}_1\|_{L_2[0,\infty)}^2$$

$$= \int_0^\infty (\mathbf{C}_1\mathbf{x} + \mathbf{D}_{12}\mathbf{u}_2)^T (\mathbf{C}_1\mathbf{x} + \mathbf{D}_{12}\mathbf{u}_2) \, dt \tag{8.184}$$

$$= \int_0^\infty \left[\mathbf{x}^T \mathbf{C}_1^T \mathbf{C}_1 \mathbf{x} + \mathbf{u}_2^T \mathbf{u}_2 \right] dt \quad (\text{since } \mathbf{D}_{12}^T \mathbf{D}_{12} = \mathbf{I} \text{ and } \mathbf{D}_{12}^T \mathbf{C}_1 = 0).$$

Let us consider the finite-time interval cost

$$J_T (\mathbf{x}_0, \mathbf{u}_2) \triangleq \int_0^T L(\mathbf{x}(t), \mathbf{u}_2(t)) \, dt, \tag{8.185}$$

where

$$L(\mathbf{x}, \mathbf{u}_2) = \mathbf{x}^T \mathbf{C}_1^T \mathbf{C}_1 \mathbf{x} + \mathbf{u}_2^T \mathbf{u}_2. \tag{8.186}$$

Clearly, for J to be well-defined, we must have

$$J = \lim_{T \to \infty} J_T. \tag{8.187}$$

We will now show that J_T can be somewhat simplified via completion of squares. Let $\mathbf{P}(t) = \mathbf{P}^T(t)$ be an arbitrary symmetric matrix and consider the function

$$\mathbf{x}^T(T)\mathbf{P}(T)\mathbf{x}(T) - \mathbf{x}_0^T \mathbf{P}(0)\mathbf{x}_0 + J_T(\mathbf{x}_0, \mathbf{u}_2)$$

$$= \int_0^T \frac{d}{dt}\left(\mathbf{x}^T \mathbf{P}\mathbf{x}\right) dt + \int_0^T L(\mathbf{x}, \mathbf{u}_2) \, dt$$

$$= \int_0^T \left[\dot{\mathbf{x}}^T \mathbf{P}\mathbf{x} + \mathbf{x}^T \mathbf{P}\dot{\mathbf{x}} + \mathbf{x}^T \dot{\mathbf{P}}\mathbf{x} + \mathbf{x}^T \mathbf{C}_1^T \mathbf{C}_1 \mathbf{x} + \mathbf{u}_2^T \mathbf{u}_2 \right] dt$$

$$= \int_0^T \Big[(\mathbf{A}\mathbf{x} + \mathbf{B}_1\mathbf{u}_1 + \mathbf{B}_2\mathbf{u}_2)^T \mathbf{P}\mathbf{x} + \mathbf{x}^T \dot{\mathbf{P}}\mathbf{x} \tag{8.188}$$

$$\qquad + \mathbf{x}^T \mathbf{P}(\mathbf{A}\mathbf{x} + \mathbf{B}_1\mathbf{u}_1 + \mathbf{B}_2\mathbf{u}_2) + \mathbf{x}^T \mathbf{C}_1^T \mathbf{C}_1 \mathbf{x} + \mathbf{u}_2^T \mathbf{u}_2 \Big] dt$$

$$= \int_0^T \Big[\mathbf{x}^T \left(\dot{\mathbf{P}} + \mathbf{A}^T \mathbf{P} + \mathbf{P}\mathbf{A} + \mathbf{P}\left(-\mathbf{B}_2\mathbf{B}_2^T\right)\mathbf{P} + \mathbf{C}_1^T \mathbf{C}_1 \right) \mathbf{x}$$

$$\qquad + \left(\mathbf{u}_2 + \mathbf{B}_2^T \mathbf{P}\mathbf{x} \right)^T \left(\mathbf{u}_2 + \mathbf{B}_2^T \mathbf{P}\mathbf{x} \right) + \mathbf{x}^T \mathbf{P}\mathbf{B}_1\mathbf{u}_1 + \mathbf{u}_1^T \mathbf{B}_1^T \mathbf{P}\mathbf{x} \Big] dt.$$

Since $\mathbf{P}(t)$ is arbitrary, to simplify the above expression, we choose $\mathbf{P}(t)$ to be the solution of the Ricatti differential equation

$$-\dot{\mathbf{P}} = \mathbf{P}\mathbf{A} + \mathbf{A}^T \mathbf{P} - \mathbf{P}\mathbf{B}_2\mathbf{B}_2^T \mathbf{P} + \mathbf{C}_1^T \mathbf{C}_1, \tag{8.189}$$

$$\mathbf{P}(T) = 0. \tag{8.190}$$

Note $\mathbf{u}_1(t) = \delta(t)$, and with the above choice of $\mathbf{P}(t)$, we obtain:

LEMMA 8.6 (Completion of squares)

$$J_T = \mathbf{x}_0^T \mathbf{P}(0)\mathbf{x}_0 + 2\mathbf{x}_0^T \mathbf{P}(0)\mathbf{B}_1 + \|\mathbf{u}_2 - \mathbf{u}_2^*\|_{L_2[0,T]}^2, \tag{8.191}$$

where $\mathbf{u}_2^* = -\mathbf{B}_2^T \mathbf{P}\mathbf{x}$.

From Lemma 8.6, it is clear that if the state $\mathbf{x}$ is available for measurement, and the initial condition $\mathbf{x}_0$ is assumed to be zero, then the cost minimizing $\mathbf{u}_2(t)$ is simply

$$\mathbf{u}_2 = \mathbf{u}_2^* = -\mathbf{B}_2^T \mathbf{P} \mathbf{x}. \tag{8.192}$$

Moreover, if $\bar{\mathbf{P}} = \lim_{T \to \infty} \mathbf{P}(t)$ exists, then it is well known that $\bar{\mathbf{P}}$ satisfies the ARE

$$0 = \mathbf{P}\mathbf{A} + \mathbf{A}^T \mathbf{P} - \mathbf{P}\mathbf{B}_2 \mathbf{B}_2^T \mathbf{P} + \mathbf{C}_1^T \mathbf{C}_1 \tag{8.193}$$

or equivalently

$$\mathbf{P} = Ric \left[\begin{array}{c|c} \mathbf{A} & -\mathbf{B}_2 \mathbf{B}_2^T \\ \hline -\mathbf{C}_1^T \mathbf{C}_1 & -\mathbf{A}^T \end{array} \right]. \tag{8.194}$$

Summarizing the above discussion, we have the following theorem.

THEOREM 8.4 (Full state feedback H_2 optimal solution)   Let

$$\mathbf{G}(s) = \left[\begin{array}{c|cc} \mathbf{A} & \mathbf{B}_1 & \mathbf{B}_2 \\ \hline \mathbf{C}_1 & 0 & \mathbf{D}_{12} \\ \mathbf{I} & 0 & 0 \end{array} \right], \tag{8.195a}$$

$$\mathbf{D}_{12}^T \mathbf{C}_1 = 0, \tag{8.195b}$$

$$\mathbf{D}_{12}^T \mathbf{D}_{12} = \mathbf{I}. \tag{8.195c}$$

Then the solution to the H_2 problem is

$$\mathbf{F}(s) = -\mathbf{B}_2^T \mathbf{P}, \tag{8.196}$$

where $\mathbf{P}$ is a solution to the ARE

$$\mathbf{P} = Ric \left[\begin{array}{c|c} \mathbf{A} & -\mathbf{B}_2 \mathbf{B}_2^T \\ \hline -\mathbf{C}_1^T \mathbf{C}_1 & -\mathbf{A}^T \end{array} \right]. \tag{8.197}$$

Now the discussion leading up to the above theorem was valid only for a scalar $\mathbf{T}_{\mathbf{y}_1 \mathbf{u}_1}$. However, as we now show, the theorem statement holds even when $\mathbf{T}_{\mathbf{y}_1 \mathbf{u}_1}$ is multi-input multi-output (MIMO). For the MIMO case, we solve the following problem:

$$\min_{\mathbf{u}_2} \sum_{i=1}^{n} \int_0^\infty \mathbf{y}_{1i}^T \mathbf{y}_{1i} dt, \tag{8.198}$$

where

$$\mathbf{y}_{1i} = \text{output with } \mathbf{u}_{1i} = \mathbf{e}_i \delta(t) \tag{8.199}$$

and $\mathbf{e}_i$ is the ith basis vector. Now, by the arguments preceding Theorem 8.4, the control input

$$\mathbf{u}_2 = -\mathbf{B}_2^T \mathbf{P} \mathbf{x} \tag{8.200}$$

minimizes each of the terms in the above summation, *independent of which basis vector*

$\mathbf{e}_i$ *is used for synthesizing the input* $\mathbf{u}_1$. Thus, the input in Equation (8.200) is also optimal for the cost in Equation (8.198). Note that Equation (8.198) is equivalent to

$$\min_{\mathbf{u}_2} \frac{1}{2\pi} \int_{-\infty}^{\infty} \text{Trace} \left[\mathbf{T}_{\mathbf{y}_1\mathbf{u}_1}^*(j\omega) \mathbf{T}_{\mathbf{y}_1\mathbf{u}_1}(j\omega) \right] d\omega = \min_{\mathbf{u}_2} \| \mathbf{T}_{\mathbf{y}_1\mathbf{u}_1}(s) \|_2. \qquad (8.201)$$

Thus, Theorem 8.4 provides the H_2 optimal full state feedback controller even for the MIMO case.

The next corollary shows that the assumption in Equation (8.195b) can be relaxed at the expense of making the formulas a little more complicated.

COROLLARY 8.3 (Full state feedback without the assumption $\mathbf{D}_{12}^T\mathbf{C}_1 = 0$) Let everything be as in Theorem 8.4, except that the condition in Equation (8.195b) does not necessarily hold, that is

$$\mathbf{D}_{12}^T\mathbf{C}_1 \neq 0. \qquad (8.202)$$

Then the solution to the H_2 optimal control problem is

$$\mathbf{F}(s) = -\mathbf{D}_{12}^T\mathbf{C}_1 - \mathbf{B}_2^T\mathbf{P}, \qquad (8.203)$$

where $\mathbf{P}$ is a solution to the ARE

$$\mathbf{P} = Ric \left[\begin{array}{c|c} \mathbf{A} - \mathbf{B}_2\mathbf{D}_{12}^T\mathbf{C}_1 & -\mathbf{B}_2\mathbf{B}_2^T \\ \hline -\mathbf{C}_1^T\left(\mathbf{I} - \mathbf{D}_{12}\mathbf{D}_{12}^T\right)\mathbf{C}_1 & -\left(\mathbf{A} - \mathbf{B}_2\mathbf{D}_{12}^T\mathbf{C}_1\right)^T \end{array} \right]. \qquad (8.204)$$

Proof Noting that the change of variables

$$\mathbf{u}_2 = \tilde{\mathbf{u}}_2 - \mathbf{D}_{12}^T\mathbf{C}_1\mathbf{x} \qquad (8.205)$$

results in

$$\tilde{\mathbf{G}}(s) = \left[\begin{array}{c|cc} \mathbf{A} - \mathbf{B}_2\mathbf{D}_{12}^T\mathbf{C}_1 & \mathbf{B}_1 & \mathbf{B}_2 \\ \hline \left(\mathbf{I} - \mathbf{D}_{12}\mathbf{D}_{12}^T\right)\mathbf{C}_1 & 0 & \mathbf{D}_{12} \\ \mathbf{I} & 0 & 0 \end{array} \right] \overset{\Delta}{=} \left[\begin{array}{c|cc} \tilde{\mathbf{A}} & \tilde{\mathbf{B}}_1 & \tilde{\mathbf{B}}_2 \\ \hline \tilde{\mathbf{C}}_1 & 0 & \tilde{\mathbf{D}}_{12} \\ \mathbf{I} & 0 & 0 \end{array} \right] \qquad (8.206)$$

and this $\tilde{\mathbf{G}}(s)$ satisfies the conditions of Theorem 8.4, it follows that

$$\tilde{\mathbf{u}}_2 = -\mathbf{B}_2^T\mathbf{P}\mathbf{x} \qquad (8.207)$$

and hence

$$\mathbf{u}_2 = -\left(\mathbf{D}_{12}^T\mathbf{C}_1 + \mathbf{B}_2^T\mathbf{P}\right)\mathbf{x}. \qquad (8.208)$$

∎

8.4.2 Observer-Reconstructed State Feedback H_2 Optimal Control

We now solve the H_2 problem when the state x is not directly measurable. As in the H_∞ case, this can be done by developing and applying a series of analogous lemmas. The first of these lemmas, which is stated next, shows that a plant $\mathbf{G}(s)$ for the H_2 problem can be replaced by an equivalent one in which $\tilde{\mathbf{G}}(s)$ has a square $\mathbf{D}_{12}$ matrix, that is $\mathbf{D}_{12} \doteq \mathbf{I}$.

LEMMA 8.7 Consider the following two plants:

$$G(s) \triangleq \left[\begin{array}{c|cc} \mathbf{A} & \mathbf{B}_1 & \mathbf{B}_2 \\ \hline \mathbf{C}_1 & 0 & \mathbf{D}_{12} \\ \mathbf{C}_2 & \mathbf{D}_{21} & 0 \end{array} \right], \tag{8.209}$$

$$\tilde{G}(s) \triangleq \left[\begin{array}{c|cc} \mathbf{A} & \mathbf{B}_1 & \mathbf{B}_2 \\ \hline \mathbf{B}_2^T \mathbf{P} & 0 & \mathbf{I} \\ \mathbf{C}_2 & \mathbf{D}_{21} & 0 \end{array} \right]. \tag{8.210}$$

Assume that the conditions of Theorem 8.4 hold, that is $\mathbf{D}_{12}^T \mathbf{C}_1 = 0$ and $\mathbf{D}_{12}^T \mathbf{D}_{12} = \mathbf{I}$. Also $\mathbf{B}_1 \mathbf{D}_{21}^T = 0$ and

$$\mathbf{P} = Ric \left[\begin{array}{c|c} \mathbf{A} & -\mathbf{B}_2 \mathbf{B}_2^T \\ \hline -\mathbf{C}_1^T \mathbf{C}_1 & -\mathbf{A}^T \end{array} \right]. \tag{8.211}$$

Then, a control law $\mathbf{u}_2 = \mathbf{F}(s)[\mathbf{y}_2]$ solves the standard H_2 problem for $G(s)$ if and only if it solves it for the "squared down plant" $\tilde{G}(s)$.

Proof The proof follows directly from Lemma 8.6 by letting

$$\tilde{\mathbf{y}}_1 = \mathbf{u}_2 + \mathbf{B}_2^T \mathbf{P} \mathbf{x}, \qquad \tilde{\mathbf{u}}_1 = \mathbf{u}_1, \tag{8.212a}$$

$$\tilde{\mathbf{y}}_2 = \mathbf{y}_2, \qquad \tilde{\mathbf{u}}_2 = \mathbf{u}_2, \tag{8.212b}$$

which when substituted into the equations for $G(s)$, yields $\tilde{G}(s)$. Note that

$$\begin{aligned} J &= \left(\|\mathbf{y}_1\|_{L_2[0,\infty)} \right)^2 \\ &= \left(\|\tilde{\mathbf{y}}_1\|_{L_2[0,\infty)} \right)^2 + \mathbf{x}_0^T \bar{\mathbf{P}} \mathbf{x}_0 + 2\mathbf{x}_0^T \bar{\mathbf{P}} \mathbf{B}_1. \end{aligned} \tag{8.213}$$

Thus, assuming $\mathbf{x}_0 = 0$, it follows that $\|\mathbf{y}_1\|_{L_2[0,\infty)}$ is minimized if and only if $\|\tilde{\mathbf{y}}_1\|_{L_2[0,\infty)}$ is minimized. ∎

Lemma 8.7 has a dual, which is the following.

LEMMA 8.8 Consider

$$G(s) = \left[\begin{array}{c|cc} \mathbf{A} & \mathbf{B}_1 & \mathbf{B}_2 \\ \hline \mathbf{C}_1 & 0 & \mathbf{D}_{12} \\ \mathbf{C}_2 & \mathbf{D}_{21} & 0 \end{array} \right] \quad \text{and} \quad \tilde{G}(s) \triangleq \left[\begin{array}{c|cc} \mathbf{A} & \mathbf{Q}\mathbf{C}_2^T & \mathbf{B}_2 \\ \hline \mathbf{C}_1 & 0 & \mathbf{D}_{12} \\ \mathbf{C}_2 & \mathbf{I} & 0 \end{array} \right], \tag{8.214}$$

where $\mathbf{B}_1 \mathbf{D}_{21}^T = 0$, $\mathbf{D}_{21} \mathbf{D}_{21}^T = \mathbf{I}$, $\mathbf{D}_{12}^T \mathbf{C}_1 = 0$, and

$$\mathbf{Q} = Ric \left[\begin{array}{c|c} \mathbf{A}^T & -\mathbf{C}_2^T \mathbf{C}_2 \\ \hline -\mathbf{B}_1 \mathbf{B}_1^T & -\mathbf{A} \end{array} \right]. \tag{8.215}$$

Then a control law $\mathbf{u}_2 = \mathbf{F}(s)[\mathbf{y}_2]$ solves the H_2 optimal control problem for $G(s)$ if and only if it solves it for the plant $\tilde{G}(s)$.

Proof Noting that $\|G(s)\|_2 = \|G^T(s)\|_2$, the result follows directly from Lemma 8.7 by taking the transpose of everything, applying Lemma 8.7, and then transposing all of the resultant equations. ∎

The following lemma shows that observer-based H_2 optimal control is possible for plants with $\mathbf{D}_{21} = \mathbf{I}$.

LEMMA 8.9 (Observer-based H_2 feedback) Consider the plant

$$\mathbf{G}(s) = \left[\begin{array}{c|cc} \mathbf{A} & \mathbf{H} & \mathbf{B}_2 \\ \hline \mathbf{C}_1 & 0 & \mathbf{D}_{12} \\ \mathbf{C}_2 & \mathbf{I} & 0 \end{array} \right] \tag{8.216}$$

and suppose that $\mathbf{A} - \mathbf{HC}_2$ is stable, and that $\mathbf{D}_{12}^T\mathbf{C}_1 = 0$ and $\mathbf{D}_{12}^T\mathbf{D}_{12} = \mathbf{I}$. Let $\mathbf{F} \triangleq -\mathbf{B}_2^T\mathbf{P}$ be the optimal full state feedback $\mathbf{u}_2 = \mathbf{Fx}$ given by Theorem 8.4. Then the observer-based feedback $\mathbf{u}_2 = \mathbf{K}(s)[\mathbf{y}_2]$ given below solves the H_2 problem for $\mathbf{G}(s)$:

$$\mathbf{u}_2 = -\mathbf{B}_2^T\mathbf{P}\hat{\mathbf{x}}, \tag{8.217}$$

$$\dot{\hat{\mathbf{x}}} = \mathbf{A}\hat{\mathbf{x}} + \mathbf{B}_2\mathbf{u}_2 + \mathbf{H}(\mathbf{y}_2 - \mathbf{C}_2\hat{\mathbf{x}}), \tag{8.218}$$

$$\mathbf{P} = Ric \left[\begin{array}{c|c} \mathbf{A} & -\mathbf{B}_2\mathbf{B}_2^T \\ \hline -\mathbf{C}_1^T\mathbf{C}_1 & -\mathbf{A}^T \end{array} \right], \tag{8.219}$$

that is

$$\mathbf{K}(s) = -\mathbf{B}_2^T\mathbf{P}\left(s\mathbf{I} - \mathbf{A} + \mathbf{B}_2\mathbf{B}_2^T\mathbf{P} + \mathbf{HC}_2\right)^{-1}\mathbf{H}. \tag{8.220}$$

Proof Let $\mathbf{e} = \mathbf{x} - \hat{\mathbf{x}}$. Then

$$\dot{\mathbf{e}} = (\mathbf{Ax} + \mathbf{Hu}_1 + \mathbf{B}_2\mathbf{u}_2) - (\mathbf{A}\hat{\mathbf{x}} + \mathbf{B}_2\mathbf{u}_2 + \mathbf{H}(\mathbf{C}_2\mathbf{x} + \mathbf{u}_1 - \mathbf{C}_2\hat{\mathbf{x}})) \tag{8.221}$$

or

$$\dot{\mathbf{e}} = (\mathbf{A} - \mathbf{HC}_2)\mathbf{e}. \tag{8.222}$$

Noting that $\mathbf{e}$ is not excited by $\mathbf{u}_1$, it follows that if $\mathbf{x}(0) = \hat{\mathbf{x}}(0)$, then $\mathbf{e}(t) \equiv 0$ for all t and hence $\mathbf{x}(t) = \hat{\mathbf{x}}(t)$ for all t. Therefore, one may substitute $\hat{\mathbf{x}}$ for $\mathbf{x}$ in the control law $\mathbf{u}_2 = -\mathbf{B}_2^T\mathbf{Px}$ without affecting $\mathbf{T}_{\mathbf{y}_1\mathbf{u}_1}$ or its H_2 norm. ∎

Combining Lemmas 8.8 and 8.9, one has a solution to the H_2 optimal control problem:

LEMMA 8.10 (A 2-Ricatti solution to the H_2 optimal control problem) Let

$$\mathbf{G}(s) = \left[\begin{array}{c|cc} \mathbf{A} & \mathbf{B}_1 & \mathbf{B}_2 \\ \hline \mathbf{C}_1 & 0 & \mathbf{D}_{12} \\ \mathbf{C}_2 & \mathbf{D}_{21} & 0 \end{array} \right] \tag{8.223}$$

with $\mathbf{D}_{12}^T\mathbf{D}_{12} = \mathbf{I}$, $\mathbf{D}_{21}\mathbf{D}_{21}^T = \mathbf{I}$, $\mathbf{D}_{12}^T\mathbf{C}_1 = 0$, and $\mathbf{B}_1\mathbf{D}_{21}^T = 0$. Define

$$\mathbf{Q} \triangleq Ric \left[\begin{array}{c|c} \mathbf{A}^T & -\mathbf{C}_2^T\mathbf{C}_2 \\ \hline -\mathbf{B}_1\mathbf{B}_1^T & -\mathbf{A} \end{array} \right]. \tag{8.224}$$

Let

$$\tilde{\mathbf{G}}(s) = \left[\begin{array}{c|cc} \mathbf{A} & \mathbf{H} & \mathbf{B}_2 \\ \hline \mathbf{C}_1 & 0 & \mathbf{D}_{12} \\ \mathbf{C}_2 & \mathbf{I} & 0 \end{array} \right] \quad \text{where} \quad \mathbf{H} \triangleq \mathbf{Q}\mathbf{C}_2^T \tag{8.225}$$

and let $\mathbf{F} = -\mathbf{B}_2^T \mathbf{P}$ where

$$\mathbf{P} = Ric \left[\begin{array}{c|c} \mathbf{A} & -\mathbf{B}_2\mathbf{B}_2^T \\ \hline -\mathbf{C}_1^T\mathbf{C}_1 & -\mathbf{A}^T \end{array} \right]. \tag{8.226}$$

Then the observer-based control law $\mathbf{u}_2 = \mathbf{K}(s)[\mathbf{y}_2]$ solves the H_2 optimal control problem with

$$\mathbf{K}(s) = -\mathbf{B}_2^T \mathbf{P} \left(s\mathbf{I} - \mathbf{A} + \mathbf{B}_2\mathbf{B}_2^T \mathbf{P} + \mathbf{H}\mathbf{C}_2 \right)^{-1} \mathbf{H} \tag{8.227}$$

or equivalently

$$\mathbf{u}_2 = -\mathbf{B}_2^T \mathbf{P}\hat{\mathbf{x}}, \tag{8.228}$$

$$\dot{\hat{\mathbf{x}}} = \mathbf{A}\hat{\mathbf{x}} + \mathbf{B}_2\mathbf{u}_2 + \mathbf{H}\left(\mathbf{y}_2 - \mathbf{C}_2\hat{\mathbf{x}} \right) \tag{8.229}$$

$$= \mathbf{A}\hat{\mathbf{x}} + \mathbf{B}_2\mathbf{u}_2 + \mathbf{Q}\mathbf{C}_2^T \left(\mathbf{y}_2 - \mathbf{C}_2\hat{\mathbf{x}} \right). \tag{8.230}$$

Proof   The proof follows directly by using Lemma 8.8 followed by Lemma 8.9. ∎

Remark 8.6   Notice from Equations (8.228) and (8.230) that in the case of H_2 optimal control, the optimal controller is obtained by just replacing the state in the full state feedback control by the reconstructed state obtained from the observer. Thus, in this case, there is a separation between the design of the stabilizing feedback (characterized by the value of $\mathbf{P}$) and the observer (characterized by the value of $\mathbf{Q}$). This is an important difference between H_2 and H_∞ optimal control.

8.5   Casting Control Problems into a Generalized Plant Framework

In this section we demonstrate how a familiar control problem can be cast in the generalized plant (GP) framework. One can then use standard software to solve the resulting H_∞ and H_2 optimal controller synthesis problems and this is illustrated in the next section. In formulating our results, several technical assumptions were made to facilitate the solution. Some of these, such as $\mathbf{D}_{12}^T\mathbf{C}_1 = 0$, can be relaxed as shown in Corollary 8.3. A similar approach can be used to relax the assumption $\mathbf{B}_1\mathbf{D}_{21}^T = 0$. Some of these adjustments are made automatically by the MATLAB software. It is also important to note that frequency shaping of various signals and introduction of additional inputs and outputs are sometimes necessary to facilitate the formulation in the GP framework, as we show below.

Consider the block diagram of a conventional control system shown in Figure 8.10. The low pass filter reflects the fact that reference inputs are usually low-frequency

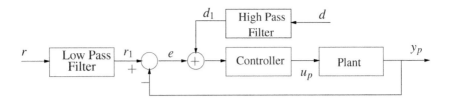

Figure 8.10 Unity feedback with low and high pass filters

signals lying in a band $[0, \omega_l]$. The high pass filter represents the fact that disturbances or noise are often higher-frequency signals in a band $[\omega_h, \infty]$.

The mathematical representation of the system in Figure 8.10 is:

$$\text{Plant: } \dot{\mathbf{x}}_p(t) = \mathbf{A}_p \mathbf{x}_p(t) + \mathbf{b}_p u_p(t), \tag{8.231a}$$

$$y_p(t) = \mathbf{c}_p \mathbf{x}_p(t). \tag{8.231b}$$

$$\text{Low pass filter: } \dot{x}_1(t) = -\omega_l x_1(t) + r(t), \tag{8.232a}$$

$$r_1(t) = \omega_l x_1(t). \tag{8.232b}$$

$$\text{High pass filter: } \dot{x}_2(t) = -\omega_h x_2(t) + d(t), \tag{8.233a}$$

$$d_1(t) = -\omega_h x_2(t) + d(t). \tag{8.233b}$$

The regulated output to be kept small is

$$\mathbf{y}_1(t) = \left[\begin{array}{c} e(t) \\ u_p(t) \end{array} \right]. \tag{8.234}$$

The measured output or input to the controller $y_2(t)$ is

$$y_2(t) = e(t) + d_1(t). \tag{8.235}$$

The exogenous input is

$$\mathbf{u}_1(t) = \left[\begin{array}{c} r(t) \\ d(t) \end{array} \right] \tag{8.236}$$

and the control input is

$$u_2(t) = u_p(t). \tag{8.237}$$

The GP representation of this problem is then

$$\left[\begin{array}{c} \dot{\mathbf{x}}_p(t) \\ \dot{x}_1(t) \\ \dot{x}_2(t) \end{array} \right] = \left[\begin{array}{ccc} \mathbf{A}_p & 0 & 0 \\ 0 & -\omega_l & 0 \\ 0 & 0 & -\omega_h \end{array} \right] \left[\begin{array}{c} \mathbf{x}_p(t) \\ x_1(t) \\ x_2(t) \end{array} \right] + \left[\begin{array}{ccc} 0 & 0 & \mathbf{b}_p \\ 1 & 0 & 0 \\ 0 & 1 & 0 \end{array} \right] \left[\begin{array}{c} r(t) \\ d(t) \\ u_p(t) \end{array} \right],$$
$$\tag{8.238}$$

$$\left[\begin{array}{c} \mathbf{y}_1(t) \\ y_2(t) \end{array} \right] = \left[\begin{array}{ccc} -\mathbf{c}_p & \omega_l & 0 \\ 0 & 0 & 0 \\ -\mathbf{c}_p & \omega_l & -\omega_h \end{array} \right] \left[\begin{array}{c} \mathbf{x}_p(t) \\ x_1(t) \\ x_2(t) \end{array} \right] + \left[\begin{array}{ccc} 0 & 0 & 0 \\ 0 & 0 & 1 \\ 0 & 1 & 0 \end{array} \right] \left[\begin{array}{c} r(t) \\ d(t) \\ u_p(t) \end{array} \right].$$

Note that the technical conditions

$$\mathbf{d}_{12}^T \mathbf{d}_{12} = 1 \quad \text{and} \quad \mathbf{d}_{21} \mathbf{d}_{21}^T = 1 \tag{8.239}$$

are satisfied.

Note that the order of the GP is $n_0 + 2$ in this simple example, where there are only two first order filters. Thus, if the original plant is of first-order, the GP is of third order and the controller is then also of third order. In general, more filters are needed and this leads to inflation of the order of the resulting controller. This is an inherent limitation of the method.

In the following, we assign numerical values in the GP model in Equation (8.238) to carry out H_∞ and H_2 controller syntheses using MATLAB.

8.5.1 H_∞ Optimal Controller

EXAMPLE 8.4   Let the plant be

$$\dot{\mathbf{x}}_p(t) = \underbrace{\begin{bmatrix} 0 & 1 & 0 \\ 0 & 0 & 1 \\ -3 & -2 & -1 \end{bmatrix}}_{\mathbf{A}_p} \mathbf{x}_p(t) + \underbrace{\begin{bmatrix} 0 & 1 \\ 1 & 1 \\ 0 & 1 \end{bmatrix}}_{\mathbf{B}_p} \mathbf{u}_p(t),$$

$$\mathbf{y}_p(t) = \underbrace{\begin{bmatrix} 1 & 0 & 1 \\ 0 & 1 & 0 \end{bmatrix}}_{\mathbf{C}_p} \mathbf{x}_p(t). \tag{8.240}$$

Following the discussion in Section 8.5, we have

$$\mathbf{A} = \begin{bmatrix} \mathbf{A}_p & 0 & 0 \\ 0 & -\omega_l \mathbf{I}_2 & 0 \\ 0 & 0 & -\omega_h \mathbf{I}_2 \end{bmatrix}, \quad \mathbf{B} = \begin{bmatrix} 0 & 0 & \mathbf{B}_p \\ \mathbf{I}_2 & 0 & 0 \\ 0 & \mathbf{I}_2 & 0 \end{bmatrix},$$

$$\mathbf{C} = \begin{bmatrix} -\mathbf{C}_p & \omega_l \mathbf{I}_2 & 0 \\ 0 & 0 & 0 \\ -\mathbf{C}_p & \omega_l \mathbf{I}_2 & -\omega_h \mathbf{I}_2 \end{bmatrix}, \quad \mathbf{D} = \begin{bmatrix} 0 & 0 & 0 \\ 0 & 0 & \mathbf{I}_2 \\ 0 & \mathbf{I}_2 & 0 \end{bmatrix}, \tag{8.241}$$

where $\omega_l = 1$ and $\omega_h = 2$. The following MATLAB script calculates the H_∞ optimal controller

$$\mathbf{A}_c = \begin{bmatrix} 7.2473 & 1.4299 & 8.7674 & -8.8825 & -1.8925 & 17.7069 & 3.2136 \\ 1.8366 & -2.1827 & 4.8050 & -4.2203 & -0.1185 & 7.4976 & 0.0148 \\ -24.3932 & -6.9536 & -20.8731 & 19.7580 & 3.4910 & -39.5741 & -7.5533 \\ 6.2823 & 1.1583 & 5.9588 & -6.7457 & -0.6085 & 12.6621 & 1.4154 \\ 11.2814 & 2.5946 & 10.4026 & -10.5084 & -3.0655 & 21.2152 & 5.5449 \\ -2.6949 & -0.6838 & -2.6949 & 2.6949 & 0.6838 & -7.3899 & -1.3676 \\ 1.3068 & 0.4968 & 1.3068 & -1.3068 & -0.4968 & 2.6136 & -1.0064 \end{bmatrix},$$

$$\mathbf{B}_c = \begin{bmatrix} 8.8535 & 1.6068 \\ 3.7488 & 0.0074 \\ -19.7870 & -3.7767 \\ 6.3310 & 0.7077 \\ 10.6076 & 2.7724 \\ -2.6949 & -0.6838 \\ 1.3068 & 0.4968 \end{bmatrix}, \tag{8.242}$$

$$\mathbf{C}_c = \begin{bmatrix} -0.3060 & -1.0131 & 0.1423 & -0.4424 & 0.1746 & 0 & 0 \\ -1.6061 & -1.1769 & -0.0861 & -0.0291 & -0.2856 & 0 & 0 \end{bmatrix},$$

and it is verified by the Simulink simulation using the standard unity feedback system. The closed-loop poles are

$$\left\{ \begin{array}{l} -2.2852 \pm j1.8643, -0.3221 \pm j1.3747, -1.1936 \pm j0.9774 \\ -1.8849, -1.2757, -1.3362, -1.0, -1.0, -2.0, -2.0, -22.9173 \end{array} \right\}, \tag{8.243}$$

$$\gamma = 0.9908, \tag{8.244}$$

and the output response driven by the initial conditions of the plant

$$\mathbf{x}_p(0) = [-0.5 \ 0.5 \ 1]^T \tag{8.245}$$

is shown in Figure 8.11.

MATLAB Solution

```
clear
wl=1; wh=2;
Ap=[0 1 0; 0 0 1; -3 -2 -1];
Bp=[0 1; 1 1; 0 1];
Cp=[1 0 1; 0 1 0];
[n,m]=size(Bp);
r=size(Cp,1);
Dp=zeros(r,m);
A=[Ap zeros(n,2*r);
   zeros(r,n) -wl*eye(r) zeros(r,r);
   zeros(r,n) zeros(r,r) -w2*eye(r)];
B=[zeros(n,2*r) Bp;
   eye(r) zeros(r,r) zeros(r,m);
   zeros(r,r) eye(r) zeros(r,m)];
C=[-Cp wl*eye(r) zeros(r,r);
   zeros(m,n) zeros(m,r) zeros(m,r);
   -Cp wl*eye(r) -w2*eye(r)];
D=[zeros(r,r) zeros(r,r) zeros(r,m);
   zeros(m,r) zeros(m,r) eye(m,m);
   zeros(r,r) eye(r,r) zeros(r,m)]; nmeas=r;
```

```
P=ss(A,B,C,D);
ncont=m;
[K,CL,gamma]=hinfsyn(P,nmeas,ncont);
[Ac,Bc,Cc,Dc]=ssdata(K);
pole(CL)
```

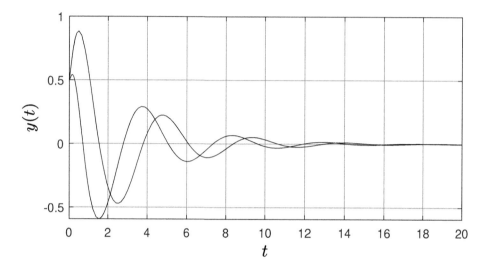

Figure 8.11 Output response (Example 8.4)

8.5.2 H_2 Controller

EXAMPLE 8.5   Considering again the plant used in Example 8.4, we now design an H_2 optimal controller. The same MATLAB script is used after replacing the line with "hinfsyn" by

```
[K,CL,gamma]=h2syn(P,nmeas,ncont);
```

The H_2 optimal controller obtained is

$$
\mathbf{A}_c = \begin{bmatrix}
-1.0823 & -0.0724 & -0.0490 & -0.0860 & -0.0236 & 0.2878 & -0.0299 \\
-1.5453 & -2.1938 & 0.8465 & -0.2156 & 0.4292 & -0.1095 & -0.4980 \\
-4.5797 & -3.0909 & -1.5465 & 0.4114 & -0.0051 & -0.7070 & -0.0669 \\
0.5068 & -0.0714 & 0.5068 & -1.5068 & 0.0714 & 1.0137 & -0.1429 \\
0.0526 & 0.4996 & 0.0526 & -0.0526 & -1.4996 & 0.1053 & 0.9993 \\
0.5075 & -0.0425 & 0.5075 & -0.5075 & 0.0425 & -0.9851 & -0.0850 \\
0.0227 & 0.4999 & 0.0227 & -0.0227 & -0.4999 & 0.0454 & -1.0002
\end{bmatrix},
$$

$$
\mathbf{B}_c = \begin{bmatrix}
0.1439 & -0.0150 \\
-0.0547 & -0.2490 \\
-0.3535 & -0.0335 \\
0.5068 & -0.0714 \\
0.0526 & 0.4996 \\
0.5075 & -0.0425 \\
0.0227 & 0.4999
\end{bmatrix},
\tag{8.246}
$$

$$
\mathbf{C}_c = \begin{bmatrix}
-0.2643 & -0.8873 & 0.0942 & -0.3283 & 0.2188 & 0 & 0 \\
-1.2262 & -1.0575 & -0.1930 & 0.0579 & -0.0386 & 0 & 0
\end{bmatrix},
$$

and the closed-loop poles are

$$
\left\{ \begin{array}{l}
-0.1378 \pm j1.5273, -1.5474 \pm j1.3038, -1.2247 \pm j0.7071 \\
-1.2247 \pm j0.7071, -1.0, -1.0, -1.2757, -1.2692, -2.0, -2.0
\end{array} \right\}.
\tag{8.247}
$$

The output response with the initial conditions of the plant in Equation (8.245) is shown in Figure 8.12.

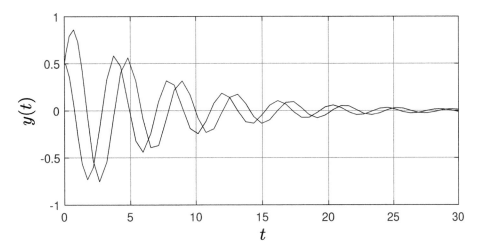

Figure 8.12 Output response (Example 8.5)

8.6   Exercises

8.1   Calculate the H_∞ norms of $G(s)$ by hand-calculation.

(a) $G(s) = \dfrac{1}{s+a}$,   for $a > 0$.

(b) $G(s) = \dfrac{s-z}{z+p}$,   for $z, p > 0$.

(c) $G(s) = \dfrac{\omega_n^2}{s^2 + 2\xi\omega_n s + \omega_n^2}$,   for $\xi, \omega_n > 0$.

(d) $G(s) = \dfrac{s-1}{s^2+3s+4}$.

8.2 Repeat Exercises 8.1(a), (c), and (d) for the H_2 norm. Does the H_2 norm exist in (b)?

8.3 Find the H_∞ and H_2 norms of $\mathbf{G}(s)$ using MATLAB.

(a)

$$\mathbf{G}(s) = \begin{bmatrix} \dfrac{1}{s+1} & \dfrac{1}{s+2} \\ \dfrac{1}{s+2} & \dfrac{1}{s+1} \end{bmatrix}. \tag{8.248}$$

(b)

$$\mathbf{G}(s) = \begin{bmatrix} \dfrac{s-1}{(s+1)^2} & \dfrac{1}{(s+1)^2} \end{bmatrix}. \tag{8.249}$$

(c)

$$\mathbf{G}(s) = \begin{bmatrix} \dfrac{1}{(s+1)^2} & \dfrac{s-1}{(s+1)^2} \\ \dfrac{1}{s+1} & \dfrac{1}{s+2} \end{bmatrix}. \tag{8.250}$$

Verify the answers by an independent check.

8.4 Find the H_∞ optimal controller for the generalized plant below using MATLAB.

$$\left[\begin{array}{ccc|cc} 0 & 1 & 0 & 1 & 0 \\ 0 & 0 & 1 & 0 & 0 \\ -1 & -1 & -1 & 0 & 1 \\ \hline 0 & 1 & 0 & 0 & 1 \\ 1 & 0 & 0 & 1 & 0 \end{array} \right]. \tag{8.251}$$

Find the optimal γ.

8.5 Find the H_2 optimal controller for the generalized plant in Exercise 8.4.

8.6 Solve the H_∞ optimal control exercise for the generalized plant

$$\left[\begin{array}{c|cc} \mathbf{A} & \mathbf{B}_1 & \mathbf{B}_2 \\ \hline \mathbf{C}_1 & \mathbf{D}_{11} & \mathbf{D}_{12} \\ \mathbf{C}_2 & \mathbf{D}_{21} & \mathbf{D}_{22} \end{array} \right] = \left[\begin{array}{c|cc} 1 & 0 & -1 \\ \hline 0 & 0 & 0.2 \\ -1 & 1 & 0 \end{array} \right] \tag{8.252}$$

with $\gamma = 1$.

8.7 Solve the H_∞ optimal control problem for the following generalized plant, and find the optimum γ:

$$
\left[\begin{array}{c|cc}
\mathbf{A} & \mathbf{B}_1 & \mathbf{B}_2 \\
\hline
\mathbf{C}_1 & \mathbf{D}_{11} & \mathbf{D}_{12} \\
\mathbf{C}_2 & \mathbf{D}_{21} & \mathbf{D}_{22}
\end{array}\right]
=
\left[\begin{array}{cc|cccccc}
1 & 0 & 1 & 2 & 0 & 0 & 1 & 0 \\
0 & 1 & 0 & 1 & 0 & 0 & 0 & 1 \\
\hline
0 & 0 & 0 & 0 & 0 & 0 & 1 & 0 \\
0 & 0 & 0 & 0 & 0 & 0 & 0 & 1 \\
1 & 0 & 0 & 0 & 0 & 0 & 0 & 0 \\
0 & 2 & 0 & 0 & 0 & 0 & 0 & 0 \\
1 & 0 & 0 & 0 & 1 & 0 & 0 & 0 \\
0 & 1 & 0 & 0 & 0 & 1 & 0 & 0
\end{array}\right].
\tag{8.253}
$$

8.8 Solve the H_∞ optimal control problem for the following generalized plant, and find the optimum γ:

$$
\left[\begin{array}{c|cc}
\mathbf{A} & \mathbf{B}_1 & \mathbf{B}_2 \\
\hline
\mathbf{C}_1 & \mathbf{D}_{11} & \mathbf{D}_{12} \\
\mathbf{C}_2 & \mathbf{D}_{21} & \mathbf{D}_{22}
\end{array}\right]
=
\left[\begin{array}{ccc|cccccc}
0 & 1 & 0 & 1 & 1 & 0 & 0 & 1 & 0 \\
0 & 2 & 1 & -1 & 0 & 0 & 0 & 0 & 1 \\
-1 & 1 & 1 & 0 & 2 & 0 & 0 & 0 & 1 \\
\hline
0 & 0 & 0 & 0 & 0 & 0 & 0 & 1 & 0 \\
0 & 0 & 0 & 0 & 0 & 0 & 0 & 0 & 1 \\
1 & 1 & 0 & 0 & 0 & 0 & 0 & 0 & 0 \\
0 & -1 & 2 & 0 & 0 & 0 & 0 & 0 & 0 \\
1 & 1 & 0 & 0 & 0 & 1 & 0 & 0 & 0 \\
0 & 0 & 1 & 0 & 0 & 0 & 1 & 0 & 0
\end{array}\right].
\tag{8.254}
$$

8.9 Repeat Exercise 8.6 for H_2 optimal control problem.

8.10 Repeat Exercise 8.7 for H_2 optimal control problem.

8.11 Repeat Exercise 8.8 for H_2 optimal control problem.

8.12 (MIMO robust stability under additive uncertainty) Let $\mathbf{P}(s) = \mathbf{P}_0(s) + \Delta\mathbf{P}(s)$, where

$$
\|\Delta\mathbf{P}(j\omega)\|_2 < |r_A(j\omega)|, \quad \text{for all } \omega
\tag{8.255}
$$

for a stable rational $r_A(s)$ and suppose that $\mathbf{P}(s)$ has an invariant number of RHP poles. Prove that $\mathbf{C}(s)$ robustly stabilizes all plants in the above family if and only if

$$
\left\| \mathbf{C}(s)\left[\mathbf{I} + \mathbf{P}_0(s)\mathbf{C}(s)\right]^{-1} r_A(s) \right\|_\infty \leq 1.
\tag{8.256}
$$

8.13 (MIMO robust stability under multiplicative uncertainty) In Exercise 8.12, let

$$
\Delta(s) = \mathbf{M}(s)\mathbf{P}_0(s),
\tag{8.257}
$$

where

$$\|\mathbf{M}(j\omega)\|_2 < |r_M(j\omega)|, \quad \text{for all } \omega \tag{8.258}$$

for a stable rational $r_M(s)$. Prove that $\mathbf{C}(s)$ robustly stabilizes all plants in the above family if and only if

$$\left\| \mathbf{P}_0(s)\mathbf{C}(s) \left[\mathbf{I} + \mathbf{P}_0(s)\mathbf{C}(s) \right]^{-1} r_M(s) \right\|_\infty \leq 1. \tag{8.259}$$

8.7   Notes and References

There is an extensive literature on H_∞ and H_2 optimal control; for a detailed treatment, see for example Zhou, Doyle, and Glover (1995). The Small Gain Theorem was first formulated by Zames (1966). Our basic presentation of H_∞ and H_2 control here is based on the results in Doyle et al. (1989) and is adapted from Bhattacharyya, Datta, and Keel (2009), where many properties of norms for linear systems are also described with proofs. The bisection algorithm for computing the H_∞ norm was derived in Boyd, Balakrishnan, and Kabamba (1989).

Sections 8.3 and 8.4 are adapted from Bhattacharyya, Datta, and Keel (2009), with permission. © 2008 Taylor & Francis LLC Books.

9 The Linear Multivariable Servomechanism

In this chapter, we discuss the tracking and disturbance rejection problem. This problem, which is also known as the *servomechanism problem*, deals with the design of systems that can impose prescribed asymptotic behavior on the outputs of a system, such as tracking a reference signal in the presence of unknown disturbance inputs and perturbations. Problems of this type arise in process control, motion control, aerospace systems, ship steering, robotics, economics, biology, medicine, and indeed in most applications of control, and the servomechanism problem is of central importance in control theory and its applications. The solution of this problem is accomplished by formulating it as a regulator or stabilization problem for the plant, augmented by a bank of signal generators placed in each error channel. The resulting regulator problem can be solved by using LQR theory, leading to optimal servomechanisms. It can also be solved using observer theory or Pearson compensators. This is illustrated by several examples.

9.1 Introduction to Servomechanisms

In this section, we informally describe servomechanisms and regulators and their interrelationship.

A *regulator* is simply an asymptotically stable feedback system. The response of every signal or system variable in a regulator to arbitrary initial conditions, converges to zero asymptotically. A regulator does not have external inputs for $t > 0$, although conceptually the initial conditions can be regarded as being the result of all past inputs prior to $t = 0$. Sometimes the initial conditions can be set up by applying an impulsive input at $t = 0$.

A *servomechanism*, on the other hand, is a control system in which the outputs asymptotically track prescribed reference signals in the presence of disturbances. In general, the reference and disturbance signals are external signals and are sometimes called *exogenous* signals. In typical applications, exogenous signals are in general persistent, such as steps and sinusoids, and can also be "unstable," such as ramps. The asymptotic tracking property called servo-action must occur

(1) for all initial conditions in the control system,
(2) for all members of a prescribed class of exogenous signals,
(3) without prior knowledge of the exact exogenous signal to be handled, and
(4) without precise knowledge of the plant model or its parameters.

In the next section, we describe how these stringent requirements can be met by a suitably designed feedback control system.

9.2   Signal Generators and the Servocontroller

9.2.1   Signal Generators

The class of exogenous signals, that is, references $r(t)$ or disturbances $d(t)$, to be handled by a control system, can be generated by a *signal generator*. Mathematically, this could be the outputs of an autonomous system described by a differential or difference equation, on which suitable initial conditions can be placed as shown below. Let

$$D^k := \frac{d^k}{dt^k}, \qquad \text{for } k = 0, 1, 2, \ldots \tag{9.1}$$

denote the kth derivative operator and $m(D)$ a prescribed polynomial in D. Consider the class of signals $z(t)$ satisfying

$$m(D)z(t) = 0, \tag{9.2}$$

called a *signal generator*, where $z(t)$ could be $r(t)$ or $d(t)$. If, for example

$$m(D) = D^2, \tag{9.3}$$

we have

$$\ddot{z}(t) = 0 \tag{9.4}$$

so that

$$z(t) = z(0) + t\dot{z}(0) \tag{9.5}$$

and Equation (9.2) along with Equation (9.3) generates all steps and ramps as $(z(0), \dot{z}(0))$ varies over $\mathbb{R}^2$. Similarly, with

$$m(D) = D^2 \left(D^2 + \omega_0^2\right) \tag{9.6}$$

in Equation (9.2) all steps, ramps, sinusoids of radian frequency ω_0, and arbitrary amplitude or phase, and their linear combinations can be generated by this signal generator. It is easy to see that in the s-plane the zeros of $m(s)$ must lie on the imaginary axis or the closed RHP, $\mathbb{C}^+$, to generate persistent and "unstable" signals.

9.2.2   The Servocontroller

In this subsection, we explain how tracking and disturbance rejection can be achieved by embedding signal generators of the exogeneous signals into a unity feedback system and stabilizing the loop. We need the following two preliminary results.

Consider the system in Figure 9.1.

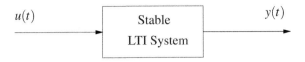

Figure 9.1 A stable LTI system

LEMMA 9.1   If $u(t)$ is any signal that satisfies

$$m(D)u(t) = 0 \tag{9.7}$$

with the zeros of $m(s)$ in $\mathbb{C}^+$, then $y(t)$ can be decomposed as

$$y(t) = y_0(t) + y_u(t) \tag{9.8}$$

where

$$\lim_{t \to \infty} y_0(t) = 0 \tag{9.9}$$

and $y_u(t)$ has only $\mathbb{C}^+$ exponents and satisfies

$$m(D)y_u(t) = 0. \tag{9.10}$$

Proof   Equation (9.9) follows from the stability of the system when $y_0(t)$ is the zero-input response of the system. For Equation (9.10), suppose the system in Figure 9.1 has transfer function $\frac{P(s)}{Q(s)}$. Then

$$Q(D)y_u(t) = P(D)u(t) \tag{9.11}$$

so that

$$\begin{aligned} m(D)Q(D)y_u(t) &= m(D)P(D)u(t) \\ &= P(D)m(D)u(t) = 0. \end{aligned} \tag{9.12}$$

Hence

$$m(D)Q(D)y_u(t) = 0 \quad \text{or} \quad Q(D)m(D)y_u(t) = 0. \tag{9.13}$$

Since the zeros of $Q(s)$ are in $\mathbb{C}^-$, the open LHP, it follows that

$$m(D)y_u(t) = 0. \tag{9.14}$$

∎

Now consider the stable system in Figure 9.2 with an external input $u(t)$ satisfying

$$m(D)u(t) = 0 \tag{9.15}$$

and the signal generator of $u(t)$ embedded within it as shown.

LEMMA 9.2   In Figure 9.2, for every $u(t)$ satisfying Equation (9.15), and for every initial condition in the system:

$$\lim_{t \to \infty} v(t) = 0. \tag{9.16}$$

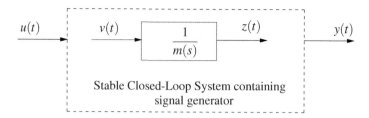

Figure 9.2 A stable system with signal generator of $u(t)$ embedded

Proof   The result follows immediately from the stability of the system and the previous lemma, since the forced response must satisfy $m(D)z(t) = v(t)$.   ∎

This result indicates how one can solve the servomechanism problem. From Equation (9.16), it is clear that if $v(t)$ is arranged to be the tracking error and $m(D)$ is chosen as the signal generator for $r(t)$ and $d(t)$, the stability of the system will guarantee that, the tracking error will be asymptotically zeroed out. This leads to the unity feedback structure given in Figure 9.3.

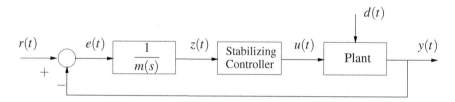

Figure 9.3 Unity feedback servomechanism

In the multivariable case, several references $r_i(t)$ and several disturbances $d_j(t)$ satisfying

$$m(D)r_i(t) = 0, \tag{9.17a}$$
$$m(D)d_j(t) = 0, \tag{9.17b}$$

may be present. The structure of Figure 9.3, however, generalizes in a straightforward manner to Figure 9.4. Note that the stabilizing controller is needed even when the plant is stable, since the signal generator or servocontroller poles are in $\mathbb{C}^+$. We discuss the single-input single-output case first, where stabilization of the plant servocontroller combination will be possible only when signal poles are disjoint from the RHP plant zeros. In the multivariable case, it will be shown that a stabilizing controller exists if and only if the (multivariable) zeros of the plant are disjoint from the signal generator poles and the number of plant inputs equals or exceeds the number of outputs n_0. In terms of the transfer function matrix $\mathbf{P}(s)$ of the plant, this will be shown to be equivalent to the condition

$$\text{rank}[\mathbf{P}(\mu)] = n_0 \tag{9.18}$$

for every zero μ of $m(s)$.

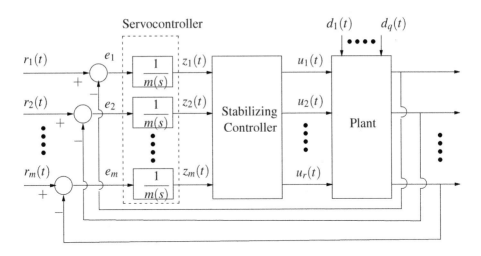

Figure 9.4 Multivariable servomechanism

9.3 SISO Pole Placement Servomechanism Controller

Consider a SISO plant in Figure 9.5 with rational, proper transfer function

$$P(s) = \frac{n_p(s)}{d_p(s)} \tag{9.19}$$

and with $d_p(s)$ monic (leading coefficient $= 1$) and of degree n.

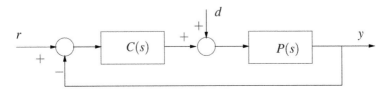

Figure 9.5 Unity feedback control system

Let $m(s)$ denote the monic characteristic polynomial of r and d. For steps, ramps, and sinusoids of frequency ω_0, for example:

$$m(s) = s^2 \left(s^2 + \omega_0^2 \right). \tag{9.20}$$

Now write the controller transfer function as

$$C(s) = \frac{n_c(s)}{d_c(s)}. \tag{9.21}$$

The controller must be proper, that is,

$$\deg [n_c] \leq \deg [d_c]. \tag{9.22}$$

Since the denominator of the controller will be in the numerator of both error transfer functions (error due to reference r and error due to disturbance d), we must have

$$d_c(s) = m(s)\beta(s) \tag{9.23}$$

for some polynomial $\beta(s)$. If

$$\deg[m(s)] = t, \quad \deg[\beta(s)] = b, \quad \deg[\alpha(s)] = a, \tag{9.24}$$

we must have

$$a \leq t + b \tag{9.25}$$

for $C(s)$ to be proper.

The closed-loop characteristic polynomial is

$$d_{cl}(s) := d_p(s)d_c(s) + n_p(s)n_c(s) \tag{9.26a}$$
$$= d_p(s)m(s)\beta(s) + n_p(s)\alpha(s) \tag{9.26b}$$

and is monic with

$$\deg[d_{cl}] = t + b + n. \tag{9.27}$$

The free parameters of the controller, available for stabilization of the closed-loop system, are represented by the coefficients of the polynomials $\beta(s)$ and $\alpha(s)$. The maximal number of free coefficients in $\alpha(s)$ is available by assuming equality in Equation (9.25). In this case, we have

$$\begin{aligned} \beta(s) &= s^b + \beta_{b-1}s^{b-1} + \cdots + \beta_1 s + \beta_0, \\ \alpha(s) &= \alpha_0 + \alpha_1 + \cdots + \alpha_{t+b}s^{t+b}. \end{aligned} \tag{9.28}$$

There are thus

$$2b + t + 1 \tag{9.29}$$

free parameters and the closed-loop characteristic polynomial has $t + b + n$ coefficients. If we choose b, so that

$$2b + t + 1 \geq t + b + n \tag{9.30}$$

or

$$t \geq n - 1, \tag{9.31}$$

it is clear that all closed-loop characteristic polynomial coefficients should be freely assignable. The only exception to this assignability will occur if $m(s)d_p(s)$ and $n_p(s)$ have a common factor. We can summarize all this as follows.

Consider the plant transfer function

$$P(s) = \frac{n_p(s)}{d_p(s)}, \quad \deg[d_p(s)] = n \tag{9.32}$$

and the controller

$$C(s) = \frac{n_c(s)}{d_c(s)} = \frac{\alpha(s)}{m(s)\beta(s)} \tag{9.33}$$

and assume that $n_p(s)$ and $m(s)d_p(s)$ are coprime, that is, have no common factor. Then by choosing the $\alpha(s)$, $\beta(s)$ as in Equation (9.28) with $\deg[\beta] = b = n - 1$, all coefficients of $d_{cl}(s)$ will be freely adjustable by choice of α_i, β_j in Equation (9.28).

EXAMPLE 9.1   Suppose

$$P(s) = \frac{s-1}{s^2+s+1} \tag{9.34}$$

and

$$m(s) = s^2, \tag{9.35}$$

that is we are to track and reject all steps and ramps. Since $n = 2$, the pole placement controller is of the form

$$C(s) = \frac{\alpha_0 + \alpha_1 s + \alpha_2 s^2 + \alpha_3 s^3}{s^2(s+\beta_0)}. \tag{9.36}$$

The closed-loop system of order 5 would have characteristic polynomial

$$\begin{aligned}
d_{cl}(s) &= s^2(s+\beta_0)(s^2+s+1) + (s-1)(\alpha_0+\alpha_1 s+\alpha_2 s^2+\alpha_3 s^3) \\
&= s^5 + (\beta_0+\alpha_3+1)s^4 + (\beta_0+1+\alpha_2-\alpha_3)s^3 + (\beta_0+\alpha_1-\alpha_2)s^2 \\
&\quad + (\alpha_0-\alpha_1)s - \alpha_0.
\end{aligned} \tag{9.37}$$

Choosing the closed-loop poles as $\{-1, -1, -1, -1 \pm j\}$, we have the corresponding *design polynomial*

$$(s+1)^3\left[(s+1)^2+1\right] = s^5 + 5s^4 + 11s^3 + 13s^2 + 8s + 2. \tag{9.38}$$

Equating coefficients in Equations (9.38) and (9.37), we have

$$\begin{aligned}
\beta_0 + \alpha_3 + 1 &= 5, \\
\beta_0 + 1 + \alpha_2 - \alpha_3 &= 11, \\
\beta_0 + \alpha_1 - \alpha_2 &= 13, \\
\alpha_0 - \alpha_1 &= 8 \quad \Rightarrow \quad (\alpha_1 = -10), \\
-\alpha_0 &= 2 \quad \Rightarrow \quad (\alpha_0 = -2),
\end{aligned} \tag{9.39}$$

or

$$\begin{bmatrix} 1 & 0 & 1 \\ 1 & 1 & -1 \\ 1 & -1 & 0 \end{bmatrix} \begin{bmatrix} \beta_0 \\ \alpha_2 \\ \alpha_3 \end{bmatrix} = \begin{bmatrix} 4 \\ 10 \\ 23 \end{bmatrix}, \tag{9.40}$$

giving $\alpha_3 = -8.33$, $\alpha_2 = -10.667$, $\beta_0 = 12.33$. Therefore, the controller transfer function is

$$C(s) = \frac{-2 - 10s - 10.667s^2 - 8.33s^3}{s^2(s+12.33)}. \tag{9.41}$$

The Simulink model shown in Figure 9.6 verifies that the controller enables the closed-loop system to track and reject steps and ramps as shown in Figure 9.7.

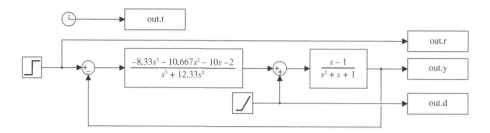

Figure 9.6 Simulink model to verify the controller (Example 9.1)

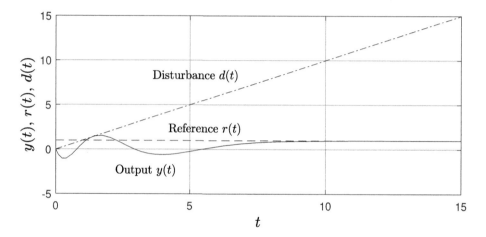

Figure 9.7 Verification of the controller with step input and ramp disturbance (Example 9.1)

The following illustrative examples display the solution in detail. They also point out that the stabilizing controller can be designed to optimize the closed-loop dynamics of the servomechanism using LQR theory. Alternatively, the closed-loop poles may be placed arbitrarily by the stabilizing controller designed by state feedback and observer theory.

EXAMPLE 9.2 (Linear quadratic servomechanism)   To illustrate the basic theory of servomechanisms and how optimal servomechanisms might be designed, consider the second-order system

$$\dot{x}_1(t) = x_2(t), \tag{9.42a}$$
$$\dot{x}_2(t) = a_1 x_1(t) + a_2 x_2(t) + u(t), \tag{9.42b}$$
$$y(t) = c_1 x_1(t) + c_s x_2(t), \tag{9.42c}$$

where $y(t)$ is to track the exponential reference signal

$$r(t) = A e^{\mu t}. \tag{9.43}$$

This signal can be generated by the signal generator, driven by the tracking error $e(t)$ as shown below:

$$\dot{z}(t) = \mu z(t) + \underbrace{(r(t) - y(t))}_{e(t)} = \mu z(t) + e(t). \tag{9.44}$$

Error convergence ($\lim_{t \to \infty} e(t) = 0$) will be achieved if Equation (9.44), called a servo-controller, is embedded in a stable unity feedback control loop as shown in Figure 9.8.

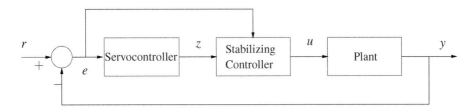

Figure 9.8 Servomechanism structure

This follows from the fact that all signals, after transients are over, are of the form of the forcing function as shown in Equation (9.43). In particular, the forced response of $z(t)$, denoted $z_f(t)$, satisfies

$$\dot{z}_f(t) = \mu z_f(t). \tag{9.45}$$

Inserting Equation (9.45) in Equation (9.44), we have

$$\dot{z}_f(t) = \mu z_f(t) + e_f(t). \tag{9.46}$$

From Equations (9.45) and (9.46):

$$\lim_{t \to \infty} e(t) = e_f(t) = 0, \tag{9.47}$$

proving that asymptotic tracking of exponentials of the type in Equation (9.43) occurs.

The stabilizing controller in Figure (9.8) can be designed by several methods. We show below how linear quadratic optimal control theory can be used to design an optimal stabilizing controller. Such a controller optimizes the closed-loop dynamics, and thus the error response.

Optimal control delivers a state feedback-generated control signal. To obtain an optimal output feedback controller, we attach an integrator

$$\dot{u}(t) = v(t) \tag{9.48}$$

and consider the augmented system:

$$\begin{bmatrix} \dot{x}_1(t) \\ \dot{x}_2(t) \\ \dot{z}(t) \\ \dot{u}(t) \end{bmatrix} = \underbrace{\begin{bmatrix} 0 & 1 & 0 & 0 \\ a_1 & a_2 & 0 & 1 \\ -c_1 & -c_2 & \mu & 0 \\ 0 & 0 & 0 & 0 \end{bmatrix}}_{\tilde{A}} \underbrace{\begin{bmatrix} x_1(t) \\ x_2(t) \\ z(t) \\ u(t) \end{bmatrix}}_{\tilde{x}(t)} + \underbrace{\begin{bmatrix} 0 \\ 0 \\ 0 \\ 1 \end{bmatrix}}_{\tilde{b}} v(t) + \begin{bmatrix} 0 \\ 0 \\ 1 \\ 0 \end{bmatrix} r(t). \tag{9.49}$$

We temporarily set $r(t) = 0$ in Equation (9.49) and consider the augmented plant

$$\dot{\bar{x}}(t) = \bar{A}\bar{x}(t) + \bar{b}v(t) \tag{9.50}$$

and a quadratic performance index

$$I = \int_0^\infty \left(\bar{x}^T(t)Q\bar{x}(t) + \rho v(t)^2 \right) dt. \tag{9.51}$$

The optimal control minimizing I in Equation (9.51) exists if

$$(\bar{A}, \bar{b}) \text{ is controllable.} \tag{9.52}$$

It is easy to verify that controllability holds if and only if

$$c_1 + c_2\mu \neq 0, \tag{9.53}$$

which means that the plant transfer function does not have a zero at $s = \mu$. Under this assumption, the optimal control minimizing the cost functional in Equation (9.51) is given by the control law:

$$v(t) = -K_1 x_1(t) - K_2 x_2(t) - K_3 z(t) - K_4 u(t). \tag{9.54}$$

From LQR theory, the control law in Equation (9.54) is stabilizing provided $(Q, \bar{A})$ is observable and this is assumed. To simplify the algebra, suppose that

$$y(t) = x_1(t), \quad \text{for } c_1 = 1, c_2 = 0. \tag{9.55}$$

Then the optimal control law can be rewritten as

$$\dot{u}(t) = -\underbrace{K_1}_{=\alpha_0} y(t) - \underbrace{K_2}_{=\alpha_1} \dot{y}(t) - \underbrace{K_3}_{\gamma} z(t) - \underbrace{K_4}_{\beta} u(t) \tag{9.56}$$

or

$$\dot{u}(t) = -\alpha_0 y(t) - \alpha_1 \dot{y}(t) - \gamma z(t) - \beta u(t). \tag{9.57}$$

This leads to the final unity feedback output feedback stabilizing controller

$$\dot{u}(t) = \alpha_0 e(t) + \alpha_1 \dot{e}(t) - \gamma z(t) - \beta u(t), \tag{9.58}$$

where $y(t)$ has now been replaced by $-e(t)$. The overall controller structure is shown in Figure 9.9.

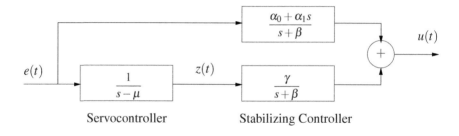

Figure 9.9 Error-driven optimal controller

MATLAB Verification

The following MATLAB script calculates and sets the parameter values for the Simulink model shown in Figure 9.10. The output response and the exponential reference input are shown in Figure 9.11.

```
% Setting SIMULINK parameters
a1=1;   a2=1;   c1=1;   c2=0;
A=[0 1; a1 a2];   b=[0 1]';   c=[c1 c2];   d=0;
mu=3;
A11=[0 0; 0 1];
Ae=[A A11; -c mu 0; 0 0 0 0];
be=[0 0 0 1]';
[K,S,CLP]=lqr(Ae,be,eye(4),1);
alpha0=K(1);   alpha1=K(2);
gamma=K(3);   beta=K(4);
```

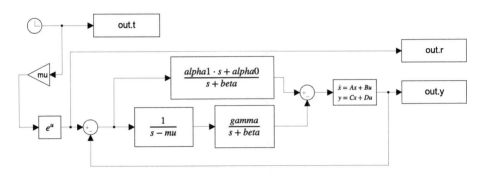

Figure 9.10 Simulink model to verify an LQ servocontroller (Example 9.2)

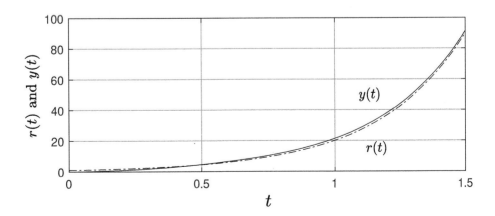

Figure 9.11 Verification of the LQ servocontroller with exponential reference (Example 9.2)

9.4   Multivariable Servomechanism: State-Space Formulation

In this section we give a general treatment of the multivariable servomechanism problem using a state-space setting. We consider the system or plant

$$\dot{\mathbf{x}}_p(t) = \mathbf{A}_p\mathbf{x}_p(t) + \mathbf{B}_p\mathbf{u}_p(t) + \mathbf{E}_p\mathbf{d}(t),$$
$$\mathbf{y}(t) = \mathbf{C}_p\mathbf{x}_p(t) + \mathbf{D}_p\mathbf{u}_p(t) + \mathbf{F}_p\mathbf{d}(t),$$
$$(9.59)$$

where $\mathbf{x}_p(\cdot)$, $\mathbf{u}_p(\cdot)$, $\mathbf{y}(\cdot)$, and $\mathbf{d}(\cdot)$ represent the n, n_i, n_0, and q-dimensional state, input, output, and disturbance signals. Let $\mathbf{r}(\cdot)$ denote a reference signal vector of dimension n_0 and

$$\mathbf{e}(t) := \mathbf{r}(t) - \mathbf{y}(t) \qquad (9.60)$$

the tracking error. The servomechanism problem aims is to find feedback-generated control signals $\mathbf{u}_p(t)$ such that $\mathbf{y}(t)$ tracks $\mathbf{r}(t)$ and rejects $\mathbf{d}(t)$, that is

$$\lim_{t \to \infty} \mathbf{e}(t) = 0. \qquad (9.61)$$

The above condition must hold for any signals from the prescribed classes of references and disturbances and arbitrary plant and controller initial conditions.

9.5   Reference and Disturbance Signal Classes

The typical servomechanism requires, for example, the tracking and disturbance rejection of arbitrary steps, ramps, and sinusoids. To represent such classes of signals, we assume that the references and disturbances satisfy the differential equations below. With

$$D := \frac{d}{dt}, \qquad (9.62)$$

consider the signals determined by the differential equations

$$\left(D^t + \beta_{t-1}D^{t-1} + \cdots + \beta_1 D + \beta_0\right) r_i(t) = 0, \qquad (9.63)$$
$$\left(D^t + \beta_{t-1}D^{t-1} + \cdots + \beta_1 D + \beta_0\right) d_j(t) = 0, \qquad (9.64)$$

for $i = 1, 2, \ldots, n_0$, $j = 1, 2, \ldots, q$. Let

$$m(D) := D^t + \beta_{t-1}D^{t-1} + \cdots + \beta_1 D + \beta_0. \qquad (9.65)$$

The classes of all reference signals to be tracked and all disturbance signals to be rejected are generated by placing all possible initial conditions in Equations (9.63) and (9.64), respectively. We make the standing assumption that the roots of

$$m(s) = 0 \qquad (9.66)$$

lie in the closed RHP, so that the reference and disturbance signals are all unstable exponentials or persistent signals such as steps, ramps, and sinusoids. In the following subsection, we show how the above problem can be solved as a regulator or stabilization problem for a suitably augmented plant.

9.6 Solution of the Servomechanism Problem

Let us note that if a linear time-invariant feedback system achieves tracking and disturbance rejection as defined by Equation (9.61), it must satisfy the following requirements:

(a) Each reference signal and each disturbance signal must be *blocked* or decoupled from any component of the error signal, asymptotically, that is, in the steady state.

(b) The closed-loop system must be asymptotically stable.

Translating (a) and (b) into specifications on transfer functions, we see that (a) is equivalent to the requirement that the reference to error and disturbance to error transfer functions possess zeros precisely at the same locations and with the same multiplicities as the roots of $m(s) = 0$. Condition (b) is, of course, the usual one of requiring the closed-loop eigenvalues to lie in the open LHP, that is, of a regulator or stabilization problem. We show below how condition (a) can be satisfied by appropriately augmenting the plant with the servocontroller described below to produce the system or "new" plant which must then be stabilized.

Introduce the servocontroller

$$\dot{\mathbf{x}}_m^i(t) = \mathbf{M}\mathbf{x}_m^i(t) + \mathbf{m}e_i(t), \tag{9.67}$$

where $e_i(t)$ is the ith component of $\mathbf{e}(t)$, $i = 1, 2, \ldots, n_0$, $(\mathbf{M}, \mathbf{m})$ is controllable, and

$$\det[s\mathbf{I} - \mathbf{M}] = m(s). \tag{9.68}$$

Then, with

$$\mathbf{x}_m(t) := \begin{bmatrix} \mathbf{x}_m^1(t) \\ \mathbf{x}_m^2(t) \\ \cdots \\ \mathbf{x}_m^{n_0}(t) \end{bmatrix}, \tag{9.69}$$

we have

$$\dot{\mathbf{x}}_m(t) = \underbrace{\begin{bmatrix} \mathbf{M} & & & \\ & \mathbf{M} & & \\ & & \ddots & \\ & & & \mathbf{M} \end{bmatrix}}_{\mathbf{A}_m} \mathbf{x}_m(t) + \underbrace{\begin{bmatrix} \mathbf{m} & & & \\ & \mathbf{m} & & \\ & & \ddots & \\ & & & \mathbf{m} \end{bmatrix}}_{\mathbf{B}_m} \mathbf{e}(t), \tag{9.70}$$

which we call the *servocontroller*

Consider now the augmented system consisting of the plant and the servocontroller:

$$
\underbrace{\begin{bmatrix} \dot{\mathbf{x}}_p(t) \\ \dot{\mathbf{x}}_m(t) \end{bmatrix}}_{\dot{\mathbf{x}}(t)} = \underbrace{\begin{bmatrix} \mathbf{A}_p & 0 \\ -\mathbf{B}_m\mathbf{C}_p & \mathbf{A}_m \end{bmatrix}}_{\mathbf{A}} \underbrace{\begin{bmatrix} \mathbf{x}_p(t) \\ \mathbf{x}_m(t) \end{bmatrix}}_{\mathbf{x}(t)} + \underbrace{\begin{bmatrix} \mathbf{B}_p \\ -\mathbf{B}_m\mathbf{D}_p \end{bmatrix}}_{\mathbf{B}} u_p(t)
$$

$$
+ \begin{bmatrix} \mathbf{E}_p \\ -\mathbf{B}_m\mathbf{F}_p \end{bmatrix} \mathbf{d}(t) + \begin{bmatrix} 0 \\ \mathbf{B}_m \end{bmatrix} \mathbf{r}(t), \tag{9.71}
$$

$$
\mathbf{e}(t) = [-\mathbf{C}_p \quad 0] \begin{bmatrix} \mathbf{x}_p(t) \\ \mathbf{x}_m(t) \end{bmatrix} - \mathbf{D}_p u_p(t) - \mathbf{F}_p \mathbf{d}(t) + \mathbf{I}\mathbf{r}(t).
$$

We can now consider a state feedback control law for Equation (9.71) of the form

$$
\mathbf{u}_p(t) = -(\mathbf{K}_p \mathbf{x}_p(t) + \mathbf{K}_m \mathbf{x}_m(t)) = -\underbrace{[\mathbf{K}_p \quad \mathbf{K}_m]}_{K} \underbrace{\begin{bmatrix} \mathbf{x}_p(t) \\ \mathbf{x}_m(t) \end{bmatrix}}_{\mathbf{x}(t)}. \tag{9.72}
$$

We will show that, provided Equation (9.72) stabilizes Equation (9.71), or equivalently stabilizes $(\mathbf{A} - \mathbf{B}K)$, asymptotic tracking and disturbance rejection will be achieved since the zero placement condition (a) will automatically be satisfied.

LEMMA 9.3 $(\mathbf{A}, \mathbf{B})$ is stabilizable if and only if

$$
\text{(i)} \quad (\mathbf{A}_p, \mathbf{B}_p) \text{ is stabilizable, and} \tag{9.73}
$$

$$
\text{(ii)} \quad \text{rank} \begin{bmatrix} \lambda I - \mathbf{A}_p & \mathbf{B}_p \\ -\mathbf{C}_p & \mathbf{D}_p \end{bmatrix} = n + n_0 \tag{9.74}
$$

for all λ satisfying $m(\lambda) = 0$.

Proof   Let

$$
\rho(\mu) := \text{rank} \begin{bmatrix} \mu I - \mathbf{A} & | & \mathbf{B} \end{bmatrix}. \tag{9.75}
$$

Since

$$
\begin{bmatrix} sI - \mathbf{A} & | & \mathbf{B} \end{bmatrix} = \begin{bmatrix} sI - \mathbf{A}_p & 0 & \mathbf{B}_p \\ \mathbf{B}_m\mathbf{C}_p & sI - \mathbf{A}_m & -\mathbf{B}_m\mathbf{D}_p \end{bmatrix}, \tag{9.76}
$$

it follows from the stabilizability of $(\mathbf{A}_p, \mathbf{B}_p)$, Equation (9.73), that when $\mu \in \mathbb{C}^+$, $\mu \notin \sigma(\mathbf{A}_m)$

$$
\rho(\mu) = n + n_0 t. \tag{9.77}
$$

Now write

$$
\begin{bmatrix} sI - \mathbf{A} & | & \mathbf{B} \end{bmatrix} = \begin{bmatrix} I_n & 0 & 0 \\ 0 & -\mathbf{B}_m & sI - \mathbf{A}_m \end{bmatrix} \begin{bmatrix} sI - \mathbf{A}_p & 0 & \mathbf{B}_p \\ -\mathbf{C}_p & 0 & \mathbf{D}_p \\ 0 & I_{n_0 t} & 0 \end{bmatrix} \tag{9.78}
$$

and consider $\mu \in \sigma(\mathbf{A}_m)$. Let

$$
x := \text{rank} \begin{bmatrix} \mu I - \mathbf{A}_p & \mathbf{B}_p \\ -\mathbf{C}_p & \mathbf{D}_p \end{bmatrix}, \tag{9.79}
$$

note that $(\mathbf{A}_m, \mathbf{B}_m)$ is controllable, and apply Sylvester's inequality for the rank of a product of matrices to Equation (9.78), to get

$$n + n_0 t + x + n_0 t - (n + n_0 + n_0 t) \le \rho(\mu) \le n + n_0 t. \tag{9.80}$$

It follows from condition (ii), Equation (9.74), that $x = n + n_0$ so that $\rho(\mu) = n + n_0 t$. Therefore

$$\rho(\mu) = n + n_0 t, \quad \text{for all } \mu \in \mathbb{C}^+, \tag{9.81}$$

proving that $(\mathbf{A}, \mathbf{B})$ is stabilizable. ■

We can now state the state feedback solution to the multivariable servomechanism problem.

THEOREM 9.1 If the plant is stabilizable, that is, $(\mathbf{A}_p, \mathbf{B}_p)$ is stabilizable and Equation (9.74) holds, then there exists a solution to the servomechanism problem of the form

$$\mathbf{u}_p(t) = -\mathbf{K}_p \mathbf{x}_p(t) - \mathbf{K}_m \mathbf{x}_m(t), \tag{9.82}$$

where

$$\dot{\mathbf{x}}_m(t) = \mathbf{A}_m \mathbf{x}_m(t) + \mathbf{B}_m \mathbf{e}(t) \tag{9.83}$$

and $(\mathbf{A} - \mathbf{B}K)$ is stable (see Figure 9.12).

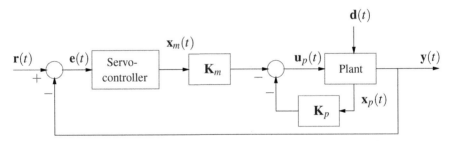

Figure 9.12 State feedback servomechanism

Proof By Lemma 9.3, there exists $\mathbf{K}$ such that $(\mathbf{A} - \mathbf{B}\mathbf{K})$ is stable. Now consider the closed-loop error system

$$\begin{bmatrix} \dot{\mathbf{x}}_p(t) \\ \dot{\mathbf{x}}_m(t) \end{bmatrix} = \underbrace{\begin{bmatrix} \mathbf{A}_p - \mathbf{B}_p \mathbf{K}_p & -\mathbf{B}_p \mathbf{K}_m \\ -\mathbf{B}_m (\mathbf{C}_p - \mathbf{D}_p \mathbf{K}_p) & \mathbf{A}_m + \mathbf{B}_m \mathbf{D}_p \mathbf{K}_m \end{bmatrix}}_{\mathbf{A}_e} \begin{bmatrix} \mathbf{x}_p(t) \\ \mathbf{x}_m(t) \end{bmatrix}$$

$$+ \underbrace{\begin{bmatrix} \mathbf{E}_p & 0 \\ -\mathbf{B}_m \mathbf{F}_p & \mathbf{B}_m \end{bmatrix}}_{\mathbf{B}_e} \begin{bmatrix} \mathbf{d}(t) \\ \mathbf{r}(t) \end{bmatrix}, \tag{9.84}$$

$$\mathbf{e}(t) = \underbrace{\begin{bmatrix} -(\mathbf{C}_p - \mathbf{D}_p \mathbf{K}_p) & \mathbf{D}_p \mathbf{K}_m \end{bmatrix}}_{\mathbf{C}_e} \begin{bmatrix} \mathbf{x}_p(t) \\ \mathbf{x}_m(t) \end{bmatrix} + \underbrace{\begin{bmatrix} -\mathbf{F}_p & \mathbf{I} \end{bmatrix}}_{\mathbf{D}_e} \begin{bmatrix} \mathbf{d}(t) \\ \mathbf{r}(t) \end{bmatrix},$$

with state $\mathbf{x}(t) = [\mathbf{x}_p(t) \ \ \mathbf{x}_m(t)]^T$, input $[\mathbf{d}(t) \ \ \mathbf{r}(t)]^T$, output $\mathbf{e}(t)$, and state-space representation $[\mathbf{A}_e, \mathbf{B}_e, \mathbf{C}_e, \mathbf{D}_e]$.

Let $h_{ij}(s)$ denote the transfer function relating $e_i(s)$ to the jth component of the input $[\mathbf{d}(t) \ \ \mathbf{r}(t)]^T$. For a matrix S, let $S_{i\cdot}$, $S_{\cdot j}$, and S_{ij} denote its ith row, jth column, and ijth element. With this notation

$$h_{ij}(s) = (\mathbf{C}_e)_{i\cdot} (s\mathbf{I} - \mathbf{A}_e)^{-1} (\mathbf{B}_e)_{\cdot j} + (\mathbf{D}_e)_{ij} = \frac{\det \begin{bmatrix} s\mathbf{I} - \mathbf{A}_e & (\mathbf{B}_e)_{\cdot j} \\ (-\mathbf{C}_e)_{i\cdot} & (\mathbf{D}_e)_{ij} \end{bmatrix}}{\det [s\mathbf{I} - \mathbf{A}_e]}. \tag{9.85}$$

The numerator of Equation (9.85) is, in expanded form

$$\det \begin{bmatrix} s\mathbf{I} - \mathbf{A}_p + \mathbf{B}_p\mathbf{K}_p & \mathbf{B}_p\mathbf{K}_m & \mathrm{X} \\[2em] \begin{array}{c} \mathbf{m}(\mathbf{C}_p - \mathbf{D}_p\mathbf{K}_p)_{1\cdot} \\ \mathbf{m}(\mathbf{C}_p - \mathbf{D}_p\mathbf{K}_p)_{2\cdot} \\ \vdots \\ \mathbf{m}(\mathbf{C}_p - \mathbf{D}_p\mathbf{K}_p)_{n_0\cdot} \end{array} & \tilde{\mathbf{A}} & \begin{array}{c} \mathbf{m}(\mathbf{D}_e)_{1j} \\ \mathbf{m}(\mathbf{D}_e)_{2j} \\ \vdots \\ \mathbf{m}(\mathbf{D}_e)_{n_0 j} \end{array} \\[2em] (\mathbf{C}_p - \mathbf{D}_p\mathbf{K}_p)_{i\cdot} & (-\mathbf{D}_p\mathbf{K}_m)_{i\cdot} & (\mathbf{D}_e)_{ij} \end{bmatrix}, \tag{9.86}$$

where

$$\tilde{\mathbf{A}} = \begin{bmatrix} s\mathbf{I} - \mathbf{M} & & & \\ & s\mathbf{I} - \mathbf{M} & & \\ & & \ddots & \\ & & & s\mathbf{I} - \mathbf{M} \end{bmatrix} - \begin{bmatrix} \mathbf{m}(\mathbf{D}_p\mathbf{K}_m)_{1\cdot} \\ \mathbf{m}(\mathbf{D}_p\mathbf{K}_m)_{2\cdot} \\ \vdots \\ \mathbf{m}(\mathbf{D}_p\mathbf{K}_m)_{n_0\cdot} \end{bmatrix}. \tag{9.87}$$

Pre-multiplying the matrix in Equation (9.86) by

$$\mathbf{T} = \begin{bmatrix} \mathbf{I}_n & & & & \\ & \mathbf{I}_t & & & \\ & & \ddots & & -\mathbf{m} \\ & & & \mathbf{I}_t & \\ & & & & 1 \end{bmatrix} \leftarrow (i+1)\text{th block}, \tag{9.88}$$

we obtain the numerator as

$$
\det
\begin{bmatrix}
s\mathbf{I} - \mathbf{A}_p + \mathbf{B}_p\mathbf{K}_p & \mathbf{B}_p\mathbf{K}_m & \mathbf{X} \\[2mm]
\hline
\begin{array}{c}
\mathbf{m}(\mathbf{C}_p - \mathbf{D}_p\mathbf{K}_p)_{1\cdot} \\
\mathbf{m}(\mathbf{C}_p - \mathbf{D}_p\mathbf{K}_p)_{2\cdot} \\
\vdots \\
0 \\
\vdots \\
\mathbf{m}(\mathbf{C}_p - \mathbf{D}_p\mathbf{K}_p)_{n_0\cdot}
\end{array} &
\bar{\mathbf{A}} &
\begin{array}{c}
\mathbf{m}(\mathbf{D}_e)_{1j} \\
\mathbf{m}(\mathbf{D}_e)_{2j} \\
\vdots \\
0 \\
\vdots \\
\mathbf{m}(\mathbf{D}_e)_{n_0 j}
\end{array} \\[2mm]
\hline
(\mathbf{C}_p - \mathbf{D}_p\mathbf{K}_p)_{i\cdot} & -(\mathbf{D}_p\mathbf{K}_m)_{i\cdot} & (\mathbf{D}_e)_{ij}
\end{bmatrix},
\tag{9.89}
$$

where

$$
\bar{\mathbf{A}} =
\begin{bmatrix}
s\mathbf{I} - \mathbf{M} & & & & \\
& s\mathbf{I} - \mathbf{M} & & & \\
& & \ddots & & \\
& & & s\mathbf{I} - \mathbf{M} & \\
& & & & \ddots & \\
& & & & & s\mathbf{I} - \mathbf{M}
\end{bmatrix}
-
\begin{bmatrix}
\mathbf{m}(\mathbf{D}_p\mathbf{K}_m)_{1\cdot} \\
\mathbf{m}(\mathbf{D}_p\mathbf{K}_m)_{2\cdot} \\
\vdots \\
0 \\
\vdots \\
\mathbf{m}(\mathbf{D}_p\mathbf{K}_m)_{n_0\cdot}
\end{bmatrix}.
\tag{9.90}
$$

Therefore, finally

$$
h_{ij}(s) = \frac{\det[s\mathbf{I} - \mathbf{M}] \cdot n_{ij}(s)}{\det[s\mathbf{I} - \mathbf{A}_e]} = \frac{m(s)n_{ij}(s)}{\det[s\mathbf{I} - \mathbf{A}_e]}
\tag{9.91}
$$

for some polynomial $n_{ij}(s)$. It follows that each entry of the error transfer matrix has $m(s)$ as a factor of the numerator and the stable polynomial

$$
\det[s\mathbf{I} - \mathbf{A}_e] = \det[s\mathbf{I} - \mathbf{A} + \mathbf{B}K]
\tag{9.92}
$$

as the denominator. Thus, the condition in Equation (9.61) is satisfied for every reference input and every disturbance input satisfying Equations (9.63) and (9.64), respectively. ∎

Remark 9.1 The special property that $m(s)$ is a factor of each and every entry of the error transfer function is described by labeling the zeros of $m(s)$ as *blocking zeros* of the closed-loop error transfer function.

Remark 9.2 The condition in Equation (9.74) requires that $n_i \geq n_0$, that is, there are at least as many control inputs as outputs to be controlled. When $n_i \geq n_0$, it is equivalent to the condition that the plant transfer function evaluated at $s = \mu$

$$
\mathbf{G}_p(\mu) := \mathbf{C}_p(\mu\mathbf{I} - \mathbf{A}_p)^{-1}\mathbf{B}_p + \mathbf{D}_p
\tag{9.93}
$$

has rank n_0.

Remark 9.3   It is easy to see that under small perturbations of the plant model $[\mathbf{A}_p, \mathbf{B}_p, \mathbf{C}_p, \mathbf{D}_p]$ and $[K_p \ K_m]$, the property that $m(s)$ is a factor of the numerator of each entry of the error transfer function is *preserved*. The latter is determined only by the servocontroller, which must be perturbation-free. Thus, the system functions as a servomechanism as long as $\mathbf{A}_e$ remains stable and the servocontroller is accurate and fixed.

Output Feedback Solution

The state feedback control law in Equation (9.82) may be unimplementable if $\mathbf{x}_p(t)$ is inaccessible as a measurement signal or otherwise unavailable for feedback. The state $\mathbf{x}_m(t)$ of the servocontroller will most often be available since the servocontroller is a synthetic device or signal generator specifically built for control.

In the case that Equation (9.82) is unimplementable because the states are unavailable for measurement, we can introduce the *output feedback* stabilizing controller

$$\dot{\mathbf{x}}_s(t) = \mathbf{A}_s \mathbf{x}_s(t) + \mathbf{B}_s \begin{bmatrix} \mathbf{e}(t) \\ \mathbf{x}_m(t) \end{bmatrix},$$
$$\mathbf{u}_p(t) = \mathbf{C}_s \mathbf{x}_s(t) + \mathbf{D}_s \begin{bmatrix} \mathbf{e}(t) \\ \mathbf{x}_m(t) \end{bmatrix}. \tag{9.94}$$

The ability of the controller in Equation (9.94) to stabilize the closed-loop system is determined by the stabilizability and detectability of $(\mathbf{A}, \mathbf{B}, \bar{\mathbf{C}})$, where

$$\bar{\mathbf{C}} = \begin{bmatrix} \mathbf{C}_p & 0 \\ 0 & \mathbf{I}_{n_0 t} \end{bmatrix}. \tag{9.95}$$

Lemma 9.3 has already given conditions for stabilizability. For detectability, observe that

$$\begin{bmatrix} s\mathbf{I} - \mathbf{A} \\ \bar{\mathbf{C}} \end{bmatrix} = \begin{bmatrix} s\mathbf{I} - \mathbf{A}_p & 0 \\ \mathbf{B}_m \mathbf{C}_p & s\mathbf{I} - \mathbf{A}_m \\ \mathbf{C}_p & 0 \\ 0 & \mathbf{I}_{n_0 t} \end{bmatrix} \tag{9.96}$$

so that $(\bar{\mathbf{C}}, \mathbf{A})$ is detectable if and only if $(\mathbf{C}_p, \mathbf{A}_p)$ is detectable. The proof that the controller in Equation (9.94) also assigns $|s\mathbf{I} - \mathbf{M}| = m(s)$ as a numerator polynomial to each entry of the error transfer function is similar to the state feedback case and is left to the reader (Exercise 7.14). Observe that this zero assignment property is only determined by the servocontroller. The output feedback servomechanism is shown in Figure 9.13. We have arrived at the following important result.

THEOREM 9.2   If the plant $(\mathbf{A}_p, \mathbf{B}_p, \mathbf{C}_p, \mathbf{D}_p)$ is stabilizable and detectable and the condition in Equation (9.74) is satisfied, then a solution to the servomechanism problem exists and is of the form shown in Figure 9.13. The solution is insensitive to perturbations of the plant and stabilizing controller parameters as long as they do not destabilize the closed loop. Tracking and disturbance rejection are not preserved if the servocontroller part of the controller is perturbed by even a vanishingly small amount.

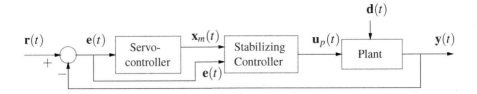

Figure 9.13 Output feedback servomechanism

Remark 9.4   As a final remark, we point out that the design of the stabilizing part of the controller can always by carried out using observer theory, optimal LQR theory, and the technique for designing optimal output feedback controllers described in Chapter 7.

9.7   Multivariable Servomechanisms: Transfer Function Analysis

In this section, we provide a transfer function analysis of the multivariable servomechanism problem. The main result to be obtained is the demonstration that the servocontroller assigns the signal poles as zeros to each of the entries of the error transfer function matrix. Zeros of this type are called blocking zeros as they prevent the "leakage" of any disturbance or reference signals to any component of the error. If the signal generator is perturbation-free, these zeros are "hard" zeros and provide robust servo action.

The block diagram of the multivariable servomechanism has the general form shown in Figure 9.14.

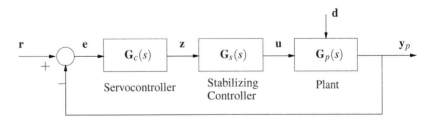

Figure 9.14 Unity feedback servomechanism

The servocontroller is described by

$$\mathbf{G}_c(s) = \begin{bmatrix} \frac{1}{m(s)} & & & \\ & \frac{1}{m(s)} & & \\ & & \ddots & \\ & & & \frac{1}{m(s)} \end{bmatrix} \tag{9.97}$$

and (by slight abuse of notation)

$$\mathbf{G}_c(s) := \frac{1}{m(s)}\mathbf{I}_{n_0}, \tag{9.98}$$

where $m(D)$ is the annihilating polynomial for the exogenous signal, that is

$$m(D)r_i(t) = 0, \quad \text{for } i \in \mathbf{n}_0, \tag{9.99}$$

$$m(D)d_j(t) = 0, \quad \text{for } j \in \{1, 2, \dots, q\}. \tag{9.100}$$

The plant is described by

$$\mathbf{y}_p(s) = \mathbf{G}_u(s)\mathbf{u}_p(s) + \mathbf{G}_d(s)\mathbf{d}(s). \tag{9.101}$$

If a more general structure is used for the stabilizing controller such as the one in Figure 9.15, it can be recast into the form of Figure 9.14 with

$$\mathbf{G}_s(s) := \mathbf{G}_2(s)\mathbf{G}_c(s)^{-1} + \mathbf{G}_1(s). \tag{9.102}$$

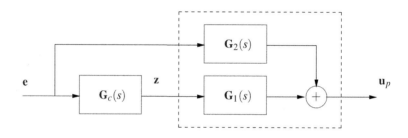

Figure 9.15 Stabilizing controller

It follows from Figure 9.14 that

$$\mathbf{e}(s) = \underbrace{\left[\mathbf{I} + \mathbf{G}_u(s)\mathbf{G}_s(s)\mathbf{G}_c(s)\right]}_{\mathbf{L}(s)}^{-1} \mathbf{r}(s) - \left[\mathbf{I} + \mathbf{G}_u(s)\mathbf{G}_s(s)\mathbf{G}_c(s)\right]^{-1}\mathbf{G}_d(s)\mathbf{d}(s). \tag{9.103}$$

Write

$$\mathbf{E}(s) := \left[\mathbf{I} + \mathbf{L}(s)\right]^{-1} \tag{9.104}$$

so that

$$\mathbf{e}(s) = \mathbf{E}(s)\mathbf{r}(s) - \mathbf{E}(s)\mathbf{G}_d(s)\mathbf{d}(s). \tag{9.105}$$

Closed-loop stability implies that $\mathbf{E}(s)$ and $\mathbf{E}(s)\mathbf{G}_d(s)$ are stable.

LEMMA 9.4

$$\mathbf{I} + \mathbf{G}_u(s)\mathbf{G}_s(s)\mathbf{G}_c(s) = \mathbf{I} + \mathbf{G}_c(s)\mathbf{G}_u(s)\mathbf{G}_s(s). \tag{9.106}$$

Proof   The proof follows from the structure of $\mathbf{G}_c(s)$ given in Equation (9.98). ∎

LEMMA 9.5

$$\left[\mathbf{I} + \mathbf{L}(s)\right]^{-1} = \underbrace{\left[\mathbf{G}_c(s)^{-1} + \mathbf{G}_u(s)\mathbf{G}_s(s)\right]^{-1}}_{\mathbf{H}(s)}\mathbf{G}_c(s)^{-1}$$

$$= \mathbf{H}\mathbf{G}_c(s)^{-1}. \tag{9.107}$$

Proof From Lemma 9.4:

$$[\mathbf{I}+\mathbf{G}_u(s)\mathbf{G}_s(s)\mathbf{G}_c(s)]^{-1} = [\mathbf{I}+\mathbf{G}_c(s)\mathbf{G}_u(s)\mathbf{G}_s(s)]^{-1}$$

$$= \left[\mathbf{G}_c(s)\mathbf{G}_c(s)^{-1}+\mathbf{G}_c(s)\mathbf{G}_u(s)\mathbf{G}s(s)\right]^{-1}$$

$$= \underbrace{\left[\mathbf{G}_c(s)^{-1}+\mathbf{G}_u(s)\mathbf{G}_s(s)\right]^{-1}}_{\mathbf{H}(s)}\mathbf{G}_c(s)^{-1} \qquad (9.108)$$

$$= \mathbf{H}(s)\mathbf{G}_c(s)^{-1}.$$

∎

Now Equation (9.105) can be rewritten as

$$\mathbf{e}(s) = \mathbf{H}(s)\mathbf{G}_c(s)^{-1}\mathbf{r}(s) - \mathbf{H}(s)\mathbf{G}_c(s)^{-1}\mathbf{G}_d(s)\mathbf{d}(s) \qquad (9.109)$$

and again, using the structure of $\mathbf{G}_c(s)$ in Equation (9.98) as

$$\mathbf{e}(s) = \mathbf{H}(s)\mathbf{G}_c(s)^{-1}\mathbf{r}(s) - \mathbf{H}(s)\mathbf{G}_d(s)\left(\mathbf{I}_q m(s)\right)\mathbf{d}(s)$$

$$= \mathbf{H}(s)\left(\mathbf{I}_{n_0} m(s)\right)\mathbf{r}(s) - \mathbf{H}(s)\mathbf{G}_d(s)\left(\mathbf{I}_q m(s)\right)\mathbf{d}(s). \qquad (9.110)$$

Equation (9.110) proves the following important result.

LEMMA 9.6   Every element of the transfer function matrix with $\mathbf{e}(s)$ as output and $(\mathbf{r}(s), \mathbf{d}(s))$ as input has $m(s)$ as a factor of the numerator.

Thus, the servocontroller assigns the zeros of $m(s)$ to each and every element of the error transfer function matrix. Zeros of this type have been called *blocking zeros*.

Existence of the Stabilizing Controller

In the state-space formulation of the multivariable servomechanism problem, we derived the condition:

$$\text{rank}\begin{bmatrix} \mu\mathbf{I}-\mathbf{A}_p & \mathbf{B}_p \\ -\mathbf{C}_p & \mathbf{D}_p \end{bmatrix} = n+n_0, \quad \text{for all } \mu \text{ such that } m(\mu)=0 \qquad (9.111)$$

as a sufficient condition for solvability of the multivariable servomechanism problem. The condition in Equation (9.111) has the following interpretation in terms of the plant transfer function.

LEMMA 9.7   If $\mu \notin \sigma(\mathbf{A}_p)$ and $m(\mu) = 0$, Equation (9.111) is equivalent to

$$\text{rank}[\mathbf{G}_u(\mu)] = n_0. \qquad (9.112)$$

Proof From

$$\begin{bmatrix} s\mathbf{I}-\mathbf{A}_p & \mathbf{B}_p \\ -\mathbf{C}_p & \mathbf{D}_p \end{bmatrix}\begin{bmatrix} (s\mathbf{I}-\mathbf{A}_p)^{-1} & -(s\mathbf{I}-\mathbf{A}_p)^{-1}\mathbf{B}_p \\ 0 & \mathbf{I} \end{bmatrix}$$

$$= \begin{bmatrix} \mathbf{I}_n & 0 \\ -\mathbf{C}_p(s\mathbf{I}-\mathbf{A}_p)^{-1} & \mathbf{G}_u(s) \end{bmatrix}, \qquad (9.113)$$

we see that, when $\mu \notin \sigma(\mathbf{A}_p)$:

$$\text{rank} \begin{bmatrix} \mu \mathbf{I} - \mathbf{A}_p & \mathbf{B}_p \\ -\mathbf{C}_p & \mathbf{D}_p \end{bmatrix} = n + \text{rank}\,[\mathbf{G}_u(\mu)] \tag{9.114}$$

and thus Equation (9.112) follows from Equation (9.111). ∎

Remark 9.5 Note that Equation (9.112) implies that $n_i \geq n_0$, that is the number of plant inputs is no less than the number of outputs and the plant transfer function matrix evaluated at every zero of $m(s)$ is right invertible.

9.8   Examples

EXAMPLE 9.3 (LQ optimal servomechanism using output feedback dynamic compensator)   Consider the example below, where the control objective is to make the plant outputs track arbitrary step reference inputs and reject step disturbances. The equations of the plant are

$$\begin{aligned}\dot{\mathbf{x}}_p(t) &= \mathbf{A}_p\mathbf{x}_p(t) + \mathbf{B}_p\mathbf{u}_p(t) + \mathbf{B}_d\mathbf{d}(t), \\ \mathbf{y}_p(t) &= \mathbf{C}_p\mathbf{x}_p(t) + \mathbf{D}_p\mathbf{u}_p(t),\end{aligned} \tag{9.115}$$

where

$$\mathbf{A}_p = \begin{bmatrix} 0 & 1 & 0 & 0 \\ 0 & 0 & 1 & 0 \\ 1 & 0 & 0 & 1 \\ -4 & -10 & -10 & -5 \end{bmatrix}, \quad \mathbf{B}_p = \begin{bmatrix} 0 & 1 \\ 1 & 1 \\ 1 & 0 \\ 0 & 0 \end{bmatrix}, \quad \mathbf{B}_d = \begin{bmatrix} 1 \\ 0 \\ 0 \\ 0 \end{bmatrix}, \tag{9.116}$$

$$\mathbf{C}_p = \begin{bmatrix} 0 & 0 & 1 & -1 \\ 0 & 1 & 0 & 1 \end{bmatrix}, \quad \mathbf{D}_p = \begin{bmatrix} 0 & 0 \\ 0 & 0 \end{bmatrix}.$$

To proceed, we first check the condition for the existence of a stabilizing controller given in Equation (9.111). Note that $n = 4$, $n_0 = 2$, and $\mu = 0$ from $m(\mu) = 0$. It is easy to see that

$$\text{rank} \begin{bmatrix} \mu \mathbf{I}_4 - \mathbf{A}_p & \mathbf{B}_p \\ -\mathbf{C}_p & 0 \end{bmatrix} = 6 \tag{9.117}$$

and thus the problem is solvable.

The servocontroller is thus described by

$$\begin{aligned}\dot{\mathbf{x}}_m(t) &= \mathbf{A}_m\mathbf{x}_m(t) + \mathbf{B}_m\mathbf{e}(t), \\ \mathbf{y}_m(t) &= \mathbf{C}_m\mathbf{x}_m(t) = \mathbf{x}_m(t), \\ \mathbf{e}(t) &= \mathbf{r}(t) - \mathbf{y}_p(t),\end{aligned} \tag{9.118}$$

where

$$\mathbf{A}_m = \begin{bmatrix} 0 & 0 \\ 0 & 0 \end{bmatrix}, \quad \mathbf{B}_m = \begin{bmatrix} 1 & 0 \\ 0 & 1 \end{bmatrix}, \quad \mathbf{C}_m = \begin{bmatrix} 1 & 0 \\ 0 & 1 \end{bmatrix}. \tag{9.119}$$

In order to render this into a "regulator problem," we redraw Figure 9.13 having inputs $\mathbf{r}(t), \mathbf{d}(t), \mathbf{u}_p(t)$ and outputs $\mathbf{e}(t)$ and $\mathbf{x}_m(t)$ for the new plant in Figure 9.16.

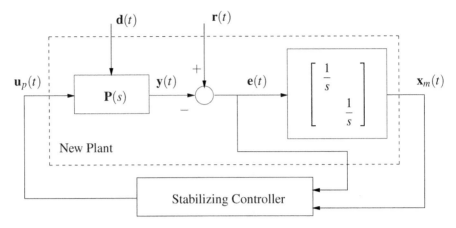

Figure 9.16 Stabilizing controller (Example 9.3)

Thus, the "new" plant (augmented by signal generator) has the following state-variable equations:

$$\begin{bmatrix} \dot{\mathbf{x}}_p(t) \\ \dot{\mathbf{x}}_m(t) \end{bmatrix} = \underbrace{\begin{bmatrix} \mathbf{A}_p & 0 \\ -\mathbf{B}_m\mathbf{C}_p & \mathbf{A}_m \end{bmatrix}}_{\mathbf{A}} \underbrace{\begin{bmatrix} \mathbf{x}_p(t) \\ \mathbf{x}_m(t) \end{bmatrix}}_{\mathbf{x}(t)} + \underbrace{\begin{bmatrix} \mathbf{B}_p \\ -\mathbf{B}_m\mathbf{D}_p \end{bmatrix}}_{\mathbf{B}} \mathbf{u}_p(t) + \begin{bmatrix} 0 \\ \mathbf{B}_m \end{bmatrix} \mathbf{r}(t), \qquad (9.120)$$

$$\underbrace{\begin{bmatrix} \mathbf{e}(t) \\ \mathbf{x}_m(t) \end{bmatrix}}_{\mathbf{y}(t)} = \begin{bmatrix} \mathbf{r}(t) - \mathbf{y}_p(t) \\ \mathbf{x}_m(t) \end{bmatrix} = \underbrace{\begin{bmatrix} -\mathbf{C}_p & 0 \\ 0 & \mathbf{C}_m \end{bmatrix}}_{\mathbf{C}} \begin{bmatrix} \mathbf{x}_p(t) \\ \mathbf{x}_m(t) \end{bmatrix} + \begin{bmatrix} \mathbf{I} \\ 0 \end{bmatrix} \mathbf{r}(t). \qquad (9.121)$$

For the design of the stabilizing controller, we assume for the moment that $\mathbf{r}(t) = 0$ so that

$$\begin{bmatrix} \dot{\mathbf{x}}_p(t) \\ \dot{\mathbf{x}}_m(t) \end{bmatrix} = \mathbf{A} \underbrace{\begin{bmatrix} \mathbf{x}_p(t) \\ \mathbf{x}_m(t) \end{bmatrix}}_{\mathbf{x}(t)} + \mathbf{B}\mathbf{u}_p(t), \qquad (9.122)$$

$$\underbrace{\begin{bmatrix} \mathbf{e}(t) \\ \mathbf{x}_m(t) \end{bmatrix}}_{\mathbf{y}(t)} = \mathbf{C} \begin{bmatrix} \mathbf{x}_p(t) \\ \mathbf{x}_m(t) \end{bmatrix}. \qquad (9.123)$$

Now we follow steps similar to those in Example 7.4 to design the stabilizing controller. First, observe that

$$\text{Rank} \begin{bmatrix} \mathbf{C} \\ \mathbf{C}\mathbf{A} \end{bmatrix} = 6. \qquad (9.124)$$

Therefore $q = 1$ and so we introduce

$$\dot{\mathbf{u}}_p(t) = \mathbf{v}(t) \qquad (9.125)$$

and define the augmented system with state $\mathbf{x}_a(t)$, input $\mathbf{v}(t)$, and corresponding matrices $\mathbf{A}_a$ and $\mathbf{B}_a$:

$$\mathbf{x}_a(t) = \begin{bmatrix} \mathbf{x}(t) \\ \mathbf{u}_p(t) \end{bmatrix}, \quad \mathbf{A}_a = \begin{bmatrix} \mathbf{A} & \mathbf{B} \\ 0 & 0 \end{bmatrix}, \quad \mathbf{B}_a = \begin{bmatrix} 0 \\ \mathbf{I} \end{bmatrix}, \tag{9.126}$$

where

$$\mathbf{x}(t) = \begin{bmatrix} \mathbf{x}_p(t) \\ \mathbf{x}_m(t) \end{bmatrix}. \tag{9.127}$$

Now let

$$\mathbf{C}_a = \begin{bmatrix} -\mathbf{C}_p & 0 & 0 \\ 0 & \mathbf{C}_m & 0 \\ 0 & 0 & \mathbf{R}_a \end{bmatrix} = \begin{bmatrix} 0 & 0 & -1 & 1 & 0 & 0 & 0 & 0 \\ 0 & -1 & 0 & -1 & 0 & 0 & 0 & 0 \\ 0 & 0 & 0 & 0 & 1 & 0 & 0 & 0 \\ 0 & 0 & 0 & 0 & 0 & 1 & 0 & 0 \\ 0 & 0 & 0 & 0 & 0 & 0 & 1 & 0 \\ 0 & 0 & 0 & 0 & 0 & 0 & 0 & 1 \end{bmatrix} \tag{9.128}$$

so that $(\mathbf{C}_a, \mathbf{A}_a)$ is observable and let $\mathbf{Q}_a := \mathbf{C}_a^T \mathbf{C}_a$ and $\mathbf{R}_a = \mathbf{I}$ in an LQR performance index for the augmented system

$$J_a = \int \mathbf{x}_a^T(t) \mathbf{Q}_a \mathbf{x}_a(t) + \mathbf{v}^T(t) \mathbf{R}_a \mathbf{v}(t) dt \tag{9.129}$$

$$= \int \mathbf{x}_p^T(t) \mathbf{C}_p^T \mathbf{C}_p \mathbf{x}_p(t + \mathbf{x}_m^T(t) \mathbf{C}_m^T \mathbf{C}_m \mathbf{x}_m(t) + \mathbf{u}_p^T(t) \mathbf{R}^T \mathbf{R} \mathbf{u}_p(t) + \dot{\mathbf{u}}^T(t) \mathbf{R}_a \dot{\mathbf{u}}(t)) dt.$$

Using LQR theory and MATLAB we obtain the optimal state feedback matrix for the augmented system

$$\mathbf{K}_a = \begin{bmatrix} \mathbf{K}_0 & \mathbf{L}_0 \end{bmatrix}, \tag{9.130}$$

where

$$\mathbf{K}_0 = \begin{bmatrix} 0.4402 & 0.4610 & 2.9376 & 0.2011 & -0.8766 & 0.4812 \\ 1.1574 & 1.2506 & 0.9188 & 0.0349 & 0.4812 & 0.8766 \end{bmatrix}, \tag{9.131}$$

$$\mathbf{L}_0 = \begin{bmatrix} 2.7256 & 0.6069 \\ 0.6069 & 2.3340 \end{bmatrix}. \tag{9.132}$$

From

$$\begin{bmatrix} \mathbf{P}_0 & \mathbf{P}_1 \end{bmatrix} \begin{bmatrix} \mathbf{C} \\ \mathbf{CA} \end{bmatrix} = \mathbf{K}_0, \tag{9.133}$$

we have

$$\mathbf{P}_0 = \begin{bmatrix} -0.4142 & 0.3197 & -0.8766 & 0.4812 \\ 0.4791 & 0.8215 & 0.4812 & 0.8766 \end{bmatrix}, \tag{9.134}$$

$$\mathbf{P}_1 = \begin{bmatrix} -1.1187 & -1.0086 & -0.4142 & 0.3197 \\ 0.7272 & 1.0165 & 0.4791 & 0.8215 \end{bmatrix}, \tag{9.135}$$

and then

$$Q_0 = L_0 - P_1 CB = \begin{bmatrix} 0.5982 & -0.4018 \\ 2.3506 & 3.3506 \end{bmatrix}. \tag{9.136}$$

To further clarify the structure of the controller in detail, define W_0, W_1, H_0, H_1 such that

$$P_0 - Q_0 P_1 = \begin{bmatrix} W_0 & W_1 \end{bmatrix}, \quad P_1 = \begin{bmatrix} H_0 & H_1 \end{bmatrix}, \tag{9.137}$$

where $(W_0, W_1, H_0, H_1 \in \mathbb{R}^{2\times 2})$. A state-space realization of the dynamic compensator is then

$$\begin{aligned} \dot{z}(t) &= -Q_0 z(t) + (P_0 - Q_0 P_1)(-y(t)) \\ &= -Q_0 z(t) - W_0 e(t) - W_1 x_m(t), \end{aligned} \tag{9.138}$$

$$\begin{aligned} u_p(t) &= z(t) + P_1(-y(t)) \\ &= z(t) - H_0 e(t) - H_1 x_m(t). \end{aligned} \tag{9.139}$$

We can now realize the stabilizing compensator and draw the block diagram of the unity feedback loop as shown in Figure 9.17.

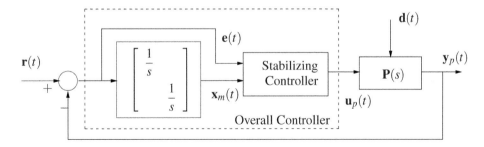

Figure 9.17 Unity feedback control loop (Example 9.3)

The state-space representation for the "overall controller" is

$$\begin{bmatrix} \dot{x}_m(t) \\ \dot{z}(t) \end{bmatrix} = \underbrace{\begin{bmatrix} A_m & 0 \\ -W_1 & -Q_0 \end{bmatrix}}_{A_c} \underbrace{\begin{bmatrix} x_m(t) \\ z(t) \end{bmatrix}}_{x_c(t)} + \underbrace{\begin{bmatrix} B_m \\ -W_0 \end{bmatrix}}_{B_c} e(t),$$

$$u_p(t) = \underbrace{\begin{bmatrix} -H_1 & I \end{bmatrix}}_{C_c} \begin{bmatrix} x_m(t) \\ z(t) \end{bmatrix} + \underbrace{(-H_0)}_{D_c} e(t). \tag{9.140}$$

Thus, the closed-loop system is described by

$$\begin{bmatrix} \dot{x}_p(t) \\ \dot{x}_c(t) \end{bmatrix} = \begin{bmatrix} A_p - B_p D_c C_p & B_p C_c \\ -B_c C_p & A_c \end{bmatrix} \begin{bmatrix} x_p(t) \\ x_c(t) \end{bmatrix} + \begin{bmatrix} B_p D_c \\ B_c \end{bmatrix} r(t),$$

$$y_p(t) = \begin{bmatrix} C_p & 0 \end{bmatrix} \begin{bmatrix} x_p(t) \\ x_c(t) \end{bmatrix}, \tag{9.141}$$

where $\mathbf{r}(t)$ is the external reference input. The eigenvalues of the system matrix are

$$\{-1.7541 \pm j2.5573, -3.4405, -0.3387 \pm j0.4052, -0.4334, -1.0, -1.0\}, \quad (9.142)$$

which coincides with the eigenvalues of $(\mathbf{A}_a - \mathbf{B}_a K_a^T)$ as expected.

MATLAB/Simulink Simulation

To verify the solution obtained, a Simulink simulation is performed. The following MATLAB script sets up the parameters needed for the Simulink model shown in Figure 9.18.

```
clear
Ap=[0 1 0 0; 0 0 1 0; 0 0 0 1; -4 -10 -10 -5];
Bp=[0 1; 1 1; 1 0; 0 0];
Bd=[1 0 0 0]';
Cp=[0 0 1 -1; 0 1 0 1];
Dp=[0 0; 0 0];
Dd=[0 0]';
B1=[Bp Bd];    D1=[Dp Dd];
%

Am=zeros(2,2);    Bm=eye(2);    Cm=eye(2); %
A=[Ap zeros(4,2); -Bm*Cp Am];
B=[Bp; -Bm*Dp];
C=[-Cp zeros(2,2); zeros(2,4) Cm];
%
Aa=[A B; zeros(2,8)];
Ba=[zeros(6,2); eye(2)];
Ca=[-Cp zeros(2,2) zeros(2,2); zeros(2,4) Cm zeros(2,2);
zeros(2,6) eye(2)];
Qa=Ca'*Ca;    Ra=eye(2);
Ka=lqr(Aa,Ba,Qa,Ra);    K0=Ka(:,1:6);    L0=Ka(:,7:8);
%
CCA=[C;C*A];
P=K0*pinv(CCA);    P0=P(:,1:4);    P1=P(:,5:8);
Q0=L0-P1*C*B;
W=P0-Q0*P1;    W0=W(:,1:2);    W1=W(:,3:4);
H0=P1(:,1:2);    H1=P1(:,3:4);
%
Ac=[Am zeros(2,2); -W1 -Q0];
Bc=[Bm; -W0];    Cc=[-H1 eye(2)];    Dc=-H0;
%
Acl=[Ap-Bp*Dc*Cp Bp*Cc; -Bc*Cp Ac];
eig(Acl);
```

The output response depicted in Figure 9.19 shows that both outputs asymptotically track unit step reference inputs and reject a unit step disturbance, as expected.

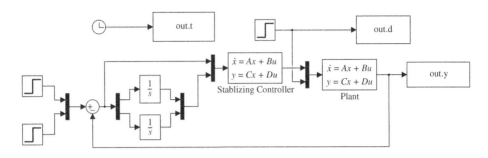

Figure 9.18 Simulink model (Example 9.3)

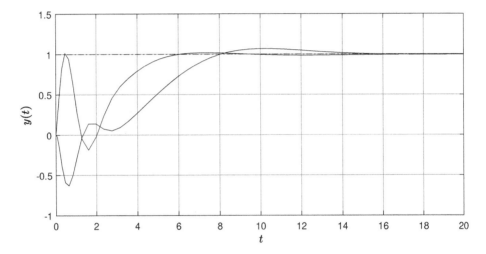

Figure 9.19 Both outputs track unit step references and reject a unit step disturbance (Example 9.3)

EXAMPLE 9.4 (LQ optimal servomechanism using output feedback dynamic compensator with a "bad" plant)   In this example, we attempt to design a servomechanism using an output feedback dynamic compensator for an unstable plant with non-minimum phase zeros, which is in general difficult to control. We introduce such a plant by replacing $\mathbf{A}_p$ in Equation (9.116) with the following, while other matrices remain the same:

$$\mathbf{A}_p = \begin{bmatrix} 1 & 0 & 1 & 0 \\ 0 & 1 & 0 & 1 \\ 1 & 0 & 0 & 0 \\ 0 & 1 & 0 & 0 \end{bmatrix}. \tag{9.143}$$

The MATLAB simulation by using the model given in Figure 9.18 shows the output response displayed in Figure 9.20.

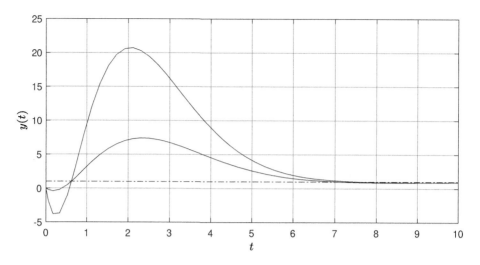

Figure 9.20 Both outputs track unit step references and reject a unit step disturbance (Example 9.4)

The next example illustrates the use of an observer to solve the servomechanism problem.

EXAMPLE 9.5 (Reduced-order observer-based servomechanism)   Consider a third-order plant.

$$\dot{\mathbf{x}}_p(t) = \mathbf{A}_p\mathbf{x}_p(t) + \mathbf{B}_p\mathbf{u}_p(t),$$
$$\mathbf{y}_p(t) = \mathbf{C}_p\mathbf{x}_p(t), \tag{9.144}$$

where

$$\mathbf{A}_p = \begin{bmatrix} -1 & 0 & 1 \\ 0 & -1 & 1 \\ 0 & 0 & -2 \end{bmatrix}, \quad \mathbf{B}_p = \begin{bmatrix} 1 & -1 \\ 0 & 1 \\ 1 & -6 \end{bmatrix}, \quad \mathbf{C}_p = \begin{bmatrix} -1 & 0 & -1 \\ 0 & 1 & 0 \end{bmatrix}. \tag{9.145}$$

The plant outputs are required to track independent step inputs. Therefore, the servo-controller is

$$\dot{\mathbf{x}}_m(t) = \mathbf{A}_m\mathbf{x}_m(t) + \mathbf{B}_m\mathbf{e}(t),$$
$$\mathbf{e}(t) = \mathbf{r}(t) - \mathbf{y}_p(t), \tag{9.146}$$
$$\mathbf{y}_m(t) = \mathbf{C}_m\mathbf{x}_m(t) = \mathbf{x}_m(t),$$

where

$$\mathbf{A}_m = \begin{bmatrix} 0 & 0 \\ 0 & 0 \end{bmatrix}, \quad \mathbf{B}_m = \begin{bmatrix} 1 & 0 \\ 0 & 1 \end{bmatrix}, \quad \mathbf{C}_m = \begin{bmatrix} 1 & 0 \\ 0 & 1 \end{bmatrix}. \tag{9.147}$$

We define the new plant as before and set $\mathbf{r}(t) = 0$. To design a reduced-order observer we construct the nonsingular transformation matrix $\mathbf{T}$ such that

$$\mathbf{T}^{-1} = \begin{bmatrix} \mathbf{C} \\ * \end{bmatrix} = \begin{bmatrix} -\mathbf{C}_p & 0 \\ 0 & \mathbf{C}_m \\ * & * \end{bmatrix} = \begin{bmatrix} 1 & 0 & 1 & 0 & 0 \\ 0 & -1 & 0 & 0 & 0 \\ 0 & 0 & 0 & 1 & 0 \\ 0 & 0 & 0 & 0 & 1 \\ 0 & 0 & 1 & 0 & 0 \end{bmatrix}. \tag{9.148}$$

The last row of $\mathbf{T}^{-1}$ is filled in by an arbitrary row vector so that it is of full rank. Then

$$\mathbf{A}_t = \mathbf{T}^{-1}\mathbf{A}\mathbf{T} = \begin{bmatrix} -1 & 0 & 0 & 0 & 0 \\ 0 & -1 & 0 & 0 & 1 \\ 1 & 0 & 0 & 0 & 0 \\ 0 & 1 & 0 & 0 & 0 \\ 0 & 0 & 0 & 0 & -2 \end{bmatrix},$$

$$\mathbf{B}_t = \mathbf{T}^{-1}\mathbf{B} = \begin{bmatrix} 2 & -7 \\ 0 & -1 \\ 0 & 0 \\ 0 & 0 \\ 1 & -6 \end{bmatrix}, \quad \mathbf{C}_t = \mathbf{C}\mathbf{T} = \begin{bmatrix} 1 & 0 & 0 & 0 & 0 \\ 0 & 1 & 0 & 0 & 0 \\ 0 & 0 & 1 & 0 & 0 \\ 0 & 0 & 0 & 1 & 0 \end{bmatrix}. \tag{9.149}$$

Next we find the state feedback gain such that

$$\sigma(\mathbf{A}_t - \mathbf{B}_t\mathbf{K}_t) = \{-1, -1 \pm j1, -1 \pm j2\} \tag{9.150}$$

and obtained

$$\mathbf{K}_t = \begin{bmatrix} \mathbf{K}_{t1} & \mathbf{K}_{t2} \end{bmatrix}$$
$$= \begin{bmatrix} 5.4716 & 28.0140 & 7.6273 & 28.1781 & -12.3018 \\ 1.2631 & 7.2739 & 1.6214 & 7.3012 & -3.0790 \end{bmatrix}. \tag{9.151}$$

To design the reduced-order observer, the matrix $\mathbf{A}_t$ is partitioned into four sub-matrices

$$\mathbf{A}_{t11} = \begin{bmatrix} -1 & 0 & 0 & 0 \\ 0 & -1 & 0 & 0 \\ 1 & 0 & 0 & 0 \\ 0 & 1 & 0 & 0 \end{bmatrix}, \quad \mathbf{A}_{t12} = \begin{bmatrix} 0 \\ -1 \\ 0 \\ 0 \end{bmatrix}, \tag{9.152}$$

$$\mathbf{A}_{t21} = \begin{bmatrix} 0 & 0 & 0 & 0 \end{bmatrix}, \quad \mathbf{A}_{t22} = -2.$$

Now find the observer gain $\mathbf{L}_{t2}$ such that

$$\sigma(\mathbf{A}_{t22} - \mathbf{L}_{t2}\mathbf{A}_{t12}) = -4, \quad \mathbf{L}_{t2} = \begin{bmatrix} 0 & -2 & 0 & 0 \end{bmatrix}. \tag{9.153}$$

Then the reduced-order observer is described by

$$\dot{w}_2(t) = \underbrace{(\mathbf{A}_{t22} - \mathbf{L}_{t2}\mathbf{A}_{t12})}_{\mathbf{P}} w_2(t)$$

$$+ \underbrace{\left[(\mathbf{A}_{t22} - \mathbf{L}_{t2}\mathbf{A}_{t12})\,\mathbf{L}_{t2} + (\mathbf{A}_{t21} - \mathbf{L}_{t2}\mathbf{A}_{t11})\right]}_{\mathbf{Q}} \begin{bmatrix} \mathbf{e}(t) \\ \mathbf{x}_m(t) \end{bmatrix} \quad (9.154)$$

$$+ \underbrace{(\mathbf{B}_{t2} - \mathbf{L}_{t2}\mathbf{B}_{t1})}_{\mathbf{R}} \mathbf{u}_p(t),$$

$$\mathbf{u}_p(t) = \underbrace{\mathbf{K}_{t2}}_{\mathbf{V}} w_2(t) + \underbrace{(\mathbf{K}_{t1} + \mathbf{K}_{t2}\mathbf{L}_{t2})}_{\mathbf{S}} \begin{bmatrix} \mathbf{e}(t) \\ \mathbf{x}_m(t) \end{bmatrix}. \quad (9.155)$$

Hence the equivalent controller transfer function is

$$\mathbf{C}(s) = \mathbf{V}(s\mathbf{I} - (\mathbf{P} + \mathbf{RV}))^{-1}(\mathbf{Q} + \mathbf{RS}) + \mathbf{S}. \quad (9.156)$$

A state-space realization of the stabilizing controller is given by

$$\mathbf{A}_s = \mathbf{P} + \mathbf{RV} = -16.3304,$$

$$\mathbf{B}_s = \mathbf{Q} + \mathbf{RS} = \begin{bmatrix} 4.6332 & 60.8380 & 5.3439 & 30.2312 \end{bmatrix},$$

$$\mathbf{C}_s = \mathbf{V} = \begin{bmatrix} 12.3018 \\ 3.0790 \end{bmatrix}, \quad (9.157)$$

$$\mathbf{D}_s = \mathbf{S} = \begin{bmatrix} -5.4716 & -52.6176 & -7.6273 & -28.1781 \\ -1.2631 & -13.4319 & -1.6214 & -7.3012 \end{bmatrix}.$$

To describe the controller structure in greater detail, define $\mathbf{B}_0, \mathbf{B}_1, \mathbf{D}_0, \mathbf{D}_1$ such that

$$\mathbf{B}_s = \begin{bmatrix} \mathbf{B}_0 & \mathbf{B}_1 \end{bmatrix}, \quad \mathbf{D}_s = \begin{bmatrix} \mathbf{D}_0 & \mathbf{D}_1 \end{bmatrix}, \quad (9.158)$$

where $\mathbf{B}_0, \mathbf{B}_1 \in \mathbb{R}^{1 \times 2}$ and $\mathbf{D}_0, \mathbf{D}_1 \in \mathbb{R}^{2 \times 2}$. Then

$$\dot{\mathbf{x}}_s(t) = \mathbf{A}_s\mathbf{x}_s(t) + \mathbf{B}_s \begin{bmatrix} \mathbf{e}(t) \\ \mathbf{x}_m(t) \end{bmatrix} = \mathbf{A}_s\mathbf{x}_s(t) + \mathbf{B}_0\mathbf{e}(t) + \mathbf{B}_1\mathbf{x}_m(t),$$

$$\mathbf{u}_p(t) = \mathbf{C}_s\mathbf{x}_s(t) + \mathbf{D}_s \begin{bmatrix} \mathbf{e}(t) \\ \mathbf{x}_m(t) \end{bmatrix} = \mathbf{C}_s\mathbf{x}_s(t) + \mathbf{D}_0\mathbf{e}(t) + \mathbf{D}_1\mathbf{x}_m(t). \quad (9.159)$$

The overall closed-loop system is described by

$$\begin{bmatrix} \dot{\mathbf{x}}_p(t) \\ \dot{\mathbf{x}}_m(t) \\ \dot{\mathbf{x}}_s(t) \end{bmatrix} = \underbrace{\begin{bmatrix} \mathbf{A}_p - \mathbf{B}_p\mathbf{D}_0\mathbf{C}_p & \mathbf{B}_p\mathbf{D}_1 & \mathbf{B}_p\mathbf{C}_s \\ -\mathbf{B}_m\mathbf{C}_p & \mathbf{A}_m & 0 \\ -\mathbf{B}_0\mathbf{C}_p & \mathbf{B}_1 & \mathbf{A}_s \end{bmatrix}}_{=:\mathbf{A}_{cl}} \begin{bmatrix} \mathbf{x}_p(t) \\ \mathbf{x}_m(t) \\ \mathbf{x}_s(t) \end{bmatrix} + \begin{bmatrix} \mathbf{B}_p\mathbf{D}_0 \\ \mathbf{B}_m \\ \mathbf{B}_0 \end{bmatrix} \mathbf{r}(t),$$

$$(9.160)$$

$$\mathbf{y}_p(t) = \begin{bmatrix} \mathbf{C}_p & 0 & 0 \end{bmatrix} \begin{bmatrix} \mathbf{x}_p(t) \\ \mathbf{x}_m(t) \\ \mathbf{x}_s(t) \end{bmatrix}.$$

Finally, the eigenvalues of $\mathbf{A}_{cl}$, the closed-loop system, are verified to be

$$\{-1, \quad -1 \pm j, \quad -1 \pm j2, \quad -4\} \quad (9.161)$$

as predicted by the theory.

MATLAB/Simulink Solution
Using the Simulink model in Figure 9.18 with the plant in Equation (9.145) and the controller in Equation (9.159), the output responses are obtained as in Figure 9.21, verifying asymptotic tracking and disturbance rejection.

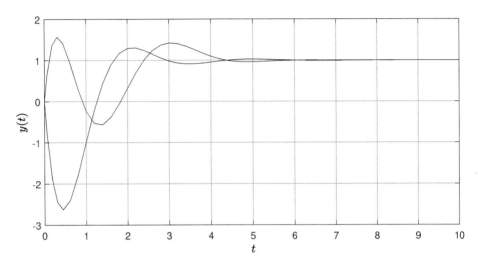

Figure 9.21 Output response of reduced-order observer-based servomechanism (Example 9.5)

9.9   Exercises

9.1   Consider the plant

$$P(s) = \frac{1}{s-1}. \tag{9.162}$$

Design a pole placement controller so that the closed-loop system tracks and rejects steps, ramps, and sinusoids with frequency of $\omega_0 = 1$ rad/s.

9.2   Consider the plant

$$\dot{\mathbf{x}}(t) = \begin{bmatrix} 0 & 1 & 0 & 1 \\ 1 & 0 & 1 & 1 \\ 0 & 0 & 0 & 1 \\ 0 & 0 & 0 & 0 \end{bmatrix} \mathbf{x}(t) + \begin{bmatrix} 1 & 0 \\ 0 & 0 \\ 0 & 0 \\ 0 & 1 \end{bmatrix} \mathbf{u}(t),$$

$$\mathbf{y}(t) = \begin{bmatrix} 1 & 0 & 0 & 0 \\ 0 & 0 & 0 & 1 \end{bmatrix} \mathbf{x}(t). \tag{9.163}$$

Design an LQR optimal servocontroller to make $y_i(t)$ track arbitrary steps $r_i(t)$, for $t \in 2$. Simulate and improve your design over the free parameters.

9.3   The transfer function of a plant **P** is

$$\mathbf{P}(s) = \begin{bmatrix} \dfrac{1}{s-1} & \dfrac{1}{s+1} \\ \dfrac{1}{(s-1)(s+1)} & \dfrac{1}{s+1} \end{bmatrix}. \tag{9.164}$$

Design an LQR optimal servocontroller to make each output track independent steps. Simulate the performance of your controller and improve the performance by varying the parameters **Q** and **R** of the performance index.

9.4   In Exercises 9.2 and 9.3, first fix **Q** and **R** and design the nominal controller. Then replace **R** by $\rho\mathbf{R}$ and call the corresponding controller $\mathbf{C}(\rho)$. Find the gain margin $g(\rho)$ and phase margin $\phi(\rho)$ and construct a graph of $g(\rho)$, $\phi(\rho)$ vs. ρ for $\rho \in (0,1]$. Select a prescribed pair of feasible margins and recover the corresponding controller.

9.5   The plant transfer function is

$$\mathbf{y}_p(s) = \begin{bmatrix} \dfrac{1}{s+1} & \dfrac{2}{s+1} \\ \dfrac{2}{s+1} & \dfrac{1}{s+1} \end{bmatrix} \mathbf{u}_p(s). \tag{9.165}$$

Design a controller to make each plant output track independent steps and reject step disturbances:

(a)   using LQR theory;

(b)   using eigenvalue assignment.

In each case, improve the error response by choosing various weights in the LQR performance index in (a), and choosing various closed-loop eigenvalues in (b). Plot the error responses using MATLAB and Simulink.

9.6   Consider a discrete-time plant with transfer function

$$y_p(z) = \frac{1}{z^2 - 0.25} u_p(z). \tag{9.166}$$

Design a controller to make the output track step inputs using:

(a)   an observer-based servocontroller;

(b)   a dead-beat controller, that is one where the error is zeroed in a finite number of steps.

9.7 The state-space model of a plant is

$$
\dot{\mathbf{x}}_p(t) = \begin{bmatrix} 0 & 0 & 1 \\ 1 & 0 & 0 \\ 0 & 1 & 1 \end{bmatrix} \mathbf{x}_p(t) + \begin{bmatrix} 0 & 0 \\ 0 & 1 \\ 1 & 0 \end{bmatrix} \mathbf{u}_p(t),
$$

$$
\mathbf{y}_p(t) = \begin{bmatrix} 1 & 0 & 0 \\ 0 & 1 & 0 \end{bmatrix} \mathbf{x}_p(t). \tag{9.167}
$$

Design a controller to make $\mathbf{y}_p(t)$ track independent step inputs using:
(a) an optimal Pearson compensator;
(b) observer theory.

9.8 Repeat Exercise 9.5 except that each plant output is now required to track steps and ramps.

9.9 Repeat Exercise 9.7 with the requirement that step inputs must be tracked and sinusoidal disturbances of radian frequency $\omega_0 = 1$ rad/s rejected. Use MATLAB for your calculations and Simulink to verify your design.

9.10   Notes and References

A special case of the servomechanism problem, of practical importance, occurs when the exogenous signals are steps. This has been intensively studied under the label proportional integral derivative (PID) control. A recent monograph by Diaz-Rodriguez, Han, and Bhattacharyya (2019) has many new results on PID control. See also Chapter 11 of this book. The servomechanism problem was initially solved for single-output multi-input plants in Bhattacharyya and Pearson (1970). It was first studied using geometric control theory as a problem of zeroing the output in Bhattacharyya, Pearson, and Wonham (1972) and Bhattacharyya and Pearson (1972). The problem of zeroing the output using bounded control energy was solved in Bhattacharyya (1973). A non-unity feedback servomechanism which is nevertheless plant-robust was studied in Howze and Bhattacharyya (1997). A clear treatment of the multivariable servomechanism using state feedback was given in Desoer and Wang (1980). The output feedback solution of the multivariable servomechanism problem was given in Davison (1976) and Francis and Wonham (1975). The servocontroller or signal generator in our treatment was called an internal model in the latter reference. The notion of blocking zeros was introduced in Ferreira and Bhattacharyya (1977), where their role in servomechanisms was clarified. Section 9.6 is adapted from Bhattacharyya, Datta, and Keel (2009), with permission. © 2008 Taylor & Francis LLC Books.

10 Robustness and Fragility of Control Systems

In this chapter we provide an introduction to the main results on robust control as developed intensively since about 1980. The main concern here is the preservation of stability of a control system under perturbations of its parameters and model descriptions. Two distinct lines of research have emerged. The first of these is based on quantifying transfer function perturbations using, for example, the H_∞ norm. This class of uncertainties are called unstructured perturbations. The second approach is based on the study of stability of a system under real parameter perturbations. In this area there are several elegant extremal results, such as the Generalized Kharitonov Theorem, the Edge Theorem, and the Mapping Theorem, which are described.

In each of the above approaches, the objective is to determine the stability margin, that is the largest ball of perturbations, in the appropriate space, that preserves stability. We show how these margins can be determined. These results are ultimately, generalizations of the notions of classical stability margins, namely gain and phase margins and similarly to these margins can be aids to the selection or design of controllers. As one important application it is shown that controllers of high order relative to the plant, as widely proposed since the 1980s, are inherently *fragile* and possess miniscule margins with respect to their own parameters, rendering them dysfunctional. Indeed, this result validates the conventional wisdom, adopted by industry, of using low-order controllers such as PIDs whenever possible, as they can be inherently more robust.

The goal of this chapter is to provide an introduction to the main results, and rigorous proofs are avoided in favor of intuitive ones.

10.1   Introduction

The robustness of a system is its ability to remain functional under perturbations. Control systems employ feedback and are thus prone to instability. Therefore, robustness of a control system most often refers to its ability to remain stable under perturbations. This property of a system is called *robust stability*. Since perturbations are unknown and unpredictable, it is important to know a priori a class of perturbations under which the system remains stable. The stability margin is a quantitative description of the largest such class of perturbations which preserves stability. Being the largest, it is a legitimate quantitative measure that may be used to compare the robustness of two or more systems, or controller designs.

10.1.1 Classical and H_∞ Stability Margins

In single-input single-output (SISO) classical control theory, we consider the closed-loop system in Figure 10.1 with $g_0(s)$, the nominal transfer function, being real rational and proper.

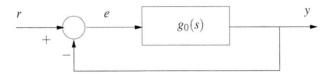

Figure 10.1 Nominal SISO control system

The stability of the nominal closed-loop system is indicated by the zeros of the characteristic equation:

$$1 + g_0(s) = 0, \tag{10.1}$$

which must lie in the open left-half plane $\mathbb{C}^-$ for stability, in the case of continuous-time systems. This is referred to as *Hurwitz stability*. Assuming nominal system stability ($1 + g_0(s)$ is Hurwitz), we consider additive perturbations of $g_0(s)$:

$$g_0(s) \quad \longrightarrow \quad g_0(s) + \Delta(s), \tag{10.2}$$

where $\Delta(s)$ is typically allowed to be real, rational, proper, and stable. Then robust stability is determined by the roots of

$$1 + g_0(s) + \Delta(s) = 0 \tag{10.3}$$

or

$$(1 + g_0(s)) \left(1 + (1 + g_0(s))^{-1} \Delta(s) \right) = 0. \tag{10.4}$$

As the nominal system is stable ($1 + g_0(s)$ is Hurwitz), robust stability requires that:

$$1 + (1 + g_0(s))^{-1} \Delta(s) \tag{10.5}$$

be Hurwitz, that is, have zeros or characteristic roots in $\mathbb{C}^-$. Let

$$h_0(s) := (1 + g_0(s))^{-1} \tag{10.6}$$

so that Equation (10.5) can be rewritten as

$$1 + h_0(s)\Delta(s) = 0. \tag{10.7}$$

Since $h_0(s)$ and $\Delta(s)$ are stable, robust stability is equivalent to the requirement that the Nyquist plots of $h_0(s)$ and $h_0(s)\Delta(s)$ have the same number of encirclements of the $-1 + j0$ point.

The above requirement holds if and only if

$$1 + h_0(j\omega)\Delta(j\omega) \neq 0, \qquad \text{for all } \omega \in [-\infty, +\infty] \tag{10.8}$$

or

$$|\Delta(j\omega)| < \frac{1}{|h_0(j\omega)|}, \qquad \text{for all } \omega \in \mathbb{R}. \tag{10.9}$$

Let ω_0 denote the frequency where

$$|h_0(j\omega)| = |1 + g_0(j\omega)|^{-1} \tag{10.10}$$

attains its supremum or maximum value, that is, equivalently

$$\text{Arg}_\omega \sup |1 + g_0(j\omega)| = \omega_0 \tag{10.11}$$

or $|1 + g_0(j\omega)|$ attains its infimum or minimum value at $\omega = \omega_0$. Thus the closed-loop is robustly stable for all perturbations $\Delta(s)$ that are rational, stable, proper, and satisfy

$$|\Delta(j\omega)| < |1 + g(j\omega_0)|. \tag{10.12}$$

This is illustrated in Figure 10.2 in the Nyquist plane where the minimum "distance" of the Nyquist plot from $-1 + j0$ is the unstructured stability margin.

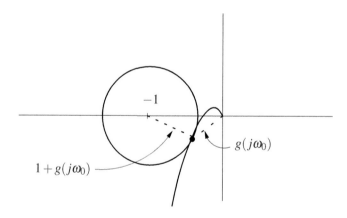

Figure 10.2 SISO H_∞ stability margin

This stability margin in some sense extends the classical notions of gain and phase margins and protects against the possibility that a system could be "close" to instability despite possessing good gain and phase margins.

In subsequent sections, this result will be extended to multivariable systems using the H_∞ norm.

10.1.2   Parametric Stability Margins

In this approach, one considers a control system containing a vector of real parameters $\mathbf{p}$, with nominal value $\mathbf{p}^0$, which is subject to perturbations. The stability of the control system is characterized by a parameter-dependent polynomial $\delta(s, \mathbf{p})$ with $\delta(s, \mathbf{p}^0)$, the nominal polynomial, assumed to be stable. Robust stability amounts to the condition that $\delta(s, \mathbf{p})$ remains stable for a ball of parameters $\mathbf{p}$, satisfying

$$\left\| \mathbf{p} - \mathbf{p}^0 \right\| < \rho \qquad (10.13)$$

in a chosen norm $\| \cdot \|$. The parametric stability margin is the largest value of ρ for which robust stability holds. There is a substantial body of useful results in the parametric uncertainty area, and these will be described in later sections of this chapter.

10.2 Unstructured Perturbations

In this section, we introduce some tools to measure the size or norm of transfer functions. Suppose that $h(s)$ is a real, rational, proper, stable (poles in $\mathbb{C}^-$) scalar transfer function. Then the ∞-norm of $h(s)$ is

$$\|h\|_\infty := \sup_{s \in \mathbb{C}^+} |h(s)| \qquad (10.14a)$$

$$= \sup_{\omega \in \mathbb{R}} |h(j\omega)|, \qquad (10.14b)$$

where the second equation follows from the Maximum Modulus Theorem for functions which are analytic over the closed right-half plane (denoted $\mathbb{C}^+$).

To extend the definition of the ∞-norm to transfer function matrices, we need the concept of singular value decomposition (SVD).

10.2.1 Singular Value Decomposition

The SVD of a complex $m \times n$ matrix $\mathbf{A}$ of rank r is the representation

$$\mathbf{A} = \mathbf{U} \Sigma \mathbf{V}^*, \qquad (10.15)$$

where $\mathbf{U}$ and $\mathbf{V}$ are $m \times m$ and $n \times n$ complex, unitary matrices ($\mathbf{U}^*\mathbf{U} = \mathbf{I}$, $\mathbf{V}^*\mathbf{V} = \mathbf{I}$, $*$ denotes conjugate transpose)

$$\Sigma = \begin{bmatrix} \sigma_1 & & & \vdots & 0 \\ & \ddots & & & \vdots \\ & & \sigma_r & \vdots & 0 \\ \hdashline 0 & \cdots & 0 & \vdots & 0 \end{bmatrix}, \qquad (10.16)$$

where σ_i are real, unique, and arranged in the order

$$\sigma_1 \geq \sigma_2 \geq \cdots \geq \sigma_r \geq 0. \qquad (10.17)$$

The representation in Equation (10.15) always exists and the notation

$$\sigma_1 := \sigma_{\max}(\mathbf{A}) =: \bar{\sigma}(\mathbf{A}), \qquad (10.18a)$$

$$\sigma_r := \sigma_{\min}(\mathbf{A}) =: \underline{\sigma}(\mathbf{A}), \qquad (10.18b)$$

will be freely used. It is easy to show that σ_1 is the induced 2-norm of $\mathbf{A}$, that is

$$\|\mathbf{A}\|_2 := \sup_{\|\mathbf{x}\|_2=1} \|\mathbf{Ax}\|_2 = \sigma_1. \tag{10.19}$$

We now turn to transfer function matrices with entries that are real, rational, proper, and stable (poles in $\mathbb{C}^-$). The ∞-norm of such a matrix $\mathbf{H}(s)$ is defined to be

$$\|\mathbf{H}(s)\|_\infty := \sup_{s \in \mathbb{C}^+} \bar{\sigma}[\mathbf{H}(s)] \tag{10.20}$$

$$= \sup_{\omega \in \mathbb{R}} \bar{\sigma}[\mathbf{H}(j\omega)], \tag{10.21}$$

where the second equality follows from an extension of the Maximum Modulus Theorem to matrices of functions which are analytic over $\mathbb{C}^+$. If the supremum in Equation (10.21) is attained at $\omega = \omega_0$, we have using Equation (10.15)

$$\mathbf{H}(j\omega_0) = \sum_{i=1}^{r} \sigma_i \mathbf{U}_u(j\omega_0) \mathbf{V}^*(j\omega_0) \tag{10.22}$$

where

$$\sigma_1 = \bar{\sigma}\left(\mathbf{H}(j\omega_0)\right) =: \|\mathbf{H}(s)\|_\infty. \tag{10.23}$$

With these preliminaries, we can now develop some basic results on robust stability of multivariable control systems subject to unstructured perturbations.

10.3　Stability Margins Against Unstructured Perturbations

The main results on robust stability can easily be derived from a fundamental theorem called the Small Gain Theorem.

10.3.1　Small Gain Theorem

Suppose $\mathbf{H}(s)$, a square matrix with entries that are real, rational, proper, and stable transfer functions, is perturbed by arbitrary real, rational, stable proper feedback matrices $\Delta(s)$ as shown in Figure 10.3.

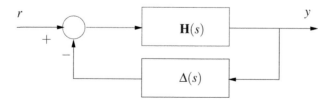

Figure 10.3 Small Gain Theorem

The question addressed by the Small Gain Theorem is: How large can $\Delta(s)$ be without destabilizing the closed-loop system of Figure 10.3? Clearly, the size of $\Delta(s)$ can be measured by $\|\Delta(s)\|_\infty$ and

$$B_\rho := \{\Delta(s) \ : \ \|\Delta(s)\|_\infty < \rho\} \tag{10.24}$$

is a ball of perturbations of radius ρ.

THEOREM 10.1 (Small Gain Theorem)   The closed-loop system of Figure 10.3 is robustly stable if

$$\|\Delta(s)\|_\infty < \frac{1}{\|\mathbf{H}(s)\|_\infty} =: \rho^* \tag{10.25}$$

and B_{ρ^*} is the largest ball of perturbations preserving robust stability of the feedback system.

Proof   The characteristic equation of the closed-loop system in Figure 10.3 is

$$\det[\mathbf{I} + \mathbf{H}(s)\Delta(s)] = 0. \tag{10.26}$$

To prove that Equation (10.26) has no $\mathbb{C}^+$ zeros, it suffices to show that

$$\inf_{s \in \mathbb{C}^+} \underline{\sigma}[\mathbf{I} + \mathbf{H}(s)\Delta(s)] > 0. \tag{10.27}$$

For robust stability, it will suffice to show that Equation (10.27) holds for all $\Delta(s)$ satisfying Equation (10.25). Since

$$\inf_{s \in \mathbb{C}^+} \underline{\sigma}[\mathbf{I} + \mathbf{H}(s)\Delta(s)] \geq 1 - \sup_{s \in \mathbb{C}^+} \bar{\sigma}[\mathbf{H}(s)\Delta(s)] \tag{10.28a}$$

$$\geq 1 - \sup_{\omega} \bar{\sigma}[\mathbf{H}(j\omega)\Delta(j\omega)] \tag{10.28b}$$

$$\geq 1 - \|\mathbf{H}(s)\Delta(s)\|_\infty \tag{10.28c}$$

$$\geq 1 - \|\mathbf{H}(s)\|_\infty \|\Delta(s)\|_\infty \tag{10.28d}$$

$$> 0 \quad \text{(by Equation (10.25))}, \tag{10.28e}$$

Equation (10.27) holds for all $\Delta(s)$ satisfying Equation (10.25), proving robust stability.
    To prove that B_{ρ^*} is the largest ball of perturbations, it is enough to construct a destabilizing perturbation $\Delta_0(s)$ with

$$\|\Delta_0(s)\|_\infty = \rho^* = \frac{1}{\|\mathbf{H}(s)\|_\infty}. \tag{10.29}$$

To construct such a $\Delta_0(s)$, suppose that

$$\sup_{\omega \in \mathbb{R}} \bar{\sigma}(\mathbf{H}(j\omega)) \tag{10.30}$$

is achieved at $\omega = \omega_0 \in [0, \infty]$, and let the SVD of $\mathbf{H}(j\omega_0)$ be

$$\mathbf{H}(j\omega_0) = \mathbf{U}(j\omega_0)\Sigma\mathbf{V}^*(j\omega_0), \tag{10.31}$$

$$\Sigma = \begin{bmatrix} \sigma_1 & & & \\ & \sigma_2 & & \\ & & \ddots & \end{bmatrix}, \tag{10.32}$$

$$\mathbf{U}(j\omega_0) = [\mathbf{u}_1(j\omega_0) \ \mathbf{u}_2(j\omega_0) \ \cdots], \tag{10.33}$$

$$\mathbf{V}(j\omega_0) = [\mathbf{v}_1(j\omega_0) \ \mathbf{v}_2(j\omega_0) \ \cdots], \tag{10.34}$$

$$\sigma_1 = \frac{1}{\rho^*}. \tag{10.35}$$

Then choose $\Delta_0(s)$ so that Equation (10.29) is satisfied and

$$\Delta_0(j\omega_0) = -\rho^*\mathbf{v}_1(j\omega_0)\mathbf{u}_1^*(j\omega_0). \tag{10.36}$$

If $\omega_0 = 0$ or ∞, then $\mathbf{U}(j\omega_0)$ and $\mathbf{V}(j\omega_0)$ in Equation (10.31) are real and with

$$\Delta_0(s) = -\rho^*\mathbf{v}_1\mathbf{u}_1^T \tag{10.37}$$

satisfy Equation (10.29) and

$$\det[\mathbf{I} + \mathbf{H}(0)\Delta_0(0)] = 0 \tag{10.38a}$$

or

$$\det[\mathbf{I} + \mathbf{H}(\infty)\Delta_0(\infty)] = 0. \tag{10.38b}$$

So $\Delta_0(s)$ is destabilizing.

If $\omega_0 \in (0, \infty)$, then Equation (10.37) would be complex. To construct a real destabilizing matrix $\Delta_0(s)$, write

$$\mathbf{v}_1(j\omega_0) = \begin{bmatrix} v_{11}e^{j\theta_1} \\ v_{12}e^{j\theta_2} \\ \vdots \\ v_{1p}e^{j\theta_p} \end{bmatrix}, \quad \mathbf{u}_1^*(j\omega_0) = \begin{bmatrix} u_{11}e^{j\phi_1} & u_{12}e^{j\phi_2} & \cdots & u_{1q}e^{j\phi_q} \end{bmatrix}, \tag{10.39}$$

so that

$$\theta_i \in [-\pi, 0] \quad \text{and} \quad \phi_j \in [-\pi, 0] \tag{10.40}$$

and v_{1i} and u_{1j} are real and positive or negative.

Next, determine real $\alpha_i > 0$ and $\beta_j > 0$ so that

$$\theta_i = \frac{\alpha_i - j\omega_0}{\alpha_i + j\omega_0}, \quad \phi_j = \frac{\beta_j - j\omega_0}{\beta_j + j\omega_0}, \tag{10.41}$$

which is always possible by Equation (10.40). Now let

$$
\mathbf{v}_1(s) :=
\begin{bmatrix}
v_{11}\left(\frac{\alpha_1 - s}{\alpha_1 + s}\right) \\[6pt]
v_{12}\left(\frac{\alpha_2 - s}{\alpha_2 + s}\right) \\[6pt]
\vdots \\[6pt]
v_{1p}\left(\frac{\alpha_p - s}{\alpha_p + s}\right)
\end{bmatrix},
\tag{10.42}
$$

$$
\mathbf{u}_1^T(s) := \begin{bmatrix} u_{11}\left(\frac{\beta_1 - s}{\beta_1 + s}\right) & u_{12}\left(\frac{\beta_2 - s}{\beta_2 + s}\right) & \cdots & u_{1q}\left(\frac{\beta_q - s}{\beta_q + s}\right) \end{bmatrix},
\tag{10.43}
$$

and choose

$$
\Delta_0(s) = -\rho^* \mathbf{v}_1(s) \mathbf{u}_1^T(s).
\tag{10.44}
$$

With this choice, we have

$$
\det\left[\mathbf{I} + \mathbf{H}(j\omega_0)\Delta_0(j\omega_0)\right] = 0
\tag{10.45}
$$

so that $\Delta_0(s)$ is real and $\|\Delta_0\|_\infty = \rho^*$. This proves that B_{ρ^*} is the largest open ball preserving robust stability. ∎

10.3.2   Additive Perturbations

The Small Gain Theorem can be used to derive stability margins of control systems under unstructured matrix perturbations.

Consider the system in Figure 10.4 and assume that the nominal closed loop is stable.

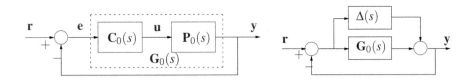

Figure 10.4 Nominal and perturbed systems

Now consider an additive matrix perturbation $\Delta_a(s)$ of $\mathbf{G}_0(s)$, with $\Delta_a(s)$ being real, rational, stable, and proper, resulting in

$$
\mathbf{G}(s) = \mathbf{G}_0(s) + \Delta_a(s).
\tag{10.46}
$$

The perturbed system is stable if and only if

$$
\det\left[\mathbf{I} + \mathbf{G}_0(s) + \Delta(s)\right]
\tag{10.47}
$$

is Hurwitz (has all zeros in $\mathbb{C}^-$).

Write Equation (10.47) as

$$\det\left[\mathbf{I}+\mathbf{G}_0(s)\right]\det\left[\mathbf{I}+\underbrace{(\mathbf{I}+\mathbf{G}_0(s))^{-1}}_{\mathbf{H}_a(s)}\Delta_a(s)\right] \qquad (10.48)$$

which shows, since $\det\left[\mathbf{I}+\mathbf{G}_0(s)\right]$ is Hurwitz, that for robust stability

$$\det\left[\mathbf{I}+\mathbf{H}_a(s)\Delta_a(s)\right] \qquad (10.49)$$

must be Hurwitz, with $\mathbf{H}_a(s)$ and $\Delta_a(s)$ stable.

By the Small Gain Theorem, it follows that the additively perturbed system is robustly stable if and only if

$$\|\Delta_a(s)\|_\infty < \frac{1}{\|\mathbf{H}_a(s)\|_\infty} =: \rho_a^*. \qquad (10.50)$$

Clearly, ρ_a^* is the stability margin for additive perturbations since there does exist a real, rational, stable $\Delta_a^0(s)$ with

$$\|\Delta_a^0(s)\|_\infty = \rho_a^* \qquad (10.51)$$

which destabilizes the system.

10.3.3 Multiplicative Perturbations

Consider now the question of robust stability of a control system under norm-bounded multiplicative perturbations. Here the nominal forward transfer function $\mathbf{G}_0(s)$ perturbs to

$$\mathbf{G}(s) = (\mathbf{I}+\Delta_m(s))\,\mathbf{G}_0(s) \qquad (10.52)$$

and the corresponding block diagram is shown in Figure 10.5.

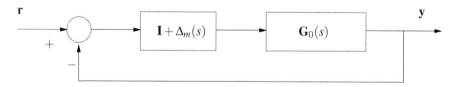

Figure 10.5 Multiplicative perturbations

For robust stability

$$\det\left[\mathbf{I}+\mathbf{G}_0(s)\,(\mathbf{I}+\Delta_m(s))\right] \qquad (10.53)$$

must be Hurwitz. Since $\det\left[\mathbf{I}+\mathbf{G}_0(s)\right]$ is Hurwitz, this is equivalent to

$$\det\left[\mathbf{I}+(\mathbf{I}+\mathbf{G}_0(s))^{-1}\mathbf{G}_0(s)\Delta_m(s)\right] \qquad (10.54)$$

being Hurwitz. Since

$$\mathbf{H}_m(s) := (\mathbf{I} + \mathbf{G}_0(s))^{-1}\mathbf{G}_0(s) \tag{10.55}$$

is stable, it follows from the Small Gain Theorem that robust stability will hold for all stable, real, rational, proper $\Delta_m(s)$ if and only if

$$\|\Delta_m(s)\|_\infty < \frac{1}{\|\mathbf{H}_m(s)\|_\infty} =: \rho_m^*. \tag{10.56}$$

As before, ρ_m^* is a stability margin under multiplicative perturbations.

EXAMPLE 10.1

Additive Perturbations

Consider the system shown in Figure 10.4 and let

$$\mathbf{G}_0(s) = \begin{bmatrix} \dfrac{1}{s+1} & \dfrac{1}{s+2} \\[2mm] \dfrac{1}{s+3} & \dfrac{1}{s+4} \end{bmatrix}. \tag{10.57}$$

Since $\det[\mathbf{I} + \mathbf{G}_0(s)]$ is Hurwitz, the closed-loop system is stable. We can now calculate the maximum additive perturbation described by its H_∞ norm that guarantees closed-loop stability by using Equations (10.50) and (10.51). Figure 10.6 generated by the following MATLAB script illustrates this.

Multiplicative Perturbations

We now calculate the maximum multiplicative perturbation described by its H_∞ norm that guarantees the closed-loop stability by using Equations (10.55) and (10.56).

MATLAB Solution

```
clear
w=logspace(-2,2);
G11d=[1 1];    G12d=[1 2];    G21d=[1 3];    G22d=[1 4];
for k=1:length(w)
  G0=[1/polyval(G11d,1i*w(k))  1/polyval(G12d,1i*w(k));
        1/polyval(G21d,1i*w(k))  1/polyval(G22d,1i*w(k))];
  Ha=inv(eye(2)+G0);
  rho_a(k)=1/max(svd(Ha));
  Hm=inv(eye(2)+G0)*G0;
  rho_m(k)=1/max(svd(Hm));
end
subplot(1,2,1), semilogx(w,rho_a)
xlabel('\omega'),    ylabel('\rho_a')
rhoastar=min(rho_a)
subplot(1,2,2), semilog(w,rho_m)
xlabel('\omega'),    ylabel('\rho_m')
rhomstar=min(rho_m)
```

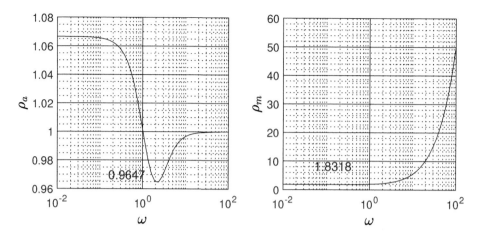

Figure 10.6 Stability margin: additive perturbation (left), multiplicative perturbation (right)

10.3.4   Remarks

In our analysis of robust stability, we assumed that the controller C was given. In synthesis problems, there may be a choice of several admissible controllers or more commonly a continuum of controllers that stabilizes the nominal system. The stability margins calculated above can then be a guide to design. We also point out that our analysis assumes that the plant-controller cascade is subject to uncertainty. This corresponds to the classical control scenario. In many modern treatments, the plant only is considered to be prone to perturbation. The Small Gain Theorem can also be employed in such situations in an obvious way. Finally, it is worth noting that the Small Gain Theorem may be considered a robustification of the classical Nyquist criterion.

10.4   The Stability Ball in Parameter Space

In this section we consider uncertainty in physical parameters, and give a useful characterization of the parametric stability margin in the general case. This can be done by finding the largest stability ball in parameter space, centered at a "stable" nominal parameter value $\mathbf{p}^0$. Let $\mathscr{S} \subset \mathscr{C}$ denote as usual an open set which is symmetric with respect to the real axis. $\mathscr{S}$ denotes the stability region of interest. In continuous-time systems $\mathscr{S}$ may be the open left half-plane or subsets thereof. For discrete-time systems $\mathscr{S}$ is the open unit circle or a subset of it. Now let $\mathbf{p}$ be a vector of real parameters,

$$\mathbf{p} = \begin{bmatrix} p_1 & p_2 & \cdots & p_l \end{bmatrix}^T. \tag{10.58}$$

The characteristic polynomial of the system is denoted by

$$\delta(s, \mathbf{p}) = \delta_n(\mathbf{p})s^n + \delta_{n-1}(\mathbf{p})s^{n-1} + \cdots + \delta_0(\mathbf{p}). \tag{10.59}$$

The polynomial $\delta(s, \mathbf{p})$ is assumed to be a real polynomial with coefficients that depend continuously on the real parameter vector $\mathbf{p}$. We suppose that for the *nominal parameter* $\mathbf{p} = \mathbf{p}^0$, $\delta(s, \mathbf{p}^0) := \delta^0(s)$ is stable with respect to $\mathscr{S}$ (has its roots in $\mathscr{S}$). Write

$$\triangle\mathbf{p} := \mathbf{p} - \mathbf{p}^0 = \begin{bmatrix} p_1 - p_1^0 & p_2 - p_2^0 & \cdots & p_l - p_l^0 \end{bmatrix}^T \tag{10.60}$$

to denote the perturbation in the parameter $\mathbf{p}$ from its nominal value $\mathbf{p}^0$. Now let us introduce a norm $\|\cdot\|$ in the space of the parameters $\mathbf{p}$ and introduce the open ball of radius ρ in this norm:

$$\mathscr{B}(\rho, \mathbf{p}^0) = \{\mathbf{p} : \|\mathbf{p} - \mathbf{p}^0\| < \rho\}. \tag{10.61}$$

The hypersphere of radius ρ is defined by

$$\mathscr{S}(\rho, \mathbf{p}^0) = \{\mathbf{p} : \|\mathbf{p} - \mathbf{p}^0\| = \rho\}. \tag{10.62}$$

With the ball $\mathscr{B}(\rho, \mathbf{p}^0)$ we associate the family of uncertain polynomials:

$$\Delta_\rho(s) := \{\delta(s, \mathbf{p}^0 + \Delta\mathbf{p}) : \|\Delta\mathbf{p}\| < \rho\}. \tag{10.63}$$

DEFINITION 10.1   The *real parametric stability margin* real parametric stability margin in parameter space is defined as the radius, denoted $\rho^*(\mathbf{p}^0)$, of the *largest* ball centered at $\mathbf{p}^0$ for which $\delta(s, \mathbf{p})$ remains stable whenever $\mathbf{p} \in \mathscr{B}(\rho^*(\mathbf{p}^0), \mathbf{p}^0)$.

This stability margin then tells us how much we can perturb the original parameter $\mathbf{p}^0$ and yet remain stable. Our first result is a characterization of this maximal stability ball. To simplify the notation, we write ρ^* instead of $\rho^*(\mathbf{p}^0)$.

THEOREM 10.2   With the assumptions as above, the parametric stability margin ρ^* is characterized by:

(a) There exists a largest stability ball $\mathscr{B}(\rho^*, \mathbf{p}^0)$ centered at $\mathbf{p}^0$, with the property that:
1. for every $\mathbf{p}'$ within the ball, the characteristic polynomial $\delta(s, \mathbf{p}')$ is stable and of degree n;
2. at least one point $\mathbf{p}''$ on the hypersphere $\mathscr{S}(\rho^*, \mathbf{p}^0)$ itself is such that $\delta(s, \mathbf{p}'')$ is unstable or of degree less than n.
(b) Moreover, if $\mathbf{p}''$ is any point on the hypersphere $\mathscr{S}(\rho^*, \mathbf{p}^0)$ such that $\delta(s, \mathbf{p}'')$ is unstable, then the unstable roots of $\delta(s, \mathbf{p}'')$ can only be on the stability boundary.

The proof is based on continuity of the roots on the parameter $\mathbf{p}$, which holds under the degree-invariant assumption, and is omitted.[1] This theorem gives the first simplification for the calculation of the parametric stability margin ρ^*. It states that to determine ρ^* it suffices to calculate the minimum "distance" of $\mathbf{p}^0$ from the set of those points $\mathbf{p}$ which endow the characteristic polynomial with a root on the stability boundary, or which cause loss of degree. This last calculation can be carried out using the complex plane image of the family of polynomials $\Delta_\rho(s)$ evaluated along the stability boundary. We will describe this in the next section.

---

[1] Readers interested in the proof of Theorem 10.2 should refer to Theorem 3.1 of Bhattacharyya, Chapellat, and Keel (1995).

The parametric stability margin or distance to instability is measured in the norm $\| \cdot \|$, and therefore the numerical value of ρ^* will depend on the specific norm chosen. We will consider, in particular, the weighted ℓ_p norms. These are defined as follows. Let $w = [w_1, w_2, \ldots, w_l]$ with $w_i > 0$ be a set of positive *weights*

$$\ell_1 \text{ norm} : \quad \|\Delta \mathbf{p}\|_1^w := \sum_{i=1}^{l} w_i |\Delta p_i|,$$

$$\ell_2 \text{ norm} : \quad \|\Delta \mathbf{p}\|_2^w := \sqrt{\sum_{i=1}^{l} w_i^2 \Delta p_i^2},$$

$$\ell_p \text{ norm} : \quad \|\Delta \mathbf{p}\|_p^w := \left[\sum_{i=1}^{l} |w_i \Delta p_i|^p \right]^{\frac{1}{p}}, \tag{10.64}$$

$$\ell_\infty \text{ norm} : \quad \|\Delta \mathbf{p}\|_\infty^w := \max_i w_i |\Delta p_i|.$$

We will write $\|\triangle \mathbf{p}\|$ to refer to a generic weighted norm when the weight and type of norm are unimportant.

10.4.1   The Image Set Approach: Zero Exclusion

The parametric stability margin may be calculated by using the complex plane image of the polynomial family $\Delta_\rho(s)$ evaluated at each point on the stability boundary $\partial \mathscr{S}$. This is based on the following idea. Suppose that the family has constant degree n and $\delta(s, \mathbf{p}^0)$ is stable but $\Delta_\rho(s)$ contains an unstable polynomial. Then the continuous dependence of the roots on $\mathbf{p}$ implies that there must also exist a polynomial in $\Delta_\rho(s)$ such that it has a root, at a point s^*, on the stability boundary $\partial \mathscr{S}$. In this case, *the complex plane image set $\Delta_\rho(s^*)$ must contain the origin of the complex plane*. This suggests that to detect the presence of instability in a family of polynomials of constant degree, we can generate the image set of the family at each point of the stability boundary and determine if the origin is included in or excluded from this set. This fact is known as the Zero Exclusion Principle.

THEOREM 10.3 (Zero Exclusion Principle)   For given $\rho \geq 0$ and $\mathbf{p}^0$, suppose that the family of polynomials $\Delta_\rho(s)$ is of constant degree and $\delta(s, \mathbf{p}^0)$ is stable. Then every polynomial in the family $\Delta_\rho(s)$ is stable with respect to $\mathscr{S}$ if and only if the complex plane image set $\Delta_\rho(s^*)$ excludes the origin for every $s^* \in \partial \mathscr{S}$.

Proof   As stated earlier, this is simply a consequence of the continuity of the roots of $\delta(s, \mathbf{p})$ on $\mathbf{p}$ and the Boundary Crossing Theorem (Theorem 2.10 in Chapter 2).   ∎

In fact, the above can be used as a computational tool to determine the maximum value ρ^* of ρ for which the family is stable. If $\delta(s, \mathbf{p}^0)$ is stable, it follows that there always exists an open stability ball around $\mathbf{p}^0$, since the stability region $\mathscr{S}$ is itself open. Therefore, for small values of ρ, the image set $\Delta_\rho(s^*)$ will exclude the origin for every point $s^* \in \partial \mathscr{S}$. As ρ is increased from zero, a *limiting* value ρ_0 may be reached where some polynomial in the corresponding family $\Delta_{\rho_0}(s)$ loses degree or a polynomial in the family acquires a root s^* on the stability boundary. From Theorem 10.2 it is clear that this value ρ_0 is equal to ρ^*, the stability margin. In case the limiting value ρ_0 is never achieved, the stability margin ρ^* is infinity.

An alternative way to determine ρ^* is as follows. Fixing s^* at a point in the boundary of $\mathscr{S}$, let $\rho_0(s^*)$ denote the limiting value of ρ such that $0 \in \Delta_\rho(s^*)$:

$$\rho_0(s^*) := \inf\left\{\rho \ : \ 0 \in \Delta_\rho(s^*)\right\}. \tag{10.65}$$

Then, we define

$$\rho_b = \inf_{s^* \in \partial\mathscr{S}} \rho_0(s^*). \tag{10.66}$$

In other words, ρ_b is the limiting value of ρ for which some polynomial in the family $\Delta_\rho(s)$ *acquires a root on the stability boundary* $\partial\mathscr{S}$. Also let the limiting value of ρ for which some polynomial in $\Delta_\rho(s)$ *loses degree* be denoted by ρ_d:

$$\rho_d := \inf\left\{\rho \ : \ \delta_n(\mathbf{p}^0 + \Delta\mathbf{p}) = 0, \ \ \|\Delta\mathbf{p}\| < \rho\right\}. \tag{10.67}$$

We have established the following theorem.

THEOREM 10.4   The parametric stability margin

$$\rho^* = \min\left\{\rho_b, \rho_d\right\}. \tag{10.68}$$

Remark 10.1   We note that this theorem remains valid even when the stability region $\mathscr{S}$ is not connected. For instance, one may construct the stability region as a union of disconnected regions $\mathscr{S}_i$ surrounding each root of the nominal polynomial. In this case the stability boundary must also consist of the union of the individual boundary components $\partial\mathscr{S}_i$. The functional dependence of the coefficients δ_i on the parameters $\mathbf{p}$ is also not restricted in any way, except for the assumption of continuity.

The above theorem shows that the problem of determining ρ^* can be reduced to the following steps:

(a) determine the "local" stability margin $\rho(s^*)$ at each point s^* on the boundary of the stability region;
(b) minimize the function $\rho(s^*)$ over the entire stability boundary and thus determine ρ_b;
(c) calculate ρ_d;
(d) set $\rho^* = \min\{\rho_b, \rho_d\}$.

In general, the determination of ρ^* is a *difficult nonlinear optimization* problem. However, the breakdown of the problem into the steps described above, exploits the structure of the problem and has the advantage that the local stability margin calculation $\rho(s^*)$, with s^* frozen, can be simple. In particular, when the parameters enter linearly or affinely into the characteristic polynomial coefficients, this calculation can be done in closed form. It reduces to a least-square problem for the ℓ_2 case, and equally simple linear programming or vertex problems for the ℓ_∞ or ℓ_1 cases. The dependence of ρ_b on s^* is in general highly nonlinear, but this part of the minimization can easily be performed computationally because sweeping the boundary $\partial \mathscr{S}$ is a *one-dimensional search*. In the next section, we describe and develop this calculation in greater detail.

10.4.2   Stability Margin Computation in the Affine Case

We develop explicit formulas for the parametric stability margin in the case in which the characteristic polynomial coefficients depend affinely on the uncertain parameters. In such cases we may write without loss of generality that

$$\delta(s, \mathbf{p}) = a_1(s)p_1 + \cdots + a_l(s)p_l + b(s), \tag{10.69}$$

where $a_i(s)$ and $b(s)$ are real polynomials and the parameters p_i are real. As before, we write $\mathbf{p}$ for the vector of uncertain parameters, $\mathbf{p}^0$ denotes the nominal parameter vector, and $\triangle \mathbf{p}$ the perturbation vector. In other words:

$$\begin{aligned}
\mathbf{p} &= \begin{bmatrix} p_1 & p_2 & \cdots & p_l \end{bmatrix}^T, \\
\mathbf{p}^0 &= \begin{bmatrix} p_1^0 & p_2^0 & \cdots & p_l^0 \end{bmatrix}^T, \\
\triangle \mathbf{p} &= \begin{bmatrix} p_1 - p_1^0 & p_2 - p_2^0 & \cdots & p_l - p_l^0 \end{bmatrix}^T \\
&= \begin{bmatrix} \triangle p_1 & \triangle p_2 & \cdots & \triangle p_l \end{bmatrix}^T.
\end{aligned} \tag{10.70}$$

Then the characteristic polynomial can be written as

$$\delta(s, \mathbf{p}^0 + \triangle \mathbf{p}) = \underbrace{\delta(s, \mathbf{p}^0)}_{\delta^0(s)} + \underbrace{a_1(s)\triangle p_1 + \cdots + a_l(s)\triangle p_l}_{\triangle \delta(s, \triangle \mathbf{p})}. \tag{10.71}$$

Now let s^* denote a point on the stability boundary $\partial \mathscr{S}$. For $s^* \in \partial \mathscr{S}$ to be a root of $\delta(s, \mathbf{p}^0 + \triangle \mathbf{p})$, we must have

$$\delta(s^*, \mathbf{p}^0) + a_1(s^*)\triangle p_1 + \cdots + a_l(s^*)\triangle p_l = 0. \tag{10.72}$$

We rewrite this equation introducing the weights $w_i > 0$:

$$\delta(s^*, \mathbf{p}^0) + \frac{a_1(s^*)}{w_1} w_1 \triangle p_1 + \cdots + \frac{a_l(s^*)}{w_l} w_l \triangle p_l = 0. \tag{10.73}$$

Obviously, the minimum $\|\triangle \mathbf{p}\|^w$ norm solution of this equation gives us $\rho(s^*)$, the calculation involved in step (a) in the last section:

$$\rho(s^*) = \inf \left\{ \|\triangle \mathbf{p}\|^w : \delta(s^*, \mathbf{p}^0) + \frac{a_1(s^*)}{w_1} w_1 \triangle p_1 + \cdots + \frac{a_l(s^*)}{w_l} w_l \triangle p_l = 0 \right\}. \tag{10.74}$$

Similarly, corresponding to loss of degree, we have the equation

$$\delta_n(\mathbf{p}^0 + \triangle \mathbf{p}) = 0. \tag{10.75}$$

Letting a_{in} denote the coefficient of the nth-degree term in the polynomial $a_i(s)$, $i = 1, 2, \ldots, l$, the above equation becomes

$$\underbrace{a_{1n} p_1^0 + a_{2n} p_2^0 + \cdots + a_{ln} p_l^0}_{\delta_n(\mathbf{p}^0)} + a_{1n}\triangle p_1 + a_{2n}\triangle p_2 + \cdots + a_{ln}\triangle p_l = 0. \tag{10.76}$$

We can rewrite this after introducing the weight $w_i > 0$:

$$\underbrace{a_{1n}p_1^0 + a_{2n}p_2^0 + \cdots + a_{ln}p_l^0}_{\delta_n(\mathbf{p}^0)} + \frac{a_{1n}}{w_1}w_1\triangle p_1 + \frac{a_{2n}}{w_2}w_2\triangle p_2 + \cdots + \frac{a_{ln}}{w_l}w_l\triangle p_l = 0.$$

(10.77)

The minimum norm $\|\triangle\mathbf{p}\|^w$ solution of this equation gives us ρ_d.

We consider the above equations in some more detail. Recall that the polynomials are assumed to be real. Equation (10.77) is real and can be rewritten in the form

$$\underbrace{\left[\begin{array}{ccc} \dfrac{a_{1n}}{w_1} & \cdots & \dfrac{a_{ln}}{w_l} \end{array}\right]}_{\mathbf{A}_n} \underbrace{\left[\begin{array}{c} w_1\triangle p_1 \\ \vdots \\ w_l\triangle p_l \end{array}\right]}_{\mathbf{t}_n} = \underbrace{-\delta_n^0}_{\mathbf{b}_n}.$$

(10.78)

In Equation (10.73), two cases may occur depending on whether s^* is real or complex. If $s^* = s_r$ where s_r is real, we have the single equation

$$\underbrace{\left[\begin{array}{ccc} \dfrac{a_1(s_r)}{w_1} & \cdots & \dfrac{a_l(s_r)}{w_l} \end{array}\right]}_{\mathbf{A}(s_r)} \underbrace{\left[\begin{array}{c} w_1\triangle p_1 \\ \vdots \\ w_l\triangle p_l \end{array}\right]}_{\mathbf{t}(s_r)} = \underbrace{-\delta^0(s_r)}_{\mathbf{b}(s_r)}.$$

(10.79)

Let x_r and x_i denote the real and imaginary parts of a complex number x:

$$x = x_r + jx_i, \qquad \text{with } x_r, x_i \text{ real.}$$

(10.80)

Using this notation, we write

$$a_k(s^*) = a_{kr}(s^*) + ja_{ki}(s^*)$$

(10.81)

and

$$\delta^0(s^*) = \delta_r^0(s^*) + j\delta_i^0(s^*).$$

(10.82)

If $s^* = s_c$ where s_c is complex, Equation (10.73) is equivalent to two equations which can be written as follows:

$$\underbrace{\left[\begin{array}{ccc} \dfrac{a_{1r}(s_c)}{w_1} & \cdots & \dfrac{a_{lr}(s_c)}{w_l} \\ \dfrac{a_{1i}(s_c)}{w_1} & \cdots & \dfrac{a_{li}(s_c)}{w_l} \end{array}\right]}_{\mathbf{A}(s_c)} \underbrace{\left[\begin{array}{c} w_1\triangle p_1 \\ \vdots \\ w_l\triangle p_l \end{array}\right]}_{\mathbf{t}(s_c)} = \underbrace{\left[\begin{array}{c} -\delta_r^0(s_c) \\ -\delta_i^0(s_c) \end{array}\right]}_{\mathbf{b}(s_c)}.$$

(10.83)

These equations completely determine the parametric stability margin in any norm. Let $\mathbf{t}^*(s_c)$, $\mathbf{t}^*(s_r)$, and $\mathbf{t}_n^*$ denote the minimum norm solutions of Equations (10.83),

(10.79), and (10.78), respectively. Thus

$$\|\mathbf{t}^*(s_c)\| = \rho(s_c),$$
$$\|\mathbf{t}^*(s_r)\| = \rho(s_r),$$
$$\|\mathbf{t}_n^*\| = \rho_d.$$

(10.84)

If any of the above Equations (10.78)–(10.83) do not have a solution, the corresponding value of $\rho(\cdot)$ is set equal to infinity.

Let $\partial \mathscr{S}_r$ and $\partial \mathscr{S}_c$ denote the real and complex subsets of $\partial \mathscr{S}$:

$$\partial \mathscr{S} = \partial \mathscr{S}_r \cup \partial \mathscr{S}_c$$

(10.85)

and

$$\rho_r := \inf_{s_r \in \partial \mathscr{S}_r} \rho(s_r),$$
$$\rho_c := \inf_{s_c \in \partial \mathscr{S}_c} \rho(s_c),$$

(10.86)

Therefore,

$$\rho_b = \inf\{\rho_r, \rho_c\}.$$

(10.87)

We now consider the specific case of the ℓ_2 norm.

10.4.3 ℓ_2 Stability Margin

In this section we suppose that the length of the perturbation vector $\triangle \mathbf{p}$ is measured by a weighted ℓ_2 norm. That is, the minimum ℓ_2-norm solution of Equations (10.78), (10.79), and (10.83) is desired. Consider first Equation (10.83). Assuming that $\mathbf{A}(s_c)$ has full row rank (=2), the minimum norm solution vector $\mathbf{t}^*(s_c)$ can be calculated as follows:

$$\mathbf{t}^*(s_c) = \mathbf{A}^T(s_c) \left[\mathbf{A}(s_c)\mathbf{A}^T(s_c)\right]^{-1} \mathbf{b}(s_c).$$

(10.88)

Similarly, if Equations (10.79) and (10.78) are consistent (i.e., $\mathbf{A}(s_r)$ and $\mathbf{A}_n$ are nonzero vectors), we can calculate the solution as

$$\mathbf{t}^*(s_r) = \mathbf{A}^T(s_r) \left[\mathbf{A}(s_r)\mathbf{A}^T(s_r)\right]^{-1} \mathbf{b}(s_r),$$
$$\mathbf{t}_n^* = \mathbf{A}_n^T \left[\mathbf{A}_n\mathbf{A}_n^T\right]^{-1} \mathbf{b}_n.$$

(10.89)

If $\mathbf{A}(s_c)$ has less than full rank, the following two cases can occur.

Case 1: $\text{Rank}[\mathbf{A}(s_c)] = 0$
In this case the equation is inconsistent since $\mathbf{b}(s_c) \neq 0$ (otherwise $\delta^0(s_c) = 0$ and $\delta^0(s)$ would not be stable with respect to $\mathscr{S}$, since s_c lies on $\partial \mathscr{S}$). In this case, Equation (10.83) has no solution and we set

$$\rho(s_c) = \infty.$$

(10.90)

Case 2: $\text{Rank}[\mathbf{A}(s_c)] = 1$

In this case the equation is consistent if and only if

$$\text{rank} \begin{bmatrix} \mathbf{A}(s_c) & \mathbf{b}(s_c) \end{bmatrix} = 1. \tag{10.91}$$

If the above rank condition for consistency is satisfied, we replace the two equations in Equation (10.83) by a single equation and determine the minimum norm solution of this latter equation. If the rank condition for consistency does not hold, Equation (10.83) cannot be satisfied and we again set $\rho(s_c) = \infty$.

EXAMPLE 10.2 (ℓ_2 Schur stability margin)   Consider the discrete-time control system with the controller and plant specified, respectively, by their transfer functions:

$$C(z) = \frac{z+1}{z^2}, \qquad G(z, \mathbf{p}) = \frac{(-0.5 - 2p_0)z + (0.1 + p_0)}{z^2 - (1.4 + 0.4p_2)z + (1.6 + 10p_1 + 2p_0)}. \tag{10.92}$$

The characteristic polynomial of the closed-loop system is

$$\delta(z, \mathbf{p}) = z^4 - (1.4 + 0.4p_2)z^3 + (1.1 + 10p_1)z^2 - (0.4 + p_0)z + (0.1 + p_0). \tag{10.93}$$

The nominal value of $\mathbf{p}^0 = [p_0^0 \ p_1^0 \ p_2^0]^T = [0 \ 0.1 \ 1]^T$. The perturbation is denoted as usual by the vector

$$\Delta \mathbf{p} = \begin{bmatrix} \Delta p_0 & \Delta p_1 & \Delta p_2 \end{bmatrix}^T. \tag{10.94}$$

The polynomial is Schur stable for the nominal parameter $\mathbf{p}^0$. We compute the ℓ_2 stability margin of the polynomial with weights $w_1 = w_2 = w_3 = 1$. Rewrite

$$\delta(z, \mathbf{p}^0 + \Delta \mathbf{p}) = (-z + 1)\Delta p_0 + 10z^2 \Delta p_1 - 0.4z^3 \Delta p_2 + (z^4 - 1.4z^3 + 1.1z^2 - 0.4z + 0.1) \tag{10.95}$$

and note that the degree remains invariant (=4) for all perturbations, so that $\rho_d = \infty$. The stability region is the unit circle. For $z = 1$ to be a root of $\delta(z, \mathbf{p}^0 + \Delta \mathbf{p})$ (see Equation (10.79)), we have

$$\underbrace{\begin{bmatrix} 0 & 10 & -0.4 \end{bmatrix}}_{\mathbf{A}(1)} \underbrace{\begin{bmatrix} \Delta p_0 \\ \Delta p_1 \\ \Delta p_2 \end{bmatrix}}_{\mathbf{t}(1)} = \underbrace{-0.4}_{\mathbf{b}(1)}. \tag{10.96}$$

Thus

$$\rho(1) = \|t^*(1)\|_2 = \left\| \mathbf{A}^T(1) \left[\mathbf{A}(1)\mathbf{A}^T(1) \right]^{-1} \mathbf{b}(1) \right\|_2 = 0.04. \tag{10.97}$$

Similarly, for the case of $z = -1$ (see Equation (10.79)), we have

$$\underbrace{\begin{bmatrix} 2 & 10 & 0.4 \end{bmatrix}}_{\mathbf{A}(-1)} \underbrace{\begin{bmatrix} \Delta p_0 \\ \Delta p_1 \\ \Delta p_2 \end{bmatrix}}_{\mathbf{t}(-1)} = \underbrace{-4}_{\mathbf{b}(-1)} \tag{10.98}$$

and $\rho(-1) = \|t^*(-1)\|_2 = 0.3919$. Thus $\rho_r = 0.04$.

For the case in which $\delta(z, \mathbf{p}^0 + \Delta\mathbf{p})$ has a root at $z = e^{j\theta}, \theta \neq \pi, \theta \neq 0$, using Equation (10.83), we have

$$
\underbrace{\left[\begin{array}{ccc} -\cos\theta + 1 & 10\cos 2\theta & -0.4\cos 3\theta \\ -\sin\theta & 10\sin 2\theta & -0.4\sin 3\theta \end{array} \right]}_{\mathbf{A}(\theta)} \underbrace{\left[\begin{array}{c} \Delta p_0 \\ \Delta p_1 \\ \Delta p_2 \end{array} \right]}_{\mathbf{t}(\theta)} \tag{10.99}
$$

$$
= -\underbrace{\left[\begin{array}{c} \cos 4\theta - 1.4\cos 3\theta + 1.1\cos 2\theta - 0.4\cos\theta + 0.1 \\ \sin 4\theta - 1.4\sin 3\theta + 1.1\sin 2\theta - 0.4\sin\theta \end{array} \right]}_{\mathbf{b}(\theta)}.
$$

Thus,

$$
\rho(e^{j\theta}) = \|\mathbf{t}^*(\theta)\|_2 = \left\| \mathbf{A}^T(\theta) \left[\mathbf{A}(\theta)\mathbf{A}^T(\theta) \right]^{-1} \mathbf{b}(\theta) \right\|. \tag{10.100}
$$

Figure 10.7 shows the plot of $\rho(e^{j\theta})$.

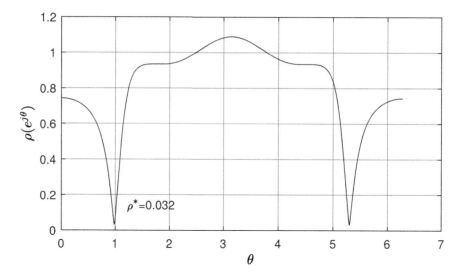

Figure 10.7 $\rho(\theta)$ vs. θ (Example 10.2)

Therefore, the ℓ_2 parametric stability margin is

$$
\min_{\theta} \rho(e^{j\theta})\rho_c = 0.032 = \rho_b = \rho^*. \tag{10.101}
$$

EXAMPLE 10.3 Consider the continuous-time control system with the plant

$$
G(s, \mathbf{p}) = \frac{2s + 3 - \frac{1}{3}p_1 - \frac{5}{3}p_2}{s^3 + (4 - p_2)s^2 + (-2 - 2p_1)s + (-9 + \frac{5}{3}p_1 + \frac{16}{3}p_2)} \tag{10.102}
$$

and the proportional integral (PI) controller

$$C(s) = 5 + \frac{3}{s}. \tag{10.103}$$

The characteristic polynomial of the closed-loop system is

$$\delta(s,\mathbf{p}) = s^4 + (4-p_2)s^3 + (8-2p_1)s^2 + (12-3p_2)s + (9-p_1-5p_2). \tag{10.104}$$

We see that the degree remains invariant under the given set of parameter variations and therefore $\rho_d = \infty$. The nominal values of the parameters are

$$\mathbf{p}^0 = [p_1^0 \ \ p_2^0]^T = [0 \ \ 0]^T. \tag{10.105}$$

Then

$$\Delta\mathbf{p} = \begin{bmatrix} \Delta p_1 & \Delta p_2 \end{bmatrix}^T = \begin{bmatrix} p_1 & p_2 \end{bmatrix}^T. \tag{10.106}$$

The polynomial is stable for the nominal parameter $\mathbf{p}^0$. Now we want to compute the ℓ_2 stability margin of this polynomial with weights $w_1 = w_2 = 1$. We first evaluate $\delta(s,\mathbf{p})$ at $s = j\omega$:

$$\delta(j\omega,\mathbf{p}^0+\Delta\mathbf{p}) = $$
$$(2\omega^2-1)\Delta p_1 + (j\omega^3 - 3j\omega - 5)\Delta p_2 + \omega^4 - 4j\omega^3 - 8\omega^2 + 12j\omega + 9. \tag{10.107}$$

For the case of a root at $s = 0$ (see Equation (10.79)), we have

$$\underbrace{\begin{bmatrix} -1 & -5 \end{bmatrix}}_{\mathbf{A}(0)} \underbrace{\begin{bmatrix} \Delta p_1 \\ \Delta p_2 \end{bmatrix}}_{\mathbf{t}(0)} = \underbrace{-9}_{\mathbf{b}(0)}. \tag{10.108}$$

Thus,

$$\rho(0) = \|\mathbf{t}^*(0)\|_2 = \left\|\mathbf{A}^T(0)\left[\mathbf{A}(0)\mathbf{A}^T(0)\right]^{-1}\mathbf{b}(0)\right\|_2 = \frac{9\sqrt{26}}{26}. \tag{10.109}$$

For a root at $s = j\omega$, $\omega > 0$, using the formula given in Equation (10.83), we have with $w_1 = w_2 = 1$:

$$\underbrace{\begin{bmatrix} (2\omega^2-1) & -5 \\ 0 & (\omega^2-3) \end{bmatrix}}_{\mathbf{A}(j\omega)} \underbrace{\begin{bmatrix} \Delta p_1 \\ \Delta p_2 \end{bmatrix}}_{\mathbf{t}(j\omega)} = \underbrace{\begin{bmatrix} -\omega^4 + 8\omega^2 - 9 \\ 4\omega^2 - 12 \end{bmatrix}}_{\mathbf{b}(j\omega)}. \tag{10.110}$$

Here, we need to determine if there exists any ω for which the rank of the matrix $A(j\omega)$ drops. It is easy to see from the matrix $\mathbf{A}(j\omega)$ that the rank drops when $\omega = \frac{1}{\sqrt{2}}$ and $\omega = \sqrt{3}$.

rank$[\mathbf{A}(j\omega)] = 1$: For $\omega = \frac{1}{\sqrt{2}}$, we have rank $\mathbf{A}(j\omega) = 1$ and rank $[\mathbf{A}(j\omega) \ \mathbf{b}(j\omega)] = 2$, and so there is no solution to Equation (10.110). Thus

$$\rho\left(j\frac{1}{\sqrt{2}}\right) = \infty. \tag{10.111}$$

For $\omega = \sqrt{3}$:

$$\text{rank}\,[\mathbf{A}(j\omega)] = \text{rank}\,[\mathbf{A}(j\omega)\ \ \mathbf{b}(j\omega)], \tag{10.112}$$

and we do have a solution to Equation (10.110). Therefore

$$\underbrace{\begin{bmatrix} 5 & -5 \end{bmatrix}}_{\mathbf{A}(j\sqrt{3})}\underbrace{\begin{bmatrix} \Delta p_1 \\ \Delta p_2 \end{bmatrix}}_{\mathbf{t}(j\sqrt{3})} = \underbrace{6}_{\mathbf{b}(j\sqrt{3})}. \tag{10.113}$$

Consequently

$$\rho(j\sqrt{3}) = \left\| \mathbf{t}^*(j\sqrt{3}) \right\|_2 = \left\| \mathbf{A}^T(j\sqrt{3})\left[\ \mathbf{A}(j\sqrt{3})\mathbf{A}^T(j\sqrt{3})\ \right]^{-1}\mathbf{b}(j\sqrt{3}) \right\|_2$$
$$= \frac{3\sqrt{2}}{5}. \tag{10.114}$$

rank$[A(j\omega)] = 2$:   In this case, Equation (10.110) has a solution (which happens to be unique) and the length of the least-square solution is found by

$$\rho(j\omega) = \left\| \mathbf{t}^*(j\omega) \right\|_2 = \left\| \mathbf{A}^T(j\omega)\left[\mathbf{A}(j\omega)\mathbf{A}^T(j\omega)\right]^{-1}\mathbf{b}(j\omega) \right\|_2$$
$$= \sqrt{\left(\frac{\omega^4 - 8\omega^2 - 11}{2\omega^2 - 1}\right)^2 + 16}. \tag{10.115}$$

Figure 10.8 shows the plot of $\rho(j\omega)$ for $\omega > 0$. The values of $\rho(0)$ and $\rho(j\sqrt{3})$ are also shown in Figure 10.8. Therefore

$$\rho(j\sqrt{3}) = \rho_b = \frac{3\sqrt{2}}{5} = \rho^* \tag{10.116}$$

is the stability margin.

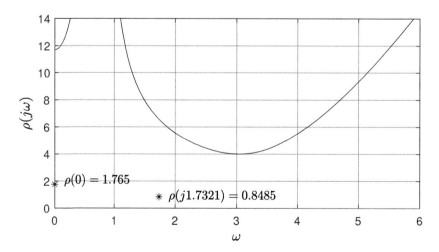

Figure 10.8 $\rho(j\omega)$ vs. ω (Example 10.2)

10.4.4 Possible Discontinuity of the Stability Margin

In the last example we notice that the function $\rho(j\omega)$ has a discontinuity at $\omega = \omega^* = \sqrt{3}$. The reason for this discontinuity is that in the neighborhood of ω^*, the minimum norm solution of Equation (10.83) is given by the formula for the rank 2 case. On the other hand, at $\omega = \omega^*$, the minimum norm solution is given by the formula for the rank 1 case. Thus, the discontinuity of the function $\rho(j\omega)$ is due to the drop of rank from 2 to 1 of the coefficient matrix $\mathbf{A}(j\omega)$ at ω^*. Furthermore, we have seen that if the rank of $\mathbf{A}(j\omega^*)$ drops from 2 to 1 but the rank of $[\mathbf{A}(j\omega^*) \ \mathbf{b}(j\omega^*)]$ does not also drop, then the equation is inconsistent at ω^* and $\rho(j\omega^*)$ is infinity. In this case, this discontinuity in $\rho(j\omega)$ does not cause any problem in finding the global minimum of $\rho(j\omega)$. Therefore, the only values of ω^* that can cause a problem are those for which the rank of $[\mathbf{A}(j\omega^*) \ \mathbf{b}(j\omega^*)]$ drops to 1. Given the problem data, the occurrence of such a situation can be *predicted* by setting all 2×2 minors of the matrix $[\mathbf{A}(j\omega) \ \mathbf{b}(j\omega)]$ equal to zero and solving for the common real roots, if any. These frequencies can then be treated separately in the calculation. Therefore, such discontinuities do not pose any problem from the computational point of view. Since the parameters for which rank dropping occurs lie on a proper algebraic variety, any slight and arbitrary perturbation of the parameters will dislodge them from this variety and restore the rank of the matrix. If the parameters correspond to physical quantities, such arbitrary perturbations are natural and hence such discontinuities should not cause any problem from a physical point of view either.

In the next section we apply the parametric stability margin calculation to determine the stability margin with respect to parameters of high-order controllers designed by various methods. The answers are somewhat surprising, as we shall see.

10.5 Fragility of High-Order Controllers

Starting in the 1980s, a number of controller synthesis methods were proposed in order to render the closed-loop system robust with respect to plant uncertainty. These methods included H_∞, H_2 l^1-optimal methods and a technique called μ. Most of these methods inflated the controller order, usually to several times that of the plant, to achieve the design objectives. In 1997 it was shown that such high-order controllers had miniscule stability margins with respect to its coefficients. They also had miniscule closed-loop stability margins, rendering them dysfunctional. This phenomena called *fragility* is described and illustrated below using examples from the literature.

EXAMPLE 10.4 (H_∞-based optimum gain margin controller)   This example uses the YJBK parametrization and the machinery of the H_∞ model matching problem to optimize the upper gain margin. The plant to be controlled is

$$P(s) = \frac{s-1}{s^2 - s - 2} \tag{10.117}$$

and the controller, designed to give an upper gain margin of 3.5 (the closed loop is stable for the gain interval $[1, 3.5]$) is obtained by optimizing the H_∞ norm of a complementary sensitivity function. The controller found is

$$C(s) = \frac{q_6^0 s^6 + q_5^0 s^5 + q_4^0 s^4 + q_3^0 s^3 + q_2^0 s^2 + q_1^0 s + q_0^0}{p_6^0 s^6 + p_5^0 s^5 + p_4^0 s^4 + p_3^0 s^3 + p_2^0 s^2 + p_1^0 s + p_0^0}, \qquad (10.118)$$

where

$$\begin{array}{llll}
q_6^0 = 379, & q_5^0 = 39383, & q_4^0 = 192306, & q_3^0 = 382993, \\
q_2^0 = 383284, & q_1^0 = 192175, & q_0^0 = 38582, & \\
p_6^0 = 3, & p_5^0 = -328, & p_4^0 = -38048, & p_3^0 = -179760, \\
p_2^0 = -314330, & p_1^0 = -239911, & p_0^0 = -67626. &
\end{array} \qquad (10.119)$$

The poles of this nominal controller are

$$174.70, \quad -65.99, \quad -1.86, \quad -1.04, \quad -0.98 \pm j0.03 \qquad (10.120)$$

and the poles of the closed-loop system are

$$\begin{aligned}
&-0.4666 \pm j14.2299, \quad -5.5334 \pm j11.3290, \quad -1.0002, \\
&-1.0000 \pm j0.0002, \quad -0.9998
\end{aligned} \qquad (10.121)$$

and this verifies that the controller is indeed stabilizing. The Nyquist plot of $P(s)C(s)$ is shown in Figure 10.9 and verifies that the desired upper gain margin is achieved.

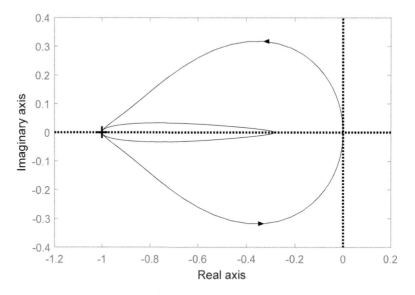

Figure 10.9 Nyquist plot of $P(s)C(s)$ (Example 10.4)

On the other hand, we see from Figure 10.9 that the lower gain margin and phase margin are

$$GM = [1, 0.9992], \qquad PM = [0°, 0.1681°]. \qquad (10.122)$$

This means roughly that a reduction in gain of one part in one thousand will destabilize the closed-loop system! Likewise, a vanishingly small phase perturbation is destabilizing. To continue with our analysis, let us consider the transfer function coefficients of the controller to be a parameter vector $\mathbf{p}$ with its nominal value being

$$\mathbf{p}^0 = \left[\; q_6^0 \quad \cdots \quad q_0^0 \quad p_6^0 \quad \cdots \quad p_0^0 \;\right]^T \tag{10.123}$$

and let $\Delta\mathbf{p}$ be the vector representing perturbations in $\mathbf{p}$. We compute the ℓ_2 parametric stability margin around the nominal point. This comes out to be

$$\rho = 0.15813903109631. \tag{10.124}$$

The normalized ratio of change in controller coefficients required to destabilize the closed loop is

$$\frac{\rho}{\|\mathbf{p}^0\|_2} = 2.103407115900516 \times 10^{-7}. \tag{10.125}$$

This shows that a change in the controller coefficients of less than 1 part in a million destabilizes the closed loop. This controller is anything but robust; in fact, we are certainly justified in labeling it as a *fragile* controller.

In order to verify this rather surprising result, we construct a closest destabilizing controller whose parameters are obtained by setting $\mathbf{p} = \mathbf{p}^0 + \Delta\mathbf{p}$:

$$
\begin{aligned}
q_6 &= 379.000285811, & p_6 &= 3.158134748, \\
q_5 &= 39382.999231141, & p_5 &= -327.999718909, \\
q_4 &= 192305.999998597, & p_4 &= -38048.000776386, \\
q_3 &= 382993.000003775, & p_3 &= -179760.000001380, \\
q_2 &= 383284.000000007, & p_2 &= -314329.999996188, \\
q_1 &= 192174.999999982, & p_1 &= -239910.999999993, \\
q_0 &= 38582.000000000, & p_0 &= -67626.000000018.
\end{aligned} \tag{10.126}
$$

The closed-loop poles of the system with this perturbed controller are

$$
\begin{aligned}
0.000 \pm j14.2717, \quad -5.5745 \pm j10.9187, \quad -1.0067 \pm j\,0.0158, \\
-1.0044, \quad -0.9820,
\end{aligned} \tag{10.127}
$$

which shows that the roots cross over to the right-half plane at $\omega = 14.27$, and the perturbed controller is indeed destabilizing.

EXAMPLE 10.5 (An arbitrary controller)   For the sake of comparison, we continue with the previous example and try a first-order controller with the same plant

$$P(s) = \frac{s-1}{s^2 - s - 2}. \tag{10.128}$$

We design a pole placement controller placing closed-loop poles on a circle of radius $\sqrt{2}$ spaced equidistantly in the left-half plane. The transfer function of this controller is:

$$C(s) = \frac{q_1^0 s + q_0^0}{s + p_0^0}, \tag{10.129}$$

where

$$q_1^0 = 11.44974739, \quad q_0^0 = 11.24264066, \quad p_0^0 = -7.03553383. \qquad (10.130)$$

Introduce the parameter vector $\mathbf{p}$ corresponding to the controller coefficients and with nominal value:

$$\mathbf{p}^0 = \begin{bmatrix} q_1^0 & q_0^0 & p_0^0 \end{bmatrix}. \qquad (10.131)$$

We compute the ℓ_2 parametric stability margin for this controller to be

$$\rho = 1.26491100827916, \quad \frac{\rho}{\|\mathbf{p}^0\|_2} = 0.07219317556675087. \qquad (10.132)$$

The normalized ratio of change in controller coefficients required to destabilize the loop is

$$\frac{\rho}{\|\mathbf{p}^0\|_2} = 0.07219317556675087, \qquad (10.133)$$

which is to be compared to the previous 10^{-7} value. This controller can tolerate a change in coefficient values of 7.2% compared to the value of 10^{-4}% for the optimum controller. The Nyquist plot of the system with this controller is shown in Figure 10.10 and gives the lower gain and phase margins

$$\text{GM} = [1, 0.79403974451346], \quad \text{PM} = [0°, -9.88729274575800°]. \qquad (10.134)$$

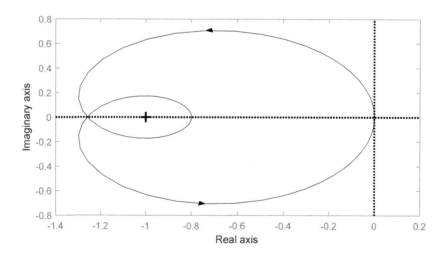

Figure 10.10 Nyquist plot of $P(s)C(s)$ (Example 10.4)

The system can tolerate gain reduction of about 21%. This is an improvement over the previous controller by a factor of about 20,000! The phase margin is improved by a factor of about 60. We have already shown the drastic improvement in the parametric stability margin. Therefore, this non-optimal controller is far less fragile than the optimal controller, on all counts.

EXAMPLE 10.6 (H_∞ robust controller)   The following example designs an optimal H_∞ robust controller that minimizes $\|W_2(s)T(s)\|_\infty$, where $T(s)$ is the complementary sensitivity function and the weight $W_2(s)$ is chosen as the high-pass function

$$W_2(s) = \frac{s+0.1}{s+1}. \tag{10.135}$$

The plant transfer function is

$$P(s) = \frac{s-1}{s^2+0.5s-0.5} \tag{10.136}$$

and the optimally robust controller found is

$$C(s) = \frac{-124.5s^3 - 364.95s^2 - 360.45s - 120}{s^3 + 227.1s^2 + 440.7s + 220}. \tag{10.137}$$

The parametric stability margin (ℓ_2 norm of the smallest destabilizing perturbation $\Delta \mathbf{p}$) and the normalized ratio of change in controller coefficients required for destabilization are

$$\rho = 8.94427190999916, \quad \frac{\rho}{\|\mathbf{p}^0\|_2} = 0.01167214151733. \tag{10.138}$$

This shows that the controller, which by design is maximally robust with respect to H_∞ perturbations, is quite fragile with respect to controller coefficient perturbations.

To continue our analysis, the Nyquist plot of the system with this controller is drawn in Figure 10.11. From this we obtain the lower gain and phase margins

$$\text{GM} = [1, \ 0.91666666666667], \quad \text{PM} = [0°, \ 12.91170327761722°], \tag{10.139}$$

which are quite poor and would probably be unacceptable in a real system.

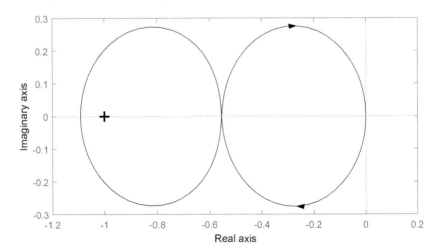

Figure 10.11 Nyquist plot of $P(s)C(s)$ (Example 10.6)

EXAMPLE 10.7 (μ-based design)   This example examines a robust controller for an electromagnetic suspension system designed by using the μ-synthesis technique. The plant transfer function is

$$P(s) = \frac{-36.27}{s^3 + 45.69s^2 - 4480.9636s - 204735.226884} \tag{10.140}$$

and the controller designed to tolerate prescribed structured plant perturbations is given as

$$C(s) = \frac{q_6^0 s^6 + q_5^0 s^5 + q_4^0 s^4 + q_3^0 s^3 + q_2^0 s^2 + q_1^0 s + q_0^0}{s^7 + p_6^0 s^6 + p_5^0 s^5 + p_4^0 s^4 + p_3^0 s^3 + p_2^0 s^2 + p_1^0 s + p_0^0}, \tag{10.141}$$

where

$$
\begin{aligned}
q_6^0 &= -5.220000000000000 \times 10^8, & p_6^0 &= 1.468170000000000 \times 10^3, \\
q_5^0 &= -1.190629800000000 \times 10^{11}, & p_5^0 &= 8.153914724000001 \times 10^5, \\
q_4^0 &= -1.089211902480000 \times 10^{13}, & p_4^0 &= 2.268680248018680 \times 10^8 \\
q_3^0 &= -5.104622252074320 \times 10^{14}, & p_3^0 &= 1.818763428483511 \times 10^{10}, \\
q_2^0 &= -1.285270261841830 \times 10^{16}, & p_2^0 &= 5.698409038920188 \times 10^{11}, \\
q_1^0 &= -1.629532689765926 \times 10^{17}, & p_1^0 &= 6.284542925855980 \times 10^{12}, \\
q_0^0 &= -7.937217972339767 \times 10^{17}, & p_0^0 &= 6.227740485023126 \times 10^{11}.
\end{aligned} \tag{10.142}
$$

This parametric stability margin and the normalized ratio of change in controller coefficients required to destabilize the closed loop are

$$\rho = 1.17938672900662 \times 10^3, \quad \frac{\rho}{\|\mathbf{p}^0\|_2} = 1.455352715525003 \times 10^{-15}. \tag{10.143}$$

This indicates that closed-loop stability is very fragile with respect to controller parameter perturbations.

To continue, in Figure 10.12 we draw the Nyquist plot of the plant $P(s)$ with this controller $C(s)$, which shows that

$$\text{GM} = [0.5745485, \ 2.585314], \quad \text{PM} = [0°, \ \pm 24.0555°]. \tag{10.144}$$

Comparing this with the previous fragility with respect to coefficient perturbations reminds us that good gain and phase margins are not necessarily reliable indicators of robustness. However, poor gain and/or phase margins *are* accurate indicators of fragileness!

EXAMPLE 10.8 (H_2 optimal design)   The plant transfer function is

$$P(s) = \frac{-s+1}{s^2+s+2} \tag{10.145}$$

and the optimal controller is determined by minimizing a weighted H_2 norm of the disturbance transfer function $\|W(s)P(s)S(s)\|_2$, where $S(s)$ is the sensitivity function. In

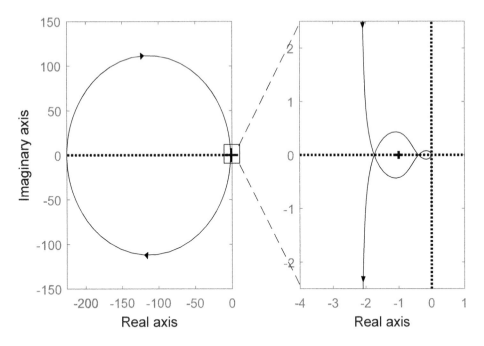

Figure 10.12 Nyquist plot of $P(s)C(s)$ (Example 10.7)

this example the optimal YJBK parameter $Q(s)$ is improper and a suboptimal controller is picked after dividing $Q(s)$ by the factor $(s\tau + 1)^k$ with $\tau = 0.01$ and $k = 2$. The controller designed is

$$C(s) = \frac{q_6 s^6 + q_5 s^5 + q_4 s^4 + q_3 s^3 + q_2 s^2 + q_1 s + q_0}{p_6 s^6 + p_5 s^5 + p_4 s^4 + p_3 s^3 + p_2 s^2 + p_1 s}, \qquad (10.146)$$

where

$$q_6 = 1.0002, \quad q_5 = 3.0406, \quad q_4 = 8.1210, \quad q_3 = 13.2010,$$
$$q_2 = 15.2004, \quad q_1 = 12.08, \quad q_0 = 4.0,$$
$$p_6 = 0.0001, \quad p_5 = 1.0205, \quad p_4 = 2.1007, \quad p_3 = 5.1403, \qquad (10.147)$$
$$p_2 = 6.06, \quad p_1 = 2.0.$$

To proceed with our analysis, we took the controller coefficients as a parameter vector and computed the l^2 parametric stability around the nominal. Here we found that

$$\rho = 1.101592322015807 \times 10^{-4}, \quad \frac{\rho}{\|\mathbf{p}^0\|_2} = 3.737066131643626 \times 10^{-6}, \quad (10.148)$$

which again shows that the controller is extremely fragile.

To continue, in Figure 10.13 we draw the Nyquist plot of the system with this controller. It shows that the gain and phase margins are

$$\text{GM} = [1, 1.02000161996515], \quad \text{PM} = [0°, 6.86646258505442°], \qquad (10.149)$$

which is again rather poor.

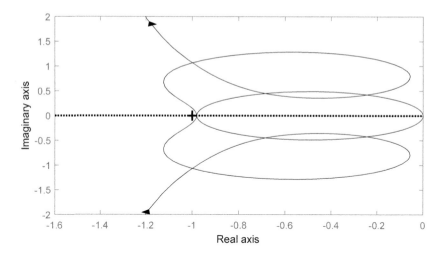

Figure 10.13 Nyquist plot of $P(s)C(s)$ (zoomed figure) (Example 10.8)

In this particular example, we found that the poles of the closed-loop system appear to be extremely sensitive with respect to controller coefficient changes. This example shows that a slight perturbation in controller coefficients of the optimal controller will result in very large perturbations of the closed-loop poles. This is due to the degree dropping of the characteristic polynomial at the perturbation in the highest coefficient.

EXAMPLE 10.9 (ℓ_1 optimal control)   The given plant which is a discrete-time model of an X-29 aircraft has transfer function:

$$P(z) = \frac{n_3 z^3 + n_2^2 + n_1 z + n_0}{z^4 + d_3 z^3 + d_2 z^2 + d_1 z + d_0}, \qquad (10.150)$$

with

$$
\begin{aligned}
n_3 &= 4.535291991023538 \times 10^{-3}, & d_3 &= -2.871701572273939, \\
n_2 &= -1.294020546545127 \times 10^{-2}, & d_2 &= 2.900568259828338, \\
n_1 &= 2.996941752535731 \times 10^{-3}, & d_1 &= -1.235443807952025, \\
n_0 &= 4.522895843871860 \times 10^{-3}, & d_0 &= 0.1888756028372526.
\end{aligned}
\qquad (10.151)
$$

The optimal controller designed to minimize the ℓ_1 norm of a disturbance transfer function is

$$C(z) = \frac{q_6 z^6 + q_5 z^5 + q_4 z^4 + q_3 z^3 + q_2 z^2 + q_1 z + q_0}{p_6 z^6 + p_5 z^5 + p_4 z^4 + p_3 z^3 + p_2 z^2 + p_1 z + p_0}, \qquad (10.152)$$

where

$$
\begin{array}{ll}
q_6 = -1.86872 \times 10^2, & p_6 = 2.5231, \\
q_5 = 8.068 \times 10, & p_5 = 1.489 \times 10^{-1}, \\
q_4 = 3.43 \times 10^2, & p_4 = -8.446 \times 10^{-1}, \\
q_3 = -1.21311 \times 10^2, & p_3 = -3.0297, \\
q_2 = -2.93168 \times 10^2, & p_2 = -1.7703, \\
q_1 = 2.18478 \times 10^2, & p_1 = 1.9728, \\
q_0 = -4.1763 \times 10, & p_0 = 1.
\end{array}
\tag{10.153}
$$

For our analysis we took the controller transfer function coefficients as a parameter vector and computed the l^2 parametric stability margin around the nominal controller. Here we found that

$$
\rho = 0.01796866210822, \quad \frac{\rho}{\|\mathbf{p}^0\|_2} = 3.231207641519448 \times 10^{-5}.
\tag{10.154}
$$

This shows how fragile the system is with respect to controller parameter perturbations: perturbations of less than 1 part in $10,000$ will destabilize the closed loop.

Next, the Nyquist plot of $P(z)C(z)$ is drawn in Figure 10.14. This gives the closed-loop gain and phase margins:

$$
\text{GM} = [0.5463916, 1.38153348], \quad \text{PM} = [0°, \pm 23.03169°].
\tag{10.155}
$$

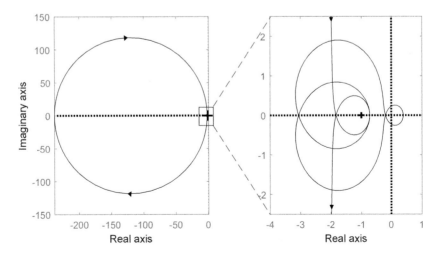

Figure 10.14 Nyquist plot of $P(z)C(z)$ (Example 10.9)

10.6   Polytopic Uncertainty: Robust Stability of a Polytope

A central problem in robust stability considers the stability of each member of a convex polytope $\mathbf{P}(s)$ of degree-invariant polynomials. The stability and instability regions $\mathscr{S}$ and $\mathscr{U}$ considered are arbitrary subject to

$$\mathscr{S} \cup \mathscr{U} = \mathbb{C}, \tag{10.156a}$$

$$\mathscr{S} \cap \mathscr{U} = \emptyset, \tag{10.156b}$$

$$\mathscr{S} \quad \text{is open.} \tag{10.156c}$$

Thus, $\mathscr{S}$ could be the open LHP (Hurwitz stability), the shifted open LHP, the open unit circle (Schur stability), and other meaningful regions representing "settling time" and damping ratio.

Let $\mathbf{Q}(s)$ denote a subset of $\mathbf{P}(s)$:

$$\mathbf{Q}(s) \subset \mathbf{P}(s), \tag{10.157}$$

with the property that its image set evaluated on the stability region boundary is identical to that of $\mathbf{P}(s)$:

$$\mathbf{Q}(s^*) = \mathbf{P}(s^*), \qquad \text{for all } s^* \in \partial\mathscr{S}. \tag{10.158}$$

Then the "smaller" set $\mathbf{Q}(s)$ can serve as a stability testing set for $\mathbf{P}(s)$, as stated below.

THEOREM 10.5 $\mathbf{P}(s)$ is robustly stable if and only if $\mathbf{Q}(s)$ is.

Proof If $\mathbf{Q}(s)$ satisfies Equation (10.158), the image set of polynomials in $\mathbf{Q}(s)$ and $\mathbf{P}(s)$ is identical. Thus, stability (instability) of a polynomial of a polynomial in $\mathbf{P}(s)$ implies the existence of a corresponding stable (unstable) polynomial in $\mathbf{Q}(s)$. ■

The theorem above can be useful to reduce the testing of a high-dimensional polytope $(\mathbf{P}(s))$ to one on a low dimension $(\mathbf{Q}(s))$, as the example below shows.

EXAMPLE 10.10   Consider the Hurwitz stability of the polytope of polynomials $\mathscr{P}(s)$:

$$\delta(s) = \delta_0 + \delta_1 s + \delta_2 s^2 + \delta_3 s^3, \tag{10.159}$$

with coefficients varying in intervals

$$\delta_0 \in [x_0, y_0] \quad \text{and} \quad \delta_2 \in [x_2, y_2]. \tag{10.160}$$

Clearly, $\mathbf{P}(s)$ corresponds to a two-dimensional polytope $\mathbf{P}$ as shown in Figure 10.15.

Let $\mathscr{Q}(s)$ be defined to be the convex combination of polynomials

$$K_1(s) = x_0 + \delta_1 s + y_2 s^2 + \delta_3 s^3, \tag{10.161a}$$

$$K_2(s) = y_0 + \delta_1 s + x_2 s^2 + \delta_3 s^3. \tag{10.161b}$$

Thus, $\mathbf{Q}(s)$ is a one-dimensional polytope

$$\mathbf{Q}(s) = \{\delta_\lambda(s) \; : \; (1-\lambda)K_1(s) + \lambda K_2(s), \; \lambda \in [0,1]\}. \tag{10.162}$$

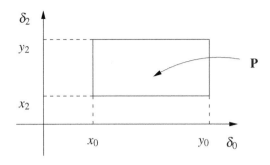

Figure 10.15 The set **P** (Example 10.10)

It is easy to verify that

$$\mathbf{Q}(j\omega) = \mathbf{P}(j\omega), \quad \text{for all } \omega \in [0, \infty) \tag{10.163}$$

and thus $\mathbf{Q}(s)$ is a Hurwitz stability testing set for $\mathbf{P}(s)$, as shown in Figure 10.16.

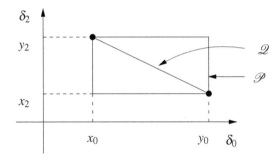

Figure 10.16 The sets **Q** and **P**

In this example, testing the stability of the two-dimensional set $\mathbf{P}(s)$ is equivalent to testing the stability of the one-dimensional set $\mathbf{Q}(s)$. The latter is a root locus problem and is thus computationally tractable. It turns out additionally, as we shall show later, that in this particular problem, stability of the line segment $\mathbf{Q}(s)$ reduces to the stability of $K_1(s)$ and $K_2(s)$.

We now state an even stronger result, namely that checking the robust stability of a convex polytope $\mathbf{P}(s)$ of degree-invariant polynomials, can be reduced to checking the robust stability of its exposed edges, $\mathbf{E}(s)$. Since $\mathbf{E}(s)$ consists of line segments, the problem reduces to a set of one-parameter root locus problems.

THEOREM 10.6 Suppose

$$0 \notin \mathscr{P}(s_0), \quad \text{for some } s_0 \in \partial\mathscr{S}, \tag{10.164}$$

then $\mathbf{P}(s)$ is robustly stable if and only if $\mathbf{E}(s)$ is robustly stable.

Proof The complex plane polygon $\mathbf{P}(s^*)$ is in the convex hull of $\mathbf{E}(s^*)$ for all $s^* \in \partial \mathscr{S}$. Due to Equation (10.164) and degree invariance, the origin can therefore enter the polygon $\mathbf{P}(s_1^*)$ if and only if it enters $\mathbf{E}(s_2^*)$ for some $s_2^* \in \partial \mathscr{S}$. Thus, $\mathbf{P}(s)$ contains an unstable polynomial if and only if $\mathbf{E}(s)$ does.   ∎

EXAMPLE 10.11   Consider the following three degree-invariant polynomials:

$$
\begin{aligned}
p_1(s) &= s^3 + 6s^2 + 10s + 4, \\
p_2(s) &= s^3 + 8s^2 + 12s + 4, \\
p_3(s) &= s^3 + 8s^2 + 12s + 5.
\end{aligned}
\tag{10.165}
$$

These polynomials and the corresponding polytopic family are depicted in the coefficient space as shown in Figure 10.17.

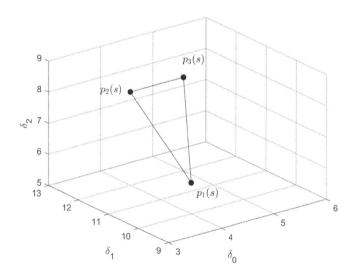

Figure 10.17 Three polynomials and their convex combinations (Example 10.11)

According to Theorem 10.6, the family of polynomials shown in Figure 10.17 is robustly stable if and only if the three exposed edges

$$
\lambda p_i(s) + (1-\lambda)p_j(s), \quad \text{for } \{(i,j) : (1,2),(2,3),(3,1)\} \text{ and } \lambda \in (0,1) \tag{10.166}
$$

are robustly stable. The robust stability of such a segment may easily be verified by a number of techniques, such as the Routh–Hurwitz criterion or the bounded phase condition in Section 10.7.1.

The above result shows that checking the robust stability of a polytope with respect to an arbitrary region reduces checking the stability of a line segment of polynomials. The next section is devoted to solving this problem.

10.7 The Segment Lemma

In this section we focus on the problem of determining the stability of a line segment joining two fixed polynomials which we refer to as the endpoints. This line segment of polynomials is a convex combination of the two endpoints and indeed is the simplest example of a polytope. This kind of problem arises in robust control problems containing a single uncertain parameter, such as a gain or a time constant, when stability of the system must be ascertained for the entire interval of uncertainty. Also we have shown in the previous section that the stability verification of an entire polytope reduces to checking that of the exposed edges.

We give some simple solutions to this problem for both the Hurwitz and Schur cases and collectively call these results the Segment Lemma. In general, the stability of the endpoints does not guarantee that of the entire segment of polynomials. For example, consider the segment joining the two polynomials

$$
\begin{aligned}
P_1(s) &= 3s^4 + 3s^3 + 5s^2 + 2s + 1, \\
P_2(s) &= s^4 + s^3 + 5s^2 + 2s + 5.
\end{aligned}
\tag{10.167}
$$

It can be checked that both $P_1(s)$ and $P_2(s)$ are Hurwitz stable and yet the polynomial at the midpoint

$$
\frac{P_1(s) + P_2(s)}{2} \qquad \text{has a root at} \quad s = j.
\tag{10.168}
$$

However, if the polynomial representing the difference of the endpoints assumes certain special forms, the stability of the endpoints does indeed guarantee stability of the entire segment. These forms, which are frequency independent, are described in the Vertex Lemma and again both Hurwitz and Schur versions of this lemma are given. The conditions specified by the Vertex Lemma are useful for reducing robust stability determinations over a continuum of polynomials to that of a discrete set of points.

The proof of the Vertex Lemma depends on certain phase relations for Hurwitz polynomials and segments, which are established here. These are also of independent interest.

10.7.1 Bounded Phase Condition

Let $\mathscr{S}$ be an open set in the complex plane representing the stability region and let $\partial\mathscr{S}$ denote its boundary. Suppose $\delta_1(s)$ and $\delta_2(s)$ are polynomials (real or complex) of degree n. Let

$$
\delta_\lambda(s) := \lambda\delta_1(s) + (1 - \lambda)\delta_2(s)
\tag{10.169}
$$

and consider the following *one-parameter family* of polynomials:

$$
[\delta_1(s), \delta_2(s)] := \{\delta_\lambda(s) \; : \; \lambda \in [0, 1]\}.
\tag{10.170}
$$

This family will be referred to as a *segment* of polynomials. We shall say that the segment is stable if and only if every polynomial on the segment is stable. This property is also referred to as *strong stability* of the pair $(\delta_1(s), \delta_2(s))$.

We begin with a lemma which follows directly from the continuity of the roots of a set of degree-invariant polynomials on its coefficients. Let $\phi_{\delta_i}(s_0)$ denote the argument of the complex number $\delta_i(s_0)$.

LEMMA 10.1 (Bounded Phase Lemma)   Let $\delta_1(s)$ and $\delta_2(s)$ be stable with respect to $\mathscr{S}$ and assume that the degree of $\delta_\lambda(s) = n$ for all $\lambda \in [0,1]$. Then the following are equivalent:

(a) the segment $[\delta_1(s), \delta_2(s)]$ is stable with respect to $\mathscr{S}$;
(b) $\delta_\lambda(s^*) \neq 0$,   for all $s^* \in \partial\mathscr{S}$, $\lambda \in [0,1]$;
(c) $|\phi_{\delta_1}(s^*) - \phi_{\delta_2}(s^*)| \neq \pi$ radians for all $s^* \in \partial\mathscr{S}$;
(d) the complex plane plot of $\dfrac{\delta_1(s^*)}{\delta_2(s^*)}$, for $s^* \in \partial\mathscr{S}$, does not cut the negative real axis.

Proof   The equivalence of (a) and (b) follows from the continuity of the roots on the coefficients. The equivalence of (b) and (c) is best illustrated geometrically in Figure 10.18.

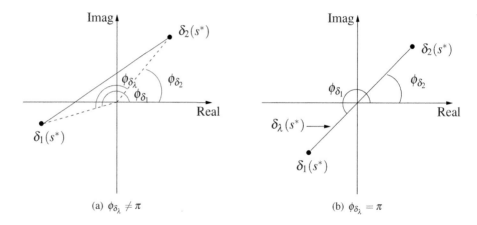

(a) $\phi_{\delta_\lambda} \neq \pi$ (b) $\phi_{\delta_\lambda} = \pi$

Figure 10.18 Image set of $\delta_\lambda(s^*)$ and ϕ_{δ_λ}

In words, this simply states that whenever $\delta_\lambda(s^*) = 0$ for some $\lambda \in [0,1]$ the phasors $\delta_1(s^*)$ and $\delta_2(s^*)$ must line up with the origin with their endpoints on opposite sides of it. This is expressed by the condition $|\phi_{\delta_1}(s^*) - \phi_{\delta_2}(s^*)| = \pi$ radians. The equivalence of (b) and (d) follows from the fact that if

$$\lambda\delta_1(s^*) + (1-\lambda)\delta_2(s^*) = 0, \tag{10.171}$$

then

$$\frac{\delta_1(s^*)}{\delta_2(s^*)} = -\left(\frac{1-\lambda}{\lambda}\right). \tag{10.172}$$

As λ varies from 0 to 1, the right-hand side of the above equation generates the negative real axis. Hence $\delta_\lambda(s^*) = 0$ for some $\lambda \in [0,1]$ if and only if $\dfrac{\delta_1(s^*)}{\delta_2(s^*)}$ is negative and real.   ∎

This lemma essentially states that the entire segment is stable provided the end points are, the degree remains invariant, and the phase difference between the endpoints evaluated along the stability boundary is bounded by π. This condition will be referred to as the *bounded phase condition*. We illustrate this result with some examples.

EXAMPLE 10.12 (Real polynomials)   Consider the feedback system shown in Figure 10.19.

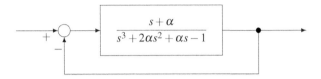

Figure 10.19 Feedback system (Example 10.12)

Suppose that we want to check the robust Hurwitz stability of the closed-loop system for $\alpha \in [2,3]$. We first examine the stability of the two endpoints of the characteristic polynomial

$$\delta(s,\alpha) = s^3 + 2\alpha s^2 + (\alpha+1)s + (\alpha-1). \tag{10.173}$$

We let

$$\begin{aligned}
\delta_1(s) &:= \delta(s,\alpha)|_{\alpha=2} = s^3 + 4s^2 + 3s + 1, \\
\delta_2(s) &:= \delta(s,\alpha)|_{\alpha=3} = s^3 + 6s^2 + 4s + 2.
\end{aligned} \tag{10.174}$$

Then $\delta_\lambda(s) = \lambda \delta_1(s) + (1-\lambda)\delta_2(s)$. We check that the endpoints $\delta_1(s)$ and $\delta_2(s)$ are stable. Then we verify the bounded phase condition, namely that the phase difference $|\phi_{\delta_1}(j\omega) - \phi_{\delta_2}(j\omega)|$ between these endpoints never reaches $180°$ as ω runs from 0 to ∞. Thus, the segment is robustly Hurwitz stable. This is shown in Figure 10.20.

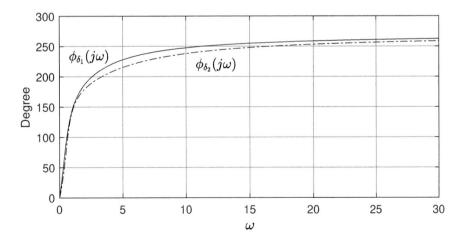

Figure 10.20 Phase difference of the endpoints of a stable segment (Example 10.12)

The condition (d) of Lemma 10.1 can also easily be verified by drawing the polar plot of $\delta_1(j\omega)/\delta_2(j\omega)$.

EXAMPLE 10.13 (Complex polynomials)   Consider the Hurwitz stability of the line segment joining the two complex polynomials:

$$
\begin{aligned}
\delta_1(s) &= s^4 + (8-j)s^3 + (28-j5)s^2 + (50-j3)s + (33+j9), \\
\delta_2(s) &= s^4 + (7+j4)s^3 + (46+j15)s^2 + (165+j168)s + (-19+j373).
\end{aligned}
\tag{10.175}
$$

We first verify that the two endpoints $\delta_1(s)$ and $\delta_2(s)$ are stable. Then we plot $\phi_{\delta_1}(j\omega)$ and $\phi_{\delta_2}(j\omega)$ with respect to ω (Figure 10.21). As we can see, the bounded phase condition is satisfied, that is the phase difference $|\phi_{\delta_1}(j\omega) - \phi_{\delta_2}(j\omega)|$ never reaches $180°$, so we conclude that the given segment $[\delta_1(s), \delta_2(s)]$ is stable.

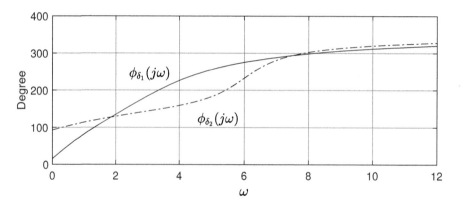

Figure 10.21 Phase difference vs. ω for a complex segment (Example 10.13)

We can also use the condition (d) of Lemma 10.1. As shown in Figure 10.22, the plot of $\delta_1(j\omega)/\delta_2(j\omega)$ does not cut the negative real axis of the complex plane.

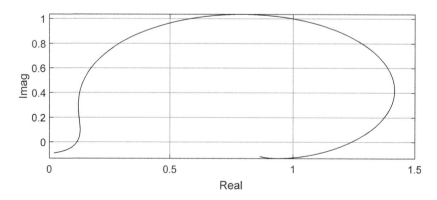

Figure 10.22 A stable segment: $\frac{\delta_1(j\omega)}{\delta_2(j\omega)} \cap \mathbb{R}^- = \emptyset$ (Example 10.13)

Therefore, the segment is stable.

EXAMPLE 10.14 (Schur stability)   Let us consider the Schur stability of the segment joining the two polynomials

$$\delta_1(z) = z^5 + 0.4z^4 - 0.33z^3 + 0.058z^2 + 0.1266z + 0.059,$$
$$\delta_2(z) = z^5 - 2.59z^4 + 2.8565z^3 - 1.4733z^2 + 0.2236z - 0.0121. \tag{10.176}$$

First we verify that the roots of both $\delta_1(z)$ and $\delta_2(z)$ lie inside the unit circle. In order to check the stability of the given segment, we simply evaluate the phases of $\delta_1(z)$ and $\delta_2(z)$ along the stability boundary, namely the unit circle. Figure 10.23 shows that the phase difference $\phi_{\delta_1}(e^{j\theta}) - \phi_{\delta_2}(e^{j\theta})$ reaches $180°$ at around $\theta = 0.81$ radians.

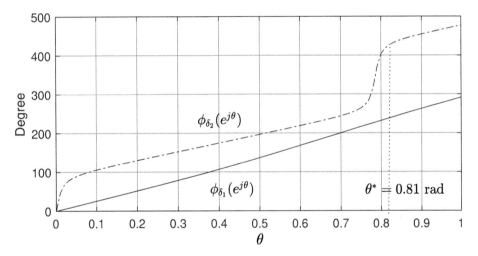

Figure 10.23 An unstable segment $(\phi_{\delta_1}(e^{j\theta^*}) - \phi_{\delta_2}(e^{j\theta^*}) \approx 182.55°)$ (Example 10.14)

Therefore, we conclude that there exists an unstable polynomial along the segment.

Lemma 10.1 can be extended to a more general class of segments. In particular, let $\delta_1(s)$, $\delta_2(s)$ be quasi-polynomials of the form

$$\delta_1(s) = as^n + \sum e^{-sT_i} a_i(s),$$
$$\delta_2(s) = bs^n + \sum e^{-sH_i} b_i(s), \tag{10.177}$$

where $T_i, H_i \geq 0$, $a_i(s)$, $b_i(s)$ have degrees less than n and a and b are arbitrary but nonzero and of the same sign. The Hurwitz stability of $\delta_1(s)$, $\delta_2(s)$ is then equivalent to their roots being in the left-half plane.

LEMMA 10.2   Lemma 10.1 holds for Hurwitz stability of the quasipolynomials of the form specified in Equation (10.177).[1]

---

[1]The proof of this lemma follows from the fact that Lemma 10.1 is equivalent to the continuity of the

EXAMPLE 10.15 (Quasi-polynomials)   Let us consider the Hurwitz stability of the line segment joining the following pair of quasi-polynomials:

$$\delta_1(s) = (s^2 + 3s + 2) + e^{-sT_1}(s+1) + e^{-sT_2}(2s+1),$$
$$\delta_2(s) = (s^2 + 5s + 3) + e^{-sT_1}(s+2) + e^{-sT_2}(2s+1),$$
(10.178)

where $T_1 = 1$ and $T_2 = 2$. We first check the stability of the endpoints by examining the frequency plots of

$$\frac{\delta_1(j\omega)}{(j\omega+1)^2} \quad \text{and} \quad \frac{\delta_2(j\omega)}{(j\omega+1)^2}.$$
(10.179)

Using the Principle of the Argument (equivalently, the Nyquist stability criterion) the condition for $\delta_1(s)$ (or $\delta_2(s)$) having all its roots in the left-half plane is simply that the plot should not encircle the origin since the denominator term $(s+1)^2$ does not have right-half plane roots. Figure 10.24 shows that both endpoints $\delta_1(s)$ and $\delta_2(s)$ are stable.

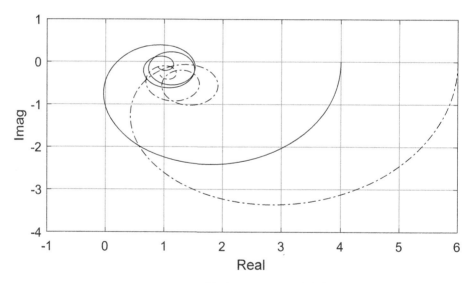

Figure 10.24 Stable quasi-polynomials $\frac{\delta_1(j\omega)}{(j\omega+1)^2}$ (solid line), $\frac{\delta_2(j\omega)}{(j\omega+1)^2}$ (dashed line) (Example 10.15)

We generate the polar plot of $\delta_1(j\omega)/\delta_2(\omega)$ (Figure 10.25). As this plot does not cut the negative real axis, the stability of the segment $[\delta_1(s), \delta_2(s)]$ is guaranteed by condition (d) of Lemma 10.1.

In the next section we focus specifically on the Hurwitz and Schur cases. We show how the frequency sweeping involved in using these results can always be *avoided* by

roots on the parameters, which applies to Hurwitz stability of quasi-polynomials $\delta_i(s)$ of the form given in Bhattacharyya, Chapellat, and Keel (1995) (see Theorem 1.14, Chapter 1).

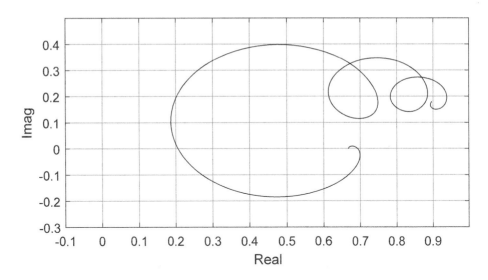

Figure 10.25 Stable segment of quasi-polynomials (Example 10.15)

isolating and checking only those frequencies where the phase difference can potentially reach 180°.

10.7.2 Hurwitz Case

In this subsection we are interested in the stability of a line segment of polynomials joining two Hurwitz polynomials. We start by introducing a simple lemma which deals with convex combinations of two real polynomials, and finds the conditions under which one of these convex combinations can have a pure imaginary root. Recall the even–odd decomposition of a real polynomial $\delta(s)$ and the notation $\delta(j\omega) = \delta^e(\omega) + j\omega\delta^o(\omega)$, where $\delta^e(\omega)$ and $\delta^o(\omega)$ are real polynomials in ω^2.

LEMMA 10.3   Let $\delta_1(\cdot)$ and $\delta_2(\cdot)$ be two arbitrary real polynomials (not necessarily stable). Then there exists $\lambda \in [0,1]$ such that $(1-\lambda)\delta_1(\cdot)+\lambda\delta_2(\cdot)$ has a pure imaginary root $j\omega$, with $\omega > 0$ if and only if

$$\delta_1^e(\omega)\delta_2^o(\omega) - \delta_2^e(\omega)\delta_1^o(\omega) = 0,$$
$$\delta_1^e(\omega)\delta_2^e(\omega) \leq 0, \tag{10.180}$$
$$\delta_1^o(\omega)\delta_2^o(\omega) \leq 0.$$

Proof   Suppose first that there exists some $\lambda \in [0,1]$ and $\omega > 0$ such that

$$(1-\lambda)\delta_1(j\omega)+\lambda\delta_2(j\omega) = 0. \tag{10.181}$$

We can write,

$$\delta_i(j\omega) = \delta_i^{\text{even}}(j\omega) + \delta_i^{\text{odd}}(j\omega)$$
$$= \delta_i^e(\omega) + j\omega\delta_i^o(\omega), \qquad \text{for } i = 1, 2. \tag{10.182}$$

Thus, taking Equation (10.182) and the fact that $\omega > 0$ into account, Equation (10.181) is equivalent to

$$
\begin{aligned}
(1-\lambda)\delta_1^e(\omega) + \lambda\delta_2^e(\omega) &= 0, \\
(1-\lambda)\delta_1^o(\omega) + \lambda\delta_2^o(\omega) &= 0.
\end{aligned}
\tag{10.183}
$$

But if Equation (10.183) holds then necessarily

$$
\delta_1^e(\omega)\delta_2^o(\omega) - \delta_2^e(\omega)\delta_1^o(\omega) = 0,
\tag{10.184}
$$

and since λ and $1-\lambda$ are both non-negative, Equation (10.183) also implies that

$$
\delta_1^e(\omega)\delta_2^e(\omega) \leq 0 \quad \text{and} \quad \delta_1^o(\omega)\delta_2^o(\omega) \leq 0,
\tag{10.185}
$$

and therefore this proves that the condition is necessary.

For the converse, there are two cases.

Case 1

Suppose that

$$
\begin{aligned}
\delta_1^e(\omega)\delta_2^o(\omega) - \delta_2^e(\omega)\delta_1^o(\omega) &= 0, \\
\delta_1^e(\omega)\delta_2^e(\omega) &\leq 0, \\
\delta_1^o(\omega)\delta_2^o(\omega) &\leq 0,
\end{aligned}
\tag{10.186}
$$

for some $\omega \geq 0$, but that we do not have $\delta_1^e(\omega) = \delta_2^e(\omega) = 0$, then

$$
\lambda = \frac{\delta_1^e(\omega)}{\delta_1^e(\omega) - \delta_2^e(\omega)}
\tag{10.187}
$$

satisfies Equation (10.183), and one can check easily that λ is in $[0,1]$.

Case 2

Suppose now that

$$
\delta_1^e(\omega)\delta_2^o(\omega) - \delta_2^e(\omega)\delta_1^o(\omega) = 0 \quad \text{and} \quad \delta_1^e(\omega) = \delta_2^e(\omega) = 0.
\tag{10.188}
$$

Then we are left with

$$
\delta_1^o(\omega)\delta_2^o(\omega) \leq 0.
\tag{10.189}
$$

Here again, if we do not have $\delta_1^o(\omega) = \delta_2^o(\omega) = 0$, then the following value of λ satisfies Equation (10.183):

$$
\lambda = \frac{\delta_1^o(\omega)}{\delta_1^o(\omega) - \delta_2^o(\omega)}.
\tag{10.190}
$$

If $\delta_1^o(\omega) = \delta_2^o(\omega) = 0$, then from Equation (10.188) we conclude that both $\lambda = 0$ and $\lambda = 1$ satisfy Equation (10.183) and this completes the proof. ∎

Based on this we may now state the Segment Lemma for the Hurwitz case.

LEMMA 10.4 (Hurwitz Segment Lemma)   Let $\delta_1(s)$, $\delta_2(s)$ be real Hurwitz polynomials of degree n with leading coefficients of the same sign. Then the line segment of polynomials $[\delta_1(s), \delta_2(s)]$ is Hurwitz stable if and only there exists no real $\omega > 0$ such that

$$\delta_1^e(\omega)\delta_2^o(\omega) - \delta_2^e(\omega)\delta_1^o(\omega) = 0,$$
$$\delta_1^e(\omega)\delta_2^e(\omega) \leq 0, \qquad\qquad (10.191)$$
$$\delta_1^o(\omega)\delta_2^o(\omega) \leq 0.$$

Proof   We first note that since the two polynomials $\delta_1(s)$ and $\delta_2(s)$ are of degree n with leading coefficients of the same sign, every polynomial on the segment is of degree n. Moreover, no polynomial on the segment has a real root at $s = 0$ because in such a case $\delta_1(0)\delta_2(0) \leq 0$, and this along with the assumption on the sign of the leading coefficients, contradicts the assumption that $\delta_1(s)$ and $\delta_2(s)$ are both Hurwitz. Therefore, an unstable polynomial can occur on the segment if and only if a segment polynomial has a root at $s = j\omega$ with $\omega > 0$. By the previous lemma this can occur if and only if the conditions in Equation (10.191) hold.   ∎

If we consider the image of the segment $[\delta_1(s), \delta_2(s)]$ evaluated at $s = j\omega$, we see that the conditions in Equation (10.191) of the Segment Lemma are the necessary and sufficient condition for the line segment $[\delta_1(j\omega), \delta_2(j\omega)]$ to pass through the origin of the complex plane. This in turn is equivalent to the phase difference condition $|\phi_{\delta_1}(j\omega) - \phi_{\delta_2}(j\omega)| = 180°$. We illustrate this in Figure 10.26.

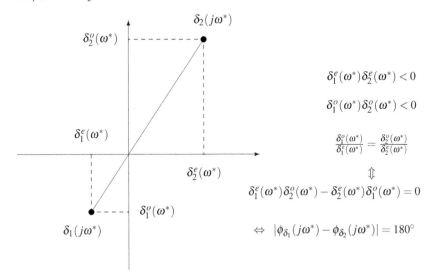

Figure 10.26 Segment Lemma: geometric interpretation

EXAMPLE 10.16   Consider the robust Hurwitz stability problem of the feedback system shown in Figure 10.27.

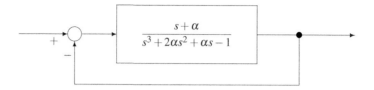

Figure 10.27 Feedback system (Example 10.16)

The characteristic polynomial is

$$\delta(s,\alpha) = s^3 + 2\alpha s^2 + (\alpha+1)s + (\alpha-1). \tag{10.192}$$

We have already verified the stability of the endpoints

$$\begin{aligned} \delta_1(s) &:= \delta(s,\alpha)|_{\alpha=2} = s^3 + 4s^2 + 3s + 1, \\ \delta_2(s) &:= \delta(s,\alpha)|_{\alpha=3} = s^3 + 6s^2 + 4s + 2. \end{aligned} \tag{10.193}$$

Robust stability of the system is equivalent to that of the segment

$$\delta_\lambda(s) = \lambda\delta_1(s) + (1-\lambda)\delta_2(s). \tag{10.194}$$

To apply the Segment Lemma we compute the real positive roots of the polynomial

$$\begin{aligned} &\delta_1^e(\omega)\delta_2^o(\omega) - \delta_2^e(\omega)\delta_1^o(\omega) \\ &= (-4\omega^2 + 1)(-\omega^2 + 4) - (-6\omega^2 + 2)(-\omega^2 + 3) = 0. \end{aligned} \tag{10.195}$$

This equation has no real root in ω and thus there is no $j\omega$ root on the line segment. Thus, from the Segment Lemma, the segment $[\delta_1(s),\delta_2(s)]$ is stable and the closed-loop system is robustly stable.

10.7.3 Schur Segment Lemma via Tchebyshev Representation

In this subsection, we use the Tchebyshev representation (see Section 11.4.1) to develop a segment lemma for Schur stability. Let $P_1(z)$, $P_2(z)$ be two real Schur stable polynomials of degree n. The question of interest is: Is the line segment of polynomials $\mathscr{S}$ joining $P_1(z)$ and $P_2(z)$ Schur stable?

By the Boundary Crossing Theorem, $\mathscr{S}$ is not Schur if and only if some polynomial in $\mathscr{S}$ has a unit circle root. To test this possibility, set $z = e^{j\theta}$ and let

$$\lambda P_1\left(e^{j\theta}\right) + (1-\lambda)P_2\left(e^{j\theta}\right) = 0 \tag{10.196}$$

for some $\theta \in [0, \pi]$.

Using the notation

$$P_{iC}(u) = R_i(u) + j\sqrt{1-u^2}T_i(u), \quad i = 1,2 \tag{10.197}$$

for the Tchebyshev representations of $P_i(z), i = 1,2$, we can write Equation (10.196) as

$$\lambda\left[R_1(u) + j\sqrt{1-u^2}T_1(u)\right] + (1-\lambda)\left[R_2(u) + j\sqrt{1-u^2}T_2(u)\right] = 0, \tag{10.198}$$

for $u \in [-1, 1]$ or, separating real and imaginary parts:

$$\lambda R_1(u) + (1 - \lambda)R_2(u) = 0, \tag{10.199a}$$

$$\lambda \sqrt{1 - u^2} T_1(u) + (1 - \lambda)\sqrt{1 - u^2} T_2(u) = 0. \tag{10.199b}$$

Since Equation (10.199b) holds at $u = \pm 1$, regardless of λ, we have, for roots to occur at $z = -1$ or $z = +1$:

$$\lambda R_1(+1) + (1 - \lambda)R_2(+1) = 0, \tag{10.200a}$$

$$\lambda R_1(-1) + (1 - \lambda)R_2(-1) = 0. \tag{10.200b}$$

Then Equation (10.200a) has a solution with $\lambda \in (0, 1)$ if and only if

$$R_1(+1)R_2(+1) \leq 0 \tag{10.201}$$

and Equation (10.200b) has a solution with $\lambda \in (0, 1)$ if and only if

$$R_1(-1)R_2(-1) \leq 0. \tag{10.202}$$

We can replace the inequalities in Equations (10.201) and (10.202) by strict inequalities because $R_i(\pm 1) \neq 0$ for $i = 1, 2$, since $P_i(z)$ are Schur. For $u \in (-1, 1)$, we can write Equations (10.199a) and (10.199b) as

$$\lambda R_1(u) + (1 - \lambda)R_2(u) = 0, \quad \text{for } \lambda \in [0, 1], \tag{10.203a}$$

$$\lambda T_1(u) + (1 - \lambda)T_2(u) = 0, \quad \text{for } u \in (-1, 1). \tag{10.203b}$$

A solution to Equations (10.203a) and (10.203b) exists if and only if

$$R_1(u)T_2(u) - R_2(u)T_1(u) = 0 \tag{10.204}$$

has at least one solution $u^* \in (-1, 1)$ such that

$$\operatorname{sgn}[R_1(u^*)]\operatorname{sgn}[R_2(u^*)] < 0. \tag{10.205}$$

To summarize, we have the following lemma.

LEMMA 10.5   The segment $\mathscr{S}$ is Schur stable if and only if

$$R_1(+1)R_2(+1) > 0, \tag{10.206a}$$

$$R_1(-1)R_2(-1) > 0, \tag{10.206b}$$

and for every real $u \in (-1, 1)$ such that

$$R_1(u)T_2(u) - R_2(u)T_1(u) = 0 \quad \text{and} \tag{10.207a}$$

$$\operatorname{sgn}[R_1(u)]\operatorname{sgn}[R_2(u)] > 0. \tag{10.207b}$$

EXAMPLE 10.17   Consider the following two Schur stable polynomials:

$$P_1(z) = z^4 + 1.3z^3 + 0.24z^2 - 0.298z - 0.136, \tag{10.208a}$$

$$P_2(z) = z^4 - 0.7z^3 + 0.42z^2 + 0.32z - 0.24. \tag{10.208b}$$

Then their respective Tchebyshev representations are as follows:

$$R_1(u) = 8u^4 - 5.2u^3 - 7.52u^2 + 4.198u + 0.624, \tag{10.209a}$$
$$T_1(u) = -8u^3 + 5.2u^2 + 3.52u - 1.598, \tag{10.209b}$$

and

$$R_2(u) = 8u^4 + 2.8u^3 - 7.1600u^2 - 2.42u + 0.3400, \tag{10.210a}$$
$$T_2(u) = -8u^3 - 2.8u^2 + 3.16u + 1.02. \tag{10.210b}$$

Using Lemma 10.5, we check the following conditions:

$$R_1(+1)R_2(+1) = 0.1591 > 0, \tag{10.211a}$$
$$R_1(-1)R_2(-1) = 1.6848 > 0, \tag{10.211b}$$
$$R_1(u)T_2(u) - R_2(u)T_1(u) = -0.832u^3 - 0.8432u^2 + 1.1898u + 1.1798. \tag{10.211c}$$

The roots of this polynomial are 1.1899, -1.25, and -0.9534. Since there is only one root in $u \in (-1,1)$, we evaluate

$$\text{sgn}\left[R_1(-0.9534)\right]\text{sgn}\left[R_2(-0.9534)\right] = (+1)(+1) > 0. \tag{10.212}$$

Since all the conditions in Lemma 10.5 are satisfied, the segment joining $P_1(z)$ and $P_2(z)$ is Schur stable.

Remark 10.2   It is easy to see that Schur stability is lost along the segment if the endpoint polynomials are of differing degrees or of equal degree with leading coefficients of opposite sign.

10.7.4   Some Fundamental Phase Relations

In this section, we develop some auxiliary results that will aid us in establishing the Vertex LemmaVertex Lemma, which deals with conditions under which vertex stability implies segment stability. The above results depend heavily on some fundamental formulas for the rate of change of phase with respect to frequency for fixed Hurwitz polynomials and for a segment of polynomials. These are derived in this section.

Phase Properties of Hurwitz Polynomials

Let $\delta(s)$ be a real or complex polynomial and write

$$\delta(j\omega) = p(\omega) + jq(\omega), \tag{10.213}$$

where $p(\omega)$ and $q(\omega)$ are real functions. Also let

$$X(\omega) := \frac{q(\omega)}{p(\omega)} \tag{10.214}$$

and

$$\varphi_\delta(\omega) := \tan^{-1}\frac{q(\omega)}{p(\omega)} = \tan^{-1}X(\omega). \tag{10.215}$$

Let $\mathrm{Im}[x]$ and $\mathrm{Re}[x]$ denote the imaginary and real parts of the complex number x.

LEMMA 10.6 If $\delta(s)$ is a real or complex Hurwitz polynomial:

$$\frac{dX(\omega)}{d\omega} > 0, \quad \text{for all } \omega \in [-\infty, +\infty]. \tag{10.216}$$

Equivalently,

$$\mathrm{Im}\left[\frac{1}{\delta(j\omega)} \frac{d\delta(j\omega)}{d\omega}\right] > 0, \quad \text{for all } \omega \in [-\infty, +\infty]. \tag{10.217}$$

Proof A Hurwitz polynomial satisfies the monotonic phase-increase property

$$\frac{d\varphi_\delta(\omega)}{d\omega} = \frac{1}{1 + X^2(\omega)} \frac{dX(\omega)}{d\omega} > 0, \quad \text{for all } \omega \in [-\infty, +\infty] \tag{10.218}$$

and this implies

$$\frac{dX(\omega)}{d\omega} > 0, \quad \text{for all } \omega \in [-\infty, +\infty]. \tag{10.219}$$

The formula

$$\frac{d\varphi_\delta(\omega)}{d\omega} = \mathrm{Im}\left[\frac{1}{\delta(j\omega)} \frac{d\delta(j\omega)}{d\omega}\right] \tag{10.220}$$

follows from the relations

$$\frac{1}{\delta(j\omega)} \frac{d\delta(j\omega)}{d\omega} = \frac{1}{p(\omega) + jq(\omega)} \left(\frac{dp(\omega)}{d\omega} + j\frac{dq(\omega)}{d\omega}\right) \tag{10.221}$$

$$= \frac{\left[p(\omega)\frac{dp(\omega)}{d\omega} + q(\omega)\frac{dq(\omega)}{d\omega}\right] + j\left[p(\omega)\frac{dq(\omega)}{d\omega} - q(\omega)\frac{dp(\omega)}{d\omega}\right]}{p^2(\omega) + q^2(\omega)}$$

and

$$\frac{d\varphi_\delta(\omega)}{d\omega} = \frac{\left[p(\omega)\frac{dq(\omega)}{d\omega} - q(\omega)\frac{dp(\omega)}{d\omega}\right]}{p^2(\omega) + q^2(\omega)}. \tag{10.222}$$

■

We shall see later that inequality in Equation (10.216) can be strengthened when $\delta(s)$ is a *real* Hurwitz polynomial.

Now let $\delta(s)$ be a *real* polynomial of degree n. We write

$$\delta(s) = \delta^{\mathrm{even}}(s) + \delta^{\mathrm{odd}}(s) = h(s^2) + sg(s^2) \tag{10.223}$$

where h and g are real polynomials in s^2. Then

$$\delta(j\omega) = h(-\omega^2) + j\omega g(-\omega^2)$$
$$= \rho_\delta(\omega)e^{j\varphi_\delta(\omega)}. \tag{10.224}$$

We associate with the real polynomial $\delta(s)$ the two auxiliary even-degree complex polynomials

$$\underline{\delta}(s) := h(s^2) + jg(s^2), \tag{10.225}$$
$$\bar{\delta}(s) := h(s^2) - js^2 g(s^2), \tag{10.226}$$

and write formulas analogous to Equation (10.224) for $\underline{\delta}(j\omega)$ and $\bar{\delta}(j\omega)$. $\delta(s)$ is *anti-Hurwitz* if and only if all its zeros lie in the open right-half plane ($\mathrm{Re}[s] > 0$). Let $t = s^2$ be a new complex variable.

LEMMA 10.7 Consider

$$h(t) + jg(t) = 0 \tag{10.227}$$

and

$$h(t) - jtg(t) = 0 \tag{10.228}$$

as equations in the complex variable t. If $\delta(s) = h(s^2) + sg(s^2)$ is Hurwitz and degree $\delta(s) \geq 2$, each of these equations has all its roots in the lower-half of the complex plane ($\mathrm{Im}[t] \leq 0$). When degree $[\delta(s)] = 1$, Equation (10.228) has all its roots in $\mathrm{Im}[t] \leq 0$.

Proof The statement regarding the case when degree$[\delta(s)] = 1$ can be directly checked. We therefore proceed with the assumption that degree$[\delta(s)] > 1$. Let

$$\delta(s) = a_0 + a_1 s + \cdots + a_n s^n. \tag{10.229}$$

Since $\delta(s)$ is Hurwitz, we can assume without loss of generality that $a_i > 0$, $i = 0, \ldots, n$. We have

$$h(-\omega^2) = a_0 + a_2(-\omega^2) + a_4(-\omega^2)^2 + \cdots,$$
$$\omega g(-\omega^2) = \omega \left[a_1 + a_3(-\omega^2) + a_5(-\omega^2)^2 + \cdots \right]. \tag{10.230}$$

As s runs from 0 to $+j\infty$, $-\omega^2$ runs from 0 to $-\infty$. We first recall the Hermite–Biehler Theorem of Chapter 2. According to this theorem if $\delta(s)$ is Hurwitz stable, all the roots of the two equations

$$h(t) = 0 \quad \text{and} \quad g(t) = 0 \tag{10.231}$$

are distinct, real, and negative. Furthermore, the interlacing property holds and the maximum of the roots is one of $h(t) = 0$. For the rest of the proof we will assume that $\delta(s)$ is of odd degree. A similar proof will hold for the case that $\delta(s)$ is of even degree. Let degree$[\delta(s)] = 2m + 1$ with $m \geq 1$. Note that the solutions of

$$h(t) + jg(t) = 0 \tag{10.232}$$

are identical with the solutions of

$$\frac{g(t)}{h(t)} = j. \tag{10.233}$$

Let us denote the roots of $h(t) = 0$ by $\lambda_1, \lambda_2, \ldots, \lambda_m$, where $\lambda_1 < \lambda_2 < \cdots < \lambda_m$. The sign of $h(t)$ changes alternately in each interval $]\lambda_i, \lambda_{i+1}[$, $(i = 1, \ldots, m-1)$.

If $\frac{g(t)}{h(t)}$ is expressed by partial fractions as

$$\frac{g(t)}{h(t)} = \frac{c_1}{t - \lambda_1} + \frac{c_2}{t - \lambda_2} + \cdots + \frac{c_m}{t - \lambda_m}, \tag{10.234}$$

then each $c_i, i = 1,\ldots,m$ should be positive. This is because when $t = -\omega^2$ passes increasingly (from left to right) through λ_i, the sign of $\frac{g(t)}{h(t)}$ changes from $-$ to $+$. This follows from the fact that $g(t)$ has just one root in each interval and $a_0 > 0$, $a_1 > 0$.

If we suppose

$$\mathrm{Im}[t] \geq 0, \tag{10.235}$$

then

$$\mathrm{Im}\left[\frac{c_i}{t - \lambda_i} \right] \leq 0 \quad i = 1, \ldots, m \tag{10.236}$$

and consequently we obtain

$$\mathrm{Im}\left[\frac{g(t)}{h(t)} \right] = \sum_{1 \leq i \leq m} \mathrm{Im}\left[\frac{c_i}{t - \lambda_i} \right] \leq 0. \tag{10.237}$$

Such a t cannot satisfy the relation in Equation (10.233). This implies that the equation

$$h(t) + jg(t) = 0 \tag{10.238}$$

in t has all its roots in the lower-half of the complex plane $\mathrm{Im}[t] \leq 0$. We can treat $[h(t) - jtg(t)]$ similarly. ∎

This lemma leads to a key monotonic phase property.

LEMMA 10.8   If $\delta(s)$ is Hurwitz and of degree ≥ 2, then

$$\frac{d\varphi_\delta}{d\omega} > 0 \tag{10.239}$$

and

$$\frac{d\varphi_{\bar{\delta}}}{d\omega} > 0. \tag{10.240}$$

Proof   From the previous lemma we have by factorizing $h(-\omega^2) + jg(-\omega^2)$:

$$h(-\omega^2) + jg(-\omega^2) = a_n(-\omega^2 - \alpha_1) \cdots (-\omega^2 - \alpha_m), \tag{10.241}$$

with some $\alpha_1, \alpha_2, \ldots, \alpha_m$ whose imaginary parts are negative. Now

$$\arg\left[h(-\omega^2) + jg(-\omega^2) \right] = \sum_{i=1}^{m} \arg(-\omega^2 - \alpha_i). \tag{10.242}$$

When $(-\omega^2)$ runs from 0 to $(-\infty)$, each component of the form $\arg(-\omega^2 - \alpha_i)$ is monotonically increasing. Consequently, $\arg\left[h(-\omega^2) + jg(-\omega^2) \right]$ is monotonically increasing as $(-\omega^2)$ runs from 0 to $(-\infty)$. In other words, $\arg\left[h(s^2) + jg(s^2) \right]$ is monotonically increasing as $s(= j\omega)$ runs from 0 to $j\infty$. This proves Equation (10.239); Equation (10.240) is proved in like manner. ∎

The dual result is given without proof.

LEMMA 10.9   If $\delta(s)$ is anti-Hurwitz, then

$$\frac{d\varphi_{\underline{\delta}}}{d\omega} < 0 \tag{10.243}$$

and

$$\frac{d\varphi_{\bar{\delta}}}{d\omega} < 0. \tag{10.244}$$

We remark that in Lemma 10.8, $\underline{\delta}(s)$ and $\bar{\delta}(s)$ are *not* Hurwitz even though they enjoy the respective monotonic phase properties in Equations (10.239) and (10.240). Similarly, in Lemma 10.9, $\underline{\delta}(s)$ and $\bar{\delta}(s)$ are not anti-Hurwitz even though they enjoy the monotonic phase properties in Equations (10.243) and (10.244), respectively.

The above results allow us to tighten the bound given in Lemma 10.6 on the rate of change of phase in the special case of a real Hurwitz polynomial.

THEOREM 10.7   For a real Hurwitz polynomial

$$\delta(s) = h(s^2) + sg(s^2), \tag{10.245}$$

the rate of change of the argument of $\delta(j\omega)$ is bounded below by

$$\frac{d\varphi_\delta(\omega)}{d\omega} \geq \left| \frac{\sin(2\varphi_\delta(\omega))}{2\omega} \right|, \qquad \text{for all } \omega > 0. \tag{10.246}$$

Equivalently, with

$$X(\omega) = \frac{\omega g\left(-\omega^2\right)}{h\left(-\omega^2\right)} \tag{10.247}$$

we have

$$\frac{dX(\omega)}{d\omega} \geq \left| \frac{X(\omega)}{\omega} \right|, \qquad \text{for all } \omega > 0. \tag{10.248}$$

In Equations (10.246) and (10.248), equality holds only when degree$[\delta(s)] = 1$.

Proof   The equivalence of the two conditions in Equations (10.246) and (10.248) follows from

$$\frac{d\varphi_\delta(\omega)}{d\omega} = \frac{1}{1 + X^2(\omega)} \frac{dX(\omega)}{d\omega} \tag{10.249}$$

and

$$\frac{1}{1+X^2(\omega)}\left|\begin{array}{c}X(\omega)\\\omega\end{array}\right|=\left|\frac{1}{1+X^2(\omega)}\frac{X(\omega)}{\omega}\right|$$

$$=\left|\frac{h^2\left(-\omega^2\right)}{h^2\left(-\omega^2\right)+\omega^2g^2\left(-\omega^2\right)}\frac{g\left(-\omega^2\right)}{h\left(-\omega^2\right)}\right|$$

$$=\left|\cos^2\left(\varphi_\delta(\omega)\right)\frac{1}{\omega}\tan\left(\varphi_\delta(\omega)\right)\right| \qquad (10.250)$$

$$=\left|\frac{1}{\omega}\cos\left(\varphi_\delta(\omega)\right)\sin\left(\varphi_\delta(\omega)\right)\right|$$

$$=\left|\frac{\sin\left(2\varphi_\delta(\omega)\right)}{2\omega}\right|.$$

We now prove Equation (10.248). The fact that equality holds in Equation (10.248) in the case where $\delta(s)$ has degree equal to 1 can easily be verified directly. We therefore proceed with the assumption that degree$[\delta(s)] \geq 2$. From Lemma 10.6, we know that

$$\frac{dX(\omega)}{d\omega} > 0 \qquad (10.251)$$

so that

$$\frac{dX(\omega)}{d\omega} = \frac{\frac{d(\omega g(-\omega^2))}{d\omega}h\left(-\omega^2\right)-\frac{d(h(-\omega^2))}{d\omega}\left(\omega g\left(-\omega^2\right)\right)}{h^2\left(-\omega^2\right)}$$

$$= \frac{g\left(-\omega^2\right)h\left(-\omega^2\right)+\omega\dot{g}\left(-\omega^2\right)h\left(-\omega^2\right)-\omega\dot{h}\left(-\omega^2\right)g\left(-\omega^2\right)}{h^2\left(-\omega^2\right)}$$

$$= \underbrace{\frac{g\left(-\omega^2\right)}{h\left(-\omega^2\right)}}_{\frac{X(\omega)}{\omega}}+\omega\underbrace{\frac{\dot{g}\left(-\omega^2\right)h\left(-\omega^2\right)-\dot{h}\left(-\omega^2\right)g\left(-\omega^2\right)}{h^2\left(-\omega^2\right)}}_{\frac{d}{d\omega}\left(\frac{g\left(-\omega^2\right)}{h\left(-\omega^2\right)}\right)} > 0. \qquad (10.252)$$

From Lemma 10.8 we have

$$\frac{d\varphi_\delta}{d\omega} > 0 \qquad \text{and} \qquad \frac{d\varphi_{\bar{\delta}}}{d\omega} > 0, \qquad (10.253)$$

where

$$\underline{\delta}(s) = h\left(s^2\right)+jg\left(s^2\right) \qquad \text{and} \qquad \bar{\delta}(s) = h\left(s^2\right)-js^2g\left(s^2\right). \qquad (10.254)$$

First consider

$$\frac{d\varphi_\delta}{d\omega} = \frac{1}{1+\left(\frac{g(-\omega^2)}{h(-\omega^2)}\right)^2}\frac{d}{d\omega}\left(\frac{g(-\omega^2)}{h(-\omega^2)}\right) > 0. \qquad (10.255)$$

Since

$$\frac{1}{1+\left(\frac{g(-\omega^2)}{h(-\omega^2)}\right)^2} > 0, \qquad (10.256)$$

we have

$$\frac{d}{d\omega}\left(\frac{g(-\omega^2)}{h(-\omega^2)}\right) > 0. \tag{10.257}$$

Thus, for $\omega > 0$, we have

$$\frac{dX(\omega)}{d\omega} = \frac{X(\omega)}{\omega} + \omega\frac{d}{d\omega}\left(\frac{g(-\omega^2)}{h(-\omega^2)}\right) > \frac{X(\omega)}{\omega}. \tag{10.258}$$

Now consider

$$\frac{d\varphi_{\tilde{\delta}}}{d\omega} = \frac{1}{1 + \left(\frac{\omega^2 g(-\omega^2)}{h(-\omega^2)}\right)^2}\frac{d}{d\omega}\left(\frac{\omega^2 g(-\omega^2)}{h(-\omega^2)}\right) > 0. \tag{10.259}$$

Here, we have

$$\frac{d}{d\omega}\left(\frac{\omega^2 g(-\omega^2)}{h(-\omega^2)}\right) = \omega\left[2\frac{g(-\omega^2)}{h(-\omega^2)} + \frac{\omega\left[h(-\omega^2)\dot{g}(-\omega^2) - g(-\omega^2)\dot{h}(-\omega^2)\right]}{h^2(-\omega^2)}\right]$$

$$= \omega\left[2\frac{X(\omega)}{\omega} + \omega\frac{d}{d\omega}\left(\frac{g(-\omega^2)}{h(-\omega^2)}\right)\right] > 0. \tag{10.260}$$

With $\omega > 0$, it follows that

$$2\frac{X(\omega)}{\omega} + \omega\frac{d}{d\omega}\left(\frac{g(-\omega^2)}{h(-\omega^2)}\right) > 0 \tag{10.261}$$

and therefore

$$\underbrace{\frac{X(\omega)}{\omega} + \omega\frac{d}{d\omega}\left(\frac{g(-\omega^2)}{h(-\omega^2)}\right)}_{\frac{dX(\omega)}{d\omega}} > -\frac{X(\omega)}{\omega}. \tag{10.262}$$

Combining Equations (10.258) and (10.262) we have, when degree$[\delta(s)] \geq 2$:

$$\frac{dX(\omega)}{d\omega} > \left|\frac{X(\omega)}{\omega}\right|, \qquad \text{for all } \omega > 0. \tag{10.263}$$

∎

10.7.5   Phase Relations for a Segment

Consider now a line segment $\lambda\delta_1(s) + (1 - \lambda)\delta_2(s)$, $\lambda \in [0, 1]$ generated by the two real polynomials $\delta_1(s)$ and $\delta_2(s)$ of degree n with leading coefficients and constant coefficients of the same sign. The necessary and sufficient condition for a polynomial in the interior of this segment to acquire a root at $s = j\omega_0$ is that

$$\lambda_0\delta_1^e(\omega_0) + (1 - \lambda_0)\delta_2^e(\omega_0) = 0, \tag{10.264a}$$

$$\lambda_0\delta_1^o(\omega_0) + (1 - \lambda_0)\delta_2^o(\omega_0) = 0, \tag{10.264b}$$

for some $\lambda_0 \in (0,1)$ and also

$$\lambda_0 \delta_1^e(\omega_0) + (1 - \lambda_0)\delta_2^e(\omega_0) = 0, \tag{10.265a}$$

$$\lambda_0 \omega_0 \delta_1^o(\omega_0) + (1 - \lambda_0)\omega_0 \delta_2^o(\omega_0) = 0, \tag{10.265b}$$

$$\lambda_0 \omega_0^2 \delta_1^o(\omega_0) + (1 - \lambda_0)\omega_0^2 \delta_2^o(\omega_0) = 0, \tag{10.265c}$$

since $\omega_0 > 0$. Noting that

$$
\begin{aligned}
\underline{\delta}_1(j\omega) &= \delta_1^e(\omega) + j\delta_1^o(\omega) = \rho_{\underline{\delta}_1}(\omega)e^{j\varphi_{\underline{\delta}_1}(\omega)}, \\
\underline{\delta}_2(j\omega) &= \delta_2^e(\omega) + j\delta_2^o(\omega) = \rho_{\underline{\delta}_2}(\omega)e^{j\varphi_{\underline{\delta}_2}(\omega)}, \\
\bar{\delta}_1(j\omega) &= \delta_1^e(\omega) + j\omega^2 \delta_1^o(\omega) = \varphi_{\bar{\delta}_1}(\omega)e^{j\varphi_{\bar{\delta}_1}(\omega)}, \\
\bar{\delta}_2(j\omega) &= \delta_2^e(\omega) + j\omega^2 \delta_2^o(\omega) = \rho_{\bar{\delta}_2}(\omega)e^{j\varphi_{\bar{\delta}_2}(\omega)},
\end{aligned}
\tag{10.266}
$$

we can write Equations (10.264b), (10.265b), and (10.265c), respectively, in the equivalent forms

$$\lambda_0 \delta_1(j\omega_0) + (1 - \lambda_0)\delta_2(j\omega_0) = 0, \tag{10.267a}$$

$$\lambda_0 \underline{\delta}_1(j\omega_0) + (1 - \lambda_0)\underline{\delta}_2(j\omega_0) = 0, \tag{10.267b}$$

$$\lambda_0 \bar{\delta}_1(j\omega_0) + (1 - \lambda_0)\bar{\delta}_2(j\omega_0) = 0. \tag{10.267c}$$

Now let

$$
\begin{aligned}
\delta_1(j\omega) &= \delta_1^e(\omega) + j\omega\delta_1^o(\omega) = \rho_{\delta_1}(\omega)e^{j\varphi_{\delta_1}(\omega)}, \\
\delta_2(j\omega) &= \delta_2^e(\omega) + j\omega\delta_2^o(\omega) = \rho_{\delta_2}(\omega)e^{j\varphi_{\delta_2}(\omega)},
\end{aligned}
\tag{10.268}
$$

$$\delta_0(s) := \delta_1(s) - \delta_2(s), \tag{10.269}$$

$$
\begin{aligned}
\underline{\delta}_0(j\omega) &:= \underline{\delta}_1(j\omega) - \underline{\delta}_2(j\omega), \\
\bar{\delta}_0(j\omega) &:= \bar{\delta}_1(j\omega) - \bar{\delta}_2(j\omega).
\end{aligned}
\tag{10.270}
$$

We now state a key technical lemma.

LEMMA 10.10 Let $\delta_1(s), \delta_2(s)$ be real polynomials of degree n with leading coefficients of the same sign and assume that $\lambda_0 \in (0,1)$ and $\omega_0 > 0$ satisfy Equations (10.265c)–(10.267c). Then

$$\left.\frac{d\varphi_{\delta_0}}{d\omega}\right|_{\omega=\omega_0} = \lambda_0 \left.\frac{d\varphi_{\delta_2}}{d\omega}\right|_{\omega=\omega_0} + (1 - \lambda_0)\left.\frac{d\varphi_{\delta_1}}{d\omega}\right|_{\omega=\omega_0}, \tag{10.271a}$$

$$\left.\frac{d\varphi_{\underline{\delta}_0}}{d\omega}\right|_{\omega=\omega_0} = \lambda_0 \left.\frac{d\varphi_{\underline{\delta}_2}}{d\omega}\right|_{\omega=\omega_0} + (1 - \lambda_0)\left.\frac{d\varphi_{\underline{\delta}_1}}{d\omega}\right|_{\omega=\omega_0}, \tag{10.271b}$$

and

$$\left.\frac{d\varphi_{\bar{\delta}_0}}{d\omega}\right|_{\omega=\omega_0} = \lambda_0 \left.\frac{d\varphi_{\bar{\delta}_2}}{d\omega}\right|_{\omega=\omega_0} + (1 - \lambda_0)\left.\frac{d\varphi_{\bar{\delta}_1}}{d\omega}\right|_{\omega=\omega_0}. \tag{10.272}$$

Proof   We prove only Equation (10.271a) in detail. If

$$\delta(j\omega) = p(\omega) + jq(\omega),\tag{10.273}$$

then

$$\tan\varphi_\delta(\omega) = \frac{q(\omega)}{p(\omega)}.\tag{10.274}$$

Let $\dot{q}(\omega) := \frac{dq(\omega)}{d\omega}$ and differentiate Equation (10.274) with respect to ω to get

$$(1+\tan^2\varphi_\delta(\omega))\frac{d\varphi_\delta}{d\omega} = \frac{p(\omega)\dot{q}(\omega) - q(\omega)\dot{p}(\omega)}{p^2(\omega)}\tag{10.275}$$

and

$$\frac{d\varphi_\delta}{d\omega} = \frac{p(\omega)\dot{q}(\omega) - q(\omega)\dot{p}(\omega)}{p^2(\omega) + q^2(\omega)}.\tag{10.276}$$

We apply the formula in Equation (10.276) to

$$\delta_0(j\omega) = (p_1(\omega) - p_2(\omega)) + j(q_1(\omega) - q_2(\omega))\tag{10.277}$$

to get

$$\frac{d\varphi_{\delta_0}}{d\omega} = \frac{(p_1-p_2)(\dot{q}_1-\dot{q}_2) - (q_1-q_2)(\dot{p}_1-\dot{p}_2)}{(p_1-p_2)^2 + (q_1-q_2)^2}.\tag{10.278}$$

Using Equation (10.267a):

$$\lambda_0 p_1(\omega_0) + (1-\lambda_0)p_2(\omega_0) = 0\tag{10.279}$$

and

$$\lambda_0 q_1(\omega_0) + (1-\lambda_0)q_2(\omega_0) = 0.\tag{10.280}$$

Since $\delta_1(s)$ and $\delta_2(s)$ are Hurwitz, $\lambda_0 \neq 0$ and $\lambda_0 \neq 1$, so that

$$p_1(\omega_0) - p_2(\omega_0) = -\frac{p_2(\omega_0)}{\lambda_0} = \frac{p_1(\omega_0)}{1-\lambda_0}\tag{10.281}$$

and

$$q_1(\omega_0) - q_2(\omega_0) = -\frac{q_2(\omega_0)}{\lambda_0} = \frac{q_1(\omega_0)}{1-\lambda_0}.\tag{10.282}$$

Substituting these relations into Equation (10.278), we have

$$
\begin{aligned}
\left.\frac{d\varphi_{\delta_0}}{d\omega}\right|_{\omega=\omega_0} &= \left.\frac{\frac{1}{1-\lambda_0}p_1\dot{q}_1 + \frac{1}{\lambda_0}p_2\dot{q}_2 - \frac{1}{1-\lambda_0}q_1\dot{p}_1 - \frac{1}{\lambda_0}q_2\dot{p}_2}{(p_1-p_2)^2 + (q_1-q_2)^2}\right|_{\omega=\omega_0}\\[2mm]
&= \left.\frac{\frac{1}{1-\lambda_0}(p_1\dot{q}_1 - q_1\dot{p}_1)}{\frac{p_1^2+q_1^2}{(1-\lambda_0)^2}}\right|_{\omega=\omega_0} + \left.\frac{\frac{1}{\lambda_0}(p_2\dot{q}_2 - q_2\dot{p}_2)}{\frac{p_2^2+q_2^2}{\lambda_0^2}}\right|_{\omega=\omega_0}\\[2mm]
&= \left.(1-\lambda_0)\frac{p_1\dot{q}_1 - q_1\dot{p}_1}{p_1^2+q_1^2}\right|_{\omega=\omega_0} + \left.\lambda_0\frac{p_2\dot{q}_2 - q_2\dot{p}_2}{p_2^2+q_2^2}\right|_{\omega=\omega_0}\\[2mm]
&= \left.(1-\lambda_0)\frac{d\varphi_{\delta_1}}{d\omega}\right|_{\omega=\omega_0} + \left.\lambda_0\frac{d\varphi_{\delta_2}}{d\omega}\right|_{\omega=\omega_0}.
\end{aligned}\tag{10.283}
$$

This proves Equation (10.271a). The proofs of Equations (10.271b) and (10.272) are identical starting from Equations (10.267b) and (10.267c), respectively.   ■

10.8   The Vertex Lemma

Using the results of the last section, it is possible to give frequency-independent conditions on $\delta_0(s) := \delta_1(s) - \delta_2(s)$ under which Hurwitz stability of the vertices implies stability of every polynomial on the segment $[\delta_1(s), \delta_2(s)]$. We first consider various special forms of the difference polynomial $\delta_0(s)$ for which this is possible. In each case we use Lemma 10.10 and Hurwitz stability of the vertices to contradict the hypothesis that the segment has unstable polynomials. We then combine the special cases to obtain the general result. This main result is presented as the Vertex Lemma.

We shall assume throughout this subsection that each polynomial on the segment $[\delta_1(s), \delta_2(s)]$ is of degree n. This will be true if and only if $\delta_1(s)$ and $\delta_2(s)$ are of degree n and their leading coefficients are of the same sign. We shall assume this without loss of generality.

We first consider real polynomials of the form

$$\delta_0(s) = s^t(as+b)P(s), \tag{10.284}$$

where t is a non-negative integer and $P(s)$ is odd or even. Suppose, arbitrarily, that t is even and $P(s) = E(s)$ an even polynomial. Then

$$\delta_0(s) = \underbrace{s^t E(s)b}_{\delta_0^{\text{even}}(s)} + \underbrace{s^{t+1}E(s)a}_{\delta_0^{\text{odd}}(s)}. \tag{10.285}$$

Defining $\underline{\delta_0}(j\omega)$ as before, we see that

$$\tan \underline{\delta_0}(\omega) = \frac{a}{b} \tag{10.286}$$

so that

$$\frac{d\varphi_{\delta_0}}{d\omega} = 0. \tag{10.287}$$

From Lemma 10.10 (i.e., Equation (10.271b)), we see that

$$\lambda_0 \frac{d\varphi_{\delta_2}}{d\omega}\bigg|_{\omega=\omega_0} + (1-\lambda_0)\frac{d\varphi_{\delta_1}}{d\omega}\bigg|_{\omega=\omega_0} = 0 \tag{10.288}$$

and from Lemma 10.8 we see that if $\delta_1(s)$ and $\delta_2(s)$ are Hurwitz, then

$$\frac{d\varphi_{\delta_2}}{d\omega} > 0 \quad \text{and} \quad \frac{d\varphi_{\delta_1}}{d\omega} > 0 \tag{10.289}$$

so that Equation (10.288) cannot be satisfied for $\lambda_0 \in [0,1]$. An identical argument works when t is odd. The case when $P(s) = O(s)$ is an odd polynomial can be handled similarly by using Equation (10.272) in Lemma 10.10. The details are left to the reader. Thus, we are led to the following result.

LEMMA 10.11   If $\delta_0(s) = s^t(as+b)P(s)$ where $t \geq 0$ is an integer, a and b are arbitrary real numbers, and $P(s)$ is an even or odd polynomial, then stability of the segment $[\delta_1(s), \delta_2(s)]$ is implied by those of the endpoints $\delta_1(s)$, $\delta_2(s)$.

We can now prove the following general and useful result.

LEMMA 10.12 (Hurwitz Vertex Lemma)

(a) Let $\delta_1(s)$ and $\delta_2(s)$ be real polynomials of degree n with leading coefficients of the same sign and let

$$\delta_0(s) = \delta_1(s) - \delta_2(s) = A(s)s^t(as+b)P(s), \tag{10.290}$$

where $A(s)$ is anti-Hurwitz, $t \geq 0$ is an integer, a, b are arbitrary real numbers, and $P(s)$ is even or odd. Then stability of the segment $[\delta_1(s), \delta_2(s)]$ is implied by that of the endpoints $\delta_1(s)$, $\delta_2(s)$.

(b) When $\delta_0(s)$ is not of the form specified in (a), stability of the endpoints is not sufficient to guarantee that of the segment.

Proof

(a) Write

$$A(s) = A^{\text{even}}(s) + A^{\text{odd}}(s) \tag{10.291}$$

and let

$$\bar{A}(s) := A^{\text{even}}(s) - A^{\text{odd}}(s). \tag{10.292}$$

Since $A(s)$ is anti-Hurwitz, $\bar{A}(s)$ is Hurwitz. Now consider the segment $[\bar{A}(s)\delta_1(s), \bar{A}(s)\delta_2(s)]$, which is Hurwitz if and only if $[\delta_1(s), \delta_2(s)]$ is Hurwitz. But

$$\bar{A}(s)\delta_0(s) = \bar{A}(s)\delta_1(s) - \bar{A}(s)\delta_2(s)$$
$$= \underbrace{[(A^{\text{even}}(s))^2 - (A^{\text{odd}}(s))^2]}_{T(s)} s^t(as+b)P(s). \tag{10.293}$$

Since $T(s)$ is an even polynomial, we may use Lemma 10.11 to conclude that the segment $[\bar{A}(s)\delta_1(s), \bar{A}(s)\delta_2(s)]$ is Hurwitz if and only if $\bar{A}(s)\delta_1(s)$ and $\bar{A}(s)\delta_2(s)$ are. Since $\bar{A}(s)$ is Hurwitz, it follows that the segment $[\delta_1(s), \delta_2(s)]$ is Hurwitz if and only if the endpoints $\delta_1(s)$ and $\delta_2(s)$ are.

(b) We prove this part by means of the following example. Consider the segment

$$\delta_\lambda(s) = (2+14\lambda)s^4 + (5+14\lambda)s^3 + (6+14\lambda)s^2 + 4s + 3.5. \tag{10.294}$$

Now set $\lambda = 0$ and $\lambda = 1$, then we have

$$\begin{aligned}\delta_\lambda|_{\lambda=0} &= \delta_1(s) = 2s^4 + 5s^3 + 6s^2 + 4s + 3.5,\\ \delta_\lambda|_{\lambda=1} &= \delta_2(s) = 16s^4 + 19s^3 + 20s^2 + 4s + 3.5,\end{aligned} \tag{10.295}$$

and consequently

$$\begin{aligned}\delta_0(s) &= \delta_2(s) - \delta_1(s)\\ &= 14s^4 + 14s^3 + 14s^2 = 14s^2(s^2+s+1).\end{aligned} \tag{10.296}$$

It can be verified that two endpoints $\delta_1(s)$ and $\delta_2(s)$ are Hurwitz. Notice that since $(s^2 + s + 1)$ is Hurwitz with a pair of complex conjugate roots, $\delta_0(s)$ cannot be partitioned into the form of Equation (10.290). Therefore, we conclude that when $\delta_0(s)$ is not of the form specified in Equation (10.290), stability of the endpoints is not sufficient to guarantee that of the segment.

■

EXAMPLE 10.18 Suppose that the transfer function of a plant containing an uncertain parameter is written in the form

$$P(s) = \frac{P_2}{P_1(s) + \lambda P_0(s)}, \tag{10.297}$$

where the uncertain parameter λ varies in $[0, 1]$, and the degree of $P_1(s)$ is greater than those of $P_0(s)$ or $P_2(s)$. Suppose that a unity feedback controller is to be designed so that the plant output follows step and ramp inputs and rejects sinusoidal disturbances of radian frequency ω_0. Let us denote the controller by

$$C(s) = \frac{Q_2(s)}{Q_1(s)}. \tag{10.298}$$

A possible choice of $Q_1(s)$ which will meet the tracking and disturbance rejection requirements is

$$Q_1(s) = s^2(s^2 + \omega_0^2)(as + b), \tag{10.299}$$

with $Q_2(s)$ being of degree 5 or less. The stability of the closed-loop requires that the segment

$$\delta_\lambda(s) = Q_2(s)P_2(s) + Q_1(s)(P_1(s) + \lambda P_0(s)) \tag{10.300}$$

be Hurwitz stable. The corresponding difference polynomial $\delta_0(s)$ is

$$\delta_0(s) = Q_1(s)P_0(s). \tag{10.301}$$

With $Q_1(s)$ of the form shown above, it follows that $\delta_0(s)$ is of the form specified in the Vertex Lemma if $P_0(s)$ is anti-Hurwitz or even or odd, or a product thereof. Thus, in such a case, robust stability of the closed loop would be equivalent to the stability of the two vertex polynomials

$$\begin{aligned} \delta_1(s) &= Q_2(s)P_2(s) + Q_1(s)P_1(s), \\ \delta_2(s) &= Q_2(s)P_2(s) + Q_1(s)P_1(s) + Q_1(s)P_0(s). \end{aligned} \tag{10.302}$$

Let $\omega_0 = 1$, $a = 1$, $b = 1$, and

$$P_1(s) = s^2 + s + 1, \quad P_0(s) = s(s-1), \quad P_2(s) = s^2 + 2s + 1,$$
$$Q_2(s) = s^5 + 5s^4 + 10s^3 + 10s^2 + 5s + 1. \tag{10.303}$$

Since $P_0(s) = s(s-1)$ is the product of an odd and an anti-Hurwitz polynomial, the conditions of the Vertex Lemma are satisfied and robust stability is equivalent to that of the two vertex polynomials

$$\delta_1(s) = 2s^7 + 9s^6 + 24s^5 + 38s^4 + 37s^3 + 22s^2 + 7s + 1,$$
$$\delta_2(s) = 3s^7 + 9s^6 + 24s^5 + 38s^4 + 36s^3 + 22s^2 + 7s + 1. \qquad (10.304)$$

Since $\delta_1(s)$ and $\delta_2(s)$ are Hurwitz, the controller

$$C(s) = \frac{Q_2(s)}{Q_1(s)} = \frac{s^5 + 5s^4 + 10s^3 + 10s^2 + 5s + 1}{s^2(s^2 + 1)(s + 1)} \qquad (10.305)$$

robustly stabilizes the closed-loop system and provides robust asymptotic tracking and disturbance rejection.

The Vertex Lemma can easily be extended to the case of Schur stability.

LEMMA 10.13 (Schur Vertex Lemma)

(a) Let $\delta_1(z)$ and $\delta_2(z)$ be polynomials of degree n with $\delta_1(1)$ and $\delta_2(1)$ nonzero and of the same sign, and with leading coefficients of the same sign. Let

$$\delta_0(z) = \delta_1(z) - \delta_2(z)$$
$$= A(z)(z - 1)^{t_1}(z + 1)^{t_2}(az + b)P(z) \qquad (10.306)$$

where $A(z)$ is anti-Schur, $t_1, t_2 \geq 0$ are integers, a, b are arbitrary real numbers, and $P(z)$ is symmetric or antisymmetric. Then Schur stability of the segment $[\delta_1(z), \delta_2(z)]$ is implied by that of the endpoints $\delta_1(z), \delta_2(z)$.

(b) When $\delta_0(z)$ is not of the form specified in (a), Schur stability of the endpoints is not sufficient to guarantee that of the segment.

Proof The proof is based on applying the bilinear transformation and using the corresponding results for the Hurwitz case. Let $P(z)$ be any polynomial and let

$$\hat{P}(s) := (s - 1)^n P\left(\frac{s + 1}{s - 1}\right). \qquad (10.307)$$

If $P(z)$ is of degree n, so is $\hat{P}(s)$ provided $P(1) \neq 0$. Now apply the bilinear transformation to the polynomials $\delta_0(z)$, $\delta_1(z)$, and $\delta_2(z)$ to get $\hat{\delta}_0(s)$, $\hat{\delta}_1(s)$, and $\hat{\delta}_2(s)$, where $\hat{\delta}_0(s) = \hat{\delta}_1(s) - \hat{\delta}_2(s)$. The proof consists of showing that under the assumption that $\delta_0(z)$ is of the form given in Equation (10.306), $\hat{\delta}_0(s)$, $\hat{\delta}_1(s)$, and $\hat{\delta}_2(s)$ satisfy the conditions of the Vertex Lemma for the Hurwitz case. Since $\delta_1(1)$ and $\delta_2(1)$ are of the same sign, $\delta_\lambda(1) = \lambda\delta_1(1) + (1 - \lambda)\delta_2(1) \neq 0$ for $\lambda \in [0, 1]$. This in turn implies that $\hat{\delta}_\lambda(s)$ is of degree n for all $\lambda \in [0, 1]$. A straightforward calculation shows that

$$\hat{\delta}_0(s) = \hat{A}(s)2^{t_1}(2s)^{t_2}(cs + d)\hat{P}(s), \qquad (10.308)$$

which is precisely the form required in the Vertex Lemma for the Hurwitz case. Thus, the segment $\hat{\delta}_\lambda(s)$ cannot have a $j\omega$ root and the segment $\delta_\lambda(z)$ cannot have a root on the unit circle. Therefore, Schur stability of $\delta_1(z)$ and $\delta_2(z)$ guarantees that of the segment. ∎

10.9   Interval Polynomials: Kharitonov's Theorem

Kharitonov's Theorem considers a special polytope of polynomials (interval polynomials) and provides a surprising and useful result for checking its robust Hurwitz stability.

Consider the polytopic family of real polynomials defined by

$$\delta(s) = \delta_0 + \delta_1 s + \cdots + \delta_i s^i + \cdots + \delta_n s^n \tag{10.309}$$

and

$$\mathscr{I}(s) := \{\delta(s) : x_i \leq \delta_i \leq y_i, \quad i = 0, 1, \ldots, n\}. \tag{10.310}$$

Kharitonov's Theorem provides a remarkably simple and effective test for the Hurwitz stability of every element of $\mathscr{I}(s)$, called an interval familyinterval family of polynomials. $\mathscr{I}(s)$ is clearly an $(n+1)$-dimensional polytope.

We assume that $\mathscr{I}(s)$ is degree invariant with positive leading coefficient, that is, $\delta_n > 0$, or equivalent:

$$0 \notin [x_n, y_n], \quad x_n > 0. \tag{10.311}$$

Introduce the four polynomials:

$$
\begin{aligned}
K_1(s) &= x_0 + x_1 s + y_2 s^2 + y_3 s^3 + x_4 s^4 + x_5 s^5 + y_6 s^6 + \cdots, \\
K_2(s) &= x_0 + y_1 s + y_2 s^2 + x_3 s^3 + x_4 s^4 + y_5 s^5 + y_6 s^6 + \cdots, \\
K_3(s) &= y_0 + y_1 s + x_2 s^2 + x_3 s^3 + y_4 s^4 + y_5 s^5 + x_6 s^6 + \cdots, \\
K_4(s) &= y_0 + x_1 s + x_2 s^2 + y_3 s^3 + y_4 s^4 + x_5 s^5 + x_6 s^6 + \cdots,
\end{aligned}
\tag{10.312}
$$

called the Kharitonov polynomials.Kharitonov polynomials

THEOREM 10.8 (Kharitonov's Theorem)   Each member of the interval polynomial $\mathscr{I}(s)$ is Hurwitz stable if and only if the four Kharitonov polynomials $K_i(s)$ in Equation (10.312) are Hurwitz stable.

The proof presented below will depend on two preliminary results. Let us define

$$
\begin{aligned}
K_{\max}^{\text{even}}(s) &:= y_o + x_2 s^2 + y_4 s^4 + x_6 s^6 + y_8 s^8 + \cdots, \\
K_{\min}^{\text{even}}(s) &:= x_o + y_2 s^2 + x_4 s^4 + y_6 s^6 + x_8 s^8 + \cdots, \\
K_{\max}^{\text{odd}}(s) &:= y_1 s + x_3 s^3 + y_5 s^5 + x_7 s^7 + y_9 s^9 + \cdots, \\
K_{\min}^{\text{odd}}(s) &:= x_1 s + y_3 s^3 + x_5 s^5 + y_7 s^7 + x_9 s^9 + \cdots,
\end{aligned}
\tag{10.313}
$$

and note that $K_i(s)$ can be rewritten as

$$
\begin{aligned}
K_1(s) &= K_{\min}^{\text{even}}(s) + K_{\min}^{\text{odd}}(s), \\
K_2(s) &= K_{\min}^{\text{even}}(s) + K_{\max}^{\text{odd}}(s), \\
K_3(s) &= K_{\max}^{\text{even}}(s) + K_{\max}^{\text{odd}}(s), \\
K_4(s) &= K_{\max}^{\text{even}}(s) + K_{\min}^{\text{odd}}(s).
\end{aligned}
\tag{10.314}
$$

An arbitrary polynomial $\delta(s) \in \mathscr{I}(s)$ can be decomposed as shown below:

$$\delta(s) = \underbrace{\left(\delta_0 + \delta_2 s^2 + \delta_4 s^4 + \cdots\right)}_{\delta_{\text{even}}(s)} + s \underbrace{\left(\delta_1 + \delta_3 s^2 + \delta_5 s^4 + \cdots\right)}_{\delta_{\text{odd}}(s)} \qquad (10.315\text{a})$$

$$= \delta_{\text{even}}(s) + \delta_{\text{odd}}(s). \qquad (10.315\text{b})$$

LEMMA 10.14   The complex plane image set $\mathscr{I}(j\omega)$ of $\mathscr{I}(s)$ is evaluated as

$$\mathscr{I}(j\omega) = \text{convex hull of } \{K_1(j\omega), K_2(j\omega), K_3(j\omega), K_4(j\omega)\}, \qquad (10.316)$$

for $\omega \in [0, \infty)$, and is an axis-parallel rectangle as shown in Figure 10.28.

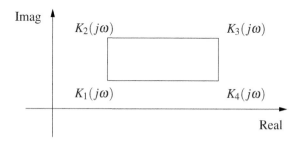

Figure 10.28 A Kharitonov rectangle

Proof   Note that

$$K_{\min}^{even}(j\omega) \leq \delta_{\text{even}}(j\omega) \leq K_{\max}^{even}(j\omega), \qquad (10.317\text{a})$$

$$\frac{1}{j} K_{\min}^{odd}(j\omega) \leq \frac{1}{j} \delta_{\text{odd}}(j\omega) \leq \frac{1}{j} K_{\max}^{odd}(j\omega). \qquad (10.317\text{b})$$

∎

To proceed, we introduce the two-parameter family of polynomials

$$\delta(\lambda, \mu, s) := (1 - \lambda) K_{\min}^{even}(s) + \lambda K_{\max}^{even}(s) + (1 - \mu) K_{\min}^{odd}(s) + \mu K_{\max}^{odd}(s) \qquad (10.318)$$

and

$$\mathscr{J}(s) := \{\delta(\lambda, \mu, s) \;:\; \lambda \in [0,1], \mu \in [0,1]\}. \qquad (10.319)$$

Clearly

$$\mathscr{J}(s) \subset \mathscr{I}(s) \qquad (10.320)$$

and $\mathscr{J}(s)$ is a two-dimensional polytope.

LEMMA 10.15 $\mathscr{J}(s)$ is robustly Hurwitz stable if and only if $\mathscr{I}(s)$ is robustly Hurwitz stable.

Proof   The implication "only of" is obvious from Equation (10.320). To prove the implication "if," note that the image sets of $\mathscr{I}(s)$ and $\mathscr{J}(s)$ are identical:

$$\mathscr{J}(j\omega) = \mathscr{I}(j\omega), \quad \text{for all } \omega \in [0, \infty). \qquad (10.321)$$

Therefore, robust Hurwitz stability of $\mathcal{J}(s)$ implies that of $\mathcal{I}(s)$.   ∎

Proof of Kharitonov's Theorem (Theorem 10.8) $\mathcal{J}(s)$ is robustly Hurwitz if and only if its exposed edges exposed edges are. From Equations (10.318) and (10.319), these exposed edges are precisely the segments

$$[K_1(s), K_2(s)], [K_2(s), K_3(s)], [K_3(s), K_4(s)], [K_4(s), K_1(s)]. \tag{10.322}$$

By the Vertex Lemma, Hurwitz stability of these segments is implied by that of the vertices, since the differences of these vertices are either even or odd. These vertices are just the Kharitonov polynomials in Equation (10.312), completing the proof.   ∎

10.10   Generalized Kharitonov Theorem

10.10.1   Motivation

For many control systems, Kharitonov's Theorem has a conservatism associated with the independence of variation of all characteristic polynomial coefficients. We illustrate this in the example below.

EXAMPLE 10.19   Consider the plant

$$G(s) = \frac{n^p(s)}{d^p(s)} = \frac{s}{1 - s + \alpha s^2 + s^3}, \quad \text{where } \alpha \in [3.4, 5] \tag{10.323}$$

and has a nominal value $\alpha^0 = 4$. It is easy to check that the controller

$$C(s) = \frac{3}{s+1} \tag{10.324}$$

stabilizes the nominal plant, yielding the nominal closed-loop characteristic polynomial

$$\delta_4(s) = 1 + 3s + 3s^2 + 5s^3 + s^4. \tag{10.325}$$

To determine whether $C(s)$ also stabilizes the family of perturbed plants, we observe that the characteristic polynomial of the system is

$$\delta_\alpha(s) = 1 + 3s + (\alpha - 1)s^2 + (\alpha + 1)s^3 + s^4. \tag{10.326}$$

In the space (δ_2, δ_3), the coefficients of s^2 and s^3 describe the segment $[R_1, R_2]$ shown in Figure 10.29.

The only way to apply Kharitonov's Theorem here is to enclose this segment in the box $\mathcal{B}$ defined by the two "real" points R_1 and R_2 and two "artificial" points A_1 and A_2 and to check the stability of the Kharitonov polynomials which correspond to the characteristic polynomial evaluated at the four corners of $\mathcal{B}$. But

$$\delta_{A_1}(s) = 1 + 3s + 2.4s^2 + 6s^3 + s^4 \tag{10.327}$$

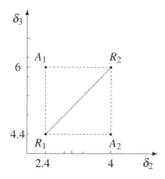

Figure 10.29 A box in parameter space is transformed into a segment in coefficient space (Example 10.19)

is unstable because its third Hurwitz determinant H_3 is

$$H_3 = \det \begin{bmatrix} 6 & 3 & 0 \\ 1 & 2.4 & 1 \\ 0 & 6 & 3 \end{bmatrix} = -1.8 < 0.$$

Therefore, using Kharitonov's Theorem here does not allow us to conclude the stability of the entire family of closed-loop systems. And yet, if one checks the values of the Hurwitz determinants along the segment $[R_1, R_2]$, one finds

$$H = \det \begin{bmatrix} 1+\alpha & 3 & 0 & 0 \\ 1 & \alpha-1 & 1 & 0 \\ 0 & 1+\alpha & 3 & 0 \\ 0 & 1 & \alpha-1 & 1 \end{bmatrix} \tag{10.328}$$

and

$$\begin{cases} H_1 = 1+\alpha, \\ H_2 = \alpha^2 - 4, \\ H_3 = 2\alpha^2 - 2\alpha - 13, \\ H_4 = H_3, \end{cases} \quad \text{all positive for } \alpha \in [3.4, 5]. \tag{10.329}$$

This example demonstrates that Kharitonov's Theorem provides only sufficient conditions, which may sometimes be too conservative for control problems.

Motivated by these considerations, we formulate the problem of generalizing Kharitonov's Theorem in the next subsection. Before proceeding to the main results, we introduce some notation and notational conventions with a view towards streamlining the presentation.

10.10.2 Problem Formulation and Notation

Motivated by applications to control systems, we will be dealing with polynomials of the form

$$\delta(s) = F_1(s)P_1(s) + F_2(s)P_2(s) + \cdots + F_m(s)P_m(s), \tag{10.330}$$

where the $F_i(s)$ represent the controller and $P_j(s)$ the plant. Write

$$\underline{F}(s) := (F_1(s), F_2(s), \ldots, F_m(s)), \tag{10.331a}$$

$$\underline{P}(s) := (P_1(s), P_2(s), \ldots, P_m(s)), \tag{10.331b}$$

and introduce the notation

$$< \underline{F}(s), \underline{P}(s) > := F_1(s)P_1(s) + F_2(s)P_2(s) + \cdots + F_m(s)P_m(s). \tag{10.332}$$

We will say that $\underline{F}(s)$ *stabilizes* $\underline{P}(s)$ if $\delta(s) = <\underline{F}(s), \underline{P}(s)>$ is *Hurwitz stable*. The polynomials $F_i(s)$ are assumed to be fixed real polynomials, whereas $P_i(s)$ are real polynomials with coefficients varying independently in prescribed intervals.

An extension of the results to the case where $F_i(s)$ are complex polynomials or quasi-polynomials will also be given in a later section.

Let $d^o(P_i)$ be the degree of $P_i(s)$:

$$P_i(s) := p_{i,0} + p_{i,1}s + \cdots + p_{i,d^o(P_i)}s^{d^o(P_i)} \tag{10.333}$$

and

$$\mathbf{p}_i := [p_{i,0}, p_{i,1}, \ldots, p_{i,d^o(P_i)}]. \tag{10.334}$$

Let $\mathbf{n} = [1, 2, \ldots, n]$. Each $P_i(s)$ belongs to an interval family $\mathbf{P}_i(s)$ specified by the intervals

$$p_{i,j} \in [\alpha_{i,j}, \beta_{i,j}], \quad \text{for } i \in \mathbf{m}, \quad j = 0, \ldots, d^o(P_i). \tag{10.335}$$

The corresponding parameter box is

$$\Pi_i := \{\mathbf{p}_i : \alpha_{i,j} \leq p_{i,j} \leq \beta_{i,j}, \quad j = 0, 1, \ldots, d^o(P_i)\}. \tag{10.336}$$

Write $\underline{P}(s) := [P_1(s), \ldots, P_m(s)]$ and introduce the family of m-tuples

$$\mathbf{P}(s) := \mathbf{P}_1(s) \times \mathbf{P}_2(s) \times \cdots \times \mathbf{P}_m(s). \tag{10.337}$$

Let

$$\mathbf{p} := [\mathbf{p}_1, \mathbf{p}_2, \ldots, \mathbf{p}_m] \tag{10.338}$$

denote the global parameter vector and let

$$\Pi := \Pi_1 \times \Pi_2 \times \cdots \times \Pi_m \tag{10.339}$$

denote the global parameter uncertainty set. Now let us consider the polynomial in Equation (10.330) and rewrite it as $\delta(s, \mathbf{p})$ or $\delta(s, \underline{P}(s))$ to emphasize its dependence on the parameter vector $\mathbf{p}$ or the m-tuple $\underline{P}(s)$. We are interested in determining the Hurwitz stability of the set of polynomials

$$\Delta(s) := \{\delta(s,\mathbf{p}) : \mathbf{p} \in \Pi\}$$
$$= \{< \underline{F}(s), \underline{P}(s) > : \underline{P}(s) \in \mathbf{P}(s)\}. \tag{10.340}$$

We call this a *linear interval polynomial* and adopt the convention

$$\Delta(s) = F_1(s)\mathbf{P}_1(s) + F_2(s)\mathbf{P}_2(s) + \cdots + F_m(s)\mathbf{P}_m(s). \tag{10.341}$$

We shall make the following standing assumptions about this family.

ASSUMPTION 10.1

(1) Elements of $\mathbf{p}$ perturb independently of each other. Equivalently, Π is an axis-parallel rectangular box.
(2) Every polynomial in $\Delta(s)$ is of the same degree.

The above assumptions will allow us to use the usual results such as Edge Theorem to develop the solution. It is also justified from a control system viewpoint, where loss of the degree of the characteristic polynomial also implies loss of bounded-input bounded-output stability. Henceforth, we will say that $\Delta(s)$ is stable if every polynomial in $\Delta(s)$ is Hurwitz stable. An equivalent statement is that $\underline{F}(s)$ stabilizes every $\underline{P}(s) \in \mathbf{P}(s)$.

The solution developed below constructs an extremal set of line segments $\Delta_E(s) \subset \Delta(s)$ with the property that the stability of $\Delta_E(s)$ implies stability of $\Delta(s)$. This solution is constructive because the stability of $\Delta_E(s)$ can be checked, for instance by a set of root locus problems. The solution will be efficient since the number of elements of $\Delta_E(s)$ will be independent of the dimension of the parameter space Π. The extremal subset $\Delta_E(s)$ will be generated by first constructing an extremal subset $\mathbf{P}_E(s)$ of the m-tuple family $\mathbf{P}(s)$. The extremal subset $\mathbf{P}_E(s)$ is constructed from the Kharitonov polynomials of $\mathbf{P}_i(s)$. We describe the construction of $\Delta_E(s)$ next.

Construction of the Extremal Subset

The Kharitonov polynomials corresponding to each $\mathbf{P}_i(s)$ are

$$\begin{aligned}
K_i^1(s) &= \alpha_{i,0} + \alpha_{i,1}s + \beta_{i,2}s^2 + \beta_{i,3}s^3 + \cdots, \\
K_i^2(s) &= \alpha_{i,0} + \beta_{i,1}s + \beta_{i,2}s^2 + \alpha_{i,3}s^3 + \cdots, \\
K_i^3(s) &= \beta_{i,0} + \alpha_{i,1}s + \alpha_{i,2}s^2 + \beta_{i,3}s^3 + \cdots, \\
K_i^4(s) &= \beta_{i,0} + \beta_{i,1}s + \alpha_{i,2}s^2 + \alpha_{i,3}s^3 + \cdots,
\end{aligned} \tag{10.342}$$

and we denote them as:

$$\mathcal{K}_i(s) := \{K_i^1(s), K_i^2(s), K_i^3(s), K_i^4(s)\}. \tag{10.343}$$

For each $\mathbf{P}_i(s)$ we introduce four line segments joining pairs of Kharitonov polynomials as defined below:

$$\mathcal{S}_i(s) := \{[K_i^1(s), K_i^2(s)], [K_i^1(s), K_i^3(s)], [K_i^2(s), K_i^4(s)], [K_i^3(s), K_i^4(s)]\}. \tag{10.344}$$

These four segments are called *Kharitonov segments*. They are illustrated in Figure 10.30 for the case of a polynomial of degree 2.

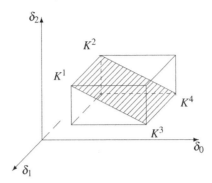

Figure 10.30 The four Kharitonov segments

For each $l \in \{1,\ldots,m\}$, let us define

$$\mathbf{P}_E^l(s) := \mathscr{K}_1(s) \times \cdots \times \mathscr{K}_{l-1}(s) \times \mathscr{S}_l(s) \times \mathscr{K}_{l+1}(s) \times \cdots \times \mathscr{K}_m(s). \qquad (10.345)$$

A typical element of $\mathbf{P}_E^l(s)$ is

$$\left(K_1^{j_1}(s), K_2^{j_2}(s), \ldots, K_{l-1}^{j_{l-1}}(s), (1-\lambda)K_l^1(s) + \lambda K_l^2(s), K_{l+1}^{j_{l+1}}(s), \ldots, K_m^{j_m}(s) \right) \qquad (10.346)$$

with $\lambda \in [0,1]$. This can be rewritten as

$$
\begin{aligned}
&(1-\lambda)\left(K_1^{j_1}(s), K_2^{j_2}(s), \ldots, K_{l-1}^{j_{l-1}}(s), K_l^1(s), K_{l+1}^{j_{l+1}}(s), \ldots, K_m^{j_m}(s) \right) \\
&+\lambda \left(K_1^{j_1}(s), K_2^{j_2}(s), \ldots, K_{l-1}^{j_{l-1}}(s), K_l^2(s), K_{l+1}^{j_{l+1}}(s), \ldots, K_m^{j_m}(s) \right).
\end{aligned}
\qquad (10.347)
$$

Corresponding to the m-tuple $\mathbf{P}_E^l(s)$, introduce the polynomial family

$$\Delta_E^l(s) := \left\{ <\underline{F}(s), \underline{P}(s)> \; : \; \underline{P}(s) \in \mathbf{P}_E^l(s) \right\}. \qquad (10.348)$$

The set $\Delta_E^l(s)$ is also described as

$$
\begin{aligned}
\Delta_E^l(s) = F_1(s)\mathscr{K}_1(s) + \cdots + F_{l-1}(s)\mathscr{K}_{l-1}(s) + F_l(s)\mathscr{S}_l(s) \\
+ F_{l+1}(s)\mathscr{K}_{l+1}(s) + \cdots + F_m(s)\mathscr{K}_m(s).
\end{aligned}
\qquad (10.349)
$$

A typical element of $\Delta_E^l(s)$ is the line segment of polynomials

$$
\begin{aligned}
&F_1(s)K_1^{j_1}(s) + F_2(s)K_2^{j_2}(s) + \cdots + F_{l-1}(s)K_{l-1}^{j_{l-1}}(s) \\
&+ F_l(s)\left[(1-\lambda)K_l^1(s) + \lambda K_l^2(s) \right] + F_{l+1}(s)K_{l+1}^{j_{l+1}}(s) + \cdots + F_m(s)K_m^{j_m}(s)
\end{aligned}
\qquad (10.350)
$$

with $\lambda \in [0,1]$.

The *extremal subset* of $\mathbf{P}(s)$ is defined by

$$\mathbf{P}_E(s) := \cup_{l=1}^m \mathbf{P}_E^l(s). \qquad (10.351)$$

The corresponding *generalized Kharitonov segment* polynomials are

$$\Delta_E(s) := \cup_{l=1}^{m} \Delta_E^l(s)$$
$$= \{< \underline{F}(s), \underline{P}(s) > \; : \; \underline{P}(s) \in \mathbf{P}_E(s)\}. \tag{10.352}$$

The set of m-tuples of Kharitonov polynomials are denoted $\mathbf{P}_K(s)$ and referred to as the *Kharitonov vertices* of $\mathbf{P}(s)$:

$$\mathbf{P}_K(s) := \mathscr{K}_1(s) \times \mathscr{K}_2(s) \times \cdots \times \mathscr{K}_m(s) \subset \mathbf{P}_E(s). \tag{10.353}$$

The corresponding set of *Kharitonov vertex* polynomials is

$$\Delta_K(s) := \{< \underline{F}(s), \underline{P}(s) > \; : \; \underline{P}(s) \in \mathbf{P}_K(s)\}. \tag{10.354}$$

A typical element of $\Delta_K(s)$ is

$$F_1(s)K_1^{j_1}(s) + F_2(s)K_2^{j_2}(s) + \cdots + F_m(s)K_m^{j_m}(s). \tag{10.355}$$

The set $\mathbf{P}_E(s)$ is made up of one-parameter families of polynomial vectors. It is easy to see that there are $m4^m$ such segments in the most general case, where there are four distinct Kharitonov polynomials for each $\mathbf{P}_i(s)$. The parameter-space subsets corresponding to $\mathbf{P}_E^l(s)$ and $\mathbf{P}_E(s)$ are denoted by Π_l and

$$\Pi_E := \cup_{l=1}^{m} \Pi_l, \tag{10.356}$$

respectively. Similarly, let Π_K denote the vertices of Π corresponding to the Kharitonov polynomials. Then, we also have

$$\Delta_E(s) := \{\delta(s, \mathbf{p}) : \mathbf{p} \in \Pi_E\}, \tag{10.357a}$$
$$\Delta_K(s) := \{\delta(s, \mathbf{p}) : \mathbf{p} \in \Pi_K\}. \tag{10.357b}$$

The set $\mathbf{P}_K(s)$ in general has 4^m distinct elements when each $\mathbf{P}_i(s)$ has four distinct Kharitonov polynomials. Thus, $\Delta_K(s)$ is a discrete set of polynomials, $\Delta_E(s)$ is a set of line segments of polynomials, $\Delta(s)$ is a polytope of polynomials, and

$$\Delta_K(s) \subset \Delta_E(s) \subset \Delta(s). \tag{10.358}$$

With these preliminaries, we are ready to state the Generalized Kharitonov Theorem (GKT).

10.10.3 The Generalized Kharitonov Theorem

Let us say that $\underline{F}(s)$ stabilizes a set of m-tuples if it stabilizes every element in the set. We can now enunciate the Generalized Kharitonov Theorem.

THEOREM 10.9 (Generalized Kharitonov Theorem)   For a given m-tuple $\underline{F}(s) = (F_1(s), \ldots, F_m(s))$ of real polynomials:

(1) $\underline{F}(s)$ stabilizes the entire family $\mathbf{P}(s)$ of m-tuples if and only if $\underline{F}$ stabilizes every m-tuple segment in $\mathbf{P}_E(s)$. Equivalently, $\Delta(s)$ is stable if and only if $\Delta_E(s)$ is stable.

(2) Moreover, if the polynomials $F_i(s)$ are of the form

$$F_i(s) = s^{t_i}(a_i s + b_i)U_i(s)Q_i(s), \qquad (10.359)$$

where $t_i \geq 0$ is an arbitrary integer, a_i and b_i are arbitrary real numbers, $U_i(s)$ is an anti-Hurwitz polynomial, and $Q_i(s)$ is an even or odd polynomial, then it is enough that $\underline{F}(s)$ stabilizes the finite set of m-tuples $\mathbf{P}_K(s)$, or equivalently, that the set of Kharitonov vertex polynomials $\Delta_K(s)$ are stable.

(3) Finally, stabilizing the finite set $\mathbf{P}_K(s)$ is not sufficient to stabilize $\mathbf{P}(s)$ when the polynomials $F_i(s)$ do not satisfy the conditions in (2). Equivalently, stability of $\Delta_K(s)$ does not imply stability of $\Delta(s)$ when $F_i(s)$ do not satisfy the conditions in (2).

The GKT has an appealing geometric interpretation in terms of the image set $\Delta(j\omega)$ of $\Delta(s)$. Recall that in step 1 of the proof, the stability of $\Delta(s)$ was reduced to that of the $2m$ parameter polytope $\Delta_I(s)$. It is easy to see that even though $\Delta_I(s)$ is in general a proper subset of $\Delta(s)$, the image sets are in fact equal:

$$\Delta(j\omega) = \Delta_I(j\omega). \qquad (10.360)$$

This follows from the fact, established in this chapter, that each of the m interval polynomials $\mathbf{P}_i(s)$ in $\Delta(s)$ can be replaced by a two-parameter family as far as its $j\omega$ evaluation is concerned. This proves that regardless of the dimension of the parameter space Π, a linear interval problem with m terms can always be replaced by a $2m$-parameter problem. Of course, in the rest of the theorem we show that this $2m$-parameter problem can be further reduced to a set of one-parameter problems.

In fact, $\Delta(j\omega)$ is a convex polygon in the complex plane and it may be described in terms of its vertices or exposed edges. Let $\partial\Delta(j\omega)$ denote the exposed edges of $\Delta(j\omega)$ and $\Delta_V(j\omega)$ denote its vertices. Then it is easy to establish the following.

LEMMA 10.16

$$(1) \ \partial\Delta(j\omega) \subset \Delta_E(j\omega), \qquad (2) \ \Delta_V(j\omega) \subset \Delta_K(j\omega). \qquad (10.361)$$

Proof   Observe that $\Delta(j\omega)$ is the sum of complex plane sets as follows:

$$\Delta(j\omega) = F_1(j\omega)\mathbf{P}_1(j\omega) + F_2(j\omega)\mathbf{P}_2(j\omega) + \cdots + F_m(j\omega)\mathbf{P}_m(j\omega). \qquad (10.362)$$

Each polygon $F_i(j\omega)\mathbf{P}_i(j\omega)$ is a rectangle with vertex set $F_i(j\omega)\mathcal{K}_i(j\omega)$ and edge set $F_i(j\omega)\mathcal{S}_i(j\omega)$. Since the vertices of $\Delta(j\omega)$ can only be generated by the vertices of $F_i(j\omega)\mathbf{P}_i(j\omega)$, we immediately have (2). To establish (1), we note that the boundary of the sum of two complex plane polygons can be generated by summing over all vertex–edge pairs with the vertices belonging to one and the edges belonging to the other. This fact used recursively to add m polygons shows that one has to sum vertices from $m - 1$ of the sets to edges of the remaining set and repeat this over all possibilities. This leads to (1).   ∎

The GKT is a powerful result applicable to robust Hurwitz stability determination of the specific type of polytope considered. In the following section we develop another powerful result that allows us to determine the root locations of an entire polytope of polynomials.

10.11   Edge Theorem: Capturing the Root Space of a Polytope

10.11.1   Edge Theorem

The Edge Theorem provides a remarkably efficient approach to determining the set of roots of an entire polytopic family of polynomials. It shows that the boundary of the root space of this family is contained in the root set of the exposed edges of the polytope.

Let us consider a family of nth-degree real polynomials whose typical element is given by

$$\delta(s) = \delta_0 + \delta_1 s + \cdots + \delta_{n-1} s^{n-1} + \delta_n s^n. \tag{10.363}$$

As usual, we identify $\mathscr{P}_n$ as the vector space of all real polynomials of degree less than or equal to n with $\mathbb{R}^{n+1}$, and we will identify the polynomial in Equation (10.363) with the vector

$$\underline{\delta} := [\delta_n, \delta_{n-1}, \dots, \delta_1, \delta_0]^T. \tag{10.364}$$

Let $\Omega \subset \mathbb{R}^{n+1}$ be an m-dimensional *polytope*, that is, the convex hull of a finite number of points. As a polytope, Ω is a closed bounded set and therefore it is compact. We make the assumption that all polynomials in Ω have the same degree.

ASSUMPTION 10.2   The sign of δ_n is constant over Ω, either always positive or always negative.

Assuming for example that this sign is always positive, and using the fact that Ω is compact, it is always possible to find $\Delta > 0$ such that,

$$\delta_n > \Delta, \text{ for every } \underline{\delta} \in \Omega. \tag{10.365}$$

A *supporting hyperplane* H is an affine set of dimension n such that $\Omega \cap H \neq \emptyset$, and such that every point of Ω lies on just one side of H. The *exposed sets* of Ω are those (convex) sets $\Omega \cap H$ where H is a supporting hyperplane. The one-dimensional exposed sets are called *exposed edges*, whereas the two-dimensional exposed sets are the *exposed faces*.

Before proceeding, we need to introduce the notion of root space. Consider any $W \subset \Omega$. Then $R(W)$ is said to be the root space of W if

$$R(W) = \{s : \delta(s) = 0, \text{ for some } \underline{\delta} \in W\}. \tag{10.366}$$

Finally, recall that the boundary of an arbitrary set S of the complex plane is designated by ∂S. We can now enunciate and prove the Edge Theorem.

THEOREM 10.10 (Edge Theorem)   Let $\Omega \subset \mathbb{R}^{n+1}$ be a polytope of polynomials which satisfies Assumption 10.2. Then the boundary of $R(\Omega)$ is contained in the root space of the exposed edges of Ω.

To prove the theorem we need two lemmas.

LEMMA 10.17 If a real s_r belongs to $R(\Omega)$, then there exists an exposed edge E of Ω such that $s_r \in R(E)$, and if a complex number s_c belongs to $R(\Omega)$, then there exists an *exposed face* F of Ω such that $s_c \in R(F)$.

Proof Consider an arbitrary $\underline{\delta}$ in Ω, and suppose that s_r is a real root of $\delta(s)$. We know that the set of all polynomials having s_r among their roots is a vector space $\mathscr{P}_{s_r}$ of dimension n. Let $\text{aff}(\Omega)$ denote the affine hull of Ω, that is, the smallest affine subspace containing Ω. Now, assume that $m = \dim[\text{aff}(\Omega)] \geq 2$. Then we have that

$$\dim\left[\mathscr{P}_{s_r} \cap \text{aff}(\Omega)\right] \geq 1, \tag{10.367}$$

and this implies that this set $\mathscr{P}_{s_r} \cap \text{aff}(\Omega)$ must pierce the relative boundary of Ω. This relative boundary, however, is the union of some $(m-1)$-dimensional polytopes, which are all exposed sets of Ω. Therefore, at least one of these boundary polytopes Ω_{m-1} satisfies

$$s_r \in R(\Omega_{m-1}). \tag{10.368}$$

If $\dim[\text{aff}(\Omega_{m-1})] \geq 2$, we see that we can repeat the preceding argument and ultimately we will find a one-dimensional boundary polytope Ω_1 for which $s_r \in R(\Omega_1)$. But Ω_1 is just an exposed edge of Ω, so that s_r does indeed belong to the root space of the exposed edges of Ω. For the case of a complex root s_c, it suffices to know that the set of all real polynomials having s_c among their roots is a vector space $\mathscr{P}_{s_c}$ of dimension $n-1$. As a consequence the same reasoning as above holds, yielding eventually an exposed face Ω_2 of Ω for which $s_c \in R(\Omega_2)$. ∎

We illustrate this lemma in Figures 10.31(a), 10.31(b), and 10.31(c) with a three-dimensional polytope Ω (see Figure 10.31(a)). Here $\mathscr{P}_{s_r}$ is a subspace of dimension two and cuts the edges of Ω (see Figure 10.31(b)). $\mathscr{P}_{s_c}$ is of dimension one and must penetrate a face of Ω (see Figure 10.31(c)).

The conclusion of this first lemma is that if p_F is the number of exposed faces, then

$$R(\Omega) = \bigcup_{i=1}^{p_F} R(F_i). \tag{10.369}$$

The next lemma focuses now on an exposed face. Let F be an exposed face of Ω and let us denote by ∂F its relative boundary. Since F is a compact set and because of Assumption 10.2 on Ω, we know from Chapter 2 that $R(F)$ is itself a closed set. We have the following.

LEMMA 10.18 $\partial R(F) \subset R(\partial F)$.

Proof Let s^* be an arbitrary element of $\partial R(F)$, we want to show that s^* is also an element of $R(\partial F)$. Since ∂F is the union of exposed edges of Ω, it follows from Lemma 10.17 that if s^* is real then $s^* \in R(\partial F)$.

Now assume that s^* is complex. Since $R(F)$ is a closed set, $\partial R(F) \subset R(F)$, so that it

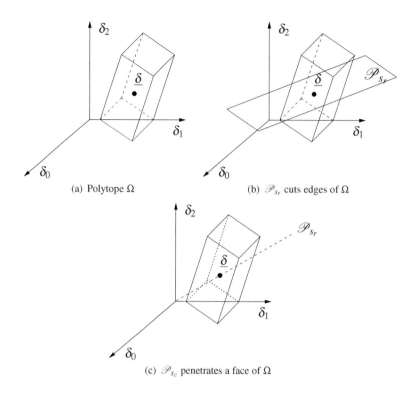

(a) Polytope Ω

(b) $\mathscr{P}_{s_r}$ cuts edges of Ω

(c) $\mathscr{P}_{s_c}$ penetrates a face of Ω

Figure 10.31 Illustration of Edge Theorem

is possible to find $\underline{\delta}^* \in F$ with $\delta^*(s^*) = 0$. We can write

$$\delta^*(s) = (s^2 + \alpha s + \beta)(d_{n-2}s^{n-2} + \cdots + d_1 s + d_0), \qquad (10.370)$$

where $\alpha = -2\mathrm{Re}(s^*)$ and $\beta = |s^*|^2$. Let $\mathrm{aff}(F)$ be the affine hull of F. Since F is two-dimensional it is possible to write $\mathrm{aff}(F) = \{\underline{\delta}^* + V\lambda; \ \lambda \in \mathbb{R}^2\}$, where V is some full-rank $(n+1) \times 2$ matrix. On the other hand, an arbitrary element of the vector space of real polynomials with a root at s^* can be written as

$$P^*(s) = (s^2 + \alpha s + \beta)\left[((\mu_{n-2} + d_{n-2})s^{n-2} + \cdots + (\mu_1 + d_1)s + (\mu_0 + d_0)\right], \quad (10.371)$$

or more generally we can write

$$\mathscr{P}_{s^*} = \{\underline{\delta}^* + W\mu : \mu = [\mu_{n-2}, \ldots, \mu_1, \mu_0]^T \in \mathbb{R}^{n-2}\}, \qquad (10.372)$$

where W is the $(n+1) \times (n-1)$ matrix

$$W = \begin{bmatrix} 1 & 0 & \cdots & 0 \\ \alpha & 1 & \cdots & 0 \\ \beta & \alpha & \cdots & 0 \\ 0 & \beta & \cdots & 0 \\ \vdots & \vdots & \ddots & \vdots \\ 0 & 0 & \cdots & 1 \\ 0 & 0 & \cdots & \alpha \\ 0 & 0 & \cdots & \beta \end{bmatrix}. \tag{10.373}$$

The intersection between $\mathrm{aff}(F)$ and $\mathscr{P}_{s^*}$ contains all λ, μ satisfying

$$\underline{\delta}^* + V\lambda = \underline{\delta}^* + W\mu, \text{ or equivalently } [V, \ -W] \begin{bmatrix} \lambda \\ \mu \end{bmatrix} = 0. \tag{10.374}$$

Two possibilities have to be considered.

(A) $[V, -W]$ does not have full rank

In this case, the space of solutions to Equation (10.374) is either of dimension one or two. If it is of dimension one, then the intersection $\mathrm{aff}(F) \cap \mathscr{P}_{s^*}$ is a straight line which must intersect ∂F at a point $\underline{\hat{\delta}}$. Since $\underline{\hat{\delta}} \in \mathscr{P}_{s^*}$, $\hat{\delta}(s^*) = 0$, which implies that $s^* \in R(\partial F)$. If the dimension is two, then $\mathrm{aff}(F) \subset \mathscr{P}_{s^*}$ and for any $\underline{\hat{\delta}} \in \partial F$ we have $\hat{\delta}(s^*) = 0$, so that clearly $s^* \in R(\partial F)$.

(B) $[V, -W]$ has full rank

In this case the intersection $\mathrm{aff}(F) \cap \mathscr{P}_{s^*}$ is reduced to $\underline{\delta}^*$. We now prove that $\underline{\delta}^* \in \partial F$, and this is where the fact that $s^* \in \partial R(F)$ is utilized. Indeed, $s^* \in \partial R(F)$ implies the existence of a sequence of complex numbers s_n such that $s_n \notin R(F)$ for all n and such that $s_n \longrightarrow s^*$ as $n \to +\infty$. In particular, this implies that

$$-2\mathrm{Re}(s_n) \longrightarrow \alpha \text{ and } |s_n|^2 \longrightarrow \beta \text{ as } n \to +\infty. \tag{10.375}$$

As usual, let $\mathscr{P}_{s_n}$ be the vector space of all real polynomials with a root at s_n. An arbitrary element of $\mathscr{P}_{s_n}$ can be expressed as

$$\begin{aligned} P(s) = \underline{\delta}^*(s) &+ \left((s^2 - 2\mathrm{Re}(s_n)s + |s_n|^2\right) \left(\mu_{n-2}s^{n2} + \cdots + \mu_1 s + \mu_0\right) \\ &+ \left(-(2\mathrm{Re}(s_n) + \alpha)s + (|s_n^2| - \beta)\right) \left(d_{n-2}s^{n-2} + \cdots + d_1 s + d_0\right), \end{aligned} \tag{10.376}$$

or similarly

$$\mathscr{P}_{s_n} = \left\{\underline{\delta}^* + W_n\mu + v_n : \mu = [\mu_{n-2}, \ldots, \mu_1, \mu_0] \in \mathbb{R}^{n-1}\right\}, \tag{10.377}$$

where

$$
W_n =
\begin{bmatrix}
1 & 0 & \cdots & 0 \\
-2\mathrm{Re}(s_n) & 1 & \cdots & 0 \\
|s_n|^2 & -2\mathrm{Re}(s_n) & \cdots & 0 \\
0 & |s_n|^2 & \cdots & 0 \\
\vdots & \vdots & \ddots & \vdots \\
0 & 0 & \cdots & 1 \\
0 & 0 & \cdots & -2\mathrm{Re}(s_n) \\
0 & 0 & \cdots & |s_n|^2
\end{bmatrix}
\tag{10.378}
$$

and

$$
V_n =
\begin{bmatrix}
d_{n-2} & 0 \\
d_{n-3} & d_{n-2} \\
d_{n-4} & d_{n-3} \\
\vdots & \vdots \\
d_0 & d_1 \\
0 & d_0
\end{bmatrix}
\begin{bmatrix}
-(2\mathrm{Re}(s_n) + \alpha) \\
|s_n|^2 - \beta
\end{bmatrix}.
\tag{10.379}
$$

Clearly

$$
W_n \longrightarrow W \quad \text{and} \quad v_n \longrightarrow 0 \quad \text{as } n \to +\infty.
\tag{10.380}
$$

Now, since $\det(\cdot)$ is a continuous function and since $\det[V, -W] \neq 0$, there must exist n_1 such that $\det[V - W_n] \neq 0$ for $n \geq n_1$. Also, for every n, the intersection between $\mathscr{P}_{s_n}$ and $\mathrm{aff}(F)$ consists of all λ, μ that satisfy

$$
\underline{\delta}^* + W_n \mu + v_n = \underline{\delta}^* + V\lambda,
\tag{10.381}
$$

or equivalently

$$
[V, -W_n]
\begin{bmatrix}
\lambda \\
\mu
\end{bmatrix}
= v_n.
\tag{10.382}
$$

For $n \geq n_1$, the system in Equation (10.382) has a unique solution,

$$
\begin{bmatrix}
\lambda_n \\
\mu_n
\end{bmatrix}
= [V, -W_n]^{-1} v_n.
\tag{10.383}
$$

From Equation (10.383), we deduce that $[\lambda_n^T, \mu_n^T] \longrightarrow 0$ when $n \to +\infty$.

We now show that $\underline{\delta}^*$ belongs to ∂F. Let us consider an arbitrary open neighborhood in $\mathrm{aff}(F)$:

$$
B_F(\underline{\delta}^*, \varepsilon) = \{\underline{\delta} \in \mathrm{aff}(F) : \|\underline{\delta} - \underline{\delta}^*\| < \varepsilon\}.
\tag{10.384}
$$

We must show that $B_F(\underline{\delta}^*, \varepsilon)$ contains at least one vector not contained in F.

To do so, consider the intersection between $\mathscr{P}_{s_n}$ and $\mathrm{aff}(F)$, that is the vector $\underline{\delta}_n = \underline{\delta}^* + V\lambda_n$. This vector belongs to $\mathrm{aff}(F)$ and, since λ_n goes to 0, it belongs to $B_F(\underline{\delta}^*, \varepsilon)$ for n sufficiently large. Moreover, the polynomial corresponding to this vector has a root at s_n and we know that s_n does not belong to $R(F)$. Hence it must be the case that $\underline{\delta}_n$ does not belong to F, and this completes the proof of the lemma. ∎

Figures 10.32(a) and 10.32(b) illustrate this lemma. The sequence s_n converges to $s^* \in R(F)$ from outside of $R(F)$. The corresponding subspaces $\mathscr{P}_{s_n}$ converge to $\mathscr{P}_{s^*}$ from outside F. Thus, $\mathscr{P}_{s^*}$ must touch an edge of F.

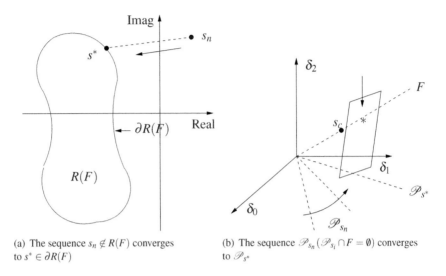

(a) The sequence $s_n \notin R(F)$ converges to $s^* \in \partial R(F)$

(b) The sequence $\mathscr{P}_{s_n}(\mathscr{P}_{s_i} \cap F = \emptyset)$ converges to $\mathscr{P}_{s^*}$

Figure 10.32 Proof of Edge Theorem

Proof of Theorem 10.10   From Equation (10.369) and Lemma 10.18, we have

$$\partial R(\Omega) = \partial \bigcup_{i=1}^{p_F} R(F_i) = \bigcup_{i=1}^{p_F} \partial R(F_i) \subset \bigcup_{i=1}^{p_F} R(\partial F_i). \tag{10.385}$$

The ∂F_i are precisely the exposed edges of Ω and this proves the theorem. ∎

Let us now consider an arbitrary simply connected domain of the complex plane, that is, a subset of the complex plane in which every simple (i.e., without self-crossings) closed contour encloses only points of the set. We can state the following corollary.

COROLLARY 10.1   If $\Gamma \subset C$ is a simply connected domain, then for any polytope satisfying Assumption 10.2, $R(\Omega)$ is contained in Γ if and only if the root space of all the exposed edges of Ω is contained in Γ.

10.11.2 Exposed Edges

In general, a polytope is defined by its vertices and it is not immediately clear how to determine the exposed edges of Ω. However, it is clear that those exposed edges are part of all pairwise convex combinations of the vertices of Ω, and therefore it is enough to check those. In the representation

$$\mathscr{P} := \{P(s) : P(s) = a_1 Q_1(s) + a_2 Q_2(s) + \cdots + a_m Q_m(s), \ \mathbf{a} \in \mathbf{A}\}, \qquad (10.386)$$

where $\mathbf{a} = [a_1, a_2, \ldots, a_m]$, the exposed edges of the polytope $\mathscr{P}$ are obtained from the exposed edges of the hypercube $\mathbf{A}$ to which $\mathbf{a}$ belongs. This can be done by fixing all a_i except one, say a_k, at a vertex $\underline{a}_i$ or $\bar{a}_i$, and letting a_k vary in the interval $[\underline{a}_k, \bar{a}_k]$, and repeating this for $k = 1, \ldots, m$. In general, the number of line segments in the coefficient space generated by this exceeds the number of exposed edges of $\mathscr{P}$. Nevertheless, this procedure captures all the exposed edges. We note that within the assumptions required by this result, stability verification amounts to checking the root location of line segments of polynomials of the form

$$P_\lambda(s) = (1 - \lambda)P_1(s) + \lambda P_2(s), \qquad \lambda \in [0, 1]. \qquad (10.387)$$

The root-locus technique can be used for this purpose. Alternatively, the Segment Lemma (Lemma 10.4) can also be used when the boundary of the domain Γ of interest can be parametrized easily. To reiterate, consider the following simple polytope consisting of the segment joining the two points

$$P_1(s) = 3s^4 + 3s^3 + 5s^2 + 2s + 1 \quad \text{and} \quad P_2(s) = s^4 + s^3 + 5s^2 + 2s + 5. \qquad (10.388)$$

It can be checked that both $P_1(s)$ and $P_2(s)$ are Hurwitz stable and yet the polynomial

$$\frac{P_1(s) + P_2(s)}{2} \qquad \text{has a root at } s = j. \qquad (10.389)$$

Remark 10.3 In light of the Edge Theorem, it is noted that the result given in Theorem 10.6 can be viewed as a weak version of the Edge Theorem.

We illustrate the Edge Theorem with some examples.

10.11.3 Examples

EXAMPLE 10.20 Consider the interval control system in Figure 10.33.

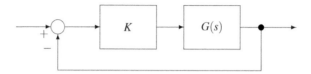

Figure 10.33 A gain feedback system (Example 10.20)

Let

$$G(s) = \frac{\delta_2 s^2 + \delta_0}{s(s^2 + \delta_1)} \qquad (10.390)$$

and assume that $K = 1$. Then the characteristic polynomial of this family of systems is the interval polynomial

$$\delta(s) = s^3 + \delta_2 s^2 + \delta_1 s + \delta_0, \tag{10.391}$$

where

$$\delta_2 \in [6,8], \quad \delta_1 \in [14,18], \quad \delta_0 \in [9.5, 10.5]. \tag{10.392}$$

The three variable coefficients form a box with 12 edges in the coefficient space. By the Edge Theorem, the boundary of the root space of the interval polynomial family can be obtained by plotting the root loci along the exposed edges of the box. The root loci of the edges is shown in Figure 10.34. Since the entire root space of the set of characteristic polynomials is found to be in the LHP, the family of feedback systems is robustly stable.

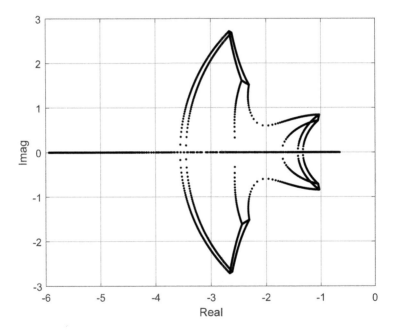

Figure 10.34 Root space for $K = 1$ (Example 10.20)

We remark that the robust stability of this system could have been checked by determining whether the Kharitonov polynomials are stable or not. However, the Edge Theorem has given us considerably more information by generating the entire root set. From this set, depicted in Figure 10.34, we can evaluate the performance of the system in terms of such useful quantities as the worst-case damping ratio, stability degree (minimum distance of the root set to the imaginary axis), largest damped and undamped natural frequencies, etc.

The movement of the entire root space with respect to the gain K can be studied systematically by repeatedly applying the Edge Theorem for each K. Figure 10.35 shows the movement of the root space with respect to various gains K.

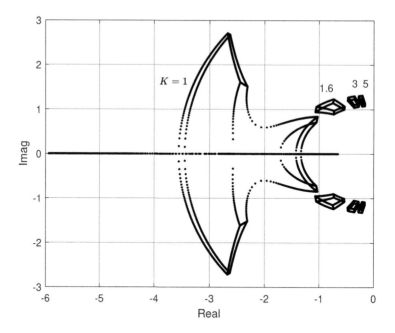

Figure 10.35 Root spaces for various $K = 1, 1.6, 3, 5$ (Example 10.20)

It shows that the root space approaches the imaginary axis as the gain K approaches the value 5. The root sets of the Kharitonov polynomials are properly contained in the root space for small values of K. However, as K approaches the value where the family is just about to become unstable, the roots of the Kharitonov polynomials move out to the right-hand boundary of the root set. These roots are therefore the "first" set of roots of the system to cross the imaginary axis.

EXAMPLE 10.21   Let us consider the unity feedback discrete-time control system with forward transfer function

$$G(z) = \frac{\delta_1 z + \delta_0}{z^2(z + \delta_2)}. \tag{10.393}$$

The characteristic polynomial is

$$\delta(z) = z^3 + \delta_2 z^2 + \delta_1 z + \delta_0. \tag{10.394}$$

Suppose that the coefficients vary in the intervals

$$\delta_2 \in [0.042, 0.158], \quad \delta_1 \in [-0.058, 0.058], \quad \delta_0 \in [-0.06, 0.056]. \tag{10.395}$$

The boundary of the root space of the family can be generated by drawing the root loci along the 12 exposed edges of the box in coefficient space. The root space is inside the unit disc as shown in Figure 10.36. Hence, the entire family is Schur stable.

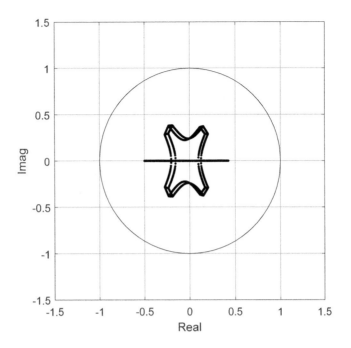

Figure 10.36 Root space of $\delta(z)$ (Example 10.21)

EXAMPLE 10.22 Consider the interval plant

$$G(s) = \frac{s+a}{s^2 + bs + c}, \tag{10.396}$$

where

$$a \in [1,2], \quad b \in [9,11], \quad c \in [15,18]. \tag{10.397}$$

The controller is

$$C(s) = \frac{3s+2}{s+5}. \tag{10.398}$$

The closed-loop characteristic polynomial is

$$
\begin{aligned}
\delta(s) &= \left(s^2 + bs + c\right)(s+5) + (s+a)(3s+2) \\
&= a(3s+2) + b\left(s^2 + 5s\right) + c(s+5) + \left(s^3 + 8s^2 + 2s\right).
\end{aligned} \tag{10.399}
$$

The boundary of the root space of $\delta(s)$ can be obtained by plotting the root loci along the 12 exposed edges. It can be seen from Figure 10.37 that the family $\delta(s)$ is stable since the root space is in the left-half plane. Hence the given compensator robustly stabilizes the interval plant. From the root set generated, we can evaluate the performance of the controller in terms of the worst-case damping ratio, the minimum stability degree, and the maximum frequency of oscillation.

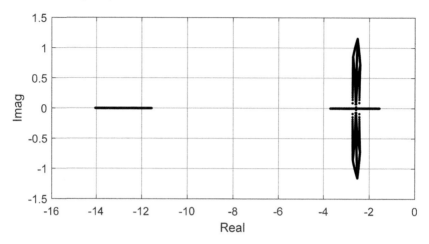

Figure 10.37 Root loci of the edges (Example 10.22)

10.12   The Mapping Theorem

10.12.1   Multilinear Dependency

The Mapping Theorem deals with a family of polynomials which depend multilinearly on a set of interval parameters. We refer to such a family as a multilinear interval polynomial. The Mapping Theorem shows us that the image set of such a family is *contained* in the convex hull of the image of the vertices. We state and prove this below.

Let $\mathbf{p} = [p_1, p_2, \dots, p_l]$ denote a vector of real parameters. Consider the polynomial

$$\delta(s, \mathbf{p}) := \delta_0(\mathbf{p}) + \delta_1(\mathbf{p})s + \delta_2(\mathbf{p})s^2 + \cdots + \delta_n(\mathbf{p})s^n, \qquad (10.400)$$

where the coefficients $\delta_i(\mathbf{p})$ are *multilinear* functions of $\mathbf{p}$, $i = 0, 1, \dots, n$. The vector $\mathbf{p}$ lies in an uncertainty set

$$\Pi := \left\{ \mathbf{p} : p_i^- \leq p_i \leq p_i^+, \ i = 1, 2, \dots, l \right\}. \qquad (10.401)$$

The corresponding set of *multilinear interval polynomials* is denoted by

$$\Delta(s) := \{ \delta(s, \mathbf{p}) : \mathbf{p} \in \Pi \}. \qquad (10.402)$$

Let $\mathbf{V}$ denote the vertices of Π:

$$\mathbf{V} := \left\{ \mathbf{p} : p_i = p_i^+ \text{ or } p_i = p_i^-, \ i = 1, 2, \dots, l \right\}, \qquad (10.403)$$

and

$$\Delta_V(s) := \{ \delta(s, \mathbf{p}) : \mathbf{p} \in \mathbf{V} \} := \{ v_1(s), v_2(s), \dots, v_k(s) \} \qquad (10.404)$$

denote the set of *vertex polynomials*. Let $\bar{\Delta}(s)$ denote the *convex hull* of the vertex polynomials $\{v_1(s), v_2(s), \ldots, v_k(s)\}$:

$$\bar{\Delta}(s) := \left\{ \sum_{i=1}^{i=k} \lambda_i v_i(s) : 0 \le \lambda_i \le 1, i = 1, 2, \ldots, k \right\}. \qquad (10.405)$$

The intersection of the sets $\Delta(s)$ and $\bar{\Delta}(s)$ contains the vertex polynomials $\Delta_V(s)$.

The Mapping Theorem deals with the image of $\Delta(s)$ at $s = s^*$. Let $\mathbf{T}(s^*)$ denote the complex plane image of the set $\mathbf{T}(s)$ evaluated at $s = s^*$ and let $co\, \mathscr{P}$ denote the convex hull of a set of points $\mathscr{P}$ in the complex plane.

THEOREM 10.11 (Mapping Theorem) Under the assumption that $\delta_i(\mathbf{p})$ are multilinear functions of $\mathbf{p}$:

$$\Delta(s^*) \subset co\, \Delta_V(s^*) = \bar{\Delta}(s^*) \qquad (10.406)$$

for each $s^* \in \mathbb{C}$.

Proof For convenience, we suppose that there are two uncertain parameters $\mathbf{p} := [p_1, p_2]$ and the uncertainty set Π is the rectangle $ABCD$ shown in Figure 10.38(a). Fixing $s = s^*$, we obtain $\delta(s^*, \mathbf{p})$, which maps Π to the complex plane. Let A', B', C', D' denote, respectively, the complex plane images of the vertices A, B, C, D under this mapping. Figures 10.38(b)–(e) show various configurations that can arise under this mapping. Now consider an arbitrary point I in Π and its complex plane image I'. The theorem is proved if we establish that I' is a convex combination of the complex numbers A', B', C', D'. We note that starting at an arbitrary vertex, say A, of Π we can reach I by moving along straight lines which are either edges of Π or are parallel to an edge of Π. Thus, as shown in Figure 10.38(a), we move from A to E along the edge AB, and then from E to I along EF, which is parallel to the edge AD. Because $\delta(s^*, \mathbf{p})$ is multilinear in the p_i, it follows that the complex plane images of AB, EF, and CD, which are parallel to edges of Π, are straight lines, respectively $A'B'$, $E'F'$, $C'D'$. Moreover, E' lies on the straight line $A'B'$, F' lies on the straight line $C'D'$, and I' lies on the straight line $E'F'$. Therefore, I' lies in the convex hull of $A'B'C'D'$.

The same reasoning works in higher dimensions. Any point in Π can be reached from a vertex by moving one coordinate at a time. In the image set, because of multilinearity, this corresponds to moving along straight lines joining vertices or convex combinations of vertices. By such movements we can never escape the convex hull of the vertices. The second equality holds by definition. ∎

We point out that the Mapping Theorem does not hold if Π is not an axis-parallel box, since the image of the edge of an arbitrary polytope under a multilinear mapping is in general not a straight line. The Mapping Theorem will also not hold when the dependency on the parameters is polynomial rather than multilinear, because of the same reason.

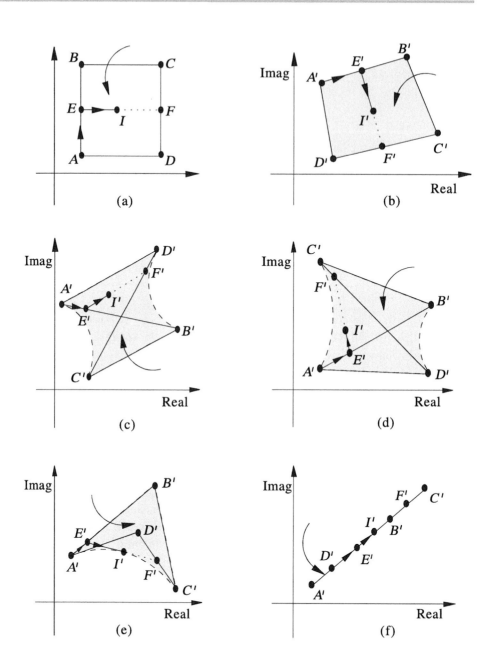

Figure 10.38 Proof of the Mapping Theorem

EXAMPLE 10.23   Consider the multilinear interval polynomial

$$\delta(s,\mathbf{p}) = p_1 Q_1(s) + p_2 Q_2(s) + p_1 p_2 Q_3(s) + Q_4(s) \tag{10.407}$$

with

$$Q_1(s) = -6s + 2, \quad Q_2(s) = -5s - 1, \quad Q_3(s) = 10s + 3, \quad Q_4(s) = 7s + 5. \quad (10.408)$$

The parameter set $\mathbf{p}$ varies inside box Π, as depicted in Figure 10.39(a). The image set $\Delta(s^*, \mathbf{p})$ of the family at $s^* = j1$ is shown in Figure 10.39(b). The convex hull of the image set $\Delta(s^*)$ is also shown. This shows that

$$\Delta(s^*, \mathbf{p}) \subset co \, \Delta_V(s^*, \mathbf{p}) = \bar{\Delta}(s^*). \quad (10.409)$$

The four corners of the polygon in Figure 10.39(b) are the vertices of the image.

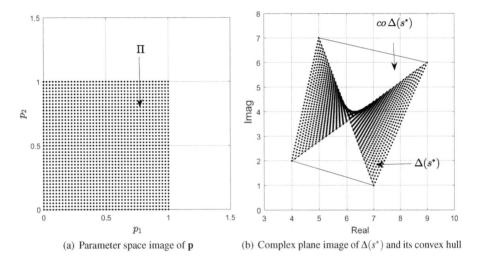

(a) Parameter space image of $\mathbf{p}$ (b) Complex plane image of $\Delta(s^*)$ and its convex hull

Figure 10.39 Parameter space, image set, and convex hull (Example 10.23)

This theorem shows us that the image set of the multilinear family $\Delta(s)$ can always be approximated by overbounding it with the image of the polytopic family $\bar{\Delta}(s)$. This approximation is extremely useful. In the next section we show how it allows us to determine robust stability and stability margins quantitatively.

10.12.2 Robust Stability via the Mapping Theorem

As we have seen in earlier chapters, the robust stability of a parametrized family of polynomials can be determined by verifying that the image set evaluated at each point of the stability boundary excludes the origin. The Mapping Theorem shows us that the image set of a multilinear interval polynomial family is contained in the convex hull of the vertices. Obviously a sufficient condition for the entire image set to exclude zero is that the convex hull exclude zero. Since the image set $\Delta(s^*)$ is overbounded by the convex hull of $\Delta_V(s^*)$, this suggests that the stability of the *multilinear* set $\Delta(s)$ can be guaranteed by solving the easier problem of verifying the stability of the *polytopic* set $\bar{\Delta}(s)$. We develop this idea in this section.

As usual, let $\mathscr{S}$, an open set in the complex plane, be the stability region. Introduce the set of edges $\mathbf{E}(s)$ of the polytope $\bar{\Delta}(s)$:

$$\mathbf{E}(s) := \left\{ \lambda v_i(s) + (1 - \lambda)v_j(s) \ : \ v_i(s), v_j(s) \in \Delta_V(s) \right\}. \tag{10.410}$$

To proceed, we make some standing assumptions about the families $\Delta(s)$ and $\bar{\Delta}(s)$.

ASSUMPTION 10.3

(1) Every polynomial in $\Delta(s)$ and $\bar{\Delta}(s)$ is of the same degree.
(2) $0 \notin \bar{\Delta}(s_0)$ for some $s_0 \in \partial\mathscr{S}$.
(3) At least one polynomial in $\Delta(s)$ is stable.

THEOREM 10.12   Under the above assumptions, $\Delta(s)$ is stable with respect to $\mathscr{S}$ if $\bar{\Delta}(s)$ and equivalently $\mathbf{E}(s)$ is stable with respect to $\mathscr{S}$.

Proof   Since the degree remains invariant, we may apply the Boundary Crossing Theorem (Chapter 1). Thus, $\Delta(s)$ can be unstable if and only if $0 \in \Delta(s^*)$ for some $s^* \in \partial\mathscr{S}$. By assumption, there exist $s_0 \in \partial\mathscr{S}$ such that

$$0 \notin \bar{\Delta}(s_0) = co\ \mathbf{E}(s_0). \tag{10.411}$$

By the Mapping Theorem

$$\Delta(s^*) \subset \bar{\Delta}(s^*) = co\ \mathbf{E}(s^*). \tag{10.412}$$

Therefore, by continuity of the image set on s, $0 \in \Delta(s^*)$ must imply the existence of $\bar{s} \in \partial\mathscr{S}$ such that $0 \in \mathbf{E}(\bar{s})$. This contradicts the stability of $\bar{\Delta}(s)$ and of $\mathbf{E}(s)$.   ∎

This theorem is just a statement of the fact that the origin can enter the image set $\Delta(s^*)$ only by piercing one of the edges $\mathbf{E}(\bar{s})$. Nevertheless, the result is rather remarkable in view of the fact that the set $\bar{\Delta}(s)$ does not necessarily contain, nor is it contained in, the set $\Delta(s)$. In fact, the inverse image of $\mathbf{E}(s)$ in the parameter space will in general include points outside Π. The theorem works because the *image set $\Delta(s^*)$ is overbounded* by $\bar{\Delta}(s)$ at every point $s^* \in \partial\mathscr{S}$. This in turn happens because the Mapping Theorem guarantees that $\Delta(s^*)$ is "concave" or bulges inward.

The test set $\bar{\Delta}(s)$ or its edges $\mathbf{E}(s)$ are linearly parametrized families of polynomials. Thus, the above theorem allows us to test a multilinear interval family using all the techniques available in the linear case.

EXAMPLE 10.24   Consider the multilinear interval polynomial family $\Delta(s)$:

$$\delta(s, \mathbf{p}) = p_1 Q_1(s) + p_2 Q_2(s) + p_1 p_2 Q_3(s), \tag{10.413}$$

where

$$p_1 \in [1, 2], \quad p_2 \in [3, 4] \tag{10.414}$$

and

$$Q_1(s) = s^4 + 4.3s^3 + 6.2s^2 + 3.5s + 0.6,$$
$$Q_2(s) = s^4 + s^3 + 0.32s^2 + 0.038s + 0.0015, \qquad (10.415)$$
$$Q_3(s) = s^4 + 3.5s^3 + 3.56s^2 + 1.18s + 0.12.$$

The edges of the polytopic set $\bar{\Delta}(s)$ are

$$\mathbf{E}(s) = \{E_i(s), \quad i = 1,2,3,4,5,6\}, \qquad (10.416)$$

with

$$
\begin{aligned}
E_1(s) &= \lambda \left[p_1^- Q_1(s) + p_2^- Q_2(s) + p_1^- p_2^- Q_3(s) \right] \\
&\quad + (1-\lambda) \left[p_1^+ Q_1(s) + p_2^- Q_2(s) + p_1^+ p_2^- Q_3(s) \right], \\
E_2(s) &= \lambda \left[p_1^- Q_1(s) + p_2^+ Q_2(s) + p_1^- p_2^+ Q_3(s) \right] \\
&\quad + (1-\lambda) \left[p_1^+ Q_1(s) + p_2^+ Q_2(s) + p_1^+ p_2^+ Q_3(s) \right], \\
E_3(s) &= \lambda \left[p_1^- Q_1(s) + p_2^- Q_2(s) + p_1^- p_2^- Q_3(s) \right] \\
&\quad + (1-\lambda) \left[p_1^- Q_1(s) + p_2^+ Q_2(s) + p_1^- p_2^+ Q_3(s) \right], \\
E_4(s) &= \lambda \left[p_1^+ Q_1(s) + p_2^- Q_2(s) + p_1^+ p_2^- Q_3(s) \right] \\
&\quad + (1-\lambda) \left[p_1^+ Q_1(s) + p_2^+ Q_2(s) + p_1^+ p_2^+ Q_3(s) \right], \\
E_5(s) &= \lambda \left[p_1^+ Q_1(s) + p_2^- Q_2(s) + p_1^+ p_2^- Q_3(s) \right] \\
&\quad + (1-\lambda) \left[p_1^- Q_1(s) + p_2^+ Q_2(s) + p_1^- p_2^+ Q_3(s) \right], \\
E_6(s) &= \lambda \left[p_1^+ Q_1(s) + p_2^+ Q_2(s) + p_1^+ p_2^+ Q_3(s) \right] \\
&\quad + (1-\lambda) \left[p_1^- Q_1(s) + p_2^- Q_2(s) + p_1^- p_2^- Q_3(s) \right].
\end{aligned}
\qquad (10.417)
$$

The stability of $\mathbf{E}(s)$ can be tested simply by applying the Segment Lemma (Lemma 10.3). Here, all edges in $\mathbf{E}(s)$ are found to be stable.

Bounded Phase Condition

An alternative way of stating Theorem 10.12 is to use the bounded phase condition. Let $\Phi_{\Delta_V}(s^*)$ equal the angle subtended at the origin by the points $\Delta_V(s^*)$. Obviously, the stability of $\mathbf{E}(s)$ is equivalent, under Assumption 10.3, to the origin being excluded from $\bar{\Delta}(s^*)$ for all $s^* \in \partial \mathscr{S}$, and this in turn is equivalent to the condition that the phase spread be less than π radians. Accordingly, we have the following theorem.

THEOREM 10.13   Under Assumptions 10.3, $\Delta(s)$ is stable with respect to $\mathscr{S}$ if

$$\Phi_{\bar{\Delta}_V}(s^*) < \pi, \qquad \text{for all } s^* \in \partial \mathscr{S}. \qquad (10.418)$$

EXAMPLE 10.25   Let us return to Example 10.24 and determine the robust stability of this multilinear interval polynomial by applying the bounded phase condition. We first construct the set of vertex polynomials corresponding to the vertices of the parameter space box $\Pi = [p_1, \ p_2]$:

$$v_1(s) = 7s^4 + 17.8s^3 + 17.84s^2 + 7.154s + 0.9645,$$
$$v_2(s) = 11s^4 + 32.6s^3 + 34.72s^2 + 14.194s + 1.9245,$$
$$v_3(s) = 9s^4 + 22.3s^3 + 21.72s^2 + 8.372s + 1.086,$$
$$v_4(s) = 14s^4 + 40.6s^3 + 42.16s^2 + 16.592s + 2.166.$$
(10.419)

The maximum phase difference over the vertex set at each frequency ω is computed as follows:

$$\Phi_{\tilde{\Delta}_V} = \sup_{i=2,3,4} \arg \frac{v_i(j\omega)}{v_1(j\omega)} - \inf_{i=2,3,4} \arg \frac{v_i(j\omega)}{v_1(j\omega)}.$$
(10.420)

The plot of the above phase function for all ω is shown in Figure 10.40. We find that the maximum phase difference over the vertex set does not reach $180°$ for any ω. Thus we conclude that the given multilinear polynomial family is Hurwitz stable.

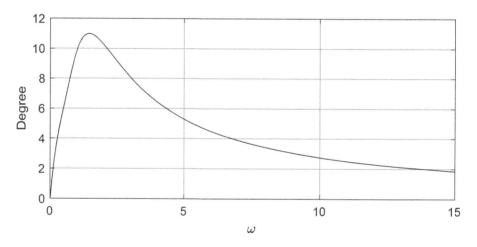

Figure 10.40 Maximum phase difference (Example 10.25)

10.13 Exercises

10.1 Consider the standard unity feedback control system given in Figure 10.41, where

$$G(s) := \frac{s+1}{s^2(s+p)}, \qquad F(s) = \frac{(s-1)}{s(s+3)(s^2 - 2s + 1.25)}$$
(10.421)

and the parameter p varies in the interval $[1, 5]$.

(a) Verify the robust stability of the closed-loop system. Is the Vertex Lemma applicable to this problem?

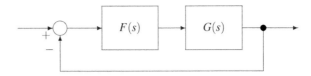

Figure 10.41 A unity feedback system (Exercise 10.1)

(b)   Verify your answer by the s-plane root locus (or Routh–Hurwitz Criterion).

10.2   For the feedback system shown in Figure 10.42, where

$$n(s,\alpha) = s^2 + (3-\alpha)s + 1,$$
$$d(s,\alpha) = s^3 + (4+\alpha)s^2 + 6s + 4 + \alpha,$$

(10.422)

partition the (K,α) plane into stable and unstable regions.

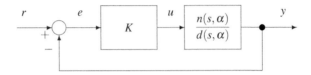

Figure 10.42 Feedback control system (Exercise 10.2)

Show that the stable region is bounded by

$$5 + K(3-\alpha) > 0$$

(10.423)

and

$$K + \alpha + 4 > 0.$$

(10.424)

10.3   Consider the feedback control system shown in Figure 10.43, where

$$n(s,\alpha) = s + \alpha,$$
$$d(s,\alpha) = s^3 + 2\alpha s^2 + \alpha s - 1,$$

(10.425)

and $\alpha \in [2,3]$.

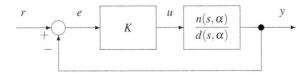

Figure 10.43 Feedback control system (Exercise 10.3)

Partition the K axis into robustly stabilizing and nonstabilizing regions.

Answer: Stabilizing for $K \in (-4.5, 0.33)$ only.

10.4   Consider the feedback system shown in Figure 10.44.

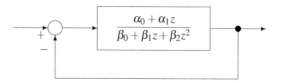

Figure 10.44 Feedback control system (Exercise 10.4)

The characteristic polynomial of the closed-loop system is

$$\delta(z) = (\alpha_0 + \beta_0) + (\alpha_1 + \beta_1)z + \beta_2 z^2$$
$$= \delta_0 + \delta_1 z + \delta_2 z^2 \qquad (10.426)$$

with nominal value

$$\delta^0(z) = \frac{1}{2} - z + z^2. \qquad (10.427)$$

Find ε_{max} so that with $\delta_i \in [\delta_i^0 - \varepsilon, \delta_i^0 + \varepsilon]$, the closed-loop system is robustly Schur stable.

Answer: $\varepsilon_{max} = 0.17$.

10.5   In a standard unity feedback control system, the plant is

$$G(s) = \frac{a_1 s - 1}{s^2 + b_1 s + b_0}. \qquad (10.428)$$

The parameters a_1, b_0, and b_1 have nominal values $a_1^0 = 1$, $b_1^0 = -1$, $b_0^0 = 2$ and are subject to perturbation. Consider the constant controller $C(s) = K$. Determine the range of values S_k of K for which the nominal system remains Hurwitz stable. For each value of $K \in S_k$, find the ℓ_2 stability margin $\rho(K)$ in the space of the three parameters subject to perturbation. Determine the optimally robust compensator value $K \in S_k$.

10.6   The transfer function of a plant in a unity feedback control system is given by

$$G(s) = \frac{1}{s^3 + d_2(\mathbf{p})s^2 + d_1(\mathbf{p})s + d_0(\mathbf{p})}, \qquad (10.429)$$

where

$$\mathbf{p} = [p_1, p_2, p_3] \qquad (10.430)$$

and

$$d_2(\mathbf{p}) = p_1 + p_2, \quad d_1(\mathbf{p}) = 2p_1 - p_3, \quad d_0(\mathbf{p}) = 1 - p_2 + 2p_3. \qquad (10.431)$$

The nominal parameter vector is $\mathbf{p}^0 = [3, 3, 1]$ and the controller proposed is

$$C(s) = K_c \frac{1 + 0.667s}{1 + 0.0667s} \qquad (10.432)$$

with the nominal value of the gain being $K_c = 15$. Compute the weighted ℓ_∞ Hurwitz stability margin with weights $[w_1, w_2, w_3] = [1, 1, 0.1]$.

10.7 In Exercise 10.6, assume that K_c is the only adjustable design parameter and let $\rho(K_c)$ denote the weighted ℓ_∞ Hurwitz stability margin with weights as in Exercise 10.6. Determine the range of values of K_c for which all perturbations within the unit-weighted ℓ_∞ ball are stabilized, i.e., for which $\rho(K_c) > 1$.

10.8 In a discrete-time control system the transfer function of the plant is:

$$G(z) = \frac{z - 2}{(z+1)(z+2)}. \tag{10.433}$$

Determine the transfer function of a first-order controller

$$C(z) = \frac{\alpha_0 + \alpha_1 z}{z + \beta_0} \tag{10.434}$$

which results in a closed-loop system that is deadbeat, that is, the characteristic roots are all at the origin. Determine the largest ℓ_2 and ℓ_∞ stability balls centered at this controller in the space of controller parameters α_0, α_1, β_0 for which closed-loop Schur stability is preserved.

10.9 Consider the Hurwitz stability of the family of polynomials

$$\delta(s, \mathbf{p}) = p_1 s(s+9.5)(s-1) - p_2(6.5s+0.5)(s-2), \tag{10.435}$$

where $p_1 \in [1, 1.1]$ and $p_2 \in [1.2, 1.25]$. Show that the family is robustly stable. Determine the worst-case stability margins over the given uncertainty set using the ℓ_2 and then the ℓ_∞ norm.

10.10 Consider the control system shown in Figure 10.45.

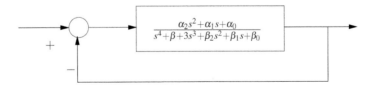

Figure 10.45 A feedback system (Exercise 10.10)

The parameters α_i, β_j vary in the following ranges:

$$\alpha_0 \in [2, 6], \quad \alpha_1 \in [0, 2], \quad \alpha_2 \in [-1, 3] \tag{10.436}$$

and

$$\beta_0 \in [4, 8], \quad \beta_1 \in [0.5, 1.5], \quad \beta_2 \in [2, 6], \quad \beta_3 \in [6, 14]. \tag{10.437}$$

Determine if the closed-loop system is Hurwitz stable or not for this class of perturbations.

10.11   Consider the Hurwitz stability of an interval family where the only coefficient subject to perturbation is δ_k, the coefficient of s^k for an arbitrary $k \in [0, 1, 2, \ldots, n]$. Show that the δ_k axis is partitioned into at most one stable segment and one or two unstable segments.

10.12   Apply the result of Exercise 10.11 to the Hurwitz polynomial

$$\delta(s) = s^4 + \delta_3 s^3 + \delta_2 s^2 + \delta_1 s + \delta_0 \tag{10.438}$$

with nominal parameters

$$\delta_3^0 = 4, \quad \delta_2^0 = 10, \quad \delta_1^0 = 12, \quad \delta_0^0 = 5. \tag{10.439}$$

Suppose that all coefficients except δ_3 remain fixed and δ_3 varies as follows:

$$\delta_3^- \le \delta_3 \le \delta_3^+. \tag{10.440}$$

Determine the largest interval (δ_3^-, δ_3^+) for which $\delta(s)$ remains Hurwitz. Repeat this for each of the coefficients δ_2, δ_1, and δ_0.

10.13   For the interval polynomial

$$\delta(z) = \delta_3 z^3 + \delta_2 z^2 + \delta_1 z + \delta_0 \tag{10.441}$$

where

$$\delta_3 \in [1 - \varepsilon, 1 + \varepsilon], \qquad \delta_2 \in \left[-\frac{1}{4} - \varepsilon, -\frac{1}{4} + \varepsilon \right],$$

$$\delta_1 \in \left[-\frac{3}{4} - \varepsilon, -\frac{3}{4} + \varepsilon \right], \qquad \delta_0 \in \left[\frac{3}{16} - \varepsilon, \frac{3}{16} + \varepsilon \right], \tag{10.442}$$

determine the maximal value of ε for which the family is Schur stable.[1]

10.14   Using the Edge Theorem, check the robust Hurwitz stability of the following family of polynomials. Also show the root cluster of the family:

$$\delta(s) := s^3 + (a + 3b)s^2 + cs + d, \tag{10.443}$$

where $a \in [1, 2]$, $b \in [0, 3]$, $c \in [10, 15]$, and $d \in [9, 14]$.

10.15   Consider the plant $G(s)$ and the controller $C(s)$:

$$G(s) := \frac{s+1}{s^2 - s - 1}, \qquad C(s) := \frac{as + b}{s + c}. \tag{10.444}$$

First, choose the controller parameter $\{a^0, b^0, c^0\}$ so that the closed-loop system has its characteristic roots at

$$-1 \pm j1 \quad \text{and} \quad -10. \tag{10.445}$$

---

[1] Refer to Theorem 5.13 of Bhattacharyya, Chapellat, and Keel (1995) for Exercise 10.13.

Now for

$$a \in \left[a^0 - \frac{\varepsilon}{2}, a^0 + \frac{\varepsilon}{2}\right], \quad b \in \left[b^0 - \frac{\varepsilon}{2}, b^0 + \frac{\varepsilon}{2}\right], \quad c \in \left[c^0 - \frac{\varepsilon}{2}, c^0 + \frac{\varepsilon}{2}\right], \quad (10.446)$$

find the maximum value ε_{max} of ε that robustly maintains closed-loop stability. Find the root set of the system when the parameters range over a box with sides $\frac{\varepsilon_{max}}{2}$.

10.16 Consider the discrete-time plant $G(z)$ and the controller $C(z)$:

$$G(z) := \frac{z-1}{z^2 + 2z + 3}, \qquad C(z) := \frac{az+b}{z+c}. \qquad (10.447)$$

Choose the controller parameter $\{a^0, b^0, c^0\}$ so that deadbeat control is achieved, namely all the closed-loop poles are placed at $z = 0$. Using the Edge Theorem, find the maximum range of the controller parameters so that the closed-loop poles remain inside the circle of radius 0.5 centered at the origin. Assume that the controller parameters are bounded by the same amount:

$$a \in [a^0 - \varepsilon, a^0 + \varepsilon], \qquad b \in [b^0 - \varepsilon, b^0 + \varepsilon], \qquad c \in [c^0 - \varepsilon, c^0 + \varepsilon]. \qquad (10.448)$$

Find the root set of the system for the parameters $\{a, b, c\}$ varying in a box

$$a \in \left[a^0 - \frac{\varepsilon}{2}, a^0 + \frac{\varepsilon}{2}\right], \quad b \in \left[b^0 - \frac{\varepsilon}{2}, b^0 + \frac{\varepsilon}{2}\right], \quad c \in \left[c^0 - \frac{\varepsilon}{2}, c^0 + \frac{\varepsilon}{2}\right]. \qquad (10.449)$$

10.17 Consider a unity feedback system with the plant $G(s)$ and the controller $C(s)$:

$$G(s) = \frac{s+b_0}{s^2 + a_1 s + a_0}, \qquad C(s) = \frac{s+1}{s+2}. \qquad (10.450)$$

Assume that the plant parameters vary independently as

$$a_0 \in [2,4], \quad a_1 \in [2,4], \quad b_0 \in [1,3]. \qquad (10.451)$$

Determine the root space of the family of closed-loop polynomials using the Edge Theorem.

10.18 Consider the polynomial $s^2 + a_1 s + a_0$ and let the coefficients (a_1, a_0) vary in the convex hull of the points

$$(0,0), \quad (0,R), \quad (R^2,0), \quad (R^2, 2R). \qquad (10.452)$$

Show that the root space of this set is the intersection with the left-half plane of the circle of radius R centered at the origin. Describe also the root space of the convex hull of the points

$$(0,0), \quad (0,2R), \quad (R^2,0), \quad (R^2, 2R). \qquad (10.453)$$

10.19 Consider the unity feedback system shown in Figure 10.46, where

$$\frac{F_1(s)}{F_2(s)} = \frac{2s^2 + 4s + 3}{s^2 + 3s + 4} \quad \text{and} \quad \frac{P_1(s)}{P_2(s)} = \frac{s^2 + a_1 s + a_0}{s(s^2 + b_1 s + b_0)}, \qquad (10.454)$$

with

$$a_1^0 = -2, \quad a_0^0 = 1, \quad b_0^0 = 2, \quad b_1^0 = 1. \tag{10.455}$$

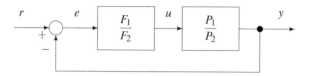

Figure 10.46 Feedback control system (Exercise 10.19)

Now let

$$a_0 \in [1 - \varepsilon, 1 + \varepsilon], \quad b_0 \in [2 - \varepsilon, 2 + \varepsilon],$$
$$a_1 \in [-2 - \varepsilon, -2 + \varepsilon], \quad b_1 \in [1 - \varepsilon, 1 + \varepsilon]. \tag{10.456}$$

Find ε_{max} for which the system is robustly stable.

Answer: $\varepsilon_{max} = 0.175$

10.20   Referring to the system given in Figure 10.46 with

$$\frac{F_1(s)}{F_2(s)} = \frac{2s^2 + 4s + 3}{s^2 + 3s + 4} \quad \text{and} \quad \frac{P_1(s)}{P_2(s)} = \frac{s^2 + a_1 s + a_0}{s(s^2 + b_1 s + b_0)}, \tag{10.457}$$

let the nominal system be

$$\frac{P_1^0(s)}{P_2^0(s)} = \frac{s^2 - 2s + 7}{s(s^2 + 8s - 0.25)}. \tag{10.458}$$

Suppose that the parameters vary within intervals

$$a_1 \in [-2 - \varepsilon, -2 + \varepsilon], \qquad a_0 \in [7 - \varepsilon, 7 + \varepsilon],$$
$$b_1 \in [8 - \varepsilon, 8 + \varepsilon], \qquad b_0 \in [-0.25 - \varepsilon, -0.25 + \varepsilon]. \tag{10.459}$$

Find the maximum value of ε for robust stability of the family using GKT.

Answer: $\varepsilon_{max} = 0.23$

10.21   Consider the plant

$$p(s) = \frac{1}{s - 1} \tag{10.460}$$

and the feedback controller

$$C(s) = K_p + \frac{K_i}{s}. \tag{10.461}$$

Using the Edge Theorem, determine the root space of the closed-loop system for all controllers satisfying

$$K_p \in [2,4], \qquad K_i \in [2,4]. \tag{10.462}$$

10.14   Notes and References

The Small Gain Theorem was formulated by Zames (1966). Kharitonov's Theorem, which sparked many developments in the parametric approach to robust control, was first published in Kharitonov (1978) and later introduced in the Western literature in Barmish (1984). The result was later extended to the complex case in Kharitonov (1979). Mikhailov's criterion was first published in Mikhailov (1938). It is equivalent to the Hermite–Bieler Theorem. The Hermite–Biehler Theorem for Hurwitz polynomials can be found in the book of Guillemin (1949). The Generalized Kharitonov Theorem was developed in Chapellat and Bhattacharyya (1989) to overcome the conservatism inherent in Kharitonov's Theorem. The Edge Theorem was discovered by Bartlett, Hollot, and Lin (1988). The problem of calculating the ℓ_2 stability margin in parameter space for Hurwitz stability was solved in Biernacki, Hwang, and Bhattacharyya (1987). The Segment Lemma, Vertex Lemma, Mapping Theorem, and other results in robustness under parametric uncertainty are proved in detail in Bhattacharyya, Chapellat, and Keel (1995). The results on fragility were first reported in Keel and Bhattacharyya (1997). The plant–controller pairs used in Examples 10.4 and 10.6 are taken from Doyle, Francis, and Tannenbaum (1992). Example 10.7 is from Fujita et al. (1995) and Example 10.9 is from Dahleh and Doaz-Bobillo (1995), respectively. The Mapping Theorem first appeared in Zadeh and Desoer (1963).

11 Analytical Design of PID Controllers

11.1 Introduction

In an earlier chapter we discussed the design of servomechanisms and demonstrated that the controller consisted of two parts: (1) a servocontroller consisting of a signal generator of the signals to be tracked or rejected and (2) a stabilizing controller for the plant augmented by the servocontroller.

In the majority of applications, it suffices to design a servomechanism to handle constant references and disturbances. In this case the signal generator is an integrator and the stabilizing controller can be a proportional derivative (PI) controller or proportional integral derivative (PID) controller.

The PI and PID controllers account for 99% of controllers in use in the world across all industries, including process control, motion control, aerospace systems, power electronics, UAVs, driverless cars, and autonomous robots. However, it is only recently that analytical approaches to their design have been developed. This was possible mainly due to the development of computationally efficient methods for determining stabilizing sets developed since 1997. In this chapter we describe these results for single-input single-output systems. In the next chapter we show how these can be extended to multi-input multi-output systems.

We begin this chapter by describing the computation of the stabilizing set of PID controllers for continuous-time and discrete-time systems, using the signature method. With this set in hand, we describe how prescribed gain margin, phase margin, H_∞ norm, and settling time specifications can be achieved over this set. This enables the mapping of the stabilizing set to the space of achievable design specifications. It allows the designer to pick a set of specifications from this achievable space and to determine a specific PI or PID controller attaining these specifications. The procedure is illustrated by examples of continuous and discrete-time systems.

11.2 Principle of Operation

The principle of operation of integral control is based on the following facts. A stable dynamic system driven by constant external signals, such as disturbances and references, reaches a steady state where: (a) every signal in the system is a constant and (b) the input to every integrator or block containing an integrator is zero. These properties

are easily proved for linear time-invariant continuous and discrete-time systems. However, they hold for a much larger class of systems, including nonlinear, distributed, and
time-varying systems. Based on (a) and (b) it is clear that a stabilizing controller driven
by the tracking error and containing an integrator will zero out the error asymptotically.
This is the basis on which the unity feedback PI or PID controller operates. Moreover,
even if the plant is nonlinear, under high gain feedback such as PID control, the closed-
loop system, when stable, behaves as a linear system. While the integrator is fixed, the
remaining two or three gains are to be adjusted to stabilize the closed-loop system and
provide satisfactory stability margins and transient performance. In the remaining sections we describe how these objectives can be attained. We begin by describing how the
stabilizing set can be computed.

11.3 Computation of Stabilizing Sets for Continuous-Time Systems

A crucial first step in our development will be the computation of the stabilizing set of
PID controllers for a given plant with transfer function $P(s)$. This is described below.

Consider the general linear time-invariant feedback system shown in Figure 11.1,
where $r(t)$ is the command signal, $e(t)$ the error signal, $u(t)$ the control input signal,
and $y(t)$ the output. $P(s)$ is the transfer function of the plant to be controlled and $C(s)$ is
the controller transfer function to be designed.

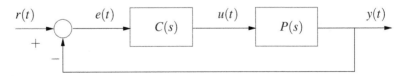

Figure 11.1 Feedback control system

The controller $C(s)$ is assumed to be of PID type:

$$C(s) = k_p + \frac{k_i}{s} + k_d s, \tag{11.1}$$

where k_p, k_i, and k_d are the proportional, integral, and derivative gains, respectively.
The pure derivative term $k_d s$ is not allowed if the measured error $e(t)$ is a noisy signal.
In such cases, we may consider a PI controller or the PID controller with a transfer
function

$$C(s) = \frac{s k_p + k_i + k_d s^2}{s(1 + sT)}, \quad \text{for } T > 0, \tag{11.2}$$

where T is usually fixed a priori at a small positive value.

The plant transfer function $P(s)$ is assumed to be rational and proper:

$$P(s) = \frac{N(s)}{D(s)}, \tag{11.3}$$

where $N(s)$, $D(s)$ are polynomials in the Laplace variable s with real coefficients. Thus, the closed-loop characteristic polynomial with $C(s)$ in Equation (11.1) becomes

$$\delta(s, k_p, k_i, k_d) = sD(s) + (k_i + k_p s + k_d s^2)N(s), \qquad (11.4)$$

or with $C(s)$ in Equation (11.2):

$$\delta(s, k_p, k_i, k_d) = s(1 + sT)D(s) + (k_i + k_p s + k_d s^2)N(s). \qquad (11.5)$$

To proceed, we introduce some facts about the relationship between the change of phase of a polynomial or rational function evaluated along a stability boundary, and its root distribution. These will be used subsequently to compute the stabilizing set.

11.3.1   Signature Formulas

Let $\delta(s)$ denote a polynomial of degree n with real coefficients without zeros on the $j\omega$ axis. Write

$$\delta(s) := \underbrace{\delta_0 + \delta_2 s^2 + \cdots}_{\delta_{\text{even}}(s^2)} + s\underbrace{\left(\delta_1 + \delta_3 s^2 + \cdots\right)}_{\delta_{\text{odd}}(s^2)} \qquad (11.6)$$

so that

$$\delta(j\omega) = \delta_r(\omega) + j\delta_i(\omega), \qquad (11.7)$$

where $\delta_r(\omega)$, $\delta_i(\omega)$ are polynomials in ω with real coefficients with

$$\delta_r(\omega) = \delta_{\text{even}}(-\omega^2), \qquad (11.8a)$$
$$\delta_i(\omega) = \omega\delta_{\text{odd}}(-\omega^2). \qquad (11.8b)$$

DEFINITION 11.1   The standard signum function $\text{sgn} : \mathbb{R} \to \{-1, 0, 1\}$ is defined by

$$\text{sgn}[x] = \begin{cases} -1 & \text{if } x < 0, \\ 0 & \text{if } x = 0, \\ 1 & \text{if } x > 0, \end{cases} \qquad (11.9)$$

where $\mathbb{R}$ represents the set of real numbers.

DEFINITION 11.2   Let $\mathbb{C}$ denote the complex plane, $\mathbb{C}^-$ the open left-half plane (LHP), $\mathbb{C}^+$ the open right-half plane (RHP), and l and r the numbers of roots of $\delta(s)$ in $\mathbb{C}^-$ and $\mathbb{C}^+$, respectively. Let $\angle\delta(j\omega)$ denote the phase or angle of $\delta(j\omega)$ and $\Delta_{\omega_1}^{\omega_2}\angle p(j\omega)$ the net change in the phase of $\delta(j\omega)$ as ω runs from ω_1 to ω_2, $(\omega_2 \geq \omega_1)$ measured in radians.

LEMMA 11.1

$$\Delta_0^\infty \angle\delta(j\omega) = \frac{\pi}{2}(l - r). \qquad (11.10)$$

Proof   Each LHP root contributes $+\pi$ and each RHP root contributes $-\pi$ to the net change in phase of $\delta(j\omega)$ as ω runs from $-\infty$ to ∞, and Equation (11.10) follows from the symmetry about the real axis of the roots since $\delta(s)$ has real coefficients. ∎

We call $l - r$, the Hurwitz *signature* of $\delta(s)$, and denote it as

$$\sigma(\delta) := l - r. \qquad (11.11)$$

Computation of $\sigma(\delta)$

By Lemma 11.1, the computation of $\sigma(\delta)$ amounts to a determination of the total phase change of $\delta(j\omega)$. To see how the total phase change may be calculated, consider typical plots of $\delta(j\omega)$ where ω runs from 0 to $+\infty$ as in Figure 11.2. We note that the frequencies $0, \omega_1, \omega_2, \omega_3, \omega_4$ are the points where the plot cuts or touches the real axis. In Figure 11.2(a), ω_3 is a point where the plot *touches* but does not cut the real axis.

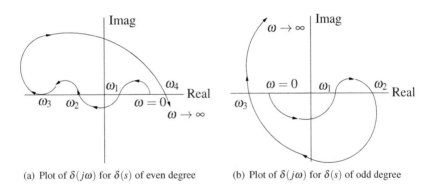

(a) Plot of $\delta(j\omega)$ for $\delta(s)$ of even degree   (b) Plot of $\delta(j\omega)$ for $\delta(s)$ of odd degree

Figure 11.2 $\delta(j\omega)$ plots

In Figure 11.2(a), we have

$$\Delta_0^\infty \angle \delta(j\omega) = \underbrace{\Delta_0^{\omega_1} \angle \delta(j\omega)}_{0} + \underbrace{\Delta_{\omega_1}^{\omega_2} \angle \delta(j\omega)}_{-\pi} + \underbrace{\Delta_{\omega_2}^{\omega_3} \angle \delta(j\omega)}_{0} + \underbrace{\Delta_{\omega_3}^{\omega_4} \angle \delta(j\omega)}_{-\pi}$$
$$+ \underbrace{\Delta_{\omega_4}^\infty \angle \delta(j\omega)}_{0}. \qquad (11.12)$$

Observe that

$$\Delta_0^{\omega_1} \angle \delta(j\omega) = \operatorname{sgn}[\delta_i(0^+)]\left(\operatorname{sgn}[\delta_r(0)] - \operatorname{sgn}[\delta_r(\omega_1)]\right)\frac{\pi}{2},$$

$$\Delta_{\omega_1}^{\omega_2} \angle \delta(j\omega) = \operatorname{sgn}[\delta_i(\omega_1^+)]\left(\operatorname{sgn}[\delta_r(\omega_1)] - \operatorname{sgn}[\delta_r(\omega_2)]\right)\frac{\pi}{2},$$

$$\Delta_{\omega_2}^{\omega_3} \angle \delta(j\omega) = \operatorname{sgn}[\delta_i(\omega_2^+)]\left(\operatorname{sgn}[\delta_r(\omega_2)] - \operatorname{sgn}[\delta_r(\omega_3)]\right)\frac{\pi}{2}, \qquad (11.13)$$

$$\Delta_{\omega_3}^{\omega_4} \angle \delta(j\omega) = \operatorname{sgn}[\delta_i(\omega_3^+)]\left(\operatorname{sgn}[\delta_r(\omega_3)] - \operatorname{sgn}[\delta_r(\omega_4)]\right)\frac{\pi}{2},$$

$$\Delta_{\omega_4}^{+\infty} \angle \delta(j\omega) = \operatorname{sgn}[\delta_i(\omega_4^+)]\left(\operatorname{sgn}[\delta_r(\omega_4)] - \operatorname{sgn}[\delta_r(\infty)]\right)\frac{\pi}{2},$$

and

$$
\begin{aligned}
\operatorname{sgn}[\delta_i(\omega_1^+)] &= -\operatorname{sgn}[\delta_i(0^+)], \\
\operatorname{sgn}[\delta_i(\omega_2^+)] &= -\operatorname{sgn}[\delta_i(\omega_1^+)] = +\operatorname{sgn}[\delta_i(0^+)], \\
\operatorname{sgn}[\delta_i(\omega_3^+)] &= +\operatorname{sgn}[\delta_i(\omega_2^+)] = +\operatorname{sgn}[\delta_i(0^+)], \\
\operatorname{sgn}[\delta_i(\omega_4^+)] &= -\operatorname{sgn}[\delta_i(\omega_3^+)] = -\operatorname{sgn}[\delta_i(0^+)],
\end{aligned}
\tag{11.14}
$$

and note also that 0, ω_1, ω_2, ω_4 are the real zeros of $\delta_i(\omega)$ of *odd* multiplicities whereas ω_3 is a real zero of *even* multiplicity. From these relations, it is evident that Equation (11.12) may be rewritten, skipping the terms involving ω_3 the root of even multiplicity so that

$$
\begin{aligned}
\Delta_0^\infty \angle \delta(j\omega) &= \Delta_0^{\omega_1} \angle \delta(j\omega) + \Delta_{\omega_1}^{\omega_2} \angle \delta(j\omega) + \Delta_{\omega_2}^{\omega_4} \angle \delta(j\omega) + \Delta_{\omega_4}^{\infty} \angle \delta(j\omega) \\
&= \frac{\pi}{2} \Big(\operatorname{sgn}[\delta_i(0^+)]\big(\operatorname{sgn}[\delta_r(0)] - \operatorname{sgn}[\delta_r(\omega_1)]\big) \\
&\quad - \operatorname{sgn}[\delta_i(0^+)]\big(\operatorname{sgn}[\delta_r(\omega_1)] - \operatorname{sgn}[\delta_r(\omega_2)]\big) \\
&\quad + \operatorname{sgn}[\delta_i(0^+)]\big(\operatorname{sgn}[\delta_r(\omega_2)] - \operatorname{sgn}[\delta_r(\omega_4)]\big) \\
&\quad - \operatorname{sgn}[\delta_i(0^+)]\big(\operatorname{sgn}[\delta_r(\omega_4)] - \operatorname{sgn}[\delta_r(\infty)]\big) \Big).
\end{aligned}
\tag{11.15}
$$

Equation (11.15) can be rewritten as

$$
\begin{aligned}
\Delta_0^\infty \angle \delta(j\omega) &= \frac{\pi}{2} \operatorname{sgn}[\delta_i(0^+)]\big(\operatorname{sgn}[\delta_r(0)] - 2\operatorname{sgn}[\delta_r(\omega_1)] + 2\operatorname{sgn}[\delta_r(\omega_2)] \\
&\quad - 2\operatorname{sgn}[\delta_r(\omega_4)] + \operatorname{sgn}[\delta_r(\infty)]\big).
\end{aligned}
\tag{11.16}
$$

In the case of Figure 11.2(b), that is, when $\delta(s)$ is of odd degree, we have

$$
\Delta_0^\infty \angle \delta(j\omega) = \underbrace{\Delta_0^{\omega_1} \angle \delta(j\omega)}_{=+\pi} + \underbrace{\Delta_{\omega_1}^{\omega_2} \angle \delta(j\omega)}_{=0} + \underbrace{\Delta_{\omega_2}^{\omega_3} \angle \delta(j\omega)}_{=-\pi} + \underbrace{\Delta_{\omega_3}^{+\infty} \angle \delta(j\omega)}_{=-\frac{\pi}{2}}
\tag{11.17}
$$

and $\Delta_0^{\omega_1} \angle \delta(j\omega)$, $\Delta_{\omega_1}^{\omega_2} \angle \delta(j\omega)$, $\Delta_{\omega_2}^{\omega_3} \angle \delta(j\omega)$ are as in Equation (11.13), whereas

$$
\Delta_{\omega_3}^\infty \angle \delta(j\omega) = \frac{\pi}{2} \operatorname{sgn}[\delta_i(\omega_3^+)]\operatorname{sgn}[\delta_r(\omega_3)].
\tag{11.18}
$$

Similarly to Equation (11.14), we have

$$
\operatorname{sgn}[\delta_i(\omega_j^+)] = (-1)^j \operatorname{sgn}[\delta_i(0^+)], \quad \text{for } j = 1,2,3.
\tag{11.19}
$$

Combining Equations (11.17), (11.18), and (11.19) for Figure 11.2(b):

$$
\begin{aligned}
\Delta_0^\infty \angle \delta(j\omega) &= \frac{\pi}{2} \operatorname{sgn}[\delta_i(0^+)]\big(\operatorname{sgn}[\delta_r(0)] - 2\operatorname{sgn}[\delta_r(\omega_1)] + 2\operatorname{sgn}[\delta_r(\omega_2)] \\
&\quad - 2\operatorname{sgn}[\delta_r(\omega_3)]\big).
\end{aligned}
\tag{11.20}
$$

We can now easily generalize the above formulas for the signature, based on Lemma 11.1.

THEOREM 11.1   Let $\delta(s)$ be a polynomial of degree n with real coefficients, without zeros on the imaginary axis. Write

$$
\delta(j\omega) = \delta_r(\omega) + j\delta_i(\omega)
\tag{11.21}
$$

and let ω_0, ω_1, ω_3, ..., ω_{l-1} denote the real non-negative zeros of $\delta_i(\omega)$ with odd multiplicities with $\omega_0 = 0$.

If n is even:

$$\sigma(\delta) = \text{sgn}[\delta_i(0^+)] \left(\text{sgn}[\delta_r(0)] + 2 \sum_{j=1}^{l-1} (-1)^j \text{sgn}[\delta_r(\omega_j)] + (-1)^l \text{sgn}[\delta_r(\infty)] \right).$$

(11.22)

If n is odd:

$$\sigma(\delta) = \text{sgn}[\delta_i(0^+)] \left(\text{sgn}[\delta_r(0)] + 2 \sum_{j=1}^{l-1} (-1)^j \text{sgn}[\delta_r(\omega_j)] \right).$$

(11.23)

11.3.2   Computation of the Stabilizing Set with PID Controllers

Consider the system configuration in Figure 11.1 with a plant $P(s)$ and a PID controller

$$C(s) = k_p + \frac{k_i}{s} + k_d s.$$

(11.24)

It will be convenient to convert the stabilization problem to an equivalent signature assignment problem. The latter problem will be easier to solve as it will lead to a nested linear programming problem in two variables with a sweeping parameter whose range can be specified a priori. This will be described as a procedure below.

1. The characteristic equation of the closed-loop system derived from Figure 11.1 is

$$\delta(s) = sD(s) + \left(k_d s^2 + k_p s + k_i \right) N(s).$$

(11.25)

    Let n denote the degree of $D(s)$.

2. Form the new polynomial

$$v(s) := \delta(s)N(-s).$$

(11.26)

    It is clear that $\delta(s)$ is Hurwitz if and only if $v(s)$ has signature $n + 1 + \sigma(N(-s))$. This will be exploited to compute the stabilizing set below.

3. The even and odd decomposition of the polynomial $v(s)$ in Equation (11.26) exhibits a separation of parameters and is denoted as

$$v(s) = v_{\text{even}}(s^2, k_i, k_d) + s v_{\text{odd}}(s^2, k_p).$$

(11.27)

4. Fix $k_p = k_p^*$ and let $0 < \omega_1 < \omega_2 < \cdots < \omega_{l-1}$ be the finite frequencies that are real and positive roots of

$$v_{\text{odd}}(-\omega^2, k_p^*) = 0$$

(11.28)

    of odd multiplicities. Define $\omega_0 := 0$ and $\omega_l := \infty$.

5. Write

$$j = \text{sgn} \left[v_{\text{odd}}(0^+, k_p^*) \right].$$

(11.29)

Let $\deg[D(s)] := n$, $\deg[N(s)] : m \le n$, and let z^+ and z^- denote the numbers of zeros (roots of $N(s) = 0$) in $\mathbb{C}^+$ and $\mathbb{C}^-$ of the plant, respectively. Let $i_0, \ldots, i_1, \ldots$ be integers $i_t \in (+1, -1)$, for all $t \in (0, \ldots, l)$. If $n + m$ is odd, the signature requirement for stability is

$$j \left(i_0 - 2i_1 + 2i_2 + \cdots + (-1)^{l-1} 2i_{l-1} + (-1)^l i_l \right) = n - m + 1 + 2z^+; \quad (11.30)$$

if $n + m$ is even, the signature requirement for stability is

$$j \left(i_0 - 2i_1 + 2i_2 + \cdots + (-1)^{l-1} 2i_{l-1} \right) = n - m + 1 + 2z^+. \quad (11.31)$$

6. Let $I_1, I_2, I_3, \ldots$ be the distinct strings of $i_0, i_1, \ldots$ that satisfy the signature condition in Equation (11.30) or (11.31). For fixed $k_p = k_p^*$, the stabilizing set in the space of (k_i, k_d) can be computed by solving the set of linear inequalities

$$v_{\text{even}} \left(-\omega_t^2, k_i, k_d \right) i_t > 0 \quad (11.32)$$

for all $\omega_t, t \in (0, \ldots, l)$ if $n + m$ is odd or for all $t \in (0, \ldots, l-1)$ if $n + m$ is even.

7. Every string I_j in the previous step creates a stability region $\mathcal{S}_j^o(k_p^*)$ and the region is a convex set since it is the intersection of solutions of linear (or affine) inequalities. Thus, the stabilizing region for the fixed $k_p = k_p^*$ is a union of convex sets:

$$\mathcal{S}^o \left(k_p^* \right) = \cup_j \mathcal{S}_j^o \left(k_p^* \right). \quad (11.33)$$

8. The stabilizing set in the space of (k_p, k_i, k_d), is computed by sweeping k_p over the real axis and by repeating the steps above.

EXAMPLE 11.1   Let us consider the plant

$$P(s) = \frac{s - 4}{s^3 + 7s^2 + 16s + 12} \quad (11.34)$$

and the controller

$$C(s) = \frac{k_d s^2 + k_p s + k_i}{s}. \quad (11.35)$$

The closed-loop characteristic polynomial is

$$\delta(s, k_p, k_i, k_d) = s^4 + (k_d + 7)s^3 + (k_p - 4k_d + 16)s^2 + (k_i - 4k_p + 12)s - 4k_i. \quad (11.36)$$

Here $n = 3$, $m = 1$, and $N(-s) = -s - 4$. Therefore, we obtain

$$\begin{aligned}
v(s) &= \delta(s, k_p, k_i, k_d) N(-s) \\
&= -s^5 + (-k_d - 11)s^4 + (-k_p - 44)s^3 + (16k_d - k_i - 76)s^2 \\
&\quad + (16k_p - 48)s + 16k_i
\end{aligned} \quad (11.37)$$

with the even and odd decomposition

$$\begin{aligned}
v(j\omega, k_p, k_i, k_d) &= (-k_d - 11)\omega^4 + (k_i - 16k_d + 76)\omega^2 + 9k_i \\
&\quad + j\left[-\omega^5 + (k_p + 44)\omega^3 + (16k_p - 48)\omega \right] \\
&= p(\omega) + jq(\omega).
\end{aligned} \quad (11.38)$$

We find that $z^+ = 1$, so that the signature requirement of Equation (11.11) on $v(s)$ for stability is

$$n - m + 1 + 2z^+ = 5. \tag{11.39}$$

Since the degree of $v(s)$ is odd, we need to use Equation (11.31). We see from the signature formulas that $q(\omega)$ must have at least two positive real roots of odd multiplicity. The range of k_p such that $q(\omega, k_p)$ has at least two real, positive, distinct, finite zeros with odd multiplicity roots was determined to be $k_p \in (-12.5, 3)$, which is then the allowable range for k_p shown in Figure 11.3. From the expression for $q(\omega)$ in Equation (11.38), we can represent k_p in terms of ω as

$$k_p(\omega) = \frac{\omega^4 - 44\omega^2 + 48}{\omega^2 + 16}. \tag{11.40}$$

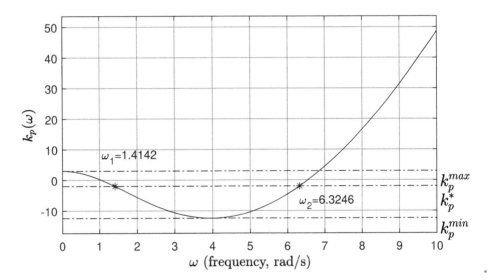

Figure 11.3 Range for k_p with at least two positive real roots (Example 11.1). © 2009 Taylor & Francis Group, LLC. Reproduced from Bhattacharyya, Datta, and Keel (2009) with permission.

For illustration, let us fix $k_p^* = -2$, which is contained in the allowable range for k_p. Now, we can find the roots ω_t from Figure 11.3. The roots are $\omega_1 = 1.4142$ and $\omega_2 = 6.3246$, as shown. Consider $\omega_0 = 0$, $\omega_1 = 1.4142$, and $\omega_2 = 6.3246$. For a small value of 0^+, we consider 0.001. Then

$$j = \text{sgn}[-(0.001)^4 + (-1 + 44)(0.001)^2 + (16(-1) - 48)] = -1. \tag{11.41}$$

Then signature expression in Equation (11.31) is given by

$$-(i_0 - 2i_1 + 2i_1) = 5 \tag{11.42}$$

and $I_1 = \{-1, +1, -1\}$ is the only string of integers i_0, i_1, i_2 that satisfies the signature condition in Equation (11.42). The stabilizing set in the space of (k_p, k_i, k_d) for a fixed $k_p^* = -2$ is thus given by the set of inequalities

$$k_i < 0, \qquad k_i - 2k_d + 6 > 0, \qquad k_i - 40k_d - 260 < 0. \tag{11.43}$$

Therefore, the stabilizing set for $k_p^* = -2$ is shown in Figure 11.4.

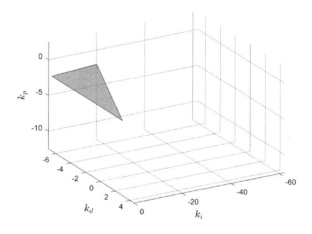

Figure 11.4 2D stabilizing set with $k_p^* = -2$ (Example 11.1) © 2009 Taylor & Francis Group, LLC. Reproduced from Bhattacharyya, Datta, and Keel (2009) with permission.

By sweeping over different k_p values within the interval $(-12.5, 3)$ we can generate the set of stabilizing (k_p, k_i, k_d) values. This set is shown in Figure 11.5.

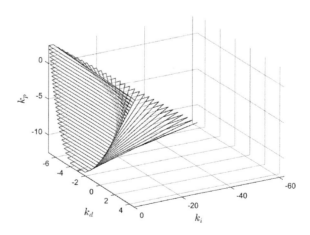

Figure 11.5 Complete 3D stabilizing set (Example 11.1) © 2009 Taylor & Francis Group, LLC. Reproduced from Bhattacharyya, Datta, and Keel (2009) with permission.

Remark 11.1   The signature method for calculating the PID stabilizing set $\mathscr{S}$ can easily be adapted to compute the stabilizing sets for controllers of the types

$$C_1(s) = \frac{k_2 s^2 + k_1 s + k_0}{s^2}, \qquad C_2(s) = \frac{k_2 s^2 + k_1 s + k_0}{s^2 + \omega_r^2}. \qquad (11.44)$$

$C_1(s)$ makes the output $y(t)$ track step and ramp inputs $r(t)$ and reject step and ramp disturbances in the closed loop. $C_2(s)$ is called a *resonant controller* and makes the output track a sinusoidal input $r(t)$ and reject a sinusoidal disturbance at the radian frequency ω_r.

The main difference in the computation of the stabilizing set is that in the characteristic polynomial $\delta(s)$, $sD(s)$, for the PID controller is replaced by $s^2 D(s)$ and $(s^2 + \omega_r^2)D(s)$ for $C_1(s)$ and $C_2(s)$, respectively.

The signature requirement for stability with $C_1(s)$ or $C_2(s)$ is

$$\sigma(v) = n - m + 2 + 2z^+. \qquad (11.45)$$

The plant $P(s)$, controlled by the resonant controller $C_2(s)$, is assumed to have no $j\omega$ axis zeros at $\omega = \omega_r$, since the closed loop cannot be stabilized otherwise.

11.4   Computation of PID Stabilizing Sets for Discrete-Time Plants

In this section we develop results on the problem of determining the PID stabilizing set for a discrete-time control system. Consider a SISO plant described by its z-domain transfer function $P(z)$ and the unity feedback controller $C(z)$ in Figure 11.6.

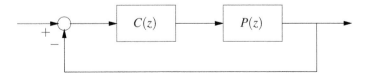

Figure 11.6 A unity feedback system

The transfer functions shown in Figure 11.6 are rational functions and we write them as ratios of polynomials:

$$P(z) = \frac{N(z)}{D(z)}, \qquad C(z) = \frac{N_C(z)}{D_C(z)}. \qquad (11.46)$$

The *characteristic polynomial* of the closed-loop system is

$$\delta(z) := D_C(z)D(z) + N_C(z)N(z) \qquad (11.47)$$

and a necessary and sufficient condition for stability of the closed-loop control system is that the characteristic roots, namely the zeros of $\delta(z)$, have magnitude less than unity. This condition is commonly referred to as *Schur stability* of $\delta(z)$.

The stabilization problem can be stated as the problem of determining $C(z)$ so that for the given $P(z)$, the closed-loop characteristic polynomial $\delta(z)$ is rendered Schur. For a fixed structure controller, such as a discrete-time PID controller, $C(z)$ is characterized by a set of gains $\underline{\mathbf{k}}$ (three gains in case of PID) and these gains must be chosen to stabilize $\delta(z)$ if possible. An important problem in multi-objective design is the determination of the entire set $\mathscr{S}$ of stabilizing gains in a constructive way. A useful characterization of $\mathscr{S}$ should allow the designer to test the feasibility of imposing various performance constraints and checking their attainability with the controller parameters ranging over the stabilizing set.

11.4.1 Tchebyshev Representation and Root Clustering

The stabilization results require us to determine the complex plane image of polynomials and rational functions evaluated on a circle of radius ρ centered at the origin. This image is used for certain root-counting formulas which we need to develop. These formulas in turn depend on the Tchebyshev representation discussed here.

Tchebyshev Representation of Real Polynomials

Let us consider a polynomial $P(z) = a_n z^n + \cdots + a_0$ with real coefficients. The image of $P(z)$ evaluated on the circle $\mathscr{C}_\rho$ of radius ρ, centered at the origin, is

$$\left\{ P(z) \ : \ z = \rho e^{j\theta}, \ \ 0 \leq \theta \leq 2\pi \right\}. \tag{11.48}$$

Since the a_is are real for all $i \in 0, \ldots, n$, $P\left(\rho e^{j\theta}\right)$, and $P\left(\rho e^{-j\theta}\right)$ are complex conjugate. Thus, it suffices to determine the image of the upper half of the circle:

$$\left\{ P(z) \ : \ z = \rho e^{j\theta}, \ \ 0 \leq \theta \leq \pi \right\}. \tag{11.49}$$

Since

$$z^k \Big|_{z=\rho e^{j\theta}} = \rho^k (\cos k\theta + j \sin k\theta), \tag{11.50}$$

we have

$$P\left(\rho e^{j\theta}\right) = \underbrace{(a_n \rho^n \cos n\theta + \cdots + a_1 \rho \cos \theta + a_0)}_{\bar{R}(\rho,\theta)} + j \underbrace{(a_n \rho^n \sin n\theta + \cdots + a_1 \rho \sin \theta)}_{\bar{I}(\rho,\theta)}$$

$$= \bar{R}(\rho,\theta) + j\bar{I}(\rho,\theta). \tag{11.51}$$

It is well known that $\cos k\theta$ and $\frac{\sin k\theta}{\sin \theta}$ can be written as polynomials in $\cos \theta$ using Tchebyshev polynomials. Write $u := -\cos \theta$. Then as θ runs increasingly from 0 to π, u runs increasingly from -1 to $+1$. Now

$$e^{j\theta} = \cos \theta + j \sin \theta = -u + j\sqrt{1-u^2} \tag{11.52}$$

and we have

$$c_k(u) := \cos k\theta \quad \text{and} \quad s_k(u) := \frac{\sin k\theta}{\sin \theta}, \tag{11.53}$$

where $c_k(u)$ and $s_k(u)$ are real polynomials in u and are known as the Tchebyshev polynomials of the first and second kind, respectively. It is easy to show that

$$s_k(u) = -\frac{c_k'(u)}{k}, \quad \text{for } k = 1, 2, \ldots \tag{11.54}$$

and that the Tchebyshev polynomials satisfy the recursive relation

$$c_{k+1}(u) = -uc_k(u) - \left(1 - u^2\right) s_k(u), \quad \text{for } k = 1, 2, \ldots \tag{11.55}$$

From Equations (11.53), (11.54), and (11.55), we can determine $c_k(u)$ and $s_k(u)$ for $k = 1, 2, 3, \ldots$. Now

$$\left(\rho e^{j\theta}\right)^k = \rho^k \cos k\theta + j\rho^k \sin k\theta \tag{11.56}$$

and so we define the generalized Tchebyshev polynomials as follows:

$$c_k(u, \rho) := \rho^k c_k(u), \quad s_k(u, \rho) := \rho^k s_k(u), \quad \text{for } k = 0, 1, 2 \ldots \tag{11.57}$$

and note that

$$s_k(u, \rho) = -\frac{1}{k} \cdot \frac{d\left[c_k(u, \rho)\right]}{du}, \quad \text{for } k = 1, 2, \ldots, \tag{11.58a}$$

$$c_{k+1}(u, \rho) = -\rho u c_k(u, \rho) - \left(1 - u^2\right) \rho s_k(u, \rho), \quad \text{for } k = 1, 2, \ldots \tag{11.58b}$$

The generalized Tchebyshev polynomials are displayed in Table 11.1 for $k = 1, \ldots, 5$.

Table 11.1 Tchebyshev polynomials of the first and second kind

k	$c_k(u, \rho)$	$s_k(u, \rho)$
1	$-\rho u$	ρ
2	$\rho^2 \left(2u^2 - 1\right)$	$-2\rho^2 u$
3	$\rho^3 \left(-4u^3 + 3u\right)$	$\rho^3 \left(4u^2 - 1\right)$
4	$\rho^4 \left(8u^4 - 8u^2 + 1\right)$	$\rho^4 \left(-8u^3 + 4u\right)$
5	$\rho^5 \left(-16u^5 + 20u^3 - 5u\right)$	$\rho^5 \left(16u^4 - 12u^2 + 1\right)$
⋮	⋮	⋮

Using the notation in Equation (11.57):

$$p\left(\rho e^{j\theta}\right) = R(u, \rho) + j\sqrt{1 - u^2}\, T(u, \rho) =: p_c(u, \rho), \tag{11.59}$$

where

$$R(u, \rho) = a_n c_n(u, \rho) + a_{n-1} c_{n-1}(u, \rho) + \cdots + a_1 c_1(u, \rho) + a_0, \tag{11.60a}$$

$$T(u, \rho) = a_n s_n(u, \rho) + a_{n-1} s_{n-1}(u, \rho) + \cdots + a_1 s_1(u, \rho). \tag{11.60b}$$

$R(u, \rho)$ and $T(u, \rho)$ are polynomials in u and ρ. The complex plane image of $p(z)$ as z traverses the upper half of the circle $\mathscr{C}_\rho$ can be obtained by evaluating $p_c(u, \rho)$ as u runs from -1 to $+1$.

LEMMA 11.2 For a fixed $\rho > 0$:

(a) if $P(z)$ has no roots on the circle of radius $\rho > 0$, $(R(u,\rho), T(u,\rho))$ have no common roots for $u \in [-1,1]$ and $R(\pm 1, \rho) \neq 0$;

(b) if $P(z)$ has $2m$ roots at $z = -\rho$ $(z = +\rho)$, then $R(u,\rho)$ and $T(u,\rho)$ have m roots each at $u = +1$ $(u = -1)$;

(c) if $P(z)$ has $2m-1$ roots at $z = -\rho$ $(z = +\rho)$, then $R(u,\rho)$ and $T(u,\rho)$ have m and $m-1$ roots, respectively, at $u = +1$ $(u = -1)$;

(d) if $P(z)$ has q_i pairs of complex conjugate roots at $z = -\rho u_i \pm j\rho\sqrt{1-u_i^2}$, for $u_i \neq \pm 1$, then $R(u,\rho)$ and $T(u,\rho)$ each have q_i real roots at $u = u_i$.

Proof To prove (a), note that $P(\rho e^{j\theta}) \neq 0$ for $\theta \in [0,\pi]$ and therefore $p_c(u,\rho) \neq 0$ for $u \in [-1,+1]$; hence the result. The statements in (b)–(d) may be verified by direct calculation. ∎

When the circle of interest is the unit circle, that is $\rho = 1$, we write $P_c(u,1) = P_c(u)$ and also

$$R(u,1) =: R(u), \qquad T(u,1) =: T(u) \tag{11.61}$$

for notational simplicity.

Interlacing Conditions for Root Clustering and Schur Stability

The formulas of the last section can be used to derive conditions for root clustering in circular regions, that is for the roots to lie strictly within a circle of radius ρ. For Schur stability we simply take $\rho = 1$. As before, let $P(z)$ be a real polynomial of degree n and

$$
\begin{aligned}
p\left(\rho e^{j\theta}\right) &= \bar{R}(\theta,\rho) + j\bar{I}(\theta,\rho), \qquad \text{where } u = -\cos\theta \\
&= R(u,\rho) + j\sqrt{1-u^2}\,T(u,\rho),
\end{aligned}
\tag{11.62}
$$

where $R(u,\rho)$ and $T(u,\rho)$ are real polynomials of degree n and $n-1$, respectively, in u, for fixed ρ.

THEOREM 11.2 $p(z)$ has all its zeros strictly within $\mathscr{C}_\rho$ if and only if

(a) $R(u,\rho)$ has n real distinct zeros r_i, $i = 1,2,\ldots,n$ in $(-1,1)$.

(b) $T(u,\rho)$ has $n-1$ real distinct zeros t_j, $j = 1,2,\ldots,n-1$ in $(-1,1)$.

(c) The zeros r_i and t_j interlace:

$$-1 < r_1 < t_1 < r_2 < t_2 < \cdots < t_{n-1} < r_n < +1. \tag{11.63}$$

Proof Let

$$t_j = -\cos\alpha_j, \qquad \alpha_j \in (0,\pi), \qquad \text{for } j = 1,2,\ldots,n-1 \tag{11.64}$$

or

$$
\begin{aligned}
\alpha_j &= -\cos^{-1} t_j, \qquad \text{for } j = 1,2,\ldots,n-1 \\
\alpha_0 &= 0, \\
\alpha_n &= \pi,
\end{aligned}
\tag{11.65}
$$

and let

$$\beta_i = -\cos^{-1} r_i, \qquad \text{for } i = 1, 2, \ldots, n, \quad \beta_i \in (0, \pi). \tag{11.66}$$

Then $(\alpha_0, \alpha_1, \ldots, \alpha_n)$ are the $n + 1$ zeros of $\bar{I}(\theta, \rho) = 0$ and $(\beta_1, \beta_2, \ldots, \beta_{n-1})$ are the n zeros of $\bar{R}(\theta, \rho) = 0$. The condition (c) means that α_i and β_j satisfy

$$0 = \alpha_0 < \beta_1 < \alpha_1 < \beta_2 < \cdots < \beta_{n-1} < \alpha_n = \pi. \tag{11.67}$$

The conditions (a)–(c) imply that the plot of $p\left(\rho e^{j\theta}\right)$ for $\theta \in [0, \pi]$ turns counterclockwise through exactly $2n$ quadrants and this condition is equivalent to $p(z)$ having n zeros inside the circle $\mathscr{C}_\rho$. ∎

Remark 11.2   The conditions (a), (b), and (c) given in Theorem 11.2 may be referred to as interlacing conditions on $R(u, \rho)$ and $T(u, \rho)$. By setting $\rho = 1$, we obtain conditions for Schur stability in terms of interlacing of the zeros of $R(u)$ and $T(u)$. This constitutes a Hermite–Biehler-like theorem for Schur stability.

Tchebyshev Representation of Rational Functions

Let $Q(z)$ be a ratio of two real polynomials $P_1(z)$ and $P_2(z)$. We compute the image of $Q(z)$ on $\mathscr{C}_\rho$ and write it as the corresponding Tchebyshev representation $Q_c(u, \rho)$. Let

$$P_i(z)\big|_{z = -\rho u + j\rho\sqrt{1-u^2}} = R_i(u, \rho) + j\sqrt{1-u^2}\, T_i(u, \rho), \qquad \text{for } i = 1, 2. \tag{11.68}$$

Then

$$
\begin{aligned}
Q(z)\big|_{z = -\rho u + j\rho\sqrt{1-u^2}} &= \left.\frac{P_1(z)}{P_2(z)}\right|_{z = -\rho u + j\rho\sqrt{1-u^2}} \\[2mm]
&= \left.\frac{P_1(z)P_2\left(z^{-1}\right)}{P_2(z)P_2\left(z^{-1}\right)}\right|_{z = -\rho u + j\rho\sqrt{1-u^2}} \\[2mm]
&= \frac{\left(R_1(u,\rho) + j\sqrt{1-u^2}\,T_1(u,\rho)\right)\left(R_2(u,\rho) - j\sqrt{1-u^2}\,T_2(u,\rho)\right)}{\left(R_2(u,\rho) + j\sqrt{1-u^2}\,T_2(u,\rho)\right)\left(R_2(u,\rho) - j\sqrt{1-u^2}\,T_2(u,\rho)\right)} \\[2mm]
&= \underbrace{\left(\frac{R_1(u,\rho)R_2(u,\rho) + \left(1-u^2\right)T_1(u,\rho)T_2(u,\rho)}{R_2^2(u,\rho) + \left(1-u^2\right)T_2^2(u,\rho)}\right)}_{=:R(u,\rho)} \\[2mm]
&\quad + j\sqrt{1-u^2}\underbrace{\left(\frac{T_1(u,\rho)R_2(u,\rho) - R_1(u,\rho)T_2(u,\rho)}{R_2^2(u,\rho) + \left(1-u^2\right)T_2^2(u,\rho)}\right)}_{=:T(u,\rho)}
\end{aligned} \tag{11.69}
$$

and we denote

$$Q_c(u, \rho) := R(u, \rho) + j\sqrt{1-u^2}\,T(u, \rho). \tag{11.70}$$

Note that $R(u, \rho)$, $T(u, \rho)$ are rational functions of the real variable u which runs from -1 to $+1$.

11.4.2   Root-Counting Formulas

In this section, we develop some formulas for counting the root distribution with respect to the circle $\mathscr{C}_\rho$, for real polynomials and real rational functions. These formulas will be necessary for our solution of the stabilization problem but are also of independent interest. We begin by relating root distribution to phase unwrapping.

Phase Unwrapping and Root Distribution

Let $\phi_p(\theta) := \angle p(\rho e^{j\theta})$ denote the *phase* of the polynomial $p(z)$ evaluated at $z = \rho e^{j\theta}$ and let $\Delta_{\theta_1}^{\theta_2}[\phi_P(\theta)]$ denote the net change in or *unwrapped phase* of $P(\rho e^{j\theta})$ as θ increases from θ_1 to θ_2. Similar notation applies to the rational function $Q(z)$ with Tchebyshev representation $Q_C(u, \rho)$. Let $\phi_{Q_C}(u) = \angle Q_C(u, \rho)$ denote the phase of $Q_C(u, \rho)$ and $\Delta_{u_1}^{u_2}[\phi_{Q_C}(u)]$ the net change in or unwrapped phase of $Q_C(u, \rho)$ as u increases from u_1 to u_2.

LEMMA 11.3   Let the real polynomial $P(z)$ have i roots in the interior of the circle $\mathscr{C}_\rho$ and no roots on the circle. Then

$$\Delta_0^\pi[\phi_P(\theta)] = \pi i. \tag{11.71}$$

Proof   From geometric considerations it is easily seen that each interior root contributes 2π to $\Delta_0^{2\pi}[\phi_p(\theta)]$ and therefore by the symmetry of roots about the real axis the interior roots contribute πi to $\Delta_0^\pi[\phi_p(\theta)]$. The exterior roots do not contribute the net change of phase $\Delta_0^{2\pi}[\phi_p(\theta)]$.   ∎

We state the corresponding result for a rational function. The proof is similar to the previous lemma and is omitted.

LEMMA 11.4   Let $Q(z) = \frac{p_1(z)}{p_2(z)}$, where the real polynomials $p_1(z)$ and $p_2(z)$ have i_1 and i_2 roots, respectively, in the interior of the circle $\mathscr{C}_\rho$ and no roots on the circle. Then

$$\Delta_0^\pi[\phi_Q(\theta)] = \pi(i_1 - i_2) = \Delta_{-1}^{+1}[\phi_{Q_C}(u)]. \tag{11.72}$$

Root Counting and Tchebyshev Representation

In this section, we first develop formulas to determine the number of roots of a real polynomial, inside a circle $\mathscr{C}_\rho$, to its Tchebyshev representation. Let us begin with a real polynomial $P(z)$ and its Tchebyshev representation

$$P_C(u, \rho) = R(u, \rho) + \sqrt{1 - u^2} T(u, \rho) \tag{11.73}$$

as developed before. Henceforth, let $t_1, \dots, t_k$ denote the real distinct zeros of $T(u, \rho)$ of odd multiplicity, for $u \in (-1, 1)$, ordered as follows:

$$-1 < t_1 < t_2 < \cdots < t_k < +1. \tag{11.74}$$

Suppose also that $T(u,\rho)$ has p zeros at $u = -1$ and let $f^{(i)}(x_0)$ denote the ith derivative of $f(x)$ evaluated at $x = x_0$. Let us also define

$$\text{sgn}[x] = \begin{cases} -1 & \text{if } x < 0, \\ 0 & \text{if } x = 0, \\ 1 & \text{if } x > 0. \end{cases} \tag{11.75}$$

THEOREM 11.3 Let $P(z)$ be a real polynomial with no roots on the circle $\mathscr{C}_\rho$ and suppose that $T(u,\rho)$ has p zeros at $u = -1$. Then the number of roots i of $P(z)$ in the interior of the circle $\mathscr{C}_\rho$ is given by

$$\begin{aligned} i = \frac{1}{2}\text{sgn}\left[T^{(p)}(-1,\rho)\right] &\left(\text{sgn}[R(-1,\rho)] + 2\sum_{j=1}^{k}(-1)^j \text{sgn}[R(t_j,\rho)] \right. \\ &\left. + (-1)^{k+1}\text{sgn}[R(+1,\rho)] \right). \end{aligned} \tag{11.76}$$

Proof Recall that

$$P(\rho e^{j\theta}) = \bar{R}(\theta,\rho) + j\bar{I}(\theta,\rho) \tag{11.77}$$

and define $\theta_i, i = 1,\ldots,k$ such that

$$t_i = -\cos\theta_i, \qquad \text{for } \theta_i \in [0,\pi], \tag{11.78}$$

where t_is are the zeros of $T(u,\rho)$ of odd multiplicity, for $u \in (-1,1)$. Let $\theta_0 := 0, t_0 := -1$ and $\theta_{k+1} := \pi$, and note that $\theta_i, i = 0,1,\ldots,k+1$ are zeros of $\bar{I}(\theta,\rho)$. The proof depends on the following elementary and easily verified facts which are first stated. (In the following, θ_i^+ denotes the point immediately to the right of θ_i.)

$$\Delta_0^\pi[\phi(\theta)] = \pi i, \tag{11.79a}$$

$$\Delta_0^\pi[\phi(\theta)] = \Delta_0^{\theta_1}[\phi(\theta)] + \Delta_{\theta_1}^{\theta_2}[\phi(\theta)] + \cdots + \Delta_{\theta_k}^{\pi}[\phi(\theta)], \tag{11.79b}$$

$$\Delta_{\theta_i}^{\theta_{i+1}}[\phi(\theta)] = \frac{\pi}{2}\text{sgn}\left[\bar{I}\left(\theta_i^+,\rho\right)\right]\left(\text{sgn}[\bar{R}(\theta_i,\rho)] - \text{sgn}[\bar{R}(\theta_{i+1},\rho)]\right),$$
$$\text{for } i = 0,1,\ldots,k, \tag{11.79c}$$

$$\text{sgn}\left[\bar{I}\left(\theta_i^+,\rho\right)\right] = -\text{sgn}\left[\bar{I}\left(\theta_{i+1}^+,\rho\right)\right], \quad \text{for } i = 0,1,\ldots,k, \tag{11.79d}$$

$$\text{sgn}\left[\bar{I}(0^+,\rho)\right] = \text{sgn}\left[T^{(p)}(-1,\rho)\right], \tag{11.79e}$$

$$\text{sgn}[\bar{R}(\theta_i,\rho)] = \text{sgn}[R(t_i,\rho)], \quad \text{for } i = 0,1,\ldots,k. \tag{11.79f}$$

Using Equations (11.79a)–(11.79f), we have

$$\pi i = \Delta_0^\pi [\phi(\theta)]$$

$$= \Delta_0^{\theta_1}[\phi(\theta)] + \cdots + \Delta_{\theta_k}^\pi[\phi(\theta)] \qquad \text{(by Equations (11.79a) and (11.79b))}$$

$$= \frac{\pi}{2}\left\{ \operatorname{sgn}\left[\bar{I}(0^+,\rho)\right] \left(\operatorname{sgn}\left[\bar{R}(0,\rho)\right] - \operatorname{sgn}\left[\bar{R}(\theta_1,\rho)\right]\right) + \right.$$

$$\left. \cdots + \operatorname{sgn}\left[\bar{I}\left(\theta_k^+,\rho\right)\right] \left(\operatorname{sgn}\left[\bar{R}(\theta_k,\rho)\right] - \operatorname{sgn}\left[\bar{R}(\pi,\rho)\right]\right)\right\} \quad \text{(by Equation (11.79c))}$$

$$= \frac{\pi}{2}\operatorname{sgn}\left[\bar{I}\left(0^+,\rho\right)\right]\left\{\left(\operatorname{sgn}\left[\bar{R}(0,\rho)\right] - \operatorname{sgn}\left[\bar{R}(\theta_1,\rho)\right]\right)\right.$$

$$\left. - \left(\operatorname{sgn}\left[\bar{R}(\theta_1,\rho)\right] - \operatorname{sgn}\left[\bar{R}(\theta_2,\rho)\right]\right) + \right. \tag{11.80}$$

$$\left. \cdots + (-1)^k \left(\operatorname{sgn}\left[\bar{R}(\theta_k,\rho)\right] - \operatorname{sgn}\left[\bar{R}(\pi,\rho)\right]\right)\right\} \quad \text{(by Equation (11.79d))}$$

$$= \frac{\pi}{2}\operatorname{sgn}[T^{(p)}(-1,\rho)]\left\{\operatorname{sgn}\left[\bar{R}(0,\rho)\right] - 2\operatorname{sgn}\left[\bar{R}(\theta_1,\rho)\right] + 2\operatorname{sgn}\left[\bar{R}(\theta_2,\rho)\right] + \right.$$

$$\left. \cdots + (-1)^k\operatorname{sgn}\left[\bar{R}(\theta_k,\rho)\right] + (-1)^{k+1}\operatorname{sgn}\left[\bar{R}(\pi,\rho)\right]\right\} \quad \text{(by Equation (11.79e))}$$

$$= \frac{\pi}{2}\operatorname{sgn}[T^{(p)}(-1,\rho)]\left\{\operatorname{sgn}[R(-1,\rho)] - 2\operatorname{sgn}\left[R\left(t_1,\rho\right)\right] + 2\operatorname{sgn}\left[R\left(t_2,\rho\right)\right] + \right.$$

$$\left. \cdots + (-1)^k2\operatorname{sgn}\left[R\left(t_k,\rho\right)\right] + (-1)^{k+1}\operatorname{sgn}[R(+1,\rho)]\right\} \quad \text{(by Equation (11.79f))},$$

from which Equation (11.76) follows.   ∎

The result derived above can now be extended to the case of rational functions. Let $Q(z) = \frac{p_1(z)}{p_2(z)}$, where $p_i(z)$ for $i = 1, 2$ are polynomials with real coefficients. Let

$$R_i(u,\rho) + j\sqrt{1-u^2}T_i(u,\rho), \qquad \text{for } i = 1,2 \tag{11.81}$$

denote the Tchebyshev representations of $p_i(z), i = 1,2$ and $Q_C(u,\rho)$ denote the Tchebyshev representation of $Q(z)$ on the circle $\mathscr{C}_\rho$. Let $R(u,\rho), T(u,\rho)$ be defined by

$$R(u,\rho) = R_1(u,\rho)R_2(u,\rho) + (1-u^2)T_1(u,\rho)T_2(u,\rho), \tag{11.82a}$$

$$T(u,\rho) = T_1(u,\rho)R_2(u,\rho) - R_1(u,\rho)T_2(u,\rho). \tag{11.82b}$$

Suppose that $T(u,\rho)$ has p zeros at $u = -1$ and let $t_1,\ldots,t_k$ denote the real distinct zeros of $T(u,\rho)$ of odd multiplicity, ordered as follows:

$$-1 < t_1 < t_2 < \cdots < t_k < +1. \tag{11.83}$$

THEOREM 11.4   Let $Q(z) = \frac{p_1(z)}{p_2(z)}$, where $p_i(z)$ for $i = 1,2$ are real polynomials with

i_1 and i_2 zeros, respectively, inside the circle $\mathscr{C}_\rho$ and no zeros on it. Then

$$i_1 - i_2 = \frac{1}{2}\mathrm{sgn}\left[T^{(p)}(-1,\rho)\right]\left(\mathrm{sgn}\left[R(-1,\rho)\right] + 2\sum_{j=1}^{k}(-1)^j\mathrm{sgn}\left[R(t_j,\rho)\right]\right. \tag{11.84}$$
$$\left. + (-1)^{k+1}\mathrm{sgn}\left[R(+1,\rho)\right]\right).$$

Proof The proof is based on the representation of $Q_C(u,\rho)$ developed in Equation (11.70). Since the denominator of Equation (11.70) is strictly positive for $u \in [-1,+1]$ it follows that the phase unwrapping can be computed from the numerator. The rest of the proof is similar to the proof for the polynomial case and is omitted. ∎

11.4.3 Stabilizing Sets of Discrete-Time PID Controllers

In this section, we first give general parametrizations of digital PI, PD, and PID controllers, respectively, in terms of parameters K_0, K_1, K_2. These will be used in the sequel to compute the stabilizing sets. T denotes the sampling period.

1. For PI controllers, we have

$$C(z) = k_p + k_iT \cdot \frac{z}{z-1} = \frac{(k_p + k_iT)z - k_p}{z-1}$$
$$= \frac{(k_p + k_iT)\left(z - \frac{k_p}{k_iT + k_p}\right)}{z - 1}. \tag{11.85}$$

Thus, we rewrite

$$C(z) = \frac{K_1z + K_0}{z - 1}, \tag{11.86}$$

where

$$k_p = -K_0 \quad \text{and} \quad k_i = \frac{K_1 + K_0}{T}. \tag{11.87}$$

2. For PD controllers, we have

$$C(z) = k_p + \frac{k_d}{T} \cdot \frac{z-1}{z} = \frac{(k_pT + k_d)z - k_d}{Tz}$$
$$= \frac{\left(k_p + \frac{k_d}{T}\right)\left(z - \frac{\frac{k_d}{T}}{k_p + \frac{k_d}{T}}\right)}{z}. \tag{11.88}$$

We rewrite

$$C(z) = \frac{K_1z + K_0}{z}, \tag{11.89}$$

where

$$k_p = K_0 + K_1 \quad \text{and} \quad k_d = -K_0T. \tag{11.90}$$

3. The general formula of a discrete-time PID controller, using backward differences for differentiation to preserve causality, is

$$
\begin{aligned}
C(z) &= k_p + k_i T \cdot \frac{z}{z-1} + \frac{k_d}{T} \cdot \frac{z-1}{z} \\
&= \frac{\left(k_p + k_i T + \frac{k_d}{T}\right) z^2 + \left(-k_p - \frac{2k_d}{T}\right) z + \frac{k_d}{T}}{z(z-1)}.
\end{aligned}
\tag{11.91}
$$

We use the general representation

$$
C(z) = \frac{K_2 z^2 + K_1 z + K_0}{z(z-1)},
\tag{11.92}
$$

where

$$
k_p = -K_1 - 2K_0 \quad \text{and} \quad k_i = \frac{K_0 + K_1 + K_2}{T},
\tag{11.93}
$$

and

$$
k_d = K_0 T.
\tag{11.94}
$$

Computation of the Stabilizing Set with PID Controllers

The results of the previous sections on Tchebyshev representations, root counting, and the representations of digital PID controllers are used here to develop constructive techniques for computing the stabilizing set. As in the continuous-time case, the main idea is to construct a polynomial or rational function such that the controller parameters are *separated* as much as possible into the real and imaginary parts. By this we mean the stabilization problem is converted into an equivalent signature assignment problem for this rational function. By applying the root-counting formulas to this function we can often "linearize" the problem. We emphasize that other root-counting formulas such as Jury's test applied to these problems result in difficult nonlinear problems, which are often impossible to solve. In this section we outline how the technique works for PI and PID controllers, and present a complete development along with an example in the next section.

Consider the control system in Figure 11.6 with a rational and proper plant transfer function

$$
P(z) := \frac{N(z)}{D(z)},
\tag{11.95}
$$

with $\deg D(z) = n$ and $\deg N(z) \le n$. The PID controller is of the form:

$$
C(z) = \frac{K_0 + K_1 z + K_2 z^2}{z(z-1)}.
\tag{11.96}
$$

The procedure to compute the PID stabilizing set can now be stated as the following steps:

1. Form the characteristic polynomial of the closed-loop system in Figure 11.6 of degree $n + 2$:

$$\delta(z) = z(z-1)D(z) + (K_0 + K_1 z + K_2 z^2)N(z). \tag{11.97}$$

2. Form the new function denoted $\nu(z)$:

$$z^{-1}\delta(z)N(z^{-1}) = (z-1)D(z)N(z^{-1}) + (K_0 z^{-1} + K_1 + K_2 z)N(z)N(z^{-1}). \tag{11.98}$$

Clearly $\delta(z)$ is Schur if and only if this new function $\nu(z)$ has Schur signature $n + 2 + \sigma\left(z^{-1}N(z^{-1})\right)$. This will be exploited below to compute the stabilizing set.

3. Use the Tchebyshev representations to calculate the image of $\nu(z)$ over the unit circle:

$$\begin{aligned} z^{-1}\delta(z)N(z^{-1})_{z=e^{j\theta}} = &-(u+1)P_1(u) - \left(1-u^2\right)P_2(u) \\ &- \left((K_0 + K_2)u - K_1\right)P_3(u) \\ &+ j\sqrt{1-u^2}\left(-(u+1)P_2(u) + P_1(u) + (K_2 - K_0)P_3(u)\right), \end{aligned} \tag{11.99}$$

where

$$\begin{aligned} P_1(u) &= R_D(u)R_N(u) + (1-u^2)T_D(u)T_N(u), \\ P_2(u) &= R_N(u)T_D(u) - T_N(u)R_D(u), \\ P_3(u) &= R_N^2(u) + (1-u^2)T_N^2(u), \end{aligned} \tag{11.100}$$

and $R_N(u)$, $R_D(u)$, $T_N(u)$, and $T_D(u)$ are calculated as in Equation (11.60). Now, let $K_3 := K_2 - K_0$. Rewriting Equation (11.99), we have

$$\begin{aligned} z^{-1}\delta(z)N(z^{-1}) &= -(u+1)P_1(u) - (1-u^2)P_2 - ((2K_2 - K_3)u - K_1)P_3(u) \\ &\quad + j\sqrt{1-u^2}\left(-(u+1)P_2(u) + P_1(u) + K_3 P_3(u)\right) \\ &= R(u, K_1, K_2, K_3) + j\sqrt{1-u^2}T(u, K_3). \end{aligned} \tag{11.101}$$

4. Fixing a specific value of K_3, we can calculate the zeros of t_i of $T(u, K_3)$, which are real, distinct, and of odd multiplicity for $u \in (-1, +1)$:

$$-1 < t_1 < t_2 < \cdots < t_k < +1. \tag{11.102}$$

5. For this fixed K_3, calculate the set of strings of sign patterns for the real part $R(t_j, K_1, K_2, K_3)$, corresponding to stability, using

$$\begin{aligned} (n+2) + i_{N_r} - (l+1) = \frac{1}{2}\mathrm{sgn}\left[T^{(p)}(-1)\right]\Big(\mathrm{sgn}[R(-1, K_1, K_2, K_3)] \\ +2\sum_{j=1}^{k}(-1)^j\mathrm{sgn}[R(t_j, K_1, K_2, K_3)] + (-1)^{k+1}\mathrm{sgn}[R(+1, K_1, K_2, K_3)]\Big), \end{aligned} \tag{11.103}$$

where i_{N_r} is the number of zeros of $N_r(z)$ inside the unit circle. For fixed K_3, this leads to linear inequalities in (K_1, K_2).

6. Sweep over the K_3 range for which an adequate number of real roots t_k exist in $(-1, 1)$ for $T(u, K_3) = 0$, and repeat the above steps.

The algorithm is illustrated in the following example.

EXAMPLE 11.2   Consider the plant $P(z)$:

$$P(z) = \frac{1}{z^2 - 0.25}. \tag{11.104}$$

Then

$$R_D(u) = 2u^2 - 1.25, \quad T_D(u) = -2u, \quad R_N(u) = 1, \quad T_N(u) = 0 \tag{11.105}$$

and

$$P_1(u) = 2u^2 - 1.25, \quad P_2(u) = -2u, \quad P_3(u) = 1. \tag{11.106}$$

Since $P(z)$ is of order 2 and $C(z)$, the PID controller, is of order 2, the number of roots of $\delta(z)$ inside the unit circle is required to be four, for stability. From Theorem 11.3:

$$i_i - i_2 = \underbrace{(i_\delta + i_{N_r})}_{i_1} - \underbrace{(l+1)}_{i_2}, \tag{11.107}$$

where i_δ and i_{N_r} are the numbers of roots of $\delta(z)$ and the reverse polynomial of $N(z)$ inside the unit circle, respectively, and l is the degree of $N(z)$. Since the required i_δ is 4, $i_{N_r} = 0$, and $l = 0$, $i_1 - i_2$ is required to be 3. To illustrate the example in detail, we first fix $K_3 = 1.3$. Then the real roots of $T(u, K_3)$ in $(-1, 1)$ are -0.4736 and -0.0264. Furthermore, $\mathrm{sgn}[T(-1)] = 1$, and from Theorem 11.3, $i_1 - i_2 = 3$ requires that

$$\frac{1}{2}\mathrm{sgn}[T(-1)]\Big(\mathrm{sgn}[R(-1, K_1, K_2)] - 2\mathrm{sgn}[R(-0.4736, K_1, K_2)]$$

$$+ 2\mathrm{sgn}[R(-0.0264, K_1, K_2)] - \mathrm{sgn}[R(1, K_1, K_2)] \Big) = 3. \tag{11.108}$$

We have only one admissible string satisfying the above equation, namely

$$\begin{aligned}
\mathrm{sgn}[R(-1, K_1, K_2)] &= 1, \\
\mathrm{sgn}[R(-0.4736, K_1, K_2)] &= -1, \\
\mathrm{sgn}[R(-0.0264, K_1, K_2)] &= 1, \\
\mathrm{sgn}[R(1, K_1, K_2)] &= -1.
\end{aligned} \tag{11.109}$$

Corresponding to this string, we have the following set of linear inequalities:

$$\begin{aligned}
-1.3 + K_1 + 2K_2 &> 0, \\
-0.9286 + K_1 + 0.9472 &< 0, \\
1.1286 + K_1 + 0.0528K_2 &> 0, \\
-0.2 + K_1 - 2K_2 &< 0.
\end{aligned} \tag{11.110}$$

This set of inequalities characterizes the stability region in (K_1, K_2) space for the fixed

$K_3 = 1.3$. By repeating this procedure for the range of K_3 for which $T(u, K_3)$ has at least two real roots, we obtain the stability region shown in the left of Figure 11.7. Using

$$
\begin{bmatrix} k_p \\ k_i \\ k_d \end{bmatrix} = \begin{bmatrix} -2 & -1 & 0 \\ \frac{1}{T} & \frac{1}{T} & \frac{1}{T} \\ T & 0 & 0 \end{bmatrix} \begin{bmatrix} K_0 \\ K_1 \\ K_2 \end{bmatrix} = \begin{bmatrix} -2 & -1 & 0 \\ \frac{1}{T} & \frac{1}{T} & \frac{1}{T} \\ T & 0 & 0 \end{bmatrix} \begin{bmatrix} 0 & 1 & -1 \\ 1 & 0 & 0 \\ 0 & 1 & 0 \end{bmatrix} \begin{bmatrix} K_1 \\ K_2 \\ K_3 \end{bmatrix}
$$

$$
= \begin{bmatrix} -1 & -2 & 2 \\ \frac{1}{T} & \frac{2}{T} & -\frac{1}{T} \\ 0 & T & -T \end{bmatrix} \begin{bmatrix} K_1 \\ K_2 \\ K_3 \end{bmatrix}, \tag{11.111}
$$

we can determine, for a fixed sampling period T, the stabilizing region in (k_p, k_i, k_d) space as shown in Figure 11.7.

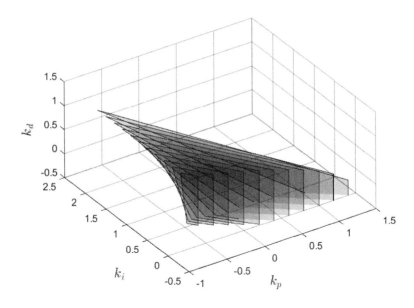

Figure 11.7 Stabilizing set in (k_p, k_i, k_d) space (Example 11.2) © 2009 Taylor & Francis Group, LLC. Reproduced from Bhattacharyya, Datta, and Keel (2009) with permission.

11.5 Stabilizing Sets from Frequency Response Data

In this section, we develop methods to compute PID stabilizing sets for continuous-time systems from frequency response data. This data could be measured data obtained from frequency response measurements and given in tabular or graphical form, such as Bode plots. First consider a real rational function

$$R(s) = \frac{A(s)}{B(s)}, \tag{11.112}$$

where $A(s)$ and $B(s)$ are polynomials with real coefficients and of degrees m and n, respectively. We assume that $A(s)$ and $B(s)$ have no zeros on the $j\omega$ axis. Let z_R^+, p_R^+ (z_L^-, p_L^-) denote the numbers of open right-half plane (RHP) and open left-half plane (LHP) zeros and poles of $R(s)$. Also let $\Delta_0^\infty \angle R(j\omega)$ denote the net change in phase of $R(j\omega)$ as ω runs from 0 to $+\infty$. Then we have

$$\Delta_0^\infty \angle R(j\omega) = \frac{\pi}{2} \left[z_L^- - z_R^+ - \left(p_L^- - p_R^+ \right) \right]. \tag{11.113}$$

This formula follows from the fact that each LHP zero and each RHP pole contribute $+\frac{\pi}{2}$ to the net phase change, whereas each RHP zero and LHP pole contribute $-\frac{\pi}{2}$ to the net phase change.

We define the (Hurwitz) signature of $R(s)$ to be

$$\sigma(R) := z_L^- - z_R^+ - \left(p_L^- - p_R^+ \right). \tag{11.114}$$

Write

$$R(j\omega) = R_r(\omega) + jR_i(\omega), \tag{11.115}$$

where $R_r(\omega)$ and $R_i(\omega)$ are rational functions in ω with real coefficients. It is easy to see that $R_r(\omega)$ and $R_i(\omega)$ have no real poles for $\omega \in (-\infty, +\infty)$, since $R(s)$ has no imaginary axis poles. To compute the net change in phase, that is, the left-hand side of Equation (11.113), it is convenient to develop formulas in terms of $R_r(\omega)$ and $R_i(\omega)$. Note that $\omega_0 = 0$ is always a zero of $R_i(\omega)$, since $R(s)$ is real. Let

$$0 = \omega_0 < \omega_1 < \omega_2 < \cdots < \omega_{l-1} \tag{11.116}$$

denote the real, finite non-negative zeros of $R_i(\omega) = 0$ of odd multiplicities, let

$$\mathrm{sgn}[x] = \begin{cases} +1, & \text{if } x > 0, \\ 0, & \text{if } x = 0, \\ -1, & \text{if } x < 0, \end{cases} \tag{11.117}$$

and define $\omega_l = \infty^-$. Define, for a real function $f(t)$:

$$f\left(t_0^-\right) := \lim_{t \to t_0, t < t_0} f(t), \qquad f\left(t_0^+\right) := \lim_{t \to t_0, t > t_0} f(t). \tag{11.118}$$

LEMMA 11.5 (Real Hurwitz Signature Lemma)
For $n - m$ even:

$$\sigma(R) = \left(\mathrm{sgn}\left[R_r(\omega_0)\right] + 2 \sum_{j=1}^{l-1} (-1)^j \mathrm{sgn}\left[R_r(\omega_j)\right] + (-1)^l \mathrm{sgn}\left[R_r(\omega_l)\right] \right) \tag{11.119}$$

$$\cdot (-1)^{l-1} \mathrm{sgn}\left[R_i(\infty^-)\right].$$

For $n - m$ odd:

$$\sigma(R) = \left(\operatorname{sgn}[R_r(\omega_0)] + 2 \sum_{j=1}^{l-1} (-1)^j \operatorname{sgn}[R_r(\omega_j)] \right) \cdot (-1)^{l-1} \operatorname{sgn}[R_i(\infty^-)]. \quad (11.120)$$

Proof The calculation of the signature is based on the total phase change of a frequency-dependent function as the frequency ranges from 0 to ∞. The latter in turn can be broken up into a summation of phase changes over a disjoint partition of the frequency axis. This is similar to what we did in the case of polynomials.

Note first that

$$\Delta_0^\infty \angle R(j\omega) = \frac{\pi}{2} \sigma(R) \quad (11.121)$$

and

$$\Delta_0^\infty \angle R(j\omega) = \Delta_{\omega_0=0}^{\omega_1} \angle R(j\omega) + \cdots + \Delta_{\omega_{l-1}}^{\omega_l=\infty^-} \angle R(j\omega). \quad (11.122)$$

For the case when $n - m$ is even, the plot of $R(j\omega)$ approaches the negative or positive real axis as ω approaches ∞. Thus, we have

$$\Delta_{\omega_k}^{\omega_{k+1}} \angle R(j\omega) = \frac{\pi}{2} \left(\operatorname{sgn}[R_r(\omega_k)] - \operatorname{sgn}[R_r(\omega_{k+1})] \right) \cdot \operatorname{sgn}[R_i(\omega_{k+1}^-)], \quad (11.123)$$

for $k = 0, \ldots, l - 2$ and

$$\Delta_{\omega_{l-1}}^{\omega_l=\infty^-} \angle R(j\omega) = \frac{\pi}{2} \left(\operatorname{sgn}[R_r(\omega_{l-1})] - \operatorname{sgn}[R_r(\omega_l)] \right) \cdot \operatorname{sgn}[R_i(\infty^-)]. \quad (11.124)$$

From Equations (11.121) and (11.122):

$$\begin{aligned}
\sigma(R) = {} & \left(\operatorname{sgn}[R_r(\omega_0)] - \operatorname{sgn}[R_r(\omega_1)] \right) \operatorname{sgn}[R_i(\omega_1^-)] \\
& + \left(\operatorname{sgn}[R_r(\omega_1)] - \operatorname{sgn}[R_r(\omega_2)] \right) \operatorname{sgn}[R_i(\omega_2^-)] + \cdots \\
& + \left(\operatorname{sgn}[R_r(\omega_{l-2})] - \operatorname{sgn}[R_r(\omega_{l-1})] \right) \operatorname{sgn}[R_i(\omega_{l-1}^-)] \\
& + \left(\operatorname{sgn}[R_r(\omega_{l-1})] - \operatorname{sgn}[R_r(\omega_l)] \right) \operatorname{sgn}[R_i(\infty^-)].
\end{aligned} \quad (11.125)$$

Since

$$\operatorname{sgn}[R_i(\omega_i^-)] = -\operatorname{sgn}[R_i(\omega_{i+1}^-)], \qquad \operatorname{sgn}[R_i(\omega_{l-1}^-)] = -\operatorname{sgn}[R_i(\infty^-)], \quad (11.126)$$

we have

$$\begin{aligned}
\operatorname{sgn}[R_i(\omega_{l-2}^-)] &= (-1)^2 R_i(\infty^-), \\
\operatorname{sgn}[R_i(\omega_{l-3}^-)] &= (-1)^3 R_i(\infty^-), \\
&\vdots \\
\operatorname{sgn}[R_i(\omega_2^-)] &= (-1)^{l-2} R_i(\infty^-), \\
\operatorname{sgn}[R_i(\omega_1^-)] &= (-1)^{l-1} R_i(\infty^-),
\end{aligned} \quad (11.127)$$

so that

$$
\begin{aligned}
\sigma(R) =& \left(\mathrm{sgn}[R_r(\omega_0)] - 2\mathrm{sgn}[R_r(\omega_1)] + 2\mathrm{sgn}[R_r(\omega_2)] + \cdots \right. \\
& \left. + (-1)^{l-1}\mathrm{sgn}[R_r(\omega_{l-1})] + (-1)^l\mathrm{sgn}[R_r(\omega_l)] \right) \cdot (-1)^{l-1}\mathrm{sgn}[R_i(\infty^-)] \\
=& \left(\mathrm{sgn}[R_r(\omega_0)] + 2\sum_{j=1}^{l-1}(-1)^j\mathrm{sgn}[R_r(\omega_j)] + (-1)^l\mathrm{sgn}[R_r(\omega_l)] \right) \quad (11.128) \\
& (-1)^{l-1}\mathrm{sgn}[R_i(\infty^-)].
\end{aligned}
$$

For the case when $n - m$ is odd, the plot of $R(j\omega)$ approaches the negative or positive imaginary axis as ω approaches ∞. Thus, we have

$$
\Delta_{\omega_k}^{\omega_{k+1}} \angle R(j\omega) = \frac{\pi}{2} \left(\mathrm{sgn}[R_r(\omega_k)] - \mathrm{sgn}[R_r(\omega_{k+1})] \right) \cdot \mathrm{sgn}[R_i(\omega_{k+1}^-)],
$$

$$
\text{for } k = 0, \dots, l-2, \tag{11.129}
$$

$$
\Delta_{\omega_{l-1}}^{\omega_l = \infty^-} \angle R(j\omega) = \frac{\pi}{2}\mathrm{sgn}[R_r(\omega_{l-1})] \cdot \mathrm{sgn}[R_i(\infty^-)]. \tag{11.130}
$$

Then, using Equation (11.127), we have

$$
\begin{aligned}
\sigma(R) =& \left(\mathrm{sgn}[R_r(\omega_0)] - \mathrm{sgn}[R_r(\omega_1)] \right)\mathrm{sgn}[R_i(\omega_1^-)] \\
& + \left(\mathrm{sgn}[R_r(\omega_1)] - \mathrm{sgn}[R_r(\omega_2)] \right)\mathrm{sgn}[R_i(\omega_2^-)] + \cdots \\
& + \left(\mathrm{sgn}[R_r(\omega_{l-2})] - \mathrm{sgn}[R_r(\omega_{l-1})] \right) \cdot \mathrm{sgn}[R_i(\omega_{l-1}^-)] \\
& + (-1)^{l-1}\mathrm{sgn}[R_r(\omega_{l-1})]\mathrm{sgn}[R_r(\infty^-)] \tag{11.131} \\
=& \left(\mathrm{sgn}[R_r(\omega_0)] - 2\mathrm{sgn}[R_r(\omega_1)] + \cdots \right. \\
& \left. + (-1)^{l-1}2\mathrm{sgn}[R_r(\omega_{l-1})] \right)(-1)^{l-1}\mathrm{sgn}[R_i(\infty^-)] \\
=& \left(\mathrm{sgn}[R_r(\omega_0)] + 2\sum_{j=1}^{l-1}(-1)^j\mathrm{sgn}[R_r(\omega_j)] \right) \cdot (-1)^{l-1}\mathrm{sgn}[R_i(\infty^-)].
\end{aligned}
$$

Similar results may be derived for complex rational functions.

11.5.1   Phase, Signature, Poles, Zeros, and Bode Plots

Let P denote an LTI plant and $P(s)$ its real, rational transfer function without imaginary axis poles or zeros and with z^+, p^+ (z^-, p^-) denoting the numbers of open RHP (open LHP) zeros and poles, and $n(m)$ the denominator (numerator) degrees. Let the relative degree be denoted r_P:

$$r_P := n - m. \tag{11.132}$$

As defined earlier, the signature of P is

$$\sigma(P) = (z^- - z^+) - (p^- - p^+). \tag{11.133}$$

LEMMA 11.6

$$r_P = -\frac{1}{20} \cdot \frac{dP_{db}(\omega)}{d(\log_{10}\omega)}\bigg|_{\omega\to\infty}, \tag{11.134}$$

$$\sigma(P) = \frac{2}{\pi}\Delta_0^\infty \angle P(j\omega), \tag{11.135}$$

where

$$P_{db}(\omega) := 20\log_{10}|P(j\omega)|. \tag{11.136}$$

Proof Equation (11.134) states that the relative degree is the high-frequency slope of the Bode magnitude plot and Equation (11.135) states that the signature can be found from the net change in phase from the phase plot. ∎

Since $P(s)$ has no $j\omega$ axis poles and zeros, we can also write

$$\sigma(P) = -(n-m) - 2\left(z^+ - p^+\right) \tag{11.137a}$$

$$= -r_P - 2\left(z^+ - p^+\right). \tag{11.137b}$$

Therefore, using Lemma 11.6 and Equation (11.137b), $z^+ - p^+$ can be determined from the Bode plot of P. In particular, if $P(s)$ is stable, the Bode plot can often be obtained experimentally by measuring the frequency response of the system. Then the above relations with $p^+ = 0$ determine z^+ from the Bode plot data.

Unstable Plant
Now suppose that G is an *unstable* LTI plant with a rational transfer function *unknown* to us (see Figure 11.8) and assume that $P(s)$ does not have imaginary axis poles and zeros. We assume, however, that a known feedback controller $C(s)$ stabilizes G and the closed-loop frequency response can be *measured* and is denoted by $G(j\omega)$ for $\omega \in [0,\infty)$.

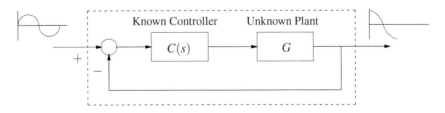

Figure 11.8 Frequency response measurement on an unstable plant

Then

$$P(j\omega) = \frac{G(j\omega)}{C(j\omega)(1 - G(j\omega))} \tag{11.138}$$

is the *computed* frequency response of the unstable plant. The next result shows that knowledge of $C(s)$ and $G(j\omega)$ is sufficient to determine the numbers z^+ and p^+, that is the numbers of RHP zeros and poles of the plant. Let z_c^+ denote the number of RHP zeros of $C(s)$.

THEOREM 11.5

$$z^+ = \frac{1}{2}\left[-r_P - r_C - 2z_c^+ - \sigma(G)\right], \tag{11.139a}$$

$$p^+ = \frac{1}{2}\left[\sigma(P) - \sigma(G) - r_C\right] - 2z_c^+. \tag{11.139b}$$

Proof We have

$$G(s) = \frac{P(s)C(s)}{1 + P(s)C(s)} \tag{11.140}$$

and, since $G(s)$ is stable:

$$\begin{aligned}
\sigma(G) &= \left(z^- + z_c^-\right) - \left(z^+ + z_c^+\right) - (n + n_c) \\
&= -r_P - r_C - 2z_c^+ - 2z^+,
\end{aligned} \tag{11.141}$$

which implies Equation (11.139a). From Equation (11.133) applied to $P(s)$, we have

$$p^+ = z^+ + \frac{\sigma(P)}{2} + \frac{r_P}{2}. \tag{11.142}$$

Substituting Equation (11.139a) into Equation (11.142), we have Equation (11.139b). ∎

Remark 11.3 In the above theorem, we assume that $C(s)$, a stabilizing controller, is known, and the corresponding closed-loop frequency response $G(j\omega)$ for $\omega \in [0,\infty)$ can be *measured*. Thus, $P(j\omega)$ can be computed from Equation (11.138). Now r_P and $\sigma(G)$ can be found by applying the results of Lemma 11.6 to $P(j\omega)$ and $G(j\omega)$, respectively. r_C and z_c^+ are known as $C(s)$ is known. Therefore, z^+ and p^+ can be found from frequency response data.

Remark 11.4 In the above analysis we have assumed for simplicity that the plant is devoid of imaginary axis poles and zeros. When such poles and zeros are present, their numbers may be known from physical considerations or their numbers and locations may be ascertained from the experimentally determined or computed Bode plots. Once identified, we can lump these poles and zeros with the controller and proceed with the design procedure. At an imaginary axis zero (pole) of multiplicity k, away from the origin, the magnitude plot is zero (infinity) and the phase plot undergoes an instantaneous change of phase $k\pi$. If such poles or zeros occur at the origin, there is a corresponding phase shift of $\frac{k\pi}{2}$ at zero frequency. It is straightforward to establish that, in this case, the relations in Equations (11.139a) and (11.139b) need to be modified to the following:

$$z^+ = \frac{1}{2}\left[-r_P - r_C - 2z_c^+ - z_c^i - \sigma(G)\right], \tag{11.143a}$$

$$p^+ = \frac{1}{2}\left[\sigma(P) - \sigma(G) - r_C\right] - 2z_c^+ - z_c^i + z^i - p^i, \tag{11.143b}$$

where z^i, z_c^i, p^i, p_c^i denote the numbers of imaginary axis zeros and poles of plant and controller.

11.5.2 PID Controller Synthesis for Continuous-Time Systems

In this section, we consider the synthesis and design of PID controllers for a continuous-time LTI plant, with underlying transfer function $P(s)$ with $n(m)$ poles (zeros) (see Figure 11.9.)

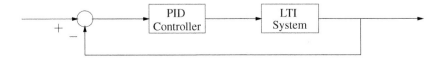

Figure 11.9 A unity feedback system with a PID controller

We assume that the *only* information available to the designer is the following.

1. Knowledge of the frequency response magnitude and phase, equivalently, $P(j\omega)$, $\omega \in [0, \infty)$ if the plant is stable.
2. Knowledge of a known stabilizing controller and the corresponding closed-loop frequency response $G(j\omega)$, if the plant is unstable.

Such assumptions are reasonable for many systems. We also make the technical assumption that the plant has no $j\omega$ poles or zeros so that the magnitude, its inverse, and phase are well-defined for all $\omega \geq 0$. As we have seen from the discussion in the last section, the numbers and locations of RHP poles and zeros can be found from the above data for any LTI plant and either "divided out" or lumped with the controller.
Write

$$P(j\omega) = |P(j\omega)|e^{j\phi(\omega)} = P_r(\omega) + jP_i(\omega), \qquad (11.144)$$

where $|P(j\omega)|$ denotes the *magnitude* and $\phi(\omega)$ the *phase* of the plant, at the frequency ω. Let the PID controller be of the form

$$C(s) = \frac{k_i + k_p s + k_d s^2}{s(1 + sT)}, \quad \text{for } T > 0, \qquad (11.145)$$

where T is assumed to be fixed and small. We now present results for determining the stabilizing set. The approach adopted is to convert the stabilization problem to an equivalent signature assignment problem, as shown below.

LEMMA 11.7   Let

$$F(s) := s(1 + sT) + \left(k_i + k_p s + k_d s^2\right) P(s) \qquad (11.146)$$

and

$$\bar{F}(s) := F(s)P(-s). \qquad (11.147)$$

Then closed-loop stability is equivalent to

$$\sigma\left(\bar{F}(s)\right) = n - m + 2z^+ + 2. \tag{11.148}$$

Proof Closed-loop stability is equivalent to the condition that all zeros of $F(s)$ lie in the LHP. This in turn is equivalent to the condition

$$\sigma\left(F(s)\right) = n + 2 - \left(p^- - p^+\right). \tag{11.149}$$

Now consider the rational function

$$\bar{F}(s) = F(s)P(-s). \tag{11.150}$$

Note that

$$\sigma\left(\bar{F}(s)\right) = \sigma(F(s)) + \sigma(P(-s)). \tag{11.151}$$

Therefore, the stability condition becomes

$$\sigma\left(\bar{F}(s)\right) = n + 2 - \left(p^- - p^+\right) + \left(z^+ - z^-\right) - \left(p^+ - p^-\right)$$
$$= n + 2 + z^+ - z^- = n - m + 2z^+ + 2. \tag{11.152}$$

$\blacksquare$

Write

$$\bar{F}(j\omega) = j\omega(1 + j\omega T)P(-j\omega) + \left(k_i + j\omega k_p - \omega^2 k_d\right) P(j\omega)P(-j\omega)$$
$$= \underbrace{\left(k_i - k_d\omega^2\right)|P(j\omega)|^2 - \omega^2 TP_r(\omega) + \omega P_i(\omega)}_{\bar{F}_r(\omega,k_i,k_d)}$$
$$+ j\omega \underbrace{\left(k_p|P(j\omega)|^2 + P_r(\omega) + \omega TP_i(\omega)\right)}_{\bar{F}_i(\omega,k_p)} \tag{11.153}$$
$$= \bar{F}_r(\omega, k_i, k_d) + j\omega\bar{F}_i(\omega, k_p).$$

THEOREM 11.6 Fix $k_p = k_p^*$ and let $\omega_1 < \omega_2 < \cdots < \omega_{l-1}$ denote the distinct frequencies of odd multiplicities which are solutions of

$$\bar{F}_i(\omega, k_p^*) = 0, \tag{11.154}$$

or, equivalently, of,

$$k_p^* = -\frac{P_r(\omega) + \omega TP_i(\omega)}{|P(j\omega)|^2} = -\frac{\cos\phi(\omega) + \omega T\sin\phi(\omega)}{|P(j\omega)|} =: g(\omega) \tag{11.155}$$

for fixed $k_p = k_p^*$. Let $\omega_0 = 0$ and $\omega_l = \infty$, and $j := \text{sgn}\left[\bar{F}_i(\infty^-, k_p^*)\right]$. Determine strings of integers

$$I = [i_0, i_1, i_2, \ldots, i_l], \tag{11.156}$$

with $i_l \in \{+1, -1\}$ such that for $n - m$ even:

$$\left[i_0 - 2i_1 + 2i_2 + \cdots + (-1)^{l-1}2i_{l-1} + (-1)^l i_l\right](-1)^{l-1}j = n - m + 2z^+ + 2; \tag{11.157}$$

for $n - m$ odd:

$$\left[i_0 - 2i_1 + 2i_2 + \cdots + (-1)^{l-1} 2i_{l-1} \right] (-1)^{l-1} j = n - m + 2z^+ + 2. \qquad (11.158)$$

Then for the fixed $k_p = k_p^*$, the (k_i, k_d) values corresponding to closed-loop stability are given by

$$\bar{F}_r(\omega_t, k_i, k_d) i_t > 0, \qquad (11.159)$$

where the i_ts are taken from strings satisfying Equations (11.157) and (11.158), and ω_ts are taken from the solutions of Equation (11.154).

Proof By Lemma 11.7, the stability condition has been reduced to the signature condition in Equation (11.148). The theorem follows from applying Lemma 11.5 to compute the signature of $\bar{F}(s)$. ∎

The following result clarifies what range the parameters k_p must be swept over.

THEOREM 11.7 For the given function $g(\omega)$ in Equation (11.155) determined completely by the plant data $P(j\omega)$ and T:

(1) A necessary condition for PID stabilization is that there exists k_p such that the function

$$k_p = g(\omega) \qquad (11.160)$$

has at least k distinct roots of odd multiplicities, where

$$
\begin{aligned}
k &\geq \frac{n - m + 2z^+ + 2}{2} - 1, \quad \text{if } n - m \text{ even,} \\
k &\geq \frac{n - m + 2z^+ + 3}{2} - 1, \quad \text{if } n - m \text{ odd.}
\end{aligned}
\qquad (11.161)
$$

(2) There exists a unique range of ω, $\underline{\omega} = (\omega_{\min}, \omega_{\max})$, over which the condition (1) occurs and this determines the range of ω to be swept.
(3) Every k_p belonging to the stabilizing set of PID parameters lies in the range

$$k_p \in \left(k_p^{\min}, k_p^{\max} \right), \qquad (11.162)$$

where

$$k_p^{\min} := \min_{\omega \in \underline{\omega}} g(\omega), \qquad k_p^{\max} := \max_{\omega \in \underline{\omega}} g(\omega). \qquad (11.163)$$

where $\underline{\omega} = (\omega_{\min}, \omega_{\max})$.

Remark 11.5 The function $g(\omega)$ is well-defined due to the assumption that the plant either has no $j\omega$ zeros or these have been lumped with the controller. The frequency $\omega_{\max}$ can be selected as any frequency after which $g(\omega)$ continues to grow monotonically. This determines the range of frequencies over which the $P(j\omega)$ data of the plant should be known accurately. Note that the ranges $(\omega_{\min}, \omega_{\max})$ and $(k_p^{\min}, k_p^{\max})$ can consist of multiple segments.

The computation of the stabilizing set implied by the above results is summarized as the following procedure.

Computation of PID Stabilizing Sets from Frequency Response Data

The complete set of stabilizing PID gains can be computed by the following procedure. For stable systems, the available data is the frequency response of the plant $P(j\omega)$.

1. Determine the relative degree of the plant $r_P = n - m$ from the high-frequency slope of the Bode magnitude plot of $P(j\omega)$.
2. Let $\Delta_0^\infty[\phi(\omega)]$ denote the net change of phase of $P(j\omega)$ for $\omega \in [0, \infty)$. Determine z^+ from

$$\Delta_0^\infty[\phi(\omega)] = -\left[(n-m) + 2z^+\right]\frac{\pi}{2}, \tag{11.164}$$

which follows from Equation (11.137a) with $p^+ = 0$.

For unstable systems, the available data are a stabilizing controller transfer function $C(s)$ and the frequency response of the corresponding stable closed-loop system $G(j\omega)$.

1. Compute the frequency response $P(j\omega)$ from Equation (11.138).
2. Determine the relative degree of the plant r_P from the high-frequency slope of the Bode magnitude plot of $P(j\omega)$.
3. Determine z_c^+ and r_C from $C(s)$.
4. Compute $\sigma(G)$ from Equation (11.135) applied to $G(j\omega)$.
5. Compute z^+ using Equation (11.139a) in Theorem 11.5.
6. Compute $g(\omega)$ using Equation (11.155) and the frequency response measured data.

(1) Fix $k_p = k_p^*$, solve Equation (11.155) and let $\omega_1 < \omega_2 < \cdots < \omega_{l-1}$ denote the distinct frequencies of odd multiplicities which are solutions of Equation (11.155).
(2) Set $\omega_0 = 0$, $\omega_l = \infty$, and $j = \operatorname{sgn}\bar{F}_i(-\infty^-, k_p^*)$. Determine all strings of integers $i_t \in \{+1, -1\}$ such that for $n - m$ even:

$$\left[i_0 - 2i_1 + 2i_2 + \cdots + (-1)^{i-1}2i_{l-1} + (-1)^l i_l\right](-1)^{l-1}j = n - m + 2z^+ + 2; \tag{11.165}$$

for $n - m$ odd:

$$\left[i_0 - 2i_1 + 2i_2 + \cdots + (-1)^{i-1}2i_{l-1}\right](-1)^{l-1}j = n - m + 2z^+ + 2. \tag{11.166}$$

(3) For the fixed $k_p = k_p^*$ chosen in step (1), solve for the stabilizing (k_i, k_d) values from

$$\left[k_i - k_d\omega_t^2 + \frac{\omega_t\sin\phi(\omega_t) - \omega_t^2 T\cos\phi(\omega_t)}{|P(j\omega_t)|}\right]i_t > 0, \quad \text{for } t = 0, 1, \ldots, l. \tag{11.167}$$

(4) Repeat the previous three steps by updating k_p over prescribed ranges. The ranges over which k_p must be swept is determined from the requirements that Equation (11.165) or (11.166) is satisfied for at least one string of integers as in Theorem 11.7.

We emphasize that all computations are based on the data $P(j\omega)$ and knowledge of the transfer function $P(s)$ is *not required*. For well-posedness of the loop, it is necessary that

$$k_d \neq -\frac{T}{P(\infty)}. \tag{11.168}$$

For strictly proper plants, $P(\infty) = 0$ and this constraint vanishes. The next example illustrates these calculations.

EXAMPLE 11.3 (PID synthesis without analytical model)   To illustrate the previous results, we take a set of frequency response data points for a *stable* delay-free plant:

$$\mathbf{P}(j\omega) := \{P(j\omega), \quad \omega \in (0,10) \text{ sampled every } 0.01\}. \tag{11.169}$$

The Nyquist and Bode plots are shown in Figures 11.10 and 11.11.

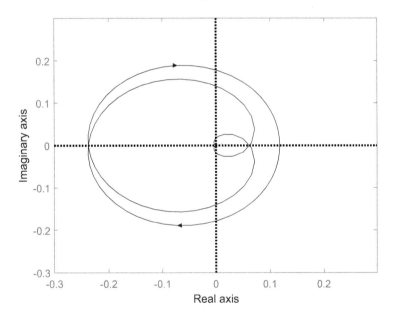

Figure 11.10 Nyquist plot of the plant (Example 11.3) © 2009 Taylor & Francis Group, LLC. Reproduced from Bhattacharyya, Datta, and Keel (2009) with permission.

The high-frequency slope of the Bode magnitude plot is -40 dB/decade and thus $n - m = 2$. The total change of phase is $-540°$ and so

$$-6\frac{\pi}{2} = -\left((n-m) - 2(p^+ - z^+)\right)\frac{\pi}{2}, \tag{11.170}$$

and since the plant is stable, $p^+ = 0$, giving $z^+ = 2$. The required signature for stability can now be determined and is

$$\sigma(\bar{F}(s)) = (n-m) + 2z^+ + 2 = (2) + 2(2) + 2 = 8. \tag{11.171}$$

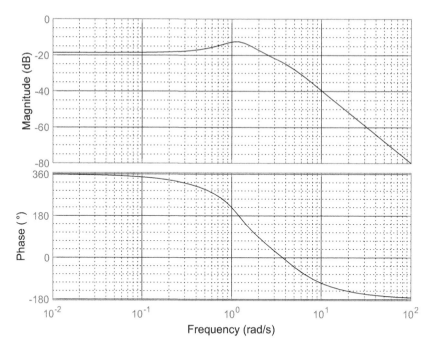

Figure 11.11 Bode plot of the plant (Example 11.3) © 2009 Taylor & Francis Group, LLC. Reproduced from Bhattacharyya, Datta, and Keel (2009) with permission.

Since $n - m$ is even, we must have

$$\left[i_0 - 2i_1 + 2i_2 - 2i_3 + 2i_4 - \cdots + (-1)^l i_l\right](-1)^{l-1} j = 8, \qquad (11.172)$$

where

$$j = \text{sgn}\left[\bar{F}_i(\infty^-, k_p)\right] = -\text{sgn}\left[\lim_{\omega \to \infty} g(\omega)\right] = -1 \qquad (11.173)$$

and it is clear that at least five terms are required to satisfy the above. In other words, $l \geq 4$. From Figure 11.12 it is easy to see that Equation (11.155) has at most three positive frequencies as solutions and therefore we have $i_0 - 2i_1 + 2i_2 - 2i_3 + i_4 = 8$. Also $i_4 = \text{sgn}\left[\bar{F}_r(\infty^-, k_i, k_d)\right] = 1$, independent of k_i and k_d. This means that k_p must be chosen so that $\bar{F}_i(\omega, k_p^*) = 0$ has exactly three positive real zeros. This gives the feasible range of k_p values as shown in Figure 11.12, which depicts the function:

$$g(\omega) = -\frac{\cos\phi(\omega) + \omega T \sin\phi(\omega)}{|P(j\omega)|}. \qquad (11.174)$$

The feasible range of k_p is such that k_p intersects the graph of $g(\omega)$ three times. This feasible range is shown in Figure 11.12. In Figure 11.12, we also observe that the *frequency range over which plant data must accurately be known for PID control* is $[0, 8.2]$. We now fix $k_p^* = 1$ and compute the set of ωs that satisfies

$$-\frac{\cos\phi(\omega) + \omega T \sin\phi(\omega)}{|P(j\omega)|} = 1. \qquad (11.175)$$

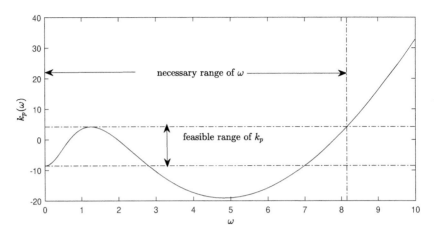

Figure 11.12 Graph of the function $g(\omega)$ in Equation (11.174) (Example (11.3)) © 2009 Taylor & Francis Group, LLC. Reproduced from Bhattacharyya, Datta, and Keel (2009) with permission.

To find the set of ωs satisfying the above, we plot the function $g(\omega)$ as shown in Figure 11.13.

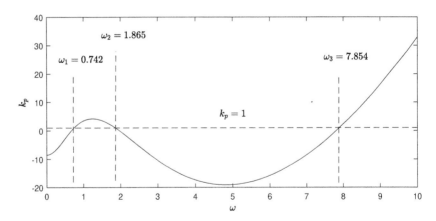

Figure 11.13 Finding the set of ωs satisfying Equation (11.155) with $k_p = 1$ (Example 11.3) © 2009 Taylor & Francis, Group, LLC. Reproduced from Bhattacharyya, Datta, and Keel (2009) with permission.

The three frequencies $\omega_1, \omega_2, \omega_3$ are required to compute the stability set in (k_i, k_d) space.

From this we found the solutions $\{\omega_1, \omega_2, \omega_3\} = \{0, 0.742, 1.865, 7.854\}$. This leads to the requirement $i_0 - 2i_1 + 2i_2 - 2i_3 = 7$, giving the only feasible string

$$\mathscr{F} = \{i_0, i_1, i_2, i_3\} = \{1, -1, 1, -1\}. \tag{11.176}$$

Thus, we have the following set of linear inequalities for stability:

$$k_i > 0,$$
$$-3.8114 + k_i - 0.5506k_d < 0,$$
$$12.2106 + k_i - 3.4782k_d > 0, \tag{11.177}$$
$$-457.0235 + k_i - 61.6853k_d < 0.$$

The complete set of stabilizing PID gains for $k_p^* = 1$ is given in Figure 11.14. This set is determined by finding the string of integers $\{i_0, i_1, i_2, i_3\}$ satisfying the stability conditions in Equations (11.165) and (11.166). The corresponding linear inequalities in Equation (11.167) determine the stabilizing set shown in the (k_i, k_d) space. The point marked "*" was used as a test point to verify stability.

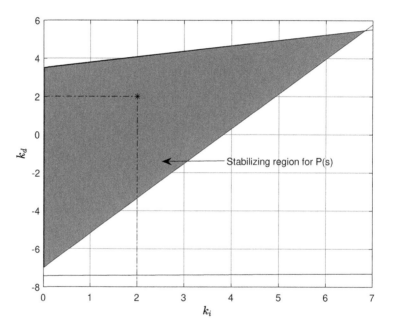

Figure 11.14 The set of stabilizing PID gains in (k_i, k_d) space when $k_p = 1$ (Example 11.3) © 2009 Taylor & Francis Group, LLC. Reproduced from Bhattacharyya, Datta, and Keel (2009) with permission.

By sweeping k_p we have the entire set of stabilizing PID gains as shown in Figure 11.15. This set is determined by sweeping over the feasible range of k_p determined in Figure 11.13, and solving the corresponding linear inequalities given by Equation (11.167) in (k_i, k_d) space for a fixed k_p^* in this range.

Note that the range of k_p over which the search needs to be carried out is also obvious from Figure 11.12 as discussed, and it is $k_p \in [-8.5, 4.2]$.

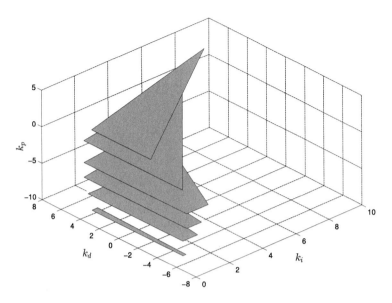

Figure 11.15 Entire set of PID gains (Example 11.3) © 2009 Taylor & Francis Group, LLC. Reproduced from Bhattacharyya, Datta, and Keel (2009) with permission.

11.5.3   Data-Based Design versus Model-Based Design

In this section, we discuss some differences between model-based design and the data-based designs described here. In model-based design, mathematical models are obtained from the laws of physics that describe the dynamic behavior of a system to be controlled. On the other hand, the most common way of obtaining mathematical models in engineering systems is through a system identification process. Let us assume that the frequency domain data is obtained by exciting a linear time-invariant plant by sinusoidal signals. In theory, a system identification procedure should exactly be able to determine the unknown rational transfer function. In this ideal situation there should be no distinction between model-based and data-based synthesis methods. However, typical system identification procedures can fail to find an exact rational function even if exact (or perfect) data is available. This is especially true when the order of the plant is high. The following example illustrates that this in turn can lead to drastic differences in control synthesis and design.

EXAMPLE 11.4   Let us assume that the frequency domain data $P(j\omega)$ shown below is obtained from a 20th-order plant. Note that the plant is unstable with two RHP poles. Then mathematical models of 20th, 10th, 7th, and 4th orders were obtained by a system identification process applied to $P(j\omega)$. Figure 11.16 shows the Bode plots of the four identified models along with the frequency domain data collected from the 20th-order plant considered here. It is observed that the Bode plots of these are almost identical, except that the 4th-order identified model is relatively crude. We now compute the stabilizing PID parameter regions of each of these systems. Figure 11.17 shows the stabilizing regions in the PID controller parameter space.

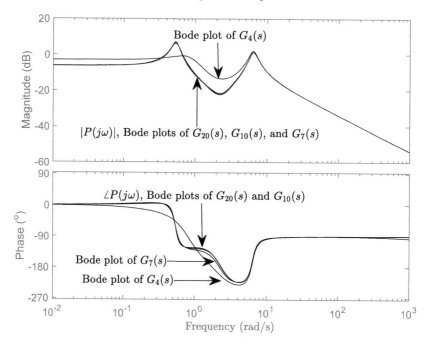

Figure 11.16 Frequency domain data and the Bode plots of 20th, 10th, 7th, 4th-order identified models (Example 11.4) © 2009 Taylor & Francis Group, LLC. Reproduced from Bhattacharyya, Datta, and Keel (2009) with permission.

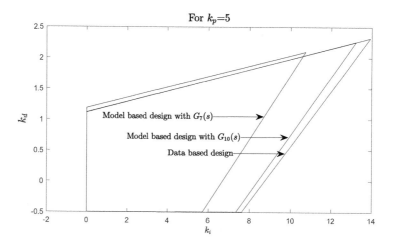

Figure 11.17 Stabilizing regions (Example 11.4) © 2009 Taylor & Francis Group, LLC. Reproduced from Bhattacharyya, Datta, and Keel (2009) with permission.

We can make the following observations. For convenience, let us denote by $G_{20}(s)$,

$G_{10}(s)$, $G_7(s)$, and $G_4(s)$ the 20th, 10th, 7th, and 4th-order models identified, respectively.

1. The models $G_{20}(s)$ and $G_4(s)$ are found to not be PID stabilizable for the chosen $k_p = 5$. In other words, the stabilizing region in PID parameter space for the given plant is empty with $k_p = 5$. This is consistent with the data-driven case, as is evident from Figure 11.18. The signature condition requires that the line representing $k_p = 5$ must intersect the $g(\omega)$ graph a minimum of three times for stability. Figure 11.18 shows that such a necessary condition cannot be satisfied with $k_p > 4.5$.

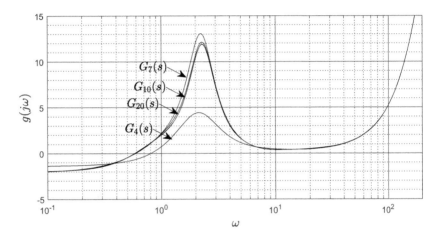

Figure 11.18 k_p vs ω plot for the identified systems (Example 11.4) © 2009 Taylor & Francis Group, LLC. Reproduced from Bhattacharyya, Datta, and Keel (2009) with permission.

This is not unexpected for the model $G_4(s)$, since there is some difference between the identified and actual Bode (frequency response) plots.

2. We have verified that $G_{20}(s)$ is also not PID stabilizable for any value of k_p. This may seem surprising since the Bode magnitude and phase plot of $G_{20}(s)$ are indistinguishably close to the data $P(j\omega)$. In fact, $G_{20}(s)$ has additional RHP poles and zeros over those in the original plant model, which makes stabilization difficult.

3. The stabilizing regions are found for $G_{10}(s)$ and $G_7(s)$. These regions overlap, but are not the same. This suggests that selection of controllers might be done inside the intersection of the stabilizing regions for $G_{10}(s)$ and $G_7(s)$.

4. The stabilizing region obtained from the data-based method given here differs from those for $G_{10}(s)$ and $G_7(s)$. Thus, a reasonable selection of controller may be done inside the intersection of the stabilizing regions for $G_{10}(s)$, $G_7(s)$, and the region obtained from the data-based method.

5. In general, the accuracy of system identification depends on the accuracy of the data considered. On the other hand, the data-based approach will work effectively as long as the roots of $g(\omega) = k_p^*$ are found reliably.

In practice, experimentally obtained frequency domain data always contains noise and measurement errors. As discussed above, the stability regions determined by the model-based design and the data-based design will generally be different. Although the accuracy of the regions depends on each particular case, the data-based design gives useful alternatives to model-based design methods and in general the two complement each other. In particular, it gives new guidelines for identification when it is to be used for controller design.

In the next section we discuss the design of PID controllers under classical gain margin and phase margin specifications, based on the computed stabilizing set.

11.6   Design with Gain and Phase Margin Requirements

In this section, we introduce an approach to the robust design of PID controllers for continuous-time plants based on gain and phase margin specifications. This design is based on a simple parametrization of constant-magnitude and constant-phase loci of the PID controller. This parametrization produces ellipses and straight lines for first-order and PI controllers, and cylinders and planes for PID controllers. These geometric figures are computed by considering a prescribed but arbitrary gain crossover frequency and prescribed but arbitrary phase margin for the closed-loop system with the given plant. These graphical representations enable the retrieval of PI, PID, and first-order controller designs with simultaneous specifications on gain and phase margins.

11.6.1   Magnitude and Phase Loci

Consider the general feedback system shown in Figure 11.19.

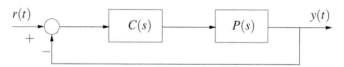

Figure 11.19 Feedback control system

Here $P(s)$ is the transfer function of the plant to be controlled and $C(s)$ is the transfer function of the controller to be designed. The plant $P(s)$ will be assumed to be rational and strictly proper. Therefore

$$P(s) = \frac{N(s)}{D(s)}, \tag{11.178}$$

where $N(s)$, $D(s)$ are polynomials in the Laplace variable s with degree $N(s)$ < degree $D(s)$. Let us assume that the rational proper controller $C(s)$ stabilizes $P(s)$. We denote by $P(j\omega)$ and $C(j\omega)$ the frequency responses of the plant and controller, respectively, where ω, the frequency in radians, runs from 0 to ∞. The Nyquist and Bode plots of

$$G(j\omega) := C(j\omega)P(j\omega) \tag{11.179}$$

are shown in Figures 11.20 and 11.21, respectively.

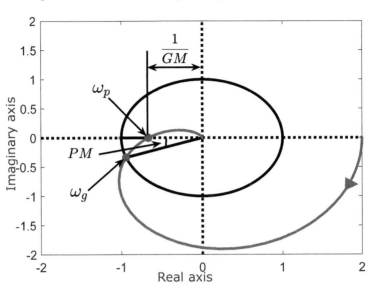

Figure 11.20 Nyquist plot of $G(s)$ in Equation (11.179) © Springer Nature Switzerland AG 2019. Reproduced from Diaz-Rodriguez, Han, and Bhattacharyya (2019) with permission.

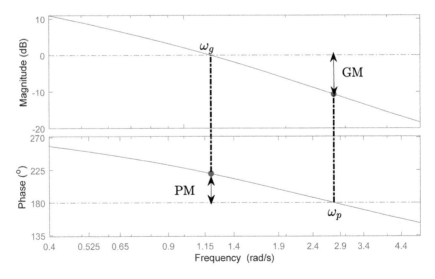

Figure 11.21 Bode plot of $G(s)$ in Equation (11.179) © Springer Nature Switzerland AG 2019. Reproduced from Diaz-Rodriguez, Han, and Bhattacharyya (2019) with permission.

Let ω_g denote a proposed gain crossover frequency, ω_p the phase crossover frequency, GM the gain margin, and PM the phase margin of the closed-loop system. As explained in the previous chapter, gain and phase margins determine how stable or robust the control system is and they are the basis of classical control designs. From the frequency responses shown in Figures 11.20 and 11.21, we have:

$$|C(j\omega_g)P(j\omega_g)| = 1, \tag{11.180a}$$

$$\angle(C(j\omega_p)P(j\omega_p)) = -n\pi, \qquad \text{for } n = 1, 3, 5, \ldots, \tag{11.180b}$$

$$\text{GM} := \frac{1}{|C(j\omega_p)P(j\omega_p)|}, \tag{11.180c}$$

$$\text{PM} := \angle(C(j\omega_g)P(j\omega_g)) - \pi. \tag{11.180d}$$

For a prescribed value of ω_g, Equations (11.180a) and (11.180d) represent the requirements on the magnitude and phase of the controller:

$$|C(j\omega_g)| = \frac{1}{|P(j\omega_g)|}, \tag{11.181}$$

$$\angle(C(j\omega_g)) = n\pi + \text{PM} - \angle P(j\omega_g)), \qquad \text{for } n \text{ odd.} \tag{11.182}$$

Equations (11.181) and (11.182) will lead to a geometrical representation in the space of controller parameters as an ellipse and a straight line for PI and first-order controllers, and a cylinder and a plane for continuous-time PID controllers. In the following sections, we describe this parametrization.

We remark that ω_g is sometimes considered the bandwidth (BW) of the closed loop system and can also represent a design specification.

11.6.2   PI and PID Controllers

For a PI controller

$$C(s) = \frac{k_p s + k_i}{s}, \tag{11.183}$$

where k_p and k_i are the design parameters. Then, setting $s = j\omega$:

$$C(j\omega) = \frac{k_p(j\omega) + k_i}{j\omega}, \tag{11.184}$$

so that

$$|C(j\omega)|^2 = k_p^2 + \frac{k_i^2}{\omega^2} =: M^2, \tag{11.185}$$

$$\angle C(j\omega) = \arctan\left(\frac{-k_i}{\omega k_p}\right) := \Phi. \tag{11.186}$$

Equations (11.185) and (11.186) can be rewritten as

$$\frac{k_p^2}{a^2} + \frac{k_i^2}{b^2} = 1, \tag{11.187a}$$

$$k_i = c k_p, \tag{11.187b}$$

where

$$a^2 = M^2, \tag{11.188a}$$

$$b^2 = M^2 \omega^2, \tag{11.188b}$$

$$c = \omega \tan \Phi. \tag{11.188c}$$

Thus, for given ω, constant M loci are ellipses and constant-phase loci are straight lines in k_p, k_i space. The major and minor axes of the ellipse are given by a and b (see Figure 11.22). The slope of the line is represented by c (see Figure 11.22).

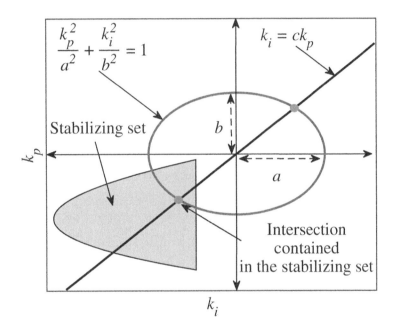

Figure 11.22 Ellipse and straight line intersecting with a stabilizing set © Springer Nature Switzerland AG 2019. Reproduced from Diaz-Rodriguez, Han, and Bhattacharyya (2019) with permission.

Suppose ω_g is the prescribed closed-loop gain crossover frequency. Then

$$M = M_g := \frac{1}{|P(j\omega_g)|}. \tag{11.189}$$

If ϕ_g^* is the desired closed-loop phase margin in radians:

$$\Phi = \Phi_g := \pi + \phi_g^* - \angle P(j\omega_g). \tag{11.190}$$

From Equations (11.185) and (11.186) we obtain the ellipse and straight line corresponding to $M = M_g$ and $\Phi = \Phi_g$, giving the design point (k_p^*, k_i^*). If this intersection point lies in the stabilizing set S, the design is feasible, otherwise the specifications are unattainable and have to be revised.

For a PID controller

$$C(s) = \frac{k_d s^2 + k_p s + k_i}{s}, \tag{11.191}$$

where k_p, k_i, and k_d are the design parameters. Then, setting $s = j\omega$:

$$C(j\omega) = \frac{k_d(j\omega)^2 + k_p(j\omega) + k_i}{j\omega}. \tag{11.192}$$

From Equation (11.192), we have

$$|C(j\omega)|^2 = k_p^2 + \left(k_d\omega - \frac{k_i}{\omega}\right)^2 := M^2, \tag{11.193}$$

$$\angle C(j\omega) = \arctan\left(\frac{k_d\omega - \frac{k_i}{\omega}}{k_p}\right) := \Phi. \tag{11.194}$$

Equation (11.193) represents an ellipse in (k_p, k_d) space with center at $(0, \frac{k_i}{\omega^2})$ and in (k_p, k_i, k_d) space is an elliptical cylinder. From Equations (11.193) and (11.194) we have

$$k_p^2 = M^2 - \left(k_d\omega - \frac{k_i}{\omega}\right)^2 = \frac{\left(k_d\omega - \frac{k_i}{\omega}\right)^2}{\tan^2\Phi}, \tag{11.195}$$

which leads to the following expressions:

$$k_i = k_d\omega^2 \pm k_p\omega\tan\Phi, \tag{11.196a}$$

$$k_p = \pm\sqrt{\frac{M^2}{1 + \tan^2\Phi}}. \tag{11.196b}$$

Suppose ω_g is the prescribed closed-loop gain crossover frequency. Then

$$M = \frac{1}{|P(j\omega_g)|}. \tag{11.197}$$

If ϕ_g^* is the desired phase margin in radians:

$$\Phi = \pi + \phi_g^* - \angle P(j\omega_g). \tag{11.198}$$

Thus, from Equations (11.193) and (11.194) with $M = M_g$, $\Phi = \Phi_g$, we have a cylinder and a plane (see Figure 11.23) in (k_p, k_i, k_d) space. The plane is represented by Equation (11.196a) and the cylinder by Equation (11.193).

11.6.3   Design with Classical Performance Specifications

Achievable Gain-Phase Margin Design Curves

In this section we show how simultaneous specifications of gain and phase margins can be obtained over the stabilizing set of PID controllers. The idea is to construct, using the stabilizing set, the set of achievable gain and phase margin specifications. The gain-phase margin design curves represent the set of achievable gain margin (GM), phase margin (PM), and gain crossover frequencies (ω_g) for our plant with a PI or PID controller. The procedure for constructing these design curves is the following.

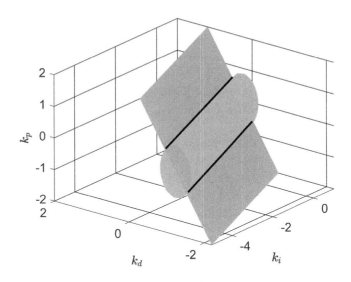

Figure 11.23 Cylinder and plane intersecting in the (k_p, k_i, k_d) space © Springer Nature Switzerland AG 2019. Reproduced from Diaz-Rodriguez, Han, and Bhattacharyya (2019) with permission.

1. Set a range of PM $\phi_g^* \in [\phi_g^-, \phi_g^+]$ and gain crossover frequency $\omega_g \in [\omega_g^-, \omega_g^+]$.
2. For prescribed values of ϕ_g^* and ω_g, plot the corresponding ellipse and straight line.
3. If the intersection point of the ellipse and straight line lies outside of the stabilizing set, then this point is rejected and we go to step 2.
4. If the intersection of the ellipse and straight line is contained in the stabilizing set, it represents a feasible design point with the PI or PID controller gains (k_p^*, k_i^*) or (k_p^*, k_i^*, k_d^*) that satisfies the prescribed ϕ_g^* and ω_g.
5. Given the selected PI or PID controller gains (k_p^*, k_i^*) or (k_p^*, k_i^*, k_d^*), the upper and lower GM of the system are given by

$$\text{GM}_{\text{upper}} = \frac{k_p^{ub}}{k_p^*} \quad \text{and} \quad \text{GM}_{\text{lower}} = \frac{k_p^{lb}}{k_p^*}, \tag{11.199}$$

where k_p^{ub} and k_p^{lb} are the controller gains at the further and closer boundary, respectively, of the stabilizing set following the straight line intersecting the ellipse.
6. Go to step 2 and repeat for all values of ϕ_g^* and ω_g in the ranges.

Time-Delay Tolerance

The time-delay tolerance design curves represent the actual time-delay tolerance achievable with a given PI or PID controller. This set of design curves is an extension of the previous gain-phase design curves because we can use the information calculated before to create this new time-delay tolerance design set. The time-delay tolerance for a chosen design point can be calculated by

$$\tau := \frac{\text{PM}}{\omega_g}, \tag{11.200}$$

where PM is the phase margin in radians and ω_g is the gain crossover frequency in radians per second. Then, taking all the points calculated from the gain-phase margin design set, we can find the values of time-delay tolerance and express the new plot with the x-axis as phase margin and the y-axis as time-delay tolerance. Similar to gain-phase margin design curves, these time-delay tolerance curves are indexed by a prescribed value of gain crossover frequency.

Simultaneous Performance Specifications and Retrieval of Controller Gains

The designer can select a desired design point from the achievable gain-phase margin curves and retrieve the controller gains corresponding to this simultaneous specification of desired GM, PM, and ω_g. The controller gain retrieval process is the following.

1. Select desired GM, PM, and ω_g from the achievable gain-phase margin curves.
2. For the specified point, construct the ellipse and straight line for a PI controller and cylinder and plane for a PID controller by using the selected PM and ω_g from the constant-gain and constant-phase loci.
3. Select the intersection of the ellipse and straight line or cylinder and plane contained in the stabilizing set. This will provide the gains (k_p^*, k_i^*) or (k_p^*, k_i^*, k_d^*).
4. The controller that satisfies the prescribed gain and phase margins and ω_g specifications is $C(s) = \frac{k_p^* s + k_i^*}{s}$ or $C(s) = \frac{k_d^* s^2 + k_p^* s + k_i^*}{s}$.

The above procedure can also be used for retrieving the controller gains using the time-delay tolerance design curves. We can select a point from the time-delay tolerance design curves and use the value of phase margin and gain crossover frequency to compute the ellipse and straight line and find the intersection contained in the stabilizing set, if it exists.

Gain-Phase Margin Based Controller Design

The PI, PID, or first-order controller design approach to satisfy simultaneous specifications of gain margin, phase margin, and gain crossover frequency for a linear time-invariant SISO plant can be summarized as follows.

1. Compute the PI, first-order, or PID stabilizing set.
2. Parametrize constant gain and phase loci.
3. Construct the gain-phase margin design curves.
4. Select simultaneous gain margin, phase margin, and gain crossover frequency design specifications from the achievable gain-phase margin design curves.
5. Retrieve the PI or PID controller gains satisfying the design specifications from the intersection of gain and phase loci points, lying in the stabilizing set.

The calculations described above are illustrated in the examples below.

EXAMPLE 11.5 (Continuous-time PI controller design)   Consider Figure 11.19 with an unstable plant with transfer function

$$P(s) = \frac{9}{s-2} \tag{11.201}$$

and the PI controller in Equation (11.183). In this example, we show how to find the controller gains that satisfy the desired performance by computing an ellipse and a straight line and superimposing on the stabilizing set. For this example, the stabilizing set can be found from the characteristic equation

$$\delta(s) = s^2 + (9k_p - 2)s + 9k_i. \tag{11.202}$$

We find the stabilizing region is given by

$$k_p > \frac{2}{9} \quad \text{and} \quad k_i > 0. \tag{11.203}$$

This means the stabilizing region is unbounded and contained in the first quadrant in the (k_p, k_i) plane. Now, consider a design value for a specific gain crossover frequency and phase margin. For illustration, consider $\omega_g^* = 15$ rad/s and $\phi_g^* = 60°$. Using Equations (11.187a) and (11.187b) representing the constant magnitude and constant phase of the controller, we can compute the corresponding ellipse and straight line for the prescribed values of ω_g^* and ϕ_g^*. In Figure 11.24 we can see the intersection of the ellipse and straight line superimposed on the stabilizing set.

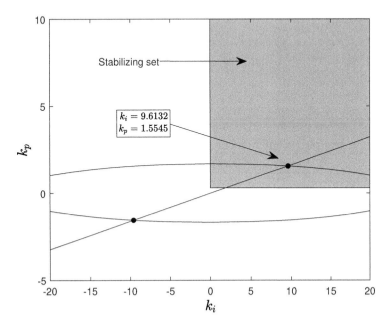

Figure 11.24 Intersection of ellipse and straight line superimposed on the stabilizing set (Example 11.5) © Springer Nature Switzerland AG 2019. Reproduced from Diaz-Rodriguez, Han, and Bhattacharyya (2019) with permission.

The intersection of the ellipse and straight line that is contained in the stabilizing set provides the desired values of ω_g^* and ϕ_g^*. The corresponding PI controller gains are

$$k_p = 1.5545 \qquad \text{and} \qquad k_i = 9.6132. \tag{11.204}$$

Considering the controller gains in Equation (11.204) and the plant in Equation (11.201), we can inspect the Nyquist plot and verify stability, phase margin, and gain crossover frequency. In Figure 11.25, we can see the Nyquist plot using a logarithmic scale. In this figure, we notice that we have one counterclockwise encirclement of $-1 + j0$. Therefore, the closed-loop system is asymptotically stable.

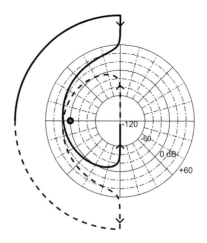

Figure 11.25 Nyquist plot in logarithmic scale with the controller gains in Equation (11.204) and the plant in Equation (11.201) (Example 11.5) © Springer Nature Switzerland AG 2019. Reproduced from Diaz-Rodriguez, Han, and Bhattacharyya (2019) with permission.

In Figure 11.26, we can verify that the phase margin of the system is 60° and the gain crossover frequency is 10 rad/s.

EXAMPLE 11.6 (Continuous-time PI controller design)   As another example, consider the continuous-time system represented in Figure 11.19 using the plant and the controller

$$P(s) = \frac{s-5}{s^2 + 1.6s + 0.2} \quad \text{and} \quad C(s) = \frac{k_p s + k_i}{s}. \tag{11.205}$$

(A) Computation of the Stabilizing Set
The first step in the controller design procedure is to obtain the stabilizing set of PI controllers for the given plant. The closed-loop characteristic polynomial is

$$\delta(s, k_p, k_i) = s^3 + (k_p + 1.6)s^2 + (k_i - 5k_p + 0.2)s - 5k_i. \tag{11.206}$$

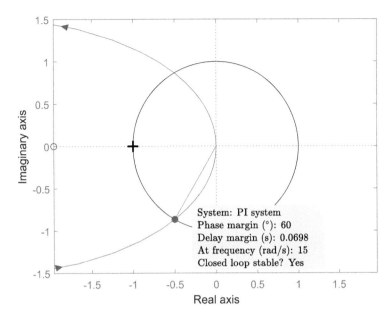

Figure 11.26 Nyquist plot with the controller gains in Equation (11.204) and the plant in Equation (11.201) (Example 11.5) © Springer Nature Switzerland AG 2019. Reproduced from Diaz-Rodriguez, Han, and Bhattacharyya (2019) with permission.

Here $n = 2$, $m = 1$, and $N(-s) = -5 - s$. Therefore, we obtain

$$v(s) = \delta(s, k_p, k_i)N(-s)$$
$$= -s^4 - (6.6 + k_p)s^3 - (8.2 + k_i)s^2 + (25k_p - 1)s + 25k_i, \tag{11.207}$$

so that

$$v(j\omega, k_p, k_i) = (-\omega^4 + (k_i + 8.2)\omega^2 + 25k_i) + j[(k_p + 6.6)\omega^3 + (25k_p - 1)\omega]$$
$$= p(\omega) + jq(\omega). \tag{11.208}$$

We find that $z^+ = 1$, so that the signature requirement on $v(s)$ for stability is

$$n - m + 1 + 2z^+ = 4. \tag{11.209}$$

Since the degree of $v(s)$ is even, we see from the signature formulas that $q(\omega)$ must have at least one positive real root of odd multiplicity. The range of k_p such that $q(\omega, k_p)$ has at least one real, positive, distinct, finite zero with odd multiplicity was determined to be $k_p \in (-1.6, 0.04)$, which is the allowable range for k_p. By sweeping over different k_p values within the interval $(-1.6, 0.04)$, we can generate the set of stabilizing (k_p, k_i) values. This set is shown in Figure 11.27.

(B) Construction of Achievable Gain-Phase Margin Design Curves
For a range of desired phase margins and gain crossover frequencies, we superimpose the corresponding ellipses and straight lines on the stabilizing set (see Figure 11.28).

We can see in Figure 11.28 the intersection points of ellipses and straight lines for

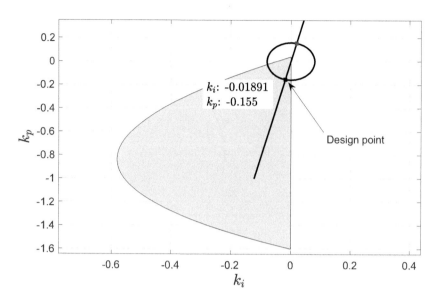

Figure 11.27 PI stabilizing set and intersection of ellipse and straight line for the final design point (Example 11.6) © 2016 IEEE. Reproduced from Diaz-Rodriguez and Bhattacharyya (2016) with permission.

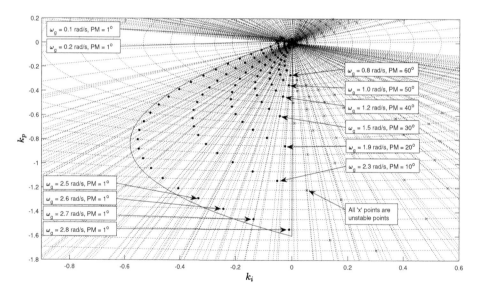

Figure 11.28 Construction of the gain-phase margin design curves for PI controller design by intersection points of ellipses and straight lines (Example 11.6) © 2016 IEEE. Reproduced from Diaz-Rodriguez and Bhattacharyya (2016) with permission.

different values of phase margin and gain crossover frequencies. We notice how for different values of phase margin, the gain crossover frequency limit is different. In this way, we obtain the maximum achievable values for the gain crossover frequency. For example, the maximum value of gain crossover frequency is 2.3 rad/s when requiring a PM = 10° and the maximum gain crossover frequency is 0.8 rad/s when requiring PM = 60°. All the intersection points, contained in the stabilizing set, determine the corresponding PI controller gains and are used to construct the achievable gain-phase design curves shown in Figure 11.29.

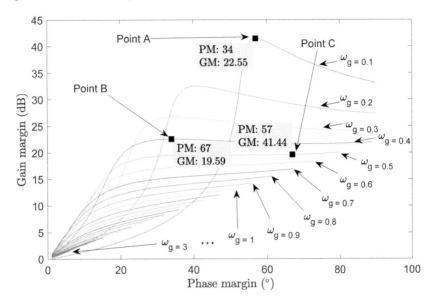

Figure 11.29 Achievable gain-phase margin design curves in the gain-phase plane for PI controller design (Example 11.6) © 2016 IEEE. Reproduced from Diaz-Rodriguez and Bhattacharyya (2016) with permission.

For this example, the evaluated range for the phase margin is from 1° to 90° and the range for the gain crossover frequency is from 0.1 to 3 rad/s.

(C) Simultaneous Performance Specifications and Retrieval of Controller Gains
As shown in Figure 11.29, we can clearly see the achievable performance for Example 11.6. In this case, the maximum gain margin that we can get is 41.44 dB. The phase margin, corresponding to this maximum gain margin, is 57° with a gain crossover frequency of 0.1 rad/s. We represent this as Point A in Figure 11.29. We can see how increasing the gain crossover frequency leads to a decrease in our achievable values for gain and phase margins. For example, for a gain crossover frequency of 0.4 rad/s, the corresponding maximum gain margin we can get is 22.55 dB, and the corresponding phase margin is 34°. We call this Point B in Figure 11.29. In Figure 11.29, we have chosen a candidate design (Point C). The controller corresponding to this design specification can be recovered by constructing the straight line and ellipse corresponding to these specifications (see Figure 11.27). The PI controller gains for these specifications are

$$k_p^* = -0.1556, \quad k_i^* = -0.0189. \tag{11.210}$$

The step response for this controller is given in Figure 11.30.

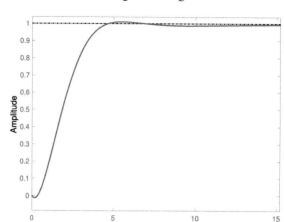

Figure 11.30 Step response for the system using the PI controller design $C(s)^* = \frac{k_p^* s + k_i^*}{s}$ (Example 11.6) © 2016 IEEE. Reproduced from Diaz-Rodriguez and Bhattacharyya (2016) with permission.

These controller gains correspond to the point of $\omega_g = 0.5$, PM $= 67°$, and GM $= 19.6$ dB in the gain-phase margin design plane (see Point C in Figure 11.29). In Figure 11.31, we can see the Nyquist plot for the controller gains selected.

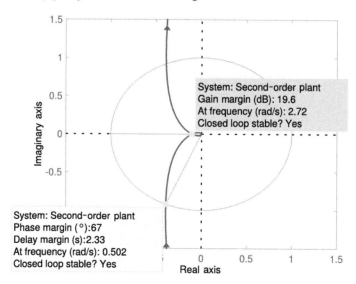

Figure 11.31 Nyquist plot for $k_p = -0.1556$, $k_i = -0.0189$ in the PI controller design (Example 11.6) © 2016 IEEE. Reproduced from Diaz-Rodriguez and Bhattacharyya (2016) with permission.

Here, we can see that those controller gains satisfy the desired performance specifications, PM $= 67°$, GM $= 19.6$ dB. We can also compute the time-delay tolerance design curves. Following Equation (11.200) and taking the values from Figure 11.29, we get Figure 11.32. In Figure 11.32, we can see the achievable time-delay tolerances for the system using the proposed controller.

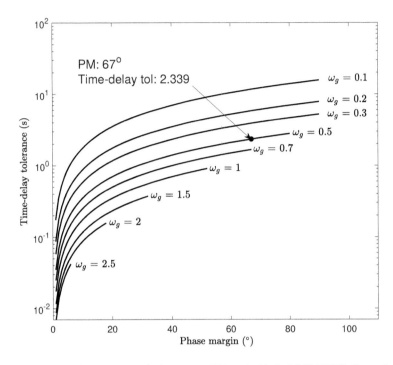

Figure 11.32 Time-delay tolerance design curves (Example 11.6) © 2016 IEEE. Reproduced from Diaz-Rodriguez and Bhattacharyya (2016) with permission.

We can select any point from the curves and retrieve the controller gains following the same procedure as selecting a point from the gain-phase margin design curves. In this case, we selected the same PM $= 67°$ and $\omega_g = 0.5$ rad/s. The time-delay tolerance is $\tau = 2.339$ s.

EXAMPLE 11.7 (Continuous-time PI controller design)   Let us consider the continuous-time LTI plant and the controller

$$P(s) = \frac{s^3 - 4s^2 + s + 2}{s^5 + 8s^4 + 32s^3 + 46s^2 + 46s + 17} \quad \text{and} \quad C(s) = \frac{k_p s + k_i}{s}. \tag{11.211}$$

(A) Computation of the Stabilizing Set
The closed-loop characteristic polynomial is

$$\begin{aligned}
\delta(s, k_p, k_i) = {}& s^6 + 8s^5 + (k_p + 32)s^4 + (k_i - 4k_p + 46)s^3 \\
& + (k_p - 4k_i + 46)s^2 + (k_i + 2k_p + 17)s + 2k_i.
\end{aligned} \tag{11.212}$$

Here $n = 5$, $m = 3$, and $N(-s) = -s^3 - 4s^2 - s + 2$. Therefore, we obtain

$$
\begin{aligned}
v(s) &= \delta(s, k_p, k_i)N(-s) \\
&= -s^9 - 12s^8 + (-k_p - 65)s^7 + (-k_i - 180)s^6 \\
&\quad + (14k_p - 246)s^5 + (14k_i - 183)s^4 + (-17k_p - 22)s^3 \\
&\quad + (75 - 17k_i)s^2 + (4k_p + 34)s + 4k_i,
\end{aligned} \tag{11.213}
$$

so that

$$
\begin{aligned}
v(j\omega, k_p, k_i) &= -12\omega^8 + (k_i + 180)\omega^6 + (14k_i - 183)\omega^4 + (17k_i - 75)\omega^2 \\
&\quad + 4k_i + j[-\omega^9 + (k_p + 65)\omega^7 + (14k_p - 246)\omega^5 \\
&\quad + (17k_p + 22)\omega^3 + (4k_p + 34)\omega] = p(\omega) + jq(\omega).
\end{aligned} \tag{11.214}
$$

We find that $z^+ = 2$, so that the signature requirement on $v(s)$ for stability is

$$
n - m + 1 + 2z^+ = 7. \tag{11.215}
$$

Since the degree of $v(s)$ is odd, we see from the signature formulas that $q(\omega)$ must have at least three positive real zeros of odd multiplicity. The range of k_p such that $q(\omega, k_p)$ has at least one real, positive zero with odd multiplicity was determined to be $k_p \in (-8.5, 4.2)$, which is the allowable range for k_p. By sweeping over different k_p values within the interval $(-8.5, 4.2)$, we can generate the set of stabilizing (k_p, k_i) values. This set is shown in Figure 11.33.

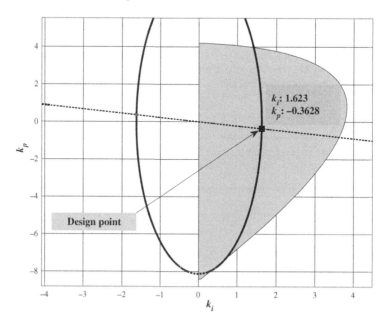

Figure 11.33 PI stabilizing set and intersection of ellipse and straight line for the final design point (Example 11.7) © 2016 IEEE. Reproduced from Diaz-Rodriguez and Bhattacharyya (2016) with permission.

(B) Construction of Achievable Gain-Phase Margin Design Curves
As in Example 11.6, for a prescribed range of phase margins and gain crossover frequencies, we superimpose ellipses and straight lines on the stabilizing set. The intersection points, contained in the stabilizing set, determine the PI controller gains. For this example, the evaluated range for the achievable phase margin is from 1° to 90° and the range for the achievable gain crossover frequency is from 0.1 to 1 rad/s.

(C) Simultaneous Specifications and Retrieval of Controller Gains
Figure 11.34 displays the achievable performance for Example 11.7. In this case, the maximum gain margin that we can get is 13.14 dB. The phase margin, corresponding to this maximum gain margin, is 79° with a gain crossover frequency of 0.1 rad/s as shown in Figure 11.34, marked as Point A.

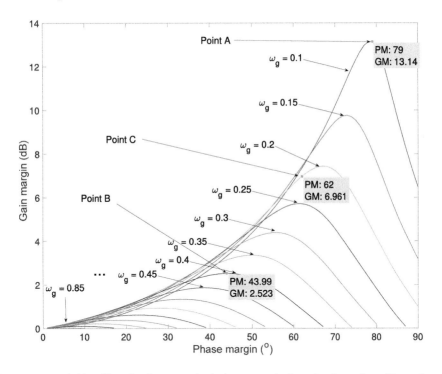

Figure 11.34 Achievable gain-phase margin design curves in the gain-phase plane (Example 11.7) © 2016 IEEE. Reproduced from Diaz-Rodriguez and Bhattacharyya (2016) with permission.

We can see how, when increasing the gain crossover frequency, our values for achievable gain and phase margins decrease. For example, for a gain crossover frequency of 0.4 rad/s, the maximum gain margin we can get is 2.522 dB, and the corresponding phase margin is 44°, marked as Point B. In Figure 11.34, we have chosen a candidate design labeled as Point C. The controller corresponding to this design specification can be recovered by constructing the straight line and ellipse corresponding to these specifications (see Figure 11.33). The PI controller gains for these specifications are

$$k_p^* = -0.36283, \qquad k_i^* = 1.6228. \tag{11.216}$$

The step response for this controller is given in Figure 11.35.

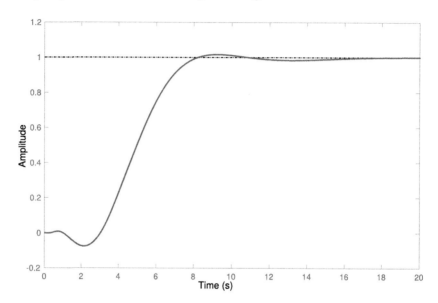

Figure 11.35 Step response for the system in Example 11.7 using $C^*(s) = \frac{k_p^* s + k_i^*}{s}$ © 2016 IEEE. Reproduced from Diaz-Rodriguez and Bhattacharyya (2016) with permission.

These controller gains correspond to the point $\omega_g = 0.2$, PM $= 62°$, and GM $= 6.96$ dB in the gain-phase margin design plane (see Point C in Figure 11.34). In Figure 11.36, we display the Nyquist plot for the controller gains selected.

Here, we can see that those controller gains satisfy the desired performance specifications, PM $= 62°$, GM $= 6.96$ dB. We can also compute the time-delay tolerance design curves. Following Equation (11.200) and taking the values from Figure 11.34, we get Figure 11.37, where we can see the achievable time-delay tolerance for the system using the proposed controller. We can select any point from the curves and retrieve the controller gains following the same procedure as taking a point from the gain-phase margin design curves. In this case, we selected PM $= 62°$ and $\omega_g = 0.2$ rad/s. The time-delay tolerance is $\tau = 5.411$ s.

EXAMPLE 11.8 (Continuous-time PID controller design)   Consider the continuous-time LTI plant represented in Figure 11.19 with the plant and the controller

$$P(s) = \frac{s - 3}{s^3 + 4s^2 + 5s + 2} \quad \text{and} \quad C(s) = \frac{k_d s^2 + k_p s + k_i}{s}. \tag{11.217}$$

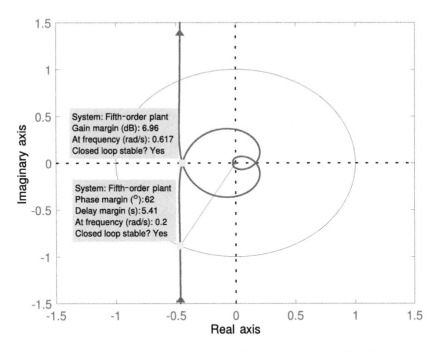

Figure 11.36 Nyquist plot for $k_p = -0.36283$, $k_i = 1.6228$ in the PI controller design (Example 11.7) © Springer Nature Switzerland AG 2019. Reproduced from Diaz-Rodriguez, Han, and Bhattacharyya (2019) with permission.

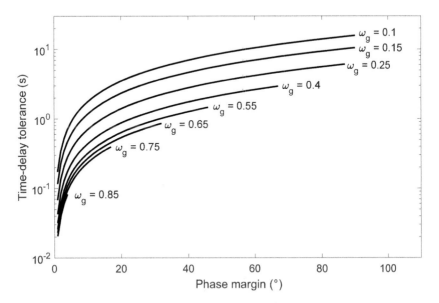

Figure 11.37 Time-delay tolerance design curves (Example 11.7) © Springer Nature Switzerland AG 2019. Reproduced from Diaz-Rodriguez, Han, and Bhattacharyya (2019) with permission.

(A) Computation of the Stabilizing Set

The closed-loop characteristic polynomial is

$$\delta(s,k_p,k_i) = s^4 + (k_d+4)s^3 + (k_p-3k_d+5)s^2 + (k_i-3k_p+2)s - 3k_i. \quad (11.218)$$

Here $n=3$, $m=1$, and $N(-s) = -s-3$. Therefore, we obtain

$$\begin{aligned}
\nu(s) &= \delta(s,k_p,k_i)N(-s) \\
&= -s^5 + (-k_d-7)s^4 + (-k_p-17)s^3 + (9k_d-k_i-17)s^2 \quad (11.219)\\
&\quad + (9k_p-6)s + 9k_i,
\end{aligned}$$

so that

$$\begin{aligned}
\nu(j\omega,k_p,k_i) &= (-k_d-7)\omega^4 + (k_i-9k_d+17)\omega^2 + 9k_i \\
&\quad + j[\omega^5 + (k_p+17)\omega^3 + (9k_p-6)\omega] = p(\omega)+jq(\omega).
\end{aligned} \quad (11.220)$$

We find that $z^+ = 1$, so that the signature requirement on $\nu(s)$ for stability is

$$\sigma(\nu) = n-m+1+2z^+ = 5. \quad (11.221)$$

Since the degree of $\nu(s)$ is odd, we see from the signature formulas that $q(\omega)$ must have at least two positive real roots of odd multiplicity. The range of k_p such that $q(\omega,k_p)$ has at least two real, positive, distinct, finite zeros with odd multiplicity was determined to be $k_p \in (-4,0.65)$, which is the allowable range for k_p. By sweeping over different k_p values within the interval $(-4,0.65)$, we can generate the set of stabilizing (k_p,k_i) values. This set is shown in Figure 11.38.

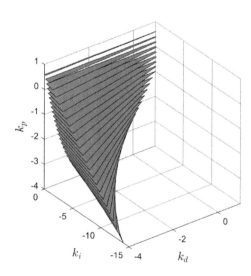

Figure 11.38 PID stabilizing set (Example 11.8) © Springer Nature Switzerland AG 2019. Reproduced from Diaz-Rodriguez, Han, and Bhattacharyya (2019) with permission.

(B) Construction of Achievable Gain-Phase Margin Design Curves

For the construction of the achievable gain-phase curves for the PID controller, the evaluated range of ω_g is $[0.1, 1.2]$ and the range for PM is from $1°$ to $100°$. Using the constant-gain and constant-phase loci in Equations (11.193) and (11.194), we now get a cylinder and a plane in the (k_p, k_i, k_d) space. The cylinder and the plane, superimposed on the stabilizing set (see Figure 11.39 below), will have two intersection line segments in the (k_i, k_d) plane.

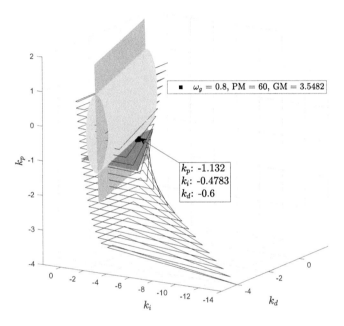

Figure 11.39 Intersection of a cylinder and a plane superimposed on the PID stabilizing set and PID controller design (Example 11.8) © Springer Nature Switzerland AG 2019. Reproduced from Diaz-Rodriguez, Han, and Bhattacharyya (2019) with permission.

The specific value where the intersection occurs can be obtained using Equation (11.196b). Equation (11.196b) will give us two values for k_p, but only one is contained in the stabilizing set. The intersection line segment in (k_p, k_i, k_d) represents the PID controller gains that satisfy the PM and ω_g. Evaluating for a range of PM and ω_g, we can construct the achievable gain-phase margin set represented in 3D in Figure 11.40.

If we fix $\omega_g = 0.8$ rad/s, we can see the achievable performance in 2D in Figure 11.41.

Here we can see that the maximum GM we can get is 6.269 with a PM of $9°$ and for a PM of $60°$ the GM is 3.548.

(C) Simultaneous Specifications and Retrieval of Controller Gains

In Figure 11.40, we can see the achievable gain-phase margin set of curves indexed by ω_g in different colors. Notice that we can get more GM and PM for lower values of ω_g. For example, for $\omega_g = 0.1$ rad/s, the maximum GM that we can get is 38.44 with a PM

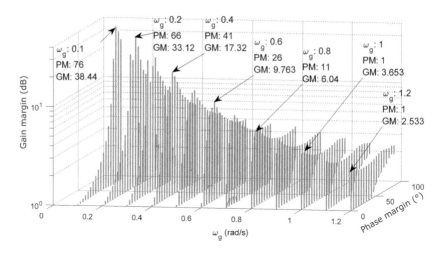

Figure 11.40 Achievable performance in terms of GM, PM, and ω_g for PID controller design (Example 11.8) © Springer Nature Switzerland AG 2019. Reproduced from Diaz-Rodriguez, Han, and Bhattacharyya (2019) with permission.

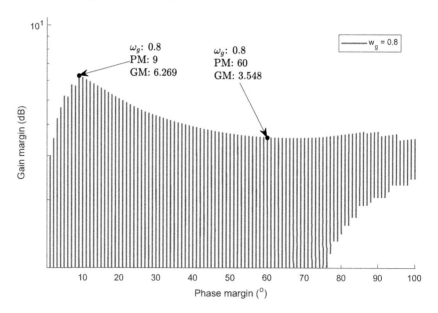

Figure 11.41 Achievable gain-phase margin set for $\omega_g = 0.8$ rad/s for PID controller design (Example 11.8) © Springer Nature Switzerland AG 2019. Reproduced from Diaz-Rodriguez, Han, and Bhattacharyya (2019) with permission.

of 76°. For $\omega_g = 0.2$ rad/s, the maximum GM is 33.12 with a PM = 66°. For a larger value of ω_g, we get lower values for GM and PM. For example, for $\omega_g = 1.2$ rad/s we get a maximum GM = 2.533 and PM = 1°. The designer has the liberty of using Figure 11.40 to choose values for GM, PM, and ω_g that best suits his design needs.

After the selection of simultaneous GM, PM, and ω_g from the achievable gain-phase margin set, the designer can retrieve the controller gains corresponding to this performance point. For illustration purposes, let us say that the desired performance values chosen for this example are PM $= 60°$, GM $= 3.548$, and $\omega_g = 0.8$ rad/s (see Figure 11.41). Then, using these values and the constant-gain and constant-phase loci for PID controllers, we can find the intersection of the cylinder and the plane in (k_p, k_i, k_d) in 3D space, shown in Figure 11.39. The controller gains are $k_p^* = -1.1317$, $k_i^* = -0.4783$, and $k_d^* = -0.6$. In Figure 11.42 we can see the Nyquist plot for the controller gains selected.

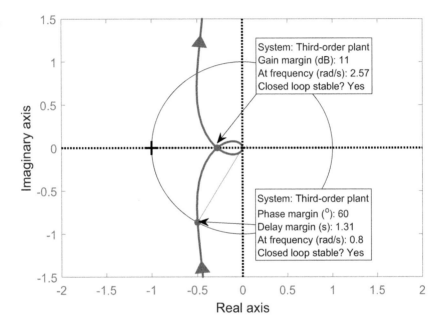

Figure 11.42 Nyquist plot for $k_p^* = -1.1317$, $k_i^* = -0.4783$, and $k_d^* = -0.6$ in the PID controller design (Example 11.8) © Springer Nature Switzerland AG 2019. Reproduced from Diaz-Rodriguez, Han, and Bhattacharyya (2019) with permission.

Here, we can see that those controller gains satisfy the desired performance specifications, PM $= 60°$, GM $= 3.5482$ (11 dB).

11.7   Design with H_∞ Norm Constraints

The H_∞ norm has proved itself to be a useful criterion for control system design. This section presents a constructive determination of a set of stabilizing PI and PID controllers, for a given plant, achieving an H_∞ norm bound of γ on the error transfer function. The results in this chapter utilize the computation of the complete stabilizing set $\mathscr{S}$. We point out connections between the H_∞ design and gain and phase margin designs.

We show that the design criterion is expressed as the intersection of the stabilizing set and the exterior of a family of ellipses in controller parameter space.

11.7.1 H_∞ Optimal Control and Stability Margins

Consider the unity feedback system in Figure 11.43 with the error transfer function

$$\frac{e(s)}{r(s)} = \frac{1}{1 + G(s)}. \tag{11.222}$$

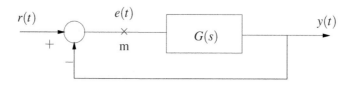

Figure 11.43 Unity feedback control loop

Suppose that $G(s)$ includes a controller designed to make the H_∞ norm of Equation (11.222) less than γ, a prescribed real positive number. Then

$$\frac{1}{|1 + G(j\omega)|} < \gamma, \qquad \text{for all } \omega \geq 0 \tag{11.223}$$

and Equation (11.223) is equivalent to

$$|1 + G(j\omega)| > \frac{1}{\gamma}, \qquad \text{for all } \omega \in [0, \infty). \tag{11.224}$$

We will now establish that Equation (11.224) implies guaranteed gain and phase margins at the loop breaking point "m" in Figure 11.43.

Remark 11.6   Let γ^* denote the infimum value of γ satisfying Equation (11.224). When $G(s)$ is strictly proper, $\gamma^* \geq 1$. When $G(s)$ is proper, $\gamma^* > 1/|1 + G(j\infty)|$.

Case 1: $\gamma > 1$
The condition in Equation (11.224) implies that the Nyquist plot $G(j\omega)$ stays out of the circle CEDB centered at $-1 + j0$ and of radius $1/\gamma$. In Figure 11.44, we have the limiting case in which $G(j\omega)$ passes through B, the phase margin is ϕ, and

$$G(j\omega) = \overrightarrow{OB}, \tag{11.225a}$$

$$-1 + j0 = \overrightarrow{OA}, \tag{11.225b}$$

$$1 + G(j\omega) = \overrightarrow{AB}. \tag{11.225c}$$

Since $\overrightarrow{OA} + \overrightarrow{AB} = \overrightarrow{OB}$, we have

$$-1 + j0 + \frac{1}{\gamma}e^{-j\theta} = -1e^{j\phi}. \tag{11.226}$$

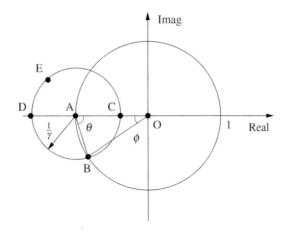

Figure 11.44 $\gamma > 1$

Also

$$2\theta + \phi = \pi \tag{11.227}$$

from the triangle $\overrightarrow{OAB}$.

From Equations (11.226) and (11.227):

$$-1 + \frac{1}{\gamma}\sin\left(\frac{\phi}{2}\right) = -\cos\phi, \tag{11.228a}$$

$$\sin\phi = \frac{1}{\gamma}\cos\left(\frac{\phi}{2}\right). \tag{11.228b}$$

From Equation (11.228b):

$$\phi = 2\sin^{-1}\left(\frac{1}{2\gamma}\right), \tag{11.229}$$

which is the guaranteed minimum phase margin for the H_∞ controller with norm less than γ.

The guaranteed gain margin is the interval

$$\left[\frac{1}{OD}, \frac{1}{OC}\right] = \left[\frac{\gamma}{\gamma+1}, \frac{\gamma}{\gamma-1}\right]. \tag{11.230}$$

Case 2: $\gamma = 1$
In this case, Figure 11.44 is replaced by Figure 11.45.

It is easy to see that the guaranteed phase margin is $\phi = \pi/3$ and the guaranteed gain margin is $\left[\frac{1}{2}, \infty\right]$. These also follow from formulas in Equations (11.229) and (11.230) evaluated at $\gamma = 1$.

Case 3: $\gamma < 1$
The geometry corresponding to this case is shown in Figure 11.46.

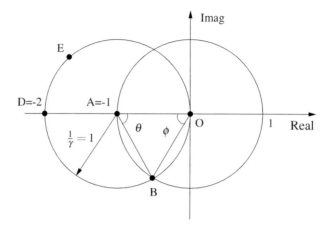

Figure 11.45 $\gamma = 1$

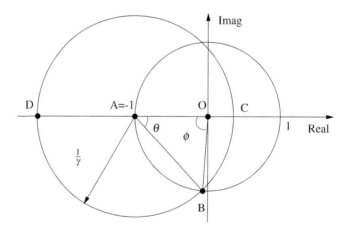

Figure 11.46 $\gamma < 1$

In this case, it also follows that the guaranteed phase margin is

$$\phi = 2\sin^{-1}\left(\frac{1}{2\gamma}\right) \tag{11.231}$$

and the guaranteed gain margin is

$$\left[\frac{1}{OD}, \infty\right] = \left[\frac{\gamma}{1+\gamma}, \infty\right]. \tag{11.232}$$

Combining the above cases, we have the following result.

THEOREM 11.8 Consider the unity feedback system in Figure 11.43. If the H_∞ norm of the error transfer function is less than γ:

$$\left\|\frac{1}{1+G(s)}\right\|_\infty < \gamma, \tag{11.233}$$

then the guaranteed phase margin at the loop breaking point "m" is

$$\phi = 2\sin^{-1}\left(\frac{1}{2\gamma}\right). \tag{11.234}$$

The guaranteed gain margin is:

$$g_m = \begin{cases} \left[\frac{\gamma}{\gamma+1}, \frac{\gamma}{\gamma-1}\right], & \text{for} \quad \gamma > 0, \\[2ex] \left[\frac{\gamma}{\gamma+1}, \infty\right], & \text{for} \quad \gamma \leq 0. \end{cases} \tag{11.235}$$

Now consider the control system in Figure 11.47 where $r(t)$ is the reference signal, $e(t)$ the error signal, $u(t)$ the input signal (to the plant), $y(t)$ the output signal, $P(s)$ the plant transfer function, and $C(s)$ the controller transfer function which we will consider to be either PI or PID.

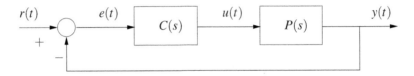

Figure 11.47 Unity feedback control loop

The problem to be solved in this section is: Find the set $\mathscr{S}_\gamma$ of all stabilizing PI or PID controllers satisfying

$$\left\|\frac{1}{1+P(s)C(s)}\right\|_\infty < \gamma. \tag{11.236}$$

In the following two subsections we develop the computation of $\mathscr{S}_\gamma$ for PID controllers. Note that Equation (11.236) is equivalent to

$$|1+P(j\omega)C(j\omega)| > \frac{1}{\gamma}, \qquad \text{for all } \omega \in [0,\infty). \tag{11.237}$$

11.7.2   Computation of $\mathscr{S}_\gamma$ for PI Controllers

PI controllers have the form

$$C(s) = k_p + \frac{k_i}{s}. \tag{11.238}$$

Write

$$P(j\omega) = P_r(\omega) + j\omega P_i(\omega), \tag{11.239}$$

$$C(j\omega) = k_p - j\frac{k_i}{\omega}. \tag{11.240}$$

Substituting Equations (11.239) and (11.240) into Equation (11.237), we get

$$\left| 1 + \underbrace{k_p P_r(\omega) + k_i P_i(\omega)}_{L_0(\omega)} + j\underbrace{(\omega k_p P_i(\omega) - \frac{k_i}{\omega} P_r(\omega))}_{L_1(\omega)} \right| > \frac{1}{\gamma}, \tag{11.241}$$

which can be rewritten as

$$(1 + L_0(\omega))^2 + L_1^2(\omega) > \frac{1}{\gamma^2}, \tag{11.242}$$

$$\begin{bmatrix} P_r(\omega) & P_i(\omega) \\ \omega P_i(\omega) & -\frac{P_r(\omega)}{\omega} \end{bmatrix} \begin{bmatrix} k_p \\ k_i \end{bmatrix} = \begin{bmatrix} L_0(\omega) \\ L_1(\omega) \end{bmatrix}. \tag{11.243}$$

Equation (11.243) has a unique solution if

$$|P(j\omega)| \neq 0, \tag{11.244}$$

that is the plant has no $j\omega$-axis zeros.

Assuming Equation (11.244), Equation (11.243) can be solved:

$$\begin{bmatrix} k_p \\ k_i \end{bmatrix} = \underbrace{\frac{1}{|P(j\omega)|^2} \begin{bmatrix} P_r(\omega) & \omega P_i(\omega) \\ -\omega^2 P_i(\omega) & -\omega P_r(\omega) \end{bmatrix}}_{T(\omega)} \begin{bmatrix} L_0(\omega) \\ L_1(\omega) \end{bmatrix}. \tag{11.245}$$

Equation (11.242) represents the outside of a circle C_γ of radius $\frac{1}{\gamma}$ in the (L_0, L_1) plane centered at $(-1, 0)$ (Figure 11.48).

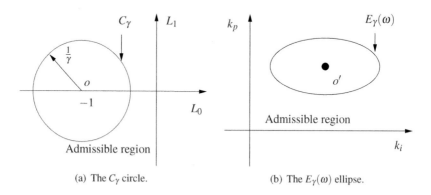

(a) The C_γ circle.　　　　　(b) The $E_\gamma(\omega)$ ellipse.

Figure 11.48 The circle and the ellipse

LEMMA 11.8　The condition in Equation (11.237) at a fixed ω is equivalent to k_p, k_i lying in the exterior of the axis-parallel ellipse $E_\gamma(\omega)$ with center o' at $\left(-\frac{\omega^2 P_i(\omega)}{|P(j\omega)|^2}, -\frac{P_r(\omega)}{|P(j\omega)|^2} \right)$, and major and minor axes of lengths $\left(\frac{2}{\gamma |P(j\omega)|}, \frac{2\omega}{\gamma |P(j\omega)|} \right)$.

Proof   For each $\omega \geq 0$, Equation (11.241) is

$$\left| 1 + (P_r(j\omega) + j\omega P_i(j\omega))(k_p - j\frac{k_i}{\omega}) \right| > \frac{1}{\gamma}$$

$$\Leftrightarrow \; (1 + P_r(j\omega)k_p + P_i(j\omega)k_i)^2 + \left(\omega P_i(j\omega)k_p - P_r(j\omega)\frac{k_i}{\omega} \right)^2 > \frac{1}{\gamma^2} \qquad (11.246)$$

$$\Leftrightarrow \; \frac{(k_i - c_1)^2}{a^2} + \frac{(k_p - c_2)^2}{b^2} > 1,$$

where

$$c_1 = \frac{-\omega^2 P_i(\omega)}{|P(j\omega)|^2}, \quad c_2 = \frac{-P_r(\omega)'}{|P(j\omega)|^2}, \quad a = \frac{\omega/\gamma}{|P(j\omega)|}, \quad b = \frac{1/\gamma}{|P(j\omega)|}. \qquad (11.247)$$

■

For a fixed ω, let $\mathscr{S}_\gamma(\omega)$ denote the intersection of the stabilizing set S with the exterior of the ellipse $E_\gamma(\omega)$ as shown in Figure 11.49(a). In other words:

$$\mathscr{S}_\gamma(\omega) = \mathscr{S} \setminus E_\gamma(\omega), \qquad \text{for all } \omega \in [0, \infty). \qquad (11.248)$$

Since Equation (11.237) must hold for all ω:

$$\mathscr{S}_\gamma = \bigcap_{\omega=0}^{\infty} \mathscr{S}_\gamma(\omega), \qquad (11.249)$$

as shown in Figure (11.49(b)).

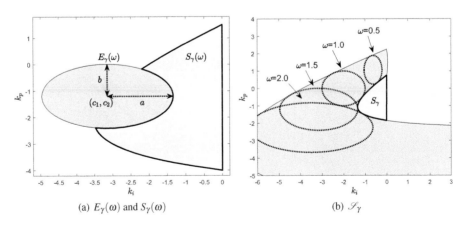

(a) $E_\gamma(\omega)$ and $S_\gamma(\omega)$ (b) $\mathscr{S}_\gamma$

Figure 11.49 $E_\gamma(\omega)$ and $S_\gamma(\omega)$, and $\mathscr{S}_\gamma$ © 2018 IFAC. Reproduced from Han, Keel, and Bhattacharyya (2018) with permission.

We state this result in the following theorem.

THEOREM 11.9   In the unity feedback control loop, suppose that the plant $P(s)$ has no $j\omega$-axis zeros. All stabilizing PI controllers $C(s)$ satisfying the H_∞ norm bound of γ

on the error transfer function form the set $\mathscr{S}_\gamma$:

$$\mathscr{S}_\gamma = \bigcap_{\omega=0}^{\infty} \mathscr{S}_\gamma(\omega). \tag{11.250}$$

Proof $\mathscr{S}_\gamma(\omega)$ is the admissible set for each ω and the controller must satisfy the H_∞ norm for all frequencies. Hence we have the set $\mathscr{S}_\gamma$ by intersecting the admissible sets $\mathscr{S}_\gamma(\omega)$ for all ω.   ∎

Note that $\mathscr{S}$ can be determined using the concept of signature developed in Section 11.3. If $E_\gamma(\omega)$ is outside of $\mathscr{S}$, then $\mathscr{S}_\gamma(\omega) = S$. If $\mathscr{S} \subset E_\gamma(\omega)$, then $\mathscr{S}_\gamma$ is empty.

Remark 11.7 We can determine the minimum achievable γ for a given plant under PI or PID control. The minimum γ, denoted γ^*, is the value for which the union of the family of ellipses eclipses the stabilizing set $\mathscr{S}$.

Remark 11.8 The computation of $\mathscr{S}_\gamma$ would not be possible without knowing the stabilizing set $\mathscr{S}$.

EXAMPLE 11.9 Consider the second-order plant and the PI controller:

$$P(s) = \frac{s-2}{s^2+4s+3} \quad \text{and} \quad C(s) = k_p + \frac{k_i}{s}. \tag{11.251}$$

The stabilizing set was first computed for the plant and the PI controller given in Equation (11.251). The family of ellipses $E_\gamma(\omega)$ were drawn by sweeping over ω and $\mathscr{S}_\gamma$ was found accordingly for $\gamma = 1.6$, 2.0, 4.0, and 8.0. In Figure 11.50 we observe that $\mathscr{S}_\gamma$ is contained in the stabilizing set $\mathscr{S}$ and $\mathscr{S}_{\gamma_1} \subset \mathscr{S}_{\gamma_2}$ if $\gamma_1 < \gamma_2$.

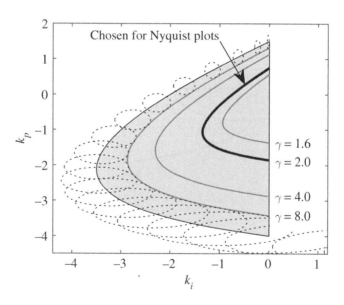

Figure 11.50 $\mathscr{S}_\gamma$ for $\gamma = 1.6$, 2.0, 4.0, 8.0 with the stabilizing set (Example 11.9) © 2018 IFAC. Reproduced from Han, Keel, and Bhattacharyya (2018) with permission.

So, $\mathscr{S}_\gamma$ for $\gamma \in [1, \infty)$ is the telescoping series of sets shown. If k_p, k_i were chosen from sets $\mathscr{S}_\gamma$, the Nyquist plot must stay outside of a circle centered at the critical point $-1 + j0$ with radius $1/\gamma$. We chose some boundary points in $\mathscr{S}_\gamma$ that were inside the stabilizing set $\mathscr{S}$, where $\gamma = 2$, and drew the Nyquist plots in Figure 11.51. Each Nyquist plot was at least 0.5 away from the critical point.

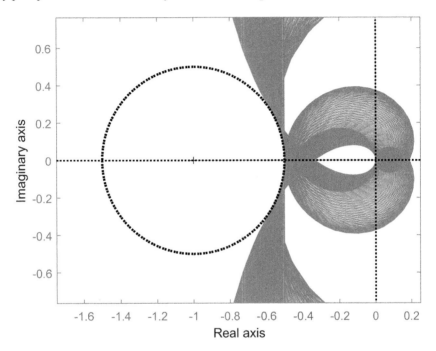

Figure 11.51 Nyquist plots with k_p, k_i along the curve of $\gamma = 2$ (Example 11.9) © 2018 IFAC. Reproduced from Han, Keel, and Bhattacharyya (2018) with permission.

Following Theorem 11.8, the guaranteed gain margin was

$$\left[\frac{\gamma}{\gamma+1}, \ \frac{\gamma}{\gamma-1} \right] = \left[\frac{2}{3}, \ 2 \right], \tag{11.252}$$

and the guaranteed phase margin ϕ was

$$\phi = 2\sin^{-1}\left(\frac{1}{2\gamma} \right) = 28.955° \tag{11.253}$$

for $\gamma = 2$. Figure 11.52 shows the guaranteed gain and phase margins when we choose k_p and k_i from S_γ for $\gamma = 2$.

For all controllers achieving the same H_∞ norm at the boundary of S_γ, there is a trade off between gain and phase margins. When a higher gain margin is desired, one should sacrifice some phase margin and vice versa. Nevertheless with the H_∞ norm we get the guaranteed gain and phase margins calculated in Equations (11.252) and (11.253).

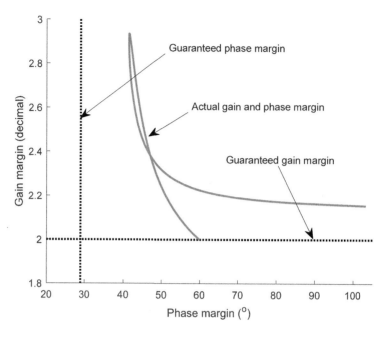

Figure 11.52 Guaranteed gain and phase margin of the boundary points of $\mathscr{S}_\gamma$ for $\gamma = 2$ (Example 11.9) © 2018 IFAC. Reproduced from Han, Keel, and Bhattacharyya (2018) with permission.

11.7.3   Computation of $\mathscr{S}_\gamma$ for PID Controllers

PID controllers are of the form

$$C(s) = k_p + \frac{k_i}{s} + k_d s. \tag{11.254}$$

Substituting $s = j\omega$, we have

$$C(j\omega) = k_p - j\frac{1}{\omega}\left(k_i - \omega^2 k_d\right). \tag{11.255}$$

It is easy to show that Equation (11.237) implies that the controller parameters k_p, k_i, k_d must lie in the exterior of $E_\gamma(\omega)$, described by

$$\frac{\left(k_i - \omega^2 k_d - c_1\right)^2}{a^2} + \frac{\left(k_p - c_2\right)^2}{b^2} > 1, \tag{11.256}$$

where $E_\gamma(\omega)$ is an elliptic cylinder with the center lying on the line

$$\begin{cases} k_i - \omega^2 k_d &= \dfrac{-\omega^2 P_i(\omega)}{|P(j\omega)|^2}, \\ k_p &= \dfrac{-P_r(\omega)}{|P(j\omega)|^2}, \end{cases} \tag{11.257}$$

and major and minor axes $\dfrac{2}{\gamma|P(j\omega)|}$ and $\dfrac{2\omega}{\gamma\sqrt{\omega^4+1}|P(j\omega)|}$ (Figure 11.53).

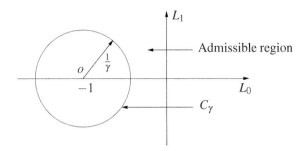

Figure 11.53 The $E_\gamma(\omega)$ elliptic cylinder

As before:

$$\mathscr{S}_\gamma(\omega) = \mathscr{S} \setminus E_\gamma(\omega) \ \forall \ \omega \in [0, \infty) \qquad (11.258)$$

and

$$\mathscr{S}_\gamma = \bigcap_{\omega=0}^{\infty} \mathscr{S}_\gamma(\omega). \qquad (11.259)$$

Remark 11.9   We can also consider the H_∞ norm with a weighting function $W(s)$ multiplied by the error transfer function in Equation (11.236). In this case, we may replace γ by γ', where $\gamma' = \frac{\gamma}{|W(j\omega)|}$. Then, the major and minor axes of the axis-parallel ellipse $E_\gamma(\omega)$ are subject to change with ω according to the frequency response of the weighting function. However, the rest of the derivation of the equations in this section remains the same.

Remark 11.10   If $C(s)$ is replaced by

$$C_\tau(s) = \frac{k_p s + k_i + k_d s^2}{s(\tau s + 1)}, \qquad (11.260)$$

then

$$C_\tau(s)P(s) = C(s)\frac{1}{\tau s + 1}P(s). \qquad (11.261)$$

Since τ can be fixed a priori, replace $P_r(j\omega)$ and $P_i(j\omega)$ by

$$P_r'(j\omega) = \frac{P_r(j\omega) + \tau\omega^2 P_i(j\omega)}{1 + \tau^2\omega^2}, \qquad (11.262a)$$

$$P_i'(j\omega) = \frac{P_i(j\omega) - \tau P_r(j\omega)}{1 + \tau^2\omega^2}. \qquad (11.262b)$$

Then the controller design can be carried out as before. We illustrate these calculations with an example below.

EXAMPLE 11.10   Consider a rational plant transfer function and the PID controller

$$P(s) = \frac{s^3 + s^2 + 18s + 2}{s^5 + 2s^4 + 35s^3 + 40s^2 + 120s + 90}, \qquad C(s) = k_p + \frac{k_i}{s} + k_d s. \quad (11.263)$$

The stabilizing set was computed using the signature method and is shown in Figure 11.54.

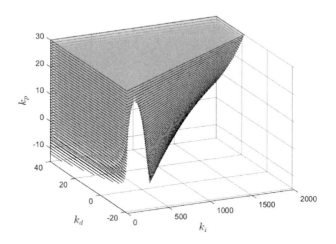

Figure 11.54 The stabilizing set in k_p, k_i, k_d space using the signature method (Example 11.10) © 2018 IFAC. Reproduced from Han, Keel, and Bhattacharyya (2018) with permission.

We chose $k_d = 20$ and computed $\mathscr{S}_\gamma$ for $\gamma = 1$ in the k_p, k_i plane. Figure 11.55 shows S_γ and the family of ellipses, $E_\gamma(\omega)$.

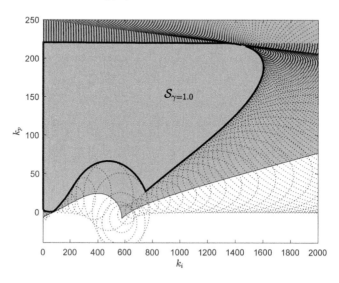

Figure 11.55 S_γ and family of ellipses for $\gamma = 1$ in k_p, k_i plane with $k_d = 20$ (Example 11.10) © 2018 IFAC. Reproduced from Han, Keel, and Bhattacharyya (2018) with permission.

We observe that the stabilizing set with $k_d = 10$ is unbounded in the k_p, k_i plane. However, $\mathscr{S}_\gamma$ for $\gamma = 1$ in the same plane is bounded. For high values of ω the major and minor axes of the ellipses grow as the centers c_1 and c_2 in Equation (11.247) go away from the origin. So we suggest that the family of ellipses be computed for high enough values of ω to get the exact set $\mathscr{S}_\gamma$.

Clearly, in this case, $\mathscr{S}_\gamma$ is not empty and the H_∞ norm condition less than $\gamma = 1$ provides very good robustness, namely $[0.5, \infty]$ gain margin and $60°$ phase margin. All of the points in $\mathscr{S}_\gamma$ guarantee such good robustness. In fact, since the open-loop transfer function $P(s)C(s)$ is strictly proper, the Nyquist plot of $P(j\omega)C(j\omega)$ goes to 0 as $\omega \to \infty$ and so every point in $\mathscr{S}_\gamma$ achieves the same H_∞ norm.

Time Response Considerations
So far we have discussed stability and robustness. However, the design of a controller should include consideration of the time response. To illustrate this, we chose the following three design points:

$$C_1(s) = 40 + \frac{245}{s} + 9s \quad \text{and} \quad C_2(s) = 188 + \frac{1601}{s} + 9s. \tag{11.264}$$

The first point is an arbitrary point from the boundary of $\mathbf{S}_\gamma$ and the second point has the maximum k_i value in $\mathscr{S}_\gamma$. The Nyquist plots in Figure 11.56 confirm that both design points satisfy the robustness condition.

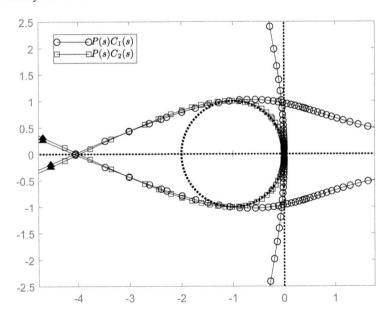

Figure 11.56 Nyquist diagram for $P(s)C_1(s)$ and $P(s)C_2(s)$ (Example 11.10)

The step responses in Figure 11.57 shows that both controller designs result in different time responses in terms of overshoot and settling time.

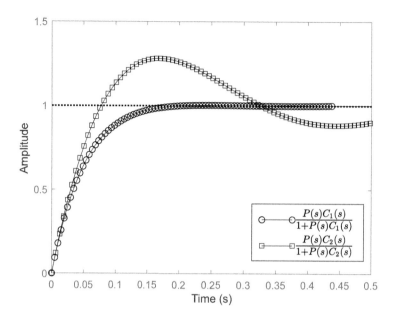

Figure 11.57 Step responses for the closed-loop systems of $P(s)C_1(s)$ and $P(s)C_2(s)$ (Example 11.10)

While $C_2(s)$ has the highest integral gain, $C_1(s)$ provides much shorter settling time and lower overshoot than $C_2(s)$.

The integrator in the controller provided zero steady-state error and we found all stabilizing controllers achieving the prescribed H_∞ norm of the error transfer function. While the robustness and zero steady-state error could be achieved by the proposed method, one should also consider the quality of the transient response when tuning the PID parameters within the set $\mathscr{S}_\gamma$. It is difficult to have analytical approaches to time response design. However, settling time can be handled by shifting the imaginary axis to the left by σ and repeating the methods of this chapter with respect to the shifted Hurwitz region. It is also known empirically that overshoot can, roughly speaking, be reduced by obtaining good phase margin and small H_∞ norm for the error transfer function. In general though transient response-based design remains an active and open area of research.

11.8   Exercises

11.1   Carry out a Ziegler–Nichols step response design for the plant

$$\frac{K}{1+sT} \cdot e^{-sL} \tag{11.265}$$

where $K = 1, T = 1, L = 1$. Find the gain and phase margins of the system.

11.2   Repeat Exercise 11.1 with $K = 1$ and
(a) $T = 1$ and $L \in [1, 10]$,
(b) $L = 1$ and $T \in [1, 10]$.
In each case, determine the gain and phase margins and their variations with respect to T and L.

11.3   Prove that any strictly proper first-order plant with transfer function

$$P(s) = \frac{K}{1 + sT} \qquad (11.266)$$

can be stabilized by the PI controller

$$C(s) = K_p + \frac{K_i}{s}. \qquad (11.267)$$

(a)   Find the stabilizing sets $\mathscr{S}^+$ and $\mathscr{S}^-$ for $T < 0$ (unstable plant) and $T > 0$ (stable plant), respectively, in (K_p, K_i) space, and show that

$$\mathscr{S}^+ \cap \mathscr{S}^- = \emptyset \quad \text{and} \quad \mathscr{S}^+ \cup \mathscr{S}^- = \mathbb{R}^2. \qquad (11.268)$$

(b)   Determine the subsets of $\mathscr{S}^+$ and $\mathscr{S}^-$ for which the closed-loop characteristic roots are (i) real, (ii) complex.
(c)   Show that the steady-state error to a ramp input can be made arbitrarily small.

11.4   Consider the second-order plant

$$P(s) = \frac{K(s - z)}{s^2 + a_1 s + a_0} = \frac{K(s - z)}{(s - p_1)(s - p_2)}, \qquad \text{for } K, z > 0 \qquad (11.269)$$

with the feedback controller $C(s) = K_p$.
(a)   Prove that stabilization by constant gain is possible if and only if

$$-a_1 < \frac{a_0}{z} \qquad (11.270)$$

and if Equation (11.270) is true the stabilizing set is given by

$$K_p \in \left(-\frac{a_1}{K}, \frac{a_0}{Kz} \right). \qquad (11.271)$$

(b)   Prove that the necessary and sufficient condition for constant gain stabilizability is

$$p_1 + p_2 < \frac{p_1 p_2}{z}. \qquad (11.272)$$

(c)   Show that the steady-state error to a unit step is at least

$$\frac{p_1 p_2}{p_1 p_2 - (p_1 + p_2)z}. \qquad (11.273)$$

11.5   Consider the unstable first-order plant with time delay

$$P(s) = \left[\frac{K}{1 - sT} \right] e^{-sL}, \qquad \text{for } K, T, L > 0 \qquad (11.274)$$

with the constant gain feedback controller

$$C(s) = K_p. \tag{11.275}$$

Use the first-order Padé approximation

$$e^{-sL} = \frac{1 - s\frac{L}{2}}{1 + s\frac{L}{2}} \tag{11.276}$$

and
(a)   prove that stabilization is possible if and only if $L < T$;
(b)   find the stabilizing range for K_p;
(c)   prove that the step response has undershoot;
(d)   prove that the minimum steady-state error to a unit step input is

$$\frac{1}{2}\left(\frac{L}{T - L}\right). \tag{11.277}$$

11.6   Prove that the PID controller

$$C(s) = \frac{K_i + K_p s + K_d s^2}{s(1 + K_0 s)} \tag{11.278}$$

can stabilize any strictly proper second-order plant.

11.7   Consider the PID controller

$$C(s) = \frac{K_i + K_p s + K_d s^2}{s} \tag{11.279}$$

and the plant with transfer function

$$P(s) = \frac{b_1 s + b_0}{s^2 + a_1 s + a_0}. \tag{11.280}$$

Use the Routh–Hurwitz criterion to obtain an explicit description of the stabilizing set $\mathscr{S}$. Prove that the subsets of $\mathscr{S}$ with constant K_p are described by linear inequalities.

11.8   Determine the PID stabilizing sets for the following plants using the controller

$$\frac{K_p s + K_i + K_d s^2}{s}. \tag{11.281}$$

(a) $P(s) = \dfrac{1}{s^2 + 3s + 4}$.

(b) $P(s) = \dfrac{s - 1}{(s + 1)^2}$.

(c) $P(s) = \dfrac{s - 1}{(s - 2)(s + 3)}$.

(d) $P(s) = \dfrac{s - 1}{s^4 + 3s^2 + 2s + 1}$.

(e) $P(s) = \dfrac{s - 1}{s^4 + 3s^2 - 2s + 1}$.

In each case, determine if stabilization is possible, the range of admissible k_p, and some typical stabilizing sets in k_i, k_d space.

11.9 Repeat Exercise 11.8 with the controller

$$\frac{K_p s + K_i + K_d s^2}{s(1 + sT)} \tag{11.282}$$

where $T = 0.1$.

11.10 Repeat Exercises 11.8 and 11.9 with the following performance specifications:
(a) gain margin of 2;
(b) phase margin of $45°$;
(c) H_∞ norm of the closed-loop transfer function less than or equal to 1.4.

11.11 Consider the PID controller with the transfer function

$$C(s) = \frac{K_i + K_p s + K_d s^2}{s} \tag{11.283}$$

and the plant with transfer function

$$P(s) = \frac{K(s - z)}{s^2 + a_1 s + a_0}. \tag{11.284}$$

Use signature methods to develop explicit conditions describing the stabilizing set $\mathscr{S}$. Consider the cases when $z > 0$ and $z < 0$.

11.12 The settling time T_s of a control system can be defined to be the time required for the step response of the system to reach and remain within 2% of its steady-state value. Show that this specification corresponds to shifting the closed-loop poles to the left of the line

$$s = -\sigma = -\frac{5}{T_s} \tag{11.285}$$

and the PID controllers attaining this specification can be obtained by Hurwitz stabilization of the closed-loop system in the

$$\hat{s} = s + \sigma \tag{11.286}$$

plane. Apply this to the plant

$$P(s) = \frac{s - 15}{s^2 + s - 1} \tag{11.287}$$

to determine all PID controllers attaining a settling time of 5 s or less.

11.13 (Extension of the parameter separation property) Consider the controller parametrized as

$$C(s) = \frac{N_e(s^2, x_1) + s N_o(s^2, x_2)}{D_e(s^2, x_3) + s D_o(s^2, x_4)} \tag{11.288}$$

connected to an LTI plant, and assume that the parameter vectors x_i, $i = 1, 2, 3, 4$ appear linearly in the polynomial coefficients. Prove that the stabilizing set can be described by linear inequalities in x_i for fixed parameters $x_j = \mathbf{x}_j^*$, $j \neq i$ for $i = 1, 2, 3, 4$.

Hint: Multiply the characteristic polynomial by $N(-s)$, $D(-s)$, $sN(-s)$, or $sD(-s)$, as needed to achieve parameter separation.

11.14   Determine the stabilizing set of controllers

$$C(s) = K_p + \frac{K_i}{s} + K_d s \qquad (11.289)$$

for the plants

$$G(s) = \frac{K}{1 + sT} e^{-sL} \qquad (11.290)$$

where
(a) $K = 1$, $T = 1$, $L \in [0, 0.5]$,
(b) $K = 1$, $T = -1$, $L \in [0, 0.5]$.

11.15   Repeat the previous problem with $K \in [0.5, 1.5]$ and $T \in [0.5, 1.5]$ in part (a) and $T \in [-1.5, -0.5]$ in part (b).

11.16   Consider the following plants with $L \in [0, 1]$:
(a) $P(s) = \dfrac{1}{s^2 + 3s + 4} e^{-sL}$,
(b) $P(s) = \dfrac{s - 1}{s^2 + 3s + 4} e^{-sL}$,
(c) $P(s) = \dfrac{s - 1}{(s + 1)^2} e^{-sL}$,
(d) $P(s) = \dfrac{s - 1}{s^4 + 3s^2 + 2s + 1} e^{-sL}$,
(e) $P(s) = \dfrac{s - 1}{s^4 + 3s^2 - 2s + 1} e^{-sL}$.
In each case, find the set of stabilizing controllers of the form

$$C(s) = K_p + \frac{K_i}{s} + K_d s \qquad (11.291)$$

using the methods of this and the last chapter. In particular, find the admissible ranges of k_p. For distinct values in this range, calculate the stability sets in K_i, K_d space.

11.17   For the plants with transfer functions
(a) $P(z) = \dfrac{z}{z^2 - z + 1}$,
(b) $P(z) = \dfrac{z - 2}{z^2 - z + 1}$,
(c) $P(z) = \dfrac{z - 0.2}{2z^3 - z^2 + z - 1}$,
determine the complete PI and PID controller stabilizing sets using the methods of this chapter.

11.18   Solve the above problem with the additional requirement that the gain margin be at least 3 dB.

11.19   To the problem above now add the requirement that the phase margin be at least $25°$ and compute the controller sets achieving the gain margin and phase margin specifications simultaneously.

11.20   To the gain margin and phase margin specifications already stated above add the requirements that (i) the closed-loop system tolerates at least two units of delay and (ii) the unit ramp response has a steady-state error that is less than 10% in magnitude, and recompute the controller sets achieving all specifications.

11.21   The frequency response $P(j\omega)$ of a plant with one RHP zero is given and its Bode and Nyquist plots are shown in Figures 11.58 and 11.59.

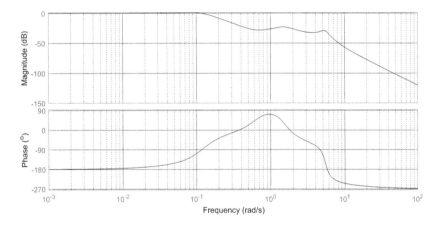

Figure 11.58 Bode plot of the plant to be controlled (Exercise 11.21)

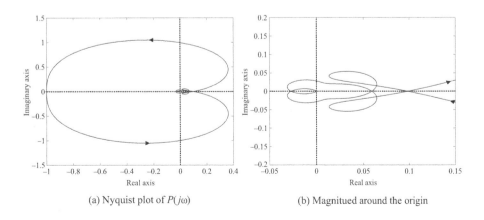

(a) Nyquist plot of $P(j\omega)$ (b) Magnitued around the origin

Figure 11.59 Nyquist plot of the plant $P(j\omega)$ (Exercise 11.21)

11.22   Find the relative degree $n - m$ of the plant.

11.23   Suppose that we want to determine the set of PID controllers with $T = 0$ (ideal PID controllers) that stabilizes the given plant; we need to first determine the required signature for stability. What is the required signature of $\bar{F}(s)$ for stability?

11.24   For the stability region in PID parameter space to exist, what is the minimum number of solutions that must satisfy

$$g(\omega) = K_p \tag{11.292}$$

for each fixed K_p?

11.25 $g(\omega)$ is given in Figure 11.60. What is the admissible range(s) of K_p values? Also determine

$$j = \operatorname{sgn}\left[\bar{F}_i(\infty^-, K_p)\right]. \tag{11.293}$$

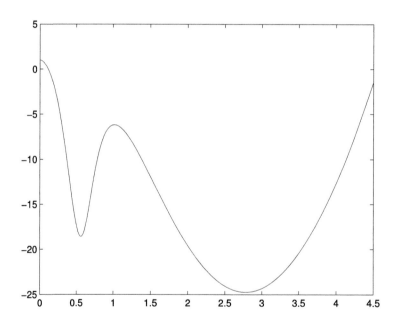

Figure 11.60 Graph of the function $g(\omega)$ (Exercise 11.25)

11.26   If possible, determine

$$i_l = \operatorname{sgn}\left[\bar{F}_r(\infty^-, K_i, K_d)\right]. \tag{11.294}$$

11.27   Write down all possible strings that satisfy the required signature.

11.28 Fix $K_p = -18$ and determine the frequency range where the frequency response must be accurately measured to determine the entire set of stabilizing PID controllers.

11.29 Using the $P(j\omega)$ data given in Table 11.2, construct the sets of linear inequalities that determine the stability region.

Table 11.2 Frequency response data $P(j\omega)$ (Exercise 11.29)

$\omega = 0:0.1:10$	$\omega = 11:0.1:20$	$\omega = 21:0.1:30$	$\omega = 31:0.1:40$
$-1.0000 + j0.0000$	$0.0147 - j0.0464$	$0.0477 + j0.0126$	$0.0171 + j0.0202$
$-0.3255 + j1.0473$	$0.0192 - j0.0509$	$0.0422 + j0.0157$	$0.0158 + j0.0203$
$0.3179 + j0.2366$	$0.0272 - j0.0540$	$0.0374 + j0.0177$	$0.0146 + j0.0204$
$0.1558 + j0.0338$	$0.0382 - j0.0538$	$0.0333 + j0.0188$	$0.0136 + j0.0205$
$0.0832 - j0.0071$	$0.0503 - j0.0484$	$0.0298 + j0.0195$	$0.0126 + j0.0207$
$0.0489 - j0.0206$	$0.0600 - j0.0376$	$0.0268 + j0.0199$	$0.0117 + j0.0209$
$0.0307 - j0.0274$	$0.0646 - j0.0238$	$0.0243 + j0.0201$	$0.0109 + j0.0212$
$0.0205 - j0.0323$	$0.0638 - j0.0105$	$0.0221 + j0.0202$	$0.0100 + j0.0215$
$0.0152 - j0.0368$	$0.0595 + j0.0001$	$0.0202 + j0.0202$	$0.0092 + j0.0220$
$0.0134 - j0.0415$	$0.0536 + j0.0076$	$0.0185 + j0.0202$	$0.0084 + j0.0225$

11.30 Using the sets of inequalities, graphically determine the stability region for $K_p = -18$. Sweep K_p over the admissible range to determine the entire 3D sets.

11.31 From the stabilizing set found, determine the subset that provides a gain margin $A_m \geq 1.5$ and a phase margin $\theta_m \geq 40°$.

11.9 Notes and References

The characterization of all stabilizing PID controllers for a given delay-free linear time-invariant plant was developed in Ho, Datta, and Bhattacharyya (1997). For systems with delays, the corresponding results were obtained in Silva, Datta, and Bhattacharyya (2004). The stabilizing set for Ziegler–Nichols plants was first calculated in Silva, Datta, and Bhattacharyya (2002). For an extensive description of these results, the reader can refer to Datta, Ho, and Bhattacharyya (2000). Stabilization sets for discrete-time PID controllers were calculated in Keel, Rego, and Bhattacharyya (2003). Tchebyshev polynomials of the first and second kind are treated in Mason and Handscomb (2002).

The frequency response data-based design for PID and first-order controllers shown in Section 11.5 was obtained in Keel and Bhattacharyya (2008). The algorithm for the discrete-time counterpart of this problem was developed in Keel, Mitra, and Bhattacharyya (2008).

The gain and phase margin-based designs discussed in Section 11.6 were reported in Diaz-Rodriguez, Han, and Bhattacharyya (2019). The algorithms for discrete-time systems as well as H_∞ optimal PID controllers are also developed therein.

The results on H_∞ optimal PID control were obtained in Han, Keel, and Bhattacharyya

(2018). A similar approach was described in Tantaris, Keel, and Bhattacharyya (2006) for first-order controllers and in this case the stability region was computed a priori. In Krajewski and Viaro (2012) it was shown that at a fixed frequency and for a fixed k_d, the derivative gain, the L_2 norm of the error transfer function being equal to γ was represented by an ellipse in (k_p, k_i) space. An H_∞ optimal PID design using a frequency loop-shaping approach was reported in Ashfaque and Tsakalis (2012) and Tsakalis and Dash (2013).

Sections 11.3.1, 11.4.1, 11.4.2, and 11.5.1–11.5.3 are adapted from Bhattacharyya, Datta, and Keel (2009) with permission. © 2009 Taylor & Francis Group, LLC. Sections 11.6 and 11.7 are adapted from Diaz-Rodriguez, Han, and Bhattacharyya (2019), with permission. © Springer Nature Switzerland AG 2019. Figures 11.27–11.30, 11.33–11.35 are reproduced from Diaz-Rodriguez and Bhattacharyya (2016) with permission. © 2016 IEEE.

12 Multivariable Control Using Single-Input Single-Output Methods

In this chapter we present some results on the analysis and design of multivariable feedback systems using single-input single-output (SISO) methods. We begin by considering the stability of a multivariable unity feedback system with a rational proper transfer function matrix $\mathbf{G}(s)$ in the forward path. The stability of this system is shown to be equivalent to that of a SISO feedback system with scalar transfer function $g(s)$ in the forward path. The transfer function $g(s)$ consists of the sum of the determinants of all leading principal minors of $\mathbf{G}(s)$. Thus, multi-input multi-output (MIMO) feedback system stability can be determined by applying the Nyquist criterion to the scalar transfer function $g(s)$. This test can even be a measurement-based test constructed from the individual frequency responses $g_{ij}(j\omega)$ of the entries of $\mathbf{G}(s)$. Finally, this avoids the necessity of constructing state-space models for $\mathbf{G}(s)$.

Next, a new method of designing multivariable controllers for linear MIMO control systems using the Smith–McMillan form and SISO theory is presented. The Smith–McMillan form of the transfer function matrix of a MIMO plant is an equivalent *diagonal* transfer function matrix using which the problem of multivariable controller design can be reduced to multiple independent SISO controller designs. This opens up a significant alternative to the state-feedback observer-based approaches and enables the application of many powerful and effective, including low order, SISO design methods to the MIMO problem. If the designed SISO controllers satisfy certain relative degree conditions, then the corresponding multivariable controller, to be connected to the MIMO plant, will be proper. We show how such multivariable servomechanisms can be designed to satisfy closed-loop stability, reference tracking, disturbance rejection, and robust stability specifications. Examples are included for illustration.

12.1 A Scalar Equivalent of a MIMO System

Consider the multivariable (MIMO) feedback system shown in Figure 12.1, where $\mathbf{G}(s)$ is a rational, proper, $r \times r$ transfer function matrix.

In this section, we introduce a scalar "equivalent" of the above system as shown in Figure 12.2.

In what follows, we make precise the sense in which the two systems are equivalent. Let $\delta_G(s)$ denote the characteristic polynomial of $\mathbf{G}(s)$, that is, $\delta_G(s)$ is the least common multiple (lcm) of the denominators of all minors of $\mathbf{G}(s)$ after all cancellations.

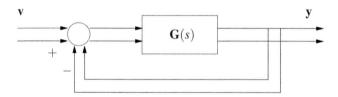

Figure 12.1 A multivariable feedback system

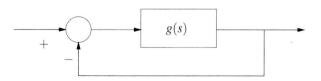

Figure 12.2 An equivalent scalar feedback system

Then the characteristic polynomial $\alpha(s)$ of the closed-loop system is

$$\alpha(s) = \delta_G(s)\det[\mathbf{I} + \mathbf{G}(s)]. \tag{12.1}$$

Next we have a key lemma.

LEMMA 12.1

$$\det[\mathbf{I} + \mathbf{G}(s)] = 1 + g(s), \tag{12.2}$$

where

$$g(s) = m_1(s) + m_2(s) + \cdots + m_r(s) \tag{12.3}$$

and

$$m_i(s) = \sum_i (\text{all } i \times i \text{ leading principal minors}). \tag{12.4}$$

Proof The proof of the lemma is similar to that of Lemma 3.1 given in Appendix A.2. ∎

EXAMPLE 12.1 If

$$\mathbf{G}(s) = \begin{bmatrix} g_{11}(s) & g_{12}(s) & g_{13}(s) \\ g_{21}(s) & g_{22}(s) & g_{23}(s) \\ g_{31}(s) & g_{32}(s) & g_{33}(s) \end{bmatrix}, \tag{12.5}$$

we have

$$g(s) = m_1(s) + m_2(s) + m_3(s),$$
$$m_1(s) = g_{11}(s) + g_{22}(s) + g_{33}(s), \tag{12.6}$$
$$m_2(s) = \det \begin{bmatrix} g_{11}(s) & g_{12}(s) \\ g_{21}(s) & g_{22}(s) \end{bmatrix} + \det \begin{bmatrix} g_{11}(s) & g_{13}(s) \\ g_{31}(s) & g_{33}(s) \end{bmatrix} + \det \begin{bmatrix} g_{22}(s) & g_{23}(s) \\ g_{32}(s) & g_{33}(s) \end{bmatrix},$$
$$m_3(s) = \det[\mathbf{G}(s)].$$

EXAMPLE 12.2   For

$$\mathbf{G}(s) = \begin{bmatrix} g_1(s) & 0 & 0 \\ 0 & g_2(s) & 0 \\ 0 & 0 & g_3(s) \end{bmatrix}, \tag{12.7}$$

$$\det[\mathbf{I} + \mathbf{G}(s)] = 1 + g_1(s) + g_2(s) + g_3(s)$$
$$+ g_1(s)g_2(s) + g_1(s)g_3(s) + g_2(s)g_3(s) + g_1(s)g_2(s)g_3(s). \tag{12.8}$$

To proceed, write

$$g(s) = \frac{v_g(s)}{\delta_g(s)}, \tag{12.9}$$

where $v_g(s)$ and $\delta_g(s)$ are coprime.

LEMMA 12.2

$$\delta_g(s) \mid \delta_G(s). \tag{12.10}$$

Proof $\delta_g(s)$ is the lcm of the denominators of the sums of all leading principal minors, whereas $\delta_G(s)$ is the lcm of the denominators of all minors of $\mathbf{G}(s)$.   ∎

From Lemma 12.2, it follows that

$$\delta_G(s) = \delta_g(s)\delta_f(s) \tag{12.11}$$

for some polynomial $\delta_f(s)$ with

$$\text{degree}\left[\delta_f(s)\right] = \text{degree}\left[\delta_G(s)\right] - \text{degree}\left[\delta_g(s)\right]. \tag{12.12}$$

THEOREM 12.1   The closed-loop characteristic roots of the MIMO system in Figure 12.1 are identical to the closed-loop characteristic roots of the SISO system in Figure 12.2, along with the zeros, if any, of $\delta_f(s)$.

Proof From Equation (12.1), Lemmas 12.1 and 12.2:

$$\alpha(s) = \delta_G(s)(1+g(s)) \tag{12.13a}$$

$$= \delta_G(s)\left(1+\frac{v_g(s)}{\delta_g(s)}\right) \tag{12.13b}$$

$$= \delta_f(s)\left(\delta_g(s)+v_g(s)\right). \tag{12.13c}$$

Therefore

$$\text{zeros of } \alpha(s) = (\text{zeros of } (1+g(s)))\cup(\text{zeros of } \delta_f(s)), \tag{12.14}$$

since $(v_g(s), \delta_g(s))$ are coprime. ∎

The following examples demonstrate the usefulness of these results for stability verification.

12.1.1   Illustrative Examples

EXAMPLE 12.3   Let

$$\mathbf{G}(s) = \begin{bmatrix} \dfrac{1}{s-1} & \dfrac{1}{(s-1)^2} \\[2mm] \dfrac{s-1}{s-2} & \dfrac{1}{s-2} \end{bmatrix}. \tag{12.15}$$

Then

$$g(s) = m_1(s) + m_2(s) \tag{12.16}$$

with

$$m_1(s) = \frac{1}{s-1} + \frac{1}{s-2} \quad \text{and} \quad m_2(s) = \det[G(s)] = 0, \tag{12.17}$$

so that

$$g(s) = \frac{(2s-3)}{(s-1)(s-2)} = \frac{v_g(s)}{\delta_g(s)}. \tag{12.18}$$

Also

$$\delta_{\mathbf{G}}(s) = (s-1)^2(s-2) \tag{12.19}$$

and therefore

$$\delta_f(s) = s-1. \tag{12.20}$$

The closed-loop characteristic polynomial is, by Theorem 12.1, predicted to be

$$\delta_{cl}(s) = [(s-1)(s-2)+2s-3](s-1) = \left(s^2-s-1\right)(s-1). \tag{12.21}$$

This result may be verified using a state-space realization of minimal order for $\mathbf{G}(s)$:

$$\dot{\mathbf{x}}(t) = \begin{bmatrix} 1 & 1 & 0 \\ 0 & 1 & 0 \\ 0 & 0 & 2 \end{bmatrix} \mathbf{x}(t) + \begin{bmatrix} 1 & 0 \\ 0 & 1 \\ 1 & 1 \end{bmatrix} \mathbf{u}(t) =: \mathbf{A}\mathbf{x}(t) + \mathbf{B}\mathbf{u}(t),$$

$$\mathbf{y}(t) = \begin{bmatrix} 1 & 0 & 0 \\ 0 & 0 & 1 \end{bmatrix} \mathbf{x}(t) + \begin{bmatrix} 0 & 0 \\ 1 & 0 \end{bmatrix} \mathbf{u}(t) =: \mathbf{C}\mathbf{x}(t) + \mathbf{D}\mathbf{u}(t). \tag{12.22}$$

The closed-loop system matrix

$$\mathbf{A} - \mathbf{B}(\mathbf{I} + \mathbf{D})^{-1}\mathbf{C} = \begin{bmatrix} 0 & 1 & 0 \\ 1 & 1 & -1 \\ 0 & 0 & 1 \end{bmatrix}, \tag{12.23}$$

and it can be seen that its characteristic polynomial coincides with Equation (12.21), verifying the theorem.

Remark 12.1 Generically, the fixed polynomial $\delta_f(s)$ will be equal to 1, that is, there will be no fixed poles if infinitesimally small perturbations are allowed to occur. For example, in Example 12.3, if the (1,1) entry of $\mathbf{G}(s)$ is perturbed to be

$$\frac{1+\varepsilon}{s-1},$$

$\delta_f(s) = 1$. In such cases, stability of a MIMO system is exactly equivalent to that of the equivalent SISO system.

EXAMPLE 12.4 Consider

$$\mathbf{G}(s) = \frac{1}{s^2 + 7s + 6} \begin{bmatrix} s+4 & 1 & s+0.5 \\ s+0.5 & s+3 & 2 \\ 1 & 2s+1 & s+5 \end{bmatrix}. \tag{12.24}$$

Then

$$g(s) = \frac{v_g(s)}{\delta_g(s)} = \frac{s+4}{s^2+7s+6} + \frac{s+3}{s^2+7s+6} + \frac{s+5}{s^2+7s+6}$$

$$+ \det\left(\frac{1}{s^2+7s+6}\begin{bmatrix} s+4 & 1 \\ s+0.5 & s+3 \end{bmatrix}\right)$$

$$+ \det\left(\frac{1}{s^2+7s+6}\begin{bmatrix} s+3 & 2 \\ 2s+1 & s+5 \end{bmatrix}\right)$$

$$+ \det\left(\frac{1}{s^2+7s+6}\begin{bmatrix} s+4 & 1 \\ 2s+1 & s+5 \end{bmatrix}\right) \tag{12.25}$$

$$+ \det\left(\frac{1}{s^2+7s+6}\begin{bmatrix} s+4 & 1 & s+0.5 \\ s+0.5 & s+3 & 2 \\ 1 & 2s+1 & s+5 \end{bmatrix}\right)$$

$$= \frac{3s^5 + 57s^4 + 393s^3 + 1181s^2 + 1554s + 746.3}{s^6 + 21s^5 + 165s^4 + 595s^3 + 990s^2 + 756s + 216}.$$

Since the degrees of $\delta_G(s)$ and $\delta_g(s)$ are the same, there are no fixed poles. Thus, the poles of $\mathbf{G}(s)$ (roots of $\delta_G(s)$ and also those of $\delta_g(s)$) are $\{-6, -1, -6, -1, -6, -1\}$. Then the closed-loop poles found from the zeros of $1 + g(s)$ are $\{-7.6737, -6.0149 \pm j1.5817, -1.5406, -1.3779 \pm j0.4533\}$ and thus the closed-loop system is stable. This is also verified from the Nyquist plot of $g(s)$, as shown in Figure 12.3.

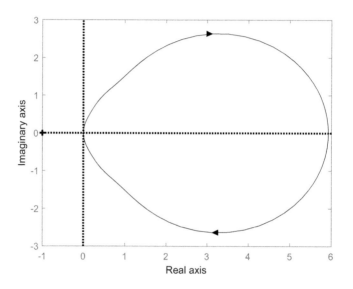

Figure 12.3 Nyquist plot of $g(s)$ (Example 12.4) © 2015 IEEE. Reproduced from Keel and Bhattacharyya (2015) with permission.

EXAMPLE 12.5 Consider

$$\mathbf{G}(s) = \frac{1}{s^2 + 5s - 6} \begin{bmatrix} s+4 & 1 & s+0.5 \\ s+0.5 & s+3 & 2 \\ s+2 & 2s-1 & s+5 \end{bmatrix}. \tag{12.26}$$

Then

$$g(s) = \frac{s+4}{s^2+7s-6} + \frac{s+3}{s^2+7s-6} + \frac{s+5}{s^2+7s-6}$$
$$+ \det\left(\frac{1}{s^2+7s-6}\begin{bmatrix} s+4 & 1 \\ s+0.5 & s+3 \end{bmatrix}\right)$$
$$+ \det\left(\frac{1}{s^2+7s-6}\begin{bmatrix} s+3 & 2 \\ 2s-1 & s+5 \end{bmatrix}\right)$$
$$+ \det\left(\frac{1}{s^2+7s-6}\begin{bmatrix} s+4 & 1 \\ 2s-1 & s+5 \end{bmatrix}\right)$$

$$+ \det \left(\frac{1}{s^2 + 7s - 6} \begin{bmatrix} s+4 & 1 & s+0.5 \\ s+0.5 & s+3 & 2 \\ s+2 & 2s-1 & s+5 \end{bmatrix} \right) = \frac{v_g(s)}{\delta_g(s)}, \tag{12.27}$$

where

$$v_g(s) = 3s^5 + 44s^4 + 187.5s^3 + 96.5s^2 - 453s + 213.3,$$
$$\delta_g(s) = s^6 + 15s^5 + 57s^4 - 55s^3 - 342s^2 + 540s - 216. \tag{12.28}$$

Since $\delta_G(s)$ and $\delta_g(s)$ have the same degree, there are no fixed poles. Thus, the poles of $G(s)$ (roots of $\delta_G(s)$ and also $\delta_g(s)$) are $\{-6, 1, -6, 1, -6, 1\}$. The closed-loop poles are, from the zeros of $1 + g(s)$, $\{-7.4801, -5.8108 \pm j1.0048, 0.0350, 0.5334 \pm j0.1325\}$ and it shows that the numbers of RHP open-loop and closed-loop poles are the same, namely 3. This is also verified from the Nyquist plot of $g(s)$, as shown in Figure 12.4.

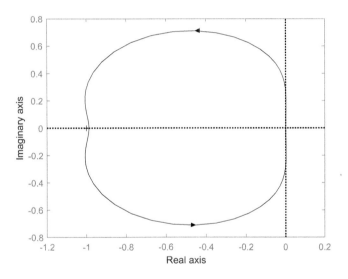

Figure 12.4 Nyquist plot of $g(s)$ (Example 12.5) © 2015 IEEE. Reproduced from Keel and Bhattacharyya (2015) with permission.

12.1.2   Stability Margin Calculations

The above result can easily be extended to calculate the gain margin, phase margin, and time-delay margin of the closed-loop system. Assume that the closed-loop system in Figure 12.1 (equivalently, Figure 12.2) is stable. Let

$$\Delta := \begin{bmatrix} \Delta & & \\ & \ddots & \\ & & \Delta \end{bmatrix} \tag{12.29}$$

with compatible dimension and consider

$$\det[I + \underline{\Delta}\mathbf{G}(s)] = 1 + g_\Delta(s) \tag{12.30}$$

where

$$g_\Delta(s) = \sum_{i=1}^{r} m_i^\Delta(s) \tag{12.31}$$

and $m_i^\Delta(s)$ is the sum of the determinants of all $i \times i$ leading principal minors of $\underline{\Delta}\mathbf{G}(s)$. The elements of $\underline{\Delta}$ may be replaced by $\Delta = k$ for gain margin, $\Delta = e^{j\theta}$ for phase margin, and $\Delta = e^{-sT}$ for time-delay margin calculations.

EXAMPLE 12.6 Consider the two-input two-output (TITO) plant $\mathbf{G}(s)$ with transfer function

$$\mathbf{G}(s) = \begin{bmatrix} g_{11}(s) & g_{12}(s) \\ g_{21}(s) & g_{22}(s) \end{bmatrix} = \frac{1}{1.25(s+1)(s+2)} \begin{bmatrix} s-1 & s \\ -6 & s-2 \end{bmatrix}. \tag{12.32}$$

Then we have

$$g(s) = g_{11}(s) + g_{22}(s) + \det \begin{bmatrix} g_{11}(s) & g_{12}(s) \\ g_{21}(s) & g_{22}(s) \end{bmatrix} = \frac{1.6s - 1.76}{s^2 + 3s + 2}. \tag{12.33}$$

The Nyquist plot of $g(s)$ is shown in Figure 12.5 and verifies that the MIMO closed-loop system is stable.

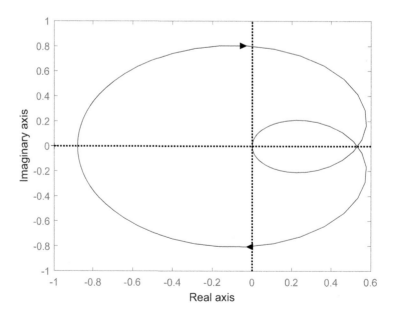

Figure 12.5 Nyquist plot of $g(s)$ (Example 12.6)

Gain Margin Calculation

Let $\Delta = k$, then

$$g_\Delta(s) = kg_{11}(s) + kg_{22}(s) + \det \begin{bmatrix} kg_{11}(s) & kg_{12}(s) \\ kg_{21}(s) & kg_{22}(s) \end{bmatrix}. \tag{12.34}$$

The ranges of the stabilizing gain calculated are:

$$-1.89 < k < 0, \quad 0 < k < 1.25, \quad 2.5 < k < \infty. \tag{12.35}$$

Nyquist plots of $g_\Delta(s)$ for $k = -1.89$, $k = 1.25$, and $k = 2.5$ are given in Figure 12.6.

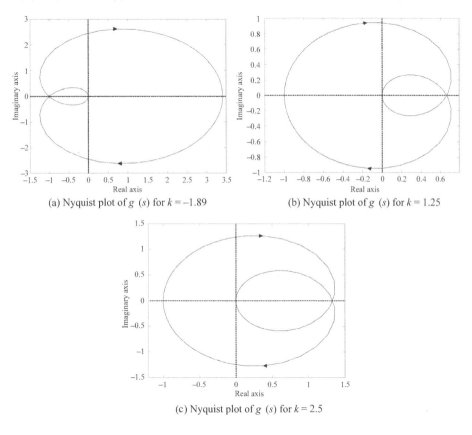

(a) Nyquist plot of g (s) for $k = -1.89$

(b) Nyquist plot of g (s) for $k = 1.25$

(c) Nyquist plot of g (s) for $k = 2.5$

Figure 12.6 Gain margin problem (Example 12.6)

Phase Margin Calculation

Let $\Delta = e^{j\theta}$, then

$$g_\Delta(s) = e^{j\theta} g_{11}(s) + e^{j\theta} g_{22}(s) + \det \begin{bmatrix} e^{j\theta} g_{11}(s) & e^{j\theta} g_{12}(s) \\ e^{j\theta} g_{21}(s) & e^{j\theta} g_{22}(s) \end{bmatrix}. \tag{12.36}$$

The phase margin is calculated to be

$$0° < \theta < 40°. \tag{12.37}$$

The Nyquist plot of $g_\Delta(s)$ for $\theta = 40°$ is given in Figure 12.7.

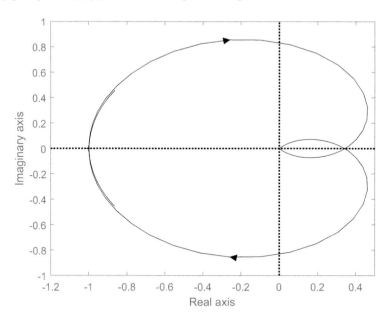

Figure 12.7 Nyquist plot of $g_\Delta(s)$ for $\theta = 40°$ (phase margin problem: Example 12.6)

Time-Delay Margin Calculation

Let $\Delta = e^{-sT}$, then

$$g_\Delta(s) = e^{-sT} g_{11}(s) + e^{-sT} g_{22}(s) + \det \begin{bmatrix} e^{-sT} g_{11}(s) & e^{-sT} g_{12}(s) \\ e^{-sT} g_{21}(s) & e^{-sT} g_{22}(s) \end{bmatrix}. \tag{12.38}$$

The time-delay margin calculated is

$$0 < T < 3.2. \tag{12.39}$$

12.1.3   Discussion

The main result, Theorem 12.1, shows that the characteristic roots of the multivariable feedback system of Figure 12.1 can be completely characterized by the scalar transfer function $g(s)$ constructed as the sum of the determinants of the leading principal minors of $\mathbf{G}(s)$. These characteristic roots coincide with those of the scalar system with unity feedback around $g(s)$, along possibly with some fixed roots which are those characteristic roots of $\mathbf{G}(s)$ counting multiplicities, which are not poles of $g(s)$. As our derivation clearly shows, these fixed roots occur in addition to the uncontrollable and unobservable modes of the open-loop system which, as is well known, continue to be poles of the closed-loop system matrix. Fixed poles will not occur if and only if $g(s)$ has all the poles of $\mathbf{G}(s)$, counting multiplicities. This translates to the condition that fixed poles do not occur if and only if the lcm of the denominator of the sum of the determinants of all leading principal minors (after cancellations) has the same degree as the lcm of the denominators of the determinants of all minors of $\mathbf{G}(s)$. It is easy to see that fixed poles will occur only in degenerate cases, that is, for $\mathbf{G}(s)$ having special fragile structure that is destroyed under vanishingly small perturbations. Since $g(s)$ completely characterizes the closed-loop stability of the multivariable system, its poles and zeros may be defined as the scalar-equivalent poles and zeros of the multivariable system.

The results given here can be applied to the feedback system in Figure 12.8 by defining $\mathbf{G}(s) = F(s)P(s)C(s)$. The scalar equivalent $g(s)$ in this case will be multilinear functions of the entries of $F(s)$, $P(s)$, and $C(s)$ (which follows from the property of determinants), and this fact could be important for controller synthesis and robustness analysis using, say, the Mapping Theorem (see Notes and References). The definition of meaningful stability margins for the multivariable system using the scalar feedback loop of Figure 12.2, as well as the development of MIMO controller synthesis using SISO synthesis methods, are promising areas of research.

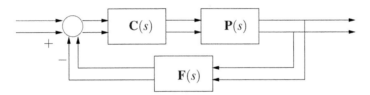

Figure 12.8 A multivariable feedback system with controller and sensor

Finally, it is worth pointing out that the scalar Nyquist criterion can be applied to $g(j\omega)$ for stability analysis of the multivariable system as shown in the examples. If $\mathbf{G}(s)$ is stable, $g(j\omega)$ can be constructed from the measured frequency responses of the individual entries of $\mathbf{G}(s)$. These simplifications are important for practical applications as well as research. Since multivariable stability analysis may now be carried out directly without the computational hurdles of constructing minimal realizations, coprime factorizations or characteristic loci, it can also simplify the teaching of linear systems.

In the next sections we show how SISO design methods can be applied to design multivariable controllers.

12.2   SISO Design Methods for MIMO Systems

Controller design for multivariable (MIMO) systems is challenging and difficult because of the coupling between inputs and outputs. Indeed, if a plant were decoupled, multivariable controller design would reduce to the design of a number of SISO systems. This would allow the powerful methods of classical control, including low-order controller designs, to be applied to the MIMO design problem. This was the motivation for the intense research into the decoupling problem in the 1970s and 1980s. However, decoupling controller designs is not always possible and moreover requires exact knowledge of plant parameters.

In this section, we exploit the Smith–McMillan diagonal form of the plant transfer function matrix, which always exists and is unique, to develop a SISO approach to MIMO controller design. We begin with the following fundamental result.

THEOREM 12.2   Let $\mathbf{P}(s)$ be a proper rational $n \times n$ transfer function matrix. Then there exist unimodular polynomial matrices $\mathbf{U}(s)$, $\mathbf{Y}(s)$ such that

$$\mathbf{P}(s) = \mathbf{Y}^{-1}(s)\mathbf{P}_d(s)\mathbf{U}^{-1}(s), \tag{12.40}$$

where

$$\mathbf{P}_d(s) = \begin{bmatrix} \frac{\varepsilon_1(s)}{\phi_1(s)} & 0 & \cdots & 0 \\ 0 & \frac{\varepsilon_2(s)}{\phi_2(s)} & & \vdots \\ \vdots & & \ddots & 0 \\ 0 & \cdots & 0 & \frac{\varepsilon_n(s)}{\phi_n(s)} \end{bmatrix} = \begin{bmatrix} P_1(s) & & 0 \\ & \ddots & \\ 0 & & P_n(s) \end{bmatrix}, \tag{12.41}$$

$\varepsilon_i(s), \phi_i(s)$ for $i = 1, 2, \ldots, n$ are monic, and

$$\varepsilon_i(s) \mid \varepsilon_{i+1}(s), \quad \text{for } i = 1, 2, \ldots, n-1, \tag{12.42a}$$
$$\phi_{j+1}(s) \mid \phi_j(s), \quad \text{for } j = 1, 2, \ldots, n-1. \tag{12.42b}$$

Proof   The proof is given in Appendix A.9.   ∎

Remark 12.2   If the rank r of $\mathbf{P}(s)$ is less than n, the last $n - r$ entries of $\mathbf{P}_d(s)$ are zero.

Now consider the control system in Figure 12.9.

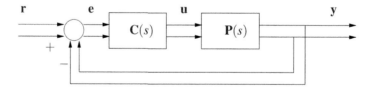

Figure 12.9 A MIMO closed-loop system with a multivariable controller

Using the representation in Equations (12.40) and (12.41) for $\mathbf{P}(s)$, we *propose* the controller structure

$$\mathbf{C}(s) = \mathbf{U}(s)\mathbf{C}_d(s)\mathbf{Y}(s) \tag{12.43}$$

where $\mathbf{C}_d(s)$ is diagonal:

$$\mathbf{C}_d(s) = \begin{bmatrix} C_1(s) & & 0 \\ & \ddots & \\ 0 & & C_n(s) \end{bmatrix}. \tag{12.44}$$

Then

$$\det[\mathbf{I}+\mathbf{P}(s)\mathbf{C}(s)] = \det\left[\mathbf{Y}^{-1}(s)\right]\det\left[\mathbf{I}+\mathbf{P}_d(s)\mathbf{C}_d(s)\right]\det[\mathbf{Y}(s)] \tag{12.45a}$$
$$= \det\left[\mathbf{I}+\mathbf{P}_d(s)\mathbf{C}_d(s)\right] \tag{12.45b}$$
$$= \prod_{i=1}^{n}(1+P_i(s)C_i(s)). \tag{12.45c}$$

Equation (12.45) shows that the characteristic equation of the MIMO system in Figure 12.9 is identical to the product of the characteristic equations of the n scalar loops shown in Figure 12.10.

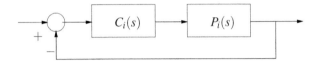

Figure 12.10 The ith SISO loop

The controller $\mathbf{C}(s)$ in Equation (12.43) must be proper in order to be realizable without differentiation. This can be ensured by choosing $C_i(s)$ to have sufficient relative degree t_i (denominator degree – numerator degree). Let r_i (s_j) denote the highest degrees among the terms in the ith row $u_i^T(s)$ of $\mathbf{U}(s)$ (jth column $y_j(s)$ of $\mathbf{Y}(s)$).

LEMMA 12.3 $\mathbf{C}(s)$ will be proper if

$$t_i \geq r_i + s_j, \quad \text{for } i, j \in \mathbf{n}. \tag{12.46}$$

Proof The conclusion follows immediately from the expression for an arbitrary element of $C(s)$:

$$C_{ij}(s) = C_i(s)u_i^T(s)y_j(s) \quad \text{for } i \in \mathbf{n}, j \in \mathbf{n}. \tag{12.47}$$

∎

Next we deal with the issue of asymptotic tracking. Consider the MIMO control system with the diagonal plant and diagonal controller shown in Figure 12.11.

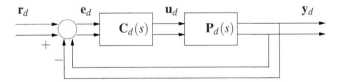

Figure 12.11 Closed-loop system with the diagonal plant and controller

Let us define

$$\mathbf{r}_d(s) := \mathbf{Y}(s)\mathbf{r}(s) \tag{12.48}$$

and

$$\mathbf{y}_d(s) := \mathbf{Y}(s)\mathbf{y}(s). \tag{12.49}$$

From the closed-loop system in Figure 12.11, we have

$$\mathbf{e}_d(s) = \mathbf{r}_d(s) - \mathbf{y}_d(s), \tag{12.50}$$

or

$$\begin{aligned}\mathbf{e}_d(s) &= \mathbf{Y}(s)\mathbf{r}(s) - \mathbf{Y}(s)\mathbf{y}(s) = \mathbf{Y}(s)(\mathbf{r}(s) - \mathbf{y}(s)) \\ &= \mathbf{Y}(s)\mathbf{e}(s).\end{aligned} \tag{12.51}$$

The inputs and outputs of $\mathbf{C}_d$ are related through

$$\mathbf{u}_d(s) = \mathbf{C}_d(s)\mathbf{e}_d(s), \tag{12.52}$$

which can be rewritten as

$$\mathbf{U}^{-1}(s)\mathbf{u}(s) = \mathbf{C}_d(s)\mathbf{Y}(s)\mathbf{e}(s). \tag{12.53}$$

The following lemmas lead us towards the design of a diagonal controller $\mathbf{C}_d$ that guarantees stability, asymptotic tracking, and properness with the corresponding multivariable controller $\mathbf{C}$.

LEMMA 12.4   The multivariable controller $\mathbf{C}(s)$ stabilizes the MIMO plant $\mathbf{P}(s)$ if and only if the designed diagonal controller $\mathbf{C}_d(s)$ stabilizes the equivalent diagonal plant $\mathbf{P}_d(s)$.

Proof   We have already proved this in the previous comments. Indeed, the characteristic polynomial of the closed-loop system attained by connecting the designed diagonal controller $\mathbf{C}_d(s)$ to the equivalent diagonal plant $\mathbf{P}_d(s)$ (see Figure 12.11) is the numerator of $\delta_d(s)$:

$$\delta_d(s) := \det\left[\mathbf{I} + \mathbf{P}_d(s)\mathbf{C}_d(s)\right], \tag{12.54}$$

or using the forms in Equations (12.41) and (12.44):

$$\delta_d(s) = \prod_{k=1}^{n}\det\left(1 + \frac{\varepsilon_k(s)}{\phi_k(s)}C_k(s)\right), \tag{12.55}$$

because $\mathbf{I} + \mathbf{P}_d(s)\mathbf{C}_d(s)$ is diagonal. Using Equations (12.40) and (12.43), we can write

$$\begin{aligned}\delta_d(s) &= \det\left[\mathbf{I} + \mathbf{P}_d(s)\mathbf{C}_d(s)\right] \\ &= \det[\mathbf{I} + \mathbf{Y}(s)\mathbf{P}(s)\underbrace{\mathbf{U}(s)\mathbf{U}^{-1}(s)}_{I}\mathbf{C}(s)\mathbf{Y}^{-1}(s)] \\ &= \det\left[\mathbf{I} + \mathbf{Y}(s)\mathbf{P}(s)\mathbf{C}(s)\mathbf{Y}^{-1}(s)\right] \\ &= \det\left[\mathbf{Y}(s)\mathbf{Y}^{-1}(s) + \mathbf{Y}(s)\mathbf{P}(s)\mathbf{C}(s)\mathbf{Y}^{-1}(s)\right] \\ &= \det\left[\mathbf{Y}(s)(\mathbf{I} + \mathbf{P}(s)\mathbf{C}(s))\mathbf{Y}^{-1}(s)\right] \\ &= \det[\mathbf{I} + \mathbf{P}(s)\mathbf{C}(s)] =: \delta(s),\end{aligned} \tag{12.56}$$

where the numerator of $\delta(s)$ is the characteristic polynomial of the original closed-loop system in Figure 12.9.   ∎

Next we have a useful technical result.

LEMMA 12.5 If $\mathbf{Y}(s)$ is a unimodular polynomial matrix, then $\mathbf{Y}(0)$ is a full-rank matrix.

Proof Let us write $\mathbf{Y}(s)$ as

$$\mathbf{Y}(s) = \mathbf{Y}_0 + \mathbf{Y}_1 s + \cdots, \tag{12.57}$$

and its inverse $\mathbf{Y}^{-1}(s)$, which exists and is a unimodular polynomial matrix as well, in the following form:

$$\mathbf{Y}^{-1}(s) = \mathbf{L}_0 + \mathbf{L}_1 s + \cdots. \tag{12.58}$$

We now have

$$
\begin{aligned}
\mathbf{I} &= \mathbf{Y}(s)\mathbf{Y}^{-1}(s) \\
&= (\mathbf{Y}_0 + \mathbf{Y}_1 s + \cdots)(\mathbf{L}_0 + \mathbf{L}_1 s + \cdots) \\
&= \mathbf{Y}_0\mathbf{L}_0 + (\mathbf{Y}_0\mathbf{L}_1 + \mathbf{Y}_1\mathbf{L}_0)s + \cdots.
\end{aligned}
\tag{12.59}
$$

Equation (12.59) is valid if and only if

$$
\begin{aligned}
\mathbf{Y}_0\mathbf{L}_0 &= \mathbf{I}, \\
\mathbf{Y}_0\mathbf{L}_1 + \mathbf{Y}_1\mathbf{L}_0 &= 0, \\
&\;\;\vdots
\end{aligned}
\tag{12.60}
$$

Therefore, the first condition in Equation (12.60), $\mathbf{Y}_0\mathbf{L}_0 = \mathbf{I}$, proves that $\mathbf{Y}(0) = \mathbf{Y}_0$ is a full-rank matrix. ∎

This leads to the following conclusion regarding asymptotic tracking.

LEMMA 12.6 The output signal $\mathbf{y}(t)$ asymptotically tracks the reference signal $\mathbf{r}(t)$ in the original system (see Figure 12.9) if and only if the output $\mathbf{y}_d(t)$ asymptotically tracks the reference $\mathbf{r}_d(t)$ (see Figure 12.11).

Proof If $\mathbf{y}_d(t)$ tracks $\mathbf{r}_d(t)$, then

$$\lim_{t\to\infty} \mathbf{e}_d(t) = 0, \tag{12.61}$$

or by the final value theorem, we have

$$\lim_{s\to 0} s\mathbf{e}_d(s) = 0. \tag{12.62}$$

Substituting $\mathbf{e}_d(s)$ with Equation (12.51) gives

$$\lim_{s\to 0} s\mathbf{Y}(s)\mathbf{e}(s) = 0. \tag{12.63}$$

According to Lemma 12.5, $\mathbf{Y}(0)$ is a full-rank matrix; thus, Equation (12.63) is valid if and only if

$$\lim_{s\to 0} s\mathbf{e}(s) = 0, \tag{12.64}$$

or

$$\lim_{t \to \infty} \mathbf{e}(t) = 0 \tag{12.65}$$

for the original closed-loop system, implying that the output $\mathbf{y}(t)$ tracks the reference $\mathbf{r}(t)$. The reverse direction of the proof follows similarly. ∎

Remark 12.3 For $y_{d_k}(t)$ to track $r_{d_k}(t)$, the controller $C_k(s)$ should be designed as

$$C_k(s) = \frac{\alpha_k(s)}{\beta_k(s)m(s)}, \tag{12.66}$$

where the characteristic polynomial

$$\phi_k(s)\beta_k(s)m(s) + \varepsilon_k(s)\alpha_k(s) \tag{12.67}$$

is Hurwitz, $m(s)$ has zeros at the poles of $r_{d_k}(s)$, and the product $\beta_k(s)m(s)$ has no RHP cancellations with $\varepsilon_k(s)$.

Based on these results, we can present in the next section our approach to the design of multivariable servomechanisms using SISO theory.

12.3   Multivariable Servomechanism Design Using SISO Methods

In this section we develop an approach to the design of multivariable servomechanisms based on SISO theory using the Smith–McMillan representation of the plant and the results of the last section. Consider the MIMO servomechanism depicted in Figure 12.12.

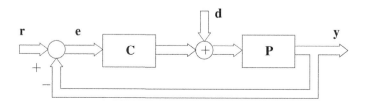

Figure 12.12 Multivariable servomechanism

First, consider the case of a square plant with the same number n of inputs and outputs. Letting $\mathbf{P}(s)$ and $\mathbf{C}(s)$ denote $n \times n$ transfer function matrices, we have

$$\mathbf{e}(s) = \underbrace{[\mathbf{I} + \mathbf{P}(s)\mathbf{C}(s)]^{-1}}_{\mathbf{E}_r(s)} \mathbf{r}(s) - \underbrace{[\mathbf{I} + \mathbf{P}(s)\mathbf{C}(s)]^{-1}\mathbf{P}(s)}_{\mathbf{E}_d(s)} \mathbf{d}(s) \tag{12.68}$$

$$= \mathbf{E}_r(s)\mathbf{r}(s) - \mathbf{E}_d(s)\mathbf{d}(s). \tag{12.69}$$

Now let $r_i(t), d_j(t)$ denote the components of the vector reference and disturbance signals and suppose they satisfy the common differential equation

$$\begin{aligned} m(D)r_i(t) &= 0, \quad \text{for } i \in \mathbf{n}, \\ m(D)d_j(t) &= 0, \quad \text{for } j \in \mathbf{n}, \end{aligned} \tag{12.70}$$

where $m(D)$ is a polynomial in the differential operator $D := \frac{d}{dt}$. It is convenient to also define the diagonal matrices

$$\mathbf{M}(D) := \begin{bmatrix} m(D) & & & \\ & m(D) & & \\ & & \ddots & \\ & & & m(D) \end{bmatrix} \tag{12.71}$$

and

$$\mathbf{M}(s) := \begin{bmatrix} m(s) & & & \\ & m(s) & & \\ & & \ddots & \\ & & & m(s) \end{bmatrix}. \tag{12.72}$$

We note that the Laplace transforms of $\mathbf{r}(t), \mathbf{d}(t)$ are of the form

$$\mathbf{r}(s) = \begin{bmatrix} \dfrac{r_1^o(s)}{m(s)} \\ \vdots \\ \dfrac{r_n^o(s)}{m(s)} \end{bmatrix}, \quad \mathbf{d}(s) = \begin{bmatrix} \dfrac{d_1^o(s)}{m(s)} \\ \vdots \\ \dfrac{d_n^o(s)}{m(s)} \end{bmatrix}, \tag{12.73}$$

where $r_i^o(s), d_j^o(s)$ are polynomials depending on the initial conditions applied to Equation (12.70). We assume that references and disturbances are persistent or unstable signals so that $m(s)$ is anti-Hurwitz, that is, the zeros of $m(s)$ are in $\mathbb{C}^+$. Now bring in the Smith–McMillan form $\mathbf{P}_d(s)$ of $\mathbf{P}(s)$. Assuming $\text{rank}[\mathbf{P}(s)] = n$, we have

$$\mathbf{P}_d(s) = \begin{bmatrix} \dfrac{\varepsilon_1(s)}{\phi_1(s)} & & \\ & \ddots & \\ & & \dfrac{\varepsilon_n(s)}{\phi_n(s)} \end{bmatrix}, \tag{12.74}$$

with

$$\mathbf{P}(s) = \mathbf{Y}(s)^{-1}\mathbf{P}_d(s)\mathbf{U}(s)^{-1} \tag{12.75}$$

for unimodular matrices $\mathbf{Y}(s), \mathbf{U}(s)$.

Consider the diagonal controller

$$\mathbf{C}_d(s) = \begin{bmatrix} C_1(s) & & \\ & \ddots & \\ & & C_n(s) \end{bmatrix}, \tag{12.76}$$

where

$$C_i(s) = \frac{\alpha_i(s)}{m(s)\beta_i(s)}, \quad \text{for } i \in \mathbf{n} \tag{12.77}$$

is designed to stabilize the SISO Smith–McMillan plants

$$P_i(s) = \frac{\varepsilon_i(s)}{\phi_i(s)}, \quad \text{for } i \in \mathbf{n} \tag{12.78}$$

in unity feedback SISO loops, which we call the Smith–McMillan loops. Let

$$\mathbf{C}(s) := \mathbf{U}(s)\mathbf{C}_d(s)\mathbf{Y}(s). \tag{12.79}$$

Then we have

$$[\mathbf{I}+\mathbf{P}(s)\mathbf{C}(s)]^{-1} = \mathbf{Y}(s)^{-1}[\mathbf{I}+\mathbf{P}_d(s)\mathbf{C}_d(s)]^{-1}\mathbf{Y}(s), \tag{12.80}$$

$$[\mathbf{I}+\mathbf{P}(s)\mathbf{C}(s)]^{-1}\mathbf{P}(s) = \mathbf{Y}(s)^{-1}[\mathbf{I}+\mathbf{P}_d(s)\mathbf{C}_d(s)]^{-1}\mathbf{P}_d(s)\mathbf{U}(s)^{-1}, \tag{12.81}$$

and

$$\mathbf{I}+\mathbf{P}_d(s)\mathbf{C}_d(s) = \begin{bmatrix} 1+\dfrac{\alpha_1(s)\varepsilon_1(s)}{m(s)\beta_1(s)\phi_1(s)} & & \\ & \ddots & \\ & & 1+\dfrac{\alpha_n(s)\varepsilon_n(s)}{m(s)\beta_n(s)\phi_n(s)} \end{bmatrix}. \tag{12.82}$$

Then let

$$\delta_i(s) := m(s)\beta_i(s)\phi_i(s) + \alpha_i(s)\varepsilon_i(s), \quad \text{for } i \in \mathbf{n} \tag{12.83}$$

denote the characteristic polynomials of the individual Smith–McMillan unity feedback SISO loops with $P_i(s), C_i(s)$ for $i \in \mathbf{n}$ in the forward paths. We have

$$[\mathbf{I}+\mathbf{P}_d(s)\mathbf{C}_d(s)]^{-1} = \text{diag}\left[\frac{m(s)\beta_i(s)\phi(s)}{\delta_i(s)}\right]$$

$$= \underbrace{\begin{bmatrix} \dfrac{\beta_1(s)\phi_1(s)}{\delta_1(s)} & & \\ & \ddots & \\ & & \dfrac{\beta_n(s)\phi_n(s)}{\delta_n(s)} \end{bmatrix}}_{\mathbf{H}(s)} \mathbf{M}(s). \tag{12.84}$$

Inserting Equation (12.84) into Equations (12.80) and (12.81), we have

$$\mathbf{E}_r(s) = [\mathbf{I}+\mathbf{P}(s)\mathbf{C}(s)]^{-1} = \mathbf{Y}(s)^{-1}\mathbf{H}(s)\mathbf{M}(s)\mathbf{Y}(s), \tag{12.85}$$

$$\mathbf{E}_d(s) = [\mathbf{I}+\mathbf{P}(s)\mathbf{C}(s)]^{-1}\mathbf{P}(s) = \mathbf{Y}(s)^{-1}\mathbf{H}(s)\mathbf{M}(s)\mathbf{P}_d(s)\mathbf{U}(s)^{-1}. \tag{12.86}$$

Since $\mathbf{M}(s) = m(s)\mathbf{I}_n$, we have

$$\mathbf{E}_r(s) = \mathbf{Y}(s)^{-1}\mathbf{H}(s)\mathbf{Y}(s)\mathbf{M}(s), \tag{12.87}$$

$$\mathbf{E}_d(s) = \mathbf{Y}(s)^{-1}\mathbf{H}(s)\mathbf{P}_d(s)\mathbf{U}(s)^{-1}\mathbf{M}(s). \tag{12.88}$$

Then

$$\begin{aligned} \mathbf{e}_r(s) &= \mathbf{Y}(s)^{-1}\mathbf{H}(s)\mathbf{Y}(s)\mathbf{M}(s)\mathbf{r}(s) \\ &= \mathbf{Y}(s)^{-1}\mathbf{H}(s)\mathbf{Y}(s)\mathbf{r}_0(s). \end{aligned} \tag{12.89}$$

Since $\mathbf{H}(s)$ is stable and $\mathbf{Y}(s), \mathbf{Y}(s)^{-1}, \mathbf{r}_0(s)$ are polynomials, we have

$$\lim_{t \to \infty} \mathbf{e}_r(t) = 0. \tag{12.90}$$

Similarly

$$\begin{aligned} \mathbf{e}_d(s) &= \mathbf{Y}(s)^{-1}\mathbf{H}(s)\mathbf{P}_d(s)\mathbf{U}(s)^{-1}\mathbf{M}(s)\mathbf{d}(s) \\ &= \mathbf{Y}(s)^{-1}\mathbf{H}(s)\mathbf{P}_d(s)\mathbf{U}(s)^{-1}\mathbf{d}_0(s) \end{aligned} \tag{12.91}$$

and since $\mathbf{H}(s)\mathbf{P}_d(s)$ is stable, $\mathbf{Y}(s)^{-1}, \mathbf{U}(s)^{-1}$, and $\mathbf{d}_0(s)$ are polynomial, we have

$$\lim_{t \to \infty} \mathbf{e}_d(t) = 0. \tag{12.92}$$

The closed-loop characteristic equation is

$$\begin{aligned} \det[\mathbf{I} + \mathbf{P}(s)\mathbf{C}(s)] &= \det\left[\mathbf{V}(s)^{-1}\right] \det\left[\mathbf{I} + \mathbf{P}_d(s)\mathbf{C}_d(s)\right] \det[\mathbf{Y}(s)] \\ &= \prod_{i=1}^{n} \left(1 + \frac{\alpha_i(s)\varepsilon_i(s)}{m(s)\beta_i(s)\phi_i(s)}\right) \\ &= \prod_{i=1}^{n} \frac{\delta_i(s)}{m(s)\beta_i(s)\phi_i(s)} \end{aligned} \tag{12.93}$$

and has zeros which are the roots of

$$\delta(s) = \prod_{i=1}^{n} \delta_i(s). \tag{12.94}$$

The polynomials $\delta_i(s)$ can be rendered stable by choice of $\alpha_i(s), \beta_i(s)$ if and only if $(m(s), \varepsilon_i(s))$ are coprime $i \in \mathbf{n}$. The controller designed above assigns $m(s)$ as the numerator of each element of the error transfer function matrices $\mathbf{E}_r(s)$ and $\mathbf{E}_d(s)$. Thus the poles of the external signals r and d are the common zeros of each error transfer function. Such zeros are called blocking zeros and were introduced in 1978. Under perturbations of the plant, these zeros remain as "hard" zeros of the error transfer function and thus the asymptotic tracking and disturbance rejection properties hold robustly as long as the perturbations do not destabilize the closed-loop system.

12.3.1 Nonsquare Plants

When the number of plant outputs, say m, differs from the number n of plant inputs, we have to consider two cases.

1. $n > m$

   The Smith–McMillan form of $\mathbf{P}(s)$ is now of the form

   $$\left[\bar{\mathbf{P}}_d(s) \mid 0\right] \tag{12.95}$$

   with $\bar{\mathbf{P}}_d(s)$ $m \times m$ and diagonal. We can then proceed as before by designing diagonal $\bar{\mathbf{C}}_d(s)$ for $\bar{\mathbf{P}}_d(s)$ and considering the controller

   $$\mathbf{C}_d(s) = \left[\begin{array}{c} \bar{\mathbf{C}}_d(s) \\ \hline 0 \end{array}\right]. \tag{12.96}$$

The corresponding controller $\mathbf{C}(s)$ for $\mathbf{P}(s)$ obtained as in the previous section provides tracking, disturbance rejection, and closed-loop stability.

2. $n < m$

In this case, the Smith–McMillan form of $\mathbf{P}(s)$ is of the form

$$\begin{bmatrix} \bar{\mathbf{P}}_d(s) \\ 0 \end{bmatrix} \tag{12.97}$$

where $\bar{\mathbf{P}}_d(s)$ is $n \times n$ and diagonal. With

$$\bar{\mathbf{C}}_d(s) = \begin{bmatrix} \bar{\mathbf{C}}_{d_1}(s) & \bar{\mathbf{C}}_{d_2}(s) \end{bmatrix} \tag{12.98}$$

we have

$$\mathbf{I}_m + \begin{bmatrix} \bar{\mathbf{P}}_d(s) \\ 0 \end{bmatrix} \begin{bmatrix} \mathbf{C}_{d_1}(s) & \mathbf{C}_{d_2}(s) \end{bmatrix} = \left[\begin{array}{c|c} \mathbf{I}_r + \bar{\mathbf{P}}_d(s)\mathbf{C}_{d_1}(s) & \bar{\mathbf{P}}_d(s)\mathbf{C}_{d_2}(s) \\ \hline 0 & \mathbf{I}_{m-r} \end{array} \right]. \tag{12.99}$$

Hence, we have

$$[\mathbf{I} + \mathbf{P}_d(s)\mathbf{C}_d(s)]^{-1} = \left[\begin{array}{c|c} [\mathbf{I}_r + \bar{\mathbf{P}}_d(s)\mathbf{C}_{d_1}(s)]^{-1} & * \\ \hline 0 & \mathbf{I}_{m-r} \end{array} \right] \tag{12.100}$$

and therefore it is impossible to assign $m(s)$ as the numerator of $m - n$ of the diagonal blocks as required. Therefore the servomechanism problem, as formulated here, has no solution.

We can now summarize our conclusions in the following theorem.

THEOREM 12.3   Given the rational proper transfer function matrix $\mathbf{P}(s)$, of a linear MIMO plant, and a diagonal controller $\mathbf{C}_d(s)$ designed such that it stabilizes the equivalent Smith–McMillan form of $\mathbf{P}(s)$, denoted by $\mathbf{P}_d(s)$, the output $\mathbf{y}_d(t)$ tracks the reference $\mathbf{r}_d(t)$, rejects the disturbances $\mathbf{d}_d(t)$, and $\mathbf{C}_d(s)$ satisfies the relative degree conditions given in Lemma 12.3, then the corresponding multivariable controller $\mathbf{C}(s)$ will be proper, will stabilize $\mathbf{P}(s)$, the output $\mathbf{y}(t)$ will asymptotically track the reference $\mathbf{r}(t)$, in the presence of the disturbance $\mathbf{d}(t)$.

Proof   The proof follows immediately from Lemmas 12.3 to 12.6.   ■

12.4   Illustrative Examples

EXAMPLE 12.7 (A TITO stable plant)   Consider the following transfer function matrix:

$$\mathbf{P}(s) = \begin{bmatrix} \dfrac{4}{(s+1)(s+2)} & \dfrac{-1}{s+1} \\[3mm] \dfrac{2}{s+1} & \dfrac{-1}{2(s+1)(s+2)} \end{bmatrix} \tag{12.101}$$

representing a TITO stable plant and suppose that a multivariable controller is to be designed so that the closed-loop system is stable and the output signals track unit step reference signals.

The lcm of the denominators of all the elements in $\mathbf{P}(s)$ is

$$d(s) = (s+1)(s+2). \tag{12.102}$$

Thus, we can write $\mathbf{P}(s)$ as

$$\mathbf{P}(s) = \frac{1}{(s+1)(s+2)} \underbrace{\begin{bmatrix} 4 & -s-2 \\ 2s+4 & \dfrac{-1}{2} \end{bmatrix}}_{\mathbf{N}(s)}. \tag{12.103}$$

The Smith form of $\mathbf{N}(s)$ can be obtained by multiplying $\mathbf{N}(s)$ by the following unimodular matrices $\mathbf{Y}(s)$ and $\mathbf{U}(s)$:

$$\mathbf{S}(s) = \underbrace{\begin{bmatrix} \dfrac{1}{4} & 0 \\ -s-2 & 2 \end{bmatrix}}_{\mathbf{Y}(s)} \underbrace{\begin{bmatrix} 4 & -s-2 \\ 2s+4 & \dfrac{-1}{2} \end{bmatrix}}_{\mathbf{N}(s)} \underbrace{\begin{bmatrix} 1 & \dfrac{s+2}{4} \\ 0 & 1 \end{bmatrix}}_{\mathbf{U}(s)}$$

$$= \begin{bmatrix} 1 & 0 \\ 0 & s^2+4s+3 \end{bmatrix}. \tag{12.104}$$

Dividing each element of $\mathbf{S}(s)$, in Equation (12.104), by $d(s)$, in Equation (12.102), and performing all possible cancellations gives the Smith–McMillan form of $\mathbf{P}(s)$ as

$$\mathbf{P}_d(s) = \begin{bmatrix} \dfrac{1}{(s+1)(s+2)} & 0 \\ 0 & \dfrac{s+3}{s+2} \end{bmatrix}. \tag{12.105}$$

Now, two SISO controllers $C_1(s)$ and $C_2(s)$ can be designed for two single loops corresponding to the diagonal elements of $\mathbf{P}_d(s)$, to satisfy closed-loop stability and reference signal tracking. Let us consider the following controller:

$$\mathbf{C}_d(s) = \begin{bmatrix} \dfrac{3}{s(s+4)} & 0 \\ 0 & \dfrac{1}{2s(s+1)} \end{bmatrix}, \tag{12.106}$$

where both relative degrees of the controllers have been chosen to be 2. It is easily verified that this choice renders the multivariable controller $\mathbf{C}(s)$ to be proper. The transfer function matrix $\mathbf{C}(s)$ is

$$\mathbf{C}(s) = \mathbf{U}(s)\mathbf{C}_d(s)\mathbf{Y}(s) = \begin{bmatrix} \dfrac{-(s^3+8s^2+14s+10)}{8s(s+1)(s+4)} & \dfrac{s+2}{4s(s+1)} \\ \dfrac{-(s+2)}{2s(s+1)} & \dfrac{1}{s(s+1)} \end{bmatrix}, \tag{12.107}$$

which is proper. Connecting $\mathbf{C}(s)$ to $\mathbf{P}(s)$ and closing the loop results in the closed-loop system transfer function matrix $\mathbf{H}(s)$:

$$\mathbf{H}(s) = [I + \mathbf{P}(s)\mathbf{C}(s)]^{-1}\mathbf{P}(s)\mathbf{C}(s) \tag{12.108}$$

$$= \begin{bmatrix} \dfrac{3}{s^4 + 7s^3 + 14s^2 + 8s + 3} & 0 \\ \dfrac{-s(s+2)(s^4 + 10s^3 + 29s^2 + 32s + 12)}{2(s^4 + 7s^3 + 14s^2 + 8s + 3)(2s^3 + 6s^2 + 5s + 3)} & \dfrac{s+3}{2s^3 + 6s^2 + 5s + 3} \end{bmatrix}.$$

The response of the closed-loop system to unit steps is plotted in Figure 12.13 and verifies asymptotic tracking of step inputs.

MATLAB Solution

```
n={4 [-1 -2]; [2 4] -0.5};
d=conv([1 1],[1 2]);
p=tf(n,d);
hn={3 0; -conv([1 2 0],[1 10 29 32 12]) [1 3]};
d=[1 7 14 8 3];
hd={d 1; 2*conv(d,[2 6 5 3]) [2 6 5 3]};
sysh=tf(hn,hd);
step(sysh,'-k'), hold on,
step(p,'--k'), grid
hold off
```

EXAMPLE 12.8 (A TITO unstable plant)   Consider the following TITO unstable transfer function matrix:

$$\mathbf{P}(s) = \begin{bmatrix} \dfrac{4}{s-1} & \dfrac{-1}{s+1} \\ \dfrac{2}{s+1} & \dfrac{1}{s-1} \end{bmatrix}. \tag{12.109}$$

Our objective is to design a multivariable controller $\mathbf{C}(s)$ to stabilize the closed-loop system and also make the outputs track unit steps. Here, we have $d(s) = (s+1)(s-1)$. Thus, $\mathbf{P}(s)$ can be written as

$$\mathbf{P}(s) = \frac{1}{(s+1)(s-1)} \underbrace{\begin{bmatrix} 4s+4 & -s+1 \\ 2s-2 & s+1 \end{bmatrix}}_{\mathbf{N}(s)}, \tag{12.110}$$

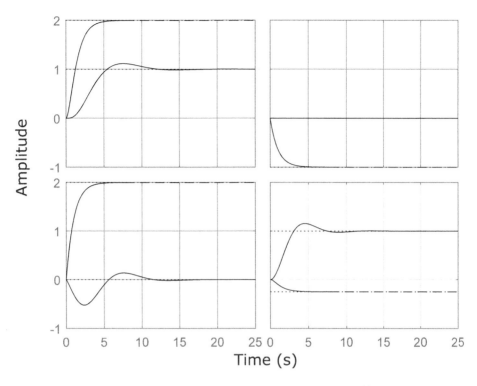

Figure 12.13 Step response of the closed-loop system after connecting $\mathbf{C}(s)$ in Equation (12.107) to $\mathbf{P}(s)$ in Equation (12.101) (solid line), and the step response of the open-loop plant $\mathbf{P}(s)$ in Equation (12.101) (dashed line) (Example 12.7) © 2015 IEEE. Reproduced from Mohsenizadeh, Keel, and Bhattacharyya (2015) with permission.

and the Smith form of $\mathbf{N}(s)$ can be calculated as follows:

$$
\mathbf{S}(s) = \underbrace{\begin{bmatrix} \dfrac{1}{8} & \dfrac{-1}{4} \\[2mm] \dfrac{-s+1}{3} & \dfrac{2s+2}{3} \end{bmatrix}}_{\mathbf{Y}(s)} \underbrace{\begin{bmatrix} 4s+4 & -s+1 \\ 2s-2 & s+1 \end{bmatrix}}_{\mathbf{N}(s)} \underbrace{\begin{bmatrix} 1 & \dfrac{3s+1}{8} \\[2mm] 0 & 1 \end{bmatrix}}_{\mathbf{U}(s)}
$$

$$
= \begin{bmatrix} 1 & 0 \\[2mm] 0 & s^2 + \dfrac{2}{3}s + 1 \end{bmatrix}.
\tag{12.111}
$$

The Smith–McMillan form of $\mathbf{P}(s)$ will be

$$
\mathbf{P}_d(s) = \begin{bmatrix} \dfrac{1}{(s^2-1)} & 0 \\[3mm] 0 & \dfrac{3s^2+2s+3}{3(s^2-1)} \end{bmatrix}.
\tag{12.112}
$$

Now, two SISO stabilizing controllers $C_1(s)$ and $C_2(s)$ are to be designed for the diagonal elements of $\mathbf{P}_d(s)$ in Equation (12.112) that also guarantee step reference tracking. Moreover, the relative degrees of these controllers need to satisfy the inequality conditions given in the statement of Lemma 12.3. Consider the following controller:

$$\mathbf{C}_d(s) = \begin{bmatrix} \dfrac{5s^2+5s+1}{s(0.1s+1)} & 0 \\ 0 & \dfrac{5}{s(0.1s+1)} \end{bmatrix}, \tag{12.113}$$

where the relative degrees of $C_1(s)$ and $C_2(s)$ meet the inequality conditions in Lemma 12.3. The multivariable controller $\mathbf{C}(s)$ becomes

$$\mathbf{C}(s) = \begin{bmatrix} \dfrac{1.05s+0.336}{s(0.1s+1)} & \dfrac{0.415s+0.166}{s(0.1s+1)} \\ \dfrac{-1.67(s-1)}{s(0.1s+1)} & \dfrac{3.33(s+1)}{s(0.1s+1)} \end{bmatrix}, \tag{12.114}$$

which is strictly proper, and the closed-loop system transfer function matrix $\mathbf{H}(s)$ is

$$\mathbf{H}(s) = \begin{bmatrix} \dfrac{h_{11}(s)}{g(s)} & \dfrac{h_{12}(s)}{g(s)} \\ \dfrac{h_{21}(s)}{g(s)} & \dfrac{h_{22}(s)}{g(s)} \end{bmatrix}, \tag{12.115}$$

where

$$\begin{aligned} g(s) &= s^8 + 20s^7 + 198s^6 + 1043s^5 + 3094s^4 + 3703s^3 + 3873s^2 + 2233s + 500, \\ h_{11}(s) &= 58s^6 + 605s^5 + 2688s^4 + 3862s^3 + 4420s^2 + 1533s + 500, \\ h_{12}(s) &= -s\left(17s^5 + 143s^4 - 290s^3 - 543s^2 + 273s + 400\right), \\ h_{21}(s) &= s\left(4s^5 + 28s^4 - 136s^3 + 73s^2 + 132s - 100\right), \\ h_{22}(s) &= 42s^6 + 478s^5 + 3105s^4 + 3988s^3 + 4020s^2 + 2533s + 500. \end{aligned} \tag{12.116}$$

The response of this closed-loop system to unit steps is plotted in Figure 12.14. Since $\mathbf{H}(s)|_{s=0} = \mathbf{I}$, the closed-loop system tracks steps.

In the next section we discuss the stability margins of the multivariable controller and their relation to the margins of the Smith–McMillan loops.

12.5   MIMO Stability Margin Calculations

If a diagonal perturbation matrix Δ with identical elements $\delta(s)$ is inserted into the multivariable feedback loop preceding the controller in Figure 12.12, the characteristic

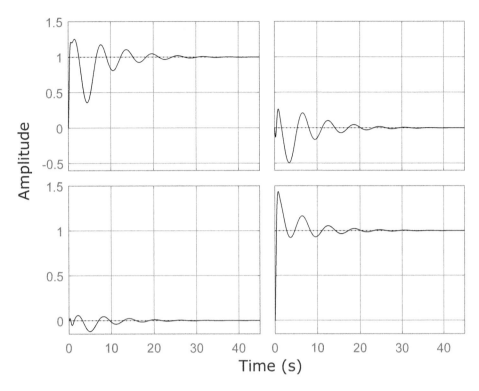

Figure 12.14 Step response of the closed-loop system after connecting $C(s)$ in Equation (12.114) to $P(s)$ in Equation (12.109) (Example 12.8) © 2015 IEEE. Reproduced from Mohsenizadeh, Keel, and Bhattacharyya (2015) with permission.

equation becomes

$$
\det[\mathbf{I} + \Delta \mathbf{P}(s)\mathbf{C}(s)] = \det[\mathbf{I} + \Delta \mathbf{P}_d(s)\mathbf{C}_d(s)]
$$
$$
= \prod_{i=1}^{n} \left(1 + \frac{\delta \alpha_i(s)\varepsilon_i(s)}{m(s)\phi_i(s)\beta_i(s)} \right). \tag{12.117}
$$

By setting $\delta = K$, $\delta = e^{-j\theta}$, or e^{-sT}, respectively, we may obtain the gain, phase, and time-delay margins of the multivariable system. In each case, we see from the above construction, that this margin equals the smallest margins, respectively of the individual Smith–McMillan loops. We can now summarize our conclusions in the following theorem.

Let us define the perturbation matrix Δ as

$$
\Delta := \begin{bmatrix} \delta & & 0 \\ & \ddots & \\ 0 & & \delta \end{bmatrix}. \tag{12.118}
$$

We insert Δ at the usual loop breaking point as shown in Figure 12.15.

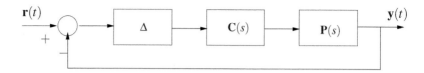

Figure 12.15 The multivariable control system with perturbation Δ

We can define the multivariable stability margins, namely gain and phase margins and time-delay tolerances for the multivariable control system.

THEOREM 12.4   Suppose $\mathbf{C}(s) = \mathbf{U}(s)\mathbf{C}_d(s)\mathbf{Y}(s)$ is a proper controller such that $C_i(s)$ stabilizes the Smith–McMillan plant $P_i(s)$ with gain margin g_i, phase margin ϕ_i, and time-delay tolerance τ_i for the ith SISO loop for each $i \in \mathbf{n}$. Then $\mathbf{C}(s)$ stabilizes $\mathbf{P}(s)$ with gain margin G, phase margin Φ, and time-delay tolerance T, where

$$G = \min_{i=1,2,\dots,n} g_i, \tag{12.119a}$$

$$\Phi = \min_{i=1,2,\dots,n} \phi_i, \tag{12.119b}$$

$$T = \min_{i=1,2,\dots,n} \tau_i. \tag{12.119c}$$

Proof   The proof of this theorem follows from the following observations. Substituting $\delta = k$ in Equation (12.118):

$$\Delta := \begin{bmatrix} k & & 0 \\ & \ddots & \\ 0 & & k \end{bmatrix}. \tag{12.120}$$

By Lemma 12.4, we obtain

$$\det\left[\mathbf{I} + \Delta\mathbf{P}(s)\mathbf{C}(s)\right] = \prod_{i=1}^{n}\left(1 + kP_i(s)C_i(s)\right). \tag{12.121}$$

Substituting $\delta = e^{j\theta}$ in Equation (12.118):

$$\Delta := \begin{bmatrix} e^{j\theta} & & 0 \\ & \ddots & \\ 0 & & e^{j\theta} \end{bmatrix}. \tag{12.122}$$

We obtain, in the same way:

$$\det\left[\mathbf{I} + \Delta\mathbf{P}(s)\mathbf{C}(s)\right] = \prod_{i=1}^{n}\det\left(1 + e^{j\theta}P_i(s)C_i(s)\right). \tag{12.123}$$

Substituting $\delta = e^{-sT}$ in Equation (12.118):

$$\Delta := \begin{bmatrix} e^{-sT} & & 0 \\ & \ddots & \\ 0 & & e^{-sT} \end{bmatrix}. \tag{12.124}$$

We obtain likewise

$$\det\left[\mathbf{I}+\Delta\mathbf{P}(s)\mathbf{C}(s)\right]=\prod_{i=1}^{n}\det\left[1+e^{-sT}P_i(s)C_i(s)\right].\qquad(12.125)$$

Evaluating Equations (12.121), (12.123), and (12.125) at $s=j\omega$, we conclude with Equations (12.119a), (12.119b), and (12.119c), respectively. ∎

Remark 12.4 This theorem shows that the gain margin, phase margin, and time-delay tolerance for the multivariable system as defined are the minimum of the gain margins, the minimum of the phase margins, and the minimum of the time-delay tolerances, respectively, of the n SISO loops in Figure 12.10. The fact that the perturbation model requires identical perturbations in each loop somewhat diminishes the significance of these margins. A true multivariable margin would allow independent perturbations in each loop and even matrix perturbations. This, however, is beyond the scope of presently available theory and is a fruitful area of research.

In the next section we show via an example how a multivariable PI controller can be designed for a multivariable plant using SISO theory and the above results.

12.6 Example: Multivariable PI Controller Design

For the closed-loop system designed to have known gain margin, phase margin, and time-delay tolerance, the following steps are needed.

1. Compute a stabilizing set for each SISO loop.
2. Parametrize constant-gain and constant-phase loci for PI or PID controllers.
3. Construct the achievable gain-phase margin design curves.
4. Select achievable gain margin, phase margin, and gain crossover frequency and retrieve the controllers $C_i(s)$ achieving them.

The methods for the steps above are developed in Chapter 11 with examples. The example below illustrates our approach to design $\mathbf{C}(s)$ with predesigned gain and phase margins and time-delay tolerance specifications for a given multivariable plant $\mathbf{P}(s)$.

EXAMPLE 12.9 Let us consider a TITO system $\mathbf{P}(s)$ in Figure 12.9:

$$\mathbf{P}(s)=\begin{bmatrix}\dfrac{4}{(s+1)(s+2)} & \dfrac{-1}{s+1}\\[3mm]\dfrac{2}{s+1} & -\dfrac{6s+7}{2(s^2+3s+2)}\end{bmatrix}.\qquad(12.126)$$

The objective is to find the controller $\mathbf{C}(s)$ such that it satisfies the predesigned gain margin, phase margin, and gain crossover frequency.

(A) The Smith–McMillan form of $\mathbf{P}(s)$ *and structure of* $\mathbf{C}_d(s)$

The lcm of the denominators in $\mathbf{P}(s)$ is

$$d(s) = (s+1)(s+2). \tag{12.127}$$

$\mathbf{P}(s)$ can be written as

$$\mathbf{P}(s) = \frac{1}{(s+1)(s+2)} \underbrace{\begin{bmatrix} 4 & -1(s+2) \\ 2(s+2) & -(3s+3.5) \end{bmatrix}}_{=:\mathbf{N}(s)}. \tag{12.128}$$

The Smith form of $\mathbf{N}(s)$ is expressed as

$$\mathbf{S}(s) = \underbrace{\begin{bmatrix} \dfrac{1}{4} & 0 \\ -(s+2) & 2 \end{bmatrix}}_{=:Y(s)} \underbrace{\begin{bmatrix} 4 & -1(s+2) \\ 2(s+2) & -(3s+3.5) \end{bmatrix}}_{\mathbf{N}(s)} \underbrace{\begin{bmatrix} 1 & \dfrac{1}{4}(s+2) \\ 0 & 1 \end{bmatrix}}_{=:U(s)} \tag{12.129}$$

$$= \begin{bmatrix} 1 & 0 \\ 0 & s^2 - 2s - 3 \end{bmatrix}.$$

Dividing $\mathbf{S}(s)$ by $d(s)$ we get the Smith–McMillan form

$$\mathbf{P}_d(s) = \begin{bmatrix} \dfrac{1}{(s+1)(s+2)} & 0 \\ 0 & -\dfrac{s-3}{s+2} \end{bmatrix}. \tag{12.130}$$

Let us consider $\mathbf{C}_d(s)$ as

$$\mathbf{C}_d(s) = \begin{bmatrix} \dfrac{k_{p_1} s + k_{i_1}}{s} & 0 \\ 0 & \dfrac{k_{p_2} s + k_{i_2}}{s(s+2)^2} \end{bmatrix}. \tag{12.131}$$

There are two additional poles included in $C_2(s)$ for the controller $C(s)$ to be proper. The location of these poles can be considered as additional design variables since it will affect the achievable performance of the system.

(B) Computation of the stabilizing sets for Smith–McMillan plants

We can find the stabilizing set and achievable performance in terms of gain margin, phase margin, and ω_g for each SISO loop. For the first SISO loop, the admissible range of k_{p_1} for stability was determined to be $k_{p_1} \in (-2, \infty)$. For the second SISO loop, the admissible range of k_{p_2} for stability was determined to be $k_{p_2} \in (-9.2702, 2.6667)$. By sweeping k_{p_1} and k_{p_2} over these intervals, we can generate the stabilizing sets in (k_{p_1}, k_{i_1}) and (k_{p_2}, k_{i_2}) space, respectively. The stabilizing sets are shown in Figures 12.16 and 12.17.

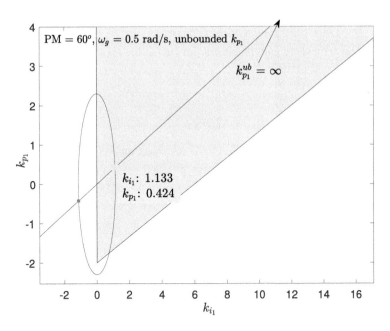

Figure 12.16 Stabilizing set for $P_1(s)$, intersection of an ellipse and a straight line and the PI controller $C_1(s)$ (inside the stabilizing set) achieving $60°$ of phase margin at $\omega_g = 0.5$ rad/s. $k_{p_1}^{ub} = \infty$ indicates that the upper gain margin is infinity (Example 12.9)

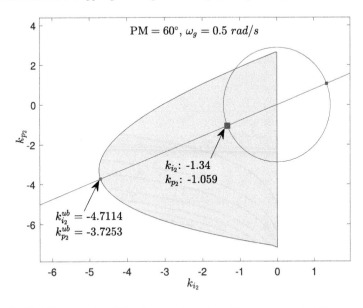

Figure 12.17 Stabilizing set for $P_2(s)$, intersection of an ellipse and a straight line and the PI controller $C_2(s)$ (inside the stabilizing set) achieving $60°$ of phase margin at $\omega_g = 0.5$ rad/s. $(k_{p_2}^{ub} = -3.7253,\ k_{i_2}^{ub} = -4.7114)$ is a boundary point in the stabilizing set for the computation of the upper gain margin (Example 12.9)

(C) Gain-phase margin design curves

The achievable gain-phase margin curves for $P_1(s)$ are generated as shown in Figures 12.18 and 12.19.

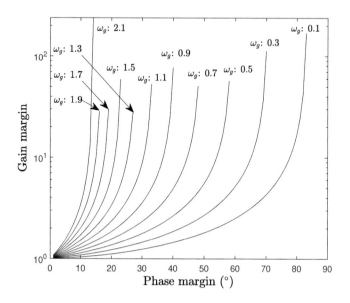

Figure 12.18 Achievable performance in terms of gain margin (GM), phase margin (PM) and gain crossover frequency ω_g for $P_1(s)$ (Example 12.9)

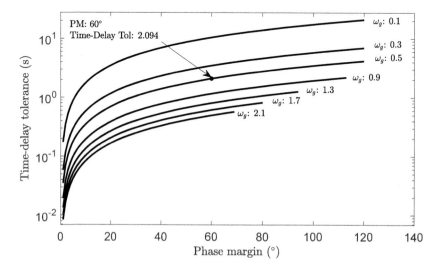

Figure 12.19 Achievable performance in terms of time-delay tolerance τ_{max}, phase margin (PM), and gain crossover frequency ω_g for $P_1(s)$ (Example 12.9)

Similarly, the achievable gain-phase margin curves for $P_2(s)$ are generated as shown in Figures 12.20 and 12.21.

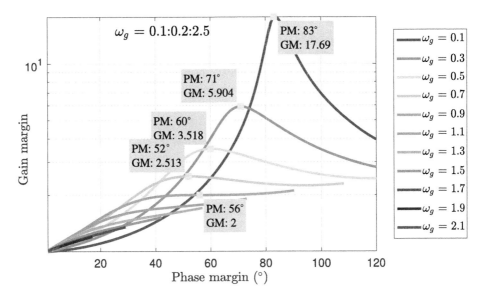

Figure 12.20 Achievable performance in terms of gain margin (GM), phase margin (PM) and gain crossover frequency ω_g for $P_2(s)$ (Example 12.9)

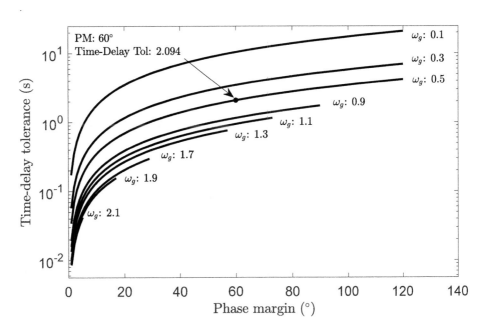

Figure 12.21 Achievable performance in terms of time-delay tolerance τ_{max}, phase margin (PM), and gain crossover frequency ω_g for $P_2(s)$ (Example 12.9)

(D) Retrieval of the PI controller gains

Figure 12.18 shows that the gain margin increases without bound as the phase margin increases. For $P_1(s)$, it can be observed from Figure 12.18 that the achievable phase margin is 83° at $\omega_g = 0.1$ rad/s.

For $P_2(s)$, we see that the achievable gain margin is bounded by 17.69 with the phase margin of 83° at $\omega_g = 0.1$ rad/s as shown in Figure 12.20. The achievable gain margins for different ω_g values are also shown in Figure 12.20. The designer has the liberty to choose different values of gain margin, phase margin, and ω_g that best suits the design needs from the generated curves. In this example, we notice, by Theorem 12.4, that the achievable gain margin for $\mathbf{P}(s)$ is equal to the minimum of the margins. For illustration purposes, we select GM $= \infty$, PM $= 60°$, and $\omega_g = 0.5$ rad/s for $P_1(s)$. For $P_2(s)$, we select GM $= 3.518$, PM $= 60°$, and $\omega_g = 0.5$ rad/s. The time-delay tolerances are $\tau_1 = 2.094$ and $\tau_2 = 2.094$, respectively. After the selection of the margins from the achievable gain-phase margin curves, the designer can retrieve the controller gains corresponding to these points. The controller gains are found to be $k_{p_1}^* = 0.424$ and $k_{i_1}^* = 1.133$ for $P_1(s)$ and $k_{p_2}^* = -1.059$ and $k_{i_2}^* = -1.34$ for $P_2(s)$.

(E) Recovering $\mathbf{C}(s)$ from $\mathbf{C}_d(s)$ and design verification

The final step is to obtain $\mathbf{C}(s)$ from $\mathbf{C}_d(s)$ using the relation

$$\mathbf{C}(s) = \mathbf{U}(s)\mathbf{C}_d(s)\mathbf{Y}(s) \tag{12.132}$$

and Equation (12.129):

$$\mathbf{C}(s) = \begin{bmatrix} \dfrac{0.371s + 0.618}{s} & \dfrac{-0.53s - 0.67}{s(s+2)} \\ \dfrac{1.061s + 1.34}{s(s+2)} & \dfrac{-2.12s - 2.68}{s(s+2)^2} \end{bmatrix}. \tag{12.133}$$

We can verify the results by computing the gain margin and phase margin of the multivariable system. The characteristic polynomial of the multivariable system is given by

$$\det[\mathbf{I} + \Delta\mathbf{P}(s)\mathbf{C}(s)], \tag{12.134}$$

where Δ is defined as in Equation (12.118). For the gain margin, substitute Equation (12.120) into Equation (12.134):

$$\begin{aligned}
\det[\mathbf{I} + \Delta\mathbf{P}(s)\mathbf{C}(s)] =\ & s^7 + 9s^6 + (32 - 0.637k)s^5 + (56 + 2.33k)s^4 \\
& + (48 - 0.4484k^2 + 19.288k)s^3 \\
& + (16 - 0.4216k^2 + 32.708k)s^2 \\
& + (3.7833k^2 + 17.096k)s + 4.5506k^2.
\end{aligned} \tag{12.135}$$

The range of k for the closed-loop stability is determined to be

$$0 < k < 3.518, \tag{12.136}$$

and the roots of Equation (12.135) at $k = 3.518$ are

$$\left\{ \begin{array}{c} 0.00030784192 \pm j1.5532363, \quad -0.38174169 \pm j1.2786136, \\ -1.2283752, \quad -2.2365012, \quad -4.7722559 \end{array} \right\}, \quad (12.137)$$

showing that two of the roots just crossed the imaginary axis. Thus, the gain margin is $k^* = 3.518$ for the MIMO control system and is the minimum value of the gain margins of the Smith–McMillan loops that we selected in the design step for the individual loops.

For the phase margin, substitute Equation (12.122) into Equation (12.134):

$$\det[\mathbf{I} + \Delta\mathbf{P}(s)\mathbf{C}(s)] = s^7 + 9s^6 + (32 - 0.637e^{-j\theta})s^5$$
$$+ (2.33e^{-j\theta} + 56.0)s^4 + (19.29e^{-j\theta} - 0.4484e^{-j2\theta} + 48.0)s^3$$
$$+ (32.7e^{-j\theta} - 0.4218e^{-j2\theta} + 16.0)s^2$$
$$+ (e^{-j\theta} + 3.784e^{-j2\theta} + 17.1)s + 4.551e^{-j2\theta}. \quad (12.138)$$

The range of θ for the closed-loop stability is determined to be

$$0 < \theta < 60° \quad (12.139)$$

and the roots of Equation (12.138) for $\theta = 60°$ are

$$\left\{ \begin{array}{c} -3.5159631 - j0.67153706, \quad -2.0790212 - j0.081079624, \\ -1.2506782 - j0.1211673, \quad -1.2333837 + j1.2928996, \\ -0.92099468 + j0.58077106, \quad 0.000015906289 - j0.49969144, \\ 0.000025099732 - j0.50019525 \end{array} \right\}, \quad (12.140)$$

showing that two of the roots just crossed the imaginary axis. Therefore, the phase margin is $\theta^* = 60°$ for the MIMO control system and is the minimum value of the phase margins that we selected in the design steps.

For the time-delay tolerance, substitute Equation (12.124) into Equation (12.134):

$$\det[\mathbf{I} + \Delta\mathbf{P}(s)\mathbf{C}(s)] = s^7 + 9s^6 + s^5(0.423e^{-Ts} + 32.0)$$
$$+ s^4(4.45e^{-Ts} + 56.0) + s^3(18.23e^{-Ts} + 48.0)$$
$$+ s^2(30.59e^{-Ts} + 0.3299e^{-2Ts} + 16.0)$$
$$+ s(17.1e^{-Ts} + 2.583e^{-2Ts}) + 4.551e^{-2Ts}. \quad (12.141)$$

The range of T for the closed-loop stability is determined to be

$$0 < T < 2.094 \text{ s.} \quad (12.142)$$

The time-delay tolerance is $T = 2.094$ s for the MIMO control system and is the minimum value of the time-delay tolerances that we selected in the design steps.

12.7   Conclusions and Summary

In this chapter we have shown that the MIMO control design task can be accomplished through multiple independent SISO controller designs. The latter designs can be executed on the Smith–McMillan plants. This enables the designer to use the full power of classical control techniques, including low-order designs, developed for SISO systems to the MIMO problem. We also determined conditions on the relative degrees of the designed SISO controllers to guarantee the properness of the resulting multivariable controller as well as stability, disturbance rejection, and reference tracking. We showed that the gain, phase, and time-delay margins of the MIMO system are the respective mimimum margins of the individual Smith–McMillan loops. These results are useful alternatives to the traditional state feedback, observer-based, or quadratic optimization approaches commonly advocated in the control literature and should have a positive impact on applications.

12.8   Exercises

12.1   Consider the plant

$$\mathbf{P}(s) = \begin{bmatrix} \dfrac{1}{s+1} & \dfrac{1}{s-1)} \\ \dfrac{1}{s+2} & \dfrac{s-1}{s+2} \end{bmatrix}. \tag{12.143}$$

Design a servocompensator using LQR theory to make the outputs of the plant track independent step reference signals.

12.2   Suppose

$$\mathbf{P}(s) = \begin{bmatrix} \dfrac{1}{(s+1)^2} & \dfrac{1}{s-1} \\ \dfrac{s-1}{s^2} & \dfrac{1}{(s-2)^2} \end{bmatrix} \tag{12.144}$$

is the open-loop transfer function in Figure 12.1. Find the equivalent scalar transfer function $g(s)$. Are there any fixed poles? Using $g(s)$, determine if the multivariable feedback system is stable. What are the characteristic roots of the multivariable closed-loop system?

12.3   Repeat Exercise 12.1, using the Smith–McMillan form of $\mathbf{P}(s)$ and the SISO design approach developed in this chapter. Find the gain, phase, and time-delay margins of the designed controller.

12.4   Consider the plant

$$\mathbf{P}(s) = \begin{bmatrix} \dfrac{1}{s+1} & \dfrac{1}{s-1} \\ \dfrac{1}{s+2} & \dfrac{s-1}{(s+2)^2} \end{bmatrix}. \tag{12.145}$$

Design a servocompensator using LQR theory to make the output of the plant track independent step reference signals.

12.5 Suppose

$$\mathbf{G}_0(s) = \begin{bmatrix} \dfrac{1}{(s+1)^2} & \dfrac{1}{s-1} \\[3mm] \dfrac{s-1}{s^2} & \dfrac{1}{(s-1)^2} \end{bmatrix}. \tag{12.146}$$

is the open-loop transfer function in Figure 12.1. Find the equivalent scalar transfer function $g(s)$. Are there any fixed poles? Using $g(s)$, determine if the multivariable feedback system is stable. What are the characteristic roots of the multivariable closed-loop system?

12.6 Repeat Exercise 12.4, using the Smith–McMillan form of $\mathbf{P}(s)$ and the SISO design approach developed in this chapter. Find the gain. phase, and time-delay margins of the designed controller.

12.7 Consider the plant transfer function

$$\mathbf{P}(s) = \begin{bmatrix} \dfrac{1}{(s+1)^2} & \dfrac{1}{s-1} \\[3mm] \dfrac{1}{s-1} & \dfrac{1}{(s+1)(s-1)} \end{bmatrix}. \tag{12.147}$$

(a) Find the Smith–McMillan form $\mathbf{P}_d(s)$ of $\mathbf{P}(s)$.

(b) Design a diagonal controller $\mathbf{C}_d(s)$ to track step inputs.

(c) Find the MIMO controller $\mathbf{C}(s)$, corresponding to $\mathbf{C}_d(s)$.

12.8 The state-space model of a plant is

$$\dot{\mathbf{x}}(t) = \begin{bmatrix} 0 & 1 & 0 & 1 \\ 1 & 0 & 1 & 1 \\ 0 & 0 & 0 & 1 \\ 0 & 0 & 0 & 0 \end{bmatrix} \mathbf{x}(t) + \begin{bmatrix} 1 & 0 \\ 0 & 0 \\ 0 & 0 \\ 0 & 1 \end{bmatrix} \mathbf{u}(t),$$

$$\mathbf{y}(t) = \begin{bmatrix} 1 & 0 & 0 & 0 \\ 0 & 0 & 0 & 1 \end{bmatrix} \mathbf{x}(t). \tag{12.148}$$

(a) Find the transfer function matrix $\mathbf{P}(s)$.

(b) Determine the Smith–McMillan form $\mathbf{P}_d(s)$ of $\mathbf{P}(s)$.

(c) Design a diagonal controller $\mathbf{C}_d(s)$ to track step inputs.

(d) Find the MIMO controller $\mathbf{C}(s)$ corresponding to $\mathbf{C}_d(s)$.

(e) Simulate the performance of the closed-loop system and improve the design iteratively.

12.9   Notes and References

The SISO equivalent of a MIMO system was first reported in Keel and Bhattacharyya (2015). The results presented in this chapter utilizing the Smith–McMillan form to enable SISO theory to be used for MIMO systems, were originally presented in Mohsenizadeh, Keel, and Bhattacharyya (2015).

Epilogue

Control theory (CT) came into existence essentially to support and advance control engineering (CE), a field that drives applications. CT and CE form a closed loop, and progress in CT can positively impact CE, have no impact at all, or even have a negative effect. In the recent past, especially post-1960, the field of CT has witnessed several upheavals, twists, and turns, and the emergence of seductively promising new theories that ultimately led to dead ends as far as applicability was concerned. On the other hand, classical control theory, which predates 1960, enjoys a solid foundation of legitimacy because of its wholehearted adoption by the control engineering community and the world of applications. The purpose of this epilogue is to make some sense of this state of affairs and to attempt to place these contrasting scenarios in perspective, from the vantage point of hindsight in 2020. In this way, the reader may obtain a better appreciation of the various topics of this book and their role in control engineering theory for design. We begin with a short historical overview.

Classical Control Theory

Classical control theory (CCT) is based on the Laplace and Fourier transforms in Lathi (2005), Black's feedback amplifier in Black (1934), the Nyquist criterion in Nyquist (1932), and frequency-domain design techniques as developed by Bode (1945) and others. The main goal of CCT was the effective design of simple, that is, low-dynamic-order, feedback controllers that would provide tracking, disturbance rejection, and robust stability based on measured frequency response data. The controllers that were the objects of interest and took center stage were the proportional integral derivative (PID) and lead–lag (LL) controllers, each of which had three adjustable coefficients and were thus sometimes referred to as "three-term controllers." The design techniques developed involved trial and error, as multiple and challenging time and frequency-domain specifications had to be satisfied in real applications.

The Rise and Fall of Optimal Control

Kalman (1960) proposed a radical departure from classical methods. He advocated the use of state-variable models to represent multi-input multi-output or multivariable plants in the time-domain and the design of control systems based on optimality with respect to a time-domain quadratic performance index. His proposal was persuasive as the state-space theory brought the power of linear algebra and matrix theory to bear on the computation of solutions, development of simulations, and analysis of stability. Moreover, he was able to derive the optimal control signal as a state feedback control law that was stabilizing and had excellent universal stability margins, specifically gain and phase margins, that held regardless of the particular performance index chosen by the designer. The only caveat in this rosy scenario was that the optimal controller was not an output feedback controller and thus could not, in general, be directly implemented based on available measurements. However, Kalman's ready response to this criticism was that the Kalman filter could optimally estimate the states from output measurements and that estimated states could replace actual states in the implementation of the optimal state feedback control law without destroying nominal closed-loop stability and with only slight deterioration in the cost.

The publication of Kalman's paper marked the birth of modern control theory, which came to dominate the field of control for the next six decades. The main underpinnings of this theory were state-variable models, state feedback, optimality, state estimation, observer theory, controllability and observability, eigenvalue assignability, and most significantly the removal of constraints on the controller order since, with the advent of computers, "poles and zeros could be sprinkled like salt and pepper" on the complex plane to implement the designed high-order controllers digitally. At the same time, that is in the 1960s and 1970s, motivated by the desire to render single-input single-output classical theory applicable to multivariable systems, a solution to the problem of decoupling multivariable systems was also being developed using a new mathematically elegant approach called "geometric theory," developed by Wonham (1985).

It would take control scientists almost 20 years to grasp the fact that while stability was necessarily preserved in transitioning from state feedback to output feedback, the guaranteed robustness results of optimal control were not. This was pointed out via an example in a paper by Doyle and Stein (1979). It generated renewed interest in and emphasis on robustness or tolerance to uncertainty as a fundamental necessity of control systems and sprouted the term "robust control," which in hindsight appears almost tautological. At the same time it became clear that many algebraic design approaches, such as decoupling, could not work in practice because they involved exact knowledge of plant parameters and exact implementation of controller parameters to zero out off-diagonal transfer functions. Geometric theory did not survive this scrutiny either, because the solvability conditions involving subspace inclusions, central to the theory, broke down under small parameter perturbations as they were based on exact pole-zero cancellations. A promising initiative that resulted in response to these developments was the formulation and solution of an optimal disturbance rejection problem that would provide an output feedback controller capable of robustly maintaining closed-loop stability under H_∞ norm-bounded perturbations of the plant transfer function. This solution derived using state-space methods came to be known as the DGKF solution in Doyle et al. (1989) and was an extension of Kalman's quadratic optimization approach. A transfer function solution was also obtained using a parametrization of all stabilizing controllers, which came to be known as the YJBK parametrization in Youla, Jabr, and Bongiorno (1976). At this point it seemed that almost all hurdles for modern control theory to be applicable to control engineering practice had been removed. The only remaining item of concern was that both DGKF and YJBK design strategies produced extremely high-order controllers, often several times the order of the plant. Certainly this approach could not design the PID or LL controllers preferred by the control industry. The celebrated YJBK parametrization of *all* stabilizing controllers could not even produce a single PID controller stabilizing a given plant. Nevertheless, the "salt and pepper" argument remained in vogue and industry was generally regarded as lagging theory. Indeed, it is not an exaggeration to say that in the 1990s, both the robust control and adaptive control theory communities freely promulgated and were completely committed to producing design theories utilizing unbridled inflation of the controller order.

Fragility: The Curse of Controller Order
This situation changed dramatically when Keel and Bhattacharyya (1997) analyzed the robustness of several high-order optimal controllers designed by various methods proposed in the literature. The surprising conclusion obtained by Keel and Bhattacharrya was that high-order controllers were dysfunctionally fragile and would destabilize the feedback loop under vanishingly small perturbations of the controller coefficients. This was true even if they were robust to plant uncertainty. The analysis presented in Keel and Bhattacharyya (1997) generated much controversy as it challenged the entire culture of modern control theory centered around the design of high-order controllers. However the results were indisputable; the numbers obtained in the examples in Keel and Bhattacharyya (1997) spoke for themselves. Prof. J. B. Pearson stated that this result "was the last nail in the coffin of optimal control."

Modern PID Control
The result on fragility of high-order controllers proved that industry resistance to adoption of the high-order controller designs of MCT was amply justified and indeed that it was theory that lagged applications. This paved the way for renewed interest and research on classical low-order controllers. The main bottleneck to progress was the lack of effective characterization of stabilizing sets of such controllers. These were not convex or even connected. Achieving multiple performance objectives over such sets was a challenging task.

Since the publication of Keel and Bhattacharyya (1997), considerable progress has been made on characterizing stabilizing sets, achievable multi-objective performance, and multivariable controller design using three-term controllers in Diaz-Rodriguez, Han, and Bhattacharyya (2019). These results are reported in the last two chapters of this book. They are applicable to traditional engineering problems of control as well as new technologies such as UAVs, driverless cars, and distributed robotics. In these applications the controller must be frequently updated in an agile and efficient manner as the environment and thus the "plant" changes dynamically. In a lane-changing environment, for example, there is no time to solve matrix Ricatti equations. As Prof. E. J. Davison remarked at the 2017 CCTA conference, it is "back to PIDs" for the field of control and old is new once again.

Appendix

A.1 Cramer's Rule

THEOREM A.1 Consider a system of n linear equations in n unknowns:

$$
\underbrace{\begin{bmatrix} a_{11} & a_{12} & \cdots & a_{1n} \\ a_{21} & a_{22} & \cdots & a_{2n} \\ \vdots & & & \vdots \\ a_{n1} & a_{n2} & \cdots & a_{nn} \end{bmatrix}}_{\mathbf{A}} \underbrace{\begin{bmatrix} x_1 \\ x_2 \\ \vdots \\ x_n \end{bmatrix}}_{\mathbf{x}} = \underbrace{\begin{bmatrix} b_1 \\ b_2 \\ \vdots \\ b_n \end{bmatrix}}_{\mathbf{b}}.
\tag{A.1}
$$

If the equations have a unique solution:

$$
x_j = \frac{\det[\mathbf{A}_j]}{\det[\mathbf{A}]},
\tag{A.2}
$$

where

$$
\mathbf{A}_j = \begin{bmatrix} a_{11} & \cdots & a_{1j-1} & b_1 & a_{1j+1} & \cdots & a_{1n} \\ \vdots & & \vdots & \vdots & \vdots & & \vdots \\ a_{n1} & \cdots & a_{nj-1} & b_n & a_{1j+1} & \cdots & a_{nn} \end{bmatrix}.
\tag{A.3}
$$

Proof Using $\mathbf{x}$ the unique solution of the linear equations, we write

$$
\det[\mathbf{A}_j] = \det \begin{bmatrix} a_{11} & \cdots & a_{1j-1} & \mathbf{a}_1\mathbf{x} & a_{1j+1} & \cdots & a_{1n} \\ \vdots & & \vdots & \vdots & \vdots & & \vdots \\ a_{n1} & \cdots & a_{nj-1} & \mathbf{a}_n\mathbf{x} & a_{1j+1} & \cdots & a_{nn} \end{bmatrix},
\tag{A.4}
$$

where $\mathbf{a}_k$ is the kth row of the matrix $\mathbf{A}$. From the property of the determinant, we write

$$\det[\mathbf{A}_j] = \det \begin{bmatrix} a_{11} & \cdots & a_{1j-1} & a_{11}x_1 & a_{1j+1} & \cdots & a_{1n} \\ \vdots & & \vdots & \vdots & & & \vdots \\ a_{n1} & \cdots & a_{nj-1} & a_{n1}x_1 & a_{nj+1} & \cdots & a_{nn} \end{bmatrix} + \cdots$$

$$+ \det \begin{bmatrix} a_{11} & \cdots & a_{1j-1} & a_{1n}x_n & a_{1j+1} & \cdots & a_{1n} \\ \vdots & & \vdots & \vdots & & & \vdots \\ a_{n1} & \cdots & a_{nj-1} & a_{nn}x_n & a_{nj+1} & \cdots & a_{nn} \end{bmatrix}$$

$$= \sum_{k=1}^{n} x_k \det \begin{bmatrix} a_{11} & \cdots & a_{1j-1} & a_{1k} & a_{1j+1} & \cdots & a_{1n} \\ \vdots & & \vdots & \vdots & & & \vdots \\ a_{n1} & \cdots & a_{nj-1} & a_{nk} & a_{nj+1} & \cdots & a_{nn} \end{bmatrix} \tag{A.5}$$

$$= x_j \det \begin{bmatrix} a_{11} & \cdots & a_{1j-1} & a_{1j} & a_{1j+1} & \cdots & a_{1n} \\ \vdots & & \vdots & \vdots & & & \vdots \\ a_{n1} & \cdots & a_{nj-1} & a_{nj} & a_{nj+1} & \cdots & a_{nn} \end{bmatrix} = x_j \det[\mathbf{A}]$$

and therefore,

$$x_j = \frac{\det[\mathbf{A}_j]}{\det[\mathbf{A}]}. \tag{A.6}$$

∎

A.2   Proof of Lemma 3.1

LEMMA A.1   If $\mathbf{T}$ is an $n \times n$ matrix:

$$\det[\mathbf{I} + \mathbf{T}] = 1 + t, \tag{A.7}$$

where

$$t = m_1(\mathbf{T}) + m_2(\mathbf{T}) + \cdots + m_n(\mathbf{T}). \tag{A.8}$$

To construct the proof, consider $\mathbf{T}$ to be an $n \times n$ matrix with real or complex elements or elements in a ring and let

$$\mathbf{t}_k = [t_{1k} \ \ t_{2k} \ \ t_{3k} \ \ \cdots \ \ t_{nk}]^T \tag{A.9}$$

denote the kth column of $\mathbf{T}$. Also denote a column vector with kth element equal to one and all other elements zero by $\mathbf{e}_k$.

LEMMA A.2   Let

$$\mathbf{T} := [\mathbf{t}_1 \ \ \mathbf{t}_2 \ \ \cdot \ \ \mathbf{t}_k \ \ \cdots \ \ \mathbf{t}_n]. \tag{A.10}$$

Then

$$\det[\mathbf{t}_1 \ \ \cdots \ \ \mathbf{t}_{k-1} \ \ \mathbf{e}_k + \mathbf{t}_k \ \ \mathbf{t}_{k+1} \ \ \cdots \ \ \mathbf{t}_n]$$
$$= \det[\mathbf{t}_1 \ \ \cdots \ \ \mathbf{t}_{k-1} \ \ \mathbf{e}_k \ \ \mathbf{t}_{k+1} \ \ \cdots \ \ \mathbf{t}_n] + \det[\mathbf{T}]. \tag{A.11}$$

Proof Let $\mathbf{T}_{ij}$ be the matrix obtained by removing the ith row and the jth column from $\mathbf{T}$. After row and column exchanges, it is easy to see that

$$\det \begin{bmatrix} \mathbf{t}_1 & \cdots & \mathbf{t}_{k-1} & \mathbf{e}_k + \mathbf{t}_k & \mathbf{t}_{k+1} & \cdots & \mathbf{t}_n \end{bmatrix}$$

$$= \det \begin{bmatrix}
1+t_{kk} & t_{k1} & \cdots & t_{k,k-1} & t_{k,k+1} & \cdots & t_{kn} \\
t_{1k} & t_{11} & & t_{1,k-1} & t_{1,k+1} & & t_{1n} \\
\vdots & & & & & & \vdots \\
t_{k-1,k} & t_{k-1,1} & \cdots & t_{k-1,k-1} & t_{k-1,k+1} & \cdots & t_{k-1,n} \\
t_{k+1,k} & t_{k+1,1} & \cdots & t_{k+1,k-1} & t_{k+1,k+1} & \cdots & t_{k+1,n} \\
\vdots & & & & & & \vdots \\
t_{nk} & t_{n1} & \cdots & t_{n,k-1} & t_{n,k+1} & \cdots & t_{nn}
\end{bmatrix} \qquad (A.12)$$

$$= (1+t_{kk}) \det\left[\mathbf{T}_{kk}\right] + \sum_{j=1,j\neq k}^{n} t_{kj} \det\left[\mathbf{T}_{kj}\right]$$

$$= \underbrace{\det\left[\mathbf{T}_{kk}\right] + \sum_{j=1}^{n} t_{kj} \det\left[\mathbf{T}_{kj}\right]}_{\det[T]}$$

$$= \det\begin{bmatrix} \mathbf{t}_1 & \cdots & \mathbf{t}_{k-1} & \mathbf{e}_k & \mathbf{t}_{k+1} & \cdots & \mathbf{t}_n \end{bmatrix} + \det[\mathbf{T}].$$

∎

Next, we have the following.

LEMMA A.3 Consider the square matrix $\mathbf{T}_r$ that is constructed by replacing columns, say $\mathbf{t}_{r_1}, \mathbf{t}_{r_2}, \ldots, \mathbf{t}_{r_r}$ by corresponding columns in $\mathbf{I}$ (i.e., $\mathbf{e}_{r_1}, \mathbf{e}_{r_2}, \ldots, \mathbf{e}_{r_r}$). Then

$$\det[\mathbf{T}_r] = \det\left[\bar{\mathbf{T}}_r\right], \qquad (A.13)$$

where the matrix $\bar{\mathbf{T}}_r$ is obtained by removing the corresponding columns and rows where 1 appears.

Proof It is easy to see that by row exchanges and column exchanges applied to $\mathbf{T}_r$, we obtain the block upper-triangular matrix of the form

$$\left[\begin{array}{c|c} \bar{\mathbf{T}}_r & \mathbf{0} \\ \hline * & \mathbf{I} \end{array}\right]. \qquad (A.14)$$

∎

LEMMA A.4 If $\mathbf{T}$ is a $n \times n$ matrix, then

$$\det[\mathbf{I} + \mathbf{T}] = 1 + \sum_{j=1}^{n} m_j(\mathbf{T}), \qquad (A.15)$$

where $m_j(\mathbf{T})$ is the sum of the determinants of all $j \times j$ leading principal minors of $\mathbf{T}$, $j \in \mathbf{n}$.

Proof From Lemma A.2, we write

$$\det[\mathbf{I} + \mathbf{T}] = \det[\mathbf{e}_1 + \mathbf{t}_1 \quad \mathbf{e}_2 + \mathbf{t}_2 \quad \cdots \quad \mathbf{e}_n + \mathbf{t}_n]$$
$$= \det[\mathbf{e}_1 \quad \mathbf{e}_2 + \mathbf{t}_2 \quad \cdots \quad \mathbf{e}_n + \mathbf{t}_n] + \det[\mathbf{t}_1 \quad \mathbf{e}_2 + \mathbf{t}_2 \quad \cdots \quad \mathbf{e}_n + \mathbf{t}_n].$$

Note that by Lemma A.2:

$$\det[\mathbf{e}_1 \quad \mathbf{e}_2 + \mathbf{t}_2 \quad \cdots \quad \mathbf{e}_n + \mathbf{t}_n] = \det[\mathbf{e}_1 \quad \mathbf{e}_2 \quad \mathbf{e}_3 + \mathbf{t}_3 \quad \cdots \quad \mathbf{e}_n + \mathbf{t}_n]$$
$$+ \det[\mathbf{e}_1 \quad \mathbf{t}_2 \quad \mathbf{e}_3 + \mathbf{t}_3 \quad \cdots \quad \mathbf{e}_n + \mathbf{t}_n] \tag{A.16}$$

and

$$\det[\mathbf{t}_1 \quad \mathbf{e}_2 + \mathbf{t}_2 \quad \cdots \quad \mathbf{e}_n + \mathbf{t}_n] = \det[\mathbf{t}_1 \quad \mathbf{e}_2 \quad \mathbf{e}_3 + \mathbf{t}_3 \quad \cdots \quad \mathbf{e}_n + \mathbf{t}_n]$$
$$+ \det[\mathbf{t}_1 \quad \mathbf{t}_2 \quad \mathbf{e}_3 + \mathbf{t}_3 \quad \cdots \quad \mathbf{e}_n + \mathbf{t}_n]. \tag{A.17}$$

Clearly, the determinant of $\mathbf{I} + \mathbf{T}$ can be expressed as the sum of determinants of the 2^n matrices obtained by the two sets of vectors $\{\mathbf{e}_1, \ldots, \mathbf{e}_n\}$ and $\{\mathbf{t}_1, \ldots, \mathbf{t}_n\}$ as shown below:

$$\det[\mathbf{I} + \mathbf{T}] = \det[\mathbf{I}] + \det[\mathbf{e}_1 \quad \mathbf{e}_2 \quad \cdots \quad \mathbf{e}_{n-1} \quad \mathbf{t}_n] + \det[\mathbf{e}_1 \quad \mathbf{e}_2 \quad \cdots \quad \mathbf{e}_{n-2} \quad \mathbf{t}_{n-1} \quad \mathbf{e}_n] + \cdots$$
$$+ \det[\mathbf{e}_1 \quad \mathbf{t}_2 \quad \mathbf{e}_3 \quad \cdots \quad \mathbf{e}_n] + \det[\mathbf{t}_1 \quad \mathbf{e}_2 \quad \cdots \quad \mathbf{e}_n]$$
$$+ \det[\mathbf{t}_1 \quad \mathbf{e}_2 \quad \cdots \quad \mathbf{e}_{n-1} \quad \mathbf{t}_n] + \cdots + \det[\mathbf{t}_1 \quad \mathbf{t}_2 \quad \mathbf{e}_3 \quad \cdots \quad \mathbf{e}_n]$$
$$+ \cdots \tag{A.18}$$
$$+ \det[\mathbf{e}_1 \quad \mathbf{t}_2 \quad \cdots \quad \mathbf{t}_n] + \cdots + \det[\mathbf{t}_1 \quad \mathbf{t}_2 \quad \cdots \quad \mathbf{t}_{n-1} \quad \mathbf{e}_n]$$
$$+ \det[\mathbf{t}_1 \quad \mathbf{t}_2 \quad \cdots \quad \mathbf{t}_n].$$

Therefore, we have

$$\det[\mathbf{I} + \mathbf{T}] = 1 + \sum_{j=1}^{n} m_j(\mathbf{T}), \tag{A.19}$$

where $m_j(\mathbf{T})$ is the sum of the determinants of all $j \times j$ principle minors of $\mathbf{T}$. ∎

A.3 Vector and Matrix Norms

DEFINITION A.1 (Vector norms) A vector norm is a real-valued function $\|\cdot\| \in \mathbb{R}$ of a vector $\mathbf{x} \in \mathbb{C}^n$ that satisfies the following properties.

(1) Positivity:

$$\|\mathbf{x}\| \geq 0, \quad \|\mathbf{x}\| = 0 \text{ if and only if } \mathbf{x} = 0. \tag{A.20}$$

(2) Homogeneity:

$$\|a\mathbf{x}\| = |a| \|\mathbf{x}\|, \quad \text{for } a \in \mathbb{C}. \tag{A.21}$$

(3) Triangle inequality:

$$\|\mathbf{x} + \mathbf{b}\| \le \|\mathbf{x}\| + \|\mathbf{y}\|, \quad \text{for } \mathbf{x}, \mathbf{y} \in \mathbb{C}^n. \tag{A.22}$$

Let

$$\mathbf{x} = \begin{bmatrix} x_1 & x_2 & \cdots & x_n \end{bmatrix} \in \mathbb{C}^n. \tag{A.23}$$

A useful class of vector norms are the p-norms defined by

$$\|\mathbf{x}\|_p := \left(\sum_{i=1}^n |x_1|^p \right)^{\frac{1}{p}}, \quad \text{for } p \ge 1. \tag{A.24}$$

Since $\mathbb{C}^{n \times m}$ is isomorphic to $\mathbb{C}^{nm}$, the definition of a matrix norm should be compatible to that of a vector norm. Therefore, a real-valued function $\|\cdot\| \in \mathbb{R}$ of a matrix $\mathbf{M}$ is a matrix norm of $\mathbf{M}$ if the following properties hold:

$$\|\mathbf{M}| \ge 0, \quad \|\mathbf{M}\| = 0 \text{ if and only if } \mathbf{M} = 0, \tag{A.25}$$

$$\|a\mathbf{M}| = |a| \|\mathbf{M}\|, \quad \text{for } a \in \mathbb{C}, \tag{A.26}$$

$$\|\mathbf{M} + \mathbf{N}\| \le \|\mathbf{M}\| + \|\mathbf{N}\|, \quad \text{for } \mathbf{M}, \mathbf{N} \in \mathbb{C}^{n \times m}. \tag{A.27}$$

Let $\mathbf{M} = [m_{ij}] \in \mathbb{C}^{n \times m}$. The p-norm is defined as

$$\|\mathbf{M}\|_p := \sum_{\mathbf{x} \ne 0} \frac{\|\mathbf{Mx}\|_p}{\|\mathbf{x}\|_p}. \tag{A.28}$$

The most widely used matrix norms are, for $p = 1, 2, \infty$:

$$\|\mathbf{M}\|_1 := \max_j \left(\sum_{i=1}^n |m_{ij}| \right), \quad \text{for } \mathbf{M} \in \mathbb{C}^{n \times m} \quad \text{(maximum column sum)}, \tag{A.29}$$

$$\|\mathbf{M}\|_2 := \sqrt{\lambda_{\max} (\mathbf{M}^* \mathbf{M})}, \quad \text{(square root of the largest eigenvalue of } \mathbf{M}^* \mathbf{M}), \tag{A.30}$$

$$\|\mathbf{M}\|_\infty := \max_i \left(\sum_{j=1}^n |m_{ij}| \right), \quad \text{for } \mathbf{M} \in \mathbb{C}^{n \times m} \quad \text{(maximum row sum)}. \tag{A.31}$$

Also the Frobenius "norm," which is not an induced norm, is defined by

$$\|\mathbf{M}\|_F := \sqrt{\sum_{i=1}^n \sum_{j=1}^m |m_{ij}|^2}. \tag{A.32}$$

A.4   Proof of the Cayley–Hamilton Theorem (Theorem 4.1)

THEOREM A.2 (Cayley–Hamilton Theorem)   Let $a(s) = \det[s\mathbf{I} - \mathbf{A}]$ denote the characteristic polynomial of $\mathbf{A}$. Then

$$a(A) = 0. \tag{A.33}$$

Proof $(s\mathbf{I} - \mathbf{A})^{-1}$ can always be written as

$$(s\mathbf{I} - \mathbf{A})^{-1} = \frac{1}{a(s)} \left(\mathbf{R}_{n-1}s^{n-1} + \mathbf{R}_{n-2}s^{n-2} + \cdots + \mathbf{R}_1 s + \mathbf{R}_0 \right), \tag{A.34}$$

where

$$a(s) = \det(s\mathbf{I} - \mathbf{A}) = s^n + a_{n-1}s^{n-1} + \cdots + a_1 s + a_0 \tag{A.35}$$

and $\mathbf{R}_i$ for $i = 0, 1, \ldots, n-1$ are the matrix coefficients of s^i in the polynomial matrix that is the adjoint of $(s\mathbf{I} - \mathbf{A})$. Write

$$
\begin{aligned}
a(s)\mathbf{I} &= (s\mathbf{I} - \mathbf{A}) \left(\mathbf{R}_{n-1}s^{n-1} + \mathbf{R}_{n-2}s^{n-2} + \cdots + \mathbf{R}_1 s + \mathbf{R}_0 \right) \\
&= (s\mathbf{I} - \mathbf{A})\mathbf{R}_{n-1}s^{n-1} + (s\mathbf{I} - \mathbf{A})\mathbf{R}_{n-2}s^{n-2} + \cdots \\
&\quad + (s\mathbf{I} - \mathbf{A})\mathbf{R}_1 s + (s\mathbf{I} - \mathbf{A})\mathbf{R}_0 \\
&= \mathbf{R}_{n-1}s^n + (\mathbf{R}_{n-2} - \mathbf{A}\mathbf{R}_{n-1}) s^{n-1} + (\mathbf{R}_{n-3} - \mathbf{A}\mathbf{R}_{n-2}) s^{n-2} + \cdots \\
&\quad + (\mathbf{R}_0 - \mathbf{A}\mathbf{R}_1) s - \mathbf{A}\mathbf{R}_0.
\end{aligned}
\tag{A.36}
$$

Since

$$a(s)\mathbf{I} = \mathbf{I}s^n + a_{n-1}\mathbf{I}s^{n-1} + \cdots + a_1\mathbf{I}s + a_0\mathbf{I}, \tag{A.37}$$

we have by matching the coefficients:

$$
\begin{aligned}
\mathbf{R}_{n-1} &= \mathbf{I}, \\
\mathbf{R}_{n-2} &= \mathbf{A}\mathbf{R}_{n-1} + a_{n-1}\mathbf{I}, \\
\mathbf{R}_{n-3} &= \mathbf{A}\mathbf{R}_{n-2} + a_{n-2}\mathbf{I}, \\
&\;\;\vdots \\
\mathbf{R}_0 &= \mathbf{A}\mathbf{R}_1 + a_1\mathbf{I}, \\
0 &= \mathbf{A}\mathbf{R}_0 + a_0\mathbf{I}.
\end{aligned}
\tag{A.38}
$$

Substituting $\mathbf{R}_i$s successively from the bottom, we have

$$
\begin{aligned}
0 &= \mathbf{A}\mathbf{R}_0 + a_0\mathbf{I} \\
&= \mathbf{A}(\mathbf{A}\mathbf{R}_1 + a_1\mathbf{I}) = \mathbf{A}^2\mathbf{R}_1 + a_1\mathbf{A} + a_0\mathbf{I} \\
&= \mathbf{A}^3\mathbf{R}_2 + a_2\mathbf{A}^2 + a_1\mathbf{A} + a_0\mathbf{I} \\
&\;\;\vdots \\
&= \mathbf{A}^n + a_{n-1}\mathbf{A}^{n-1} + a_{n-2}\mathbf{A}^{n-2} + \cdots + a_1\mathbf{A} + a_0\mathbf{I} = a(\mathbf{A}).
\end{aligned}
\tag{A.39}
$$

Therefore, $a(\mathbf{A}) = 0$. ∎

A.5   Singular Value Decomposition

THEOREM A.3 (Singular value decomposition)   (Theorem 5.9) For a matrix $\mathbf{A} \in \mathbb{C}^{n \times m}$, there exist unitary matrices $\mathbf{U} \in \mathbb{C}^{n \times n}$ and $\mathbf{V} \in \mathbb{C}^{m \times m}$ such that

$$\mathbf{U}^*\mathbf{A}\mathbf{V} = \begin{bmatrix} \sigma_1 & & & \vdots & \\ & \ddots & & 0 & \\ & & \sigma_p & \vdots & \\ \hdashline & 0 & & \vdots & 0 \end{bmatrix} \in \mathbb{R}^{n \times m}, \quad \text{for } p = \min\{n,m\}, \tag{A.40}$$

where $\sigma_1 \geq \sigma_2 \geq \cdots \geq \sigma_p \geq 0$.

Proof of Theorem 5.9   Let $\mathbf{x} \in \mathbb{C}^m$ and $\mathbf{y} \in \mathbb{C}^n$ be unit vectors satisfying

$$\mathbf{A}\mathbf{x} = \|\mathbf{A}\|_2 \mathbf{y} =: \sigma \mathbf{y}. \tag{A.41}$$

Then there exist $\mathbf{V}_2 \in \mathbb{C}^{m \times (m-1)}$ and $\mathbf{U}_2 \in \mathbb{C}^{n \times (n-1)}$ so that

$$\mathbf{V} := [\mathbf{x}\ \mathbf{V}_2] \in \mathbb{C}^{m \times m}, \tag{A.42}$$
$$\mathbf{U} := [\mathbf{y}\ \mathbf{U}_2] \in \mathbb{C}^{n \times n} \tag{A.43}$$

are unitary (that is, $\mathbf{V}^* = \mathbf{V}^{-1}$ and $\mathbf{U}^* = \mathbf{U}^{-1}$). It can be shown that

$$\mathbf{U}^*\mathbf{A}\mathbf{V} = \begin{bmatrix} \sigma & \mathbf{w}^* \\ 0 & \mathbf{B} \end{bmatrix} =: \mathbf{A}_1, \quad \text{for } \mathbf{w} \in \mathbb{C}^{m-1},\ \mathbf{B} \in \mathbb{C}^{(n-1)\times(m-1)}. \tag{A.44}$$

Since

$$\left\| \mathbf{A}_1 \begin{bmatrix} \sigma \\ \mathbf{w} \end{bmatrix} \right\|_2^2 = \left\| \begin{bmatrix} \sigma & \mathbf{w}^* \\ 0 & \mathbf{B} \end{bmatrix} \begin{bmatrix} \sigma \\ \mathbf{w} \end{bmatrix} \right\|_2^2 \geq \left(\sigma^2 + \mathbf{w}^*\mathbf{w}\right)^2, \tag{A.45}$$

we have

$$\|\mathbf{A}_1\|_2^2 \geq \left(\sigma^2 + \mathbf{w}^*\mathbf{w}\right). \tag{A.46}$$

But $\sigma^2 = \|\mathbf{A}\|_2^2 = \|\mathbf{A}_1\|_2^2$ and so we must have $\mathbf{w} = 0$. An obvious induction argument leads to

$$\mathbf{U}^*\mathbf{A}\mathbf{V} = \Sigma. \tag{A.47}$$

∎

A.6   Positive or Negative-Definite Matrices

DEFINITION A.2   A Hermitian matrix (defined by the condition $\mathbf{A} = \mathbf{A}^* \in \mathbb{C}^{n \times n}$) is said to be *positive definite* (semi-definite), denoted by $\mathbf{A} > 0$ ($\mathbf{A} \geq 0$), if

$$\mathbf{x}^*\mathbf{A}\mathbf{x} > 0\ (\geq 0) \quad \text{for all } \mathbf{x} \neq 0,\ \mathbf{x} \in \mathbb{C}^n. \tag{A.48}$$

A Hermitian matrix has the following interesting properties.

THEOREM A.4   Any Hermitian matrix

(1) has only real eigenvalues,

(2) has a diagonal Jordan form,

(3) has orthogonal eigenvectors.

Proof

(1) Let $\lambda \in \mathbb{C}$ be an eigenvalue of the Hermitian matrix $\mathbf{A}$. Then there exists $\mathbf{v} \neq 0$ such that

$$\mathbf{A}\mathbf{v} = \lambda\mathbf{v} \tag{A.49}$$

and

$$\mathbf{v}^*\mathbf{A}\mathbf{v} = \lambda\mathbf{v}^*\mathbf{v}. \tag{A.50}$$

But since $\mathbf{A}$ is Hermitian, we have

$$\lambda\mathbf{v}^*\mathbf{v} = \mathbf{v}^*\mathbf{A}\mathbf{v} = (\mathbf{v}^*\mathbf{A}\mathbf{v})^* = \bar{\lambda}\mathbf{v}^*\mathbf{v}, \tag{A.51}$$

where $\bar{\lambda}$ is the complex conjugate of λ. Therefore, λ must be equal to $\bar{\lambda}$.

(2) If the symmetric matrix $\mathbf{A}$ is not diagonalizable, then it must have generalized eigenvalues of order 2 or higher. That is, for some repeated eigenvalue λ_i, there exists $\mathbf{v} \neq 0$ such that

$$(\mathbf{A} - \lambda_i\mathbf{I})^2\mathbf{v} = 0, \quad (\mathbf{A} - \lambda_i\mathbf{I})\mathbf{v} \neq 0. \tag{A.52}$$

But note that

$$0 = \mathbf{v}^*(\mathbf{A} - \lambda_i\mathbf{I})^2\mathbf{v} = \mathbf{v}^*(\mathbf{A} - \lambda_i\mathbf{I})(\mathbf{A} - \lambda_i\mathbf{I}) \neq 0, \tag{A.53}$$

which is a contradiction. Therefore, as there exists no generalized eigenvectors of order 2 or higher, $\mathbf{A}$ must be diagonalizable.

(3) As $\mathbf{A}$ can have no generalized eigenvector of order 2 or higher:

$$\begin{aligned} \mathbf{A}\mathbf{T} &= \mathbf{A}\begin{bmatrix} \mathbf{v}_1 & \cdots & \mathbf{v}_n \end{bmatrix} \\ &= \begin{bmatrix} \mathbf{v}_1 & \cdots & \mathbf{v}_n \end{bmatrix}\Lambda = \mathbf{T}\Lambda, \quad \text{for } \det[\mathbf{T}] \neq 0. \end{aligned} \tag{A.54}$$

That is $\mathbf{A} = \mathbf{T}^{-1}\Lambda\mathbf{T}$. But since $\mathbf{A}$ is Hermitian:

$$\mathbf{T}^{-1}\Lambda\mathbf{T} = \mathbf{A} = \mathbf{A}^* = (\mathbf{T}^{-1}\Lambda\mathbf{T})^* = \mathbf{T}^*\Lambda(\mathbf{T}^*)^{-1} \quad \Rightarrow \quad \mathbf{T}^* = \mathbf{T}^{-1} \tag{A.55}$$

or

$$\mathbf{T}^*\mathbf{T} = \mathbf{I} \quad \Rightarrow \quad \mathbf{v}_i^*\mathbf{v}_i = 1, \; \mathbf{v}_i^*\mathbf{v}_j = 0, \quad \text{for all } i \neq j. \tag{A.56}$$

A.7   Orthogonal, Orthonormal, and Unitary Matrices

DEFINITION A.3 (Orthogonal matrices)   A matrix $\mathbf{A} \in \mathbb{C}^{n \times n}$ is an orthogonal matrix if

$$\mathbf{A}\mathbf{A}^* = \mathrm{diag} \begin{bmatrix} d_1 & d_2 & \cdots & d_n \end{bmatrix}, \qquad (A.57)$$

where the d_is are real.

DEFINITION A.4 (Orthonormal matrices)   A matrix $\mathbf{A} \in \mathbb{C}^{n \times n}$ is an orthonormal matrix if

$$\mathbf{A}\mathbf{A}^* = \mathbf{I} \qquad (A.58)$$

and this leads to the equivalent characterization

$$\mathbf{A}^{-1} = \mathbf{A}^*. \qquad (A.59)$$

The column vectors of an orthonormal matrix are necessarily orthogonal to each other and each vector has unit norm. An orthonormal matrix is also called a *unitary matrix*.

A.8   Polynomial Matrices: Basic Theory

In this section, we develop some basic results on polynomial matrices, that is, matrices whose elements are polynomials with real or complex coefficients. This theory is useful in its own right and the Jordan form may also be derived using the Smith canonical form, invariant polynomials, and elementary divisors of a polynomial matrix. This useful connection is developed in the present section.

A.8.1   Elementary Operations on Polynomial Matrices

The following *elementary operations* are defined for polynomial matrices:

1. Interchange rows (columns) i and j.
2. Replace row (column) i by c (row i) (c (column i)) where c is a nonzero real or complex constant.
3. Replace row i (column i) by (row i) $+ \alpha(s)$(row j) ((column i) $+ \alpha(s)$(column j)) where $\alpha(s)$ is a polynomial.

Elementary row or column operations applied to the identity matrix result in matrices called *elementary matrices*. It is easy to see that elementary row (column) operations on a given matrix correspond to left (right) multiplication of the given matrix by elementary matrices. Elementary matrices are characterized by the fact that their determinants are nonzero constants. They are also characterized by the fact that their inverses exist and are polynomial matrices. Polynomial matrices which are square and whose inverses are polynomial are said to be *unimodular*. In fact, it is easy to show that any unimodular matrix is a product of elementary matrices.

A.8.2 Equivalence of Polynomial Matrices

Two polynomial matrices $\mathbf{A}(s)$ and $\mathbf{B}(s)$ are said to be *left equivalent* if

$$\mathbf{A}(s) = \mathbf{P}(s)\mathbf{B}(s) \tag{A.60}$$

for some unimodular matrix $\mathbf{P}(s)$ and *right equivalent* if

$$\mathbf{A}(s) = \mathbf{B}(s)\mathbf{Q}(s) \tag{A.61}$$

for some unimodular matrix $\mathbf{Q}(s)$. They are said to be *equivalent* if

$$\mathbf{A}(s) = \mathbf{P}(s)\mathbf{B}(s)\mathbf{Q}(s) \tag{A.62}$$

for some unimodular matrices $\mathbf{P}(s), \mathbf{Q}(s)$. The relations of left equivalence, right equivalence, and equivalence are reflexive, symmetric, and transitive, and thus constitute *bona fide* equivalence relations. Thus they partition the set of all polynomial matrices of the same size into disjoint *equivalence classes*. Each of these classes can be characterized by a *canonical form* that is unique to that class. In the sections below, we describe such a canonical form, called the *Smith canonical form*.

A.8.3 Hermite Forms: Row (Column) Compression Algorithm

Consider the polynomial matrix $\mathbf{A}(s)$ of rank v, that is one where the determinants of all minors of order $v+1$ or higher are identically zero and at least one of order v is nonzero. The Hermite form of $\mathbf{A}(s)$ can be obtained by the following *row compression algorithm*, carried out by successive left multiplications of $\mathbf{A}(s)$ by unimodular matrices.

(1) By row interchanges, bring the lowest-degree term in column 1 to the $(1,1)$ position. Call the corresponding elements of the new matrix $a_{ij}(s)$.

(2) By Euclidean division, write all the elements of column 1 by $a_{11}(s)$:

$$a_{j1}(s) = \gamma_{j1}(s)a_{11}(s) + r_{j1}(s), \tag{A.63}$$

where $\deg\left[\gamma_{j1}(s)\right] < \deg\left[a_{11}(s)\right]$.

(3) Replace $a_{j1}(s)$ by $a_{j1}(s) - \gamma_{j1}(s)a_{11}(s) = r_{j1}(s)$ which is of lower degree, by performing the row operation of replacing row j by (row j) $- q_{j1}(s)$(row 1), $j = 2, 3, \ldots$.

(4) By repeating step (1), bring the lowest-degree $r_{j1}(s)$, $j = 2, 3, \ldots$ to the $(1,1)$ position.

(5) Repeat steps (2)–(4) until all terms in column 1, below the $(1,1)$ element, are zero.

(6) Repeat steps (1)–(5) on the second through last row of the new matrix.

(7) The resulting matrix, called the *Hermite form*, will be of the row-compressed form

$$
\mathbf{U}(s)\mathbf{A}(s) =
\begin{bmatrix}
a_{11}(s) & a_{12}(s) & \cdots & \cdots & \cdots & a_{1n}(s) \\
0 & a_{22}(s) & & & & \vdots \\
\vdots & 0 & \ddots & & & \vdots \\
0 & & & a_{rr}(s) & \cdots & a_{rn}(s) \\
0 & \cdots & \cdots & \cdots & \cdots & 0 \\
\vdots & & & & & \vdots \\
0 & \cdots & \cdots & \cdots & \cdots & 0
\end{bmatrix},
\tag{A.64}
$$

where $\mathbf{U}(s)$ is unimodular and $a_{ij}(s)$ are polynomials.

A similar algorithm can be applied to the columns of $\mathbf{A}(s)$ to produce the column compressed form:

$$
\mathbf{A}(s)\mathbf{V}(s) =
\begin{bmatrix}
b_{11}(s) & 0 & \cdots & 0 & 0 & \cdots & 0 \\
b_{21}(s) & b_{22}(s) & 0 & \vdots & \vdots & & \vdots \\
\vdots & & \ddots & 0 & \vdots & & \vdots \\
b_{m1}(s) & b_{n2}(s) & \cdots & b_{mr}(s) & 0 & \cdots & 0
\end{bmatrix},
\tag{A.65}
$$

where $V(s)$ is unimodular.

A.8.4   Smith Form, Invariant Polynomials, and Elementary Divisors

If we apply both elementary row and column operations to a polynomial matrix, it follows from the previous discussion that we can obtain the polynomial matrix:

$$
\begin{bmatrix}
\mathring{a}_1(s) & 0 & \cdots & 0 \\
0 & & & \\
\vdots & & \mathbf{A}_1(s) & \\
0 & & &
\end{bmatrix}.
\tag{A.66}
$$

Furthermore, we can assume without loss of generality that $\mathring{a}_1(s)$ is monic and has lower degree than all the elements of $\mathbf{A}_1(s)$, since the preceding row and column operations can be repeated until this condition is satisfied. Now we can divide each entry of $\mathbf{A}_1(s)$ by $\mathring{a}_1(s)$ and replace it by the remainder, a polynomial of lower degree, and repeat this procedure until every entry of $\mathbf{A}_1(s)$ is divisible without remainder by $\mathring{a}_1(s)$. We can then repeat this procedure on $\mathbf{A}_1(s)$ and reduce it to

$$
\begin{bmatrix}
\mathring{a}_2(s) & 0 & \cdots & 0 \\
0 & & & \\
\vdots & & \mathbf{A}_2(s) & \\
0 & & &
\end{bmatrix},
\tag{A.67}
$$

where each entry of $\mathbf{A}_2(s)$ is divisible without remainder by $\mathring{a}_2(s)$. Proceeding in this way, we finally obtain the matrix $\mathbf{S}(s)$, called the *Smith form* of $\mathbf{A}(s)$.

The Smith form is a canonical form for the equivalence class to which $\mathbf{A}(s)$ belongs. In other words, every polynomial matrix in this equivalence class has the same Smith form. The Smith form is described by

$$\mathbf{S}(s) = \left[\begin{array}{ccc:cc} i_1(s) & & & 0 & \\ & \ddots & & \vdots & \\ & & i_r(s) & 0 & \\ \hdashline 0 & \cdots & 0 & 0 & \end{array}\right], \tag{A.68}$$

where the polynomials $i_1(s), i_2(s), \ldots, i_r(s)$ are monic and each divides the succeeding one. $i_1(s), i_2(s), \ldots, i_r(s)$ are the *invariant polynomials* of $\mathbf{A}(s)$.

From the derivation of the Smith form described above, it follows that there exist unimodular $\mathbf{U}(s), \mathbf{V}(s)$ so that

$$\mathbf{U}(s)\mathbf{A}(s)\mathbf{V}(s) = \mathbf{S}(s) \tag{A.69}$$

and every matrix equivalent to $\mathbf{A}(s)$ has the same invariant polynomials and therefore the same Smith canonical form. Finally, the *elementary divisors* of $\mathbf{A}(s)$ are all the nontrivial ($\neq 1$) monic factors of $i_j(s)$, $j = 1, 2, \ldots, r$ over the real or complex field.

The invariant polynomials can also be obtained directly from $\mathbf{A}(s)$, that is without constructing unimodular transformations, as described below. The procedure is based on the fact that the minors of various orders of a polynomial matrix are unaltered by unimodular left and right multiplications.

(1) Suppose that rank $A(s) = r$ and let $m_j(s)$ denote the monic greatest common divisor (gcd) of all $j \times j$ minors of $\mathbf{A}(s)$, $j = 1.2. \ldots, r$, and let $m_0(s) := 1$. Then the invariant polynomials of $A(s)$ are

(2)

$$i_1(s) = \frac{m_1(s)}{m_0(s)}, \quad i_2(s) = \frac{m_2(s)}{m_1(s)}, \quad \ldots, \quad i_r(s) = \frac{m_r(s)}{m_{r-1}(s)}. \tag{A.70}$$

Factor the invariant polynomials of $A(s)$ as

$$\begin{aligned} i_1(s) &= \alpha_1(s)^{p_1} \alpha_2(s)^{p_2} \cdots \alpha_t(s)^{p_t}, \\ i_2(s) &= \alpha_1(s)^{q_1} \alpha_2(s)^{q_2} \cdots \alpha_t(s)^{q_t}, \\ &\vdots \\ i_r(s) &= \alpha_1(s)^{v_1} \alpha_2(s)^{v_2} \cdots \alpha_t(s)^{v_t}, \end{aligned} \tag{A.71}$$

where $\alpha_i(s), \alpha_j(s)$ are monic and coprime over $\mathbb{F}$ for $i \neq j, i = 1, 2, \ldots, t; j = 1, 2, \ldots, t$, and

$$\begin{aligned} 0 &\leq p_1 \leq q_1 \leq \cdots \leq v_1, \\ 0 &\leq p_2 \leq q_2 \leq \cdots \leq v_2, \\ &\vdots \\ 0 &\leq p_t \leq q_t \leq \cdots \leq v_t. \end{aligned} \tag{A.72}$$

Then all the nontrivial ($\neq 1$) factors

$$\alpha_1(s)^{p_1}, \ldots, \alpha_t(s)^{p_t}, \ldots, \alpha_1(s)^{q_1}, \ldots, \alpha_t(s)^{v_t} \tag{A.73}$$

are called the *elementary divisors* of $A(s)$.

EXAMPLE A.1   If

$$\begin{aligned}
i_1(s) &= (s+1)(s+2), \\
i_2(s) &= (s+1)^2(s+2)^2, \\
i_3(s) &= (s+1)^3(s+2)^2.
\end{aligned} \tag{A.74}$$

the elementary divisors are

$$(s+1),\ (s+2),\ (s+1)^2,\ (s+2)^2,\ (s+1)^3,\ (s+2)^2. \tag{A.75}$$

Note that when $\mathbb{F}$ is the complex field, each $\alpha_i(s)$ in Equation (A.71) is of the form $(s - \lambda_i)^{p_i}$.

LEMMA A.5

$$\mathbf{J} = \begin{bmatrix} \lambda_i & 1 & 0 & \cdots & 0 \\ 0 & \lambda_i & 1 & & 0 \\ \vdots & & \ddots & & \vdots \\ & & & \ddots & 1 \\ 0 & \vdots & \vdots & 0 & \lambda_i \end{bmatrix} \in \mathbb{R}^{h_i \times h_i} \tag{A.76}$$

has $(s - \lambda_i)^{h_i}$ as its only elementary divisor.

Proof   The gcds of minors of $(s\mathbf{I} - \mathbf{J})$ of various orders are

$$m_0 := 1,\ m_1 = 1,\ \ldots,\ m_{h_i - 1}(s) = 1,\ m_{h_i} = (s - \lambda_i)^{h_i} \tag{A.77}$$

and so the invariant polynomials are

$$i_1(s) = 1,\ i_2(s) = 1,\ \ldots,\ i_{h_i}(s) = (s - \lambda_i)^{h_i} \tag{A.78}$$

and therefore $(s - \lambda_i)^{h_i}$ is the only elementary divisor of $\mathbf{J}$.   ∎

The next three lemmas connect the Jordan form of a matrix A to the elementary divisors of $(s\mathbf{I} - \mathbf{A})$.

LEMMA A.6   The set of elementary divisors of

$$\mathbf{C}(s) = \begin{bmatrix} \mathbf{A}(s) & 0 \\ 0 & \mathbf{B}(s) \end{bmatrix} \tag{A.79}$$

is the union of the elementary divisors of $\mathbf{A}(s)$ with the elementary divisors of $\mathbf{B}(s)$.

Proof   The proof is omitted and can be found in the reference given in Section 4.9.   ∎

Now consider elementary divisors of $(s\mathbf{I} - \mathbf{J})$ where $\mathbf{J}$ is in Jordan form:

$$\mathbf{J} = \begin{bmatrix} \mathbf{J}_1 & & \\ & \ddots & \\ & & \mathbf{J}_t \end{bmatrix} \tag{A.80}$$

and

$$\mathbf{J}_i = \begin{bmatrix} \lambda_i & 1 & 0 & \cdots & 0 \\ 0 & \lambda_i & 1 & & 0 \\ \vdots & & \ddots & & \vdots \\ \vdots & & & \ddots & 1 \\ 0 & \vdots & \vdots & 0 & \lambda_i \end{bmatrix} \in \mathbb{R}^{n_i \times n_i}, \quad \text{for } i = 1, 2, \dots . \tag{A.81}$$

LEMMA A.7   The elementary divisors of $(s\mathbf{I} - \mathbf{J})$ are

$$(s - \lambda_i)^{n_1} (s - \lambda_2)^{n_2} \cdots (s - \lambda_t)^{n_t}. \tag{A.82}$$

Proof   Follows immediately from Lemmas A.5 and A.6.   ■

Remark A.1   Note that Lemma A.7 does not require that the λ_i be distinct.

We can now appreciate the significance of the above results by considering a square matrix $\mathbf{A}$ with real or complex coefficients. Let

$$\mathbf{A}(s) := s\mathbf{I} - \mathbf{A} \tag{A.83}$$

and let the elementary divisors of $\mathbf{A}(s)$ be:

$$(s - \lambda_1)^{n_1} (s - \lambda_2)^{n_2} \cdots (s - \lambda_t)^{n_t}. \tag{A.84}$$

LEMMA A.8   The polynomial matrices $(s\mathbf{I} - \mathbf{A})$ and $(s\mathbf{I} - \mathbf{J})$ in Equations (A.83) and (A.80) are equivalent.

Proof   The result follows from the fact that the invariant polynomials of $(s\mathbf{I} - \mathbf{A})$ and $(s\mathbf{I} - \mathbf{J})$ and hence the Smith forms of $(s\mathbf{I} - \mathbf{A})$ and $(s\mathbf{I} - \mathbf{J})$ are identical.   ■

It is now possible to constructively determine the matrix transforming a given matrix A to its Jordan form. This is shown below.

From Lemma A.8, it follows that there exist unimodular matrices $\mathbf{P}(s)$ and $\mathbf{Q}(s)$ such that

$$\mathbf{P}(s)(s\mathbf{I} - \mathbf{A})\mathbf{Q}(s) = s\mathbf{I} - \mathbf{J}. \tag{A.85}$$

Write

$$\mathbf{Q}(s) = \mathbf{Q}_0 + \mathbf{Q}_1 s + \cdots + \mathbf{Q}_v s^v, \tag{A.86a}$$

$$\mathbf{P}(s) = \mathbf{P}_0 + \mathbf{P}_1 s + \cdots + \mathbf{P}_u s^u. \tag{A.86b}$$

LEMMA A.9   With

$$\mathbf{T} := \mathbf{Q}(\mathbf{J}) = \mathbf{Q}_0 + \mathbf{Q}_1\mathbf{J} + \cdots + \mathbf{Q}_v\mathbf{J}^v, \qquad (A.87a)$$

$$\mathbf{T}^{-1} = \mathbf{P}(\mathbf{J}) = \mathbf{P}_0 + \mathbf{J}\mathbf{P}_1 + \cdots + \mathbf{J}^u\mathbf{P}_u, \qquad (A.87b)$$

Equation (A.85) reduces to

$$\mathbf{T}^{-1}\mathbf{A}\mathbf{T} = \mathbf{J} \qquad (A.88)$$

and thus $\mathbf{A}$ is similar to $\mathbf{J}$, its Jordan form.

Proof   The proof is omitted and can be found in the reference given in Section 4.9.   ■

One way to determine the unimodular matrices $\mathbf{P}(s)$ and $\mathbf{Q}(s)$ is to first determine unimodular matrices that reduce $(s\mathbf{I} - \mathbf{A})$ and $(s\mathbf{I} - \mathbf{J})$ to their common Smith form:

$$\mathbf{P}_1(s)(s\mathbf{I} - \mathbf{A})\mathbf{Q}_1(s) = \begin{bmatrix} i_1(s) & & 0 \\ & \ddots & \\ 0 & & i_n(s) \end{bmatrix} = S(s),$$

$$\mathbf{P}_2(s)(s\mathbf{I} - \mathbf{J})\mathbf{Q}_2(s) = \begin{bmatrix} i_1(s) & & 0 \\ & \ddots & \\ 0 & & i_n(s) \end{bmatrix} = S(s). \qquad (A.89)$$

Therefore,

$$s\mathbf{I} - \mathbf{J} = \underbrace{\mathbf{P}_2(s)^{-1}\mathbf{P}_1(s)}_{\mathbf{P}(s)}(s\mathbf{I} - \mathbf{A})\underbrace{\mathbf{Q}_1(s)\mathbf{Q}_2(s)^{-1}}_{\mathbf{Q}(s)} \qquad (A.90)$$
$$= \mathbf{P}(s)(s\mathbf{I} - \mathbf{A})\mathbf{Q}(s).$$

The point of the above procedure, which is clearly constructive, is that the reductions in Equation (A.89) may be easier to compute.

A.9   Smith–McMillan Form

Suppose that $\mathbf{P}(s)$ is an $n \times n$ real rational transfer function matrix:

$$\mathbf{P}(s) = \begin{bmatrix} p_{11}(s) & \cdots & p_{1n}(s) \\ \vdots & \ddots & \vdots \\ p_{n1}(s) & \cdots & p_{nn}(s) \end{bmatrix}, \qquad (A.91)$$

where each transfer function $p_{ij}(s)$, $i, j = 1, 2,,\ldots, n$ is rational and proper. $\mathbf{P}(s)$ can be written as

$$\mathbf{P}(s) = \frac{1}{d(s)}\mathbf{N}(s), \qquad (A.92)$$

where $d(s)$ is the lcm of the denominators of all the elements in $\mathbf{P}(s)$, and $\mathbf{N}(s)$ is a polynomial matrix. Pre-multiplication and post-multiplication of $\mathbf{N}(s)$ by appropriate choices of unimodular matrices results in an equivalent diagonal polynomial matrix $\mathbf{S}(s)$, the Smith form of $\mathbf{N}(s)$:

$$\mathbf{S}(s) = \underbrace{\mathbf{Y}_y(s)\ldots\mathbf{Y}_2(s)\mathbf{Y}_1(s)}_{\mathbf{Y}(s)}\,\mathbf{N}(s)\,\underbrace{\mathbf{U}_1(s)\mathbf{U}_2(s)\ldots\mathbf{U}_u(s)}_{\mathbf{U}(s)}. \tag{A.93}$$

A unimodular polynomial matrix is a square polynomial matrix whose inverse is also a polynomial matrix. A necessary and sufficient condition for the latter is that the determinant of the polynomial matrix be a constant. Therefore, as shown previously, $\mathbf{S}(s)$ can be written as

$$\mathbf{S} = \text{diag}[\varepsilon_1'(s), \varepsilon_2'(s), \ldots, \varepsilon_n'(s)], \tag{A.94}$$

where

$$\varepsilon_i'(s) \mid \varepsilon_{i+1}'(s), \quad \text{for } i = 1, 2, \ldots, n-1, \tag{A.95}$$

meaning that each polynomial $\varepsilon_i'(s)$ divides $\varepsilon_{i+1}'(s)$ for $i = 1, 2, \ldots, n-1$. Dividing each diagonal element of $S(s)$, $\varepsilon_i'(s)$, by $d(s)$ and performing all possible cancellations yields coprime polynomials $\varepsilon_i(s)$ and $\psi_i(s)$ such that

$$\frac{\varepsilon_i(s)}{\psi_i(s)} = \frac{\varepsilon_i'(s)}{d(s)}, \quad \text{for } i = 1, 2, \ldots, n \tag{A.96}$$

and

$$\varepsilon_i(s) \mid \varepsilon_{i+1}(s),$$
$$\psi_{i+1}(s) \mid \psi_i(s), \quad \text{for } i = 1, 2, \ldots, n-1. \tag{A.97}$$

The Smith–McMillan form of the transfer function matrix $P(s)$ is by definition:

$$\mathbf{P}_d(s) = \text{diag}\left[\frac{\varepsilon_1(s)}{\psi_1(s)}, \frac{\varepsilon_2(s)}{\psi_2(s)}, \ldots, \frac{\varepsilon_n(s)}{\psi_n(s)}\right]. \tag{A.98}$$

The *poles* of $\mathbf{P}(s)$ are the roots of the following polynomial, known as the *pole polynomial* $p(s)$:

$$p(s) := \psi_1(s)\psi_2(s)\cdots\psi_n(s). \tag{A.99}$$

Similarly, the roots of the *zero polynomial* $z(s)$:

$$z(s) := \varepsilon_1(s)\varepsilon_2(s)\cdots\varepsilon_n(s), \tag{A.100}$$

are the *zeros* of $\mathbf{P}(s)$. The degree of $p(s)$ is called the *McMillan degree* of $\mathbf{P}(s)$ and is the order of any minimal realization of $\mathbf{P}(s)$.

A.10 Proof of the Faddeev–LeVerrier Algorithm (Section 3.6)

Write

$$(s\mathbf{I} - \mathbf{A})^{-1} = \frac{\text{Adj}[s\mathbf{I} - \mathbf{A}]}{\det[s\mathbf{I} - \mathbf{A}]} = \frac{\mathbf{T}(s)}{a(s)}, \tag{A.101}$$

where

$$a(s) = \det[s\mathbf{I} - \mathbf{A}] = s^n + a_{n-1}s^{n-1} + \cdots + a_1 s + a_0, \tag{A.102}$$

$$\mathbf{T}(s) = \mathrm{Adj}[s\mathbf{I} - \mathbf{A}] = \mathbf{T}_{n-1}s^{n-1} + \mathbf{T}_{n-2}s^{n-2} + \cdots + \mathbf{T}_1 s + \mathbf{T}_0, \tag{A.103}$$

$$\text{for } \mathbf{T}_i \in \mathbb{R}^{n \times n}.$$

Thus, $(s\mathbf{I} - \mathbf{A})^{-1}$ is determined if the matrices $\mathbf{T}_i$ and the coefficients a_j are found. LeVerrier's algorithm does that as follows. Let

$$a(s) = (a - \lambda_1)(s - \lambda_2) \cdots (s - \lambda_n), \tag{A.104}$$

where the λ_i are complex numbers and possibly repeated. Then it is easily seen that

$$a'(s) := \frac{da(s)}{ds}$$
$$= (s - \lambda_2) \cdots (s - \lambda_n) + (s - \lambda_1)(s - \lambda_3) \cdots (s - \lambda_n) + \cdots$$
$$+ (s - \lambda_1) \cdots (s - \lambda_{n-1}). \tag{A.105}$$

This leads to the following result.

LEMMA A.10

$$\frac{a'(s)}{a(s)} = \frac{1}{s - \lambda_1} + \frac{1}{s - \lambda_2} + \cdots + \frac{1}{s - \lambda_n}. \tag{A.106}$$

To proceed, we note that

$$(s\mathbf{I} - \mathbf{A})(s\mathbf{I} - \mathbf{A})^{-1} = \mathbf{I} \tag{A.107}$$

or

$$\underbrace{s\mathbf{I}(s\mathbf{I} - \mathbf{A})^{-1}}_{s(s\mathbf{I} - \mathbf{A})^{-1}} - \mathbf{A}(s\mathbf{I} - \mathbf{A})^{-1} = \mathbf{I} \tag{A.108}$$

and taking the trace of both sides, we have

$$s\mathrm{Trace}\left[(s\mathbf{I} - \mathbf{A})^{-1}\right] - \mathrm{Trace}\left(\frac{\mathbf{A}\mathbf{T}(s)}{a(s)}\right) = n \tag{A.109}$$

or

$$s\mathrm{Trace}\left[(s\mathbf{I} - \mathbf{A})^{-1}\right] - \mathrm{Trace}\left(\frac{\mathbf{A}\mathbf{T}_0 + \mathbf{A}\mathbf{T}_1 s + \cdots + \mathbf{A}\mathbf{T}_{n-1}s^{n-1}}{a(s)}\right) = n. \tag{A.110}$$

LEMMA A.11

$$\mathrm{Trace}(s\mathbf{I} - \mathbf{A})^{-1} = \frac{a'(s)}{a(s)}. \tag{A.111}$$

Proof   If **J** denotes the Jordan form of **A**:

$$\mathbf{J} = \begin{bmatrix} \begin{array}{cccc|cccc|c} \lambda_1 & 1 & & & & & & & \\ & \ddots & & & & & & & \\ & & & 1 & & & & & \\ 0 & \cdots & \cdots & \lambda_1 & & & & & \\ \hline & & & & \lambda_2 & 1 & & & \\ & & & & & \ddots & & & \\ & & & & & & 1 & & \\ & & & & 0 & \cdots & \cdots & \lambda_2 & \\ \hline & & & & & & & & \ddots \end{array} \end{bmatrix} \tag{A.112}$$

and

$$\mathbf{A} = \mathbf{T}^{-1}\mathbf{J}\mathbf{T} \tag{A.113}$$

for some **T**. Thus

$$(s\mathbf{I} - \mathbf{A})^{-1} = \mathbf{T}(s\mathbf{I} - \mathbf{J})^{-1}\mathbf{T}^{-1} \tag{A.114}$$

and

$$\begin{aligned} \text{Trace}(s\mathbf{I} - \mathbf{A})^{-1} &= \text{Trace}\left(\mathbf{T}(s\mathbf{I} - \mathbf{J})^{-1}\mathbf{T}^{-1}\right) \\ &= \text{Trace}\left((s\mathbf{I} - \mathbf{J})^{-1}\mathbf{T}^{-1}\mathbf{T}\right) \\ &= \text{Trace}(s\mathbf{I} - \mathbf{J})^{-1} \\ &= \frac{1}{s - \lambda_1} + \frac{1}{s - \lambda_2} + \cdots + \frac{1}{s - \lambda_n} \\ &= \frac{a'(s)}{a(s)} \quad \text{(by Lemma A.10)}. \end{aligned} \tag{A.115}$$

Then Equation (A.110) becomes

$$s\left(\frac{a'(s)}{a(s)}\right) - \text{Trace}\left(\frac{\mathbf{AT}_0 + \mathbf{AT}_1 s + \cdots + \mathbf{AT}_{n-1}s^{n-1}}{a(s)}\right) = n \tag{A.116}$$

or

$$sa'(s) - \text{Trace}(\mathbf{AT}_0) - s\,\text{Trace}(\mathbf{AT}_1) - \cdots - s^{n-1}\text{Trace}(\mathbf{AT}_{n-1}) = na(s). \tag{A.117}$$

Now note that

$$sa'(s) = na(s) - a_{n-1}s^{n-1} - 2a_{n-2}s^{n-2} - \cdots - (n-2)a_2 s^2 - (n-1)a_1 s - na_0 \tag{A.118}$$

so that Equation (A.117) reduces to

$$\begin{aligned} -\,\text{Trace}\,(\mathbf{AT}_0) - s\,\text{Trace}\,(\mathbf{AT}_1) - s^2\text{Trace}\,(\mathbf{AT}_2) - \cdots - s^{n-1}\text{Trace}\,(\mathbf{AT}_{n-1}) \\ = a_0 n + sa_1(n-1) + s^2 a_2(n-2) + \cdots + s^{n-1}a_{n-1}. \end{aligned} \tag{A.119}$$

Equating coefficients in (A.119) we get

$$a_0 = -\frac{1}{n}\text{Trace}\,(\mathbf{AT}_0),$$

$$a_1 = -\frac{1}{n-1}\text{Trace}\,(\mathbf{AT}_1),$$

$$\vdots$$

$$a_k = -\frac{1}{n-k}\text{Trace}\,(\mathbf{AT}_k),$$ (A.120)

$$\vdots$$

$$a_{n-1} = -\text{Trace}\,(\mathbf{AT}_{n-1}).$$

Now from

$$(s\mathbf{I} - \mathbf{A})(s\mathbf{I} - \mathbf{A})^{-1} = \mathbf{I}$$ (A.121)

we obtain

$$a(s)\mathbf{I} = (s\mathbf{I} - \mathbf{A})\left(\mathbf{T}_0 + \mathbf{T}_1 s + \cdots + \mathbf{T}_{n-1}s^{n-1}\right)$$
$$= a_0\mathbf{I} + a_1\mathbf{I}s + \cdots + a_{n-1}\mathbf{I}s + \mathbf{I}s^n$$ (A.122)

or

$$-\mathbf{AT}_0 + (\mathbf{T}_0 - \mathbf{AT}_1)s + (\mathbf{T}_1 - \mathbf{AT}_2)s^2 + \cdots + (\mathbf{T}_{n-2} - \mathbf{AT}_{n-1})s^{n-1}$$
$$+ \mathbf{T}_{n-1}s^n = a_0\mathbf{I} + a_1\mathbf{I}s + a_2\mathbf{I}s^2 + \cdots + a_{n-1}\mathbf{I}s^{n-1} + \mathbf{I}s^n.$$ (A.123)

Equating the matrix coefficients of powers of s, we get

$$\mathbf{T}_{n-1} = \mathbf{I},$$

$$\mathbf{T}_{n-2} = \mathbf{AT}_{n-1} + a_{n-1}\mathbf{I},$$

$$\vdots$$ (A.124)

$$\mathbf{T}_0 = \mathbf{AT}_1 + a_1\mathbf{I},$$

$$\mathbf{0} = \mathbf{AT}_0 + a_0\mathbf{I}.$$

The relations in Equations (A.120) and (A.124) suggest the following sequence of calculations:

$$\mathbf{T}_{n-1} = \mathbf{I}, \qquad a_{n-1} = -\text{Trace}\,(\mathbf{AT}_{n-1}),$$

$$\mathbf{T}_{n-2} = \mathbf{AT}_{n-1} + a_{n-1}\mathbf{I}, \qquad a_{n-2} = -\frac{1}{2}\text{Trace}\,(\mathbf{AT}_{n-2}),$$

$$\vdots$$ (A.125)

$$\mathbf{T}_1 = \mathbf{AT}_2 + a_2\mathbf{I}, \qquad a_1 = -\frac{1}{n-1}\text{Trace}\,(\mathbf{AT}_1),$$

$$\mathbf{T}_0 = \mathbf{AT}_1 + a_1\mathbf{I}, \qquad a_0 = -\frac{1}{n}\text{Trace}\,(\mathbf{AT}_0).$$

References

Anderson, G. E. and Lin, P. M. (1973), 'Computer generation of symbolic network functions-A new theory and implementation', *IEEE Transactions on Circuit Theory* **20**(1), 48–56.

Antsaklis, P. J. and Michel, A. N. (2005), *Linear Systems*, Birkäuser, Boston, MA.

Ashfaque, B. S. and Tsakalis, K. (2012), 'Discrete-time PID controller tuning using frequency loop-shaping', *IFAC Proceedings* **45**(3), 613–618.

Barmish, B. R. (1984), 'Invariance of strict Hurwitz property of polynomials with perturbed coefficients', *IEEE Transactions on Automatic Control* **AC-29**(10), 935–936.

Bartlett, A. C., Hollot, C. V. and Lin, H. (1988), 'Root location of an entire polytope of polynomials: It suffices to check the edges', *Mathematics of Controls, Signals, and Systems* **1**, 61–71.

Bhattacharyya, S. P. (1973), 'Output regulation with bounded energy', *IEEE Transactions on Automatic Control* **AC-18**(4), 381–383.

Bhattacharyya, S. P. (1976), 'The structure of robust observers', *IEEE Transactions on Automatic Control* **21**(4), 581–588.

Bhattacharyya, S. P., Chapellat, H. and Keel, L. H. (1995), *Robust Control: The Parametric Approach*, Prentice Hall PTR, Englewood Cliffs, NJ.

Bhattacharyya, S. P., Datta, A. and Keel, L. H. (2009), *Linear Control Theory: Structure, Robustness, and Optimization*, CRC Press, Boca Raton, FL.

Bhattacharyya, S. P. and de Souza, E. (1982), 'Pole assignment via Sylvester's equation', *Systems and Control Letters* **1**(4), 261–263.

Bhattacharyya, S. P. and Howze, J. W. (1984), 'Transfer function conditions for stabilizability', *IEEE Transactions on Automatic Control* **AC-29**(3), 253–254.

Bhattacharyya, S. P., Keel, L. H. and Mohsenizadeh, D. N. (2014), *Linear Systems: A Measurement Based Approach*, Springer, New York, NY.

Bhattacharyya, S. P. and Pearson, J. B. (1970), 'On the linear servomechanism problem', *International Journal of Control* **12**(5), 795–806.

Bhattacharyya, S. P. and Pearson, J. B. (1972), 'On error systems and the servomechanism problem', *International Journal of Control* **15**(6), 1041–1062.

Bhattacharyya, S. P., Pearson, J. B. and Wonham, W. M. (1972), 'On zeroing the output of a linear system', *Information and Control* **20**(2), 135–142.

Biernacki, R. M., Hwang, H. and Bhattacharyya, S. P. (1987), 'Robust stabilization of plants subject to structured real parameter perturbations', *IEEE Transactions on Automatic Control* **AC-32**(6), 495–506.

Black, H. S. (1934), 'Stabilized feedback amplifiers', *The Bell System Technical Journal* **13**(1), 1–18.

Bode, H. W. (1945), *Network Analysis and Feedback Amplifier Design*, D. Van Nostrand, New York, NY.

Boyd, S., Balakrishnan, V. and Kabamba, P. (1989), 'A bisection method for computing the H_∞ norm of a transfer matrix and related problems', *Mathematics of Control, Signal and Systems* **2**, 207–219.

Brasch, F. and Pearson, J. B. (1970), 'Pole placement using dynamic compensators', *IEEE Transactions on Automatic Control* **AC-15**(1), 34–43.

Brittain, J. (1990), 'Thévenin's theorem', *IEEE Spectrum* **27**(3), 42.

Callier, F. M. and Desoer, C. A. (1982), *Multivariable Feedback Systems*, Springer, New York, NY.

Chapellat, H. and Bhattacharyya, S. P. (1989), 'A generalization of Kharitonov theorem: Robust stability of interval plants', *IEEE Transactions on Automatic Control* **AC-34**(3), 306–311.

Dahleh, M. A. and Doaz-Bobillo, I. J. (1995), *Control of Uncertain Systems: A Linear Programming Approach*, Prentice Hall PTR, Englewood Cliffs, NJ.

Datta, A., Ho, M.-T. and Bhattacharyya, S. P. (2000), *Structure and Synthesis of PID Controllers*, Springer, New York, NY.

Davison, E. J. (1976), 'The robust control of a servomechanism problem for linear time-invariant systems', *IEEE Transactions on Automatic Control* **AC-21**(1), 25–34.

de Souza, E. and Bhattacharyya, S. P. (1981), 'Controllability, observability and the solution of $AX - XB = C$', *Linear Algebra and its Applications* **39**, 167–188.

DeCarlo, R. A. and Lin, P. M. (1995), *Linear Circuit Analysis: Time Domain, Phasor, and Laplace Transform Approaches*, Prentice Hall, Englewood Cliffs, NJ.

Desoer, C. A. and Wang, Y. T. (1980), 'Linear time-invariant robust servomechanism problem: A self-contained exposition', *Control and Dynamics Systems* **16**, 81–129.

Diaz-Rodriguez, I. D. and Bhattacharyya, S. P. (2016), PI controller design in the achievable gain-phase margin plane, *in* 'Proceedings of the 55th IEEE Conference on Decision and Control'.

Diaz-Rodriguez, I. D., Han, S. and Bhattacharyya, S. P. (2019), *Analytical Design of PID Controllers*, Springer, New York, NY.

Dorato, P., Abdallah, C. and Cerone, V. (1994), *Linear Quadratic Control: An Introduction*, Macmillan, New York, NY.

Doyle, J. C., Francis, B. A. and Tannenbaum, A. R. (1992), *A Feedback Control Theory*, Macmillan, New York, NY.

Doyle, J. C., Glover, K., Khargonekar, P. P. and Francis, B. A. (1989), 'State-space solutions to standard H_2 and H_∞ control problems', *IEEE Transactions on Automatic Control* **34**(8), 831–847.

Doyle, J. C. and Stein, G. (1979), 'Robustness with observers', *IEEE transactions on automatic control* **24**(4), 607–611.

Faddeev, D. K. and Sominsky, I. S. (1972), *Sbornik zadatch po vyshej algebra (Problems in higher algebra)*, Mir publishers. Moskow-Leningrad (1949).

Ferreira, P. and Bhattacharyya, S. P. (1977), 'On blocking zeros', *IEEE Transactions on Automatic Control* **22**(2), 258–259.

Francis, B. A. and Wonham, W. M. (1975), 'The internal model principle for linear multivariable regulators', *Applied Mathematics and Optimization* **2**, 170–194.

Fujita, M., Namerikawa, T., Matsumura, F. and Uchida, K. (1995), 'μ-systhesis of an electromagnetic suspension system', *IEEE Transactions on Automatic Control* **AC-40**(3), 530–536.

Gantmacher, F. R. (1959), *The Theory of Matrices*, Vol. 1, Chelsea Publishing Company, New York, NY.

Gilbert, E. G. (1963), 'Controllability and observability in multivariable control systems', *Journal of the Society for Industrial and Applied Mathematics Series A Control* **1**, 128–151.

Giustolisi, O., Kapelan, Z. and Savic, D. (2008), 'Algorithm for automatic detection of topological changes in water distribution networks', *Journal of Hydraulic Engineering* **134**(4), 435–446.

Golub, G. H. and Van Loan, C. F. (1996), *Matrix Computation*, The Johns Hopkins University Press, Baltimore, MD.

Green, M. and Limebeer, D. (1995), *Linear Robust Control*, Prentice Hall PTR, Englewood Cliffs, NJ.

Guillemin, E. A. (1949), *The Mathematics of Circuit Analysis*, John Wiley & Sons, New York, NY.

Han, S., Keel, L. H. and Bhattacharyya, S. P. (2018), 'PID controller design with an H_∞ criterion', *IFAC-PapersOnLine* **51**(4), 400–405.

Ho, M.-T., Datta, A. and Bhattacharyya, S. P. (1997), A linear programming characterization of all stabilizing PID controllers, *in* 'Proceedings of the 1997 American Control Conference'.

Ho, M. T., Datta, A. and Bhattacharyya, S. P. (1998), 'An elementary derivation of the Routh-Hurwitz criterion', *IEEE Transactions on Automatic Control* **43**(3), 405–409.

Howze, J. W. and Bhattacharyya, S. P. (1997), 'Robust tracking, error feedback and two degree of freedom controllers', *IEEE Transactions on Automatic Control* **AC-42**(7), 980–984.

Johnson, D. H. (2003), 'Origins of the equivalent circuit concept: The voltage-source equivalent', *Proceedings of the IEEE* **91**(4), 636–640.

Kailath, T. (1980), *Linear Systems*, Prentice-Hall, Englewood Cliffs, NJ.

Kalman, R. E. (1960), 'Contribution to the theory of optimal control', *Boletin de la Sociedad Matematica Mexicana* **5**, 102–119.

Kalman, R. E. (1964), 'When is a linear control system optimal?', *ASME Transactions Series D (Journal of Basic Engineering)* **86**, 51–60.

Kalman, R. E. (1965), 'Irreducible realizations and the degree of a rational matrix', *SIAM Journal of Applied Mathematics* **13**(2), 520–544.

Keel, L. H. and Bhattacharyya, S. P. (1997), 'Robust, fragile, or optimal', *IEEE Transactions on Automatic Control* **AC-42**(8), 1098–1105.

Keel, L. H. and Bhattacharyya, S. P. (2008), 'Controller synthesis free of analytical models: Three term controllers', *IEEE Transactions on Automatic Control* **53**(6), 1353–1369.

Keel, L. H. and Bhattacharyya, S. P. (2010a), 'A Bode plot characterization of all stabilizing controllers', *IEEE Transactions on Automatic Control* **55**(11), 2650–2654.

Keel, L. H. and Bhattacharyya, S. P. (2010b), 'Structural instability and minimal realizations', *IEEE Transactions on Automatic Control* **55**(4), 1014–1017.

Keel, L. H. and Bhattacharyya, S. P. (2015), On the stability of multivariable feedback systems, *in* 'Proceedings of the 2015 IEEE Conference on Decision and Control'.

Keel, L. H. and Bhattacharyya, S. P. (2017), A new formula for the characteristic equation of a matrix with applications, *in* 'Proceedings of the 2017 IEEE Conference on Control Technology and Applications', p. 1631–1636.

Keel, L. H., Mitra, S. and Bhattacharyya, S. P. (2008), 'Data driven synthesis of three term digital controllers', *SICE Journal of Control, Measurement, and System Integration* **1**(2), 102–110.

Keel, L. H., Rego, J. I. and Bhattacharyya, S. P. (2003), 'A new approach to digital PID controller design', *IEEE Transactions on Automatic Control* **48**(4), 687–692.

Keel, L. H., Shafai, B. and Bhattacharyya, S. P. (2009), A frequency response parametrization of all stabilizing controllers for continuous time systems, *in* 'Proceedings of the 2009 American Control Conference'.

Kharitonov, V. L. (1978), 'Asymptotic stability of an equilibrium position of a family of systems of differential equations', *Differentsialnye uravneniya* **14**, 2086–2088.

Kharitonov, V. L. (1979), 'The Routh–Hurwitz problem for families of polynomials and quasi-polynomials', *Izvetiy Akademii Nauk Kazakhskoi SSR, Seria fizikomatemticheskaia* **26**, 69–79.

Kirchhoff, G. (1847), 'Ueber die auflsung der gleichungen, auf welche man bei der untersuchung der linearen vertheilung galvanischer strme gefhrt wird', *Annalen der Physik* **148**(12), 497–508.

Knap, M. J., Keel, L. H. and Bhattacharyya, S. P. (2013), 'Controller design and the Gauss–Lucas theorem', *IEEE Transactions on Automatic Control* **58**(11), 2940–2944.

Krajewski, W. and Viaro, U. (2012), On robust PID control for time-delay plants, *in* 'Proceedings of the 17th International Conference on Methods and Models in Automation and Robotics (MMAR)'.

Lathi, B. P. (2005), *Linear Systems and Signals*, Oxford University Press, Oxford, UK.

Layek, R., Datta, A. and Bhattacharyya, S. P. (2011), Linear circuits: A measurement based approach, *in* 'Proceedings of 20th European Conference on Circuit Theory and Design', Linkoping, Sweden, pp. 476–479.

Layek, R., Nounou, H., Nounou, M., Datta, A. and Bhattacharyya, S. P. (2012), A measurement based approach for linear circuit modeling and design, *in* 'Proceedings of the 51st IEEE Conference on Decision and Control'.

LeVerrier, U. (1840), 'Sur les variations séculaires des éléments des orbites pour les sept planètes principales', *Journal d'Analyse Mathématique* **1**(5), 230.

Lin, P. (1973), 'A survey of applications of symbolic network functions', *IEEE Transactions on Circuit Theory* **20**(6), 732–737.

Luenberger, D. G. (1979), *Introduction to Dynamic Systems: Theory, Models, and Applications*, John Wiley & Sons, New York, NY.

Mason, J. C. and Handscomb, D. C. (2002), *Chebyshev Polynomials*, Chapman & Hall/CRC, London, UK.

Mason, S. J. (1953), 'Feedback theory-some properties of signal flow graphs', *Proceedings of the IRE* **41**(9), 1144–1156.

Mason, S. J. (1956), 'Feedback theory-further properties of signal flow graphs', *Proceedings of the IRE* **44**(7), 920–926.

Mikhailov, A. V. (1938), 'The methods of harmonie analysis in the theory of control', *Avtomatika i Telemekhanika* **3**, 27–81.

Mohsenizadeh, D. N., Keel, L. H. and Bhattacharyya, S. P. (2015), Multivariable controller synthesis using SISO design methods, *in* 'Proceedings of the 2015 IEEE Conference on Decision and Control'.

Mohsenizadeh, D. N., Nounou, H., Nounou, M., Datta, A. and Bhattacharyya, S. P. (2012), A measurement based approach to circuit design, *in* 'Proceedings of IASTED International Conference on Engineering and Applied Science'.

Nyce, D. S. (2003), *Linear Position Sensors: Theory and Application*, John Wiley & Sons, New York, NY.

Nyquist, H. (1932), 'Regeneration theory', *Bell System Technical Journal* **11**(1), 126–147.

Pearson, J. B. (1968), 'Compensator design for dynamic optimization', *International Journal of Control* **9**(4), 473–482.

Pontryagin, L. S. (1955), 'On the zeros of some elementary transcendental functions', *American Mathematical Society Translations* **2**, 95–110.

Quanser (2012), *Rotary Double Inverted Pendulum: Laboratory Guide*, Quanser, Inc., Ontario, Canada.

Rao, S. S. (2000), *Mechanical Vibrations*, Addison-Wesley, Boston, MA.

Reddy, J. N. (2006), *An Introduction to the Finite Element Method*, McGraw-Hill, New York, NY.

Rosenbrock, M. M. (1970), *State-Space and Multivariable Theory*, John Wiley & Sons, New York, NY.

Silva, G. J., Datta, A. and Bhattacharyya, S. P. (2002), 'New results on the synthesis of PID controllers', *IEEE Transactions on Automatic Control* **AC-47**(2), 241–252.

Silva, G. J., Datta, A. and Bhattacharyya, S. P. (2004), *PID Controllers for Time-Delay Systems*, Birkhäuser, Boston, MA.

Tantaris, R. N., Keel, L. H. and Bhattacharyya, S. P. (2006), 'H_∞ design with first-order controllers', *IEEE Transactions on Automatic Control* **51**(8), 1343–1347.

Thévenin, L. (1883), 'Sur un nouveau théoréme d'électricité dynamique [on a new theorem of dynamic electricity]', *C. R. des Séances de l'Académie des Sciences* **97**, 159–161.

Tichý, J., Erhart, J., Kittinger, E. and Prívratská, J. (2010), *Fundamentals of Piezoelectric Sensorics: Mechanical, Dielectric, and Thermodynamical Properties of Piezoelectric Materials*, Springer, New York, NY.

Tsakalis, K. S. and Dash, S. (2013), 'Approximate H_∞ loop shaping in PID parameter adaptation', *International Journal of Adaptive Control and Signal Processing* **27**(1-2), 136–152.

Wolovich, W. A. (1974), *Linear Multivariable Systems*, Springer, New York, NY.

Wonham, W. M. (1985), *Linear Multivariable Control: A Geometric Approach*, Springer, New York, NY.

Youla, D. C., Jabr, H. A. and Bongiorno, J. J. (1976), 'Modern Wiener–Hopf design of optimal controllers Part I: The single-input-output case', *IEEE Transactions on Automatic Control* **AC-21**(1), 3–13.

Zadeh, L. A. and Desoer, C. A. (1963), *Linear Systems Theory: The State Space Approach*, McGraw-Hill, New York, NY.

Zames, G. (1966), 'On the input–output stability for time-varying nonlinear feedback systems: Part I, conditions derived using concepts of loop gain, conicity and positivity', *IEEE Transactions on Automatic Control* **11**(2), 228–238.

Zhou, K., Doyle, J. C. and Glover, K. (1995), *Robust and Optimal Control*, Prentica-Hall, Englewood Cliffs, NJ.

Ziegler, J. G. and Nichols, N. B. (1942), 'Optimum setting for automatic controllers', *Transactions of ASME* **64**, 759–768.

Index